PROCEEDINGS OF THE

THIRD INTERNATIONAL CONFERENCE ON

PRESSURE SURGES

Canterbury, England
March 25-27, 1980

VOLUME 1

Organised by BHRA Fluid Engineering
in conjunction with The City University, London

bhra fluid engineering

THE
CITY
UNIVERSITY

ACKNOWLEDGEMENTS

The valuable assistance of the Organising Committee and Panel of Referees is gratefully acknowledged.

ORGANISING COMMITTEE

Dr. K.J. Enever (Chairman)	Cranfield Institute of Technology
Mr. J.A. Fox	University of Leeds
Mr. A. Lupton	Binnie & Partners
Mr. M. Papworth	BHRA Fluid Engineering
Mr. H.S. Stephens	BHRA Fluid Engineering
Dr. A.R.D. Thorley	The City University, London
Prof. A.E. Vardy	University of Dundee
Mr. J.A. Hanson (Secretary)	BHRA Fluid Engineering

EDITORIAL NOTE

Volume 1 contains the papers presented at the 3 rd International Conference on Pressure Surges held at Canterbury, England, March 25-27, 1980. An edited record of the discussion and contributions to these papers will be produced in a separate volume - Volume 2.

The Organisers cannot be held responsible for statements or opinions made in the papers in Volume 1 or in the discussion recorded in Volume 2.

These papers have been reproduced by offset printing from author's original typescript to minimise delay.

When citing papers from this volume the following reference should be used:- Title, Author(s), Paper No., Pages, 3 rd International Conference on Pressure Surges, BHRA Fluid Engineering, Cranfield, Bedford, U.K. (March 1980)

Editors: H.S. Stephens
J.A. Hanson

British Library Cataloguing in Publication Data

International Conference on Pressure Surges,
3rd, Canterbury, 1980
Proceedings of the Third International
Conference on Pressure Surges, Canterbury,
March 1980.
1. Hydrodynamics - Congresses
2. Hydraulic transients - Congresses
532'.56 QC150

ISBN 0-906085-24-1

Printed and Published by:

BHRA FLUID ENGINEERING
CRANFIELD, BEDFORD, U.K.

PAPERS PRESENTED AT THE THIRD INTERNATIONAL CONFERENCE ON PRESSURE SURGES
CANTERBURY, ENGLAND
March 25-27, 1980

CONTENTS

Third International Conference on

PRESSURE SURGES

Canterbury, England
March 25-27, 1980

PAPER A1

VELOCITY PROFILES IN ACCELERATING PIPE FLOWS STARTED FROM REST

E. van de Sande, A.P. Belde, B.J.G. Hamer, and W. Hiemstra

N.V. Keuring van Electrotechnische Materialen, The Netherlands.

Summary

The mathematical description of transient pipe flows is one-dimensional. Plug flow is assumed and frictional losses are calculated with a quasi-steady approximation.

In accelerating flows the energy losses due to viscous dissipation deviate substantially from the steady-state values. This has been proven by the following study.

In a simple pipe line (plexiglass) accelerating flows, initially at rest, have been investigated experimentally.

Transient velocity signals were obtained from a Laser Doppler Aneometer (reference beam). From these local velocity data complete flow profiles could be derived. Additional flow visualisation experiments reveal the phenomena preceding the onset to turbulence.

Held at the University of Kent at Canterbury, U.K.

Conference Organised by BHRA Fluid Engineering, Cranfield, Bedford, England.
In conjunction with The City University, London.

Nomenclature

D	Diameter of tube	(m)
f	friction factor	(-)
f_i	unsteady friction factor	(-)
L	length of the pipe	(m)
Δp	pressure drop	(N/m^2)
R	radius of the pipe	(m)
r	radial coordinate	(m)
Re	Reynolds Number ($= \rho v D/\mu$)	(-)
Re_i	Reynolds Number based on actual velocity	(-)
t	time coordinate	(s)
$\underline{v}$	local velocity	(m/s)
$\bar{v}$	mean velocity	(m/s)
v^*	dimensionless velocity ($= v/v_r$)	(-)
v_r	core velocity defined by eq (5)	(m/s)
v_s	mean velocity in the steady-state	(m/s)
μ	dynamic viscosity	(Ns/m^2)
ρ	density	(kg/m^3)

Introduction

Transient pipe flows are often analysed by assuming plug flow and by calculating frictional losses with a quasi-steady approximation (Ref. 1). The value of the friction factor at every moment is determined by the Reynolds Number in the steady state, implying a momentarily establishment of the instantaneous velocity profile. Shortly after the start of a transient the inertia forces dominate and viscous losses are not important. Waterhammer calculations using the quasi-steady approach, therefore lead to a correct prediction of the first pressure peak after a change in the flow rate. After some time the frictional losses due to viscous dissipation cannot be neglected anymore and the energy content of a pressure wave decreases. The quasi-steady theory usually underestimates the contribution of the viscous losses (Ref. 2). This results in unrealistic values of the height of successive pressure waves.
A more serious consequence of the reduced energy content might be a erroneous prediction of column separation phenomena (cavities) during the passage of reflected low pressure waves.

Several papers dealing with transient viscous losses have been published. Daily (Ref. 3), Carstens (Ref. 4), Denisov (Ref. 5) and Popov (Ref. 6) experimentally determined the unsteady friction factor in turbulent flow. The lack of a theoretical background together with the variety in the measuring procedures lead to diverging results. Zielke (Ref. 7) derived an equation for the frequency dependent friction in laminar flow. His theoretical result is indirectly verified through comparison with the measured pressure fluctuations after sudden valve closure. A boundary layer approximation for transient losses in turbulent flow is given by Wood (Ref. 8). The model is based upon a constant boundary layer thickness and a complete extinction of the flow after the passage of a pressure wave. These contradictory assumptions were justified by comparison with some experimental data.
Kawamura (Ref. 9) numerically evaluated the influence upon friction losses and velocity profiles due to the unsteady motion of accelerating and decelerating turbulent flows. In the momentum equation the 'Prandtl mixing length" theory is applied to calculate the eddy diffusivity. However, the experiments of Maruyama (Ref. 10) showed a time-dependency of the turbulence intensity during transient flows. Kataoka (Ref. 11) measured velocity profiles in initial stagnant pipe flows after a step change in the flow rate. In a similar study of Letelier (Ref. 12) the unsteady flow rates were investigated after a step change in pressure. Only from this last experiment the analytical solution is known.
About 50 years ago Szymanski (Ref. 13) derived his expression describing the development of the velocity profile after a sudden pressure change. His analysis is restricted to laminar flow. Leteliers data of the flow rate show a considerable spread but the agreement with the theory is satisfactory.
An extension of the theory of Szymanski is given by Chambré (Ref. 14) who considered the case of a gradually increasing pressure at the beginning of a pipe (laminar flow).

As is clear from this concise literature review, limited information about flow field characteristics in transient flow is available. Only for some special cases in laminar flow a theory is developed (Ref. 7, 13, 14) but direct experimental verification, as for instance a measured velocity profile has, to our knowledge, never been done. For turbulent flow no theory exists.

In our laboratories an experimental programme has been set-up in which accelerating and decelerating flows can be investigated. This paper will present a part of the total study, i.e. the acceleration of pipe flows, initially at rest, after a sudden pressure step. When the final steady velocity has been reached a fully developed turbulent flow has been established. Transient velocity profiles are obtained during the entire acceleration.
Local velocities were measured with a Laser Doppler Anemometer. Additional flow visualisation experiments gave insight in the phenomena preceding the onset of turbulence.

The experimental procedures

The experimental set-up is shown in fig. 1. Two constant level tanks are connected by a straight plexiglass pipe (length 25.4 m, diameter 0.05 m). The liquid can be accelerated by opening the air-controlled valve. This valve opens almost instantaneously (within 0.1 sec.). Via the low level reservoir the water flows into the sump tank. A centrifugal pump recirculates the water to the high level reservoir without disturbing the entrance conditions in the pipe. During the entire acceleration period a constant pressure drop over the pipe is maintained.

A Laser Doppler Anemometer (LDA) is situated 400 pipe diameters downstream from the air-controlled valve. The principle of this optical measuring device is based upon the measurement of the Doppler shifted frequency of monochromatic light scattered by small particles in the liquid. Instantaneous local velocity data can be obtained without disturbing the flow field (Ref. 15).
The LDA consists of a 5 mW He-Ne Laser, an optical unit including a rotating grating and a photodetector. The reference beam mode is used and the length of the measuring volume is 0.5 mm. A frequency tracker converts the Doppler frequency into an analog voltage proportional to the local fluid velocity.
The output of the LDA is connected with an instrumentation recorder. During play-back the tape speed is reduced and the velocity signal is recorded upon a X-Y plotter. From the local velocity data a complete profile has been composed. Twenty positions across the pipe diameter were scanned under the same experimental conditions. All measurements have been duplicated to check the reproducibility.

In a similar experimental set-up the hydrogen bubble technique (Ref. 16) is used for flow visualisation experiments. A pipe diameter of 0.02 m was used.
A very fine platinum wire is stretched across the region of the flow to be investigated and by electrolysis hydrogen is introduced into the water. The anode is a cylindrical copper tube attached to the pipe wall 0.3 m behind the wire. With a stroboscope the hydrogen bubbles were illuminated. The stroboscope and the electric pulse, which is fed to the wire, are synchronized.
Flow patterns were recorded on video-tape.

Some preliminary flow visualisation experiments

The Reynolds Number based upon the final steady velocity ($2{,}500 < Re < 21{,}600$) used in the flow visualisation experiments is lower then during the comparative experiments with the LDA ($Re = 60{,}500$). It was expected that this lower range would facilitate the visualisation technique. As will be shown the essential features of the flow field remain the same.

In fig. 2 the development of the flow pattern at various moments after the pressure step is shown. The final Reynolds Number is 2,500. This situation was chosen as an example because the moderate behaviour of the fluid motions can be made clearly visible and the phenomena preceeding the onset of turbulence are more pronounced then at higher Reynolds Numbers.
The first moments after the acceleration plug flow exists. A boundary layer can be observed within a few seconds. The growing boundary layer gradually penetrates the bulk of the fluid causing a retarded acceleration. Outside this layer plug flow persists until at a certain moment a parabolic velocity profile is obtained. Now the fluid hardly accelerates anymore. About 14 seconds have been elapsed since the start of the experiment.
The small dynamic velocity changes still occuring in the flow, ultimately create a situation which is unstable. The location of maximum velocity in the flow pattern shifts out of the centre line of the pipe. The displacement reverses and some time later the point of maximum velocity can be found in the opposite part of the pipe. This "waggling" of the velocity pattern remains until inflection points in the profile occur. The events, happening before the initiation of a turbulent burst, have now become visible.

In the given example the "waggling" of the profile could be observed for at least twenty seconds before the appearance of the first turbulent burst. After this burst the turbulent motions are damped and the "waggling" starts again. Of course, this last phenomenon of re-laminarisation occurs only for Reynolds Numbers in the transition region.
At increasing Reynolds Numbers the duration time of the "waggling" decreases. Above 10,000 "waggling" itself is not observed anymore. When the acceleration has almost finished, plug flow outside the boundary layer still exists. Then the rather flat profile starts "waggling" directly followed by the creation of turbulent eddies also due to instabilities in the profile. A remarkable fact is that the onset to turbulence only takes place almost at the end of the acceleration. Until that moment streamlined flow is maintained.

Laminar accelerating flows initially at rest (theory)

The preliminary experiments show that turbulence is initiated only at the end of the acceleration. The laminar flow theory originally developed for Reynolds Numbers below 2,000 might therefore be also applicable to higher values.

The analytical solution of Szymanski (Ref. 13) is based upon the Navier-Stokes equation. By neglecting radial velocity components and the possible effects of compressibility he derived the following expression describing the flow field:

$$v = \frac{\Delta p R^2}{4\mu L} \left\{ 1 - \left(\frac{r}{R}\right)^2 + 8 \sum_{n=1}^{\infty} \frac{J_o(\alpha_n \frac{r}{R})}{\alpha_n^3 \; J_o'(\alpha_n)} \exp\left(\frac{-\alpha_n^2 \mu t}{\rho R^2} \right) \right\} \tag{1}$$

The pressure step is Δp, r and R are the radial coordinate resp. the radius of the pipe, μ is the dynamic viscosity, t the time coordinate and L the length of the accelerated liquid column. J_o and J_o' are the Bessel function of the first kind of the order zero resp. the derivative with respect to the argument. The zeros of J_o are α_n.

Straight forward calculation of the mean velocity leads to:

$$\bar{v} = \frac{\Delta p . R^2}{8 \mu L} \{ 1 - 32 \sum_{n=1}^{\infty} \frac{1}{\alpha_n^4} \exp (\frac{- \alpha_n^2 \mu t}{\rho R^2}) \} \qquad (2)$$

The unsteady frictional losses can be evaluated. By defining the Fanning friction factor:

$$f = - 8 \mu \frac{\partial v}{\partial r} / \rho \bar{v}^2 \qquad (3)$$

substitution of eq's (1) and (2) into (3) gives after some rearranging:

$$f_i Re_i = 64 \{ (1 - 4 \sum_{n=1}^{\infty} \frac{1}{\alpha_n^2} \exp(\frac{-\alpha_n^2 \mu t}{\rho R^2})) / (1 - 32 \sum_{n=1}^{\infty} \frac{1}{\alpha_n^4} \exp(\frac{-\alpha_n^2 \mu t}{\rho R^2})) \} \qquad (4)$$

Re_i is the Reynolds Number based on the actual time-dependent mean velocity.
The functional relationship expressed in eq. (4) is given in Fig. 3. As is clear from this figure a substantial deviation between unsteady and steady state values of the friction losses exist. One conclusion might be obvious. A quasi-steady approach cannot estimate the energy losses due to dissipation processes in laminar transient flows. However, a direct experimental verification of the derived equations is necessary to support this conclusion.

Flow measurements with the Laser Doppler Anemometer (LDA)

In the experimental set-up built for the visualisation tests a measurement of the local velocity with the LDA had been performed. Fig. 4 shows a recording of the frequency tracker signal. The Reynolds Number based on the final velocity is about 8,000.
As could be expected the registrated increasing velocity shows no fluctuations. At the end of the acceleration period the velocity decreases due to the changes of the profile. After that the onset of turbulence takes place.
This introductary experiment confirms the applicability of Laser Doppler Anemometry to accelerating flows. The phenomena also revealed by the visualisation tests, can be determined quantitatively and a direct verification of the analytical solution of the Szymanski theory is possible now.

The experimental set-up (Fig. 1) and the procedures during each run have been described before. The local velocity data used in the profiles are the arithmetic mean of two subsequent recordings. The reproducibility of the experiments appears to be satisfactory.
Four successive velocity profiles are given in Fig. 5. In the figure the measured velocities have been made dimensionless. The reference value is the core velocity predicted by the Szymanski equation.

$$v_r = \frac{\Delta p R^2}{4 \mu L} \{ 1 + 8 \sum_{n=1}^{\infty} \frac{1}{\alpha_n^3 J_o'(\alpha_n)} \exp(\frac{-\alpha_n^2 \mu t}{\rho R^2}) \} \qquad (5)$$

The circles and dots each represent one half of the tube diameter.

The agreement between experiment and theory is good.
In Fig. 6 the further development of the profiles is shown. Two and a half seconds after the start of the transient the theory still fits, but one second later the measured velocities are too low. The deviation can be explained by considering the assumptions of the analytical solution. The pressure drop responsible for the liquid motion remains constant during the entire acceleration. In the experiments minor (entrance) losses continuously become more important and the available pressure drop along the pipe decreases.

In the forty runs made during the experiments the onset to turbulence takes place about 4.2 seconds after the start. The reproducibility is within 0.1 seconds. Velocity profiles before (dots) and after (open circles) the transition are given in Fig. 7. The line in the figure represents the measured steady-state profile.
The LDA recordings show that total redistribution of momentum takes place after the initiation of turbulence. Almost instantaneously the shape of the final profile has been approached.

From the measured profiles the increase of the mean velocity can be calculated numerically.
The dimensionless velocity on the abscis in Fig. 8 is the ratio of the time dependent mean and the final steady state velocity. On the ordinate a dimensionless time (Fourier Number for momentum) is plotted.

The results can be compared with two theoretical models. Line 1 is calculated with the aid of the quasi-steady approach.
Then it is assumed that at a Reynolds Number of 2,300 the transition to turbulence takes place. In line 2 (Szymanski) no transition is assumed. The final steady velocity evaluated with the latter approach is therefore unrealistic and much too high. Another model which is also known in the literature is not presented here. The theoretical analysis of Zielke (Ref. 7) should give the same results as given in line 2. The calculated shear stresses are a numerical approximation of the exact solution of Szymanski.

For short time intervals both lines fit the data very well. The contribution of the dissipative losses in the two approximations is totally different. However, shortly after the start of a transient the acceleration or deceleration is mainly determined by the inertia forces and a significant deviation between the curves cannot be expected.
For longer periods the frictional losses calculated with the quasi-steady approach apparently delay the flow and the predicted velocities are too low. Line 2 matches the experimental data until minor losses become important.

The point of transition to turbulence has been marked in Fig. 8. At that moment already 95% of the final velocity has been reached. The corresponding Reynolds Number is 57,500. This value will be dependent on the diameter of the pipe.
From the experiments described it is shown that somewhere in the final stage of the acceleration the onset to turbulence takes place. For larger pipe diameters the value of this critical Reynolds Number will increase.

Conclusions

The development of the flow pattern of a liquid column during acceleration from rest has been investigated. The results prove that accelerating flows are of a stable character. The transition to turbulence is strongly delayed. Visualisation experiments show that instabilities in the shape of the velocity profile are responsible for the initiation of turbulence. The originations of turbulent eddies can be made clearly visible and accelerating flows are therefore most suitable for fundamental studies of turbulence.
After the transition to turbulence the final steady state velocity has almost been reached. The time to establish a fully developed turbulent flow pattern is shorter than could be expected from any known theory.

The frictional losses during transient flows, deduced from the measured velocity profiles deviate significantly from the steady-state values. The lack of theoretical models properly describing the unsteady dissipation processes is shown by the experiments.

References

1 Wylie, E.B. and Streeter, V.L.: "Fluid Transients". McGraw-Hill Inc. (1978).

2 Wood, D.J. and Kao, T.Y.: "Evaluation of Quasi-steady approximation for viscous effects in unsteady liquid pipe flow". ASME paper 68-FE-33 (1968).

3 Daily, J.W., Hankey, W.L., Olive, R.W. and Jordaan, J.M.: "Resistance coefficients for accelerated and decelerated flows through smooth tubes and orifices". Trans-ASME 78, pp.1071-1077 (1956).

4 Carstens, M.R. and Roller, J.E.: "Boundary-shear stress in unsteady turbulent pipe flow". J. Hydr. Division HY2, pp. 67-81 (1959).

5 Denisov, S.V. : "The friction coefficient in non-stationary flows". J. Eng. Phys. 18, part 1, pp. 88-92 (1970).

6 Popov, D.N.: "Research into unsteady fluid flow during transient processes in a short pipe". Russian Eng. J. 54, 6, pp. 7-9 (1974).

7 Zielke, W.: "Frequency-dependent friction in transient pipe flow". J. Basic Eng. 90, pp. 109-115 (March 1968).

8 Wood, D.J. and Funk, J.E.: "A boundary-layer theory for transient viscous losses in turbulent flow". J. Basic Eng. 92, pp. 865-873 (December 1970).

9 Kawamura, H.: "Transient hydraulics and heat transfer in turbulent flow". Nucl. Techn. 30, pp. 246-255 (September 1976).

10 Maruyama, T., Kuribayashi, T. and Mizushina, T.: "The structure of turbulence in transient pipe flows". J. Chem. Eng. Japan 9, 6, pp. 431-439 (1976).

11 Kataoka, K., Kawabata, T. and Miki, K.: "The start-up response of pipe flow to a step change in flow rate". J. Chem. Eng. Japan 8, 4, pp. 266-271 (1975).

12 Letelier, M.F. and Leutheusser, H.J.: "Skin friction in unsteady laminar pipe flow". J. Hydr. Division HY1, pp. 41-56 (1976).

13 Szymanski, P.: "Some exact solutions of the hydro-dynamic equations of a viscous fluid in the case of a cylindrical tube". (Quelques solutions exactes des equations de l'hydrodynamique du fluide visqueux dans le cas d'un tube cylindrique). J. Math. Pures et Appl. 11, pp. 67-107 (1932) (In French).

14 Chambré, P.L., Schrock, V.E. and Gopalakrishnan, A.: "Reversal of laminar flow in a circular pipe". Nucl. Eng. Design 47, pp. 239-250 (1978).

15 Durst, F., Melling, A and Whitelaw, J.H.: "Principles and practice of Laser-Doppler Anemometry". Acedemic Press (1976).

16 Davis, W. and Fox, R.W.: "An evaluation of the hydrogen bubble technique for the quantitative determination of fluid velocities within clear tubes". J. Basic. Eng. 89, pp. 771-781 (December 1967).

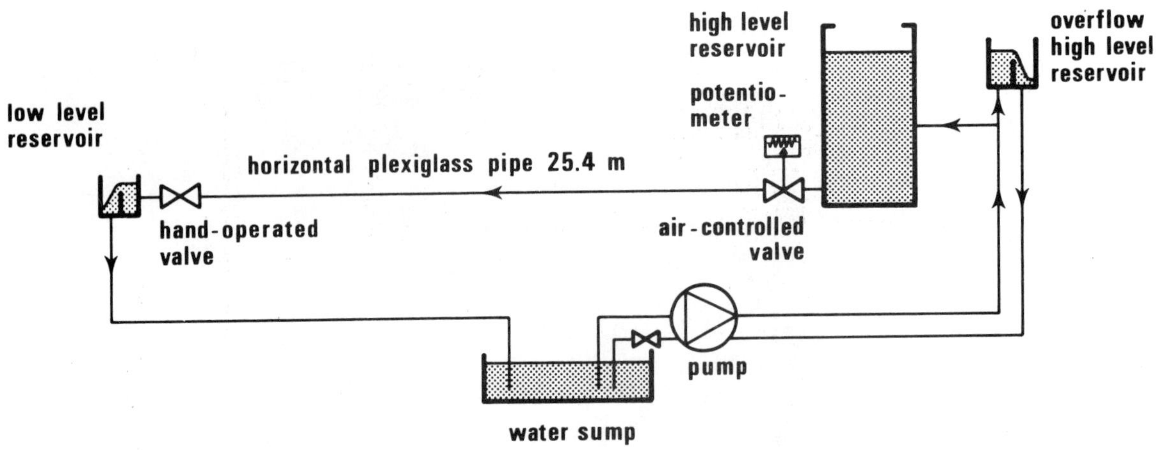

Fig.1 SCHEMATIC DIAGRAM OF THE EXPERIMENTAL SET-UP

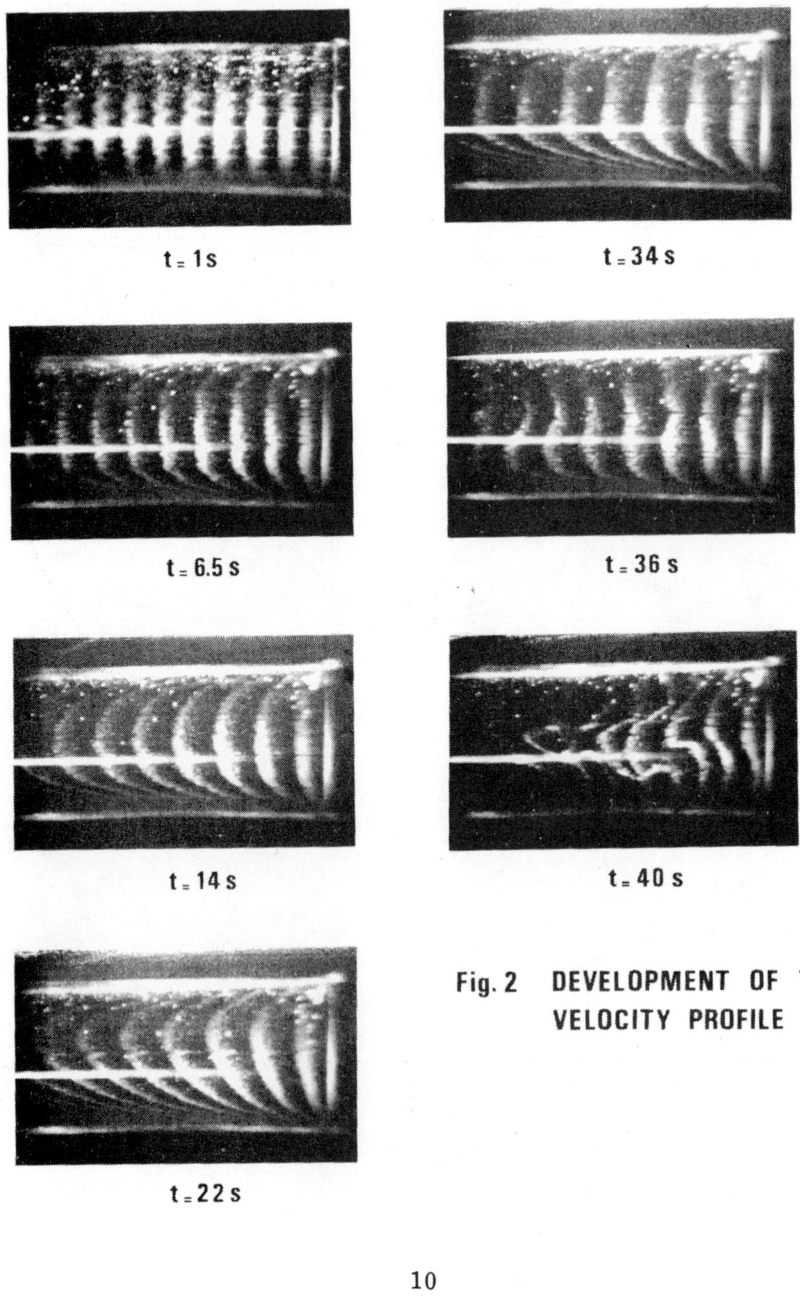

Fig. 2 DEVELOPMENT OF THE VELOCITY PROFILE

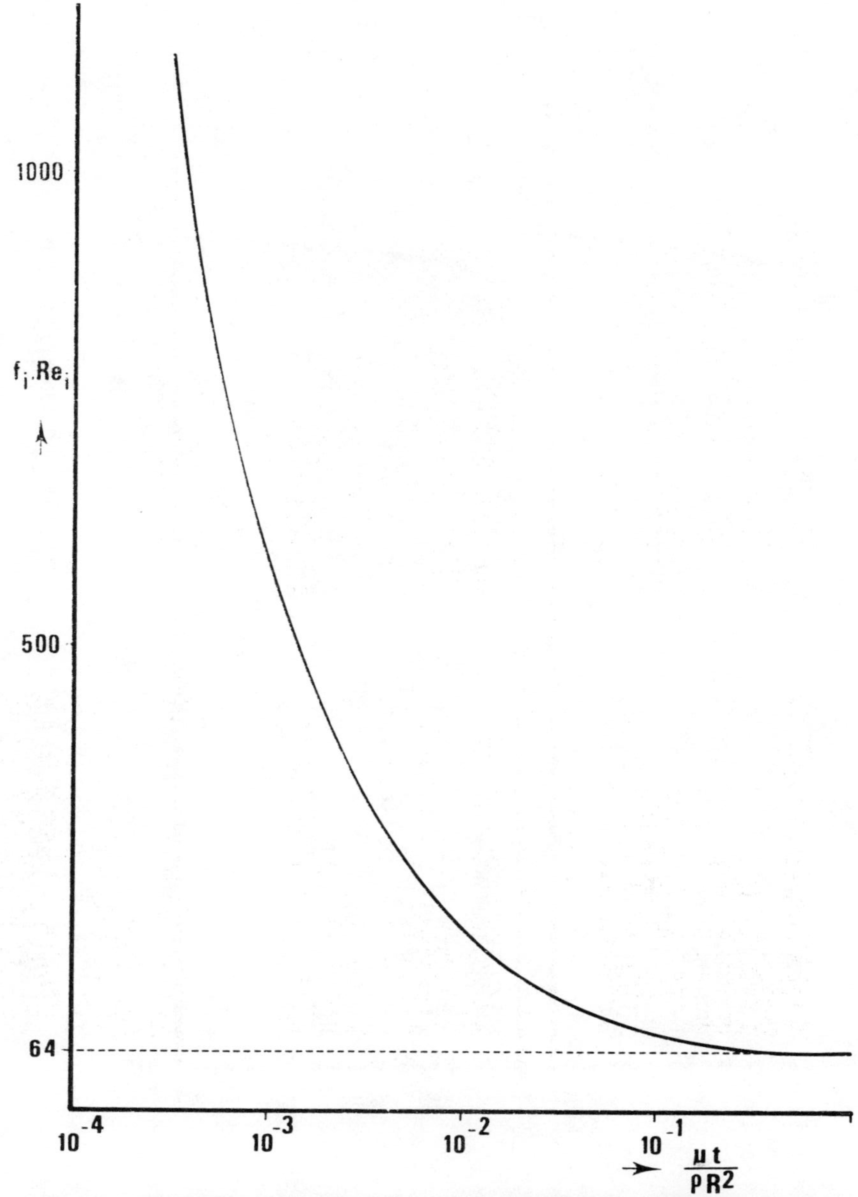

Fig.3 FRICTION LOSSES IN ACCELERATING LAMINAR FLOW

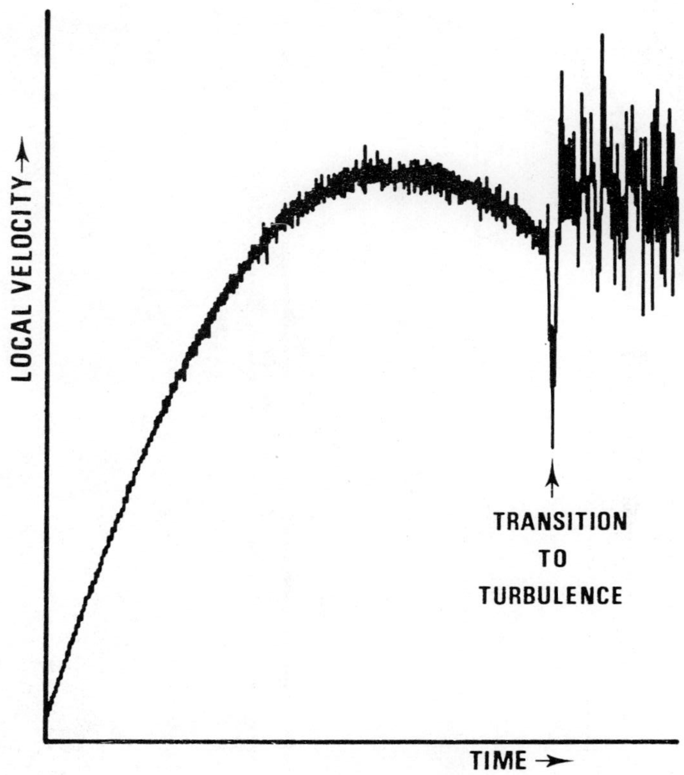

Fig.4 LDA RECORDING

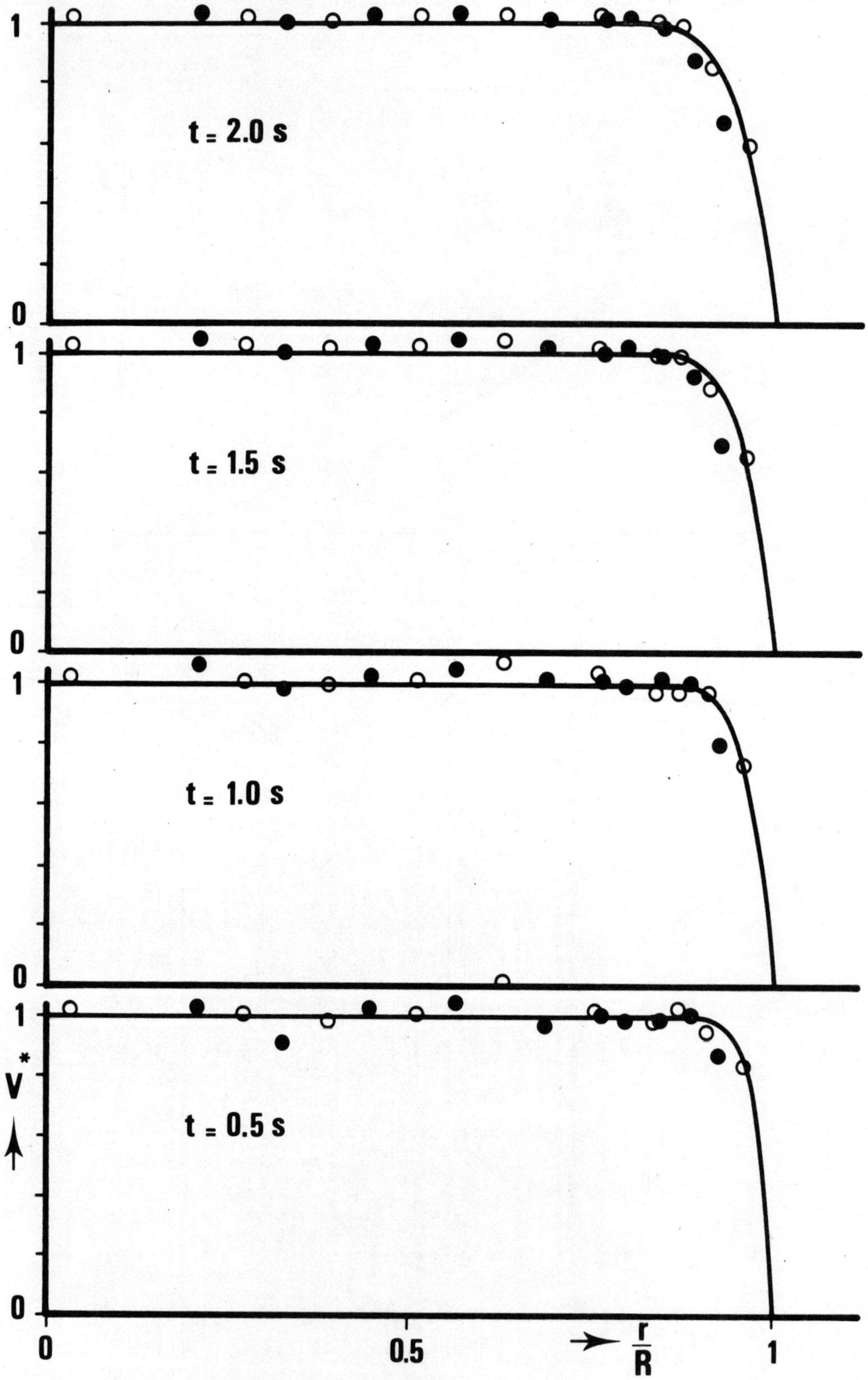

Fig.5 INSTANTANEOUS VELOCITYPROFILES

(— SOLUTION OF SZYMANSKI)

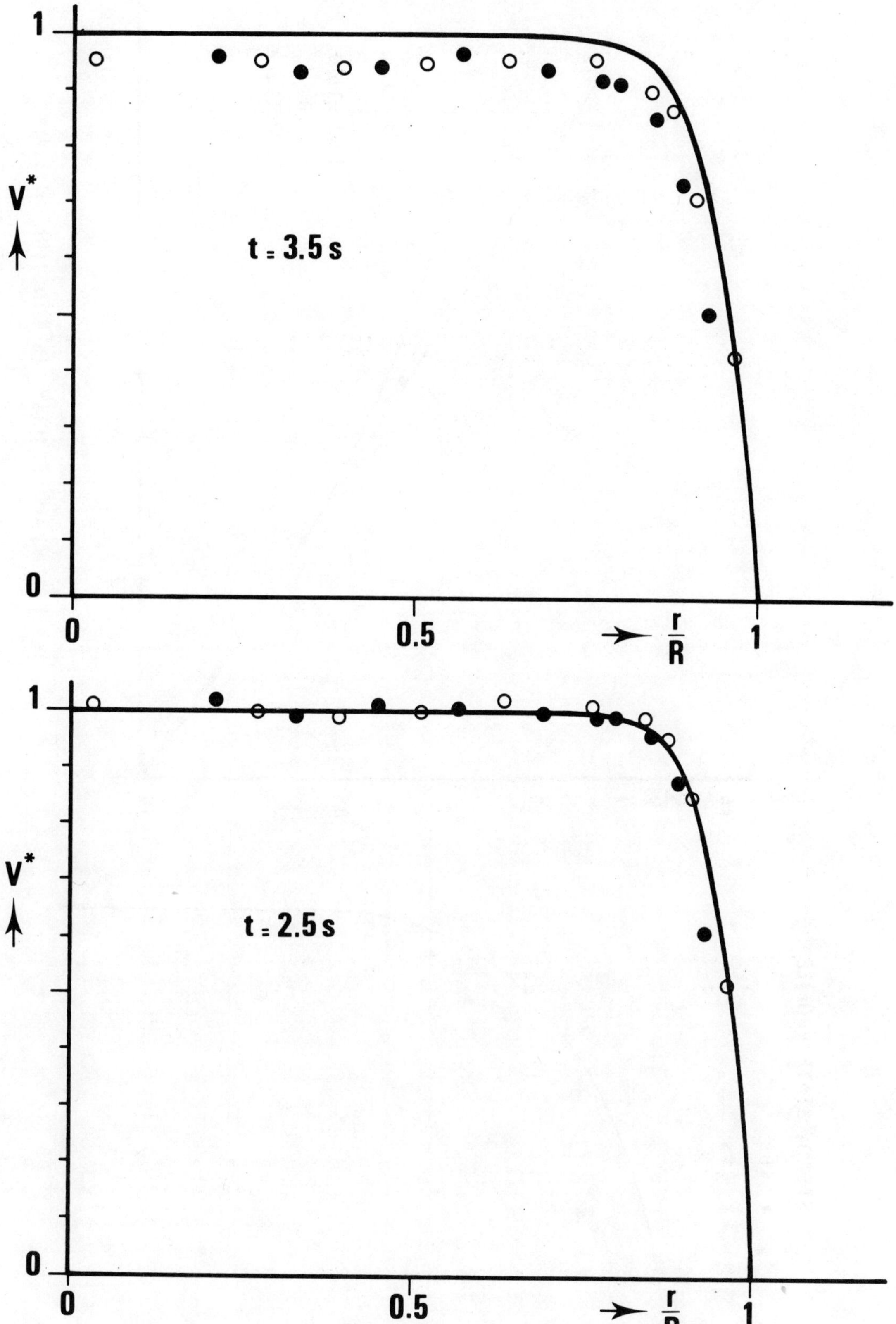

Fig.6 INSTANTANEOUS VELOCITYPROFILES

(— SOLUTION OF SZYMANSKI)

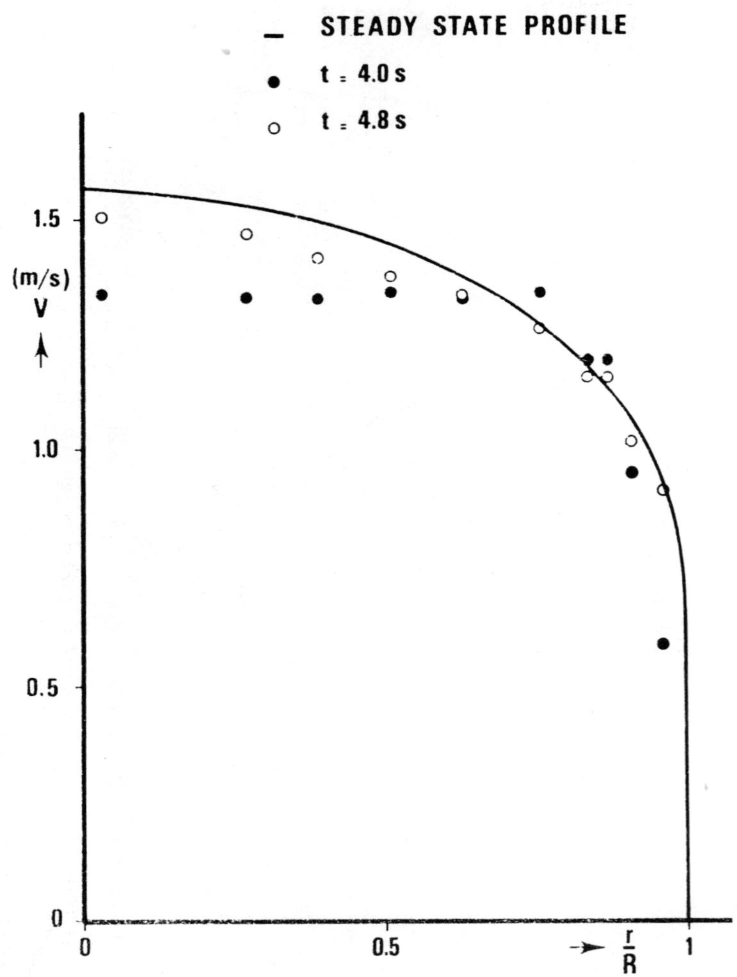

Fig. 7 INSTANTANEOUS VELOCITYPROFILES

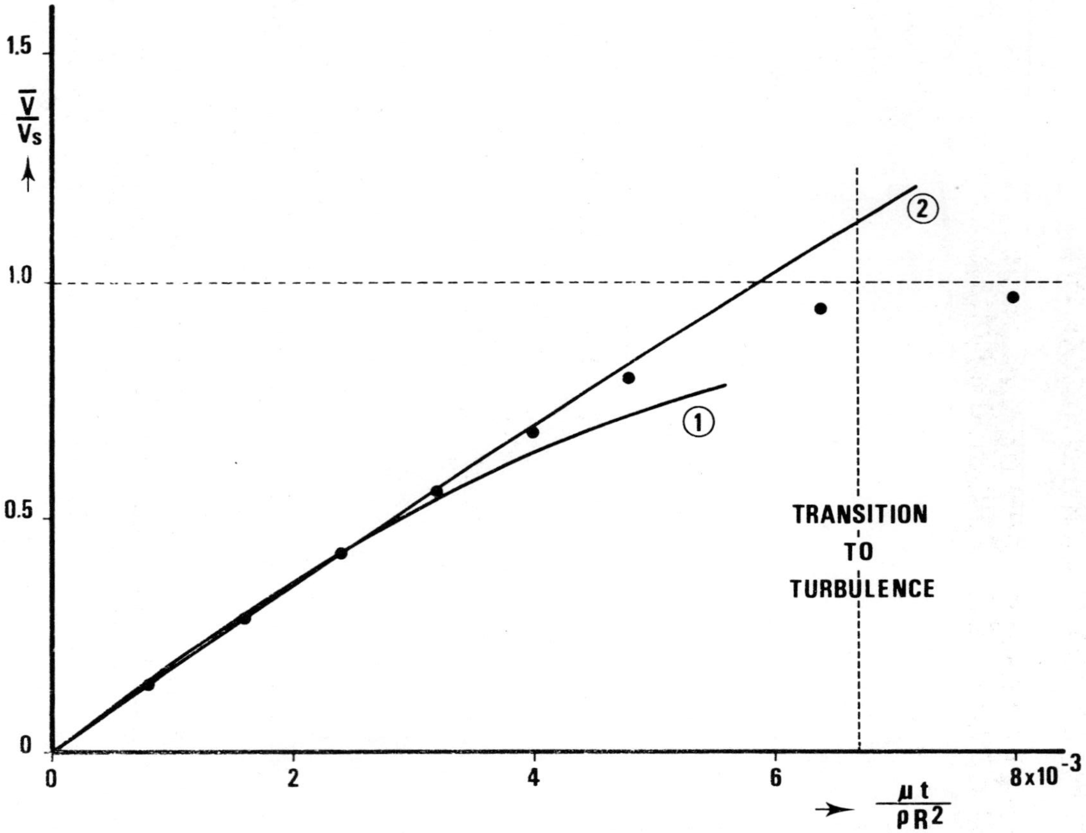

Fig. 8 MEAN VELOCITY AS FUNCTION OF TIME

Third International Conference on
PRESSURE SURGES
Canterbury, England
March 25-27, 1980

PAPER A2

UNSTEADY FLOWS: FACT AND FRICTION

A.F. Vardy

Dundee University, U.K.

Summary

Discrepancies between theory and experiment in the simulation of unsteady air-flows in railway tunnels are attributed to the derivation of the momentum equation. A possible cause of the discrepancies is the manner in which unsteady skin friction is simulated. It is argued that the representation of unsteady skin friction is more difficult in a numerical simulation than in an analytical simulation of the airflows. An approximate formula is proposed which is shown to have the correct influence on the wavespeed. It is concluded that this formula represents a possible starting point for simulating skin friction in numerical solutions of highly unsteady flows.

Held at the University of Kent at Canterbury, U.K.

Conference Organised by BHRA Fluid Engineering, Cranfield, Bedford, England.
In conjunction with The City University, London.

NOTATION

a	cross-sectional area
A,B	parameters in equations (7) and (8)
c	sound speed relative to fluid particles
C_F	skin friction coefficient for steady-state flows
K_1, K_2	coefficients in equation (10)
ℓ	cross-sectional perimeter
L	length of control volume
p	pressure
t	time co-ordinate
$\bar{U}$	mean fluid velocity at a cross-section
x	distance co-ordinate
β	momentum flux correction coefficient
γ	ratio of principal specific heats of air
ρ	fluid mass density
τ_w	wall shear stress
τ_{ws}	wall shear stress in steady flow
τ_{wu}	the difference $(\tau_w - \tau_{ws})$

1. INTRODUCTION

The class of unsteady flows is very large. It can be broadly categorised into two sections, namely slowly varied and rapidly varied flows. This paper deals primarily with the latter category. In railway tunnels the unsteadiness is caused by a piston (train) suddenly entering a long cylinder (tunnel). When the nose and tail of the train enter or leave the tunnel, strong flow disturbances are caused and pressure waves propagate along the tunnel. Authors in several countries have developed computer programs in which the unsteady airflows are simulated and a reasonable measure of agreement has been achieved. However, the analyses all fail to correctly predict the speed at which pressure waves propagate along the tunnel and they also underestimate the attenuation of the pressure waves in the system.

The discrepancies between theory and experiment are almost certainly due to theoretical inadequacies because they are consistently present for different experimental measurements, notably even when great care has been taken to simplify the experimental configuration, e.g. Gawthorpe & Pope (Ref. 3). In the past, relatively little attention has been paid to the discrepancies because the analyses have been sufficiently accurate to give good estimates of the influence of major geometrical changes to the system, for example, the provision of ventilation shafts in a tunnel. However, it is now felt that there is a need to improve the analyses sufficiently to be able to predict accurately what happens when, say, pressure waves interact with passing trains within a double-track tunnel. In particular, it is important that the speed of transmission of the pressure waves is correctly reproduced because this determines whether or not particular waves will pass alongside trains as they meet.

For economic reasons, it is important that the method used to improve the analyses should not involve great increases in complexity. For most purposes, a simple approximate method will be more valuable than a complex accurate model even though the latter is far superior from an academic viewpoint.

For many years it was assumed that the major cause of the discrepancies was the simple manner in which the energy equation is usually simulated. It has been normal to approximate the effect of the energy equation by assuming that the pressure and density changes satisfy isentropic relationships. This is clearly not exact when friction and/or heat transfers are present, as has been clearly demonstrated by Hammitt (Ref. 4). However, Woods & Pope (Ref. 8) have recently shown that the principal errors are not due to this cause alone. On the contrary, it can be inferred from their work that pseudo-homentropic analyses *should* be capable of giving improved predictions of unsteady airflows in railway tunnels. It is therefore necessary to look elsewhere for the principal source of error.

The object of this paper is to demonstrate that the error could be a result of the representation of unsteady skin friction in the momentum equation. Firstly, however, it is necessary to briefly indicate the manner in which airflows in railway tunnels are currently simulated.

2. REPRESENTATION OF AIRFLOWS IN RAILWAY TUNNELS

Figure 1 illustrates a railway train passing through a tunnel. Three main regions of flow can be identified, namely (1) in front of the train, (2) alongside the train and (3) behind the train. These regions are bounded by the tunnel portals, A & D, and by the nose and tail of the train, B & C. Within the regions 1, 2 and 3, the continuity and momentum equations may be written for horizontal flow as

$$\frac{\partial \rho}{\partial t} + \bar{U}\frac{\partial \rho}{\partial t} + \rho\frac{\partial \bar{U}}{\partial x} = 0 \qquad (1)$$

$$\frac{\partial p}{\partial x} + \tau_w \frac{\ell}{a} + \frac{\partial}{\partial t}(\rho\bar{U}) + \frac{\partial}{\partial x}(\beta\rho\bar{U}^2) = 0 \qquad (2)$$

respectively, in which

a & ℓ are the cross-sectional area and perimeter of the rigid tube (tunnel),
p, ρ, and $\bar{U}$ are the pressure, density and mean velocity of the air,
β is the momentum flux correction coefficient,
τ_w is the shear stress at the wall and
x, t are the distance and time coordinates respectively.

These equations are linked by assuming that the pressure and density fluctuations satisfy standard thermodynamic relationships for the isentropic flow of a perfect gas.

Following the usual procedure, the momentum flux correction coefficient, β, is approximated to unity. Possible errors resulting from this simplification will be discussed in a separate publication. Useful information has been given by Brown (Ref. 1). The only other approximation required in (1) and (2) is the representation of the skin friction. Usually it is assumed that the instantaneous unsteady skin friction differs negligibly from the steady state skin friction corresponding to the same local mean velocity. An expression such as

$$\tau_w = \tau_{ws} = \frac{1}{2} C_F \rho \bar{U} |\bar{U}| \quad (3)$$

is normally used in which C_F is regarded either as a constant or as some function of the Reynolds number and the ratio of the roughness size to the pipe diameter. No allowance is normally made for the influence of unsteadiness.

At the boundaries (A, B, C, D in Fig. 1), the instantaneous flow is represented by steady state continuity and Bernoulli equations, allowance being made empirically for stagnation pressure loss coefficients in the usual way.

2.1 Selection of empirical coefficients

Figure 2 shows the pressure histories at three positions inside a railway tunnel as a train passes through at constant speed. The agreement between experiment and theory is reasonable, but this is partly due to the careful choice of stagnation pressure loss coefficients. In particular, the loss coefficient at the train nose is chosen so that the pressure rise generated when the nose enters the tunnel is correctly reproduced. The tunnel skin friction coefficient is chosen to ensure that the pressures and velocities at a late stage in the journey are well represented, and the train skin friction coefficient is chosen to ensure that the gradual pressure rise shortly after nose entry is well simulated. Finally, a loss coefficient must be assumed at the tail of the train. This is normally taken to be equal to the well known expansion loss from a small duct to a larger duct.

2.2 Discrepancies between theory and experiment

Figure 2 shows at least two systematic errors between theory and experiment. Firstly, the wavespeed is incorrectly predicted; a phase difference exists between the theoretical and the experimental pressure traces. Secondly, the attenuation of the pressure waves as they pass along the tunnel is underestimated in the theoretical analysis. It will be shown that both of these effects could be at least partly due to inaccuracies in the method of representing skin friction.

3. APPROXIMATE FORMULA FOR UNSTEADY SKIN FRICTION

Suppose that a fully developed flow with a mean velocity $\bar{U}_1$ is disturbed by a wavefront which causes a single step change in velocity to a new mean velocity $\bar{U}_2$. Figure 3 depicts a control volume drawn relative to the wavefront which moves with a speed U_w relative to the tube. The section 3 is an arbitrary distance behind the wavefront. Between the sections 2 and 3, there is no change in the mean velocity (by stipulation) and so the axial pressure distribution balances the skin friction. The steady flow momentum equation for the control volume is

$$(p_3 - p_1) - \frac{\ell}{a} \int_0^L \tau_w \, dx = \rho_1 (\bar{U}_1 - U_w)(\bar{U}_1 - \bar{U}_3) \qquad (4)$$

in which L denotes the length of the control volume.

In order to make use of this equation, it is useful to express τ_w as

$$\tau_w = \tau_{ws} + \tau_{wu} \qquad (5)$$

in which τ_{ws} and τ_{wu} denote the steady and unsteady contributions respectively. The value of τ_{ws} may be obtained from (3) in the usual manner, and so the problem reduces to the evaluation of $\int \tau_{wu} dx$ over the length L. No exact integration is presented in this paper, but a useful guide can be obtained from a qualitative study of the physical behaviour of the flow.

3.1 Velocity profiles in unsteady flow

Figure 4 depicts the velocity profiles (1) in front of, (2) just behind and (3) far behind

(a) a compression wavefront accelerating the flow,
(b) a rarefaction wavefront accelerating the flow,
(c) a compression wavefront decelerating the flow, and
(d) a rarefaction wavefront decelerating the flow.

The profiles are not exact. They illustrate the notion that the wavefront initially influences only the region very close to the wall, but that inward diffusion of vorticity subsequently relaxes the profile to a new steady state condition. These simple illustrations are supported by the predictions of Wood & Funk (Ref. 6) who used a theoretical model of turbulent flows based on a laminar boundary layer surrounding an inviscid core. By inspection, wavefronts of type (a) and (b) accelerate the flow and cause a temporary increase in τ_w in the region 2 in comparison with the steady state value (N.B. τ_w is regarded as positive in the negative x-direction). Similarly, wavefronts of type (c) and (d) cause a deceleration accompanied by a temporary reduction in τ_w. This simple description leads to a clear conclusion that the *sign* of τ_{wu} is the same as the sign of the overall velocity change, irrespective of the direction of propagation of the wavefront.

The whole of the region of varying shear stress is behind the hypothesised step change in velocity. Since there is no acceleration of the mean flow in this region, it follows that the shear stress *cannot* be a function of $\bar{U}$ alone, and neither can it be a function of local derivatives of $\bar{U}$. The wall shear stress *must* be a function of the flow history. It therefore appears to be necessary to adopt an expression such as that developed by Zielke (Ref. 11) who used weighted contributions of the past history of the flow to describe the shear stress. Zielke's analysis is especially useful because it can be used in conjunction with the Method of Characteristics. However, its application to railway tunnel airflows is not straightforward because discontinuities (trains) are present within the (x-t) domain. These prevent the evaluation of some of the velocities required to describe the flow history in Zielke's expression.

In practice, a more important limitation exists. The principal disturbances to the airflows are very rapid indeed, as can be seen by inspection of Fig. 2. For example, the wavefront caused by nose entry to a tunnel is typically generated in about 0.1 seconds. For economic reasons, this is usually less than the time increment chosen for each step of the numerical integration. It follows that the whole of the velocity change associated with the most important disturbances is generated within one time step. Therefore, expressions requiring a knowledge of the flow history during the passage of the wavefront cannot be used. Instead, an expression must be found that relates the unsteady skin friction to the overall velocity change.

An alternative approach would be to analyse the airflows using sufficiently small time increments for expressions describing the flow history to be used. In the long term, this may be a viable option, but in the short term it is discarded on economic grounds. The computational cost involved is unacceptable.

3.2 Numerical integration

In the above discussion, it was shown that the *sign* of τ_{wu} is the same as the sign of the overall velocity change, $\Delta\bar{U}$. Since the *magnitude* of τ_{wu} can be expected to increase with the magnitude of the velocity change, it is reasonable to look for a formula which, on integration, will lead to a result of the form

$$\int_0^L \tau_{wu}\, dx = \rho\, A\, \Delta\, \bar{U} \tag{6}$$

in which A is an unknown parameter with the dimensions $[L^2\ T^{-1}]$. Of course, it is possible to hypothesise many different forms for (6), but this simple expression is sufficient for present purposes. It is the *influence* of unsteady skin friction, not its precise representation, that is the primary subject of this paper.

A possible expression for τ_{wu} that would yield (6) on integration if A is constant is

$$\tau_{wu} = \rho A \frac{d\bar{U}}{dx}\ . \tag{7}$$

Since this derivation has been based on flows relative to the wavefront, the corresponding expression for τ_{wu} *relative to the tube* is

$$\tau_{wu} = \rho B \frac{d\bar{U}}{dt} \tag{8}$$

in which B is a constant with the dimensions of a length.

On substituting (3) and (8) into (5), we obtain

$$\tau_w = \frac{1}{2}\rho\, C_F\, \bar{U}\,|\bar{U}| + \rho B \frac{d\bar{U}}{dt} \tag{9}$$

and it is interesting to compare this expression with the relationship used by Safwat & Polder (Ref. 5) who dealt with oscillating flows in a large U-tube. Their expression,

$$\tau_w = K_1\, \bar{U} + K_2 \frac{d\bar{U}}{dt} \tag{10}$$

was found to lead to a reasonable correlation with experiment, but the numerical values of the coefficients varied considerably with the frequency of oscillation. Indeed, even the sign of K_2 was not the same for all frequencies. Equation (10) is essentially of the same form as (5) because the outer region in the U-tube flow was dominated by viscous effects for which τ_{ws} is proportional to $\bar{U}$. Nevertheless, the work was not directly relevant to the present discussion because the dimensionless frequencies considered were all very low in comparison with the values present in the sharp wavefronts considered herein.

4. METHOD OF CHARACTERISTICS

Using (9), the equations (1) and (2) can be cast into their characteristic forms in the usual manner. By substituting $\beta = 1$ and using isentropic expressions to replace the density and pressure by the sound speed c, we obtain the compatibility relationships

$$\pm \frac{2}{\gamma - 1}\left(1 + \frac{B\ell}{a}\right)^{\frac{1}{2}} \frac{dc}{dt} + \left(1 + \frac{B\ell}{a}\right) \frac{d\bar{U}}{dt} = -\frac{1}{2}\frac{\ell}{a} C_F\, \bar{U}\,|\bar{U}| \tag{11}$$

which are applicable only in the characteristic directions

$$\frac{dx}{dt} = \bar{U} \pm c(1 + \frac{B\ell}{a})^{-\frac{1}{2}} \qquad (12)$$

respectively.

4.1 Wavespeed

The general form of the compatibility relationships (11) is similar to equation (4) which also describes flows relative to a wavefront. However, the value of U_w, which was an unknown in (4), is now seen from (12) to be dependent upon the unsteady contribution to the skin friction. This result is an inevitable consequence of the introduction of any additional function of $\partial\bar{U}/\partial t$ and/or $\partial\bar{U}/\partial x$ into either of (1) or (2). A similar procedure was used to good advantage by Yow (Ref. 10). He introduced an inertial multiplier to increase the time steps of integration in the analysis of transient flows in natural gas pipelines. No physical significance was claimed for the parameter.

5. RESULTS

The influence of the parameter B is illustrated in Fig. 5 which deals with the same train journey as that used in Fig. 2. By inspection, the magnitudes of the pressure fluctuations are not greatly changed, but there is a considerable improvement in the simulation of the phase of the peaks. To achieve this improvement, the optimum value of B has been deduced by trial and error in the same way as that used to choose the steady-state skin friction coefficients and the train nose stagnation pressure loss coefficient. The chosen value, 0.06 m, corresponds to a value of 0.05 for the product $B\ell/a$.

The magnitudes of the predicted pressures in Fig. 2 were shown to be dependent upon empirical loss coefficients which were chosen by trial and error to give the closest fit with the experimental results when B = 0. For clarity, Fig. 5 is presented with no change in any of the coefficients except B. This allows the influence of B to be inferred from a simple comparison of the two figures. However, it does not yield the closest possible agreement with experiment. For a true comparison to be made, it is necessary to choose new values for the original coefficients to give the closest fit with the experimental results when $B\ell/a = 0.05$. Figure 6 illustrates the best correlation obtained after adjusting C_F and the train nose stagnation pressure loss coefficient. The agreement with experiment is again improved.

5.1 Interpretation of the parameter B

By comparing Figs. 2 and 6, it is possible to deduce that the introduction of the parameter B leads to an improvement in the ability of the theoretical analysis to reproduce experimental results. However, it may not be acceptable to infer that equation (8) is a valid method of representing the unsteady contribution of skin friction. There are at least two important limitations that must be stressed.

Firstly, equation (6) is only one of an infinity of possible relationships that would satisfy the conditions stated in the paragraph preceding it. It is highly unlikely that the total additional shear stress due to unsteadiness is directly proportional to the velocity change. Even if this were the case, it would not be reasonable to neglect the length behind a wavefront that is needed to establish a fully developed velocity profile corresponding to the new mean velocity. (N.B. When (6) is used in a numerical solution, it is implicitly assumed that this length is less than the distance travelled by a wavefront during one time step of integration).

Secondly, equation (8) is merely a convenient expression which will yield (6) on integration between two sections a finite distance apart. However, in the development of the characteristic equations, it is assumed to be valid between sections separated by infinitesimal distances. As a result, the wavespeeds (12) are incorrectly predicted. This is an important conclusion because the value of B was sensibly chosen to give good phase agreement between theory and experiment.

Notwithstanding these limitations, the introduction of (8) has led to the possibility of improving the theoretical/experimental correlation. It may therefore be justifiable to develop this work by searching for more realistic forms of (6) and (8). One possibility is to repeat the work of, say, Wood & Funk (Ref. 8), using a wide range of solutions to develop more exact empirical expressions than (6) and (8).

Alternatively, use might be made of a more direct approach such as that currently being developed by Fox & Hunt. In a preliminary description of the type of analysis they are considering (Ref. 2), they describe an extension of earlier work by Vasiliev & Kvon (Ref. 6). A series expansion is assumed for the shear stress distribution at any cross-section, and the coefficients are functions of the rate of change of the mean flow. Fox & Hunt are using a polynomial expansion, but other series could be considered. The resulting expressions may be used in their own right, but they could also be used to develop improved forms of (8).

As previously indicated, it is important that the method chosen should not involve a large increase in computing costs. The results obtained using the simple expression (8) are already a useful improvement on the quasi-steady skin friction results. The advantages of further developments must exceed any disadvantages due to, say, increased computing costs.

6. CONCLUSIONS

Based on a qualitative discussion of the physical response of a fluid to a step wavefront, an expression (8) has been developed which appears to give an approximate description of the influence of rapid unsteadiness on skin friction.

The expression has been shown to have severe limitations, but it has nevertheless led to an improved correlation between theory and experiment for unsteady airflows in railway tunnels.

A possible future development leading to further improvements in the numerical solution of rapidly varied unsteady flows has been outlined.

7. ACKNOWLEDGMENT

The author is pleased to acknowledge the financial support given by the British Railways Board to allow this work to be carried out.

REFERENCES

1. Brown, F.T: "A quasi method of characteristics with application to fluid lines with frequency dependent wall shear and heat transfer". Trans. ASME, J. of Basic Engrg., 91, pp 217-227. (June 1969).

2. Fox, J.A. and Hunt, S.C: "A theory of unsteady friction in open channels". Proc. Eurovisc III, held at Liverpool University, U.K. (April 1979).

3. Gawthorpe, R.G. and Pope, C.W: "The measurement and interpretation of transient pressures generated by trains in tunnels". Proc. 2nd int. Symp. on the Aerodynamics and Ventilation of Vehicle Tunnels, BHRA, Cambridge. (March 1976).

4. Hammitt, A.G: "Unsteady aerodynamics of vehicles in tubes". AIAA Journal, 13 (4), pp 497-503. (April 1975).

5. Safwat, H.H. and v.d.Polder, J: "Friction-frequency dependence for oscillatory flows in circular pipes". Proc. ASCE, J. Hyd. Div., 99 (HY11), pp 1933-1945. (Nov. 1973).

6. Vasiliev, O.F. and Kvon, V.I: "Effect of nonstationarity in the motion of an open fluid flow". Zhurnal Prikladnoi Meckhaniki I, Tekhnicheskoi Fiziki, No. 1, pp 126-128. (1966).

7. Vasiliev, O.F. and Kvon, V.I: "Friction forces of unsteady flows in open channels and pipes". Proc. 14th Conf. IAHR, pp 187-196. (1971).

8. Wood, D.J. and Funk, J.E: "A boundary-layer theory for transient viscous losses in turbulent flow". Trans. ASME, J. Fluids Engrg, 92, pp 865-873. (Dec 1970). See also 96, pp 365-396. (Dec 1974).

9. Woods, W.A. and Pope, C.W: "On the range of validity of simplified one-dimensional theories for calculating unsteady flows in railway tunnels". Proc. 3rd int. Symp. on the Aerodynamics and Ventilation of Vehicle Tunnels, BHRA. Sheffield. (March 1979).

10. Yow, W: "Numerical error on natural gas transient calculations". Trans. ASME, J. Basic Engrg, 94, pp 422-428. (June 1972).

11. Zielke, W: "Frequency-dependent friction in transient pipe flow". Trans. ASME, J. Basic Engrg, 90, pp 109-115. (March 1968).

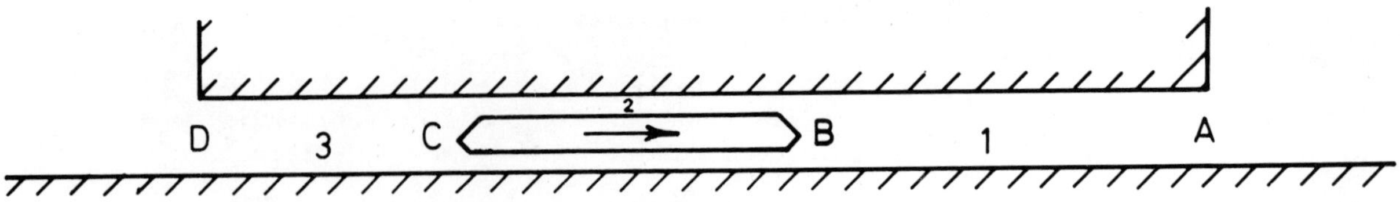

Fig. 1 Railway train in a tunnel

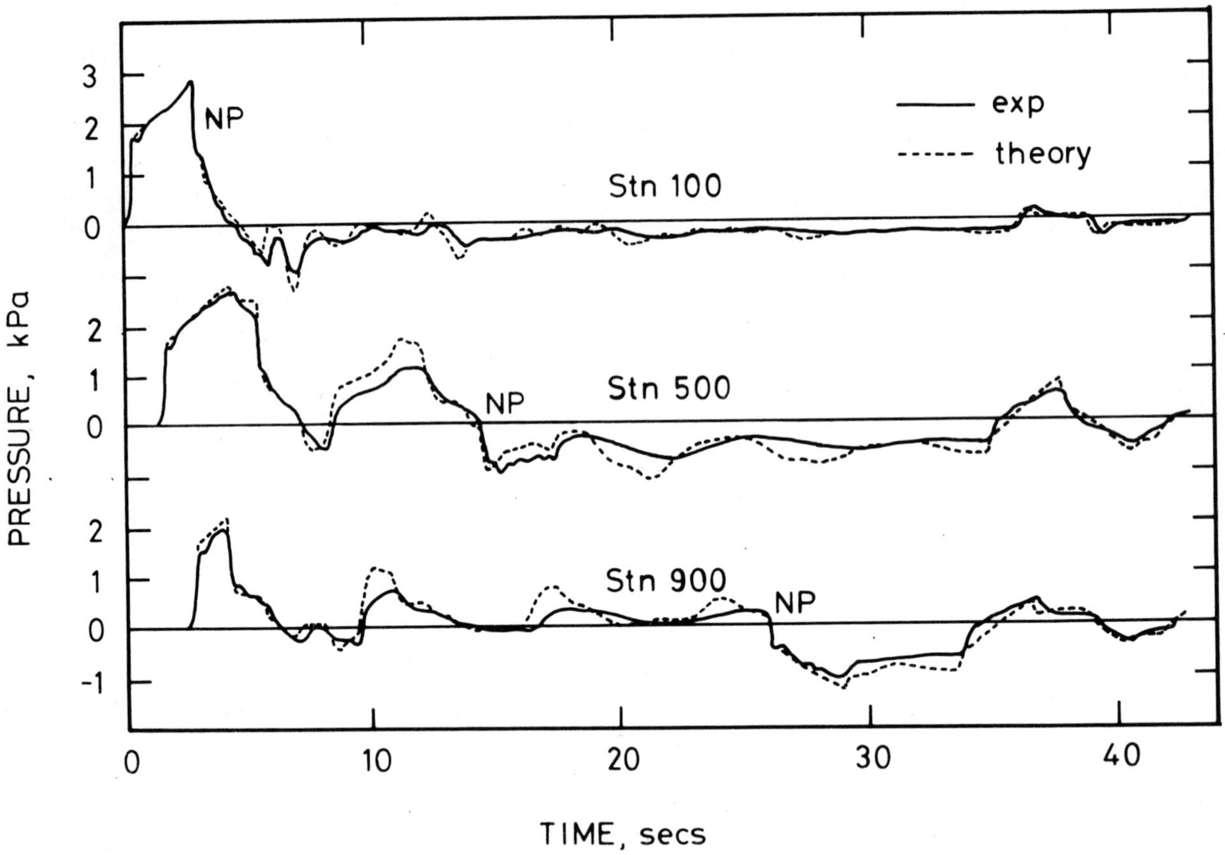

Fig. 2 Pressure histories 100m, 500m and 900m from the entrance of Patchway Tunnel when a train passes through at 34.7m/s. (After Gawthorpe & Pope, Ref. 3)

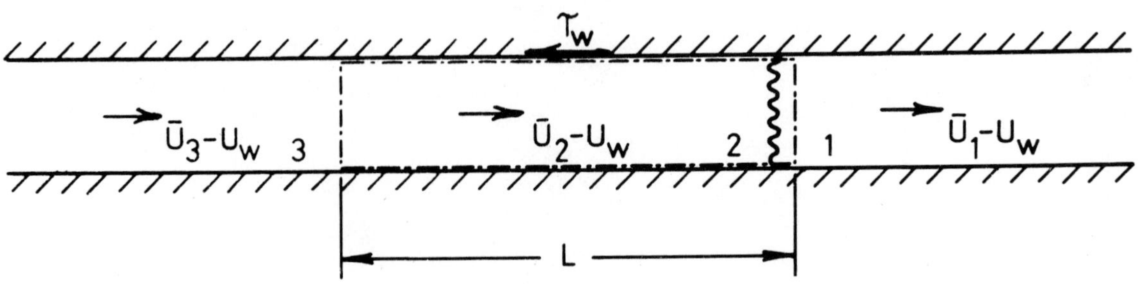

Fig. 3 Control volume relative to a wavefront causing a step change in velocity.

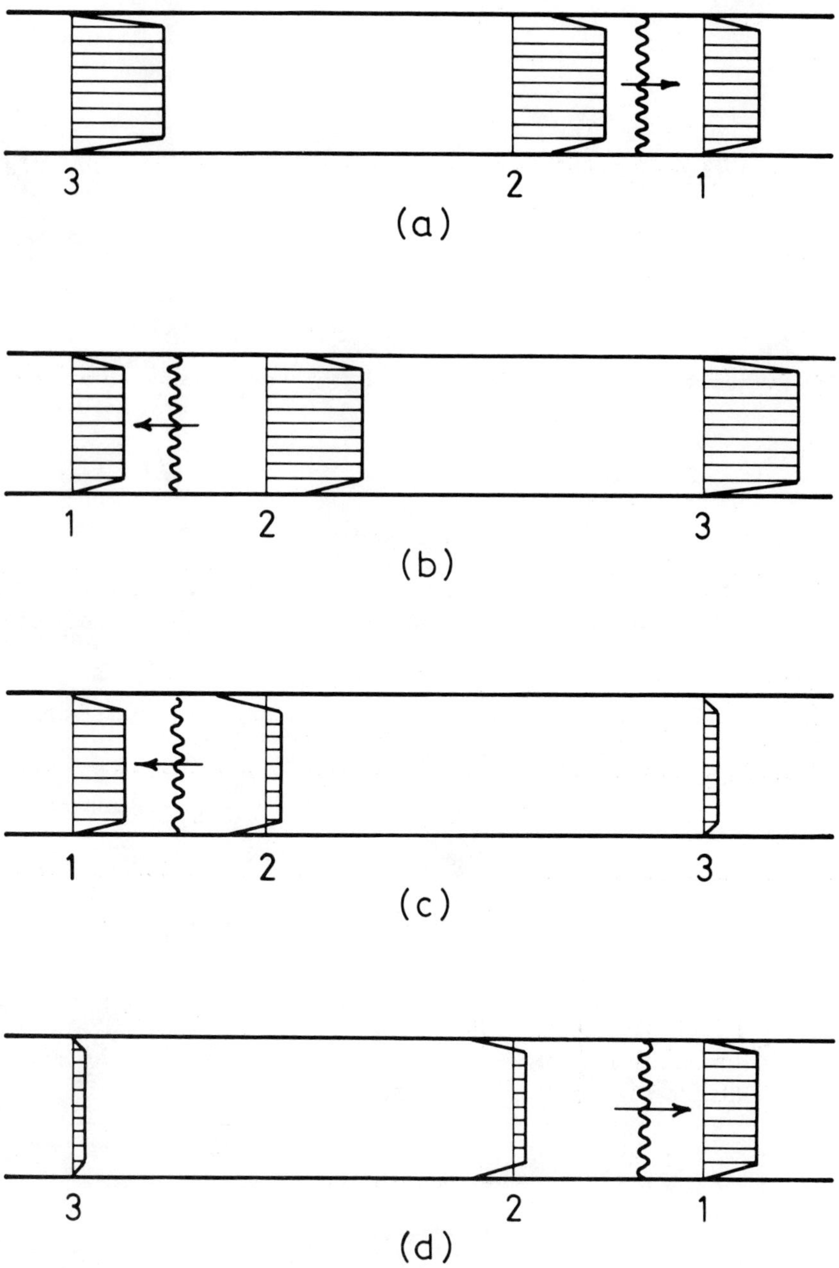

Fig. 4 Schematic representation of velocity profiles during the passage of a plane wavefront causing a step change in the mean velocity.

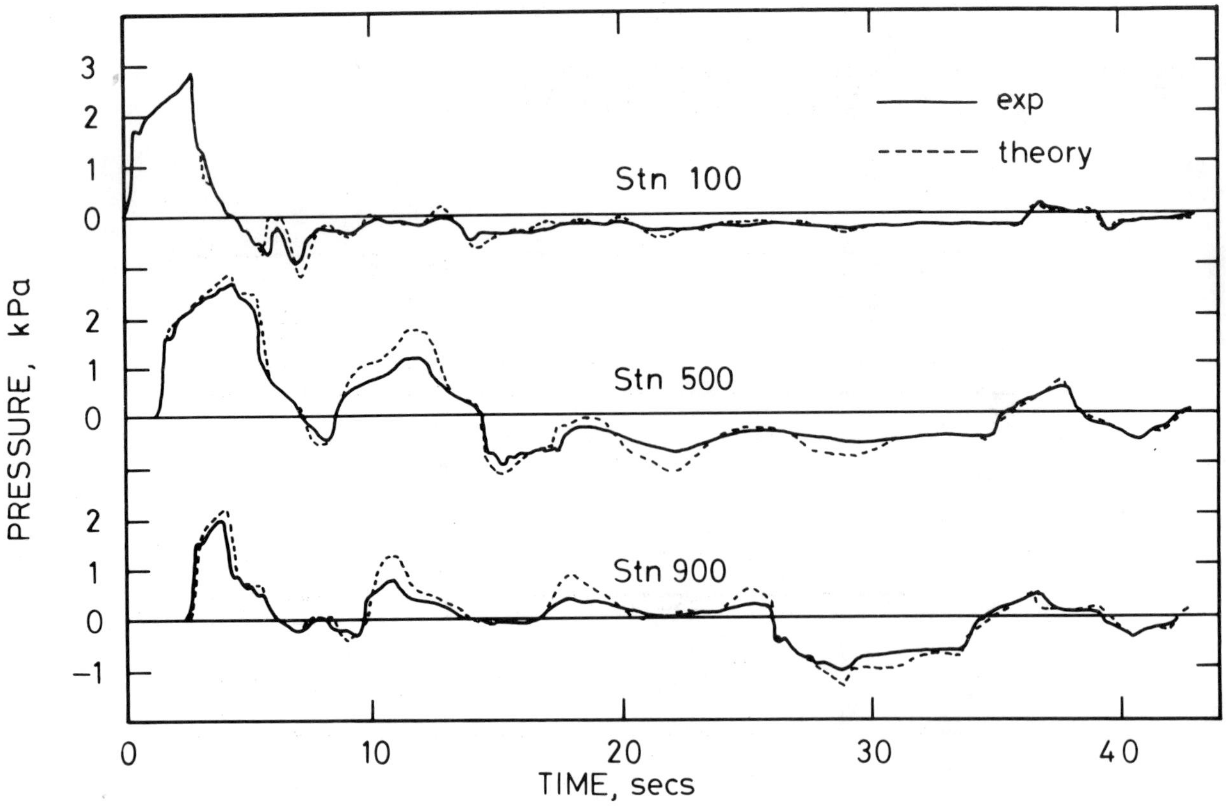

Fig. 5 Pressure histories in Patchway Tunnel. The phase correlation is greatly improved when account is taken of the dependence of skin friction on accelerations.

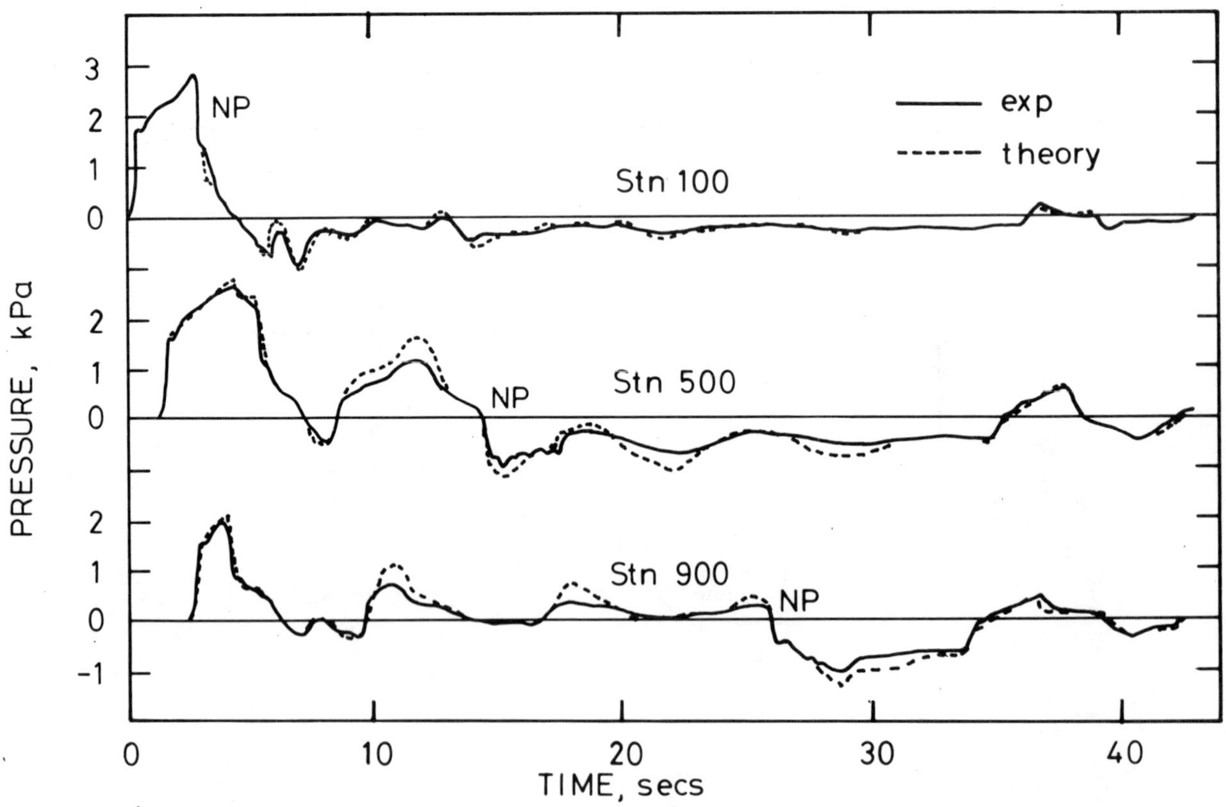

Fig. 6 Pressure histories in Patchway Tunnel. The phase correlation and the magnitudes are both improved when optimum values are obtained by trial and error for all empirical coefficients.

Third International Conference on

PRESSURE SURGES

Canterbury, England
March 25-27, 1980

PAPER B1

FREE AIR IN LIQUID TRANSIENT FLOW

E.B. Wylie

University of Michigan, U.S.A.

Summary

The presence of free air in a liquid, even in small quantities, has a profound influence on the pressure wave propagation velocity in the system, particularly at low pressures. The wave speed becomes strongly dependent on the local pressure magnitude: at high pressure it is the speed of sound in the liquid in the system, whereas at low pressure it may have a value less than the speed of sound in air.

This paper develops a numerical model for the treatment of a transient flow in a two-phase homogeneous mixture of low free-gas content, generally less than one percent at standard pressure. The approach utilizes the method of characteristics to transform the pair of one-dimensional partial differential equations to a suitable numerical form. A time-line interpolation scheme is utilized to reliably handle the highly-nonlinear fluid behaviour. A simplified analytical example demonstrates the dispersive character of the problem. Numerical examples illustrate the unusual behaviour of the positive and negative transient waves. The use of an overall continuity balance is recommended as an additional measure of computational accuracy.

Held at the University of Kent at Canterbury, U.K.

Conference Organised by BHRA Fluid Engineering, Cranfield, Bedford, England.
In conjunction with The City University, London.

NOMENCLATURE

a	accoustic wave speed in liquid system
a'	accoustic wave speed in two-phase mixture
A	cross-sectional area of flow passage
B	characteristic impedance of pipeline
C_2 C_3	constants
C_B	volume that expresses continuity unbalance
D	inside pipe diameter
e	pipe wall thickness
E	modulus of elasticity of pipe material
f	friction factor
g	gravitational acceleration
$\underline{H}$	absolute pressure head
H	barometric pressure head
K	bulk modulus of elasticity of the fluid
K'	system modulus of elasticity of the fluid
m	mass of free gas present per unit volume of mixture
p	absolute pressure
Q	volumetric flow rate
R	gas constant, also friction coefficient
t	time
T	absolute temperature, also total time of transient
V	velocity
∀	volume
x	distance along pipeline
z	elevation of pipeline
α	void fraction
θ	angle of pipeline, also angle of characteristic line
ρ	mass density

SUBSCRIPTS

g	gas
ℓ	liquid
m	integer used in summation
M	number of full Δt that characteristic extends
P	computation point where conditions are unknown
o	standard conditions, or initial conditions

INTRODUCTION

Liquids carrying free gas, even in very small volumetric proportions, have a significantly reduced wave-propagation velocity. Since the wave speed is one of the controlling parameters in a system transient, this reduced magnitude can easily be viewed to have a dominant influence on system response (Ref. 2,4). Systems carrying liquids with dissolved gases can be analyzed using the full wave speed as long as the gases remain in solution. Gas evolution occurs only after the ambient pressure drops below the gas saturation pressure. The rate of evolution is dependent upon a number of parameters including the gas solubility coefficient, degree of agitation of the fluid, pressure gradient, etc. Wiggert and Sundquist (Ref. 6) provide a good summary of the state-of-the-art relating to gas evolution and adsorption. In this study the mechanisms that control the release of dissolved gases are not addressed. It is assumed that free gas is present in the system initially and the mass of this gas remains constant throughout the transient. A gas evolution term, if introduced, would not change the basic procedure developed herein.

The free gas is assumed to be distributed throughout the pipe as a homogeneous bubbly-fluid mixture with gas bubbles and liquid moving at the same velocity. A void fraction is used to describe the ratio of volume of free gas to the mixture volume, and for a given mass of free gas it is pressure dependent.

$$\alpha = \forall_g/\forall \tag{1}$$

Only small void fractions at standard conditions are considered, generally less than one percent. At this level the bubbles expand and contract and may coalesce, but it is assumed that they remain uniformly dispersed throughout the pipe. It is also assumed that the momentum of the gas component relative to the liquid component is small and may be neglected.

The primary focus in this study is on the unique response generated during transients in fluids that behave in a highly nonlinear manner. The dispersion of pressure waves is not a common feature in most transients, particularly in cases of low viscous behavior. Similarly the development of shock waves and their recognition is of considerable interest.

The wave propagation velocity as a function of pressure in a two-phase mixture is first presented. The equations to describe the unsteady one-dimensional fluid behavior are then applied to a frictionless horizontal system where an analytic solution to the negative wave problem can be developed for comparison with numerical solutions. The positive wave is also considered in the simplified problem. The need for an iterative solution is illustrated with an example. Numerical results are used to illustrate different solution procedures, continuity checks, and positive and negative wave responses in both the simplified and complete equation problems.

BASIC EQUATIONS

The wave propagation velocity in a fluid in a pipeline, including the effects of pipe wall elasticity, is given by

$$a' = \frac{\sqrt{K/\rho}}{\sqrt{1 + KD/Ee}} \tag{2}$$

in which K = bulk modulus of elasticity of the fluid, ρ = mass density of the fluid, D = pipe inside diameter, e = pipe wall thickness, and E = modulus of elasticity of the pipe wall material (Ref. 10). The thin-walled pipeline is assumed to behave in a linear elastic manner. By considering a mass m of free gas present per unit volume of mixture and, assuming that it compresses isothermally, the void fraction is seen to vary inversely with the absolute pressure p.

$$\alpha = \frac{mRT}{p} \qquad (3)$$

The gas constant and absolute temperature are given by R and T, respectively. The dependence of K on void fraction and pressure is given by

$$K = \frac{K_\ell}{1 + \alpha\ (K_\ell/p - 1)} \qquad (4)$$

Again isothermal behavior of the gas is assumed, $K_g = p$. The subscripts ℓ and g refer to the liquid and gas components, respectively. For practical transient problems $K_\ell/p >> 1$ so the denominator of Eq. (4) becomes $1 + \alpha K_\ell/p$. The mass density of the two-component fluid also varies with void fraction but the variation is small enough to be neglected in cases of low void fraction, thus $\rho \approx \rho_\ell$. With these assumptions Eq. (2) may be expressed (Ref. 5,10)

$$a' = \frac{a}{\sqrt{1 + C_2 m/p^2}} \qquad (5)$$

in which a is the wavespeed in the liquid without free gas in the pipe and

$$C_2 = \frac{RT\ K_\ell}{1 + K_\ell D/Ee} \qquad (6)$$

Figure 1 shows a dimensionless graph of wavespeed vs absolute pressure for a range of air contents in air-water mixtures in a pipeline in which KD/Ee = 0.3 (Ref. 8). Each line represents a mixture with a constant mass of free gas, identified with void fractions at standard pressure, p_o. Inasmuch as void fraction increases with decreasing pressure, and Eq. (5) applies only to small void fractions, a lower limit line with a 2% void fraction is shown. However, Martin and Padmanabhan (Ref. 3) show that Eq. (5) is reasonably valid in two-phase bubbly and slug flows to void fractions as high as 30%.

The one-dimensional equations to describe the dynamic behavior of the bubbly mixture are the Equations of Motion and Continuity, (Ref. 10).

$$g\ H_x + V_t + g\ \sin\theta + \frac{f|V|V}{2D} = 0 \qquad (7)$$

$$H_t + \frac{a'^2}{g} V_x = 0 \qquad (8)$$

In these equations the absolute pressure head is $H = p/\rho g + \bar{H} - Z$, $\bar{H}$ = barometric pressure head, Z = the height of the pipeline above the reference datum, θ = angle of the pipeline measured up from horizontal in the positive x direction, V = average velocity of fluid, f = Moody diagram friction factor, and the subscripts refer to partial differentiation with respect to the independent variables x and t. Equation (5) written in terms of H becomes

$$a' = \frac{a\ H}{\sqrt{C_3 + H^2}} \qquad (9)$$

in which $C_3 = C_2 m\rho^2 g^2$, a constant for a given mass of gas/unit volume of mixture. It may be noted that if there is no free gas, $C_3 = 0$, and a' = a.

Wiggert and Sundquist Ref. (6) provide an alternative approach in which separate conservation of mass equations are written for the gas component and for the liquid component. The void fraction α becomes an additional dependent variable in the three-equation formulation. Herein the free gas content is included in Eq. (9) which is incorporated in Eq. (8).

The method of characteristics applied to Eqs. (7) and (8) transforms the equations to four ordinary differential equations without approximation.

$$\pm \frac{g}{a'} \frac{dH}{dt} + \frac{dV}{dt} + g \sin\theta + \frac{f V|V|}{2D} = 0 \tag{10}$$

$$\frac{dx}{dt} = \pm a' \tag{11}$$

These four equations are restricted in application such that Eq. (10) with the positive (negative)-valued first term can only be used along the positively (negatively)-sloped characteristic Eq. (11). By substitution of the wave speed Eq. (9), changing the dependent variable from velocity to volumetric flow rate Q, and multiplying by dt, the equations are in a suitable form to integrate to obtain the numerical form.

$$\pm dH + B \frac{a'}{a} dQ + a' \sin\theta \, dt + \frac{f a'}{2gDA^2} Q|Q| \, dt = 0 \tag{12}$$

$$\frac{dx}{dt} = \pm \frac{a H}{\sqrt{C_3 + H^2}} \tag{13}$$

The characteristic impedance of the fluid without free air in the pipeline is given by B = a/gA, with A = pipe cross-sectional area.

SIMPLIFIED EQUATIONS

In order to illustrate some of the unique response characteristics of a variable-wave-speed problem the equations are temporarily simplified by considering a frictionless horizontal pipeline. The dispersive character of the flow which occurs in some cases, and the focusing nature in others, are more easily visualized without the influence of viscous losses and pipeline slope. For this special case Eqs. (10) and (11) become

$$\frac{\sqrt{C_3 + H^2}}{H} dH \pm B \, dQ = 0 \tag{14}$$

$$\frac{dx}{dt} = \pm a' = \pm \frac{a H}{\sqrt{C_3 + H^2}} \tag{15}$$

Dispersive or Negative Wave. When the absolute pressure head at the upstream end of a system is reduced, the positive-valued characteristic lines increase in slope in the xt plane as shown in Fig. 2 (Ref. 1). As will be demonstrated these lines remain straight, in the absence of reflections, in this simplified case. An undisturbed region with uniform conditions lies below the first characteristic line.

An exact integration of Eq. (14) from A to P along any C^+ characteristic line yields

$$\left[\sqrt{C_3 + H^2} - \sqrt{C_3}\ \ell n\ \frac{\sqrt{C_3} + \sqrt{C_3 + H^2}}{H}\right]_{H_A}^{H_P} + B(Q_P - Q_A) = 0 \tag{16}$$

With the pressure head specified at the upstream end, the flow rate can be obtained by applying the second compatibility equation along the C^- characteristic line, from B_1 to P_1.

$$\left[\sqrt{C_3 + H^2} - \sqrt{C_3}\ \ell n\ \frac{\sqrt{C_3} + \sqrt{C_3 + H^2}}{H}\right]_{H_{B_1}}^{H_{P_1}} - B(Q_{P_1} - Q_{B_1}) = 0 \tag{17}$$

A second C^- characteristic line could extend from point B_2 to P_2 where the pressure may be the same as at point P_1, in which case the flow at P_2 would be the same as at P_1. Any number of these C^- characteristic lines could extend from the undisturbed region to this same pressure. If the pressure at P_2 is equal to the pressure at P_1, and equal to other similar points, it follows that these points lie along the C^+ characteristic line given by Eq. (15), along which H is a constant. Application of Eq. (16) confirms the existence of a common flow along the C^+ characteristic line, as well as the common pressure head. Thus an exact solution may be generated for any position in the xt plane in which the characteristic lines are parallel or fan-like. Parallel lines would exist above the characteristic issuing from point j, if the pressure head were reduced from i to j, and then held from j to k. It may be noted that points i and j may be infinitesimally close together without changing the fan-like picture since each pressure from maximum to minimum must exist in the region. The region encompasses a zone in which common values of pressure and flow are spreading both in distance and time, thus it is referred to as a dispersive region.

Shock or Positive Wave. By increasing rather than decreasing the pressure at the boundary in a system containing a pressure-sensitive fluid, such as free gas in a liquid, the characteristic lines become flatter in the xt plane and converge as shown in Fig. 2b. This is the condition that leads to the well-known shock wave. An anomaly in the method of characteristics exists at point P, Fig. 2b, where two C^+ characteristic lines intersect. This implies that two different pressure heads and flows could be calculated at the intersection of these compatibility equations when each is solved with the appropriate C^- equation. Point P represents a point of initial formation of a weak shock wave which, in Fig. 2b, is likely to strengthen in the downstream direction. The complete analysis of the shock wave will not be addressed herein, however the position of initiation of the shock front will be identified, following a procedure by Henderson (Ref. 1). From the calculus and from geometry in Fig. 2b, $\delta(\tan\theta) = \delta\theta/\cos^2\theta$

$$\sin\theta = x_s/\ell = \delta t'/\delta t$$

$$\delta\theta = \delta t \sin^2\theta/x_s$$

Therefore

$$x_s = \frac{\tan^2\theta}{\delta(\tan\theta)/\delta t} \tag{18}$$

Since the inverse slope of the C^+ characteristic is given by $\tan\theta = a'$, the numerator in Eq. (18) can be evaluated and substituted, yielding

$$x_s = \frac{a\,H^2\sqrt{C_3 + H^2}}{C_3\ \delta H/\delta t} \tag{19}$$

This equation provides the distance to the first intersection of characteristic lines from the boundary in terms of the local fluid properties, the pressure head at point A in Fig. 2b, and the time rate of increase of the pressure head. In this simplified problem it can provide an envelope of the shock front in the xt plane. In the numerical examples that follow it will be used to provide an indication of the validity of the solution, i.e., it will indicate whether the compatibility equations along characteristic lines are providing "good" numbers, or that the shock front has formed and more complete equations should be used to achieve the correct solution.

NUMERICAL METHOD

A specified time intervals grid is used, as shown in Fig. 3. For given reach lengths the time step is selected to satisfy the Courant condition when using the wave speed in the liquid without free air content. With wave speed a function of both mass of free air and the pressure level, interpolations of some form are necessary. One form of integration of Eq. (12) and (13) is provided by Wylie and Streeter (Ref. 10) with linear interpolations at R and S. Second order integration is utilized and the numerical process seems reasonable in cases of very low void fractions of free gas and in examples involving free gas combined with vapor column separation. In cases of significant wave speed variation numerical attenuation of waves is present in this method due to the interpolation procedure.

A time-line interpolation is proposed herein in which characteristics from point P extend backwards in time for a full reach length as shown in Fig. 4. Integration of the C^+ characteristic, Eq. (13), from R_M to P yields

$$x_P - x_{RM} = \frac{a'_{RM} + a'_P}{2} \Delta t = a^*_{RM} \Delta x \tag{20}$$

in which

$$a^*_{RM} = \frac{a'_{RM} + a'_P}{2a} \tag{21}$$

Repeated integration over each time step to R_o provides the result $x_P - x_{Ro} = \Delta x$ and

$$\sum_{m=0}^{M} a^*_{Rm} = 1 \tag{22}$$

The integration of Eq. (12) from R_M to P along the C^+ characteristic line yields

$$H_P - H_{RM} + B\, a^*_{RM}(Q_P - Q_{RM}) + a^*_{RM} \Delta x \sin\alpha + R\, a^*_{RM} Q_{RM}|Q_{RM}| = 0 \tag{23}$$

in which $R = f\Delta x/(2gDA^2)$. A first-order integration is used in the friction term in that it is evaluated only at point R_M and not at the unknown point P. However, the dimensionless wave speed a^*_{RM} is in terms of a'_P and is dependent upon the unknown H_P. Thus an interative solution is utilized. Repeated integration of Eq. (12) over each time step to R_o provides equations of the same form as Eq. (23) which, when combined, yield

$$H_P - H_{Ro} + B[a^*_{RM} Q_P - a^*_{Ro} Q_{Ro} + \sum_{m=1}^{M} Q_{Rm}(a^*_{Rm-1} - a^*_{Rm})] + \Delta z$$
$$+ R \sum_{m=0}^{M} a^*_{Rm} Q_{Rm}|Q_{Rm}| = 0 \tag{24}$$

The increase in elevation from R_o to P is given by Δz. Unknowns in Eq. (24) are H_P and Q_P, and a^*_{RM} since it is a function of H_P. Additionally the location of each subscripted R point in the xt grid moves as H_P changes. A linear interpolation between adjacent grid intersection points along the time-line provides conditions at point R_o. Various approximating techniques may be used to evaluate a^*_{Rm}. In this study a^*_{Rm} is based upon an average of the four pressure heads at the corners of reach i-1 through which the characteristic extends. The only exception is a^*_{RM} which is based upon H_P and an average of the two pressure heads at t - Δt in the reach. The C^- characteristic and compatibility equations when integrated are handled similarly but involve conditions at point P and the subscripted S locations in reach i, Fig. 4.

A first-order integration of characteristic and compatibility equations, in which a* is evaluated only at points R_o and S_o, is considerably simpler to implement than the procedure outlined above. It yields straight characteristic lines between R_o and P and S_o and P. As will be demonstrated in the numerical examples this procedure is inadequate in the highly nonlinear two-component fluid.

Continuity Balance. An overall continuity check can be made to provide an evaluation of the accuracy of the numerical method. If $\forall_i$ and $\forall_f$ represent the volumes stored at the initial and final times, respectively, and $\forall_{in}, \forall_{out}$ represent the volumetric inflow and outflow during the time T, then the continuity balance C_B can be evaluated at any time from

$$C_B = \forall_{in} - \forall_{out} + \forall_i - \forall_f \tag{25}$$

The number could be nondimensionalized by dividing by the average volume through-flow

$$C_B{}^* = 2C_B/(\forall_{in} + \forall_{out}) \tag{26}$$

The volume entering or leaving the system is computed from

$$\forall = \int_o^T Q\, dt \tag{27}$$

The volume stored in the system must include the variation in elasticity with pressure.

A modulus of elasticity of the system K' can be defined that encompasses the elasticity of the pipe and the compressibility of the fluid

$$K' = \frac{\Delta p}{\Delta\forall/\forall} \tag{28}$$

The change in volume stored, $\Delta\forall$, in the volume $\forall$ due to increase in pressure, includes the fluid stored due to pipe wall exapnsion, and due to fluid compressibility. The pressure-dependent wave speed can be related to K' by

$$a'^2 = \frac{K'}{\rho} = \frac{K/\rho}{1 + KD/Ee} \tag{29}$$

When these equations are combined

$$\Delta\forall = \frac{\Delta p\ \forall}{\rho a'^2} \tag{30}$$

which, on substitution of Eq. (9) becomes

$$\frac{\Delta\forall}{\forall} = \frac{g}{a^2}\left(1 + \frac{C_3}{H^2}\right)\Delta H \tag{31}$$

If the volume stored in a reach of pipeline at pressure H_o is $\forall_o$, then the volume in the same region at average pressure $\bar{H}$ is

$$\forall = \forall_o \, EXP[g(\bar{H} - H_o)(1 + C_3/\bar{H} H_o)/a^2] \quad (32)$$

The volume stored in the entire system at time t can be determined by summing the volumes stored in each reach determined from Eq. (32) with $\bar{H}$ being an average reach absolute pressure head.

NUMERICAL EXAMPLES

Two examples are presented, both involving essentially the same system: one involves the simplified equations in a horizontal, frictionless system, and the second includes the effects of a significant pressure gradient in the system. The single pipeline data are L = 3000 m, D = 0.61 m, ρ = 992 kg/m^3, a = 981.4 m/s, m = 0.00254 kg/m^3 (0.2% free air), R = 287 m N/kg K, T = 288K, Q_o = 0.89 m^3/s. A reservoir exists at the downstream end. Beginning at time zero the upstream head drops from 60 m to 10.3 m in 0.204 s and is held at that level for the duration of the transient.

In the first example f = 0, and H = 60 m throughout the pipeline. Three solutions are presented: the analytic solution up to the time of the first reflection from the downstream reservoir, the time-lime-interpolation procedure described herein, and the distance-line-interpolation method.

Figure 5 shows the detail of the slopes of the characteristic lines at the upstream end during the reduction in pressure head from 60 m to 10.3 m. Also shown are the computed discharges at the upstream end. For comparison purposes, if there were no free air in this example the flow would have been reduced to 0.745 m^3/s. If the wave speed corresponding to 0.2% free air at 60 m pressure were used and held constant the flow would be reduced to 0.707 m^3/s; and with proper accounting of the wave speed variation with pressure the flow is reduced to 0.606 m^3/s. The graph in Fig. 5 shows the reduction in Q computed analytically with Eq. (17). Also shown are (1) results from a first-order integration of the equations using the time-distance grid in Fig. 5a, and (2) results from the time-line-interpolation procedure with the same time-distance grid.

The response in the entire system is shown in Figs. 6 to 8. Pressure head as a function of time is presented at two locations along the pipeline in Fig. 6. These results are generated with two different levels of discretization, the number of reaches N = 15 and 30. The dispersion of the sharp pressure reduction is clearly evident in the graphs, in fact at 2400 m the head only reduces to about 31 m. Both numerical procedures show reasonable agreement with the analytic solution during the first dispersive wave. Thereafter the two results diverge considerably. The time-line procedure yields approximately the same pressure history with either of the discretizations. It also shows the sharp positive wave and little numerical damping. The distance-line interpolation shows a sharper response with N = 30 but still displays numerical damping and rounding at the extreme excursions. Use of Eq. (19) in the time-line interpolation program indicated the formation of a shock wave on the rising limb of the response when the pressure head was below about 30 m.

Results of the continuity check are shown in Fig. 7. Again the time-line method provided about the same error using both discretizations, and suffers the most serious deterioration during the formation of the shock waves. The distance-line method is better with N = 30 than with N = 15 but provides a continuous imbalance throughout the transient.

Figure 8 displays another view of the response by presenting the pressure gradient and flow variation along the pipeline at particular instants of time. The dispersive nature of the flow is evident in Fig. 8a with the positive wave formation evident in Fig. 8c.

In the second example the same horizontal pipeline is used but viscous losses are included, f = 0.02, and the steady-state gradient drops from 60.75 m upstream to the reservoir at 13.4 m downstream. The pressure head is again dropped to 10.4 m absolute at the upstream end in 0.204 s initiating a sharp transient that propagates into the system. In this case the pressure soon drops to absolute zero as the pressure wave moves down the hydraulic grade line. In the absence of any free air, the liquid vaporizes with extensive vaporous cavitation in the system and, after a brief delay, the water columns rejoin producing an absolute pressure head of about 37 m in the central portion of the pipeline. An analytic solution to this vaporous cavitation problem was given by Wylie and Streeter (Ref. 9).

With fairly small quantities of free air in the system the fluid is able to expand sufficiently to avoid pressure reduction all the way to vapor pressure, and a modified response is generated. Different amounts of free gas are used in this example, one to two orders of magnitude less than the first simplified problem. However the pressure drops to a lower level than in the first example so the degree of nonlinearity is still substantial.

Figure 9 shows the results of the different free air contents compared with the vaporous cavitation result. The latter forms a good basis for comparison to assist in the judgement of degree of dispersion, attenuation, etc. In each case 30 reaches were used to generate the results using the time-line-interpolation method. Figure 9 shows results with 0.002%, 0.004%, and 0.01% free air at standard conditions.

The free air content clearly influences the shape of the wave but probably the most interesting aspect is the influence of the various levels of free air on the timing of the events and on the peak magnitudes of the generated pressure waves. The smallest air contents shows the pressure rise occurring earlier than the vaporous cavitation event and it also shows lower pressure peaks. The mid-range air content provides timing quite similar to the vaporous case, and the highest void fraction shows a delayed timing but the largest pressure peaks. Wiggert and Sundquist (Ref. 7) noted a similar timing response in some of their gaseous cavitation studies.

CONCLUSIONS

The unique features in the response of highly nonlinear fluid transients have been discussed with particular emphasis on the negative-dispersive and positive-shock waves. The time-line-interpolation procedure provides a suitable numerical method to handle the variable wave speed with reasonable accuracy. An iterative (second order) procedure is necessary in dealing with nonlinear materials.

A continuity balance is suggested as an additional means of evaluating the success of the numerical process. In variable-wave-speed problems the results are influenced by the degree of discretization. If a significant continuity imbalance exists the results are likely to be suspect.

ACKNOWLEDGEMENTS

This study was supported by National Science Foundation Agreement No. 2931-2 to The University of Michigan. Taso Sophios assisted with the computer programming.

REFERENCES

1. Henderson, F. M.: "Open Channel Flow." pp. 288-297. Macmillan Co., (1966).

2. Kranenburg, C.: "Gas Release During Transient Cavitation." Jour. Hydr. Div., ASCE, 100, HY10, pp. 1383-1398, (Oct., 1974).

3. Martin, C. S. and Padmanabhan, M.: "Pressure Pulse Propagation in Two-Component Slug Flow." Jour. of Fluids Eng., ASME, 101, pp. 44-52. (Mar., 1979).

4. Swaffield, J. A.: "A Study of the Influence of Air Release on Column Separation in an Aviation Kerosene Pipeline." Proc. Inst. of Mec. Eng., 186 56/72, pp. 693-703, (1972).

5. Tullis, J. P., Streeter, V. L., and Wylie, E. B.: "Waterhammer Analysis with Air Release." Paper C3. Proc. Second International Conf. on Pressure Surges, Organized by B.H.R.A., Cranfield, (Sept. 22-24, 1976).

6. Wiggert, D. C., and Sundquist, M.J.: "The Effect of Gaseous Cavitation on Fluid Transients." Jour. of Fluids Eng., ASME, 101, pp. 79-86. (Mar. 1979).

7. Wiggert, D. C. and Sundquist, M. J.: " The Effect of Released Gases on Hydraulic Transients" Final Report to NSF., Michigan State Univ., p. 81. (Aug., 1977).

8. Wylie, E. B.: "Two-phase Flow During Liquid Transients in Pipelines." Proc. Institute on Transient Flow and Hydromachinery, Colorado State Univ., Fort Collins. (June, 1979).

9. Wylie, E. B. and Streeter, V. L.: "Column Separation in Horizontal Pipelines," 9th IAHR Symp. on Fluid Mechanics, Fort Collins. (June, 1978).

10. Wylie, E. B. and Streeter, V. L.: "Fluid Transients." McGraw Hill Book Co. (1978).

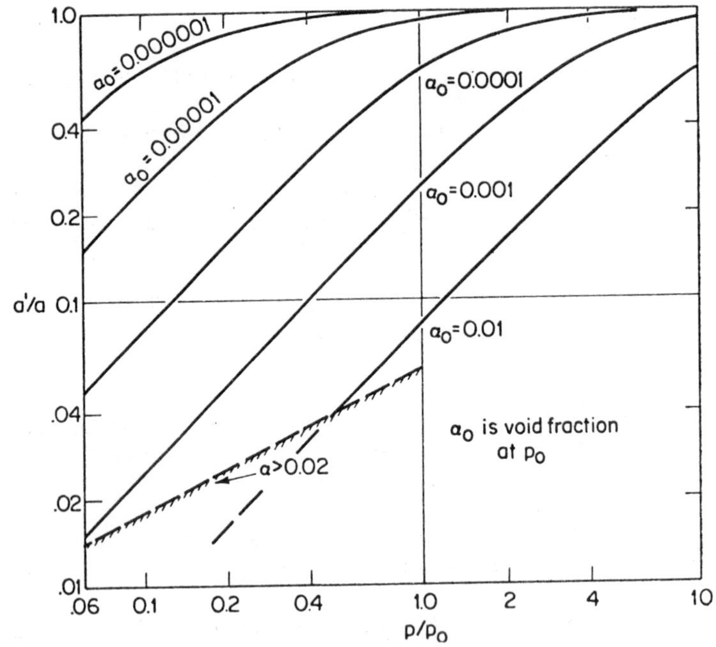

Figure 1. Wavespeed variation as a fuction of void fraction and pressure.

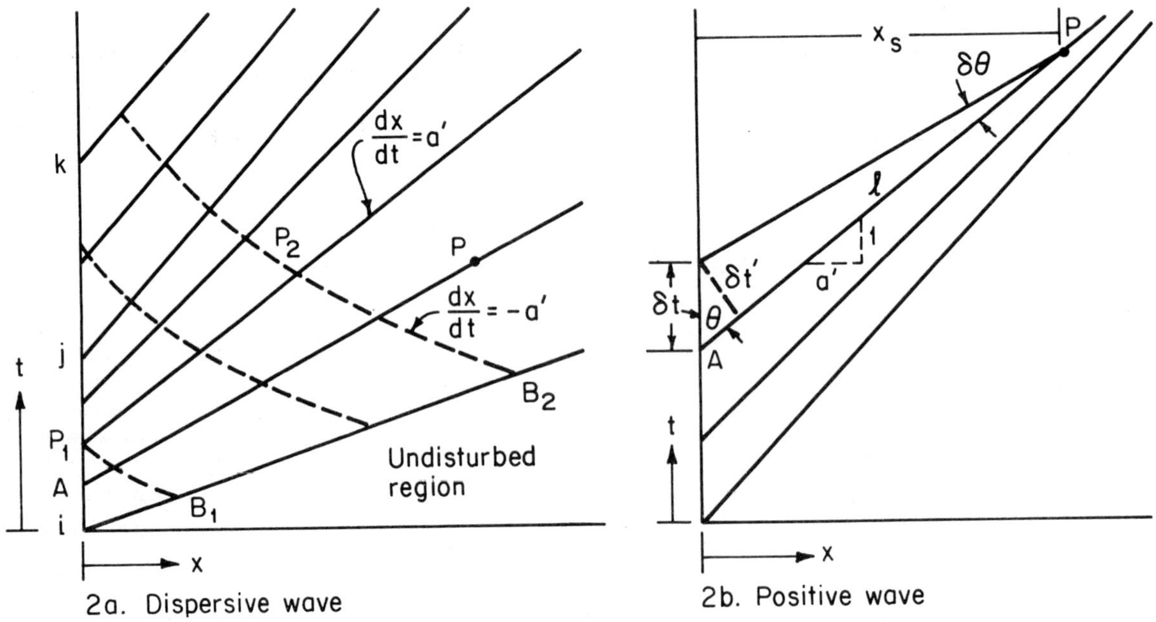

Figure 2. Simplified problem. Characteristics in xt plane.

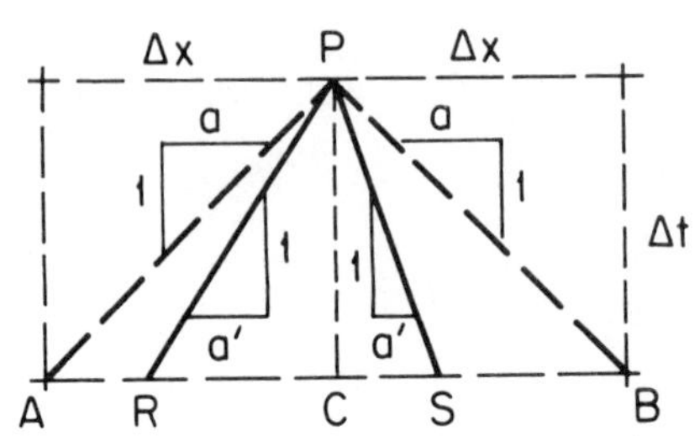

Figure 3. Specified time intervals grid.

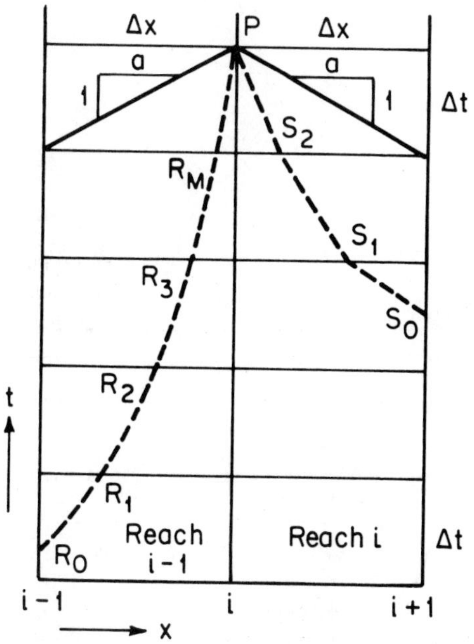

Figure 4. Time-line interpolations

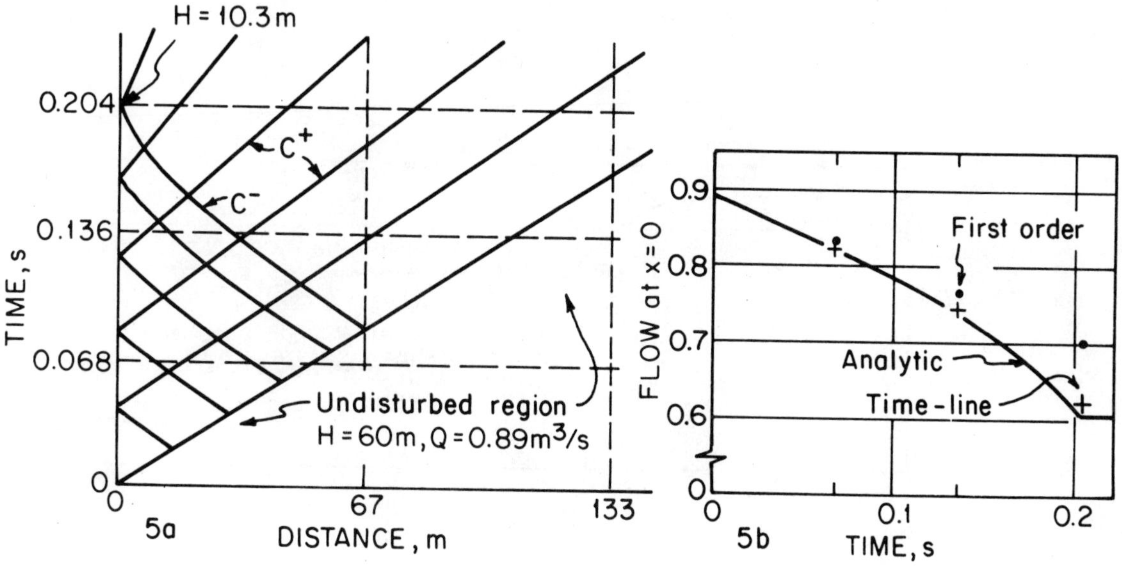

Figure 5. Typical characteristics. Flow at upstream end.

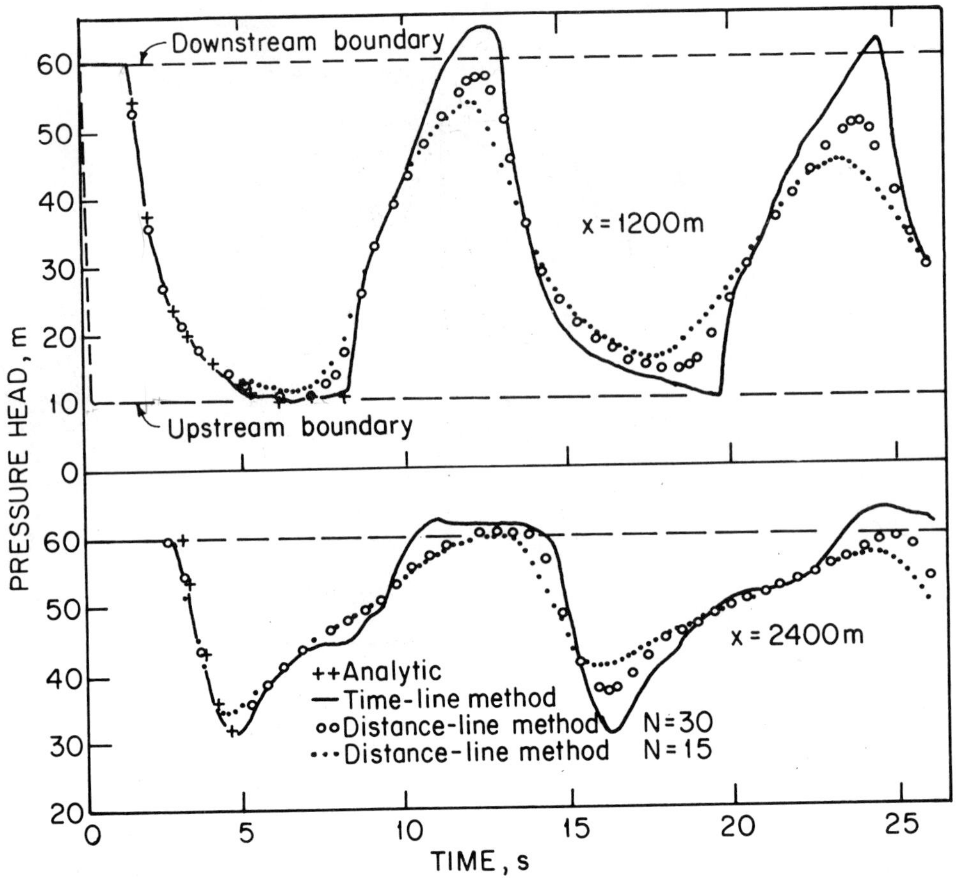

Figure 6. Pressure head vs time at 1200m and 2400m.

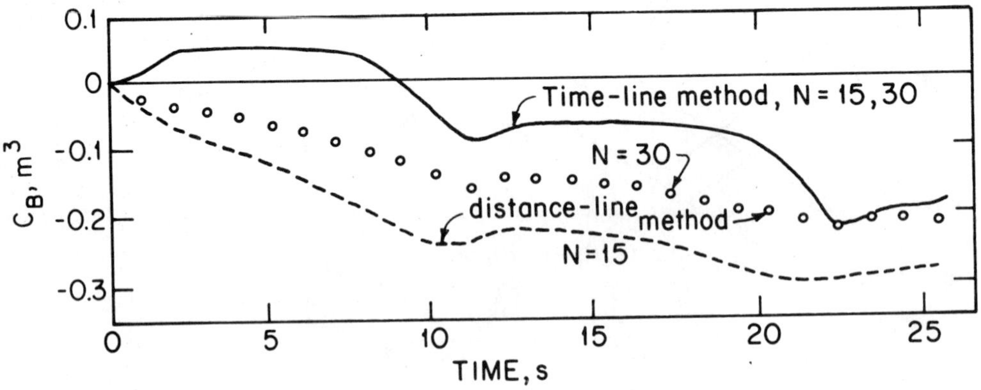

Figure 7. Continuity balance vs time.

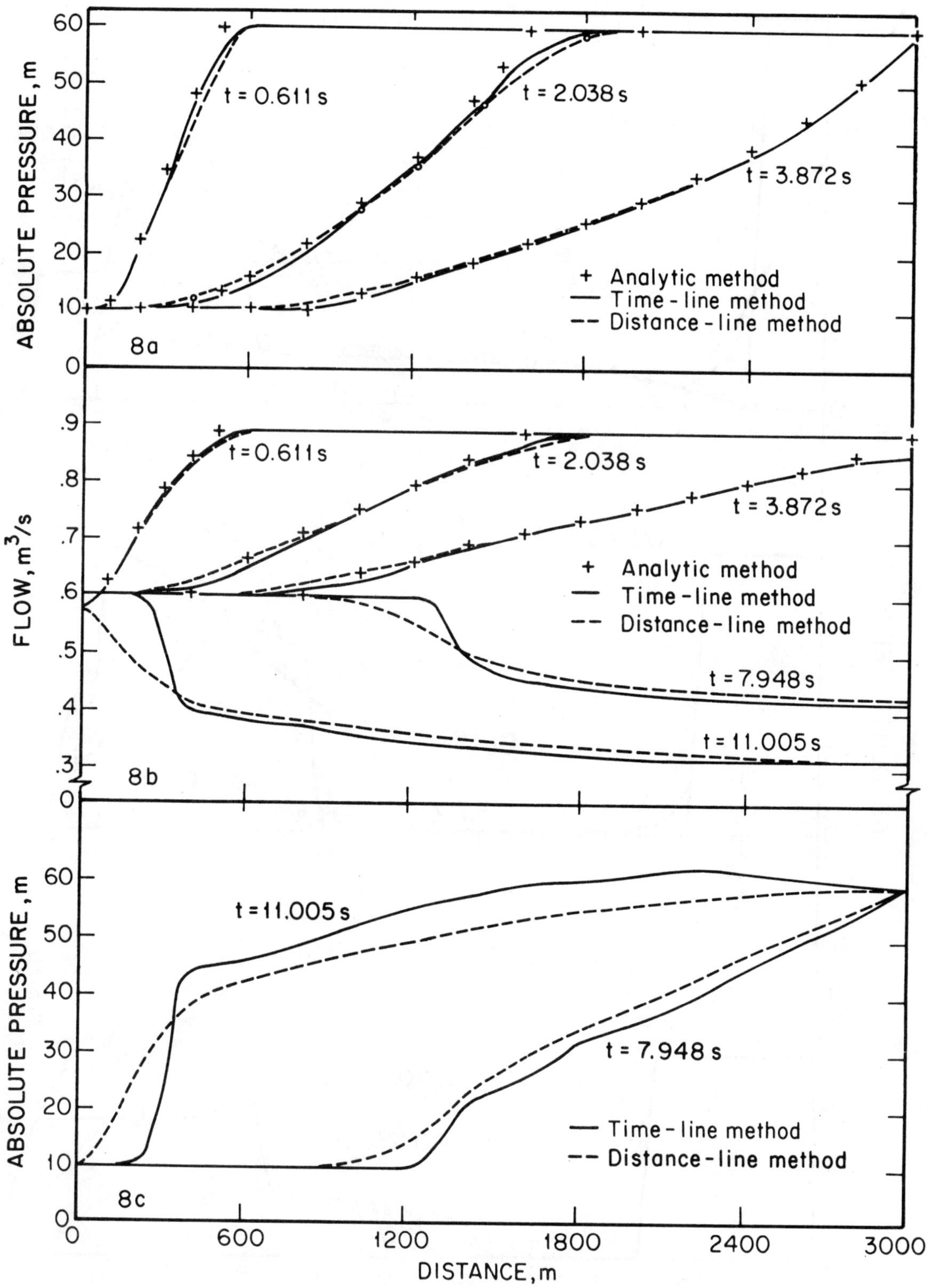

Figure 8. Pressure and flow along pipeline.

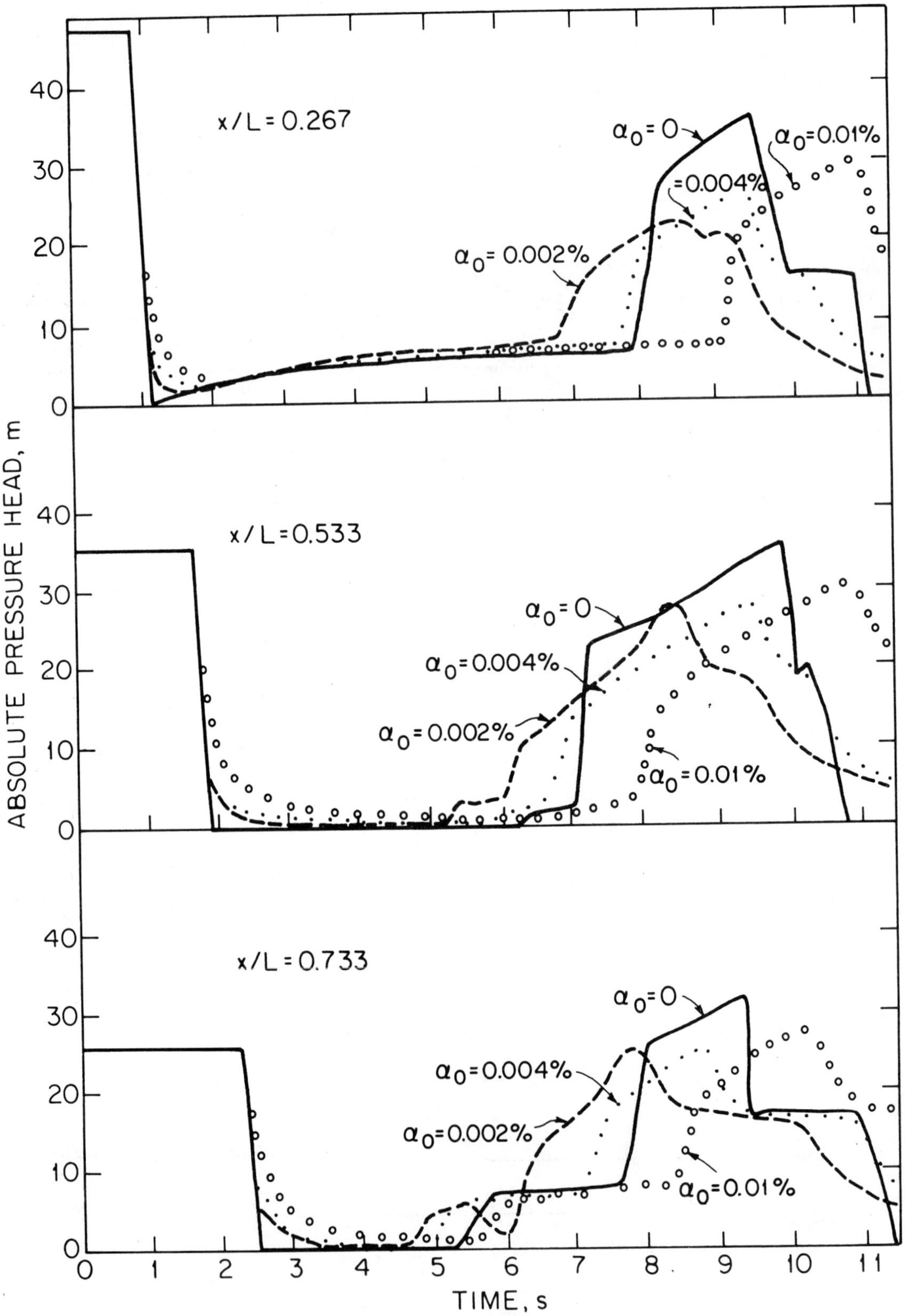

Figure 9. Example 2. Pressure response with various air contents.

Third International Conference on

PRESSURE SURGES

Canterbury, England
March 25-27, 1980

PAPER B2

COMBINATION OF FINITE DIFFERENCE AND FINITE ELEMENT TECHNIQUES IN HYDRAULIC TRANSIENT PROBLEMS

C.S. Watt

Merz and McLellan, U.K.

A.P. Boldy

Warwick University, U.K.

and

J.M. Hobbs

Thames Polytechnic, U.K.

Summary

The solution of hydraulic transient problems using the Method of Characteristics can have limitations, particularly relating to maximum permissible time step, when dealing with networks of pipes. In an attempt to overcome these difficulties a Finite Element formulation was devised and the results compared with two other numerical methods, namely an implicit finite difference scheme and the explicit finite difference Method of Characteristics. All theoretical solutions are compared with experimental data obtained from a purpose-built rig.

A modification of finite difference computer simulations is proposed involving the inclusion of a Finite Element component into such programs and the marrying-up of the two techniques. The resulting package overcomes the difficulties described above and the results compare very favourably with experimental data.

Held at the University of Kent at Canterbury, U.K.

Conference Organised by BHRA Fluid Engineering, Cranfield, Bedford, England.
In conjunction with The City University, London.

NOMENCLATURE

a	wave speed
d	internal pipe diameter
g	gravitational acceleration
t	time co-ordinate
x	distance co-ordinate of the conduit
H	piezometric head
N_i	Interpolation functions
W_i	Weighting functions
U	dimensionless form of x/L
V	velocity
α_i, β_i	nodal values of velocity, piezometric head and their first partial derivatives with respect to distance
λ	Darcy-Weisbach friction factor
ϕ	inclination of the conduit (measured anticlockwise with reference to right-hand side horizontal x-axis)
$[A]$, $[C]$, $[E]$	constant sub-element matrices used in the finite element method
$[B_n]$, $[D_n]$, $[F_n]$	time dependent sub-element matrices used in the finite element method
$[d]$	displacement vector
$[K]$	stiffness matrix
$[NU]$ $[VN]$	coefficient matrices
$[P]$	load vector
$[S_i]$, $[Z_i]$, $[\gamma_i]$	coefficient matrices

1. INTRODUCTION

It is well known that most hydraulic transient problems of internal flow can be solved satisfactorily by various numerical methods using digital computers. Among the various numerical procedures the characteristics method is by far the most popular because of its versatility in handling boundary conditions explicitly.

The implicit finite difference method, a popular modelling technique in the solution of free surface open channel flows, is known to be numerically more stable than the characteristics method and permits the use of larger time steps. Another implicit technique based on the established variational method has attracted recent attention because of its linear formulation and high accuracy.

In this paper the implicit finite difference scheme, arranged as an accurate linear nodel, and the finite element method, based on a generalized variational method, are compared with the characteristics method. Slow and rapid transient data are used to validate all the theoretical predictions.

Finally, the two implicit schemes are married-up with the characteristics method as an enhanced procedure to provide flexibilities in determining time step and mesh length.

2. BASIC DIFFERENTIAL EQUATIONS

The equation of motion and continuity, which can be found in (Ref. 1) for one-dimensional unsteady internal flow of a compressible fluid with negligible density variation, are expressed in matrix form as

$$\begin{bmatrix} V \\ H \end{bmatrix}_t + \begin{bmatrix} V & g \\ a^2/g & V \end{bmatrix} \begin{bmatrix} V \\ H \end{bmatrix}_x + \begin{bmatrix} \frac{\lambda}{2d} V|V| \\ -V \sin\phi \end{bmatrix} = 0 \qquad (2.1)$$

where the subscripts of the matrices denote partial derivatives. The system of equations is hyperbolic in the sense that the square matrix

$$\begin{bmatrix} V & g \\ a^2/g & V \end{bmatrix}$$

has distinct real eigenvalues given by V+a and V-a.

The theoretical solutions of the above equation system (Ref. 2.1) can be solved by various numerical methods. Generally they can be broadly classified into explicit and implicit solutions. The explicit finite difference scheme (see (Ref.2, 3)) and the widely used method of characteristics (Ref. 4,1) calculate the solution explicitly at any interior point within a x-t finite difference mesh. The implicit finite difference scheme (Ref. 2,5) and the variational method (Ref. 3) interlock all unknowns together and require a simultaneous solution. Advantage and disadvantage of the two classes of technique are not always obvious. But in general the explicit form is simpler and easier to apply globally because of the discrete nature of the unknowns not coupled with one another. Each unknown at new time interval is evaluated readily from the previous history of a small section of the system. Due to the confined area from which information is drawn for the calculation of unknowns the time step advance will be restricted accordingly. Thus all explicit methods are subject to a limitation on the ratio of time step to distance interval (e.g. Courant condition). Such limitation is relaxed in all implicit schemes because the previous history of the entire system is used for solving, normally in a single operation, all the activities at the new time. But the resulting procedures are much involved and rigid. Computational efforts compare unfavourably with the explicit counterpart.

3. METHOD OF CHARACTERISTICS

The use of the method of characteristics in the solution of hyperbolic type partial differential equations was well established as early as 1928 by Courant, Friedrichs and Lewy who formulated the well known necessary condition for the convergence of the scheme. The use of the method for solving water hammer problems was publicised by Gray (1953), Streeter and Lai (1962), and Evangelistic (1965) (Ref. 1, 6, 7). Nowadays the maturely developed characteristics method, despite the recent advancement of other finite difference schemes and the finite element method, is still the most widely used numerical technique for problems on pressure surges. The method is extensively documented (Ref. 1) and summarized in Fig. 1.

4. IMPLICIT FINITE DIFFERENCE METHOD

The centred implicit finite difference method is more renowned for solving unsteady free surface flow in open channels. Its application in unsteady internal flow has been limited to slow transients like gas flows (Ref. 1, 14). The method, although common enough and well established, has rarely been fully attempted in the solution of water hammer problems. Perkin et al (Ref. 2) reported that the method performed satisfactorily, in the prediction of mass oscillation in hydroelectric plants, using a time step 75 times that of the characteristics method counterpart. Streeter (Ref. 5) also advocated its use to replace interpolation in short pipe reaches among the characteristics method.

The following derivation of the implicit finite difference scheme, after Perkin, Tedrow, Eagleson and Ippen (Ref. 2) has been refined so that the resulting set of equations is nonlinear with velocity alone. Since the velocity changes are smaller and more predictable than pressure changes the linearized solution of the scheme (i.e. one pass and non-iterative procedure) has been found tolerable for engineering purposes.

The essential feature of the method is shown in Fig. 2 and the continuity equation then becomes

$$\begin{aligned}
&\left[\frac{\Delta t}{\Delta x}\left(-\frac{a^2}{g}+\frac{1}{2}\Delta x\sin\phi\right)\right][V_1] + \left[1-\frac{\Delta t}{2\Delta x}(V_1+V_2)\right][H_1] + \left[\frac{\Delta t}{\Delta x}\left(\frac{a^2}{g}+\frac{1}{2}\Delta x\sin\phi\right)\right][V_2] + \left[1+\frac{\Delta t}{2\Delta x}(V_1+V_2)\right][H_2] \\
&= \left[-\frac{\Delta t}{\Delta x}\left(-\frac{a^2}{g}+\frac{1}{2}\Delta x\sin\phi\right)\right][\bar{V}_1] + \left[1+\frac{\Delta t}{2\Delta x}(\bar{V}_1+\bar{V}_2)\right][\bar{H}_1] + \left[-\frac{\Delta t}{\Delta x}\left(\frac{a^2}{g}+\frac{1}{2}\Delta x\sin\phi\right)\right][\bar{V}_2] + \left[1-\frac{\Delta t}{2\Delta x}(\bar{V}_1+\bar{V}_2)\right][\bar{H}_2]
\end{aligned} \qquad (4.1)$$

and the momentum equation transforms to the form

$$
\begin{aligned}
&\left[\frac{\Delta x}{g\Delta t} - \frac{1}{2g}V_1 + \frac{\lambda\Delta x}{4dg}|V_1|\right]\left[V_1\right] && \left[\frac{\Delta x}{g\Delta t} + \frac{1}{2g}\bar{V}_1 - \frac{\lambda\Delta x}{4dg}|\bar{V}_1|\right]\left[\bar{V}_1\right] \\
&+\left[\quad -1 \quad\right]\left[H_1\right] && +\left[\quad 1 \quad\right]\left[\bar{H}_1\right] \\
&+\left[\frac{\Delta x}{g\Delta t} + \frac{1}{2g}V_2 + \frac{\lambda\Delta x}{4dg}|V_2|\right]\left[V_2\right] \quad = && +\left[\frac{\Delta x}{g\Delta t} - \frac{1}{2g}\bar{V}_2 - \frac{\lambda\Delta x}{4dg}|\bar{V}_2|\right]\left[\bar{V}_2\right] \\
&+\left[\quad 1 \quad\right]\left[H_2\right] && +\left[\quad -1 \quad\right]\left[\bar{H}_2\right]
\end{aligned}
\qquad (4.2)
$$

The equations (4.1) and (4.2) have four unknowns and nonlinear velocity terms (also λ). The mesh length Δx and time interval Δt can be decided arbitrarily provided the finite difference approximations of the derivatives are valid. Hence the entire length of the conduit will be used as the mesh length whenever possible. Since the scheme is neutrally stable (see Ref. 2) the size of time step Δt is only dependent on the transient response of the system (i.e. the number of points produced should be sufficient to approximate the response curve). In order to solve the equations (4.1) and (4.2) two additional boundary equations must be supplied at each end of the conduit before the four unknowns can be solved. Conduits are interlocked together through the common doundary equations. A simultaneous solution is inevitable.

The method has two major drawbacks. Since the basic unsteady flow equations (4.1) and (4.2) are nonlinear, together with nonlinear boundary equations, a two-stage global iteration is required. This may be reduced by linearizing the unsteady flow equations (4.1) and (4.2) using previous velocities for the nonlinear terms. The simultaneous solution also requires a large computer storage area for the mathematical operations. But it is less critical nowadays with the well developed out-of-core equation solvers (Ref. 9, 10) available in the literature of the finite element theory.

5. FINITE ELEMENT METHOD

The finite element concept, first used as a general method in structural analysis (e.g. see Ref. 11), has gradually evolved to a well proven approximation technique for solving differential equations. The method is powerful in solving boundary value problems in which the space domain, even in three dimensions, can be easily approximated with a wide range of selectable shape functions. However, it is still relatively new and less successful in the solution of time dependent problems because the time domain is far more difficult to model simultaneously with the space domain. Consequently most finite element applications in time dependent problems adopt a hybrid alternative; finite difference in time and finite element in space.

In pressure surge problems the finite element technique has been applied to the solution of free surface open channel flow (e.g. see (ref. 12). The established variational method (applied to gas transients by Rachford and Dupont (Ref. 3)) has been generalized in this paper as a finite element solution.

The formulation of the finite element method can be attempted by directly approximating the basic differential equations with a weight residual approach, applying the Galerkin criterion for the choice of the weighting function. The unknown velocity and piezometric head profiles are considered to be made up by a linear combination of functions of the form

$$V(x,t) = \sum_{i=1}^{k} \alpha_i(t)\, N_i(x)$$
$$H(x,t) = \sum_{i=1}^{k} \beta_i(t)\, N_i(x) \tag{5.1}$$

where $N_i(x)$ are interpolation functions to be chosen, $\alpha_i(t)$ and $\beta_i(t)$ are time dependent unknowns to be evaluated and k is an integer.

The form of the interpolation function used is described in Appendix I.

The method of Weighted Residuals then requires that

$$\int_0^L \left\{ \begin{bmatrix} V \\ H \end{bmatrix}_t + \begin{bmatrix} V & g \\ a^2/g & V \end{bmatrix} \begin{bmatrix} V \\ H \end{bmatrix}_x + \begin{bmatrix} \frac{\lambda}{2d} V|V| \\ -V\sin\phi \end{bmatrix} \right\} W_i(x)\, dx = 0 \tag{5.2}$$

for some set of weighting functions $W_i(x)$ where $[0, L]$, the domain of integration, is the length of the pipeline. The Galerkin approach requires that the weighting functions $W_i(x)$ be taken as the interpolating functions N_i, that is

$$W_i(x) = N_i(x) \; ; \quad i = 1, 2 \ldots, k \tag{5.3}$$

thus reducing the residual criterion given by (5.2) to the expression

$$\int_0^L \left\{ \begin{bmatrix} V \\ H \end{bmatrix}_t + \begin{bmatrix} V & g \\ a^2/g & V \end{bmatrix} \begin{bmatrix} V \\ H \end{bmatrix}_x + \begin{bmatrix} \frac{\lambda}{2d} V|V| \\ -V\sin\phi \end{bmatrix} \right\} N_i(x)\, dx = 0 \tag{5.4}$$

$$i = 1, 2 \ldots, k$$

Denoting the partial derivaties with respect to distance and time by subscripts x and t respectively, then substitution of equations (5.1) into equations (5.4) leads to the simultaneous differential equations

$$[A][\alpha_t] + [B][\alpha] + [C][\beta] + [D] = [0]$$
$$[A][\beta_t] + [E][\alpha] + [B][\beta] + [F] = [0] \tag{5.5}$$

where the coefficient matrices have elements given by :-

$$A_{ij} = \int_0^L N_j N_i \, dx \; ; \qquad B_{ij} = \int_0^L V N_{jx} N_i \, dx$$

$$C_{ij} = g\int_0^L N_{jx} \cdot N_i \, dx \; ; \qquad E_{ij} = \frac{a^2}{g^2} C_{ij} \tag{5.6}$$

$$D_i = \int_0^L \frac{\lambda}{2d} v|v| N_i \, dx \; ; \qquad F_i = \int_0^L - V \sin\phi \, N_i \, dx$$

with $i,j = 1, \ldots\ldots, k$. The differential equations (5.5) may be solved numerically by a time discretisation, so that if the subscript n is used to denote the nth time interval, then

$$\begin{aligned} &[A]\left[\frac{\alpha_{n+1} - \alpha_n}{\Delta t}\right] + \left[B_{n+\frac{1}{2}}\right]\left[\alpha_{n+\frac{1}{2}}\right] + [C]\left[\beta_{n+\frac{1}{2}}\right] + \left[D_{n+\frac{1}{2}}\right] = [0] \\ &[A]\left[\frac{\beta_{n+1} - \beta_n}{\Delta t}\right] + [E]\left[\alpha_{n+\frac{1}{2}}\right] + \left[B_{n+\frac{1}{2}}\right]\left[\beta_{n+\frac{1}{2}}\right] + \left[F_{n+\frac{1}{2}}\right] = [0] \end{aligned} \tag{5.7}$$

Equations (5.7) are non-linear with convective term $B_{n+\frac{1}{2}}$, $F_{n+\frac{1}{2}}$ and friction term $D_{n+\frac{1}{2}}$ evaluated at the half time step between the nth and $(n+1)^{th}$ levels. A rapid procedure to obtain a solution can be found by linearizing the discrete equations using the approximations $B_{n+\frac{1}{2}} = B_n$; $D_{n+\frac{1}{2}} = D_n$ and $F_{n+\frac{1}{2}} = F_n$ The smallness of the terms involved usually ensures a stable iterative procedure leading to an accurate solution without excessive computation time. Finally, if it is assumed that

$$\alpha_{n+\frac{1}{2}} = \frac{1}{2}(\alpha_n + \alpha_{n+1})$$

and

$$\beta_{n+\frac{1}{2}} = \frac{1}{2}(\beta_n + \beta_{n+1}) \tag{5.8}$$

then equations (5.7) reduce to a set of linear algebraic equations, with unknowns α_{n+1} and β_{n+1}, in the standard form

$$\left[\begin{array}{c|c} \frac{A}{\Delta t} + \frac{B_n}{2} & \frac{C}{2} \\ \hline \frac{E}{2} & \frac{A}{\Delta t} + \frac{B_n}{2} \end{array}\right] \left[\begin{array}{c} \alpha_{n+1} \\ \hline \beta_{n+1} \end{array}\right] = \left[\begin{array}{c|c} \frac{A}{\Delta t} - \frac{B_n}{2} & -\frac{C}{2} \\ \hline -\frac{E}{2} & \frac{A}{\Delta t} - \frac{B_n}{2} \end{array}\right] \left[\begin{array}{c} \alpha_n \\ \hline \beta_n \end{array}\right] - \left[\begin{array}{c} D_n \\ \hline F_n \end{array}\right] \tag{5.9}$$

The resultant matrices on the right-hand side of (5.9) are known and can be decomposed into a vector and the equation becomes the normal form of

$$[k]\,[d] = [P] \tag{5.10}$$

with the first matrix in the equation (5.9) as the stiffness term $[k]$, α_{n+1} and β_{n+1} as the displacement vector $[d]$ and the decomposed vector becomes the equivalent load vector $[P]$.

Applying this to the waterhammer problem α_i and β_i are selected as the nodal values of velocity, head and their first derivatives.

BOUNDARY CONDITION

The finite element formulation so far developed contains an 8 x 8 matrix with the velocity, piezometric head and their first derivatives with respect to distance as the eight unknowns. Although the system is linear and determinate no unique solution exists because the matrix is singular without the boundary conditions prescribed. The normal technique in prescribing a boundary condition is to replace a suitable row by the corresponding boundary equation. When there are several boundary equations available in a particular node they should be systematically reduced to a single equation.

6. EXPERIMENTS

Experiments were carried out using the hydraulic system shown in Fig. 4. Two different experiments were conducted. The first experiment, intended as a rapid transient, was achieved by closing a fast reacting gate valve to produce the water hammer surge. A DEC PDP 11/10 mini computer was used to record the experimental data. The fast sampling capability of the computer proved ideal for capturing the rapid transient responses. The gate valve was accurately calibrated and its movement sensed by an ultra-linear potentiometer. A good water hammer surge could be generated by closing the valve in about one second.

The second experiment, with a 100 mm diameter surge tank fitted upstream of the gate valve, was to simulate the slow mass oscillation effect of a surge tank. The gate valve was closed again to initiate the transient surge, but the valve's closure time was no longer critical. The pressure rise was considerably reduced while the transient duration was extended.

7. RESULTS

7.1 Rapid transients

The recorded fast transient, created by the rapid valve closure is shown in Fig. 5, together with the predictions by the characteristics and finite element methods. The closure was carried out in two stages; almost 90% shut in first 0.6 second then followed by closing the rest slowly in another 0.4 second. The predictions by the theoretical models, based on the recorded movements of the valve, are excellent. The disagreement after one second is due to the accuracy of the valve calibration at nearly closed setting (although the valve movement was divided into more than forty positions during the calibration). As the valve approaches the fully closed position the loss coefficient (expressed as a multiplier of the velocity head) tends to infinity. The resulting boundary equation will be ill-conditioned and the damping to the surge will be sensitive to the variation in valve loss coefficients.

The entire length of the pipe was used as mesh length in both models. The finite element solution requires four times more computer time than the characteristics method. Its step size could be chosen freely but was restricted, in order to have sufficient number of points for the curve, to about two times larger than that by the characteristics method.

7.2 Slow transients

The recorded slow transient, given by the oscillation of the surge tank level, is shown in Fig. 6 together with the predictions by the three theoretical methods. All three theoretical solutions employed the whole length as the mesh length. The characteristics method had a finite time step interval which could be increased fifty times for the other two methods without noticeable loss of accuracy. Although the experimental result agrees well in magnitude with the three theoretical solutions, the oscillation periods match less successfully. The mismatch could have been contributed by the exclusion of minor losses (e.g. one sudden contraction, two bends, one tee and one valve) in the system, inaccurate estimate of the pipe friction and the valve closure. Nevertheless, the agreements between the theoretical methods are excellent. Moreover, the large time step by the finite element method and the implicit finite difference proved competitive in computation time (see Table 1). These two methods have been found stable at a time step over two hundred times that of the characteristics method. Accuracy, however, started to deteriorate with time step increased above a hundred and the number of points became inadequate for the response curve with time step increased above two hundred.

8. MARRYING-UP BETWEEN CHARACTERISTICS METHOD AND OTHER SCHEMES

From the comparison of results between the various methods and the experimental data it becomes apparent that the implicit finite difference and the finite element methods are of comparable accuracy to the characteristics method. They both suffer from more computation time but gain in larger time step. However, the benefits from using large time steps disappear in fast transient which necessitates small time step to determine their responses.

Apart from offering larger time step the finite element method and the implicit finite difference scheme can also be used with larger mesh length Δx than the characteristics method. There is little scope to exploit such advantage if these methods are used individually since the full conduit length is already used as Δx. However, the inclusion of one or possibly two of these two schemes in the characteristics method may be suitable for relaxing some of the limitations with regard to Δx and Δt.

In a network system the time step Δt of the characteristics method is dictated by the shortest pipe in the system. In contrast the presence of an excessively long pipe can unfavourably prolong the computation due to the large number of mesh lengths required to satisfy the stability criterion. The implicit scheme and the finite element method can be used locally to replace the characteristics method for such a pipe (or pipes) and improve the efficiency of the characteristics method.

The application involves solving an element matrix (4 x 4 for implicit finite difference and 8 x 8 for finite element) for the short conduit with the appropriate boundary equations. In order to test the validity of the marrying-up principle, the previous two simulations were repeated with the pipe length now divided into three sections. The centre section was modelled by the implicit techniques, while the end sections by the characteristics method. By increasing the centre section and suitably reducing the end sections the effect of using vastly different mesh length Δx could be simulated. The interfaces of the different models in this case are simple two-way junctions. In the case of characteristics method with implicit finite difference the two implicit water hammer equations (4.1) and (4.2) were combined with the two characteristics equations (upstream and downstream characteristics from the end sections) to form a 4 x 4 matrix. If the finite element method were to be combined with the characteristics method, the corresponding characteristics equations would be used to replace two rows of the element 8 x 8 matrix.

The accuracy of the marrying-up can be seen in Fig. 7 and Fig. 8. In the slow transient case (Fig. 7) the centre section was thirteen times longer than the end sections. In the rapid transient case (Fig. 8) the centre-to-end section ratio was significantly increased to well over a hundred.

The introduction of the implicit techniques, which effectively replace a large number of mesh lengths Δx required by the characteristics method, can also cause an increase of computation time. There is no advantage to be gained by the mixed models if they do not save computation time. The comparison of computer CPU time, for two combinations employed in the rapid transient simulation, is shown in Table 2 and Fig. 9. It appears that computer time can be saved if the implicit finite difference is called upon to replace about forty Δx otherwise required by the characteristics method. The economical application of the combined finite element method and the characteristics method requires a replacement of over one hundred and fifty Δx.

9. CONCLUSIONS

(1) Comparison of results produced by the implicit finite difference scheme and the finite element method, with the experimental data, produce parallel accuracy to that obtained from the characteristics method while at the same time offer the advantages of possibly larger time step and mesh length.

(2) The computation efficiency of the characteristics method, implicit finite difference scheme and the finite element method for a single pipe can be roughly considered in the context of solving a 2x2, 4x4 and 8x8 element matrices respectively. The two implicit techniques have nearly identical characteristics; simultaneous solution necessary and algorithm independent of mesh length to time step ratio. The finite element method is more involved but higher in accuracy.

(3) The large Δt advantage of the two implicit schemes may be effectively used provided the number of points resulted is sufficient for modelling the transient response.

(4) The large Δx advantage of the implicit schemes can be achieved by combining with the characteristics method.

(5) All three methods can be employed individually to solve transients problems satisfactorily. The characteristics method is probably preferred most because of its explicit nature to deal with boundary conditions. The proposed marrying-up of the characteristics method with both implicit finite difference and finite element method adds further flexibilities and enhancements.

REFERENCES

(1) Wylie, E.B. and Streeter, V.L: "Fluid transients". McGraw-Hill Inc. (1978).

(2) Perkins, F.E., Tedrow, A.C., Eagleson, P.S. and Ippen, A.T: "Hydro-power plant transients - Part II - Response to load rejection". Hydrodunamic laboratory Report No. 71, Dept. of Civil Engineering, Massachusetts Institute of Technology. (September 1964).

(3) Rachford, H.H. Jr. and Dupont, T. "A fast, highly accurate means of modelling transient flow in gas pipeline systems by variational methods". Soc. Pet. Eng. J. (April, 1974). AIME.

(4) Streeter, V.L. and Lai, C: "Water hammer analysis including fluid friction". J. Hyd. Div., ASCE, vol.88, no. HY3, pp. 79-112. (May 1962).

(5) Streeter, V.L: "Water hammer analysis". J. Hyd. Div., ASCE, no. HY6, pp. 1959-1972. (Nov., 1969).

(6) Streeter, V.L: "Numerical methods for calculation of transient flow". Proc. 1st Int. Conf. Pressure Surges, Bedford, England BHRA. (Sept. 1972).

(7) Thorley, A.R.D: "A survey of investigations into pressure surge phenoma". Memorandum ML 83, Dept. of Mech. Eng., The City University. (April, 1976).

(8) Vardy, A.E: "On the use of the Method of Characteristics for the solution of unsteady flows in networks". Proc. 2nd Int. Conf. Pressure Surges, Bedford, England BHRA. (Sept. 1976).

(9) Meyer, C: "Solution of linear equations - state-of-the-art". J. Struct. Div., Proc. ASCE, vol. 99, no. ST7. (July, 1973).

(10) Meyer, C: "Special problems related to linear equation solvers". Proc. ASCE 6th Conf. on Electronic Computation. (August, 1974).

(11) Zienkiewicz, O.C: "The finite element method in engineering science". McGraw-Hill Publishing Co. Ltd. (1971).

(12) Jovic, V: "Non-steady flow in pipes and channels by finite element methods". 17th Congress of the Int. Association for Hydraulic Research, vol. 2, pp. 197-204. (August, 1977).

(13) Felippa, C.A. and Clough, R.W: "The finite element method in solid mechanics". Numerical solution of field problems in Continum Physics. STAM-AMS proc. vol II, pp. 210-252, by American Mathematical Society Providence, Rhode Island. (1970).

(14) Wylie, E.B., Stoner, M.A. and Streeter, V.L: "Network system transient calculations by implicit method". Soc. of Petroleum Eng. J. pp.356-363. (Dec. 1971).

APPENDIX I:

The Interpolating functions for Finite Element Technique

The interpolating functions $N_i(x)$ introduced in equation (5.1) and yet to be chosen, are one dimensional piecewise polynomials. Since the highest order derivative in the governing differential equations is only of the first order, a simple linear polynomial would be adequate to meet the compatability and completeness conditions necessary to ensure convergence to the solution (Ref. 13). There are a number of benefits to be gained, however, by the adoption of higher order polynomials, shown in Fig.3, particularly of the Hermite cubic type. The boundary conditions can be introduced in a natural way and since the Hermite polynomials have both their value and their derivative value equal to unity at each nodal point, continuity of solutions up to the first derivative in x is assured at nodal and boundary points. Thus introducing the dimensionless quantity $U = x/L$, the interpolating functions will be taken as

$$\begin{bmatrix} N_1 \\ N_2 \\ N_3 \\ N_4 \end{bmatrix} = \begin{bmatrix} 1 & 0 & -3 & 2 \\ 0 & 1 & -2 & 1 \\ 0 & 0 & 3 & -2 \\ 0 & 0 & -1 & 1 \end{bmatrix} \begin{bmatrix} 1 \\ U \\ U^2 \\ U^3 \end{bmatrix} \quad \text{or} \quad [N] = [S_1][Z_4] \qquad (A.1)$$

and thus the derivative of the function is

$$\begin{bmatrix} N_{1x} \\ N_{2x} \\ N_{3x} \\ N_{4x} \end{bmatrix} = \begin{bmatrix} 0 & -6 & 6 & 0 \\ 1 & -4 & 3 & 0 \\ 0 & 6 & -6 & 0 \\ 0 & -2 & 3 & 0 \end{bmatrix} \begin{bmatrix} 1 \\ U \\ U^2 \\ U^3 \end{bmatrix} \quad \text{or} \quad [N_x] = [S_2][Z_4] \qquad (A.2)$$

The use of Hermite polynomials enables the matrices defined in equation (5.6) to be evaluated exactly with ease. The results obtained are as follows:-

$$[A] = \begin{bmatrix} 156 & 22 & 54 & -13 \\ 22 & 4 & 13 & -3 \\ 54 & 13 & 156 & -22 \\ -13 & -3 & -22 & 4 \end{bmatrix} \times \frac{L}{420} \; ; \quad [E] = \frac{a^2}{g^2} \times [C]$$

$$(A.3)$$

$$[C] = \begin{bmatrix} 210 & 42 & 210 & -42 \\ -42 & 0 & 42 & -7 \\ -210 & -42 & 210 & 42 \\ 42 & 7 & -42 & 0 \end{bmatrix} \times \frac{g}{420}$$

Time dependent matrices like $[B_n]$, $[D_n]$ and $[F_n]$ have to be evaluated at every time step. A technique has been developed to reduce the number of repetitive integrations, by considering a general component of the form $\int_0^L N_i(x)\, U^{j-1}\, dx$ where U^{j-1} are powers of U and j ranges from one to a maximum of seven for the pressure surge problem. Thus a matrix can be introduced, with elements $NU_{ij} = \int_0^L N_i(x)\, U^{j-1}\, dx$ and given by

$$[NU] = \begin{bmatrix} 1260 & 378 & 168 & 90 & 54 & 35 & 24 \\ 210 & 84 & 42 & 24 & 15 & 10 & 7 \\ 1260 & 882 & 672 & 540 & 450 & 385 & 336 \\ -210 & -126 & -84 & -60 & -45 & -35 & -28 \end{bmatrix} \times \frac{L}{2520} \tag{A.4}$$

It can now be seen from (A.1), (A.2) and (A.4) that the constant matrices $[A]$ and $[C]$ can be obtained by premultiplying $[S_1]^T$ and $[S_2]^T$ matrices by the 4 x 4 submatrix of $[NU]$ comprising the first four columns with suitable scalar multipliers 1 and g respectively.

$[B_n]$, $[D_n]$ and $[F_n]$ matrices involve the velocity profile at the nth time interval which is approximated by $V(x,n) = \sum_{i=1}^{4} \alpha_i(n)\, N_i(x)$ This profile can be represented, at each time interval, by a third order polynomial of the type $aU^3 + bU^2 + cU + d$ by premultiplying (A.1) by the row vector $\alpha_1, \alpha_2, \alpha_3, \alpha_4$. The product of V(x,n) and $[N_x]$ will be a fifth order polynomial in U for each component $(N_x)_i$ and therefore can be represented in the form

$$[VN] = [S_3]\,[Z_6], \quad [Z_6] = \begin{bmatrix} 1 \\ U \\ U^2 \\ U^3 \\ U^4 \\ U^5 \end{bmatrix} \tag{A.5}$$

where $[S_3]$ is the resulting 4 x 6 matrix. Thus premultiplying $[S_3]^T$ with the submatrix formed by the first six columns of $[NU]$ will give the required matrix $[B_n]$. Furthermore, since the square of V(x,n) will form a sixth order polynomial of the form

$$V^2(x,n) = [\gamma_7][Z_7] \tag{A.6}$$

where $[\gamma_7]$ is a 1 x 7 row vector and $[Z_7]$ a 7 x 1 column vector, then

$$[D_n] = \frac{\lambda}{2d}\, \text{sign}(V)\, [NU]\, [\gamma_7]^T \tag{A.7}$$

where sign (V) is +1 when V is positive and -1 when V is negative. Finally, with $[\gamma_4]$ as a 1 x 4 row vector containing the basic coefficient a, b, c and d the matrix $[F_n]$ is simply given by

$$[F_n] = -\sin\phi\, [NU]\, [\gamma_4]^T \tag{A.8}$$

TABLE 1

Comparison of CPU time for slow surge tank oscillation by various methods (computer HARRIS S125)

Description of the method	CPU time for a 200 seconds surge history (sec.)	Ratio of CPU time / Transient time
Method of characteristics alone (MOC)	33.29	0.17
Implicit finite difference alone (at 50 times MOC Δt)	11.6	0.06
Finite element method alone (at 50 times MOC Δt)	4.25	0.02

TABLE 2

Comparison of CPU time for rapid valve closure by various methods (Computer HARRIS S125), unit in seconds

No. of Δx replaced	Characteristics method alone	Characteristics method with implicit finite difference scheme	Characteristics method with finite element method
1	4.34	8.22	20.50
4	6.08	12.08	32.49
13	18.75	30.11	92.93
28	50.04	55.90	184.24
58	115.81	87.08	303.58
148	839.01	239.96	916.76

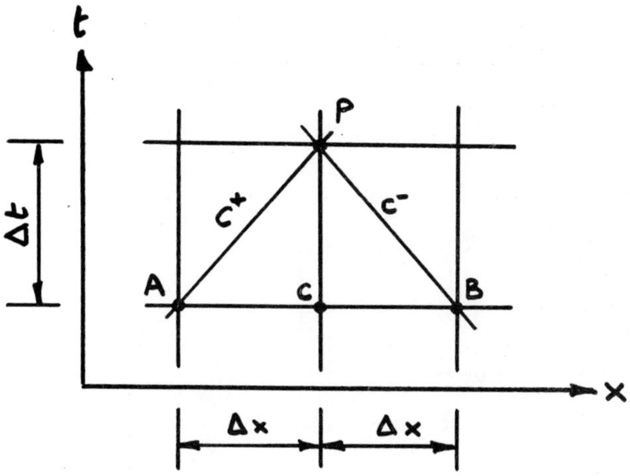

$$c^+ \quad \frac{g}{a} H_P + V_P = \psi_A \qquad (a)$$

$$c^- \quad -\frac{g}{a} H_P + V_P = \psi_B \qquad (b)$$

where ψ_A and ψ_B are functions of H and V at A and B respect'y, and (a) and (b) are valid if

$$\frac{\Delta x}{\Delta t} = \pm a$$

Fig. 1 x-t mesh grid, characteristic method when fluid velocity is small compared with wave velocity.

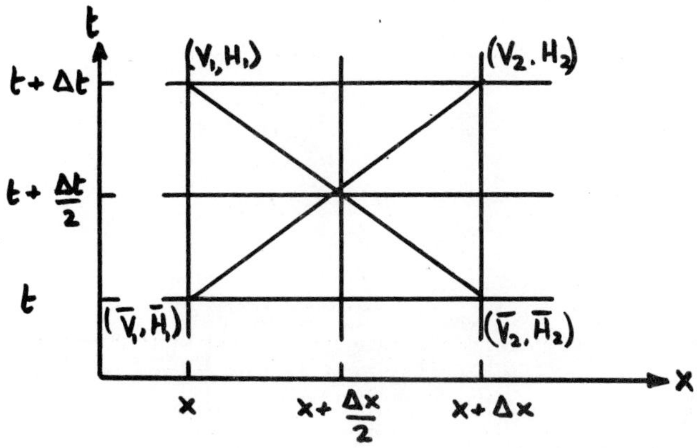

$$\frac{\partial V}{\partial t} = \frac{1}{2\Delta t}\left[V_1 + V_2 - \bar{V}_1 - \bar{V}_2\right]$$

$$\frac{\partial H}{\partial t} = \frac{1}{2\Delta t}\left[H_1 + H_2 - \bar{H}_1 - \bar{H}_2\right]$$

$$\frac{\partial V}{\partial x} = \frac{1}{2\Delta x}\left[V_2 - V_1 + \bar{V}_2 - \bar{V}_1\right]$$

$$\frac{\partial H}{\partial x} = \frac{1}{2\Delta x}\left[H_2 - H_1 + \bar{H}_2 - \bar{H}_1\right]$$

The following approximations for the derivatives in the two basic differential equations are assumed:

$$V\left[\frac{\partial H}{\partial x} + \sin\phi\right] = \frac{1}{2}\left[\frac{1}{2}(V_1+V_2)\left(\frac{H_2-H_1}{\Delta x} + \sin\phi\right) + \frac{1}{2}(\bar{V}_1+\bar{V}_2)\left(\frac{\bar{H}_2-\bar{H}_1}{\Delta x} + \sin\phi\right)\right]$$

$$\frac{1}{2}\frac{\partial}{\partial x}(V^2) = \frac{1}{4\Delta x}\left[V_2^2 + \bar{V}_2^2 - V_1^2 - \bar{V}_1^2\right]$$

$$\frac{\lambda}{2d} V|V| = \frac{1}{2}\left[\frac{\lambda}{4d}\left(V_1|V_1| + V_2|V_2|\right) + \frac{\lambda}{4d}\left(\bar{V}_1|\bar{V}_1| + \bar{V}_2|\bar{V}_2|\right)\right]$$

Fig. 2 x-t mesh grid, implicit finite difference method.

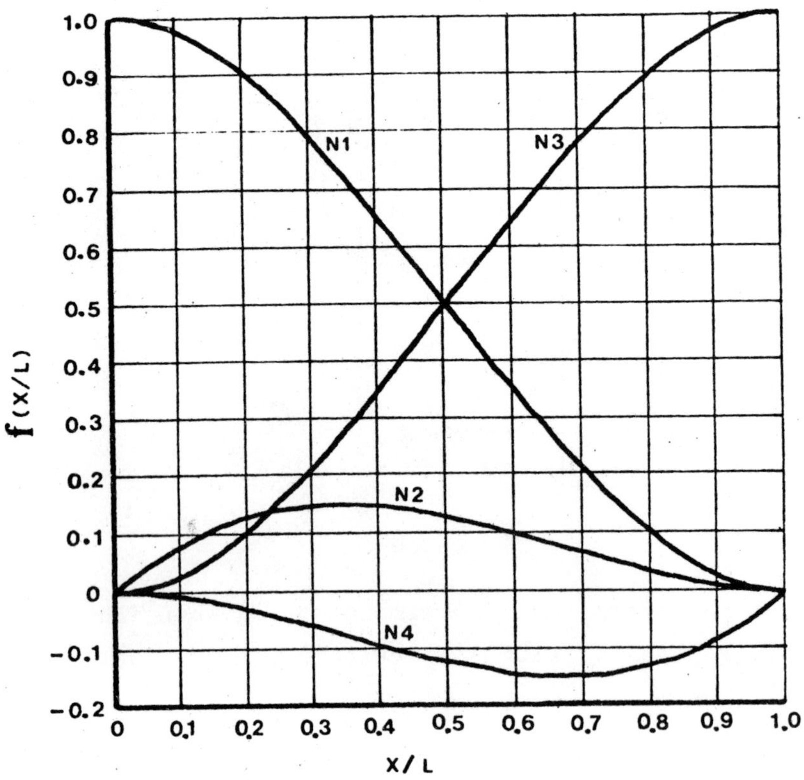

Fig. 3 Hermite polynomials

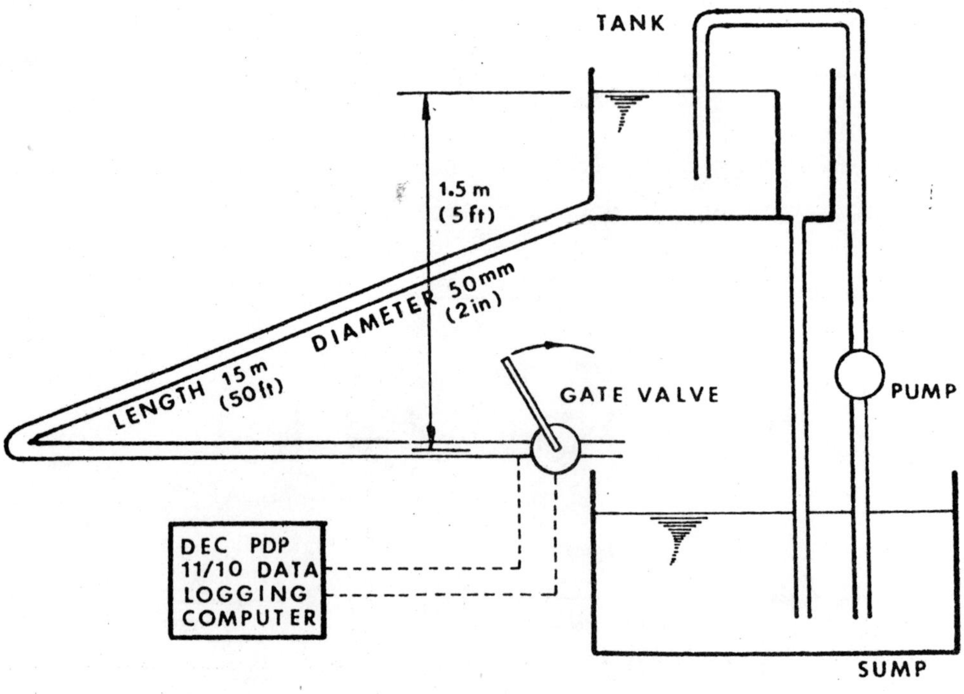

Fig. 4 Experimental apparatus.

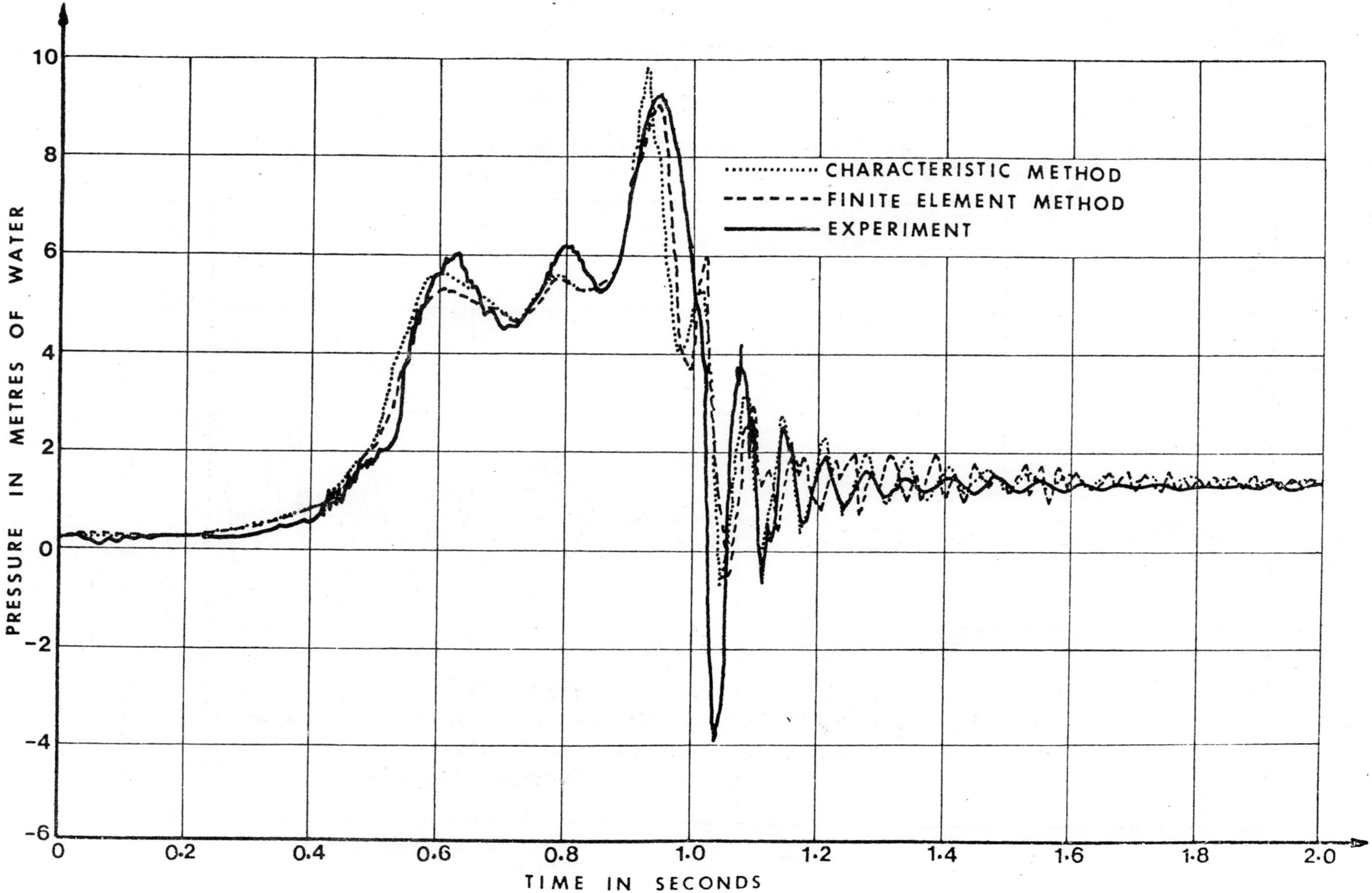

Fig. 5. Pressure history upstream of valve caused by valve closure in 1·0 second

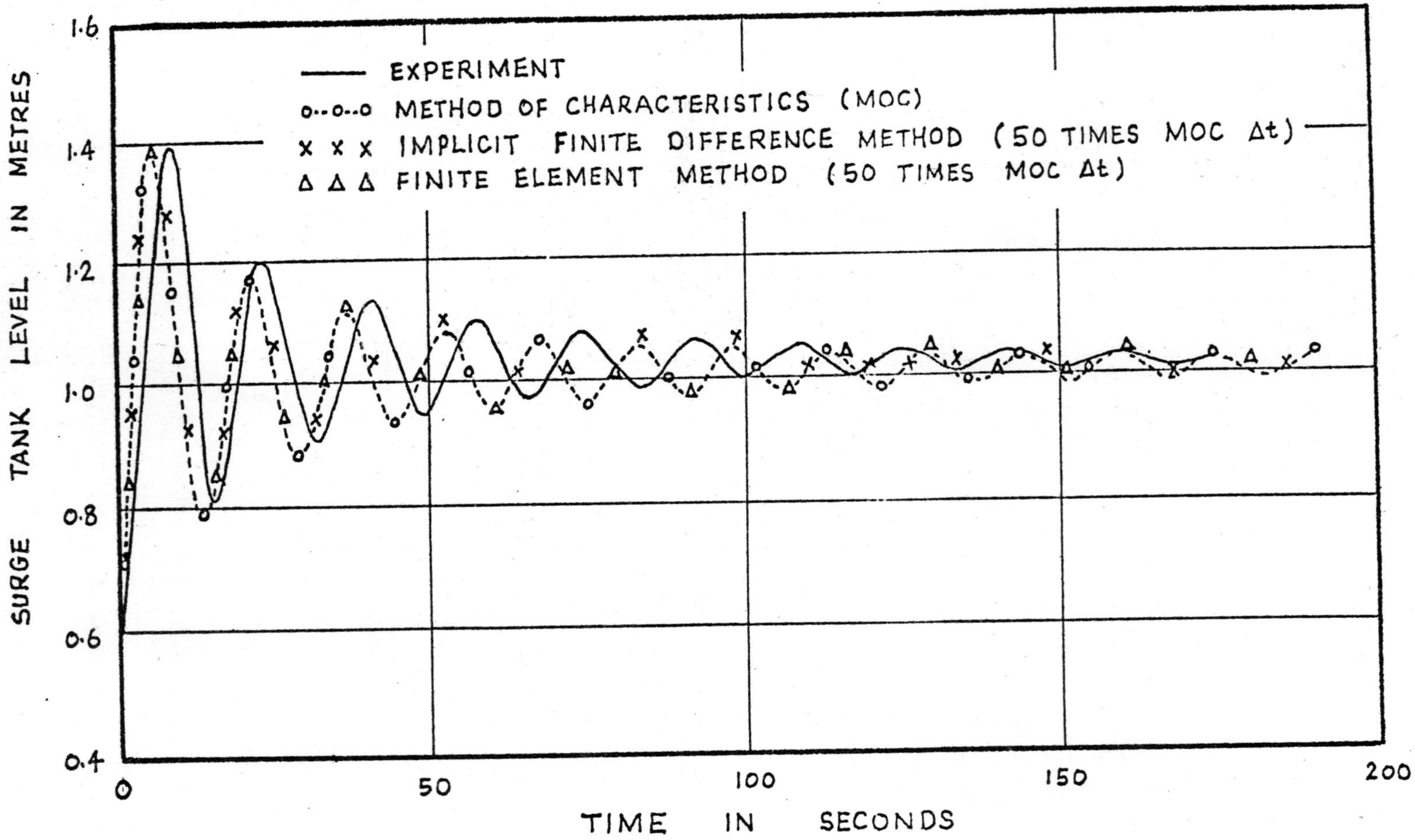

Fig. 6. Surge tank water level variation.

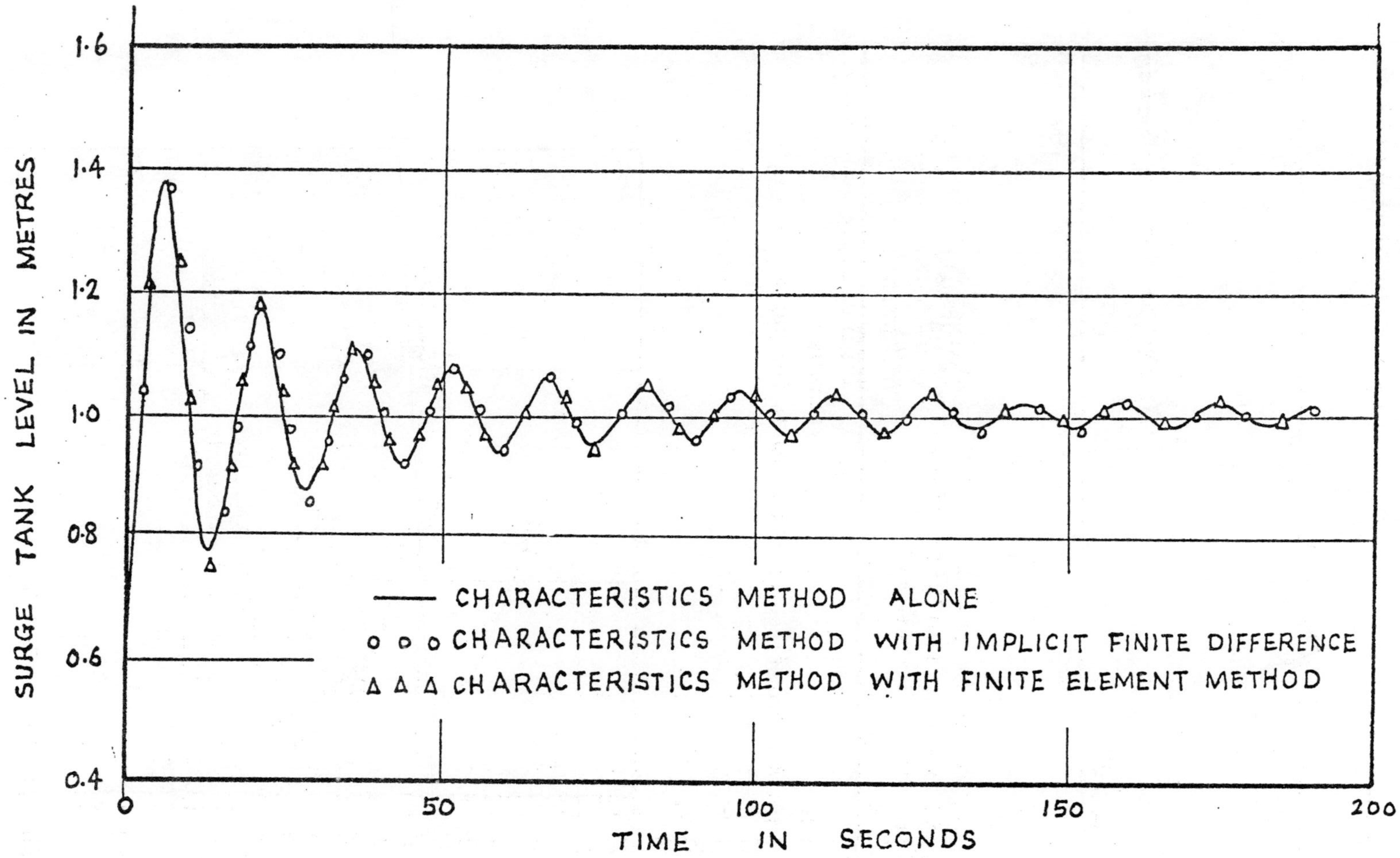

Fig. 7. Comparison of mixed models for slow transients.

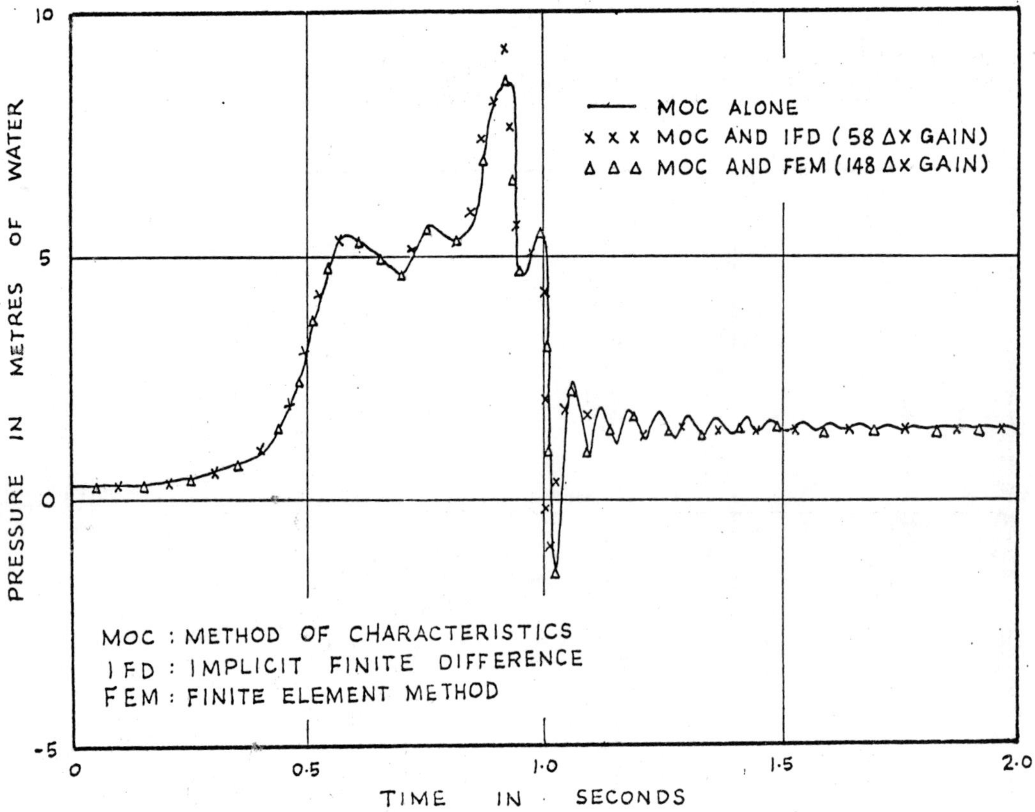

Fig. 8. Comparison of mixed models for rapid transients.

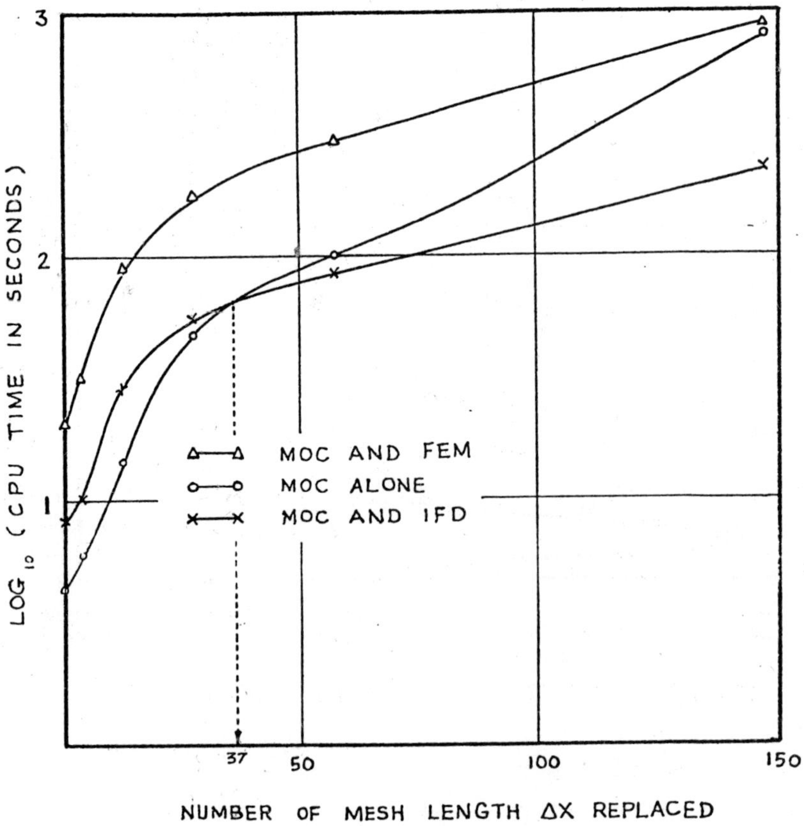

Fig. 9. Comparison of computer time.

Third International Conference on

PRESSURE SURGES

Canterbury, England
March 25-27, 1980

PAPER B3

THE CALCULATION OF PRESSURE SURGE IN COMPLEX NETWORKS

K.G. Priddin, B.Sc.

BHRA Fluid Engineering, U.K.

Summary

This paper describes a method of modelling pressure transients on complex networks. It contains details of the modelling techniques used and also problems that were encountered during the development. The methods used to eliminate these problems are discussed and a brief comparison with prototype measurements is given.

Held at the University of Kent at Canterbury, U.K.

Conference Organised by BHRA Fluid Engineering, Cranfield, Bedford, England.
In conjunction with The City University, London.

Nomenclature

a_{ijk}	coefficient in equation 3.2.2.
a	speed of sound
A	area of pipe or connection
(A)	matrix in equation 3.3.4
b_{ij}	coefficient in equation 3.2.2.
(B)	matrix in equation 3.3.4
(C)	matrix in equation 3.3.4
C_D	discharge coefficient
D	pipe diameter
(D)	matrix in equation 3.3.4.
f	pipe friction factor
$f(P_1 T_1 \frac{P2}{P1})$	function for compressible flow through a discrete loss
I	total number of nodes
J_i	number of branches at node i
K	loss coefficient
L	number of sections in pipe
m_{ij}	mass flowrate at node i branch j
M	mass of gas
p	pressure
P_1	upstream pressure in 3.5.1,3.5.2,3.6.1.
P_2	downstream pressure in 3.5.1,3.5.2,3.6.1.
(P)	pressure matrix in equation 3.3.4.
R	gas constant
T	temperature
t	time
to	time at beginning of timestep
V	volume
x	distance along pipe
γ	polytopic gas constant
ρ	density
Δx	element of length
Δt	timestep

Suffices of $_{to}$ and $_{to + t}$ indicate that the variable is evaluated at the beginning or end of the timestep.

1. Introduction

This paper describes a computer technique for modelling pressure surge in complex networks which the author has not found described in any other report. The method was adopted for a particular system containing a large number of pressure actuated control valves with many interconnecting pipes. Both gas and liquid phases were present in the system, though at no time would they ever mix. This complexity, of itself, would porbably not have necessitated the use of a new technique, but there were tow other constraints which excluded the standard methods of calculation. The main constraint was a problem concerningthe timestep. It was known that the individual valves operated in a complex manner very much faster than the overall system operation time, and so a variable timestep method was required. To use a single timestep method would have required excessive amounts of computer time. Since the method of characteristics technique for pipe flow assumes a constant timestep, it could not be used in its standard form. Even variants of this technique would not have given the required flexibility. This method was therefore ruled out as a possible approach. The other main technique is an implicit one which is usually used in resonance analysis of pipe systems. The method adopted here could be broadly classified under this heading, though it was modified to comply with the second constraint. This specified that the program had to be written in a modular form, which meant that the program was to be divided into subroutines mainly on an engineering and not mathematical or numerical basis. Each physical component would form one subroutine so that any modification to one valve, for example, would require a change to only one subroutine. The main reason for the second constraint was that there would be a considerable amount of analysis with slightly modified valves and such modification needed to be as simple and quick as possible.

The modelling technique described here met both of these requirements and it is thought that this method provided an accurate, stable and efficient model. The same approach, or a similar one, could be used on other systems where the constraints of variable timestep and ease of modification are required.

2. Modelling technique

In broad outline the technique has been stated to be modular with the emphasis on programming from an engineering point of view. Each component was modelled as a separate subroutine except for the pipes, where one subroutine could model any number of pipes.

With the system broked down into its constituent parts, there was a need for some way of joining them back together to obtain the original system again. The approach here was based on the concept of a 'node'. The various components or 'modules' were assumed to be connected together at nodes, and each node was simply an abstraction of a pipe connection. It contained no volume, and so there could be no net flow into or out of a node, on the basis of conservation of mass. This provided a fundamental equation which was used to join the various modules together. It is examined in more detail in section 2.2.

The program was required to model the whole system and following list of variables were usually sufficient to specify completely its state.

About 60 pressures
60 mass flow rates
20 valve positions
and 20 valve speeds

Section 2.1 details how these different values were calculated within the program, and all that need be mentioned here is that they were not all calculated in one large matrix equation. That would have resulted in a loss of modularity, since if one valve was changed slightly to include another piston, for example, the main matrix equation would have changed. This model was programmed so that the main program did not need to calculate anything concerning the interior state of any module. It only calculated the pressures and mass flow rates at the nodes.

This modular approach was found to be very useful in making the program easier to write and understand, which resulted in a saving of time and money. The only possible penalty was in a slightly longer run time than might otherwise have been achieved, but this was more than offset by the ease of modification. If the modification could not be implemented by a data change, it could almost certainly be done by a change to only subroutine. In short this technique was valuable in that it made the program easier to write, debug and understand.

2.1. System solution technique

The system solution is built up from a number of stages which join together to advance the system by one small timestep, Δt. By advancing from to to to+Δt, it is possible to repeat the procedure, ad infinitum, until all the required results are obtained. So here only one timestep advance will be described.

It is assumed that everything is known about the system at time to. The first stage in the calculation is to produce a number of equations relating pressures and mass flows at each of the nodes at time to+Δt. This is done by the subroutines modelling specific valves and pipes. Next a matrix equation for pressures is obtained by using continuity at the nodes. The matrix is then solved to give the pressures at the nodes at time to+Δt. The nodal mass flowrates can then be calculated. Lastly each of the subroutines calculates its own internal pressures and mass flows, and also updates piston positions and velocities. This completes one timestep advance, since all is now known about the system at time to+Δt.

2.2. Nodal connection technique

Each of the subroutines was programmed to produce a number of equations relating pressure and mass flow at each of its connecting nodes. In figure 1 a particular module is shown with three branches connected to nodes A,B and C. Three equations would therefore be produced by this subroutine, one for each branch. The equation is produced to have the form:-

$$m_{A4} = \alpha P_A + \beta P_B + \gamma P_C + \delta \qquad (2.2.1.)$$

In general this can be written as:-

$$m_{ij} = \sum_{k=1}^{I} a_{ijk} P_k + b_{ij} \qquad (2.2.2.)$$

where m_{ij} is the mass flow at node i branch j,

a_{ijk} is the coefficient of P_k for the equation at node i branch j, and

b_{ij} is a constant term.

This equation can be generated for each node i and each branch j. If node i has J_i branches and there are I nodes in all, then we have

$$1 < j < Ji \quad \text{and} \quad 1 < i < I,$$

giving a total of $\sum_{i=1}^{I} Ji$ equations.

By adding all the mass flow rates into node i, we have a set of I equations of the form:

$$\sum_{j=1}^{Ji} m_{ij} = \sum_{j=1}^{Ji} \sum_{k=1}^{I} a_{ijk} P_k + \sum_{j=1}^{Ji} b_{ij} \qquad (2.2.3.)$$

$$\text{for } i, 1 < i < I$$

But we know that the sum of the mass flowrates into any node is zero and so:

$$\sum_{j=1}^{Ji} m_{ij} = 0 \text{ for } i,\ 1 < i < I \qquad (2.2.4.)$$

Therefore $\sum_{j=1}^{Ji} \sum_{k=1}^{I} a_{ijk} P_k + \sum_{j=1}^{Ji} b_{ij} = 0 \quad 1 < i < I$ (2.2.5.)

This gives I equations in I unknowns and so the resulting matrix equation can be solved to produce $P_i, t_o+\Delta t$ for $1 < i < I$. Once all the values of P_i have been calculated, the values m_{ij} can be found using equations 2.2.2..

If the system is fairly simply connected, then the resultant matrix of coefficients is diagonally dominant. However, for more complex systems this becomes less distinct and a more general method of solution is required. A banded Gaussian elimination subroutine formed the basis of this method, which gave the required solution, even though it is not a particularly fast method. The bandwidth was found to be quite large even with an optimised node numbering system. This node optimising procedure is a separate programe developed at BHRA specifically for this sort of problem.

2.3. Modelling gas flow in pipes

It was known that the system operated at considerably different speeds during different parts of the operating cycle. Pressurising large volumes would be relatively slow, whereas a small valve operation would happen quickly. With this large variation in rates of change, a variable timestep approach was needed. Unfortunately the prime method for calculation of fluid flow in pipes is based on the method of characteristics which in its simplest form uses a constant size of steps. The method adopted here was based on an implicit approach for the partial differential equations. These are given below:-

Equation of state $$P = R\rho T \qquad (2.3.1)$$

Equation of continuity $$\frac{\partial P}{\partial t} + \frac{a^2}{A}\frac{\partial m}{\partial x} = 0 \qquad (2.3.2.)$$

Equation of motion $$\frac{\partial P}{\partial x} + \frac{1}{A}\frac{\partial m}{\partial t} + \frac{f a^2 \, m|m|}{2DA^2P} = 0 \qquad (2.3.3.)$$

By replacing the partial differential equation by finite difference equations at the point $(i + \frac{1}{2})\Delta x, t+ \frac{1}{2}\Delta t$ it is possible after much algebra to obtain a matrix equation for the internal pressures in the pipe. The algebra also yields two equations for the mass flow at the boundaries of the pipe.

This equation takes the form:-

$$(A)(P)+(B)P_o+(C)P_1 = (D) \qquad (2.3.4.)$$

where (A) is a known tridiagonal matrix of order L-1,

(P) is a column vector of order L-1, and (B),(C) and (D) are known column vectors It is assumed the pipe is split into L sections, so (P) is as shown below:-

$$(P) = \begin{pmatrix} P1 \\ P2 \\ \vdots \\ PL-1 \end{pmatrix}$$

By inverting matrix A we obtain:-

$$(P) + (A)^{-1}(B)P_o + (A)^{-1}(C)P_1 = (A)^{-1}(D) \qquad (2.3.5.)$$

This is then used to substitute back into the two mass flow boundary equations to give:-

$$m_o = \alpha p_o + \beta p_1 + \gamma \qquad (2.3.6.)$$

$$m_1 = \partial p_o + \varepsilon p_1 + \eta \qquad (2.3.7.)$$

This is now the required form for the main program, since the two flows are given in terms of the pressures at the ends of the pipe. This form is given in equation 2.2.1

When the nodal pressures have been calculated, it is possible, using equation 2.3.5., to calculate all the internal pressures in the pipe. The internal mass flowrates are also calculated at this time.

2.4. Modelling of liquid flow in a pipe.

This was a very much simpler exercise than for gas flow, since the more rigid property of water made it possible to ignore elasticity effects without loss of accuracy. Thus the flow in a pipe consisted only of an ordinary differential equation relating the inertial and frictional forces to the acceleration. With the mass flow constant along the pipe, the required form for the equation was not difficult to obtain.

For a pipe joining node i to node k we have

$$\frac{\pi D^2}{4}(P_i - P_k) - \frac{fL}{D}\frac{\rho}{2}\frac{m1}{\rho}\frac{m1}{\rho} = \rho\frac{\pi D^2 L}{4}\frac{d}{dt}\left(\frac{m}{\rho}\right) \qquad (2.4.1.)$$

and by taking the frictional term at time to a simple linearised expression can be obtained.

2.5. Modelling of gas flow through a discrete loss

It can be shown that the mass flowrate through a discrete loss can be put in the form:-

$$m = C_D A \; f\left(P_1, T_1, \frac{P_2}{P_1}\right)$$

where f is a function which depends on whether the flow is choked or not.

<u>Subsonic flows</u>

$$f\left(P_1, T_1, \frac{P_2}{P_1}\right) = C_1 \frac{P_1}{\sqrt{T_1}} \left(\frac{P_2}{P_1}\right)^{\frac{1}{\gamma}} \sqrt{1 - \left(\frac{P_2}{P_1}\right)^{\frac{\gamma - 1}{\gamma}}} \qquad (2.5.1.)$$

where $C_1 = \sqrt{\frac{2\gamma}{R(\gamma - 1)}}$

<u>Sonic flow</u>

$$f\left(P_1, T_1, \frac{P_2}{P_1}\right) = C_2 \frac{P_1}{\sqrt{T_1}} \qquad (2.5.2.)$$

$$C_2 = \sqrt{\frac{\gamma}{R\left(\frac{\gamma+1}{2}\right)^{\frac{\gamma+1}{\gamma-1}}}}$$

Both of these forms can be linearised to give the required form of equation.

2.6. Modelling of liquid flow through a discrete loss

Using incompressible theory, the equation is:-

$$P_1 - P_2 = K \frac{m1m1}{2\rho} \qquad (2.6.1.)$$

This can be linearised by taking the term m1m1 evaluated as $m_{to+\Delta t}1m_{to}1$. Thus:-

$$(P_2 - P_1) \frac{2\rho}{K1m_{to}1} = m_{to+\Delta t} \qquad (2.6.2.)$$

which is now in the required form.

2.7. Modelling of piston movement

The basic equation used for this was Newton's second law including friction. The pressure differences across piston faces provided the driving force for any motion, and the mass was simply taken as that of the piston. Piston advance was then done either in an explicit manner or a predictor-corrector method, depending on the required accuracy.

2.8. Modelling of a complete valve

In sections 2.5 to 2.7, some details of modelling specific items of a valve have been mentioned. However, to build up a complete model of a valve it was necessary to join together many of these and often variable volumes were also present. Each part of the valve would be modelled by the appropriate equations and then an intermediate matrix inversion was required to produce equations of the form 2.2.1. However, sufficient has been mentioned to show how such a model could be built up and so we shall now consider the difficulties which arose during development.

3. Problems encountered during development.

The problems can be divided into four basic areas.

1. Problems associated with obtaining data.
2. Problems associated with timestep.
3. Problems associated with boundary conditions.
4. Problems associated with the linearisation of the equations.

3.1. Data

Many of the items could be shown specifically on drawings and this gave significant amounts of data. However, certain parameters cannot be obtained from drawings.

Firstly values of piston friction are likely to vary from one prototype to another, and so to obtain good agreement, measurement of this parameter was required. In fact, only certain valves were susceptible to frictional forces since many of the valves had much larger driving forces so that friction was negligible. However, where friction was significant there was no real substitute to measurement on the prototype. This was a fairly easy matter to find and so the cost involved with this was not prohibitive.

Another value which cannot be interpreted straight from a drawing is the value of discharge coefficient for a particular discrete loss. The nominal area of the loss could be found easily, but the value of C_d depends on the geometry up and downstream of the contraction. Some physical measurements were made on small scale physical models, but this was not practicable for all the valves in the system. It was found during the tests that the discharge coefficient ranged between 0.9 for a good geometry to 0.6 for a bad one. It was possible with some engineering judgement to estimate C_d values from the drawings and this was found to be sufficient for the present study. It is noted that this was fortunate since measurement for each one would have taken a lot of time and effort.

3.2. Timestep

As with all time integration routines, if the timestep is too large then the procedure is likely to become unstable. Consider the situation shown in figure 2.

If the pressure in the large volume P_o suddenly rises to a value $2P_o$ in a time Δt, then the flow will become sonic and the equation will give:-

$$m = C_D A \sqrt{\frac{\gamma}{R\left(\frac{\gamma+1}{2}\right)^{\frac{\gamma+1}{\gamma-1}}}} \frac{2Po}{\sqrt{T_1}} \qquad (2.5.2.)$$

And so in time Δt, a mass of $m\Delta t$ will enter volume V. So the mass M suddenly increase to $M+m\Delta t$. Initially the pressure is P_o where:-

$$P_o = \frac{RTM}{V}$$

but the pressure in the volume V after the timestep is P_1 where $P_1 = \frac{RT(M+m\Delta t)}{V}$

and so if $m\Delta t>M$, then P_1 will be higher than the outside pressure of $2P_o$. Indeed if $m\Delta t>3M$ then P_1 will be larger than double the outside pressure, and so the following timestep will result in sonic flow out of the small volume V. This instability partly comes about from the nature of the sonic flow equation which only includes the upstream pressure, but it is here mainly generated by a mismatch of timestep, volume and restriction loss coefficient.

This instability can be further accentuated by a valve containing a small volume with a movable piston. Here the large predicted pressure in the small volume will accelerate the piston and so cause the volume to increase. The pressure will naturally reduce due to this effect, as well as by flow leaving the entrance port. These two effects, being in phase, make small volumes and movable pistons a rather difficult combination to model. A small timestep is the easiest solution to the problem. This limits the amount the upstream pressure rises in one timestep amd also reduces the mass flow into the volume. An alternative method is to eliminate the small volume from the model causing a little loss of accuracy for a much less sensitive program.

3.3. Boundary conditions

The main problem with boundary conditions is where to stop the model. The model must have boundaries since the technique must use a well defined system. It is not easy though to determine whether a model needs to include the air compressor, for example, or whether a pressurised volume would be adequate, or even a constant nodal pressure. All these decisions belong to the task of determining the model in the first place including what it is to be used for, and what results are required from it. The problem is not so difficult here as for those who model flow in the seas of the world. In this example almost any boundary condition will be artificial, but with all modelling there is always the problem of determing the boundaries and what conditions to impose at them.

3.4. Linearisation of equations

By far the most interesting problems (as I see them) are those which derive from the modelling technique used. Here linearisation is fundamental to the use of matrix equations for the solution technique, and so forms quite a basic area for problems.

Let us consider the simplest problem from linearisation first. This is due to the equation 2.6.2 in its linearised form:-

$$m_{t+\Delta t} = \frac{2\rho}{K|m_t|} (P_{2,t+\Delta t} - P_{1,t+\Delta t}) \qquad (2.6.2.)$$

The equation uses the mass flow rate at the previous timestep to produce the flow at the next timestep. If the initial value of m_t is too small, then the equation above will yield a value for $m_{t+\Delta t}$ which is too large. This will then produce too small a value for $m_{t+2\Delta t}$. Depending on the value of K, the oscillation may or may not be unstable. One possible method of overcoming this problem is to iterate at each timestep, so that there is no discrepancy in the equation. This does however, increase the computer time required for each step and so will require considerably more time for a complete run.

In most cases the use of iteration was rejected, in favour of a smaller timestep since this increased the accuracy of piston positions as well. It should be noted that this problem is particularly sensitive near zero flow. It has been found that while the flow was decreasing steadily, when it became close to zero, the flow suddenly increased and then started falling again. In this way the flow was never found to reverse as should have happened. This problem was eliminated by checking specifically for small flows and then not using this approach during that condition. At low flows a laminar condition might occur which does not have the quadratic behaviour, and thus eliminates this problem. There is then the problem of matching the two regions, laminar and turbulant.

A slightly unexpected problem was found when modelling two discrete losses together with no volume between them. Some prototype volumes were so small that they were ignored, and the resultant model was unstable. It is thought that this was caused by the fact that a pressure change can only cross one node each timestep. So as a pressure change passes the first loss, its pressure and flow equation changes value, while the other remains the same. Since there is no volume between the two losses, the equations are no longer compatible and can lead to some interesting results! A finite volume was found to be needed to alleviate this problem. The other method of solution was to combine the two losses into one using the well known formula for incompressible flow. Unfortunately, the 'equivalent' single loss does not result in the same flow under sonic conditions and a 'best' value for all pressure ratios was required. It was not possible to justify this method rigourously, but it seemed a reasonable engineering solution to the difficulty.

The resultant model, after all these modifications had been made, was found to agree quite well with the prototype. However, only small amounts of data for the prototype were available, and what there was agreed well with predictions.

4. Conclusions.

Outlined in this paper is a description of a technique used to model a complex flow network. The results were acceptable once the development problems had been overcome. The method has proved to be reliable, as well as easy to use and understand.

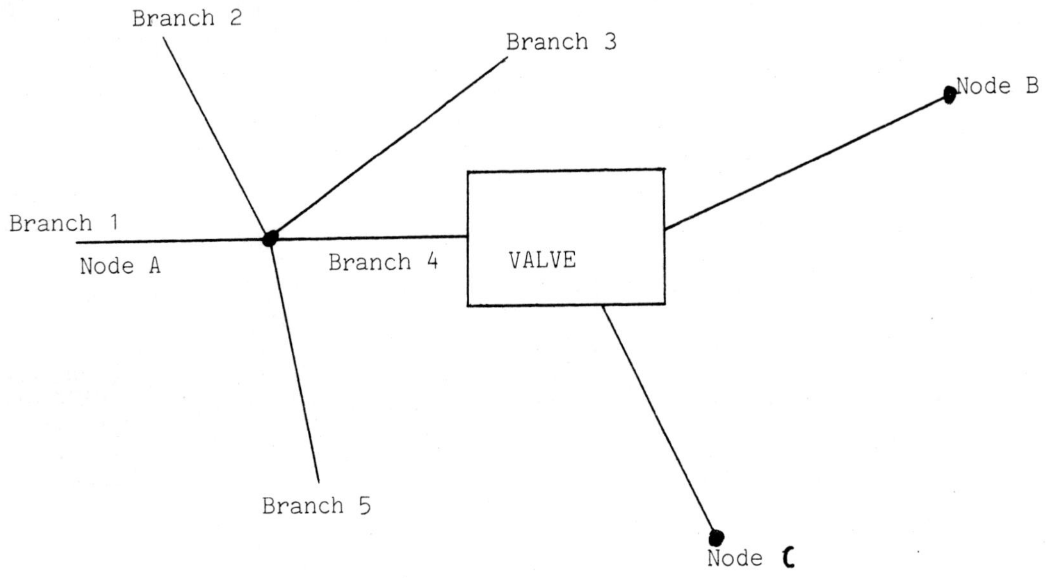

FIGURE 1.

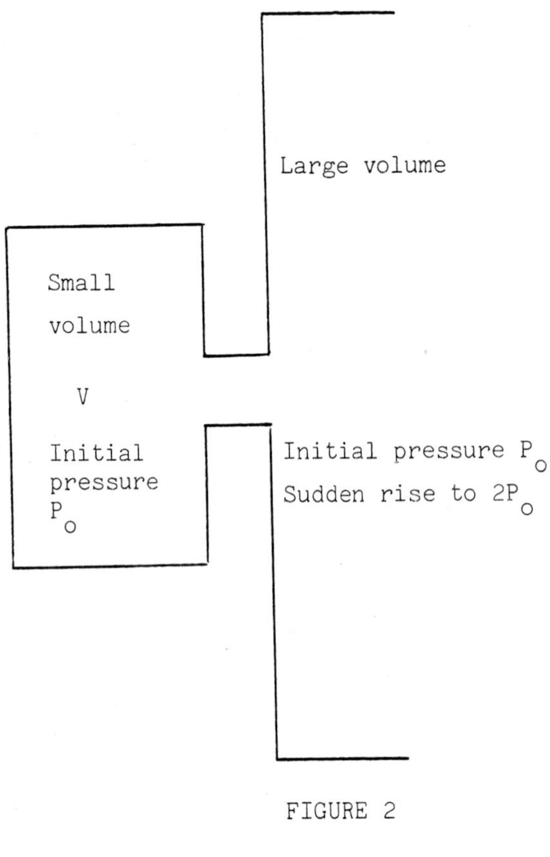

FIGURE 2

Third International Conference on

PRESSURE SURGES

Canterbury, England
March 25-27, 1980

PAPER B4

SIMULATION OF HYDRAULIC TRANSIENTS IN HEATING NETWORKS

A. Serwach

Industrial Institute of Automation and Measurements, Warsaw, Poland.

Summary

This method of simulating hydraulic transients in heating networks is based on the bond graph theory but this has been extended by the addition of new structural elements. A modular concept is used in the design of network models. Basic modules describe typical network elements: the pipe segment, the pump characteristics, the valve resistance, the resistance of the local internal installation, etc. The computer simulation is in the well known CSMP language. Programming pecularities are discussed. The selection of the pressure control system for the heating network is given as an example of the main application of the proposed method.

Held at the University of Kent at Canterbury, U.K.

Conference Organised by BHRA Fluid Engineering, Cranfield, Bedford, England.
In conjunction with The City University, London.

Nomenclature

$f(t)$ - nondimensional time function with the initial value equal one,

g - gravity constant,

H - level difference between upstream and downstream ends of the tube,

k - number of parallel coupled constant velocity pumps,

K_v - hydraulic static admittance coefficient of the tube in turbulent flow conditions

K_{vq} - flow coefficient of the local station,

K_{vs} - nominal flow coefficient of the valve,

n - rotary velocity of the variable velocity pump,

n_m - maximum value of n,

P_c - pressure of the water in the pneumo-hydraulic accumulator,

P_{cp} - pressure rise between suction and outlet ports of the pump

P_{cpm} - maximum value of P_{cp},

P_d - pressure in the downstream end of the tube,

P_{di} - temporary pressure variable,

P_{dio} - initial condition for P_{di},

P_{dit} - temporary delay pressure variable,

P_{lp} - pressure in the outlet port of the leakage compensation pump,

P_q - pressure drop in the local station,

P_u - pressure in the upstream end of the tube,

P_{ui} - temporary pressure variable,

P_{uio} - initial condition for P_{ui},

P_{uit} - temporary delay pressure variable,

P_v - valve pressure drop,

P_o - minimum gas pressure in the pneumo-hydraulic accumulator,

P_1 - maximum gas pressure in the pneumo-hydraulic accumulator,

Q_c - flow rate from the pneumo-hydraulic accumulator,

Q_{cp} - flow rate of circulation pumps,

Q_{cpm} - maximum flow rate of one pump,

Q_d - flow rate at the downstream end of the tube,

Q_{di} - temporary flow variable,

Q_{dio} - initial condition for Q_{di},

Q_{dit} - temporary delay flow variable

Q_{impl} - temporary implicit flow variable

Q_q - flow rate in the local station,

Q_u - flow rate in the upstream end of the tube,

Q_{ui} - temporary flow variable,

Q_{uio} - initial condition for Q_{ui},

Q_{uit} - temporary delay flow variable,

Q_v - valve flow rate,

t - time

T - transport delay time in the tube

V_c - volume of the water in the pneumo-hydraulic accumulator,

V_{co} - initial condition for V_c,

V_o - internal capacity of the pneumo-hydraulic accumulator,

Z - wave impedance of the tube,

$\varkappa$ - polytropic exponent,

ρ - specific density of the water,

$\varphi(h)$ - inherent flow characteristics of the valve - see eqn. (7)

DLF - module of the long hydraulic line with distributed effects of capacitance and inertia, and coupled effects of friction and gravity,

PCH - pressure-flow characteristics of the pump set,

 - first mode of the energy port - the direction of the energy flow is opposite to the causality direction,

 - second mode of the energy port - the energy flow and the causality have the same direction.

Introduction

High temperature two-pipe water district heating systems are most popular in Poland. Main elements of these are central heating plants, two-pipe networks, and local distribution stations which direct the hot water to domestic heat exchangers. The necessity to save energy and to provide living comfort requires the use of automatic temperature control in domestic heat exchangers. Such controls, from the hydraulic point of view, involve the inclusion of variable resistances in the circuit. Variable resistances cause hydraulic disturbances in the network which may be, in extreme conditions, the cause of instability. The way by which the network adjusts to the variable demand of the heating medium is by making the pump rotational speed, and hence its delivery, a function of supply and return pressures. The analysis of hydraulic process dynamics in the network is recommended for the proper selection of the controller mechanism and its parameters.

The network is a combination of hydraulic transport lines and wave processes must be considered in the dynamic analysis for this reason. This is practically impossible if traditional methods are used because the computations involved are very tedious. Here the method of computer simulation is proposed. It is an application of the general difference method originally presented in a paper by the author (ref. 3).

Modular mathematical models of the network

The basic characteristic of the proposed method is the use of modular concepts in the design of mathematical models. Typical network elements are described by mathematical models which are connected together by input and output variables. The network model is assembled by combining standard modules. Bond graphs are used for the graphical representation of mathematical models (Refs. 1 and 2).

A minimum set of modules includes mathematical models of the following network elements: long pipes, batteries of circulating pumps, leakage compensation pumps, valves, pneumo-hydraulic accumulators, and local internal installations. If it is necessary, this set may be extended by the inclusion of other elements, for example: the resistance of leakage, flow or pressure sources to simulate non included portions of the network if it should be too large for its complete inclusion in a simulation.

The pipe segment module has a major influence on the network dynamics. Wave propagation effects, steady friction in turbulent flow conditions, and gravity effects in liquid filled pipes must be included in it. These effects have been considered in the mathematical model shown in Fig. 1 a and, in the aggregated form, in Fig. 1 b, as a two port DLF module. This module has four causality forms and any of these may be described by two of the four equation sets listed below:

$$\left.\begin{aligned} &p_u(t) = p_{uit}(t) + ZQ_u(t) \\ &p_{uit}(t) = p_{ui}(t - T) \\ &\text{with initial condition } p_{uit}(t) = p_{uio} \text{ for } 0 \leqslant t \leqslant T \\ &p_{ui}(t) = p_d(t) + \rho gH + K_v^{-2} \; Q_d(t)\left|Q_d(t)\right| - ZQ_d(t) \end{aligned}\right\} \quad (1)$$

$$\left.\begin{aligned} &p_d(t) = p_{dit}(t) - K_v^{-2} \; Q_d(t)\left|Q_d(t)\right| - ZQ_d(t) - \rho gH \\ &p_{dit}(t) = p_{di}(t-T) \\ &\text{with initial condition } p_{dit}(t) = p_{dio} \text{ for } 0 \leqslant t \leqslant T \\ &p_{dit}(t) = p_u(t) + ZQ_u(t) \end{aligned}\right\} \quad (2)$$

$$\left.\begin{aligned} &Q_u(t) = Q_{uit}(t) + Z^{-1} p_u(t) \\ &Q_{uit}(t) = Q_{ui}(t - T) \\ &\text{with initial condition } Q_{uit}(t) = Q_{uio} \text{ for } 0 \leqslant t \leqslant T \\ &Q_{ui}(t) = Q_d(t) - Z^{-1}\left[p_d(t) + \rho gH + K_2^{-2} \; Q_d(t)\left|Q_d(t)\right|\right] \end{aligned}\right\} \quad (3)$$

$$
\left.
\begin{aligned}
&Q_d(t) = \frac{1}{2}\, Z\, K_v^2\left[\sqrt{1 + 4Z^{-1}K_v^{-2}\left|Q_{impl}(t)\right|} - 1\right] \operatorname{sign} Q_{impl}(t)\\
&Q_{impl}(t) = Q_{dit}(t) - Z^{-1}\left[p_d(t) + \varrho g H\right]\\
&Q_{dit}(t) = Q_{di}(t - T)\\
&\text{with initial condition } Q_{dit}(t) = Q_{dio} \text{ for } 0 \leqslant t \leqslant T\\
&Q_{di}(t) = Q_u(t) + Z^{-1}\, p_u(t)
\end{aligned}
\right\} \quad (4)
$$

The above equations are obtained by the tranformation of equations (3), (4), (15), and (18) from (Ref.3). The impedance form $\longmapsto$ DLF $\rightharpoondown\!|$ is described by equations (1) and (2), the admittance form $\rightharpoondown\!|$ DLF $\longmapsto$ by equations (3) and (4), the flow transmission form $\rightharpoondown\!|$ DLF $\rightharpoondown\!|$ by equations (1) and (4) and the pressure transmission form $\rightharpoondown\!|$ DLF $\rightharpoondown\!|$ by equations (2) and (3).

The circulating pump set is normally designed as a number of parallel coupled pumps one of which has a variable velocity drive the remainder having constant velocity drives. Such a solution gives the possibility of continuous variable velocity (capacity) control by using two stage controllers (Ref.4 and 5). The simplified mathematical model of such a pump combination may be given by the following equations:

$$
\left.
\begin{aligned}
&p_{cp} = 0 \quad \text{for} \quad \frac{Q_{cp}}{Q_{cpm}} > k + \left(\frac{n}{n_m}\right)^2\\
&p_{cp} = p_{cpm}\sqrt{\frac{k + \left(\frac{n}{n_m}\right)^2}{k+1} - \frac{Q_{cp}}{(k+1)\,Q_{cpm}}}\\
&\text{for} \quad \frac{Q_{cp}}{Q_{cpm}} \leqslant k + \left(\frac{n}{n_m}\right)^2
\end{aligned}
\right\} \quad (5)
$$

The graphical representation of this model is the special bond graph element

$\begin{matrix} n \rightarrow \\ k \rightarrow \end{matrix}$ PCH $\xrightarrow[Q_{cp}]{P_{cp}}\!|$ which includes properties of the pressure source and the bypass resistance.

Leakage compensation pumps may be simulated as a constant pressure source $S_{1p} \xrightarrow{P_{1p}}\!|$ and described by the simple equation.

$$
P_{1p} = \text{const} \qquad (6)
$$

Valves are described by the well known equation using the flow coefficient and the inherent flow characteristics

$$Q_v = K_{vs}\,\varphi(h)\sqrt{|p_v|}\ \text{sign}\ (p_v) \qquad (7)$$

The control of the valve stroke may be external, for example, automatic from the controller, or internal as in the case of a check valve where

$$\begin{aligned} \varphi(h) &= 1 \quad \text{for} \quad p_v \geqslant 0 \\ \varphi(h) &= 0 \quad \text{for} \quad p_v < 0 \end{aligned} \qquad (8)$$

The equivalent bond graph is the modulated hydraulic resistance

$\xrightarrow{h} R_v \;\overset{P_v}{\underset{Q_v}{\longleftarrow\!|}}$ in the admittance form. This form is important because the zero admittance (valve closing) may be easy simulated.

A pneumo-hydraulic accumulator may be used in the network as a pressure pulsation damper. It is described by the following equations:

$$p_c = \frac{p_o}{\left[\dfrac{V_c}{V_o} + \left(\dfrac{P_o}{P_1}\right)^{\frac{1}{\varkappa}}\right]^{\varkappa}}$$

$$\frac{dV_c}{dt} = Q_c \qquad (9)$$

with the initial condition $V_c = V_{co}$ for $t = 0$

The equivalent bond graph is a capacitance element $C_a \;\overset{P_c}{\underset{Q_c}{\longrightarrow\!|}}$

Loads of the network are local stations which deliver hot water to domestic heat exchangers. The simplest mathematical model of the local station is the variable resistance $\xrightarrow{t} R_q \;\overset{P_q}{\underset{Q_q}{|\!\longleftarrow}}$ in the admittance form which may be represented by equation

$$Q_q = K_{vq} f(t)\sqrt{|p_q|}\ \text{sign}\ p_q \qquad /10/$$

The admittance form is used for the same reason as in the valve case. The nondimensional time function $f(t)$ represents variations of the local resistance caused by temperature and pressure valve controllers in domestic heat exchangers. This function may be the combination of standard functions, step functions, or any other determinable function of time.

Simulation procedure

The simulation procedure has two main parts: the design of the mathematical model of the network and the development of the computer program.

Input data for the simulation consist of the hydraulic scheme of the network together with values of pressures, flows, hydraulic resistances, etc., which may be obtained from steady state computations or from experimental measurements.

The mathematical model of the network is designed as a bond graph which includes the elements (modules) listed above. These elements are connected by series (1) and paralled (0) junctions (Ref. 1). The linking methods for the full bond graph may be obtained by using appropriate causality forms of the DLF elements because the causality of other elements is strictly determined. For example the hydraulic scheme of a simple heating network is shown in Fig. 2 and the corresponding bond graph in Fig. 3. If the network is too big for the simulation of the whole, because of limited computer memory, the mathematical model may be partitioned and parts of the network then simulated separately. Normally, in such a case, the supply part and one or a few of the receiver parts may be isolated. The inclusion of additional bond graph elements is necessary in place of the ignored network parts. Pressure sources may replace the supply portion of the network portion and flow sinks in place of the receiver portion in a model of the supply part of a network.

A computer program may be directly written from the bond graph if equations of all graph elements are known. Very powerful in the discussed simulation method is the use of any of the continuous system simulation languages. CSMP language is used here. Properties of the CSMP language allow the definition of MACRO functions (subprograms) which describe separate modules of the model i.e. separate bond graph elements. If this is done the main program, in its DYNAMIC segment, is very short and it reduces finally to the calling of MASTER functions.

The system dynamics in the presented method are described above by difference equations of DLF elements. Initial conditions in these equations may be easily programmed by using the sum of the step function and the DELAY function in one direction. Values of initial conditions may be computed in the INITIAL segment of the program, which is practical in simple cases, or in the DYNAMIC segment when all nondimensional time functions in local station equations (10) are assumed constant and have values of unity. In the second case boundary steady state values of appropriate variables are initial condition values exactly.

Special methods should be used for the linking of algebraic loops which occur in the connection of program modules (MACRO functions). This may be done by the inclusion of the delay function with the initial condition and the zero delay time. This same function is used if the segmentation of the DYNAMIC segment is necessary and the feedback signal is included in the NOSORT section. Using such a method in the algebraic loop linkage can cause an instability in some cases. A low pass filter should be used for such a circumstance.

The main hydraulic problem in a heating network is the maintenance of semi-constant pressure drops at the supply point of local stations. This is achieved by the use of automatic control of the circulating pump delivery or rotational speed using pressure feedbacks. The most simple and economic way is to use automatic control of the pressure drop in the power plant itself. In this case the controller should have a variable setting for the compensation of pressure drops in tubes which are a function of the flow rate. Alternatively and more approximately, the number of constant speed pumps in use could be controlled. Such a pressure control system was used in the computing example for the network shown in Fig. 2. A detailed description of the control algorithm is given in the Danfoss notes (Ref. 4 and 5).

The application of a quality index is necessary for the quantity estimation of control system selection results. The relative pressure drop variations in local stations were computed in the enclosed example. The relative pressure drop variation at any network point is defined as the difference between maximum and minimum pressure drop values at the point, occurring in the selected time interval (normally a unit of computing time), divided by the initial value of this difference. Such a quality index may be computed in the TERMINAL segment of the program which gives economy of the computer memory needed for the DYNAMIC segment. The CSMP program gives the possibility of the automatic minimization of the quality index and the selection of optimal control system parameters. Some computing results are shown in Fig. 4. The general simulation results for the analyzed example are as follows:

- the best pressure stabilization in local stations is achieved if the pressure controller and the pneumo-hydraulic accumulator are used together.
- the main influence on the selection of control system parameters is the effect of the dynamics of the nearest supply pipe.

Conclusions

The simulation method presented in this paper provides the possibility of continuous prediction of significant variables but only at junctions and other pipe ends. This limitation is not important from the practical point of view in the analysis of heating networks.

The advantage of the module conception in the design of network mathematical models is the possibility of considerable simplification of the programming by the definition of the module subprograms (MACRO functions in the CSMP language). This subsequently makes possible the direct writing of the program from the bond graph. The next advantage is the possibility of easy simulation of automatic control systems which is the direct result of the use of the CSMP language.

The main domain of the application of the presented method is the selection of lay out and parameters of pressure control systems in a heating network. The method may be used too for the analysis of network damage and the design of protection systems.

The method may be applied, with some modifications, to the simulation of the dynamics of other piping systems. Directly, it may be used for the analysis and design of control systems in district water-supply networks.

Acknowledgement

The author wishes to thank the Industrial Institute of Automation and Measurements for permission to publish this paper and Mr. S. Frydliński of this Institute for computation of the enclosed example.

References

1. Karnopp, D. and Rosenberg, R.C.: "Analysis and Simulation of Multiport Systems - The Bond Graph Approach to Physical System Dynamics". The MIT Press (1968).

2. Serwach, A.: "Computer Simulation of Hydraulic System Dynamics". (Modelowanie dynamiki układów hydraulicznych za pomoca maszyn matematycznych). Archiwum Budowy Maszyn, Vol. XXI, No. 3 (September 1974) (In Polish).

3. Serwach, A.: Difference Method in the Simulation of Fluid Lines" Advances in Computer Methods for Partial Differential Equations II. pp.278-283. Proc. of the Symposium organized by IMACS/AICA/Lehigh University-Bethlehem, Pennsylvania, U.S.A. (June, 1977).

4. "Infinitely Variable Control of Differential Pressure in Plant with Circulation Pumps". Danfoss Application Note IN.OO.B1.02. (March 1976).

5. "Infinite Regulation of Pressure in Plant with Parallel-coupled Pumps." Danfoss Application Note IN.OO.C1.02. (March 1976).

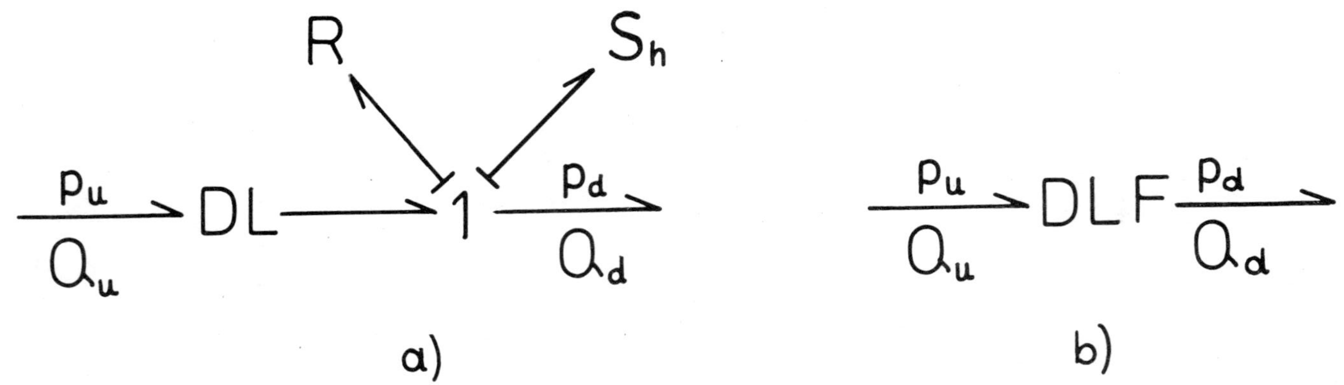

Fig. 1 The bond graph of the long hydraulic line model:
a) detailed b) aggregated

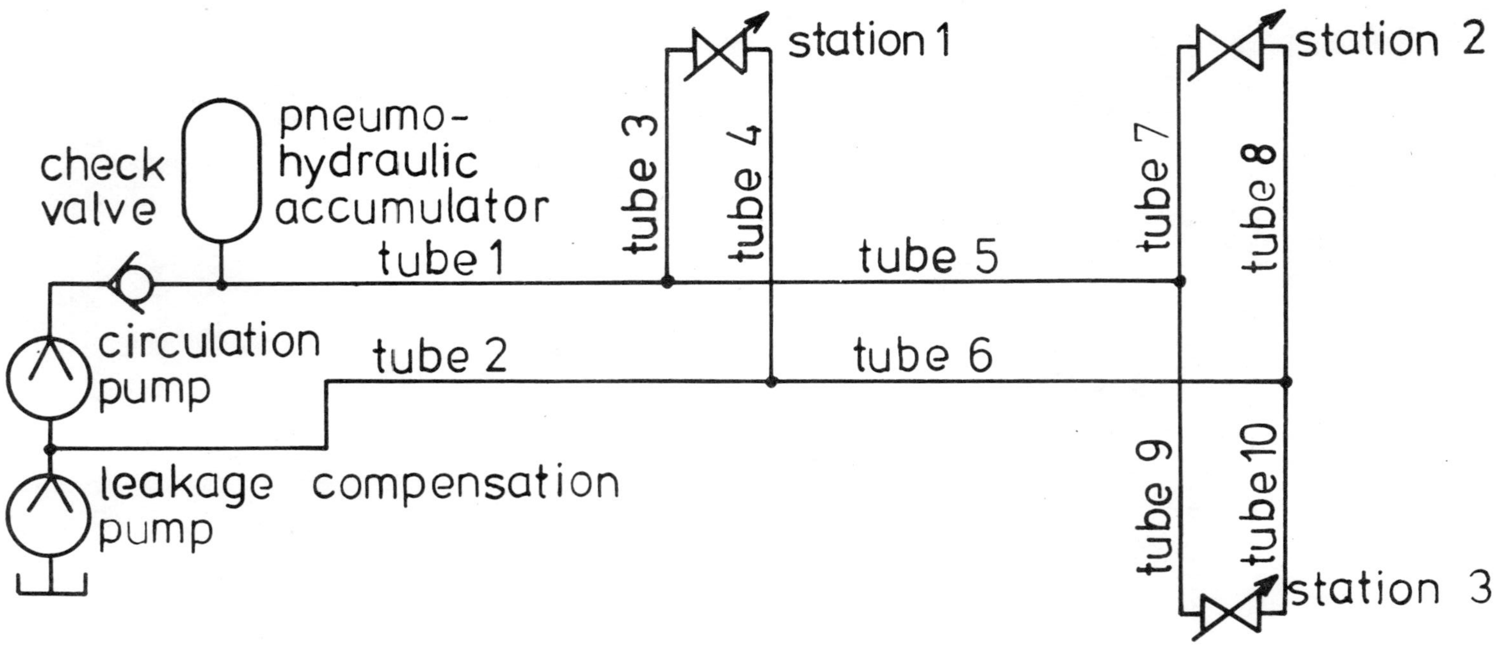

Fig. 2 The hydraulec scheme of the example heating network

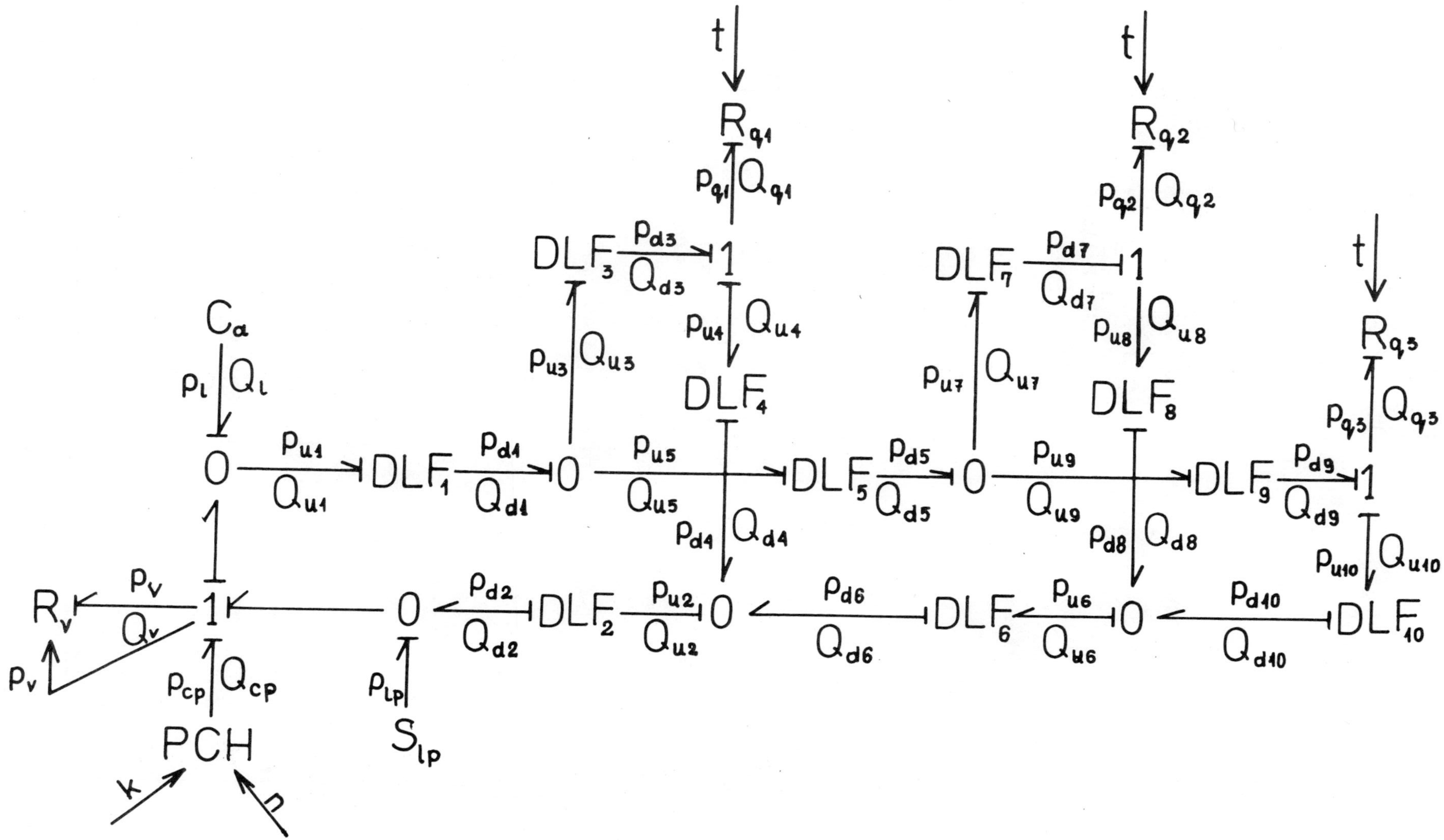

Fig. 3 The bond graph of the network from Fig. 2

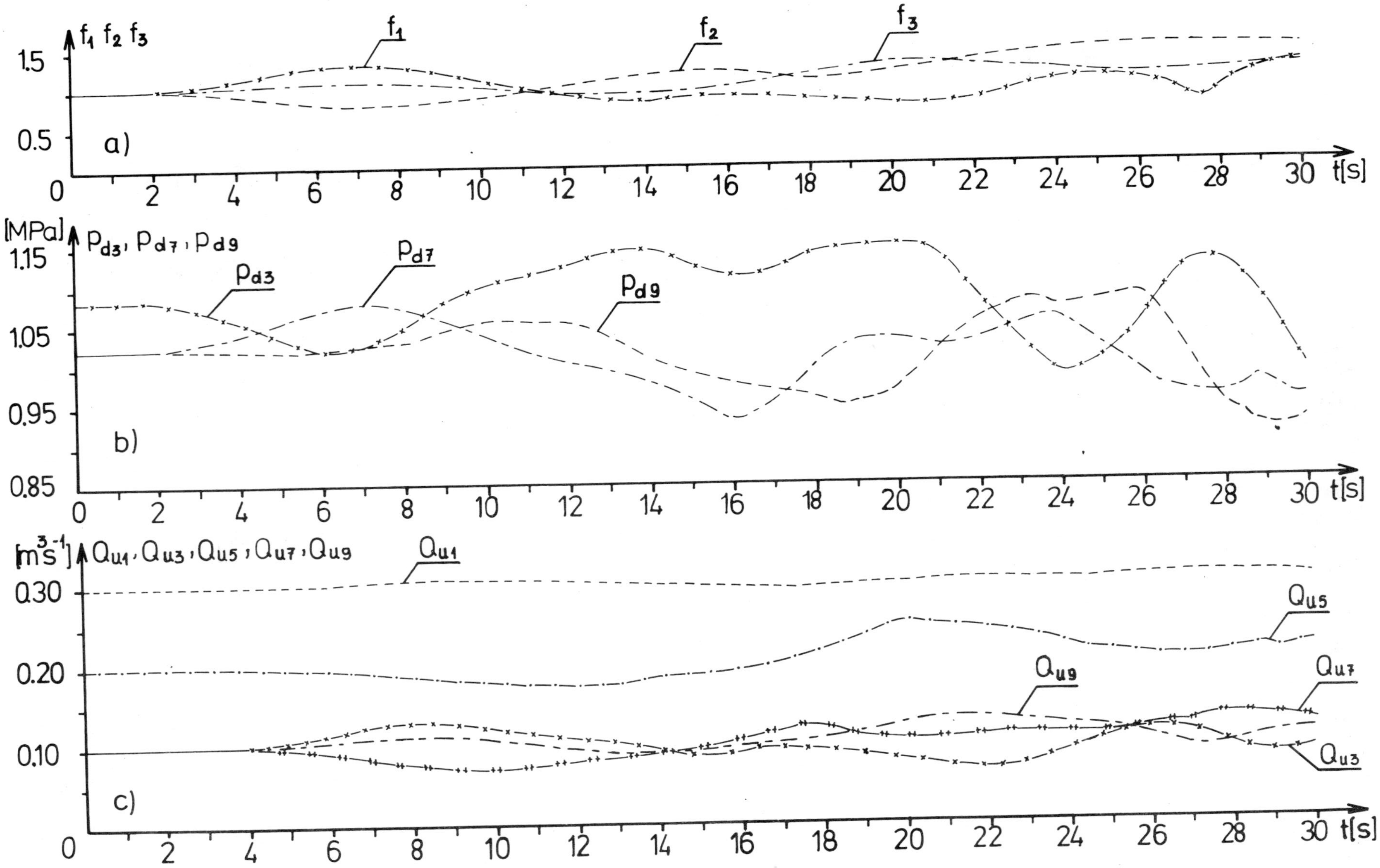

Fig. 4 Some results of the computing example:
a) input disturbances b) pressure responses c) flow responses

Third International Conference on

PRESSURE SURGES

Canterbury, England
March 25-27, 1980

PAPER B5

LIMITS OF SIMPLIFICATION OF STRONGLY LOOPED PIPE NETWORKS IN WATERHAMMER ANALYSIS

O. Füzy and L. Kullmann

Technical University, Budapest, Hungary.

Summary

The effect of the number of loops in a network on pressure transients is investigated. It is shown that the number of paths characterized by different time periods needed for the pressure wave to travel from the point of disturbance to that of detection is the substantial parameter determining the response of the network. The disturbance is a sudden pressure rise in the examples presented in this paper. The pressure of a definite point of the network is plotted as a function of time for illustration. A computer program designed for the simulation of pressure transients in pipe networks was used. It is not described here in detail, only to an extent needed for a better understanding of the results.

Held at the University of Kent at Canterbury, U.K.

Conference Organised by BHRA Fluid Engineering, Cranfield, Bedford, England.
In conjunction with The City University, London.

Nomenclature

a	velocity of pressure wave
d	diameter of a duct
E_r	reduced bulk modulus of elasticity of the fluid
F	nonlinear function
G	nonlinear function
g	gravitational acceleration
h	height above reference level
L	length of sequence of ducts
P	nondimensional pressure
p	pressure
s	distance co-ordinate along the duct
T	nondimensional time
t	time
v	flow velocity
Δ	increment
λ	friction factor in the duct
ρ	fluid density

Suffices

i	initial value
f	final value
s	partial derivative with respect to distance
t	partial derivative with respect to time

1. INTRODUCTION

In the past eight years a computer program has been made at the Hydraulic Machinery Department of the Technical University in Budapest for the simulation of pressure transients occuring in real networks. The program is the result of continual development in its present state. While developing it, we had always in mind the needs of the industry in our country. Both the boundary conditions and the flow model in a single duct has been set up in such a way that as many problems of waterworks, oil industry, heat engineering can be solved by the program as possible.

In the course of program development, following basic questions arose which are common for any digital simulation process:

- How are the parameters of the differential equations and their boundary conditions to be determined by the aid of identification methods?
- How large is the error resulting from the uncertainty of parameters of the models, that is, how sensitive is the model to inaccurate parameters?
- May the, often very complicated, network be simplified?

We consider only this third problem in this paper. The aim of simplification is not that computer programs are not suitable for the simulation of transients in complicated networks but that their simulation needs too much time because of many data and many operations. Above this, preparing the input data is very labour consuming too when simulating large networks with several junctions.

Experts for the operation of waterworks know from experience that a disturbance at one point of a simple network with only a few junctions and branches causes a well-determined change in pressure and flow velocity at an other point of the same network. Such unambiguous pressure changes in time cannot be detected in networks with many loops. The pressure history at a point far from the disturbance measured by the dimensions of loops has an impression rather stochastic than determinable /Ref. 1/. This effect is caused by the interference of waves, not by their dissipation. We do not know any exact analytical method useful in engineering practice, which would give an answer to this problem. There are models to replace the real network, borrowed from the investigation of electric networks, which have the indisputable advantage of giving mathematical formulas to calculate equivalent impedance and equivalent resistance such as described by Ewing /Ref. 2/. The application of these methods is limited by the fact that they are valid only for incompressible flow.

These reasons led us to make an experiment by the computer with the aim to obtain results of general validity.

2. MATHEMATICAL MODEL OF SIMULATION

Differential equations of flow in the tube:

The equation of motion describes the onedimensional unsteady flow of a compressible fluid:

$$v_t + v\,v_s + \frac{1}{\rho}\,p_s + g\,h_s + \frac{\lambda}{2d}\,v|v| = 0. \qquad /1/$$

Mass conservation is expressed by:

$$p_t + v p_s + \frac{\rho a^2}{d^2}\,(v\,d^2)_s = 0. \qquad /2/$$

The velocity of the pressure wave /a/ depends on the Young's modulus for the tube and can be expressed by the formula:

$$a^2 = \left(\frac{d\rho}{dp}\right)^{-1} = \frac{E_r}{\rho}. \qquad /3/$$

/See nomenclature for the meanings of symbols used in these equations./ The ducts for which these equations hold are connected in nodes characterized by prescribed boundary constraints.

Nodes of the network:

The nodes contain subsystems. They model simple junctions, pumps, surge-tanks etc. or even quite complicated pipe arrangements. There is one common characteristic feature for the subsystem, namely the fluid is considered incompressible in it. A node represents boundary conditions for each duct connected with this node. The solution of the system of generally nonlinear algebraic equations describing such subsystems happens by Newton-Raphson iteration procedure based upon the so called basic flow rates /Refs. 3 and 4/. Only the simplest types of nodes were applied in the present calculations: junction of ducts n in number, constant pressure, steady pump.

Numerical method:

Finite differences over a constant grid substituted for partial derivatives. Following formulas determine the flow velocities and pressures at the new time level /Ref. 5/:

$$p(s,t+\Delta t) + \rho a\; v(s,t+\Delta t) = p(s,t) + \rho a\; v(s,t) + F(s,s-\Delta s,t) \qquad /5/$$

$$p(s,t+\Delta t) - \rho a\; v(s,t+\Delta t) = p(s,t) - \rho a\; v(s,t) + G(s,s+\Delta s,t) \qquad /6/.$$

One of the equations of the subsystem must have the same form, it must be linear in p and v, presenting the boundary condition for the duct attached to this node.

We have tested our computer program by experiments. We found a very good agreement of measured and calculated pressure transients when we knew the velocity of the pressure wave from measurements. Comparison of analytical and numerical results gave excellent agreement.

3. STRUCTURE OF NETWORKS

The sketches of the networks are displayed in Fig. 1. There are asymmetrical /b,d,f/ and symmetrical /a,c,e/ networks. The simplest "network" is one duct /a/ following the direction of the main stream. Each network is fed from the left by a pump having steady characteristic curves. The pressure at the outlet on the right is constant in the steady state. A sudden pressure rise is imposed at the outlet at time t = 0. The pressure rise is large, it is of order half of the Allievi pressure-jump needed to stop the flow. We simulated the pressure change on the left at the inlet as a response of the network to the disturbance.

Following assumptions and limitations are made:

- The steady flow rate streaming through the network is identical in each case.
- The steady initial pressure at the inlet of the network is identical in each case.
- The steady flow velocity lies between 1 m/s and 1.5 m/s in every duct.
- In symmetrical and asymmetrical networks the steady flow velocities are identical in ducts corresponding to each other.
- The steady flow velocity is identical in every duct corresponding to each other joining the inlet or outlet nodes.
- The friction factor λ is the same /λ=0.02/ for each duct /with one exception/.
- The constant velocity of the pressure wave is identical along each duct.
- The duct diameter can be calculated from the aforesaid assumptions and may be different for different ducts.
- The total pressure drop may differ for different networks in the steady state, however, the differences are limited.

By these assumptions it was possible to construct a series of networks in which a constant pressure rise causes /or is caused by/ a decrease of flow velocity at the outlet identical for each network. After the disturbance, the initial pressure /p_i/ tends towards a new equilibrium state /p_f/. The pressure change as a function of time is represented in nondimensional form in the figures.

$$P = (p - p_i) / (p_f - p_i)$$

is the nondimensional pressure of the inlet point calculated from the actual /p/, initial /p_i/ and final /p_f/ values.

$$T = a\,t / L$$

is the nondimensional time. L marks the length of the shortest path connecting the inlet and outlet nodes. In this system of co-ordinates $P(T = 0) = 0$ and $P(T \to \infty) = 1$ hold for each network.

4. RESULTS

Figures 2, 3 and 4 show the results. P-T curves for symmetrical networks are displayed in Fig. 2. There is only one way in case "a" for the pressure wave to travel from the point of disturbance to the point of detection. There are two and three ways respectively in cases c and e along which the pressure wave needs different time intervals /characteristic time intervals/ to reach the point of detection. Accordingly the slope of the pressure curve is large at T=1 in case "a", at T=1 and T=2 in case c and at T=1, T=1.5 and T=2 in case e. For larger T values, the more different characteristic time intervals there are, the less clear is the form of the P-T curves. Interferences confuse the picture.

P-T curves for asymmetrical networks a, b, d, f are represented in Fig. 3. A disturbance starting at t=0 reaches the inlet node at T=1, 1.25, 1.5, 1.75, 2.0 along different paths with different characteristic time intervals in network f. The slope of the P-T curve rises in every such point of time resulting in a very smooth curve for $0<T<2$. It is in this case f that the pressure approaches the final value in the shortest time. However the pressure-time curve shows large oscillations with increasing time caused by interferences. To be sure that these oscillations are not consequences of numerical instabilities, the calculations were carried on as long as T=25.0 . At that time the final value /P=1/ is approached with an accuracy of $\pm$ 0.02. There are no large fluctuations in the flow velocity either /less than 10% of the new steady value along the duct/.

In Fig. 4 the P-T curves of a symmetrical and an asymmetrical network, c and b respectively, are compared with each other. Otherwise they are similar in construction. The total pressure drop differs for the two networks. It is obvious that these curves are very similar in their shape. The difference diminishes for large time periods. We investigated the effect of the total pressure drop in the network on the slope of the P-T curve at the starting point. The dots mark the altered P values for network b when the total pressure drop is equal to that of the network c. Only the friction factor λ had to be modified for this purpose. The steady pressure drop has little effect on the initial slope of the pressure curve, as can be seen. Differences between networks b and c have decreased further when the pressure drop was the same. Also the P-T curves d and e are similar in shape in the time interval $T<6$ /see figures 3 and 2 respectively/. However there are two differences between these networks. The first difference is their symmetry or asymmetry, the second one is that in three of the ducts of network d the velocity of the pressure wave has been slightly altered. It resulted in dissimilarities of the P-T curves for T >6. So the effect of interferences of waves is fundamentally influenced by the characteristic time periods along the different paths which is determined by the velocity of the pressure wave and by the length of ducts along this path.

The thin continuous line in Fig. 2 is the P-T curve of single duct "a" in case of an incompressible model of flow. It represents a very crude approximation of the pressure transient in the starting period which is just the most interesting part of the process.

5. CONCLUSIONS

The number of loops is different in the networks described in paragraph 3. One can define characteristic time intervals needed by the pressure wave caused by a disturbance to reach the point of detection.

- The presence of several paths of the same characteristic time intervals does not influence essentially the pressure history /Fig. 4/.
- The increase in number of paths of different characteristic times decreases the slope of the P-T curve at the starting time period.
- A greater number of paths of different characteristic times has the effect of a more rapid approach of the final steady value.
- A small change in the friction loss has hardly any effect on the pressure history.
- Each P-T curve goes in a rather narrow band for later time periods if making their co-ordinates dimensionless in an adequate manner.

All these effects should be taken into consideration when dimensioning surge tanks, butterfly-valves etc. for networks with many loops on the basis of simplified model networks.

6. REFERENCES

1. Jany, S: Private communication /Dr. S. Jany, Institut for Civil Engineering Design, Hungary - MÉLYÉPTERV/.
2. Ewing, D.J.F: "Analysis of pressure transients using incompressible flow theory". Proc. 2nd International Conference on Pressure Surges. Paper H1, pp. 1-14. Organized by BHRA Fluid Engineering, Cranfield, U.K. /London: Sep. 22-24, 1976/.
3. Almássy, B: Doctor's Thesis. Budapest, 1966.
4. Füzy, O. and Kullmann, L: "Rapid method for calculating the liquid flow in a pipe network".Proc. 6th Conf. on Fluid Machinery. p. 417. Organized by Scientific Society of Mech. Eng./Budapest: Sep.17-22, 1976/. - Hungary.
5. Füzy, O.-Halász, G.-Kullmann, L: "Simulation of pressure oscillations in pipelines". Proc. 5th Conf. on Fluid Machinery. p.309. Organized by Scientific Society of Mech. Eng. /Budapest: Sep.22-27, 1975/. - Hungary.

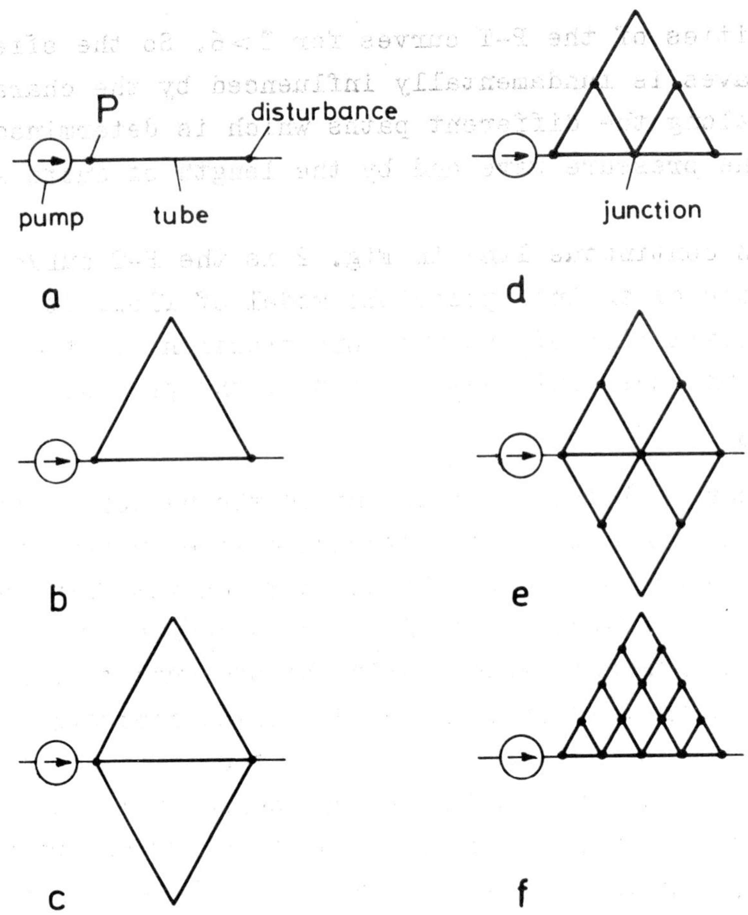

Fig. 1. Structure of the networks.

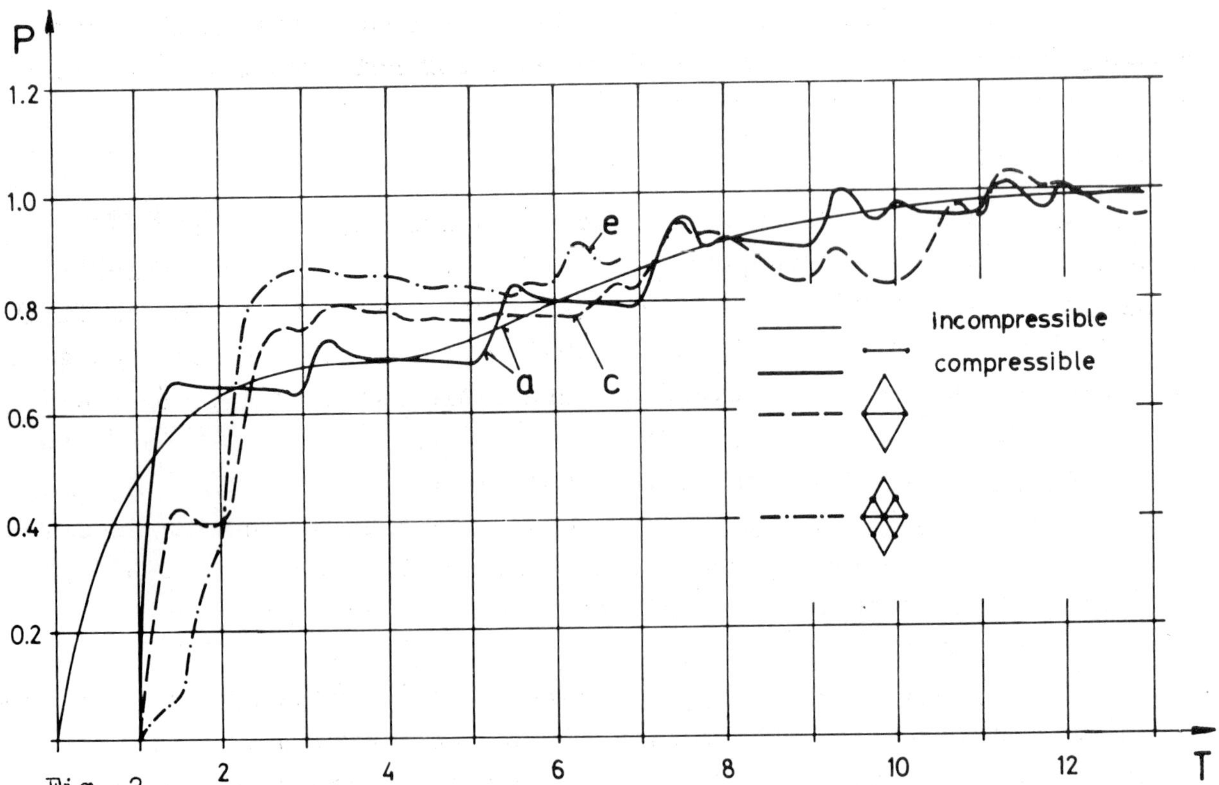

Fig. 2.
Comparison of simulated pressure transients in symmetrical networks.

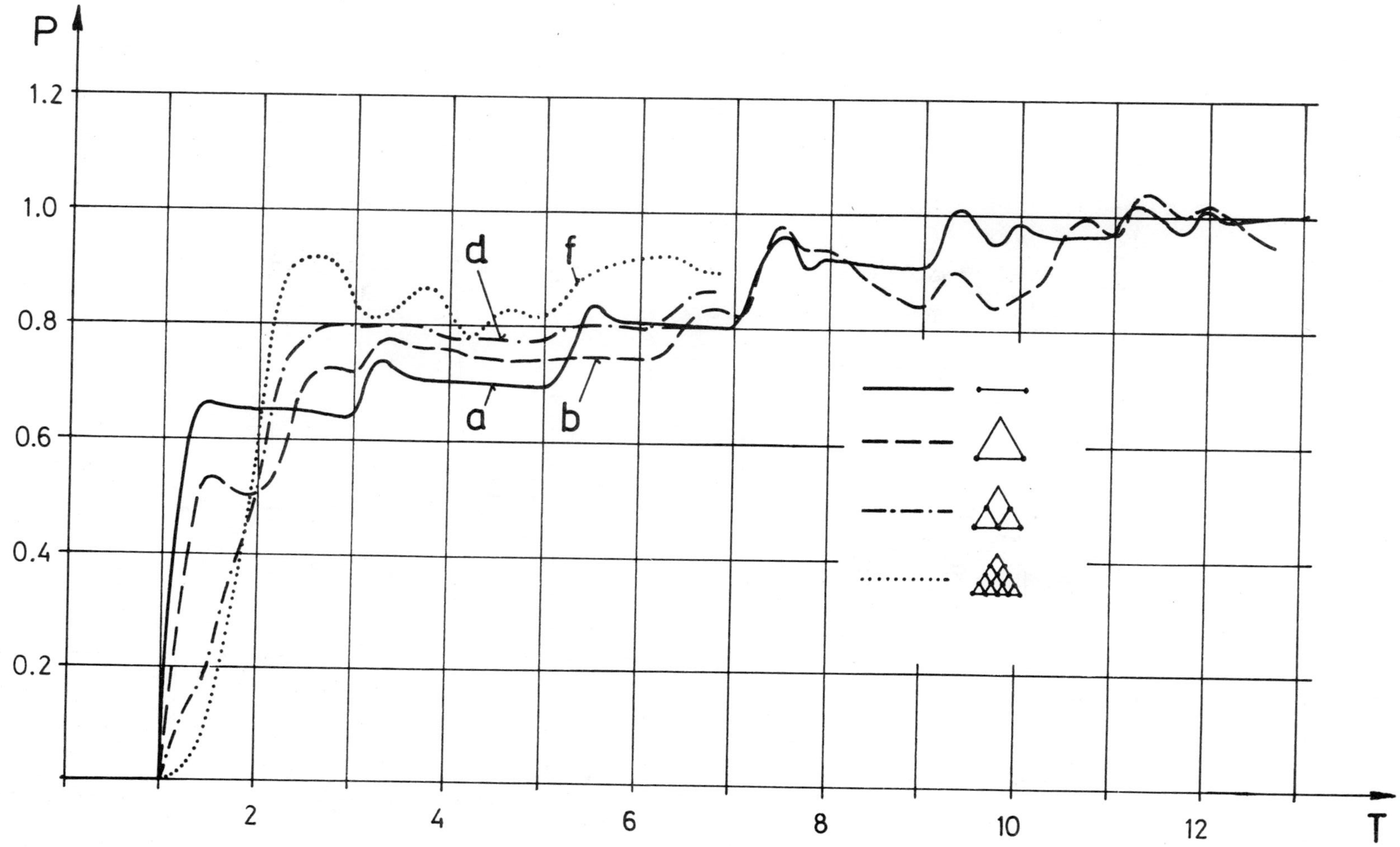

Fig. 3. Comparison of simulated pressure transients in asymmetrical networks.

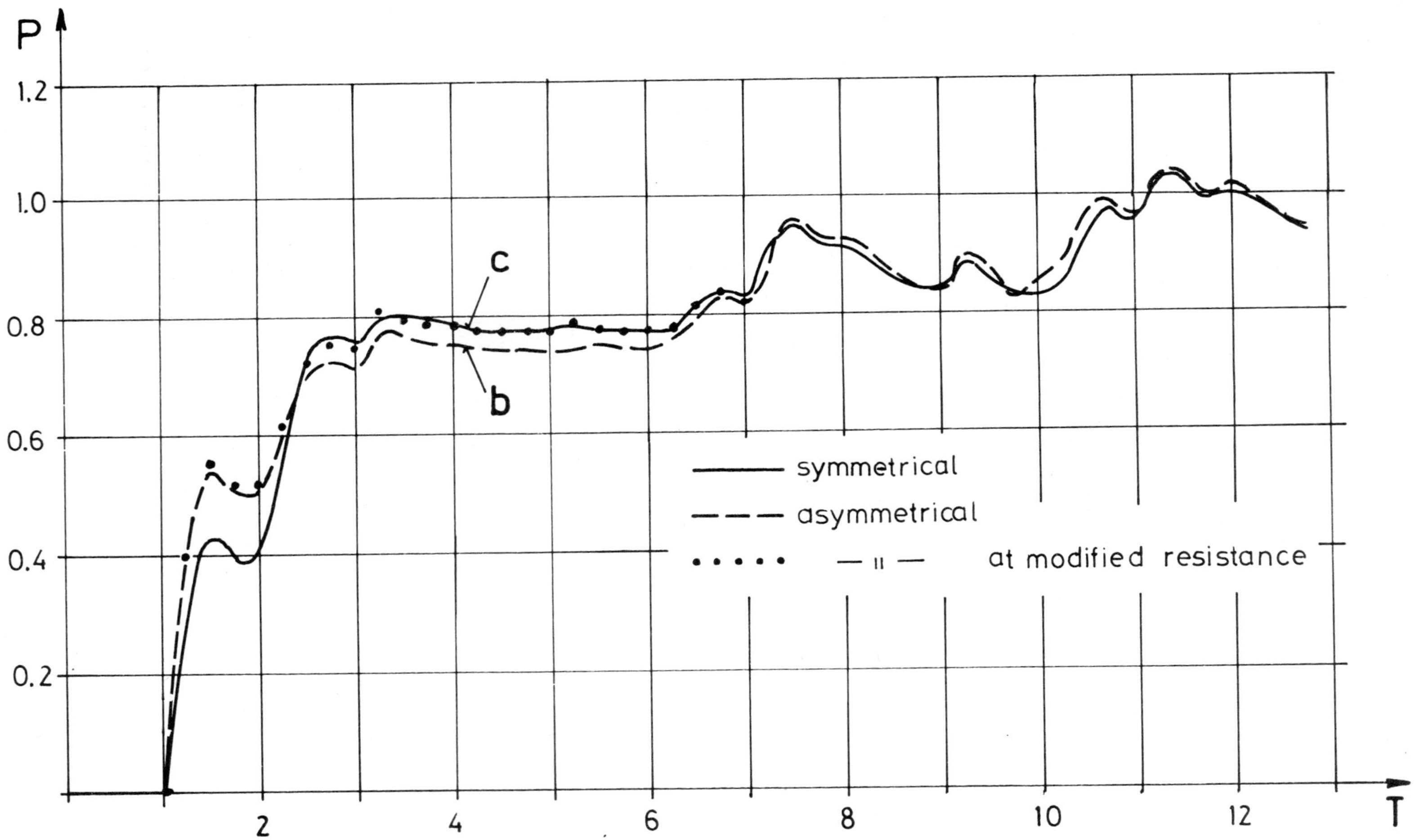

Fig. 4. Comparison of pressure histories in a symmetrical and an asymmetrical network.

Third International Conference on

PRESSURE SURGES

Canterbury, England
March 25-27, 1980

PAPER B6

COMBINED ANALYSIS OF SURGE AND WATERHAMMER PRESSURES

H.N.F. Pells and M.J. Shand

Ninham Shand Inc., S. Africa.

Summary

The traditional method of analysing hydraulic transients is to undertake separate analyses of the high pressure system using elastic theory and of the low pressure system using inelastic mass oscillation or surge theory. These analyses are easily applied where the surge shaft acts as a discontinuity between the high pressure and low pressure systems. However, where the surge shaft is partly isolated from the high pressure system by a restrictive orifice and waterhammer pressures can be transmitted into the low pressure tunnel, it is desirable to carry out the waterhammer and surge analyses as a single operation. In the case of the Steenbras Pumped Storage Scheme an existing waterhammer programme based on the method of characteristics was adapted to include a restricted orifice differential surge tank and a secondary surge device in the form of an access adit, from which overflow back into the upper reservoir can occur. The behaviour of the mathematical model was initially verified by physically modelling the system to a scale of 1 in 50. Subsequently, the predicted behaviour of the prototype was confirmed during commissioning tests of the completed pumped storage scheme.

Held at the University of Kent at Canterbury, U.K.

Conference Organised by BHRA Fluid Engineering, Cranfield, Bedford, England.
In conjunction with The City University, London.

1. INTRODUCTION

The Steenbras Pumped Storage Scheme has several features which have made the waterhammer and surge analyses (an essential part of the design of any hydro-electric scheme) particularly interesting.

Before proceeding with a description of the analyses, the hydraulic features of the scheme as illustrated in Figure 1 are described.

2. STORAGE AND FLOW

2.1 Lower Reservoir

Normal operating levels in the lower reservoir range from 75,50 m to 83,35 m giving an operating capacity of 2,16 x 10^6 m^3 which under maximum flow conditions is equivalent to 8,0 hours turbining at 75 m^3/s or 10,0 hours pumping at 60 m^3/s.

Emergency generation storage capacity between level 83,35 m and level 86,20 m gives a capacity of 1,00 x 10^6 m^3 equivalent to 3,7 hours turbining at maximum flow.

2.2 Upper Reservoir

The full supply level of the upper reservoir is at 370 m providing a total capacity of 30 x 10^6 m^3.

The operation of the pumped storage scheme causes the level in the upper reservoir, when full, to rise and fall by little more than 1 m, with the balance of the capacity being held as a reserve for water supply to the City of Cape Town. It is estimated that once in 40 years the level will be drawn below level 360 m when the upper reservoir is called upon to make up deficiencies in the water supply.

3. CONDUITS AND PUMP TURBINES

3.1 Upper Intake

The screen area of the upper intake is 19 m x 9,5 m resulting in a velocity through the screens of 0,4 m/s at full generating flow.

The intake tapers to the upper control gate (3 m x 4,5 m) through which the velocity for full operating flow is 5,6 m/s. However at time of maximum surge this velocity can increase to about 10 m/s.

In an emergency the upper control gate has been designed to close by gravity for flow velocities of up to 20 m/s.

3.2 Low Pressure Tunnel

The low pressure tunnel has an internal diameter of 4,5 m, is 850 m long and has an average fall of 2,54%. It is concrete lined throughout and the normal maximum velocity is 4,7 m/s.

Access to the low pressure tunnel is via an inclined adit (approximately 4 m x 4 m) the entrance to which is at level 371 m i.e. 1 m above the full supply level of the upper reservoir. It intersects the low pressure tunnel 250 m from the upper intake and will act as a secondary surge device discharging water into the upper reservoir on machine stoppage or reversal (see Figures 5, 6 and 9).

3.3 Surge Shaft

The surge shaft located at the junction of the low pressure and high pressure systems partially isolates the low pressure tunnel from waterhammer pressures and provides a reservoir of water to cater for rapid changes in machine demand.

The surge shaft itself is 10,0 m in diameter and 36,0 m deep. At level 371,45 m it enlarges into a tank 20 m in diameter and 8,55 m deep, which contains upsurge in the event of sudden machine stoppage during full turbining (i.e. load rejection).

3.4 High Pressure Shaft

The high pressure shaft which is 6,4 m in diameter and 118 m deep, is situated vertically below the surge shaft. A 3,0 m diameter orifice between the high pressure shaft and the surge shaft assists in damping out surge without increasing waterhammer pressures appreciably.

The orifice is shaped so as to allow minimum losses on upsurge and maximum losses on downsurge.

3.5 High Pressure Tunnel

The high pressure tunnel starts at the base of the high pressure shaft (at level 210 m). It is 408 m long and the grade is 8% for the upper 272 m which are concrete lined, and 1% in the steel lined section.

The concrete lined section of tunnel is 5,7 m in diameter. The maximum velocity in this section is 2,9 m/s.

At a distance of 136 m from the portal a debris trap has been constructed and the 3,9 m diameter steel lined section commences. The steel liner has a minimum thickness of 16 mm which increases progressively to 25 mm at the portal.

The steel lining is continuous with the surface penstock which commences at the portal.

3.6 Penstock

The 355 m long single steel penstock commences at the portal at a level of 185 m.

The first 297 m is 3,5 m in diameter and varies in thickness from 25 mm at the portal to 38 mm at Anchor Block No. 3 which is situated on the edge of the power station cutting (at level 110 m). The maximum velocity in this section is 7,6 m/s.

The last 58 m of penstock is 3,4 m in diameter and 38 to 40 mm thick. This bifurcates into two 2,3 m diameter shafts at the rear of the power station.

3.7 Penstock shafts

The two penstock shafts link the penstock at level 88 m (the power station floor level) to the bases of the two machine shafts at level 52 m, where the penstock shafts bifurcate into two branches 1,3 m in diameter.

The penstock shaft to machine shaft A is vertical and that to machine shaft B is inclined.

3.8 Reversible Pump Turbines

Each of the four pump turbines has a 1,3 m diameter spherical valve which can isolate the spiral casing from the penstock.

The machines operate at a synchronous speed of 600 rpm, both as pumps and as turbines.

The impeller of each machine is 2,39 m in diameter giving a peripheral velocity of 75 m/s (270 km/hr).

3.9 Draft Tubes and Lower Control Works

Each machine has an individual draft tube 3,0 m in diameter and approximately 50 m long, terminating in a control gate which can isolate the draft tube from the lower reservoir.

4. GENERAL APPROACH TO TRANSIENT ANALYSIS

The traditional and conventional approach to the analysis of transients in the conduits of a hydro-electric system is to regard the surge shaft as isolating the low pressure tunnel from the high pressure system. The analysis of transients in the low pressure tunnel then becomes a "surge" analysis which is regarded as a non-elastic mass oscillation. Transient pressures in the high pressure system are analysed by conventional waterhammer methods.

The conventional approach has the advantage of simplification of the analysis but where the introduction of a throttle at the base of the surge shaft allows "transmission" of waterhammer pressure into the low pressure system, these pressures cannot be easily calculated. To overcome this difficulty and to include the effects of the low pressure access adit in the calculation, the decision was taken to treat both the surge and the waterhammer as a single problem analysed by the method of characteristics (Ref. 1, 2).

5. MODELLING OF THE SYSTEM

5.1 Mathematical Model

The mathematical model of the system was derived from a standard waterhammer model that has been used and tested for many years on conventional pumping installations and rising mains and is based on the method of characteristics.

In order to adapt this mathematical model to the pumped storage concept it was necessary to introduce additional features such as the behaviour of a sophisticated surge tank and the overflow characteristics of the access adit.

It was decided that despite its previous successful use in predicting the transient behaviour of conventional pumping systems and rising mains, the mathematical model should be verified by comparing it with the behaviour of a hydraulic model of the Steenbras system.

5.2 Hydraulic Model

A hydraulic model of the pumped storage conduit system including the lower reservoir and upper intake was built and tested at the University of Cape Town. This model represented a situation which was geometrically similar to the prototype rather than a true hydraulic model, as it was not intended to predict the performance of the prototype (Ref. 3).

The primary purpose of the hydraulic model was to verify the mathematical model and particularly the mathematical predictions of the behaviour of the surge tank. Certain constants, such as the discharge coefficients for the orifice at the base of the surge tank which was designed to impose a minimum restriction on upsurge and a maximum restriction on downsurge to avoid emptying the surge shaft, were determined in the hydraulic model and checked in the mathematical model. The hydraulic model was also used to provide an initial assessment of the effects of the inclined access adit on the system.

The hydraulic model was constructed to a scale of 1 in 50. The concrete lined portions of the tunnels and the high pressure shaft were modelled with asbestos cement piping, and the steel lined portions of the tunnels, the penstocks and the draft tubes with steel pipe. The surge shaft was constructed of perspex and fibre glass. A quick closing and opening valve was inserted at the base of the high pressure system. A level recorder was mounted in the surge tank, and pressure transducers connected to high speed recorders were located at the top of the vertical shaft and at the lower end of the high pressure system.

5.3 Comparison of mathematical and hydraulic model results

The behaviour of the hydraulic and mathematical models was compared for a number of operating conditions, typical examples of which are shown in Figures 2, 3 and 4.

Figures 2 and 3 show the maximum waterhammer pressures recorded for sudden valve closure during turbining respectively at the transducer immediately upstream of the valve which simulated sudden shut down during turbining, and at a transducer below the surge shaft on the centre line of the low pressure conduit. It is evident that the maximum waterhammer pressures measured in the hydraulic model do not follow the anticipated trend of increasing pressure with increasing flow, which is caused by the superposition of the pressure wave due to sudden valve closure, and the friction losses in the system which become zero when the flow stops (Ref 4). In addition, the hydraulic model results shown in Figure 3 do not project to give the expected trend of zero pressure for no flow. The mathematical model appears to predict both these phenomena correctly.

Figure 4 shows the relationship between the maximum and minimum surge levels in the surge tank following sudden valve closure. The first maximum and first minimum have been plotted both with and without the access adit and the results from the mathematical and hydraulic models agree very well. It is evident that the access adit reduced surge amplitudes.

The results presented above and those for various other tests verified that the mathematical model would predict reasonably accurately the behaviour of the surge shaft in the hydraulic model. Agreement between measured and predicted waterhammer pressures was not good, but the inaccuracy appears to lie in the hydraulic model results. The mathematical model was therefore considered to be sufficiently reliable to predict the behaviour of the prototype.

6. DESIGN PROCEDURE

The preliminary design studies were based on conventional surge theory and their purpose was to establish the size of surge shaft required for stability considerations accepting reasonable limitations on surge levels. The diameter selected by this procedure was 14 metres, but it was realised that considerable savings could be made by reducing the diameter of the surge shaft and by introducing a throttle between the conduit system and the surge shaft. The mathematical model was therefore developed to calculate both surge and waterhammer pressures.

The mathematical model showed that the size of the surge shaft could be reduced to 10 metres by introducing an orifice with an area of 7,1 m^2 and relatively high discharge coefficients of 1,0 and 0,8 respectively for flow into and out of the surge shaft as used to correlate the mathematical and hydraulic models. This final selection of the surge shaft and orifice sizes was based on the criterion that the maximum pressure transmitted into the low pressure tunnel must not exceed the maximum anticipated upsurge. The sizes of the other components are summarised in Tables 1 and 2.

7. PREDICTED PROTOTYPE BEHAVIOUR

Some of the flow-time relationships finally selected for the design of the system and tested in the mathematical model are shown in Tables 3 to 7, together with descriptions of the machine operations causing the changes in flow. The flows quoted are those which would occur at the rated head of the machine when generating or pumping, variations due to surges being allowed for automatically in the calculations.

Figures 3 to 7 show the variations with time of the predicted surge levels in the surge chamber and access adit and of the pressures below the orifice for the following cases:

1) Load rejection of 4 units from pump operation (Fig. 5)

2) Load rejection of 4 units from turbine operation (Fig. 6)

3) Turbine start of 4 units from the spinning mode (Fig. 7)

4) Consecutive pump start of 4 units (Fig. 8)

5) Hydraulic changeover (Fig. 9)

The maximum predicted surge amplitude of 21,65 m would occur if hydraulic changeover of all sets from pumping to generating takes place when the upper reservoir level is at 370 m as shown in Figure 9. It should be noted, however, that no residual surges from the original pump start up were allowed for.

8. OBSERVED PROTOTYPE BEHAVIOUR

At time of writing only one set of observations of surge level, comparable with mathematical predicted levels had been recorded. These were for load rejection from full load in the turbining direction on three machines with the upper reservoir at a level of 368,25 m.

The results of both the observed surge tank levels and the mathematical prediction are shown in Figure 10 and give good agreement between the computed surge tank levels and observed surge tank levels. To make the mathematical predictions fit the prototype observations it was clear that lower coefficients of discharge than were determined for the hydraulic model, must be used for flow through the orifice. The values adopted were 0,76 and 0,66 respectively for flow into and out of the surge shaft.

Two sets of waterhammer results are available for comparison with mathematical predictions.

The first of these, shown in Figure 11, corresponds to full load rejection on three machines from turbining. The second set of results shown in Figure 12 gives the pressure at the spherical valve for four machine pump release.

The agreement on load rejection for three machines from the turbining mode is poor due to failure of the mathematical model to correctly simulate the behaviour of the machines.

The agreement on four machine pump release as shown in Figure 12 is good except for the initial four seconds when the observed and computed pressures do not agree. This difference is probably due to the differences between the assumed and actual discharge characteristics of the pumps.

CONCLUSION

The mathematical model developed for predicting the behaviour of the Steenbras pumped storage scheme proved to be a useful design tool which enabled considerable savings in the cost of the scheme to be effected.

Surges and waterhammer pressures predicted by this mathematical model showed reasonable agreement with measurements in a hydraulic model of the system, and comparisons with prototype behaviour are encouraging.

REFERENCES

1. Streeter, V.L. and Wylie, E.G. "Hydraulic transients" McGraw-Hill Co Inc (1967)
2. Martin, C.S. "Method of Characteristics applied to calculation of surge tank oscillations" BHRA. 1st Conf on Pressure Surges (1972)
3. Lazarus, J.H. "Experimental Investigation of the surge and waterhammer associated with the Steenbras Pumped Storage Scheme" U.C.T. (1975)
4. Stephenson, David "Pipeline Design for Water Engineers" Elsevier (1979)

TABLE 1

HYDRAULIC SYSTEM DATA

RESERVOIR LEVELS AND STATIC HEADS

UPPER RESERVOIR	
MAXIMUM FLOOD	371 m
MAXIMUM RETENTION	370 m
NORMAL MAXIMUM	370 m
NORMAL MINIMUM	369 m
EXTREME MINIMUM	360 m
LOWER RESERVOIR	
MAXIMUM FLOOD	86,5 m
MAXIMUM FOR EMERGENCY GENERATION	86,0 m
NORMAL MAXIMUM	83,35 m
NORMAL MINIMUM	75,5 m
LOWER CONTROL WORKS SILL LEVEL	72,5 m
STATIC HEADS	
EXTREME MAXIMUM	298 m
NORMAL MAXIMUM	294,5 m
NORMAL MINIMUM	286 m
EXTREME MINIMUM	274 m

TABLE 2

HYDRAULIC SYSTEM DATA

DESCRIPTION		NUMBER OF CONDUITS	L (m)	D (m)	K
LOW PRESSURE TUNNEL		1	850	4,500	0,490 $\times Q^2 \times 10^{-3}$m
HIGH PRESSURE SYSTEM					
- CONCRETE LINED SHAFT		1	118,48	6,400	0,080 $\times Q^2 \times 10^{-3}$m
- CONCRETE LINED TUNNEL		1	272,15	5,700	0,070 $\times Q^2 \times 10^{-3}$m
- STEEL LINED TUNNEL		1	135,86	3,858	0,169 $\times Q^2 \times 10^{-3}$m
- PENSTOCK		1	110,10	3,542)	0,570 $\times Q^2 \times 10^{-3}$m
			187,00	3,530)	
		1	58,16	3,422	0,135 $\times Q^2 \times 10^{-3}$m
- BIFURCATION 1		2		)	
- PENSTOCK	(SHAFT A)	1	47,7	2,300)	1,964 $\times Q^2 \times 10^{-3}$m
	(SHAFT B)	1	68,6	2,300)	
- BIFURCATION 2		2		)	
DRAFT TUBE SHAFTS	(SETS 1 + 4)	2	45,57	3,000)	1,110 $\times Q^2 \times 10^{-3}$
	(SETS 2 + 3)	2	43,15	3,000)	

L = LENGTH OF REACH; D = INTERNAL DIAMETER; K = HYDRAULIC LOSS ; Q = FLOW IN CONDUIT (m^3/s)

SURGE SHAFT

SURGE CHAMBER ORIFICE LOSSES:	
FLOW INTO SURGE CHAMBER	1,058 $\times Q^2 \times 10^{-3}$m
FLOW OUT OF SURGE CHAMBER	1,653 $\times Q^2 \times 10^{-3}$m

Q = FLOW THROUGH THE ORIFICE (m^3/s)

TABLE 3

LOAD REJECTION OF 4 UNITS
FROM PUMP OPERATION

TIME (secs)	FLOW (m^3/s)	OPERATION
0	60	ALL SETS FULLY OPEN FOLLOWED BY LOAD REJECTION
4,5	0	FLOW REVERSAL
8	-44	MAXIMUM REVERSE FLOW
21,5	0	GUIDE VANES CLOSED

FOR PLOT OF SURGES SEE FIG. 5

TABLE 4

LOAD REJECTION OF 4 UNITS
FROM TURBINE OPERATION

TIME (secs)	FLOW (m^3/s)	OPERATION
0	-75	ALL SETS FULLY OPEN FOLLOWED BY LOAD REJECTION
8,5	0	FLOW REVERSAL
10,5	8,4	MAXIMUM REVERSE FLOW
13	0	FLOW REVERSAL
16	-15	FLOW MAXIMUM
22,2	0	GUIDE VANES CLOSED

FOR PLOT OF SURGES SEE FIG. 6

TABLE 5

TURBINE START OF 4 UNITS
FROM THE SPINNING MODE

TIME (secs)	FLOW (m^3/s)	OPERATION
0	0	ALL SETS OPENED TO TURBINE
10	-70	ALL SETS FULLY OPENED
15	-75	

FOR PLOT OF SURGES SEE FIG. 7

TABLE 6 CONSECUTIVE PUMP START

4 PUMPS

TIME (secs)	FLOW (m^3/s)	OPERATION
0	0	FIRST SET OPENED TO PUMP
60	17,3	SECOND SET OPENED TO PUMP
120	32,4	THIRD SET OPENED TO PUMP
180	46,6	FOURTH SET OPENED TO PUMP
240	60	

FOR PLOT OF SURGES SEE FIG. 8

TABLE 7 HYDRAULIC CHANGEOVER

TIME (secs)	FLOW (m^3/s)	OPERATION
0	60	ALL SETS FULLY OPEN FOR PUMPING MODE FOLLOWED BY LOAD REJECTION
20	0	FLOW REVERSAL
30	-30	MAXIMUM REVERSE FLOW
60	-30	MAXIMUM REVERSE FLOW
120	- 7,5	
180	- 7,5	
185	- 75	TURBINES FULLY OPENED

FOR PLOT OF SURGES SEE FIG. 9

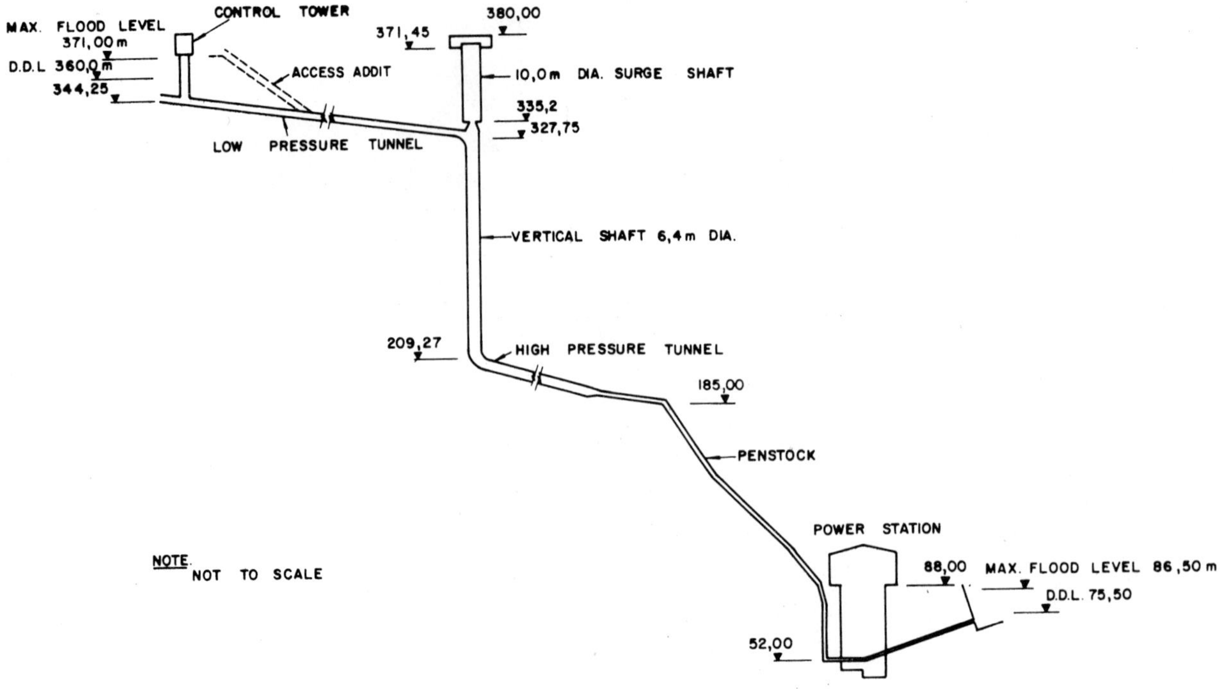

SCHEMATIC ARRANGEMENT OF SYSTEM

FIG. 1

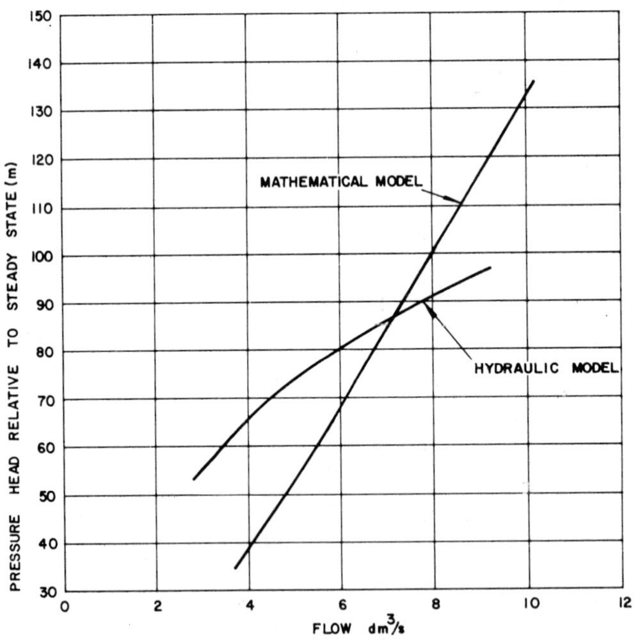

NOTE : VALVE CLOSURE IN 60 MILLISECONDS

MAXIMUM WATER HAMMER PRESSURE
AT DOWNSTREAM PRESSURE TRANSDUCER FOR
SUDDEN VALVE CLOSURE (TURBINING)
FIG 2

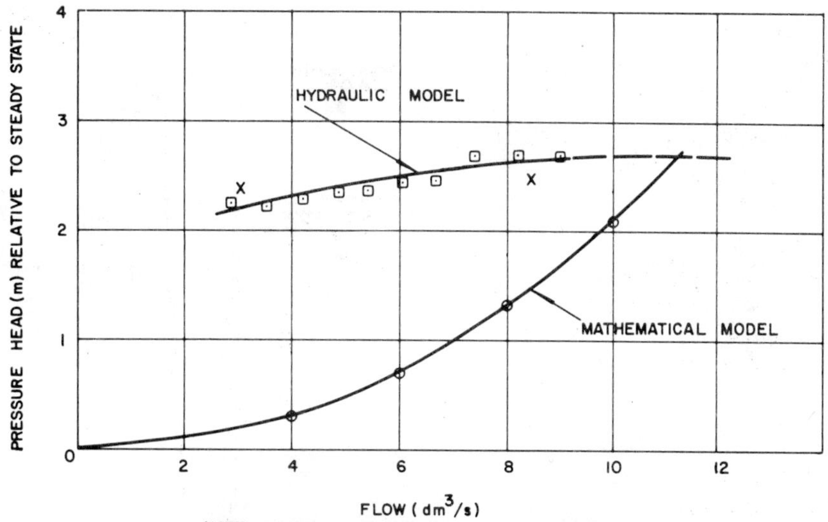

NOTE VALVE CLOSURE IN 60 MILLISECONDS

MAXIMUM WATER HAMMER PRESSURE
AT UPSTREAM PRESSURE TRANSDUCER FOR
SUDDEN VALVE CLOSURE (TURBINING)

FIG 3

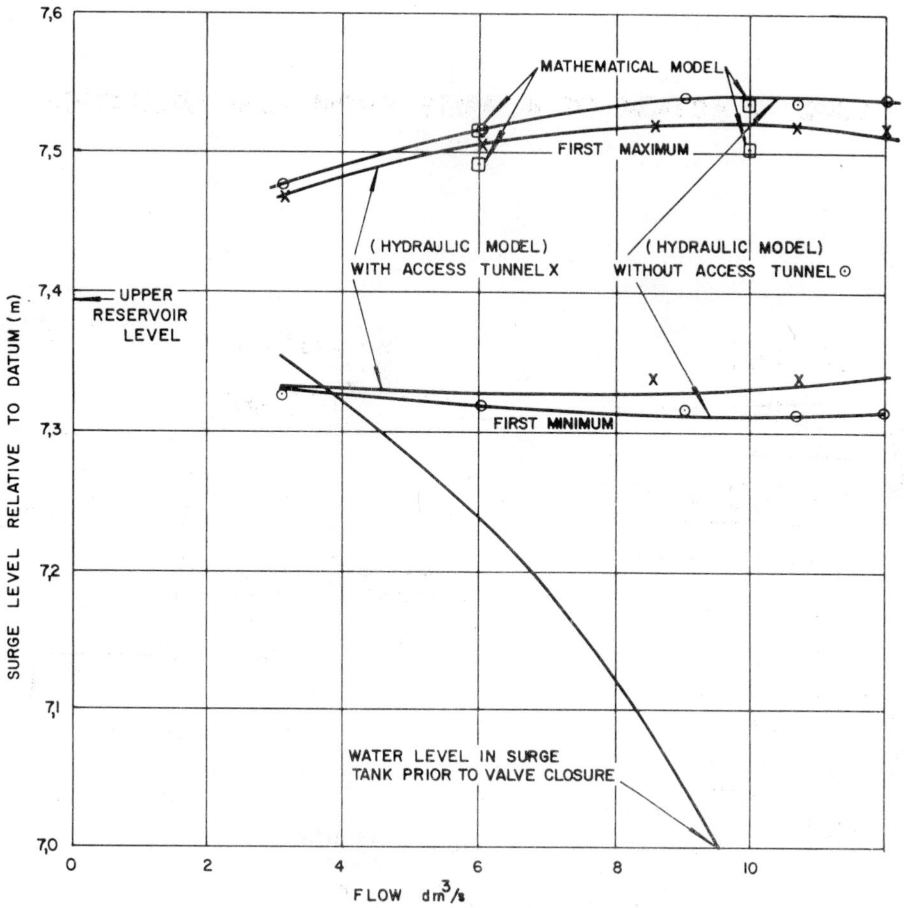

NOTE: VALVE CLOSURE IN 60 MILLISECONDS

MAXIMUM AND MINIMUM SURGE LEVELS
FOR SUDDEN VALVE CLOSURE (TURBINING)

FIG 4

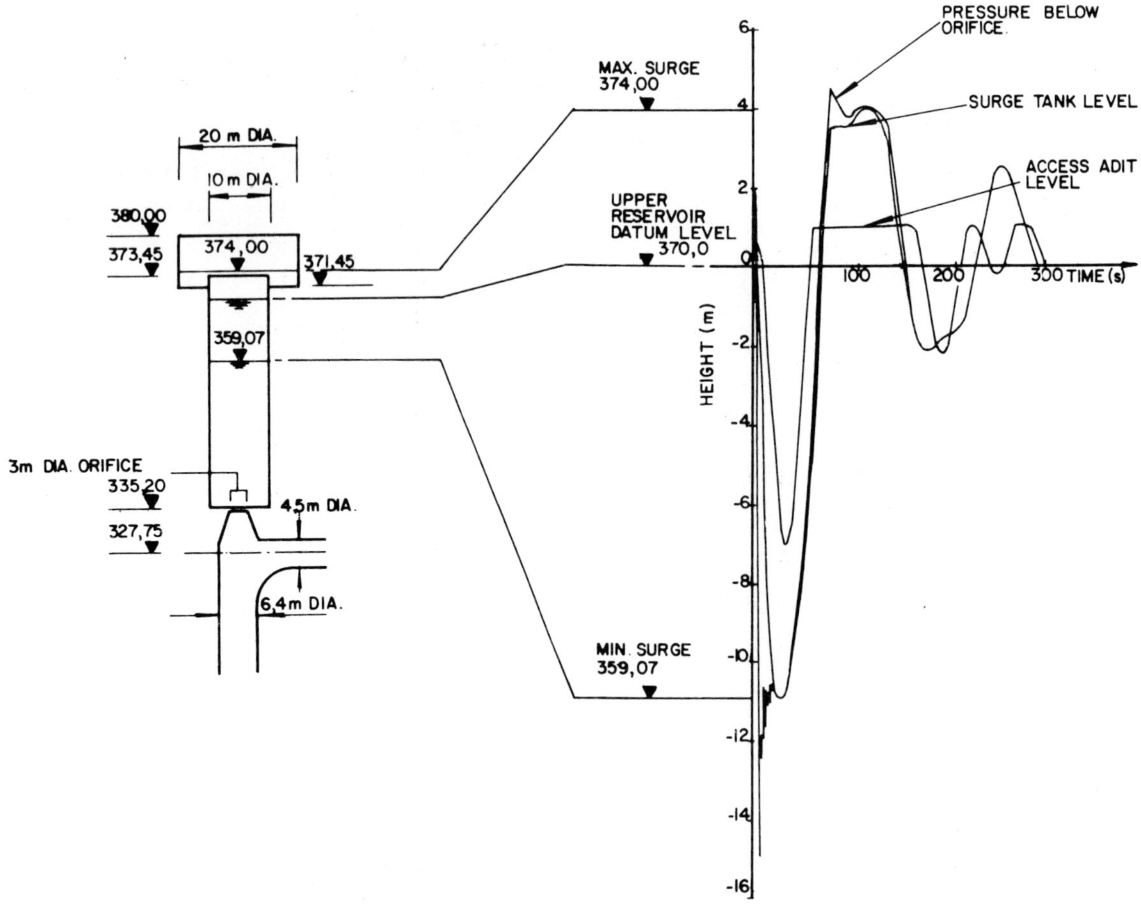

LOAD REJECTION OF 4 UNITS FROM PUMP OPERATION
Fig. 5

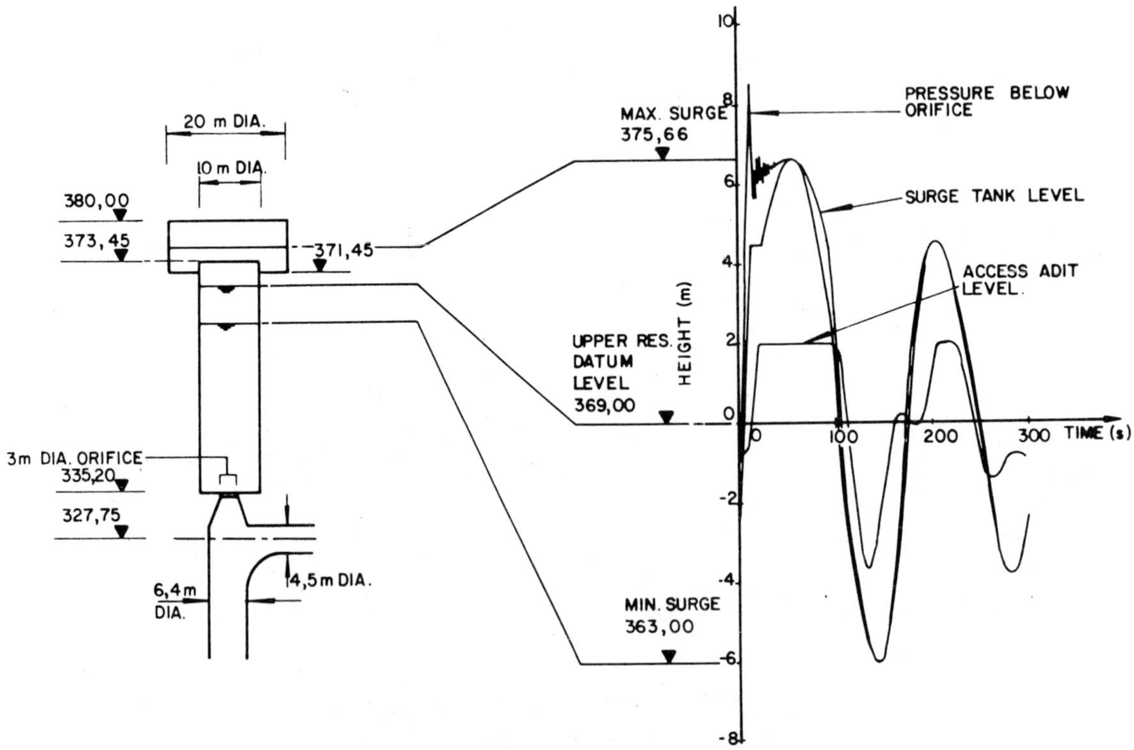

LOAD REJECTION OF 4 UNITS FROM TURBINE OPERATION
Fig. 6

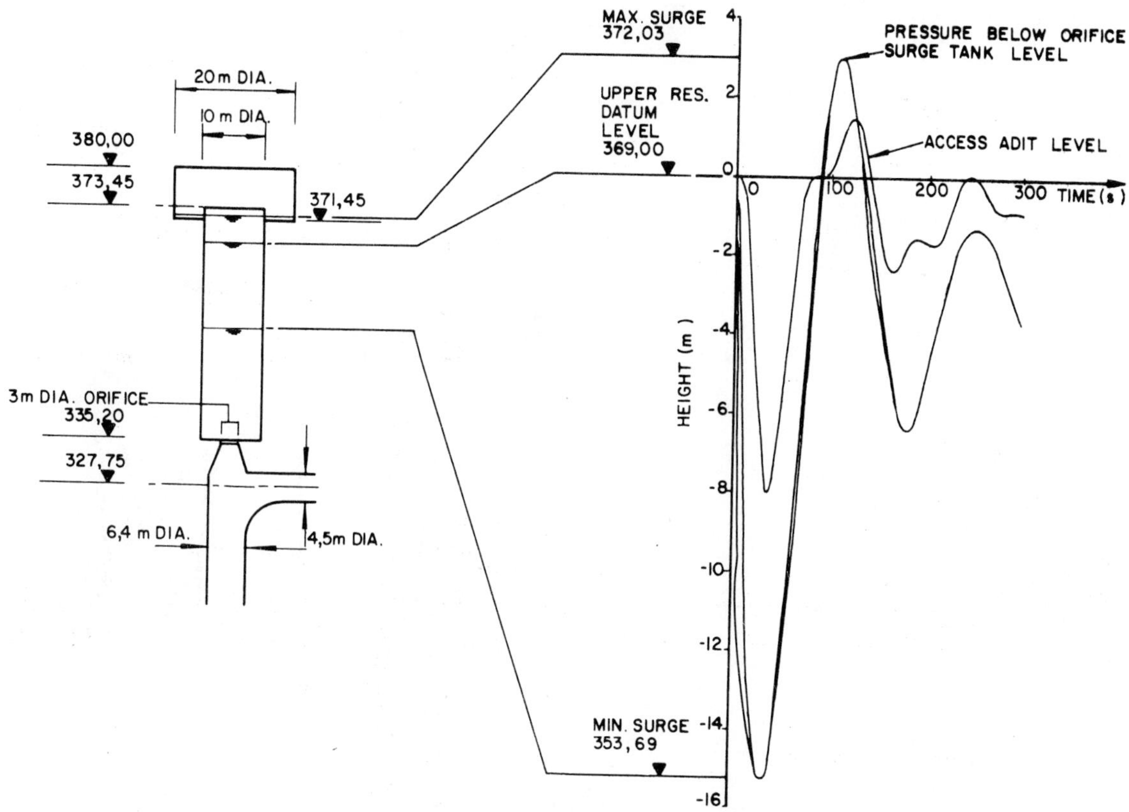

TURBINE START OF 4 UNITS
FROM SPINNING MODE
Fig. 7

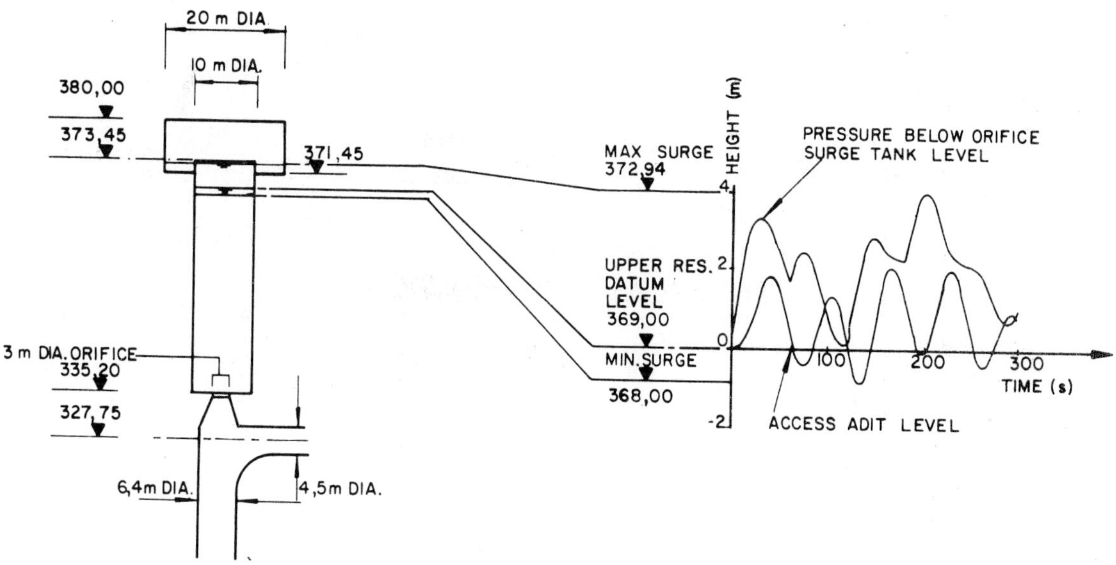

CONSECUTIVE PUMP START
OF 4 PUMPS
Fig. 8

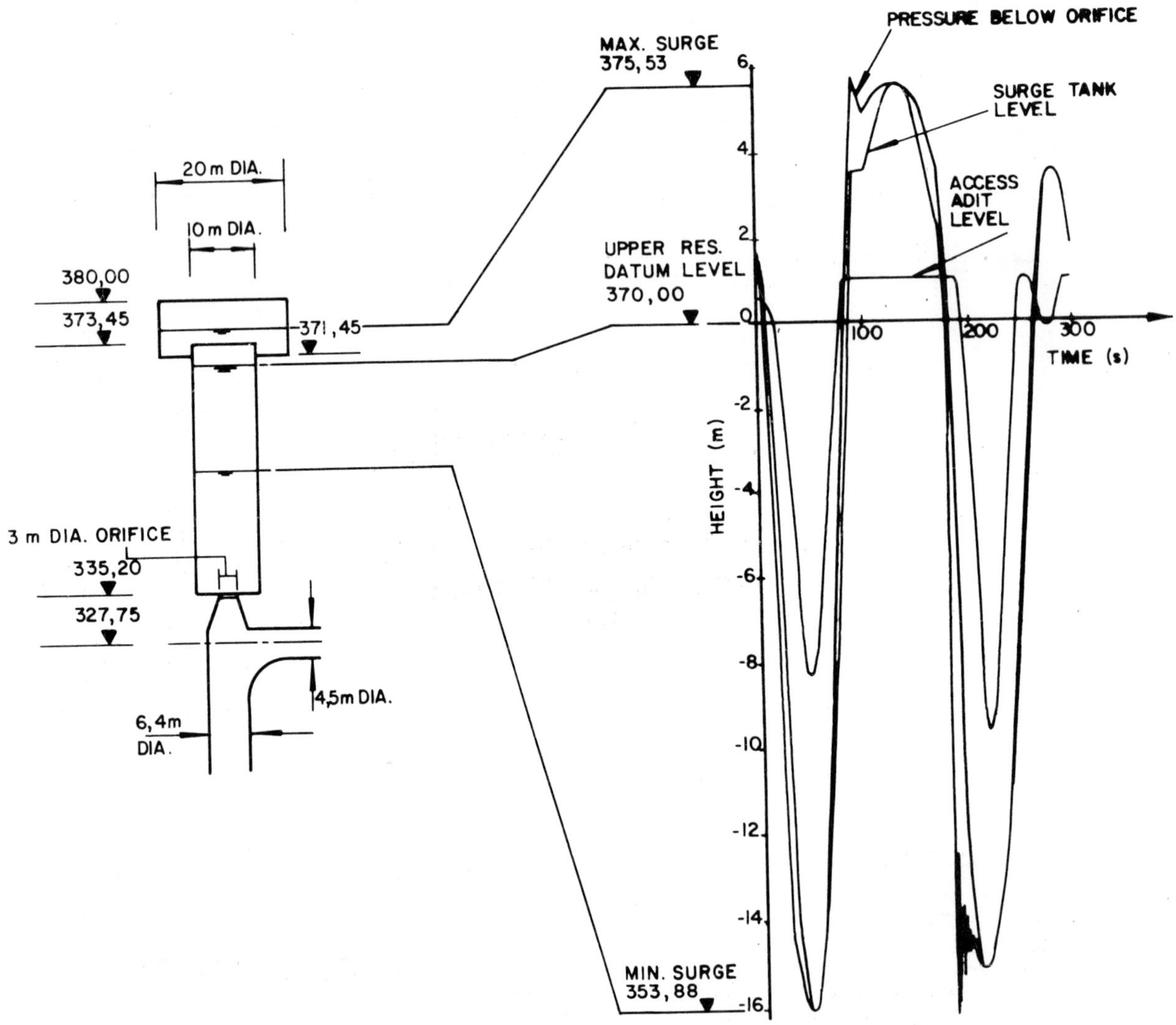

HYDRAULIC CHANGEOVER

Fig. 9

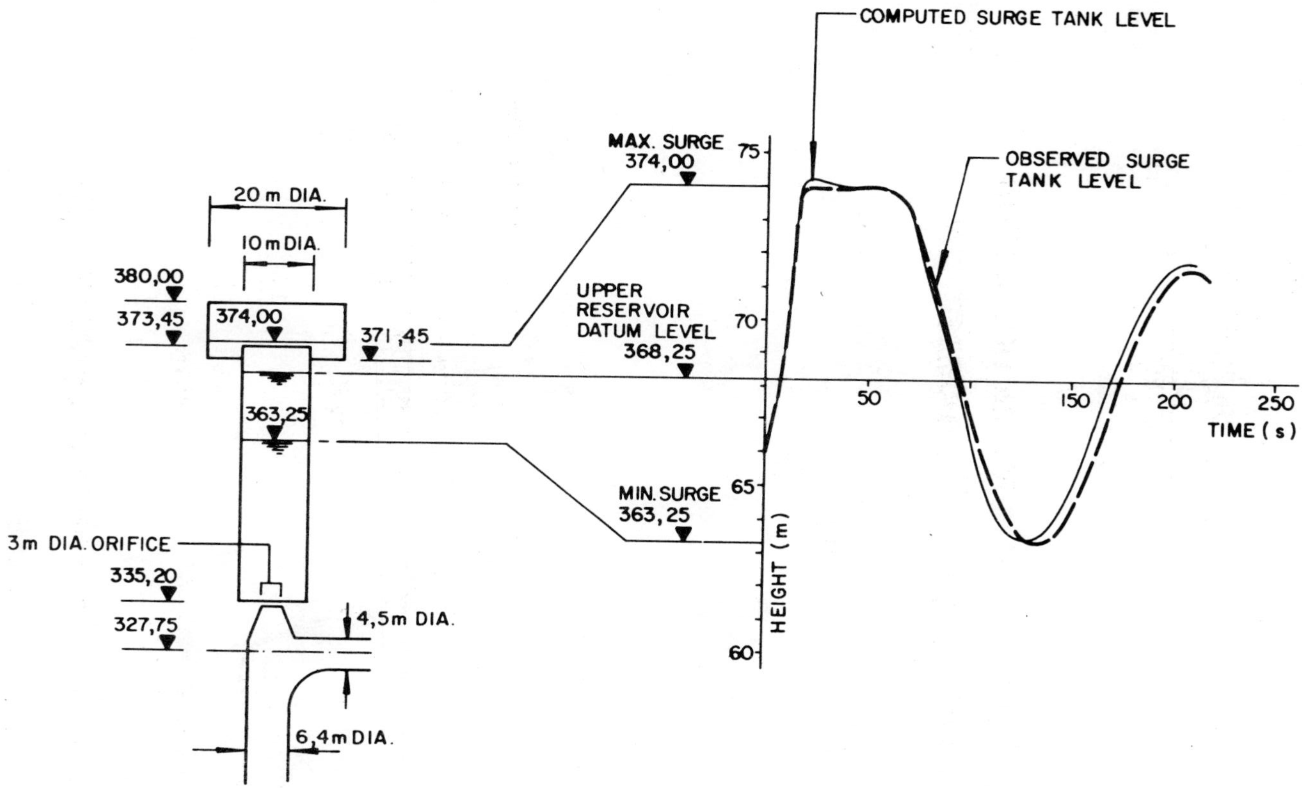

LOAD REJECTION OF 3 UNITS
FROM TURBINE OPERATION

Fig. 10

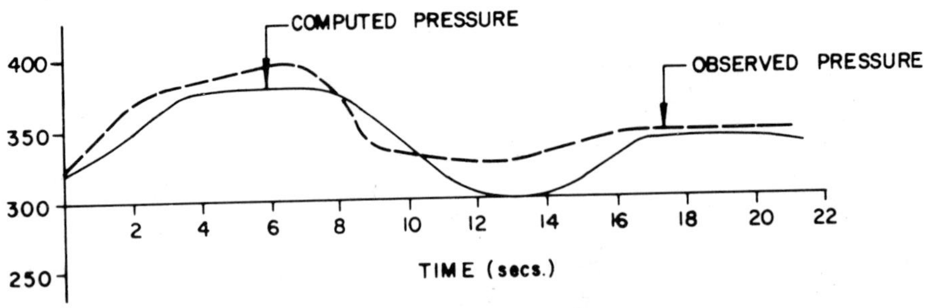

PRESSURE AT SPHERICAL VALVE FOR THREE MACHINE LOAD REJECTION

Fig . 11

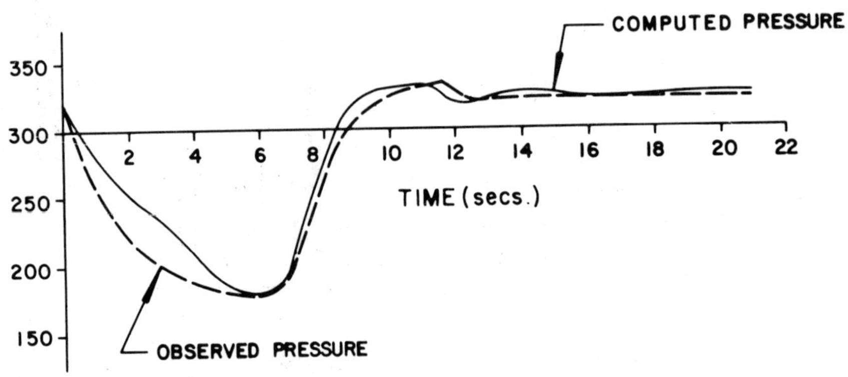

PRESSURE AT SPHERICAL VALVE FOR FOUR MACHINE PUMP RELEASE

Fig . 12

Third International Conference on

PRESSURE SURGES

Canterbury, England
March 25-27, 1980

PAPER C1

TESTING OF TRANSIENT FLOW AND COLUMN SEPARATION IN CRUDE OIL PIPELINES

J. Aga

Siv. Ing. Nybro Hansen/Norconsult, Norway.

and

T.J. Karterud and T.K. Nielsen

Norwegian Hydrodynamic Laboratories, Norway

Summary

In pipelines carrying crude oil, pressure waves and separation occur at pump shutdown and/or valve closing. The River and Harbour Laboratory is cooperating with Nybro Hansen and Kværner Brug A/S in an experimental research project to verify and develop computer programs for fluid transients and separation effects.

A test rig has been built to investigate pressure surges in pipelines, and of special interest are situations which lead to cavitation and column separation. Both water and crude oil are used in the tests. The experimental results are compared with corresponding numerical calculations.

The test rig consists of an 87 mm diameter, 250 m long steel pipeline with a total rise of 12 m. The crude oil is pumped from a 3 m^3 bottom tank up to a break-point 8.5 m above the pump, the pipeline sloping 30°. The pipeline then rises evenly up to the upper tank.

Measurements of the pressure variation and gas release, caused by pump shutdown and valve closing, are taken at the break-point.

Held at the University of Kent at Canterbury, U.K.

Conference Organised by BHRA Fluid Engineering, Cranfield, Bedford, England.
In conjunction with The City University, London.

NOMENCLATURE

p (bar)	Pressure
Q (m^3/s)	Flow
a (m/s)	Pressure wave speed
p_v (bar)	Separation pressure
V_c (m^3)	Volume of the cavity
m_c (kg)	Mass of the cavity
$\overline{M}$ (kg/mol)	Average molecular weight
R ($J/K^o mol$)	Universal gas constant
z	Compressibility factor
T (K)	Temperature
Q_i (m^3/s)	Average inflow to pipe section i
QU_i (m^3/s)	Average outflow to pipe section i
t (s)	Time
Δt (s)	Time increment
C_1, C_2	Constants
H (m)	Pressure head
I (kg m^2)	Moment of inertia
l (m)	Length
n (rpm)	Pump speed

INTRODUCTION

In pipelines carrying liquids, pressure waves and separations may occur under transient conditions.

Earlier measurements and computations, mainly with water as a fluid, have been reported by [2, 3], among others.

In general, oil pipelines are designed to allow for flow separation. However, the present pipelines may be overdimensioned as well as underdimensioned for certain operational conditions. Thus, from an economical point of view as well as safety at oil installations, it is important to study these flow processes.

The lack of data on these processes in oil pipelines led us to investigate them using crude oil from the North Sea.

A special test rig with a high-point was built to generate column separations. After a test series with water we then carried out similar tests with crude oil.

In both cases the purpose was to verify a computer program treating separation. The pressure was recorded in the expected separation region at the high-point and at the pump for various pump shutdowns and valve closures. At the same time the quality of the crude oil and the released gases was tested in order to get a basis for numerical calculations. The crude oil contains light hydrocarbons that cause, as expected, higher separation pressure than for water.

With various degrees of success we could visually observe the separation through a glass tube at the high-point.

THEORY

Numerical model

As crude oil is a multicomponent fluid the mechanism of cavitation is more complex than for water. On the basis of the model tests we have developed a simplified numerical model which seems to give relatively good agreement with the measurements. The numerical model is based on the Method of Characteristics with specified time intervals as described in [1].

The following assumptions are made:

- The wave-speed of the liquid is constant.
- The cavity is located at the internal points of the pipe element.
- There is no gas release before the pressure drops to separation pressure.
- The separation pressure is supposed to be constant, independent of the volume of the cavity.
- The gas developed during the period of separation remains in the system; i.e., the gas or vapour is not absorbed by the liquid nor condensed.
- The liquid as well as the cavity are regarded as homogeneous.
- The cavity is assumed to behave according to the equation of state: $p V = z m R/\bar{M} T$
- The temperature is constant.

Equations

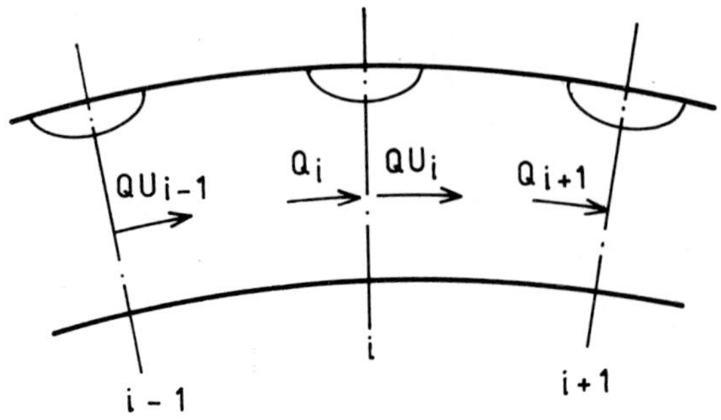

At time t we have the characteristic equations:

1) $$p_{i,t} = p_{i-1,t-\Delta t} - C_1 (Q_{i,t} - QU_{i-1,t-\Delta t}) - C_2 QU_{i-1,t-\Delta t}|QU_{i-1,t-\Delta t}|$$

2) $$p_{i,t} = p_{i+1,t-\Delta t} - C_1 (QU_{i,t} - Q_{i+1,t-\Delta t}) - C_2 Q_{i+1,t-\Delta t}|Q_{i+1,t-\Delta t}|$$

If the calculated pressure is lower than the vapour pressure then $p_{i,t} = p_v$ and the mass of the cavity, m_c, is calculated according to the equation of state:

$$m_c = \frac{p_v \cdot V_{c,t}}{z \cdot \frac{R}{M} \cdot T}$$

where $V_{c,t}$ is the volume of the cavity.

3) $$V_{c,t} = V_{c,t-\Delta t} + 0.5 \cdot \Delta t \, (QU_{i,t} + QU_{i,t-\Delta t} - Q_{i,t} - Q_{i,t-\Delta t})$$

The mass of the cavity developed through the separation period remains in the system and the pressure is later calculated as:

4) $$p_{i,t} = \frac{z \cdot m_c \cdot \frac{R}{M} \cdot T}{V_{c,t}}$$

In the actual calculations we have used the following parameter values:

Universal gas constant	:	R	=	8314 J/mol·K
Average molecular weight	:	$\overline{M}$	=	32 kg/mol
Temperature	:	T	=	285 K
Vapour pressure	:	p_v	=	-0.7 bar
Pressure propagation speed	:	a	=	900 m/s

The compressibility factor z is measured from oil probe:

Pressure	z-factor
3.0	0.916
2.0	0.951
1.5	0.960
1	0.968
0.5	0.976
0	0.984
- 0.5	0.991
- 0.8	0.997

TEST RIG

General layout

The test rig consists of an 87 mm diameter, 250 m long steel pipeline with a total rise of 12 m. The crude oil was pumped from a 3.0 m^3 bottom tank up to the break point 8.5 m above the pump, the pipeline slope ∿ 30°*. The pipeline then rised evenly up to the upper tank. (See Figure 1).

Pipeline

Perrot-pipes, snaplock couplings and rubber seals.

Inside diameter:	Φ 87 mm
Wall thickness :	∿ 1 mm
Section length :	∿ 6 m
Total length :	∿ 250 m

Tanks

Two standard petrol tanks.

Upper tank:	3.0 m^3
Lower tank:	1.5 m^3

The liquid in both tanks was exposed to atmospheric pressure. They were fitted with safety fire-screens at the top and special energy dissipators at the inlets and outlets.

Pump and valves

STORK centrifugal pump CEN65-315, with a capacity Q = 48 m^3/h when the pressure head H_c = 33 m and speed n = 1450 rpm. Moment of inertia I = 0.1325 kgm^2.

The pump was run by an 11 kW motor. The unit was fitted with an electromagnetic clutch between the pump and motor.

The pumping system had a great number of valves as shown in Figure 1. Most of them were used only to regulate the necessary stationary flow. The valve varked AV on the figure is a pneumatically operated butterfly valve. This was used for the valve closing tests. The valve marked TV is a non-return valve of disc type which was used for all the pump shutdown tests.

The vertical glass tube with length 1.0 m and diameter 0.1 m was connected to the valve V1 at the high point.

* The slope break has the same effect as a high-point and will be referred to as such.

Instrumentation

The pressures P1 and P2 in Figure 1, were measured by two Schaevitz pressure transducers, and valve position by a Schaevitz rotary variable differential transformer. An optical pulse-counter was placed on the pump shaft to measure pump speed.

Pressure and pump speed (or valve position) were recorded by a 4 channel Tandberg TIR 100 tape recorder and a Hewlett-Packard two channel ship chart recorder.

On the high-point of the test rig there is a glass for observing separation and gas release. A valve placed on the bottom of the glass can be closed when measuring pressure variation.

TEST PROCEDURE

General consideration

With the test rig described above the electromagnetic clutch between the motor and the pump makes it possible to use only 200 m of pipeline in the circuit. Thus the inertia of the motor which is 10 times greater than that for the pump rotor does not affect the pump speed in shutdowns. It should be desirable to hold the run-off time for the impeller and the valve closure time as short as possible in comparison with the pressure wave reflection time $t = 2\ l/a$.

The pressure propagation speed was calculated to a = 900 m/s. The pipeline was carefully deaerated before each measurement.

With reverse flow, i.e. stationary flow from the upper tank T2 to the lower tank T1, pressure surges were induced by means of the valve AV. Pressure recordings from test No 37 (see Figure 9) are given together with the valve angle variation.

We shall note that the pressure propagation speed decreases for low pressures more than for water because of the hydrocarbon gas release.

The test rig was situated in a sand-pit outside the laboratory and was therefore affected by variations in the air temperature and sun radiation.

A large number of valves was needed to produce flow reversal in the pipeline and for other special measurements for solving stability problems in hydroelectric power plants.

Recorded and measured parameters

There were run 7 pump shutdowns and 1 valve closure with various flow rates.

The following parameters should be useful for comparison between measurements and calculations:

Pressure p_1 (bar) on	P1 at the pump	(see Figure 1)
Pressure p_2 (bar) on	P2 at the high-point	
Pump speed n rpm		
Valve position α		

The 4 parameters were recorded on a 4 channel tape recorder.

Additional observations include:

- Oil temperature in the oil reservoir T1.
- Visual observation of the separation in the glass above the high-point.
- Gas probes were taken from the top of the glass.
- Composition of the crude oil before, under and after the tests by means of a gaschromatographic method.
- Viscosity in cPs.

The tests are tabulated as follows.

Date	Measurement No.	Type of test	Flow l/s	Oil temp °C	Remarks	Figure No.
2/7	31	Pump shutdown	10.5	12		2
"	33	"	10.3	12		3
"	38	"	9.5	13		4
"	39	"	13.7	13		5
5/7	45	"	6.2	29	V1 open. No visible separation in the glass	6
"	46	"	6.2	29		7
"	40	"	13.7	20	V1 open. Visible separation	8
2/7	37	Sudden valve closure	-3.7	13	Downwards flow	9

RESULTS

1. The oil changed its ability for gas release during the tests

The lowest separation pressure p_2 = - 0.8 bar was, as expected, much higher than for water. This is a natural consequence of the fact that crude oil contains light hydrocarbons. From the pressure recordings it can be seen that more gases are released (See Figures 2 and 3) in the first tests than later. The duration of the low pressure period after the first pressure peak reveals this. The flow rate is approximately the same in both cases. The following tests (Nos. 38 and 39) with different flows show less gas release. (See Figures 4 and 5.)

2. Influence of temperature

A comparison between tests 39 and 40 indicate that the gas release period is longer at higher temperatures. The temperature is higher in the last test but the valve V1 is also opened.

It is assumed that the valve V1 has only a small influence on the pressure history.

3. Visible separation

In the test shown in Figure 8 separation bubbles were observed in the inspection glass a moment after the shutdown.

The level in the glass oscillated near the lowest level with a = 0.23 - 0.25 m from the top and then rose to a stable level at a = 0.12 m.

CHEMICAL ANALYSIS

Composition of the cavity

A sample of dry gas was taken from the top of the inspection glass. In order to get sufficient gas filling, the probe sampling was carried out after a series of 12 shutdowns/startups within intervals of ∿20 seconds during steady state pumping at maximum flow.

Component	Formula	Vol %
Methane	CH_4	4.11
Ethylene	C_2H_6	16.06
Propane	C_3H_3	37.14
Butane	C_4H_{10}	16.69
Pentane	C_5H_{12}	5.27
Hexane	C_6H_{14}	1.65
Air	N_2O_2	19.08
Total		100.00

Accuracy ± 2%.

The occurrence of 19% air in the sample is a consequence of the fact that the oil had contact with the atmosphere during the filling and in the tanks.

Composition of dissolved gases

Chemical tests of the crude oil were made before, during and after the measuring programme was carried out.

The samples were pressurized to 100 mm Hg (0.13 bar) and it could be seen that only the contents of methane, ethylene and butane were decreased during the tests.

Viscosity

32 centipoise at $20^{\circ}C$

42 centipoise at $12^{\circ}C$

Comparison between measurements and calculations

For 4 of the tests, see Figures 3, 5, 7 and 9, calculations were made using the proposed method. The calculated pressure history shows good agreement with the measurements, but the following deviation can be noted:

- The damping effect is less in the calculations. This is a phenomenon seen often in earlier comparisons. The reason is the complex velocity distribution that causes relatively high head losses in unsteady flow.

- The deviation in the valve closure test is caused by the fact that the non-return valve opened in the low pressure period.

CONCLUSIONS

Separation caused by pump shutdowns and valve closure has been studied for both water and crude oil flows in a pipeline.

The experiments showed that the pressure history was less irregular for crude oil than for water flows. The experiments revealed further that the hydrocarbon gases released during separation in crude oil were not significantly absorbed after the separation.

A numerical method in which the separated gas mixture is not allowed to be absorbed in the fluid nor condensed after the separation, gave general agreement with the measurements.

REFERENCE

[1] Wylie, E.B. and Streeter, V.L. "Fluid transients". Mc Craw-Hill (1978).

[2] Tullis, J.P. "Water hammer analysis with air release". Paper C3 Proc. 2nd Int. Conf. on pressure surges, BHRA, 1976.

[3] Provost, G.A. "Investigation into a cavitation in a prototype pipeline caused by water hammer". Paper D2 Proc. Int. Conf. on pressure surges, BHRA, 1976.

ACKNOWLEDGEMENT

The authors wish to thank the Applied Chemistry Division at The Foundation of Scientific and Industrial Research at The Norwegian Institute of Technology for their chemical analysis and support during this work.

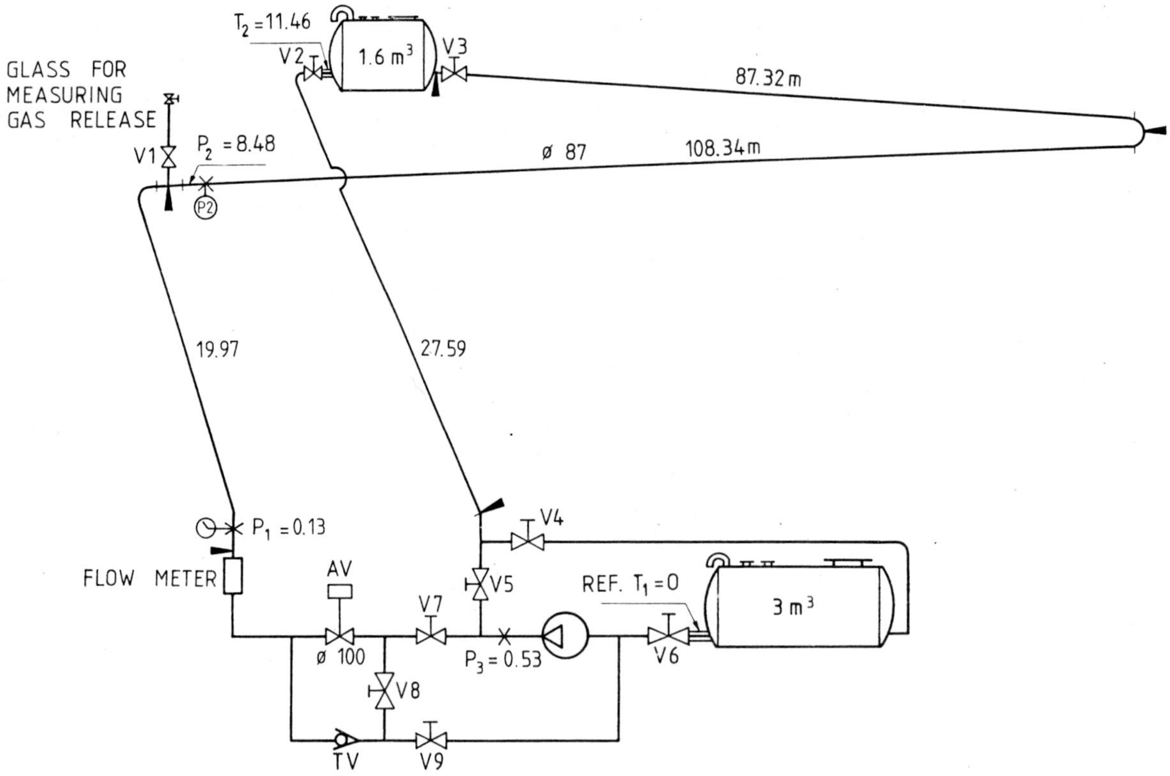

FIG. 1. TEST RIG

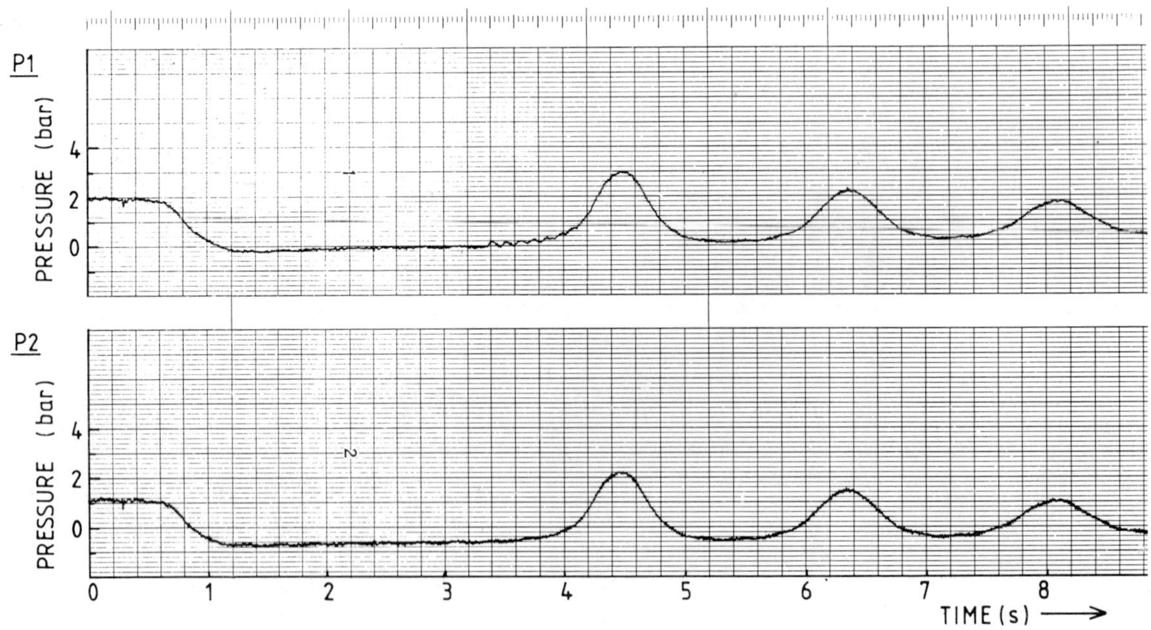

FIG. 2. PUMP SHUTDOWN AT FLOW-RATE OF $10.5 \cdot 10^{-3}$ m³/s

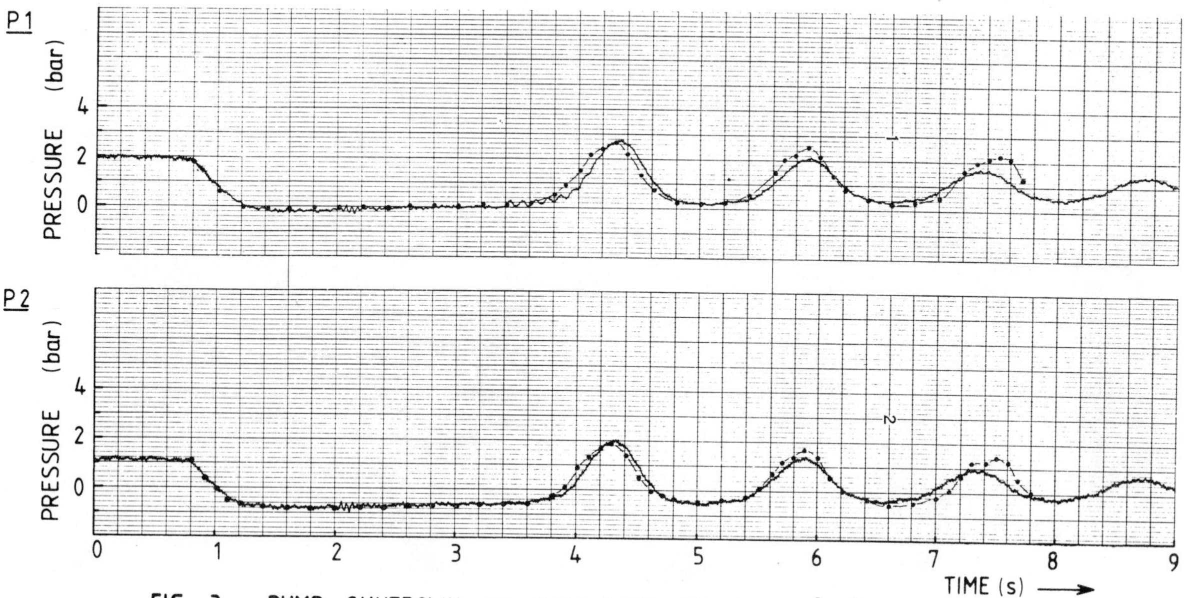

FIG. 3. **PUMP SHUTDOWN** AT FLOW-RATE OF $10.3 \cdot 10^{-3}$ m³/s

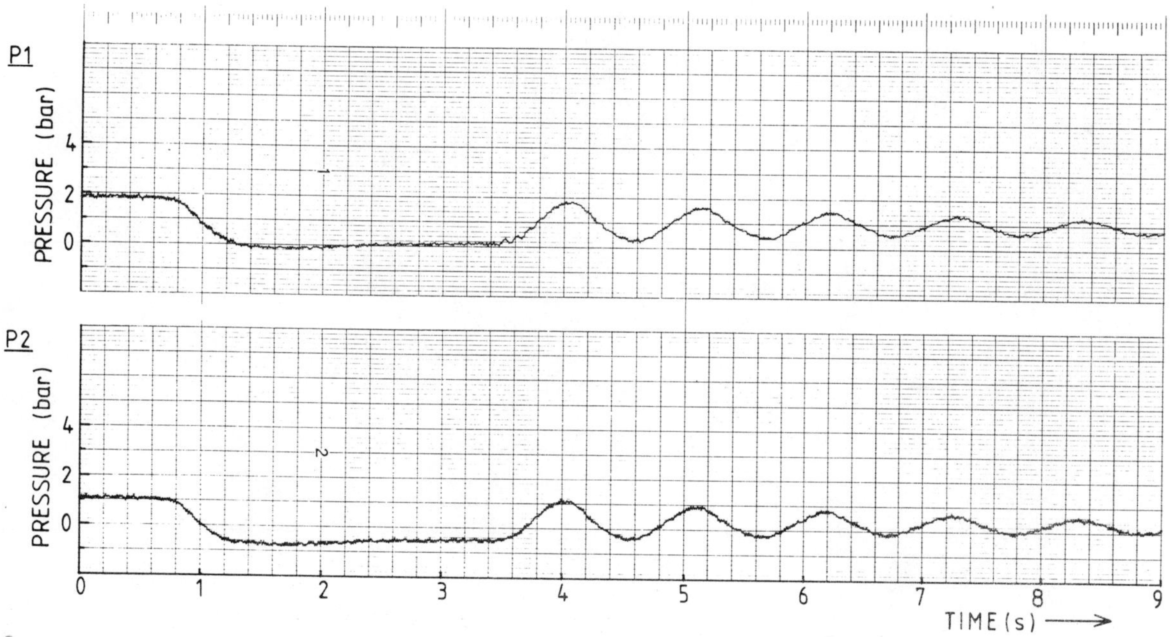

FIG. 4. **PUMP SHUTDOWN** AT FLOW-RATE OF $9.46 \cdot 10^{-3}$ m³/s

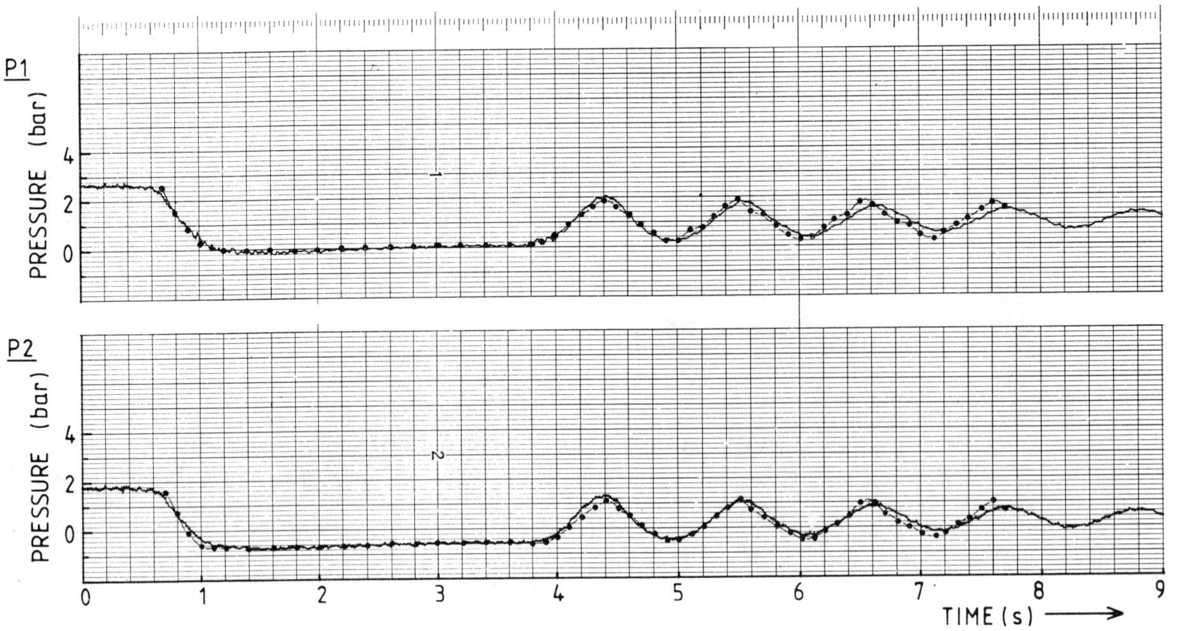

FIG. 5. **PUMP SHUTDOWN** AT FLOW-RATE OF $13.74 \cdot 10^{-3}$ m³/s

—— **TEST**
-•-•-•- **COMPUTATION**

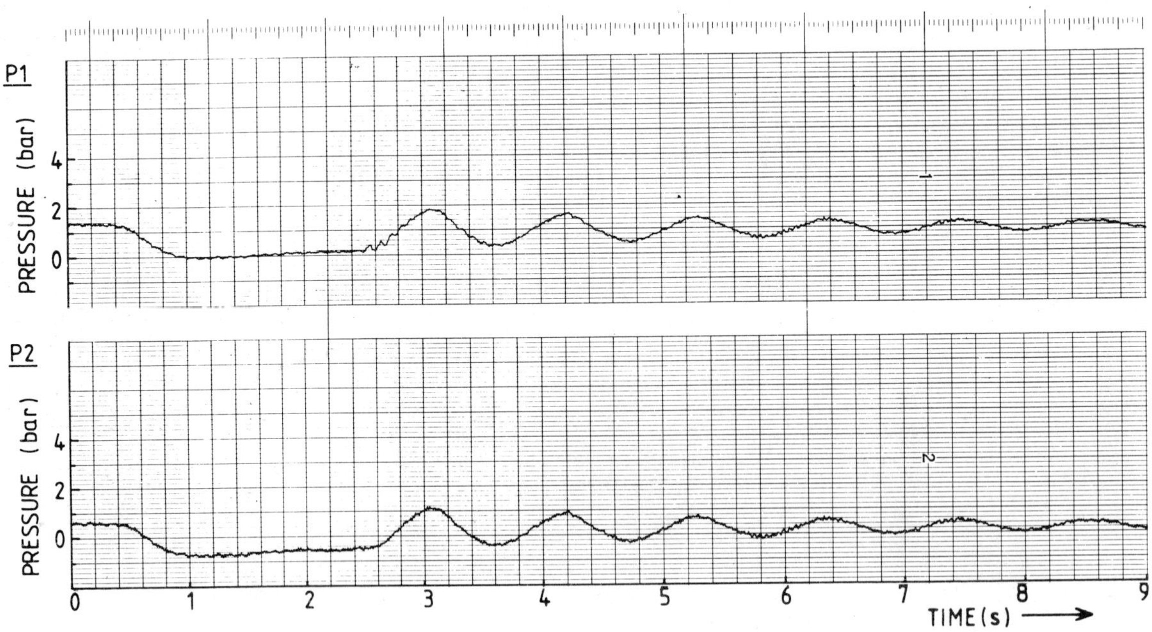

FIG. 6. **PUMP SHUTDOWN** AT FLOW-RATE OF $6.25 \cdot 10^{-3}$ m³/s
V1 OPEN

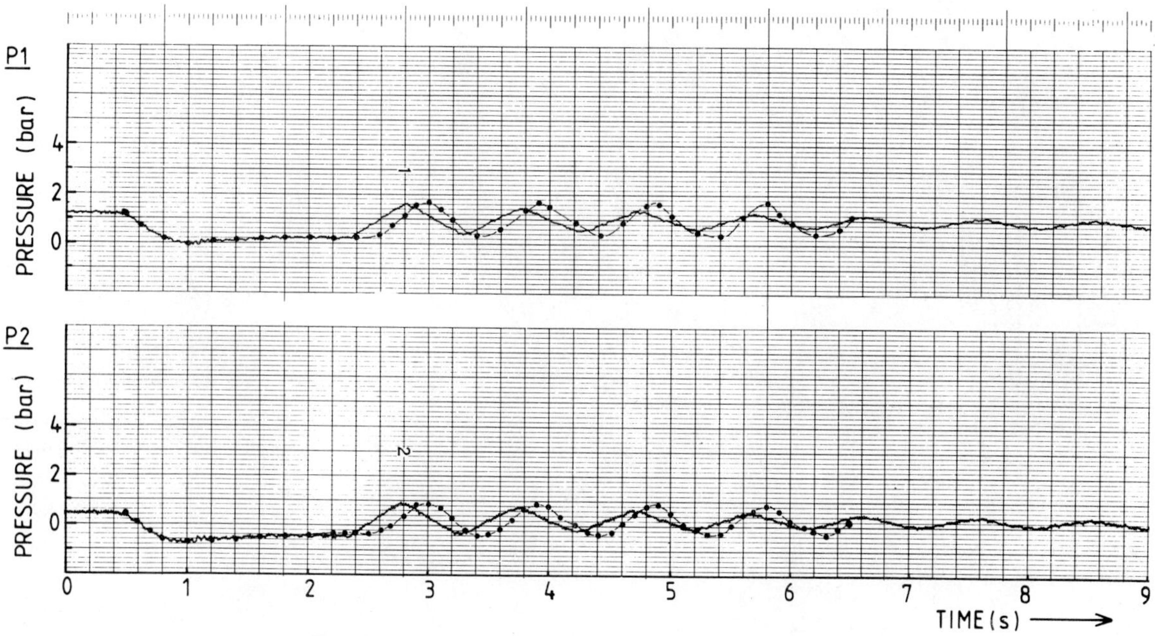

FIG. 7. PUMP SHUTDOWN AT FLOW-RATE OF $6.25 \cdot 10^{-3}$ m³/s

— TEST
-•-•-•- COMPUTATION

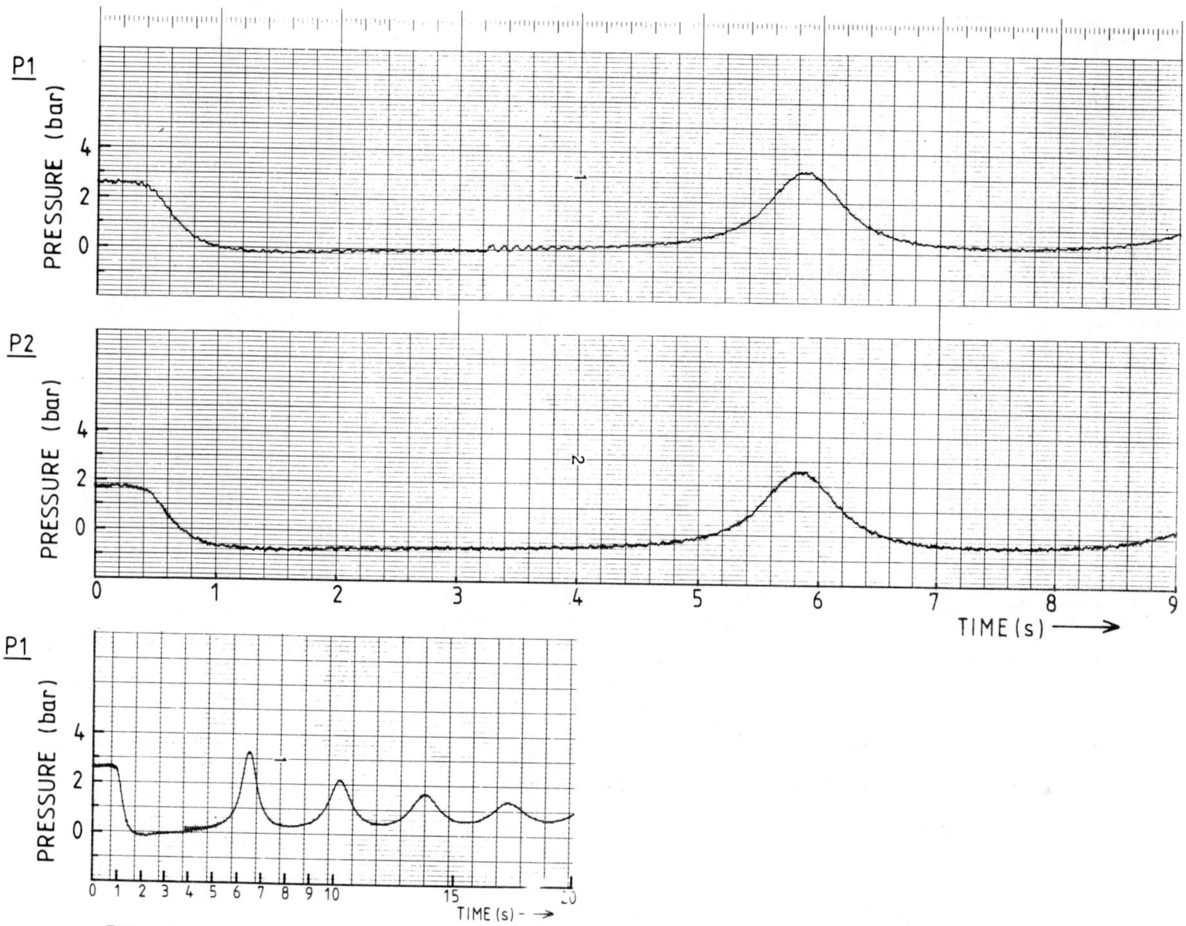

FIG. 8. PUMP SHUTDOWN AT FLOW-RATE OF $13.7 \cdot 10^{-3}$ m³/s
V1 OPEN

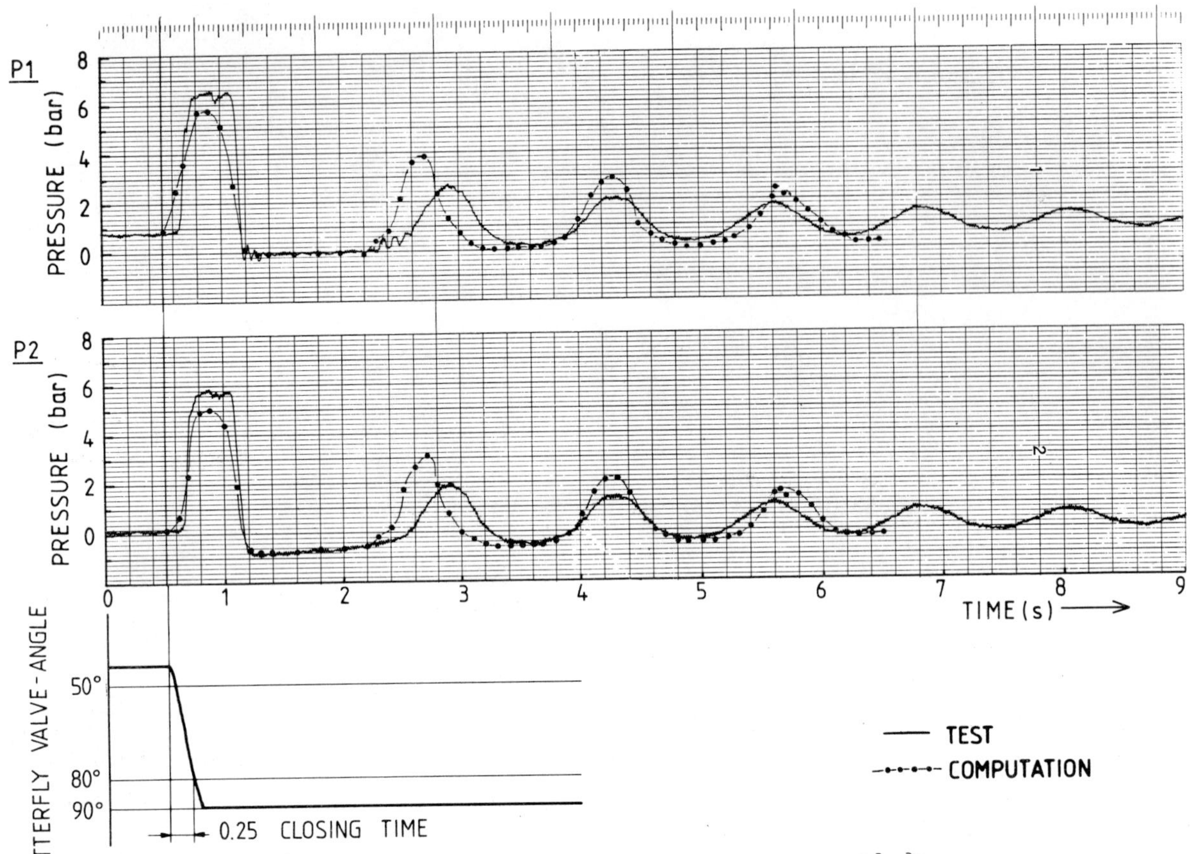

FIG. 9. SUDDEN VALVE CLOSURE AT FLOW-RATE OF $-3.71 \cdot 10^{-3}$ m³/s

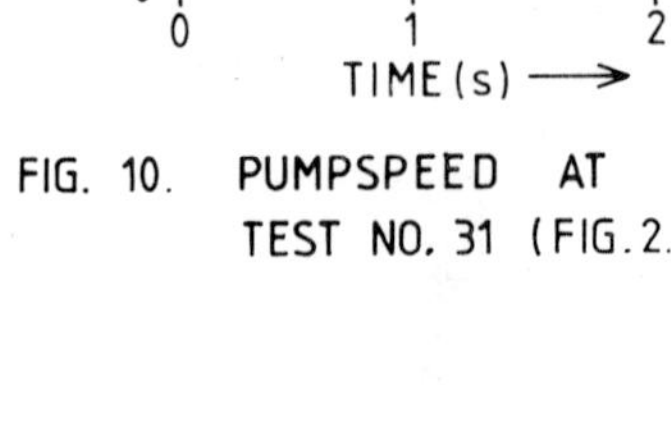

FIG. 10. PUMPSPEED AT TEST NO. 31 (FIG. 2.)

Third International Conference on

PRESSURE SURGES

Canterbury, England
March 25-27, 1980

PAPER C2

ALLOWING FOR FREE AIR IN WATERHAMMER ANALYSIS

D.J.F. Ewing

C.F.G.B., U.K.

Summary

When free, i.e. undissolved, air is present in a hydraulic system, its effect on waterhammer waves and pressure-surges is now well known. It greatly reduces the waterhammer wave speed; this speed now depends on pressure which makes all calculations nonlinear; and the free air greatly increases the surge damping.

The object of this paper is to show how one can allow for these effects approximately by elementary methods. Comparisons are made with the classical investigations of Whiteman and Pearsall (1959) and Pearsall (1966) on sewage lines, and various damping mechanisms are discussed.

Held at the University of Kent at Canterbury, U.K.

Conference Organised by BHRA Fluid Engineering, Cranfield, Bedford, England.
In conjunction with The City University, London.

NOMENCLATURE

a, a_o, a_{sh}	wave speed (m s^{-1}), wave speed with no free air, shock-wave speed
A	pipe cross-sectional area (m^2)
b=a/g	Allievi parameter (s)
c_p	specific heat of air at constant pressure (J/kg deg.K)
d	pipe diameter (m)
D	diffusion coefficient (m^2 s^{-1})
f	Darcy-Weisbach friction factor (fL/d = no. of velocity-heads lost)
F(Y)	characteristics function (equation (21))
g	acceleration due to gravity (9.81 m s^{-2})
G(Y)	lumped-air function (equation (25))
h	specific enthalpy of air (J/kg)
$H, \bar{H}$	head, usually absolute (m), mean head
k	damping factor (s^{-1})
$K_o = p_o a_o^2$	bulk modulus of water allowing for pipewall flexibility (N m^{-2})
L	pipe length (m)
$N = K_o \bar{\alpha}/\bar{p}$	air to water + pipewall elasticity-ratio
$p, \bar{p}$	absolute pressure (N m^{-2}), mean pressure
P	pressure-function defined by equation (16) (N m^{-2})
q	heat transferred to air per unit mass of air (J/kg)
r	bubble radius (m); displacement at free end of column (m) (Appendix B)
R	gas constant for air, taken as $2c_p/7$ (J/kg deg.K)
s	equivalent length of lumped-air bubble (volume ÷ pipe cross-sectional area) (m)
t	time (s)
t_h, t_s	heat-transfer relaxation time, bubble slip relaxation time (s)
T	period (s)
$U, \hat{T}$	potential energy, kinetic energy (J) (Appendix B)
v	specific volume of air (m^3 kg^{-1})
V	velocity along pipe (m s^{-1})
x	distance along pipe (m)
Y	dummy variable (usually $H/\bar{H}$)
z	height above reference level (m)
$\alpha, \bar{\alpha}$	volume-fraction of free air, mean value
β_1, β_2	amplitudes (Fig. 5) (m)
$\Delta x, \Delta t$	space and time grid steps (Fig. 1) (m)
ζ	$a\Delta t/\Delta x$ (Fig. 1)
ρ, ρ_o	mixture density (kg m^{-3}), water density (1000 kg m^{-3})
σ	air-water surface tension (73 mN m^{-1})
θ	temperature, deg. K
Suffixes:	b bubble; bv bulk viscosity; h heat-transfer; p constant-pressure; o with no free air; s slip; sh shock

1. INTRODUCTION

Air can get into large hydraulic systems, and in particular power station cooling-systems, in several ways. It can be entrained from a free surface. It can come out of solution in the water when this passes through a low-pressure region (e.g. the pump draft tubes). Finally it can be bled in deliberately, as a precautionary measure to prevent vapour cavities from forming during a low-pressure transient.

The essential effects of free air on waterhammer surges are now well known (Refs. 1,2,3): they are conveniently summarised in Ref. 4. If the air remains localised in one place it usually acts as a surge cushion, but can act as an unwanted nonlinear spring magnifying surges (Refs. 2,3,4). If the air is uniformly distributed its effects are more difficult to predict. The most noticeable effect is a large drop in the waterhammer wavespeed, which is now pressure-dependent (Refs. 1,4). This can lead to shock formation (Refs. 4,5) and can sometimes magnify the surge by comparison to the air-free case (just as concentrated air can). Another noticeable effect is the greatly increased damping of the surge waves (Refs. 1,6). This is of course beneficial; nevertheless it would be useful to be able to quantify it (for example, when interpreting site surge tests).

The purpose of the present paper is to summarise some of the writer's own research into these questions (in so far as it has not been overtaken by material in Wylie and Streeter's authoritative text, Ref. 4). Attention is mainly focussed on the problems of allowing for distributed air, which are less clear-cut than those for lumped air: fortunately, one of the conclusions found is that lumping the air provides the extreme case. Approximate techniques are presented for allowing for the effects of the air-bubbles. These methods are compared with Whiteman and Pearsall's experiments (Refs. 1,6). Various damping mechanisms are discussed.

2. REVIEW OF BASIC PHENOMENA

The basic phenomena produced by small amounts of free air (distributed as air bubbles) are by now reasonably well understood (Ref. 4). The air greatly increases the compressibility of the mixture, but has no significant effect on its inertia. The mixture can be simulated by an equivalent homogeneous fluid having a (variable) air-volume fraction α. The water acts as a heat reservoir, and so

$$\alpha p = \bar{\alpha}\,\bar{p} \qquad \text{... (1)}$$

(bar denoting mean value). The waterhammer wavespeed of the mixture satisfies

$$a = \left[(1-\alpha)\rho/K_o + \rho\alpha/p\right]^{-\frac{1}{2}} \qquad \text{... (2)}$$

where K_o is the 'effective modulus' of the air-free water allowing for pipewall elasticity in the usual way (Ref. 4) and ρ is mixture density. Unless ρ is very large or p is very small (in which case the homogeneous single-fluid approximation breaks down in any case) equation (2) simplifies to

$$a^2 = a_o^{\,2}/\left[1 + N(\bar{p}/p)^2\right] \qquad \text{... (3)}$$

where $a_o = (K_o/\rho_o)^{\frac{1}{2}}$ is the wave-speed in an air-free pipe, ρ_o is water-density, and

$$N = K_o\bar{\alpha}/\bar{p} = \rho_o a_o^{\,2}\,\bar{\alpha}/p \qquad \text{... (4)}$$

N represents the ratios of the elasticities of the air and the (water plus pipewall) combination.

The reduction in wavespeed corresponding to small amounts of free air is now well documented (Refs. 1,4). For example, taking a = 1200 m s^{-1} (representative of a thin-walled metal pipe) implies K_o = 1.44 GN m^{-2} and so N = 14.4 if the mean pressure is one bar (100 kPa) for a 0.1% mean volume-fraction of free air. This reduces a to

306 m^{-1} by equation (3). Moreover when N is large a is approximately proportional to the pressure for moderate pressure-fluctuations.

In addition bubbly mixtures have greatly increased damping: however, the precise physical cause of the damping is still not clear. The damping is broadly of two kinds: direct damping (due to direct dissipative processes) and indirect damping (due to dispersion breaking down the main long-wave surges into shorter wavelength components, which are damped out more readily).

Possible mechanisms for direct damping include:

(i) The increased bulk viscosity of the fluid due to the air bubbles (Taylor, Ref. 7).

(ii) Losses due to slip between bubbles and water (van Wijngaarden, Ref. 8)

(iii) Thermodynamic losses, e.g. due to heat transfer between bubbles and surrounding water (Ref. 9 and Appendix A).

This list is not exhaustive and others will occur to the reader. For example, if resonances can be excited in the bubbles extra losses occur: but the resonant frequencies involved are too high to be excited in prototype-sized plant.

The main mechanism for indirect damping seems to be Pearsall's reflection hypothesis (Ref. 1). If bubbly and clear stretches of water alternate - or even if more bubbly and less bubbly stretches alternate - then partial wave reflections occur at the interfaces. This breaks periodic surge waves down.

Unless the surge is very sharp, it turns out that each of the mechanisms (i-iii) above can be represented formally by adding a 'pseudoviscosity' term to the standard equation of motion. If $\rho_o D$ represents this pseudoviscosity, then D will be termed the effective diffusivity: it has units $m^2\ s^{-1}$. The calculation of D is discussed in Appendix A: it turns out that the diffusivity due to bulk-viscosity effects is below 0.01 $m^2\ s^{-1}$, that due to bubble slip is of order 10 $m^2\ s^{-1}$, whilst that due to heat transfer is of order 100 $m^2\ s^{-1}$. Thus, thermal dissipation dominates, as one might expect.

Diffusion-type damping with emphasis on the events following column separation, was also considered by Bernardinis and Siccardi (Ref. 10), but it appears that their mechanism gave diffusivities of order 0.01 $m^2\ s^{-1}$ (Ref. 3a ppX63-X64).

The standard equation of motion (Ref. 4), modified to include pseudoviscosity, is

$$\partial V/\partial t = -g\partial H/\partial x + \underbrace{D\partial^2 V/\partial x^2}_{\text{damping}} - \underbrace{fV|V|/2d}_{\text{wall friction}} \qquad \ldots (5)$$

(see Nomenclature). H is the piezometric head:

$$H = p/\rho_o g + z \qquad \ldots (6)$$

z being elevation above a reference-level. The continuity-equation is unchanged, namely:

$$g\partial H/\partial t = -a^2\partial V/\partial x \qquad \ldots (7)$$

(Ref. 4). The convection terms $V\partial V/\partial x$ and $V\partial H/\partial x$ have been ignored, and in fact H could equally well be interpreted as the total head, since changes in the velocity-head $V^2/2g$ are of the same order as the ignored convection terms.

The reader may be puzzled as to how a pseudoviscosity-type damping term can arise from a heat-transfer mechanism. In fact when heat transfer is considered (Appendix A) the continuity-equation is modified to

$$\partial H/\partial t = -(a^2/g)\partial V/\partial x + (D/a^2)\partial^2 H/\partial t^2 \qquad \ldots (8)$$

while the momentum-equation is unchanged. On defining

$$H^* = H - (D/a^2)\partial H/\partial t \qquad \ldots (9)$$

and noting that H^* is in fact equivalent to H except for very sharp shocks, the equation transforms to the standard form, equation (5). Linearised analysis is used: that is changes in D and a are ignored when differentiating.

3. NUMERICAL METHODS

The system specified by equations (3-7) is very difficult to solve accurately, even for a single reach of pipe. The difficulty arises because the wavespeed a depends upon the pressure p: this means that shocks can form (Refs. 4,5). When D is nonzero the shocks are suppressed but the equations are technically no longer hyperbolic, and strictly speaking the method of characteristics no longer applies. In practice, D is usually small compared to La, and characteristics-type methods adapt: the damping terms can be treated as small perturbations.

For a computing algorithm to be suitable for a large system, it should be

(a) numerically stable

(b) adjustable and flexible

(c) economic

and if at all possible

(d) simple to understand and to apply.

Requirement (c) makes second-order convergence highly desirable. 'Second-order convergence' means that the cumulative error at the end of a calculation is proportional to the square of the mesh size. When the wavespeed is constant, the standard 'method of specified time-intervals' (Ref. 4), Fig. 1a, meets all the above requirements: which is no doubt why it is so popular. Technically it is not second-order convergent in the friction term when applied in its usual form (Ref. 4 pp33-36), but it can easily be made so (Ref. 4 p52) and as a bonus its stability is then increased.

Unfortunately the second-order convergence is lost if interpolations are necessary (Fig. 1b). The difficulty in a nutshell is that linear interpolation between grid points is necessary to handle shocks effectively; but quadratic or higher-order interpolation is necessary to maintain the theoretical second-order convergence. Quadratic interpolation introduces spurious oscillations behind a shock.

With linear interpolation, numerical damping is introduced. The simplest case to analyse is that of constant wavespeed. Using the notation of Ref. 4, let ζ denote $a\Delta t/\Delta x$, so that $\zeta \leqslant 1$ for stability. In the frictionless case, the standard 'method of specified time-intervals' with interpolations ($\zeta < 1$) solves to first order the equations

$$\partial V/\partial t = -g\partial H/\partial x; \quad \partial H/\partial t = -(a^2/g)\partial V/\partial x \qquad \ldots (10a,10b)$$

that it is designed to solve; but that it solves to second order the equations

$$\partial V/\partial t = -g\partial H/\partial x + \tfrac{1}{2}D\partial^2 V/\partial x^2 \qquad \ldots (11a)$$

$$\partial H/\partial t = -(a^2/g)\partial V/\partial x + \tfrac{1}{2}D\partial^2 H/\partial x^2 \qquad \ldots (11b)$$

where

$$D = \zeta(1-\zeta)\Delta x^2/\Delta t = \zeta(1-\zeta)a\Delta x \qquad \ldots (12)$$

(see Appendix B). If now one defines a modified head

$$H^* = H - (D/2a^2)\partial H/\partial t \qquad \dots (13)$$

then $a^2\partial^2H/\partial x^2 = \partial^2H/\partial t^2$ plus second-order terms, and so

$$\partial H^*/\partial t = -(a^2/g)\partial V/\partial x \qquad \dots (14a)$$

$$\partial V/\partial t = -g\partial H^*/\partial x + D\partial^2V/\partial x^2 \qquad \dots (14b)$$

This is in the form of equation (5). In general, if terms $D_1\partial^2V/\partial x^2$ and $D_2\partial^2H/\partial x^2$ are added to the right-hand sides of the undamped equations (10a,10b), then the effective diffusivity is $D = D_1 + D_2$ and the equations can be transformed to standard form, equations (5,6).

In 1973 the writer tried out various numerical schemes for solving equations (3-7), ignoring both friction and damping (Ref. 11). A very idealised problem was considered: the pressure was held constant at $x = L$ whilst the velocity was assumed to vary in a known way at $x = 0$. The object was to compare various schemes against each other. The standard 'Courant algorithm', i.e. the method of specified time-intervals with interpolations, was the best one found, but of course converged rather slowly. The test problem also showed up the unexpected surge-magnification that air can sometimes produce (Refs. 3,4).

More sophisticated methods used with success by other authors have included: use of the Lax-Wendoff algorithm by Martin, Padmanabham and Wiggert (Ref. 12); an improved linear-interpolation scheme by Vardy (Ref. 13) and the air-lumping technique of Tullis, Streeter and Wylie (Ref. 14). This last involves introducing air pockets at suitable nodes and is the one the writer intends to use in future. For numerical stability, it is important to use an implicit method when integrating the equations at the air pockets; a very simple example is considered later.

4. ANALYTICAL METHODS

Because accurate all-purpose numerical solutions are so hard to obtain, it is worth looking instead at cruder analytical models. These are less accurate in principle, but should give worst-case estimates of the effects involved.

For a system made up of a single pipe and containing a given amount of free air, five principles help:

(i) Lumping the air at one point produces extreme surges.

(ii) Isothermal rather than polytropic air behaviour gives extreme surges.

(iii) Many cooling-water systems are acoustically short; that is to say, a number of full reflections take place during a transient. If so, then the surges tend to be insensitive to the wavespeed for much of the transient.

(iv) When the air remains distributed, then neglecting friction, damping and possible shock wave formation produces extreme surge estimates.

(v) Allowing for shocks in (iv) produces estimates that are very little lower.

The first principle is also valid if the true air content is not known but is estimated from the period. In (iv) it is assumed that the nonlinear method of characteristics is being used - it is not enough just to use the mean wavespeed.

For a pipe network, it is not yet known how valid these principles are, but one would expect them to carry over.

So far, principles (i) and (ii) have only been demonstrated for simple systems (Refs. 3,15 and the present paper) but there is no reason why they should not be generally true. Principle (iii) means that the variable-wavespeed effect only needs to be accounted for at the critical stages of the transient (either when things start to vary very rapidly, or when some discontinuity occurs). The writer tried to give some examples of the power of this principle earlier (Ref. 16).

Principles (iv) and (v) can be demonstrated directly as follows:

Consider a shock moving through a horizontal pipe and separating two uniform regions (p_1,V_1) and (p_2,V_2) with p_1 greater than p_2. The general analysis of this case is given by Enever (Ref. 5). His analysis simplifies if velocity-head terms are neglected in the momentum-balance (Ref. 11). This is logical whenever (as here) one ignores the convection terms in the governing equations. In the isothermal case, the end result for the shock speed a_{sh} is:

$$a_{sh}/a_o = [(p_1 - p_2)/(P_1 - P_2)]^{\frac{1}{2}} \qquad \ldots (15)$$

where

$$P = \int (a_o/a)^2 \, dp + \text{constant} \qquad \ldots (16)$$

and where

$$V_1 - V_2 = (p_1 - p_2)/\rho a_{sh} \qquad \ldots (17)$$

Equation (17) is the analogue for shocks of the Allievi-Joukowski equation

$$dp = \rho a dV \qquad \ldots (18)$$

In the present case, a(p) is given by equation (3) and so

$$P = p - N\bar{p}^2/p \qquad \ldots (19)$$

$$a_{sh}/a_o = [p_1p_2/(p_1p_2 + N\bar{p}^2)]^{\frac{1}{2}} \qquad \ldots (20)$$

Now analyse the same problem by the method of characteristics, ignoring the possibility of shock formation. Integrating equation (18) gives

$$\rho_o a_o(V_1 - V_2) = \int_{p_2}^{p_1} a_o dp/a = F(p_1') - F(p_2') \qquad \ldots (21)$$

where p' equals $p/\bar{p}$, and where

$$F(Y) = (N + Y^2)^{\frac{1}{2}} - N^{\frac{1}{2}}\ln([N^{\frac{1}{2}} + (Y^2 + N)^{\frac{1}{2}}]/Y) \qquad \ldots (22)$$

For a given velocity-jump $V_1 - V_2$ and a fixed pressure p_2 the pressure p_1 as calculated from equation (21) is slightly higher than that calculated from the shock relationship, equation (16). In other words, ignoring shock formation (and allowing characteristics of the same family to cross each other) leads to conservative calculations. Fig. 2 compares the pressure-rise across the shock with that calculated by characteristics theory for a representative special case ($p_2 = \bar{p} = 1$ bar, $a_o = 1000$ m s^{-1}, $N = 10$, $a_{sh} = 301.5$ m s^{-1}), for a range of velocity-jumps. The situation could arise if, for example, p_1 is a high-pressure region at rest, and the water in region 2 is moving to the left so that the jump $V_1 - V_2$ is positive. For p_1 less than p_2 and V_1 less than V_2, the jump does not form and we have a pure expansion wave.

From Fig. 2, the gain in realism by modelling accurately shock formation - i.e. the running together of characteristics of the same family - is quite small. In a general system, the effort involved is unlikely to be worthwhile.

Shocks cannot arise when the air is lumped, which is a further advantage of lumped-air computing methods.

5. APPLICATION TO WHITEMAN AND PEARSALL'S EXPERIMENTS

The experiments in question were carried out in 1959 at Kingston and Derby sewage-works (Ref. 1). The Kingston experiments are described in detail in Ref. 6. Their primary purpose was to check the safe working of the nonreturn valves (reflux valves) protecting the pumps; but they turned out to have very strong fundamental significance for our understanding of surge waves in prototype-sized plant when free air is present.

Fig. 3 shows schematically the Kingston layout: the main pipeline consists of a 2.7 km long rising stretch, diameter 840 mm (33 inches), with a total rise of 8.5 m (28 feet) and ending in a riser, with exit through a bellmouth. The static head on the pumps was given as 13.7 m (45 foot) gauge. The Derby pipe was approximately 200 m long, of diameter 1 m (39 inches), with a static head of 9.0 m (29.5 foot) gauge.

The tests consisted of setting running and then shutting off combinations of the available pumps. The ones of particular interest are those when all the pumps running were shut off. A typical trace of one of them (test K13, Ref. 6) is redrawn in Fig. 3.

The trace shows two characteristic features:

(i) The surges are asymmetric with respect to the static head; the first pressure-peak is in fact above that predicted by constant-wavespeed theory. The line of dots is Whiteman and Pearsall's calculation by orthodox, constant-wavespeed, waterhammer theory.

(ii) The damping is noticeably large.

At Derby, where the air content was significantly greater, this damping was even larger: it seems to be much greater than can be explained by orthodox pipewall friction (even allowing for, say, a four-fold increase in this due to the periodic reversals of the pipewall boundary-layers). So, as concluded by Whiteman and Pearsall (Ref. 6) the damping must somehow be caused by the air content.

Feature (i) is explained qualitatively by the present analysis. Graphs are available in Refs. 1 and 6 of four of the test surge results (tests K2, K4 and K13 at Kingston, and test A4 at Derby). Values of H_{min} and H_{max} have been read off these graphs and estimated to the nearest 0.5 m (Table 1 at the end of this paper). The writer appreciates that the graphs given were not necessarily intended for such a purpose, and that distortion may have arisen in the reprinting of these graphs for publication. Nevertheless, at such a long time after the original experiments, this seems to be the best that can be done.

Using the analysis of the last section, H_{max} and H_{min} should be related by

$$F(H_{min}/\bar{H}) + F(H_{max}/\bar{H}) = 2F(1) \qquad \ldots (23)$$

This equation is easily solved by Newton's rule to give H_{max} in terms of H_{min}, since the equation $F(Y) = k$ can be solved iteratively by

$$Y_{new} = Y + Y[k - F(Y)]/(N + Y^2)^{\frac{1}{2}} \qquad \ldots (24)$$

Solutions are plotted in Fig. 4 for the cases N = 0, 1 and infinity (i.e. incompressible water). For large air contents, i.e. large N, the relationship (23) simplifies

(Ref. 15) to

$$H_{min}\, H_{max} = \bar{H}^2 \qquad \ldots (25)$$

Equation (23) is derived by integrating along the C+ characteristic from the closed valve end (Fig. 3) up to the open end, and then back along the C- characteristic down to the closed end again (Ref. 11). The possibility of shocks forming because of crossing-over of C- characteristics is ignored, for reasons already explained. The equation follows because the change in velocity along the two characteristics must be equal and opposite.

Calculating H_{max} from the experimental values of H_{min} in this way gives the results shown in Table 1. One result comes as a surprise: the experimental H_{max} is about 3 m above its theoretical value for test K13.

It is difficult to understand why this should have been so. One possibility is that the air was not in fact uniformly distributed along the pipeline but is concentrated near the valve. This tends to increase the maximum surge (Ref. 15). Accordingly, a very quick investigation of this possibility was made by programmable calculator. The method and results are discussed in Appendix B: essentially the method is a one-off adaptation of the lumped-air method (Ref. 14) with friction neglected and with stable time-integration at the air pocket.

The results of this one-off calculation are shown in Fig. 3 and compared to test K13. The maximum surge has increased as expected, by about 1.3 m, but is still below the observed value. The cause of the shortfall is not clear: it could be cumulative distortion of the original results.

It is also possible to analyse the problem crudely by a lumped-parameter two-degree-of-freedom model. This gives, from Appendix B,

$$G(H_{min}/\bar{H}) = G(H_{max}/\bar{H}) \qquad \ldots (26)$$

where

$$G(Y) = \tfrac{1}{2}(1 - Y)^2 + (T^2/T_o^{\,2} - 1)(\ln Y + 1/Y) \qquad \ldots (27)$$

Like equation (23), this can be solved by Newton's rule quite easily. Results are plotted in Fig. 4, where they can be compared to the distributed-air solutions. For lumped air, N may be interpreted as $(T^2/T_o^{\,2} - 1)$ i.e. the periods of the surges are equalised.

For the case H_{min} = 12.6 m, $\bar{H}$ = 23.7 m, T_o = 0.6 T, N = 16/9 the three methods give H_{max} = 39.70 m (characteristics), 40.95 m (numerical simulation) and 42.23 m (lumped-parameter model).

The other feature of the experiments was the extremely large damping. This can be characterised by the ratio β_2/β_1 (Fig. 5). Values of β_2/β_1 are given in Ref. 1 for all the tests.

Possible damping mechanisms are internal wave reflections due to a change in air concentration (Pearsall, Ref. 1) and heat transfer. If the wave decays exponentially then β_2/β_1 represents the ratios of the energies in successive half cycles. Pearsall (Ref. 1) suggested that

$$\beta_2/\beta_1 = 4\, a\, a_o/(a + a_o)^2 \qquad \ldots (28)$$

Density-changes included in his original formula are too small to matter. The right-hand side represents the energy transmitted when a wave passes from clear water to bubbly water, or vice versa. Data from Ref. 1 have been recalculated using equation (28). Theory and experiment are compared in Fig. 5.

The basic formula (28) applies strictly speaking only to sharp-fronted step waves or to reflections between two infinite media: there are complications if the finite lengths of the waves and water stretches interact (Ref. 17). What may well happen in practice is a whole series of reflections whose general effect can be summed up by equation (28) for a one-pipe system.

Direct damping by the diffusion-type mechanisms considered in Appendix A is much too low to account for the observed damping of the fundamental surge: but these mechanisms could possibly damp out the shorter-wavelength surges produced by intermediate internal reflection. The damping due to a longitudinal diffusion coefficient D on a wave varying as cos or sin $n\pi x/2L$ ($n = 1,3,5 \ldots$) can be shown to be $\exp(-kT/2)$ per cycle, where $k = \pi^2 n^2 D/4L^2$ (Ref. 18). Take $L = 2700$ m, $T = 18.8$ s, $D = 550\ m^2\ s^{-1}$ at Kingston and $L = 200$ m, $T = 4.4$ s, $D = 75\ m^2\ s^{-1}$ (including $10\ m^2\ s^{-1}$ from bubble slip) at Derby is typical. The values of D are calculated in Appendix A assuming N = 3 and 22 respectively. Then the damping of the first harmonic is about 0.2% at Kingston and about 1% at Derby. Note that in the laboratory diffusion damping could well be very significant, because length-scales are much shorter.

The waves steepen at the crests (Fig. 3) because of the increase of wave-speed with pressure. This increases any damping due to diffusion. The second derivatives seem to have increased by about 60% (compared to a pure sine wave) for the crests in Fig. 3.

Heat transfer may be more significant as a source of damping when the air agglomerates. Suppose that the air collects as a thin layer on the roof of a horizontal stretch of pipe near the pumps. As in Appendix A, let t_h be a heat-transfer relaxation time which characterises the time needed for a small change of temperature in the air to equalise to the bulk water temperature. An analysis on the lines of Appendix A and Appendix B (iii) reveals that

$$\beta_2/\beta_1 \simeq \exp\left[-4\pi^2 t_h (T^2 - T_o^2)/7T^3\right] \qquad \ldots (29)$$

provided that t_h is small compared to $T/2\pi$. The difficulty comes in estimating t_h. If the air layer is surface-tension controlled its thickness is probably of order $2(\sigma/\rho_o g)^{\frac{1}{2}}$ (this represents the limiting thickness of a cylindrical water drop), i.e. about 5 mm, except at the leading edge of the bubble (facing the flow). This thickness of air will generally be turbulent due to the motion of the water underneath: the amplitude ΔV of the water-velocity caused by a surge ΔH is $g\Delta H/a$ which would give velocity-amplitudes around 0.17 m s^{-1} at Kingston, 0.5 m s^{-1} at Derby and thickness-based Reynolds numbers of order 1000 at Kingston, 3000 at Derby. It is not possible to estimate t_h at all accurately because of the various uncertainties and so any estimate of damping from equation (29) is purely speculative: but from equation (29) the damping might be around 20% per cycle if t_h were 5% of T.

This mechanism, like that in Appendix A, operates because of the phase lag effect: heat flow to and from the air lags the forcing pressure-change by 90°, since t_h is small. The finite amplitude effect may also cause losses: for example, when the air is distributed as small bubbles the variation of D_p with pressure and the change in surface area may introduce extra irreversibility into the cycle. Still other mechanisms have been proposed in the literature (see e.g. Refs. 4,10). All of them are difficult to quantify and it is not clear that they would predict significantly more damping than those considered here.

To sum up: direct damping by the diffusion-type mechanisms considered in detail here is not large enough to account for the observed damping. Indirect damping - caused by internal reflections where there are changes in bubble density, breaking the surge up into shorter-wavelength components - is a reasonable explanation. Other dissipative mechanisms, e.g. heat-transfer from lumped air, may be significant but are much harder to quantify.

6. DISCUSSION

The question of air release during low-pressure transients has not been discussed in this paper. A kilogram of water saturated with air at 15°C and one bar can release a mass of air equivalent to about 2 ml by volume at S.T.P. for every 1 m of head that the pressure drops below one bar. More precise statements of the law involved are found in Refs. 4,19. This effect is very significant in an engineering context. Papadakis and Hollingshead (Ref. 19) have shown how air release can be taken advantage of in power-station cooling-water circuit design, by omitting the air-admission valves usually provided. Operating such valves can lead to a loss of syphon in the condenser exit waterboxes and the need for slow repriming after a pump trip: worthwhile savings can be made by not operating such valves whenever it can be shown that it is safe to do so.

Naturally, air release contributes to surge damping (Ref. 10). However, there is no evidence that it occurred in Whiteman and Pearsall's experiments. All pressure fluctuations were above gauge and the surge periods were very regular.

The present theory may be used to correct for free air in an orthodox (constant-wavespeed) waterhammer program. Set the program wavespeed to the mean wavespeed (allowing for air) in the usual way and use either equations (23) or (26) to adjust the calculated surges. Running the program with a different wavespeed shows which surges are insensitive to the wavespeed: these need not be adjusted.

7. CONCLUSIONS

1. Five general principles help when modelling free air in pipelines. The first three are (a) that lumped air gives more extreme surges than the same amount of distributed air, (b) that isothermal surges are more extreme than polytropic surges and (c) that for acoustically short systems, results are insensitive to wavespeed for much of the transient.

The last two principles deal with shock modelling (and are not needed when the air is lumped into pockets).

2. Formulae are given for the numerical damping produced by linear interpolation, and for the speed of a shock wave: this last simplifies the normal treatment.

3. Approximate solutions are given for a single pipe containing a bubble at one end, allowing for the water's compressibility, and for the corresponding problem with the air uniformly distributed.

4. Three direct mechanisms for air damping have been analysed (extra bulk viscosity; bubble slip; and heat transfer). The last is the largest. All three can be represented by an effective longitudinal diffusion-coefficient D. Formulae for D are given: a typical combined value is around 100 $m^2\ s^{-1}$.

5. The theory is compared with Whiteman and Pearsall's experiments (1959). Agreement is moderate. The direct damping mechanisms do not account for the observed damping at the fundamental frequency, but may explain the suppression of the higher harmonics. Following Pearsall, the most likely cause of the damping observed is internal reflections at interfaces between more and less bubbly water.

6. Air release during a transient has not been analysed: it did not seem to occur during Whiteman and Pearsall's experiments.

8. ACKNOWLEDGEMENT

This work was carried out at the Central Electricity Research Laboratories and is published by permission of the Central Electricity Generating Board.

9. REFERENCES

1. Pearsall, I.S.: "The velocity of waterhammer waves", Symp. on surges in pipelines (London, Nov. 2-3, 1965), Proc. Inst. Mech. Eng. 180, 3E, 12-20 (1966)

2. Streeter, V.L. and Wylie, E.B.: "Hydraulic transients", McGraw-Hill (1967)

3. Martin, C.S.: "Entrapped air in pipelines", paper F2 of Ref. 3a (1976)

3a. Proceedings, "Second International Conference on Pressure-Surges", (London, Sept. 22-24, 1976), organised by Brit. Hydromech. Res. Assoc., Cranfield (1977)

4. Wylie, E.B. and Streeter, V.L.: "Fluid transients", McGraw-Hill (1978)

5. Enever, K.J.: "Surge pressures in a gas-liquid mixture with a low gas content", Proc. (first) Int. Conf. on Pressure-Surges, Paper C1, organised by Brit. Hydromech. Res. Assoc., Cranfield (Sept. 1972)

6. Whiteman, K.J. and Pearsall, I.S.: "Reflux valve and surge tests at Kingston pumping station", Brit. Hydromech. Res. Assoc./National Engineering Laboratory joint report No. 1 (April 1959)

7. Taylor, G.I.: "The two coefficients of viscosity for an imcompressible liquid containing air bubbles", Proc. Roy. Soc. A, 226, pp 34-39 (1954)

8. van Wijngaarden, L.: "Some problems in the formulation of the equations for gas/liquid flows", article in "Theoretical and Applied Mechanics", ed. W.T. Koiter, North-Holland Publishing Co. (1976)

9. Ewing, D.J.F.: "Additional material on air-bubble damping", discussion of Refs. 12,14, pp X21-X23 of Ref. 3a (1976)

10. Bernardinis, B. de and Siccardi, F.: "Transients in low pressure pumping systems with column separation: a simulation program", paper G2 of Ref. 3a (1976)

11. Ewing, D.J.F.: "The computer calculation of pressure-waves in cooling system pipework: Part I", internal report (1973)

12. Martin, C.S., Padmanabhan M. and Wiggert, D.C.: "Pressure-wave propagation in two-phase bubbly air-water mixtures", paper C1 of Ref. 3a (1976)

13. Vardy, A.E., "On the use of the method of characteristics for the solution of unsteady flows in networks", paper H2 of Ref. 3a (1976)

14. Tullis, J.P., Streeter, V.L. and Wylie, E.B.: "Waterhammer with air release", paper C3 of Ref. 3a (1976)

15. Ewing, D.J.F., discussion of Ref. 3, pp X56-X57 of Ref. 3a (1976)

16. Ewing, D.J.F.: "Analysis of pressure-transients using incompressible flow theory", paper H1 of Ref. 3a (1976)

17. Wood, A.B.: "A textbook of sound", 3rd Edition, Bell and Sons (1955)

18. Coulson, C.A.: "Waves", 7th edition, Oliver and Boyd (1955)

19. Papadakis, C.N. and Hollingshead, D.F.: "Air release in the transient analysis of condensers", paper G1 of Ref. 3a (1976)

APPENDIX A: DAMPING MECHANISMS

This Appendix analyses the three direct damping mechanisms mentioned in the main text. These are the bulk viscosity effect; the bubble slip effect; and heat transfer between water and air-bubbles.

It turns out that each of these three mechanisms can be represented by an effective diffusivity D (units $m^2\ s^{-1}$) which provides a good way of comparing their magnitudes. Linearised analysis is used throughout, i.e. only small fluctuations are considered.

Bulk viscosity of an air-water mixture was analysed by Taylor (Ref. 7). His analysis leads to a bulk-viscosity diffusivity D_{bv} where, putting $N = K_o\alpha/p$ as before,

$$D_{bv} = (\nu K_o/3p)(4N/(N+1)^2) \leqslant \nu K_o/3p \qquad \text{... (A1,A2)}$$

the maximum being attained for N = 1. At 20^oC D_{bv} cannot exceed 7×10^{-3} m^2 s^{-1}, its maximum value for a rigid-walled pipe: so bulk viscosity effects are negligible for prototype-scale plant. Wall flexibility causes some extra damping directly, because the walls can absorbe kinetic energy: but it can be shown, by comparing typical kinetic energies, that this effect is negligible at prototype scale.

Bubble slip occurs because the bubbles have negative inertia relative to the bulk water flow; thus relative motion occurs during acceleration. van Wijngaarden has analysed the slipping of small bubbles in various papers (e.g. Ref. 8). As pointed out to me by J.R. Ockendon, H. Ockendon and D.H. Spence, his results imply that

$$D_s \doteq 2a^2\alpha t_s \qquad \text{... (A3)}$$

where

$$t_s = r^2/18\nu \qquad \text{... (A4)}$$

and t_s is van Wijngaarden's slip relaxation time.

Superficially, equations (A3-A4) imply that D_s can become very large when the bubbles agglomerate, so that r and t_s become large. In fact the analysis only holds for spherical bubbles, i.e. bubbles stabilised by surface tension. Larger bubbles become nonspherical, the drag coefficient increases and the relaxation time decreases: so paradoxically this increase in drag coefficient lessens the slip dissipation (because the bubbles "freeze" onto the mixture faster). In practice, bubbles of up to 5 mm diameter are observed. Taking a mean bubble radius of 1 mm gives $t_s = 1/18$ s at 20^oC. Rearranging equation (A3) then gives, putting $N = K_o\alpha/p$ as before,

$$D_s = 2t_s pN/\rho_o(N+1) \qquad \text{... (A5)}$$

$$< 2t_s p/\rho_o = 11 \text{ m}^2 \text{ s}^{-1} \qquad \text{... (A6)}$$

In the typical case r = 1 mm, p = 1 bar, $K_o = 1.44$ GN m^{-2} (so that $a_o = 1200$ m/s), D_s equals 10 m^2 s^{-1} when α equals 0.1% (with a = 306 m/s and N = 14.4).

Thermal dissipation due to heat transfer has been summarised by the writer in Ref. 9. If temperature-changes in the bubbles are allowed for, the continuity-equation (7) is replaced by

$$g\partial H/\partial t = -a^2\left[\partial V/\partial x - (\alpha/\theta)\partial\theta/\partial t\right] \qquad \text{... (A7)}$$

The extra terms arise because it is $p\alpha/\theta$ which is held constant during the differentiation of the density, rather than $p\alpha$ as in the isothermal case. θ is local temperature of the air-bubbles, averaged through their volume. To estimate $\partial\theta/\partial t$, the thermodynamic equation

$$dq = dh - vdp = c_p d\theta - vdp \qquad \text{... (A8a,A8b)}$$

can be used, where h is enthalpy per unit mass, c_p is specific heat of air at constant pressure, q is heat transfer per unit mass and v is volume per unit mass (for air). $\partial q/\partial t$ can be estimated by Newton's law of cooling as

$$\partial q/\partial t = c_p(\theta - \theta_o)/t_h \qquad \text{... (A8c)}$$

where t_h is a heat-transfer relaxation time and θ_o is the bulk water-temperature. t_h turns out to be very small, of order

$$t_h = r^2/15D_p \qquad \text{... (A9)}$$

where D_p is the thermal diffusivity of air at constant pressure. The factor 15 occurs in the "well-stirred limit" of perfectly efficient supply/removal of heat from the bubble's surface by convection and mixing, and by averaging over all transients. Inside the air-bubbles, heat is supposed to transfer purely by conduction. Let θ" denote temperature a distance r" from the bubble centre. The slowest-decaying transient corresponds to θ" varying like (sin π r')/r' where r' = r"/r. If only this transient is considered the factor 15 in equation (A9) falls to 9.87.

It is hoped to give a fuller account of these calculations elsewhere. Here, the important point to note is that t_h is proportional to the bubble's surface area and, for 1 mm bubbles at 20°C, is of order 3-5 ms. D_p is 21 $mm^2\ s^{-1}$ at 20°C, so equation (A9) gives t_h = 1/300 s for 1 mm radius bubbles.

Since t_h is normally very small compared to the forcing external timescale, the term $\partial h/\partial t$ can be ignored compared to $\partial q/\partial t$. So

$$c_p(\theta - \theta_o) = (t_h R\bar{\theta}/p)\partial p/\partial t \quad \text{and} \quad \partial\theta/\partial t = (2\theta/7H)t_h \partial^2 H/\partial t^2 \quad \ldots \text{(A10,A11)}$$

Equation (A11) follows by differentiating equation (A10) and neglecting changes in the ratio $R\theta/p$, i.e. assuming bubble volume-changes are small (linearised analysis). For air, $R = 2c_p/7$. Substituting into equation (A7) now leads to equation (8) of the main text, with D equal to D_h, where

$$D_h = 2\rho_o a^4 \alpha t_h/7p \quad \ldots \text{(A12)}$$

This can be rearranged as

$$D_h = 4ND_{max}/(N + 1)^2 \leqslant D_{max} \quad \ldots \text{(A13,A14)}$$

with equality when N = 1, where

$$D_{max} = t_h K_o/14\rho_o = t_h a_o^2/14 \quad \ldots \text{(A15)}$$

With the same parameters as in the bubble slip example, this gives D_{max} = 340 $m^2\ s^{-1}$ with α = 0.007%. When α = 0.1%, then D_h = 83 $m^2\ s^{-1}$.

The temperature-changes predicted by this mechanism are minute - only about 0.03°C if the pressure is changing by 10% s^{-1}, from equation (A10). The large damping that they nevertheless create arises because they lag 90° behind the pressure-variations.

The air bubbles will also be saturated with water-vapour - they will contain about 1.4% by volume at 20°C. This produces a significant increase in the air-vapour enthalpy h, but the change in dh/dp produced is under 2%, so the net effect on the damping is negligible.

The thermal conductivity of air is approximately independent of pressure, which means that D_p is inversely proportional to pressure. This makes t_h directly proportional to pressure, assuming no change in bubble size-distribution. At Kingston and Derby, $\bar{p}$ were 2.33 and 1.87 bar respectively. Taking a_o = 1150 m s^{-1}, N = 3 at Kingston, and a_o = 940 m s^{-1}, N = 22 at Derby as typical, then D_{max} = 730 m s^{-1}, D_h = 550 $m^2\ s^{-1}$ (Kingston) and D_{max} = 390 $m^2\ s^{-1}$, D_h = 65 $m^2\ s^{-1}$ (Derby) - assuming 1 mm as the average bubble radius in both cases.

APPENDIX B: MATHEMATICAL DETAILS

(i) Analysis of Numerical Damping Produced by Interpolations

For simplicity, neglect friction and consider the constant-wavespeed case. With reference to Fig. 1b, let ζ denote XM/AM or MY/MB. Then

$$H_X = \zeta H_A + (1 - \zeta)H_M - \tfrac{1}{2}\zeta(1 - \zeta)\partial^2 H/\partial\zeta^2 + \ldots.$$

$$= H_{XL} - \tfrac{1}{2}\zeta(1 - \zeta)H'' \qquad \ldots (B1)$$

where H'' denotes $\partial^2 H/\partial\zeta^2$ and where H_{XL} is the value predicted by linear interpolation. The standard frictionless characteristic equations $dH \pm bdV$ along $dx = \pm a\,dt$ (where $b = a/g$) give

$$H_N + bV_N = H_X + bV_X, \qquad H_N - bV_N = H_Y - bV_Y$$

and so

$$H_N + bV_N = (H_{XL} + bV_{XL}) - \tfrac{1}{2}\zeta(1 - \zeta)(H'' + bV'') + \ldots. \qquad \ldots (B2)$$

Similarly

$$H_N - bV_N = (H_{YL} - bV_{YL}) - \tfrac{1}{2}\zeta(1 - \zeta)(H'' - bV'') + \ldots. \qquad \ldots (B3)$$

Adding and rearranging,

$$H_{NL} = H_N + \tfrac{1}{2}\zeta(1 - \zeta)H'' = H_N + \tfrac{1}{2}D\Delta t\partial^2 H/\partial x^2 - \ldots. \qquad \ldots (B4)$$

A similar equation holds for V_N. Now, H_N represents the true solution of the basic equations, and H_{NL} represents that calculated by the method of specified time-intervals; so H_{NL} can be regarded as solving to second order equations (11a), (11b).

(ii) Analysis of Pipe with Bubble at Closed End

The geometry of the problem is shown in Fig. 3. The air is assumed to be concentrated in a pocket at the valve end of the pipe, which is divided into sections. Three were chosen, so as to fit into the memory of the calculator used (a Hewlett-Packard HP-97 programmable calculator with twenty-six stores).

The air bubble is assumed to obey an isothermal law of form $Hs = \bar{H}\bar{s}$, where s is the equivalent length of the bubble (volume divided by pipe cross-sectional area). All heads are absolute.

The first step is to calibrate this law, i.e. to find $\bar{s}$, by reference to the observed period T and the nominal period T_o with no air present. Let T_b be the period that the system would have if the water were incompressible. The equation of motion of the system is then

$$Ld^2s/dt^2 = g(H - \bar{H}) \qquad \ldots (B5)$$

so that

$$T_b = 2\pi(L\bar{s}/g\bar{H})^{\frac{1}{2}} \qquad \ldots (B6)$$

(this is for small oscillations about equilibrium). For such oscillations, the air and water act like springs in series, whose flexibilities add. Flexibility is the reciprocal of stiffness, so period varies as the square root of flexibility. Thus, the periods satisfy the relationship

$$T_o^2 + T_b^2 = T^2 \qquad \ldots (B7)$$

Equations (B6) and (B7) determine $\bar{s}$.

The main pipe is analysed by standard methods; the only special feature is the treatment at the bubble end. To save writing unnecessary superfixes let H0,V0,H1, V1,...H3,V3 refer to quantities at the old timestep at the four nodes (0 bubble end, 3 free end), and let H and V refer to the unknown quantities at the bubble end at the new timestep. The boundary condition at the upstream end is just H3 = constant = $\bar{H}$.

The equations to be solved are

$$V = ds/dt, \quad Hs = \bar{H}\bar{s} \quad \text{and} \quad H - bV = H1 - bV1 \qquad \ldots (B8,B9,B10)$$

(wall friction is neglected). It is important to integrate equation (B8) by a stable numerical method. The implicit trapezoidal rule is very suitable:

$$s - s0 = \tfrac{1}{2}\Delta t(V + V0) \qquad \ldots (B11)$$

H and V may now be eliminated between equations (B9-B11) to give a quadratic equation for s, from which the solution can be continued in the normal way.

The simulation was started when the valve closed. The initial conditions assumed were V = 0 everywhere, $H0 = H_{min}$ and H varying linearly along the pipe. The main objective was to analyse test K13 and it was noted that the pressure at the pump end was approximately constant over the previous 20 s or so: so this approximation seemed reasonable.

Some calculated values of H0, V0, and V3 are given in Table 2. T was taken as 15.4 s and T_o as 0.6 T.

(iii) Analysis by the Lumped-Parameter Method

Consider an elastic column of water in a pipe of length L, closed at one end. Let r denote displacement at the free end. The column can be simulated by a one-degree-of-freedom mass and spring combination having stiffness AK_o/L. Its mass is $\rho_o AL$. In the simulation, uniform straining is assumed. This requires the mass to be a distance $2\pi/L$ from the closed end, in order for the system to have the correct period $(4L/a_o)$. This can be checked since the system's kinetic energy is then $(2\dot{r}/\pi)^2\rho_o AL/2$ and its potential energy is $\rho_o Aa_o^2 r^2/2L$ (since $K_o = \rho_o a_o^2$). Dots denote d/dt.

Now consider the two-degree-of-freedom system with an air pocket at the closed end. Let s denote equivalent air bubble length (as before) and let r denote displacement of the free end. The kinetic and potential energies $\hat{T}$, U are

$$\hat{T} = \tfrac{1}{2}\rho_o AL\left[\dot{s} + (2/\pi)(\dot{r} - \dot{s})\right]^2, \quad U = (\rho_o Aa_o^2/2L)(r - s)^2 - \rho_o Ag\bar{H}(\bar{s}\ \ln s' - r)$$

$$\ldots (B12)$$

where $s' = s/\bar{s}$. Isothermal behaviour is assumed $(Hs = \bar{H}\bar{s})$.

Setting up the Lagrange equations of motion for r and s and eliminating the acceleration terms gives the compatability-equation

$$(a_o^2/L)(r - s) = 2g(\bar{H} - H)/\pi - g\bar{H} \qquad \ldots (B13)$$

Substituting for r in terms of s, then putting $s = \bar{H}/Y$ where $Y = H/\bar{H}$ gives, after some algebra,

$$U/\rho_o gA\bar{H}^2 = (4gL/a_o^2)G(Y) + gL/2a_o^2 \qquad \ldots (B14)$$

where G(Y) is given by equation (28). Conservation of energy during a cycle gives $U(Y_{max}) = U(Y_{min})$ i.e. $G(Y_{max}) = G(Y_{min})$.

TABLE 1: COMPARISON WITH WHITEMAN-PEARSALL EXPERIMENTS

Test (pumps working)	Experiment					Theory (undamped heads)			
	H	H_{min}	H_{max}	a/a_o	β_2/β_1	N	H_{max}	H_{max} (linear)	β_2/β_1 (eq. 28)
K2 (1 pump)	23.7	14	35.5	0.49	0.76	3.2	37.8	33.4	0.88
K4 (2 pumps)	23.7	12	38.5	0.49	0.77	3.2	42.5	35.4	0.88
K13 (3 storm pumps)	23.7	12.5	43	0.61	0.81	1.7	39.8	34.9	0.94
A4 (1 pump) (Derby)	19.0	12	26.5	0.21	0.52	22	29.8	26.0	0.57

All heads are in m and are absolute heads (gauge + 10 m). The experimental H_{min} and H_{max} are read off graphs, and are rounded to 0.5 m. The calculated heads assume the experimental H_{min} as data.

TABLE 2: LUMPED-AIR SIMULATION RESULTS

Timestep	0	1	2	3	4	5	6	7	8	9	10	11
H(m)	12.6	12.8	13.5	14.8	16.8	19.8	24.1	29.7	35.5	39.43	40.95	40.21
aV/g(m)	0	-3.5	-6.5	-8.9	-10.6	-11.3	-10.7	-8.3	-4.9	-2.05	0.24	1.33
aV_3/g(m)	0	-3.7	-7.4	-11.1	-14.3	-16.6	-17.8	-17.5	-15.2	-10.24	-2.27	6.89

H, V denotes values at bubble end. V_3 is velocity at free end. $H = \bar{H} = 23.7$ m at the free end. Initially H varies linearly with x, V = 0. Three pipe sections were used. $T = 15.4$ s, $T_o = 0.6\ T$. The timestep is 0.77 s.

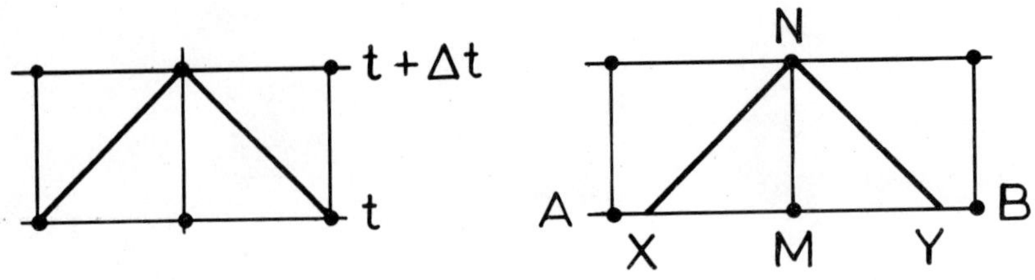

FIG. 1 Characteristic grids: (a) standard (b) with interpolations

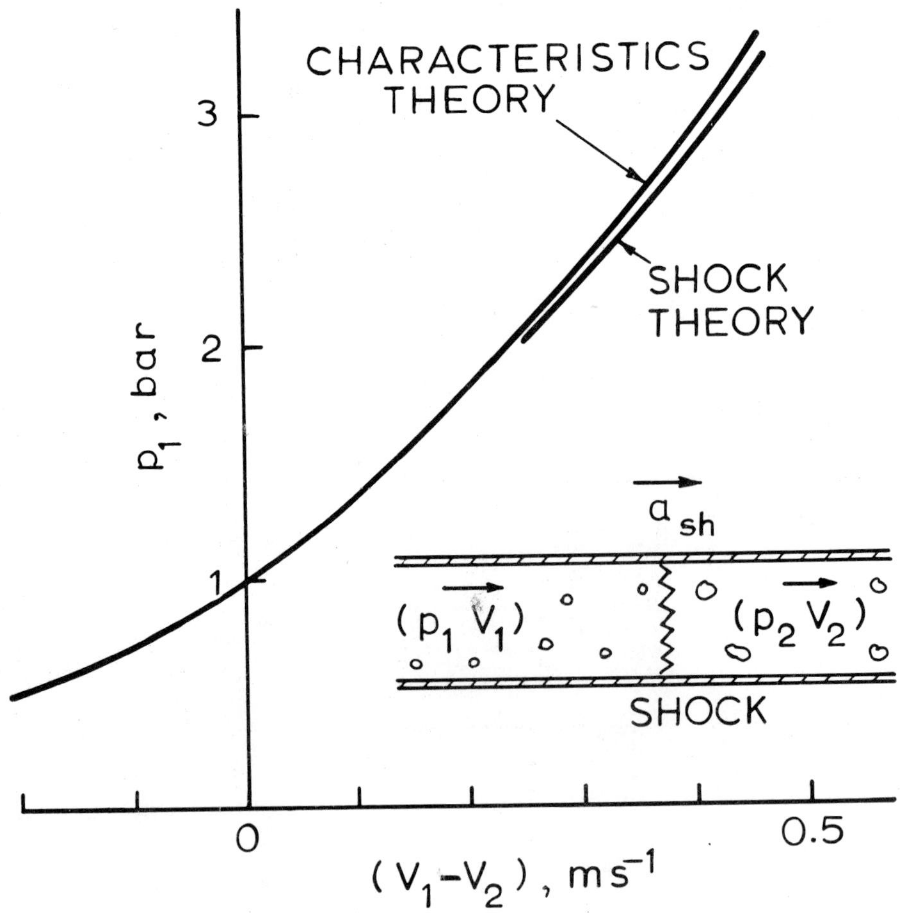

FIG. 2 Shock advancing through bubbly water ($p_2 = \bar{p} = 1$ bar)

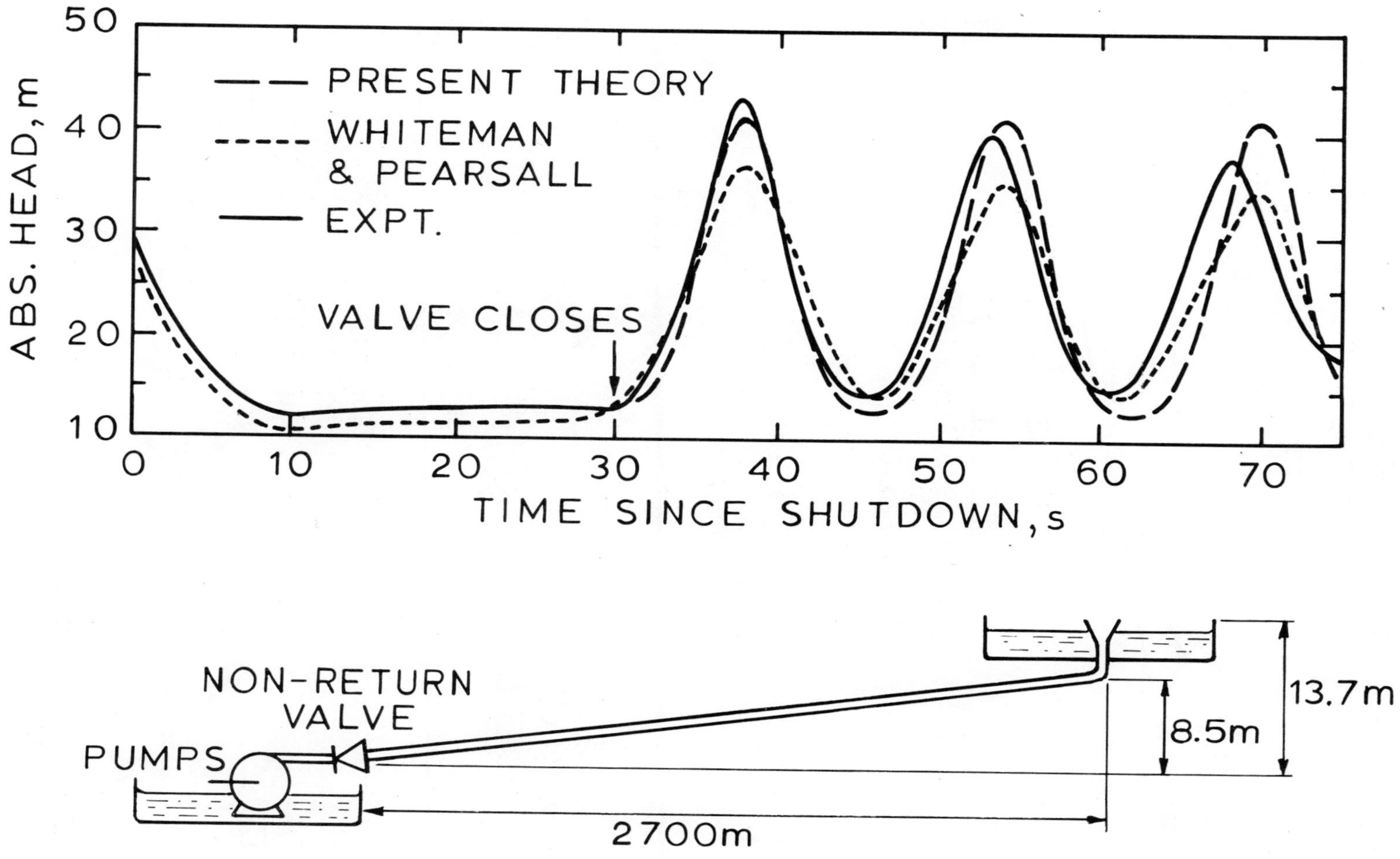

FIG. 3 Schematic layout of Kingston sewage line, and result of test K13

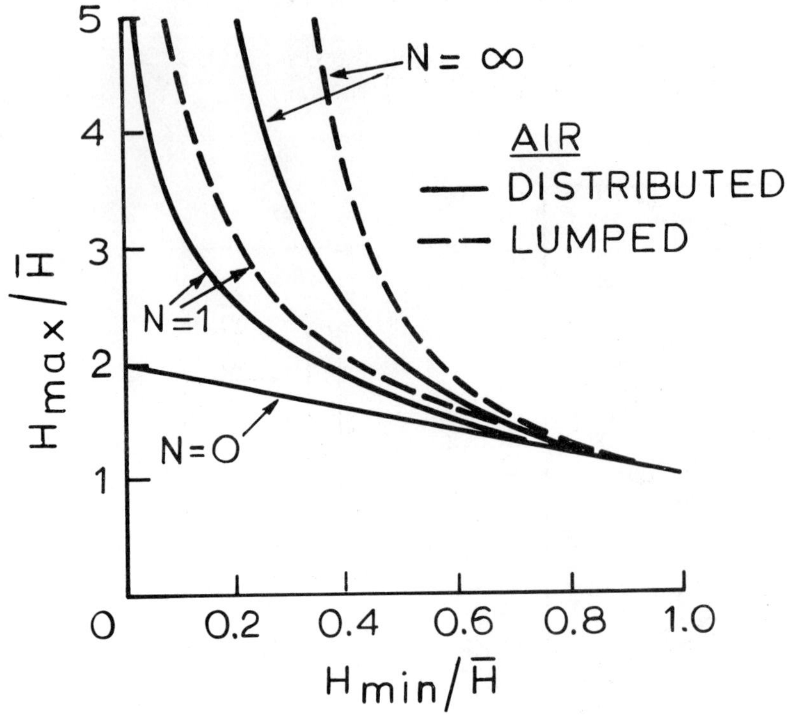

FIG. 4 H_{max} versus H_{min} for lumped and distributed air (same periods, $N = T^2/T_o^2 - 1$)

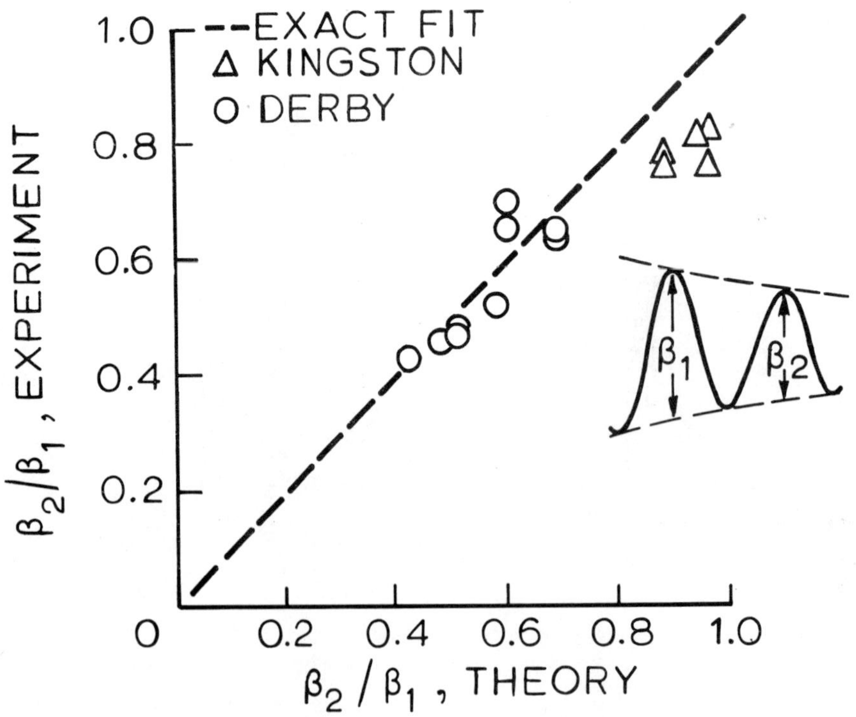

FIG. 5 Comparison of Pearsall's reflection-damping theory with the observed damping

Third International Conference on

PRESSURE SURGES

Canterbury, England
March 25-27, 1980

PAPER C3

ANALYSIS OF WATER CLEARING LOADS IN STEAM QUENCHING SYSTEMS

H.H. Safwat

Bechtel Power Corporation, U.S.A.

Summary

This paper presents a method for the analysis of water clearing transients in a steam discharge line whose end is submerged in a water tank. It covers the line transient and the induced transient flow in the suppression tank. The method has been applied to a 1/13.3 scale model Mark II BWR Loss of Coolant Accident-Vent single cell test. The calculated results compare favorably with the measurements.

Held at the University of Kent at Canterbury, U.K.

Conference Organised by BHRA Fluid Engineering, Cranfield, Bedford, England.
In conjunction with The City University, London.

NOMENCLATURE

a	= speed of propagation of a pressure wave
A	= cross sectional area
A,B,C	= x-t grid points shown in Fig. 2
C^+, C^-	= characteristic lines shown in Fig. 2
D	= pipe inside diameter
f	= wall frictional factor ($f = \tau/((\frac{1}{2})\rho V^2)$)
g_c	= constant of proportionality in Newton's second law
h	= enthalpy
H	= piezometric head
k	= ratio of specific heats of gas
K	= flow resistance of discharge device
L	= discharge device equivalent water length
N	= distance node index
P	= pressure
Q	= discharge flow rate
P,Q,R	= points shown on x-t grid, Fig. 2
t	= time
V	= velocity
x	= distance
α	= angle of inclination of flow conduit with the horizontal direction
ρ	= density
τ	= wall shear stress

Subscripts

A,B,C	= points on x-t grid, Fig. 2
eq	= equivalent
ℓ	= lower region of the tank, Fig. 4
out	= outside the discharge line in the tank
R,Q,S	= points on x-t grid, Fig. 2
T	= tank
u	= upper part of the tank, Fig. 4

INTRODUCTION

This paper deals with the analysis of water clearing loads during transients involving discharge of steam through a discharge line to a quench (suppression) tank. Two aspects are covered: i) the transient in the discharge line itself and ii) the associated transient conditions in the suppression tank. Such a problem is encountered in several systems in industrial and power plants. In boiling water reactors (BWRs), the Loss of Coolant Accident (LOCA)-Vent and the Safety Relief Valve (S/RV) discharge systems are typical systems where the problem is encountered.

Generally, the discharge line connecting between the steam source ends with a discharge device submerged in a supression tank. The transient starts as a result of steam admission into the discharge line (as a result of a pipe break in the case of a LOCA-Vent clearing or a valve opening in the case of the S/RV system in a BWR). The steam admission initially causes pressurization of the air in the discharge line, and the water slug in the submerged part of the discharge line starts to move. The water is expelled from the discharge line, and the water clearing continues until air starts entering the supression tank. It is desired to analyze flow conditions in the discharge line during water clearing to ensure acceptable back pressure (pressure buildup in the line) and to determine the dynamic forces induced by this transient on the different segments of the discharge line to ensure adequate structural support of the line. On the other hand, the water entering the suppression tank induces loads on the tank walls and any submerged structure.

ANALYSIS OF DISCHARGE LINE TRANSIENT

The application of the method of characteristics to analyze the transient in the discharge line has been the subject of several recent papers: McCready, et al (Ref. 1), Safwat (Ref. 2), Gorton (Ref. 3), and Wheeler and Moody (Ref. 4). Since Ref. 2 was published, its method of analysis has been generalized to handle a discharge line containing segments with different cross sectional areas (Fig. 1). Also, modeling of the discharge device at the end of the discharge line was given particular attention.

For a one-dimensional unsteady flow along the axis of a pipe with a uniform cross section, neglecting gravity effects and assuming adiabatic flow for an ideal gas, the mass, momentum, and energy conservation equations can be written (Refs. 1-4):

Mass Conservation

$$\frac{\partial \rho}{\partial t} + \frac{\partial}{\partial x} (\rho V) = 0 \qquad (1)$$

Momentum Conservation

$$\frac{\partial V}{\partial t} + V \frac{\partial V}{\partial x} + \frac{g_c}{\rho} \frac{\partial P}{\partial x} + \frac{f}{D} \frac{V|V|}{2} = 0 \qquad (2)$$

Energy Conservation

$$\frac{\partial h}{\partial t} + V \frac{\partial h}{\partial x} - \frac{1}{\rho} \left(\frac{\partial P}{\partial t} + V \frac{\partial P}{\partial x} \right) - \frac{f}{D} \frac{V^2 |V|}{2g_c} = 0. \qquad (3)$$

Using ideal gas properties, the enthalpy h is expressed as follows:

$$h = \frac{k}{k-1}\frac{P}{\rho} = \frac{1}{k-1}\frac{a^2}{g_c}. \qquad (4)$$

Equations (1) through (4) are solved using the method of characteristics to find the flow variables pressure P, velocity V, and density ρ as a function of time t and distance along the axis of the pipe x (for details, refer to Refs. 1-4). Numerical solution of this set of equations using constant time-constant distance increments (Fig. 2) utilizes the finite difference equations (5), (6), and (7) along the characteristic curves C^+ and C^- and fluid particle path line respectively.

The finite difference equation along C^+: $\frac{dx}{dt} = V + a$ is given by:

$$P_P - P_R + \frac{\rho_R a_R}{g_c}(V_P - V_R) = \frac{f_R}{D}\frac{\rho_R}{2g_c} V_R|V_R| \cdot [(k-1)\,V_R - a_R] \cdot \Delta t \qquad (5)$$

The finite difference equation along C^-: $\frac{dx}{dt} = V - a$ is:

$$P_P - P_S - \frac{\rho_S a_S}{g_c}(V_P - V_S) = \frac{f_S}{D}\frac{\rho_S}{2g_c} V_S|V_S| \cdot [(k-1)\,V_S + a_S] \cdot \Delta t \qquad (6)$$

Along a fluid particle path: $\frac{dx}{dt} = V$, the density change in a finite difference form is given by:

$$\rho_P - \rho_Q - \frac{g_c}{a_Q^2}(P_P - P_Q) = \frac{-1}{2}\frac{f_Q}{D}(k-1)\frac{\rho_Q}{a_Q^2} V_Q^2|V_Q|\Delta t \qquad (7)$$

The numerical solution proceeds by interpolating between known flow conditions (pressure P, velocity V, and density ρ) at time level t at grid points A, C, and B (Fig. 2) to obtain intermediate points R, S, and Q; then, using flow variables at R, S, and Q and equations (5) through (7), flow values P_P, V_P, and ρ_P at point P at time t + Δt are obtained. At the ends of a link (a segment of the discharge line bounded by two flow devices such as a valve, a change in cross section, or a discharge device), boundary conditions must be satisfied. In addition, the numerical solution allows for movements of the steam/air interface as well as the air/water interface.

Establishing where the steam/air interface is at a certain instant is accomplished by comparing the initial mass of air in the discharge line to the additive masses of air occupying Δx increments just upstream of the air/water interface. This procedure gives the node and link number where the steam/air interface is, and only influences the k value in equations (5) through (7). Downstream of the air/water interface the fluid is water, and the well known finite difference equations for hydraulic transients (Ref. 5) are used for calculation at nodes covered by water. The hydraulic transients equation (for a liquid with negligible changes in density due to change in pressure) can be obtained by using a value of k = 1 in equations (5) and (6). At the air/water interface, the pressure and velocity at both the air and water sides are the same. The logic explained in Ref. 2 for calculating conditions at the air/water interface and the nodes directly upstream and downstream of the air/water interface has been generalized to allow for moving from one link to another.

With reference to Fig. 1, the valve boundary condition at the first node of the first link is treated by using a characteristic equation along C^- (Eq. (5)) together with the quasi-steady compressible flow of gas through the throat area of the valve, allowing for either subsonic or sonic flow through the valve depending on the pressure ratio across the valve (refer to Ref. 2 for further details).

At a change in cross section, the procedures explained in Ref. 6 are utilized for modeling a change in cross section (between two links in Fig. 1) when gas is flowing between these two links. On the other hand, a junction between two links where water is flowing is treated as a hydraulic restriction (Ref. 5). At the end of the last link, the dynamic equation governing the motion of the water mass contained in the discharge device (Eq. (8)) is utilized together with a C^+ characteristic (Eq. (5)) with k = 1.0.

$$\frac{g_c}{\rho_L}(P - P_{out}) = K_{eq}\frac{V^2}{2} + L_{eq}\frac{dV}{dt} \tag{8}$$

In Eq. (8), the pressure outside the discharge device P_{out} is assumed constant, and the discharge device inertia factor L_{eq} contains added length to account for inertia effects of water just outside of the discharge device which has to be set in motion.

ANALYSIS OF TANK FLOW TRANSIENT

In this part of the analysis it is desired to predict pressure and velocity changes in the tank during the discharge line water clearing. In general, this involves a three-dimensional flow problem; however, for several practical applications the tank has a finite size with a limited pool surface area. For example, in the case of the LOCA-Vent for a BWR plant all vents are cleared simultaneously and the clearing process amounts to a clearing of a single vent in a unit cell of the suppression pool.

The first simplification of such a problem is to model axisymmetric flow in a cylindrical tank where water clears a vertical discharge line located at its axis and ends at some distance from the floor of the tank. Using the two-dimensional general purpose computer code PISCES (Ref. 7), this problem was modeled. Fig. 3 shows a sample of the flow pattern obtained from the PISCES analysis.

Examining the results of this two-dimensional analysis, one observes that the high velocities in the tank decrease sharply from the vent clearing velocity to small values at a small distance just below the vent (distance on the order of one to two vent diameters). In the lower part of the tank (below the vent exit), there are negligible velocities but increases in pressures are noticed (due to the water compressibility and the water confinement because of the tank walls). In the upper part of the tank, above the vent exit, the PISCES analysis shows vertical flow patterns, negligible variation in the velocities with the radial distance, and smaller pressure increases in comparison to those exhibited at the lower part of the tank.

These observations suggested a further simplification to model the flow in the tank using a one-dimensional approximation. This has been accomplished using two grids representing the upper and lower sections of the cylindrical tank, as shown in Fig. 4. The commonly known hydraulic transient finite difference (Eqs. (9) and (10)) (Ref. 5) is used with the additional simplifying assumption that $\Delta x = a\Delta t$ since the expected tank flow velocites are negligible in comparison to pressure wave speed in the tank's water.

Along a C^+ characteristic:

$$V_P - V_R + \frac{g_c}{a}(H_P - H_R) + \frac{g_c}{a}V_R \sin\alpha\,(\Delta t) + \frac{f}{2D}V_R|V_R|\Delta t = 0 \tag{9}$$

Along a C^- characteristic:

$$V_P - V_S - \frac{g_c}{a}(H_P - H_S) - \frac{g_c}{a} V_s \sin\alpha\,(\Delta t) - \frac{f}{2D} V_S|V_S|\,\Delta t = 0 \qquad (10)$$

With reference to Fig. 4 at node N_U, a constant pressure boundary condition is used and, at node N_D, a zero velocity is utilized. At the section representing the incoming water flow (from the discharge line) to the tank (this section is taken at the same elevation as the vent), the continuity equation is in the form

$$Q(t) = A_T\,(V_u + V_\ell) \qquad (11)$$

where Q(t) is the discharge flow rate out of the vent at any time as obtained from the discharge line transient analysis, A_T is the tank surface area, and V_u and V_ℓ are the velocities at the first node of each of the upper and lower grids respectively. Eq. (11), together with two characteristic equations connecting to the first nodes of the two grids, are solved simultaneously to obtain the three unknowns: the pressure H_P (which is the same for the first node on the upper and lower grid as they coincide), V_u, and V_ℓ.

COMPARISON OF THE ANALYSIS RESULTS WITH TEST DATA

The method described in the previous section for analysis of the water clearing transient in the discharge line has been applied extensively for the S/RV line clearing. In this paper, the results of the analysis methods described in the previous section are compared to test results from a 1/13.3 scale single cell simulating a Mark II LOCA-Vent system (Ref. 8).

Fig. 5 shows a schematic of the test setup used in Ref. 8. Prior to test initiation, the drywell and wetwell were evacuated to $7.58 \times 10^3\ Nm^{-2}$ absolute using the vacuum pump, and the supply tank was filled with compressed air to $163 \times 10^3\ Nm^{-2}$ absolute. The ball valve was then closed and the needle valve was adjusted to hold the vacuum steady. Once all instrumentation was balanced, the fast acting valve was opened and the blowdown transient was initiated. The tests presented in Ref. 8 cover simulating conditions during LOCA in the early part of the transient when flow in the downcomer is primarily airflow (this is why the test facility used compressed air). Ref. 8 studies the transient blowdown extending over the water clearing phase as well as the later pool swell phase. The present work deals only with the water clearing phase. The measurements of test #387 from Ref. 8 were chosen for comparison with the numerical results predicted using the analysis method described in the previous section. Fig. 6 shows the schemes used for modeling the discharge line as well as the water tank. The various input valves used in the numerical calculations are included in Fig. 6. The model shown in Fig. 6 only models the test apparatus downstream of the quick acting valve of Fig. 5. Ref. 8 provided the rate of air mass flow rate versus time shown in Fig. 6 during test #387. In simulating the test numerically, the valve area versus time was chosen to give the same mass flow rate-time observed during the test as reported in Ref. 8. However, a simplification was introduced in the numerical simulation; that is, the pressure upstream of the valve was assumed to remain constant and equal to the initial pressure in the supply tank. Fig. 7 shows a comparison between the calculated and measured pressure in the drywell. It also shows a comparison between the calculated and measured pressure at the bottom of the tank during the water clearing. Ref. 8 reports that, based on movie camera observations, vent clearing time is determined by averaging the time of maximum slope and the time of peak pressure of the pool pressure history. This time has been marked on Fig. 7. The time at which water clears from the vent as predicted from the analysis of the discharge line transient is also marked on Fig. 7. The calculated results agree reasonably well with the measurements. It is believed that the over-prediction in the drywell pressure comes from the approximation of the constant pressure in the supply tank throughout the analysis time while in reality, because of the finite size of the supply tank, that pressure decreased as shown in Fig. 6.

CONCLUSIONS

A comprehensive method for the analysis of the water clearing transient has been presented. The method predicts pressure, velocity, and density changes in the discharge line during water clearing. These can be used in turn to compute the dynamic forces (during water clearing) on the discharge line segments to be used for structural assessment of the piping support. The pressure, velocity changes in the quench tank during water clearing can be predicted using a simple one-dimensional approximation. The method of analysis appears to yield accurate results. Although the method presented here is quite general, it is suggested that careful evaluation be made of the unique features of each steam discharge quenching system before this technique is applied.

ACKNOWLEDGMENT

Thanks are due to Bechtel Power Corporation for supporting the work reported in this paper. The assistance of Drs. C. B. Kim and A. Arastu are acknowledged.

The author notes that the technique proposed in this paper to obtain water clearing loads on the wall of the suppression tank is inexpensive and would be very useful in evaluating various concepts of design; for final designs the multidimensional analysis may still be needed. He also notes that for BWR loss of coolant accident application, suppression pool wall loads due to subsequent air and steam clearing are larger than the water clearing loads.

REFERENCES

1. McCready, J. L., McBride, M. G., Krepps, J. G., Moody, F. J., Javid, A., and Strickland, E. L.: "Steam vent clearing phenomena and structural response of the BWR torus (Mark I containment)." General Electric Company, NEDO-10859. (April 1973).

2. Safwat, H. H.: "Analysis of the safety relief valve discharge transients of a boiling water reactor." Proceedings of the International Conference on Pressure Surges, Paper E1, organized by BHRA Fluid Engineering, Cranfield, England. (September 22-24, 1976).

3. Gorton, R. L.: "Application of finite difference methods to transient steam flow problems in power plant piping systems." Symposium of Fluid Transients and Acoustics in the Power Industry, organized by the Fluid Transients Committee of the ASME Fluid Engineering Division, San Francisco, CA. (December 10-15, 1978).

4. Wheeler, A. J., and Moody, F. J.: "A method for computing transient pressures and forces in safety relief valve discharge lines." The Third National Congress on Pressure Vessels and Piping, organized by ASME, San Francisco, CA. (June 24-29, 1979).

5. Streeter, V. L., and Wylie, E. B.: "Hydraulic Transients." McGraw-Hill. (1967).

6. Safwat, H. H.: "A generalized application of the method of characteristics for the analysis of transients in steam lines in nuclear power plants." Symposium of Fluid Transients and Acoustics in the Power Industry, organized by the Fluid Transients Committee of the ASME Fluid Engineering Division, San Francisco, CA. (December 10-15, 1978).

7. PISCES 2DELK Manuals, Physics International Company, San Leandro, CA.

8. Kiang, L. R., and Jeuck, P. R.: "A study of pool swell dynamics in a Mark II single cell model." Electric Power Research Institute. (October 1978).

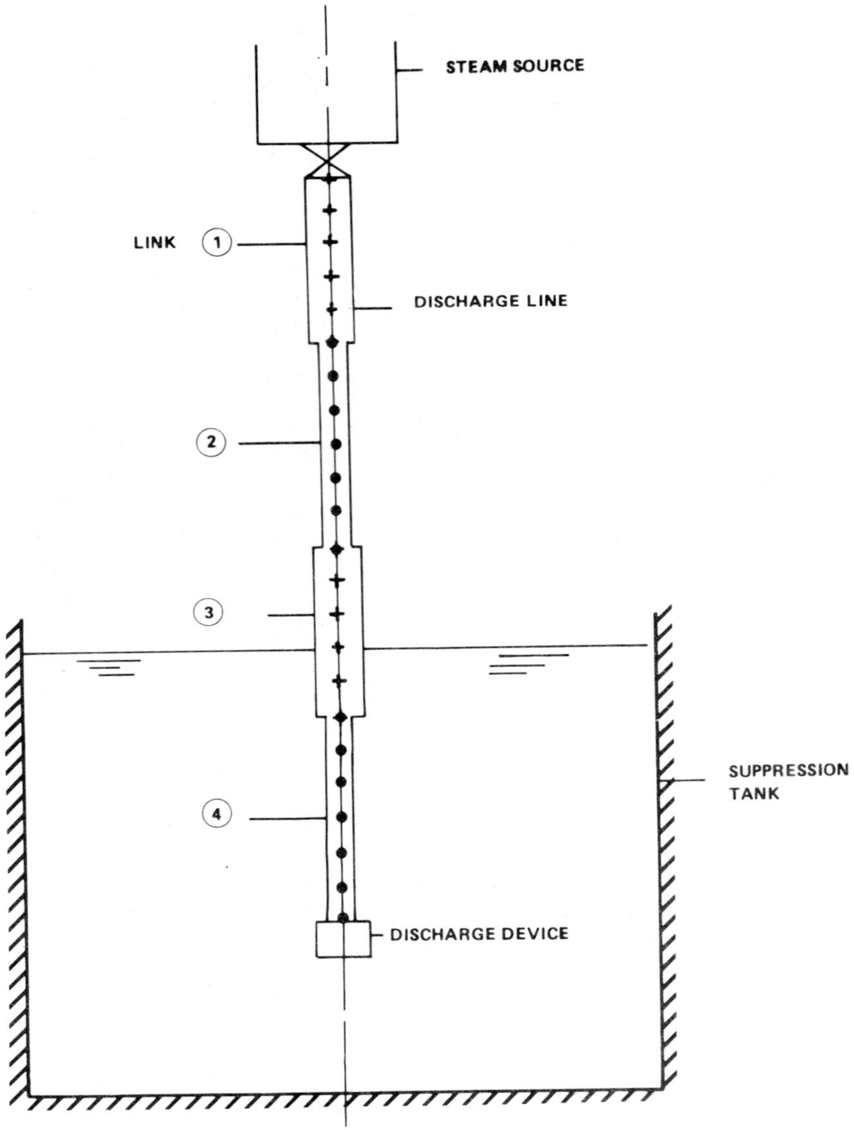

Fig. 1 SCHEME USED IN MODELING WATER CLEARING TRANSIENT FOR STEAM QUENCHING SYSTEM

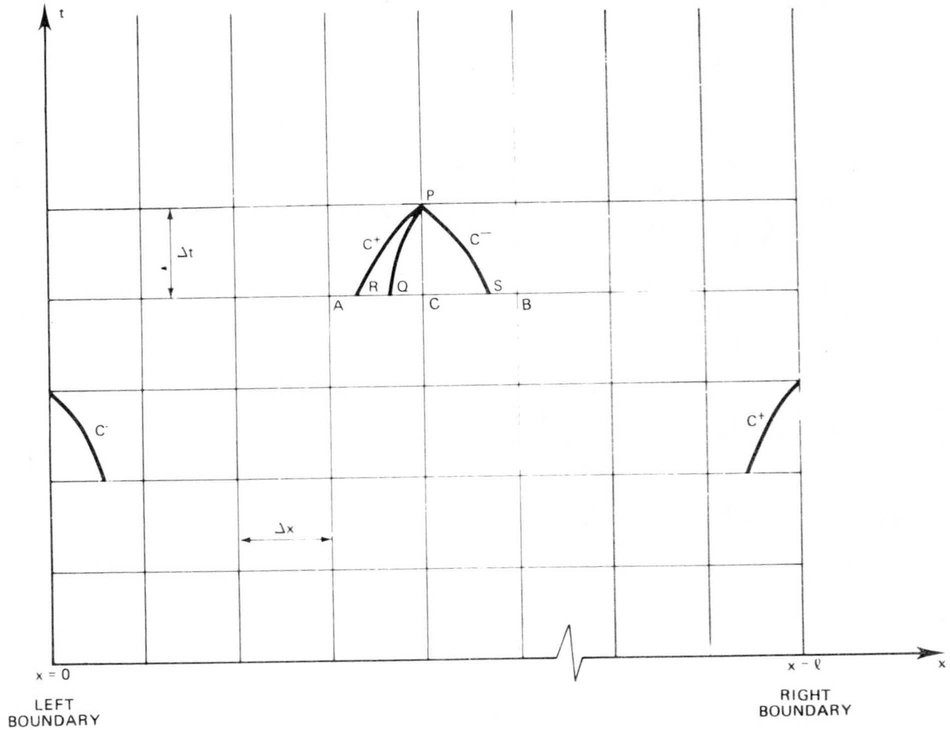

Fig. 2 CHARACTERISTIC CURVES ON x-t GRID

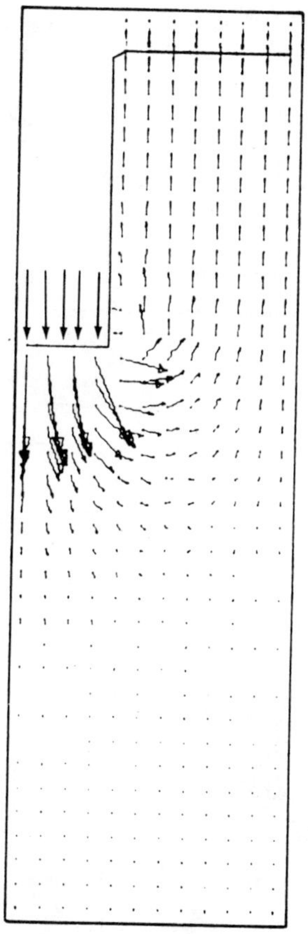

Fig. 3 VELOCITY FIELD IN TANK AT A GIVEN INSTANT AS PREDICTED BY TWO-DIMENSIONAL ANALYSIS USING THE COMPUTER PROGRAM PISCES

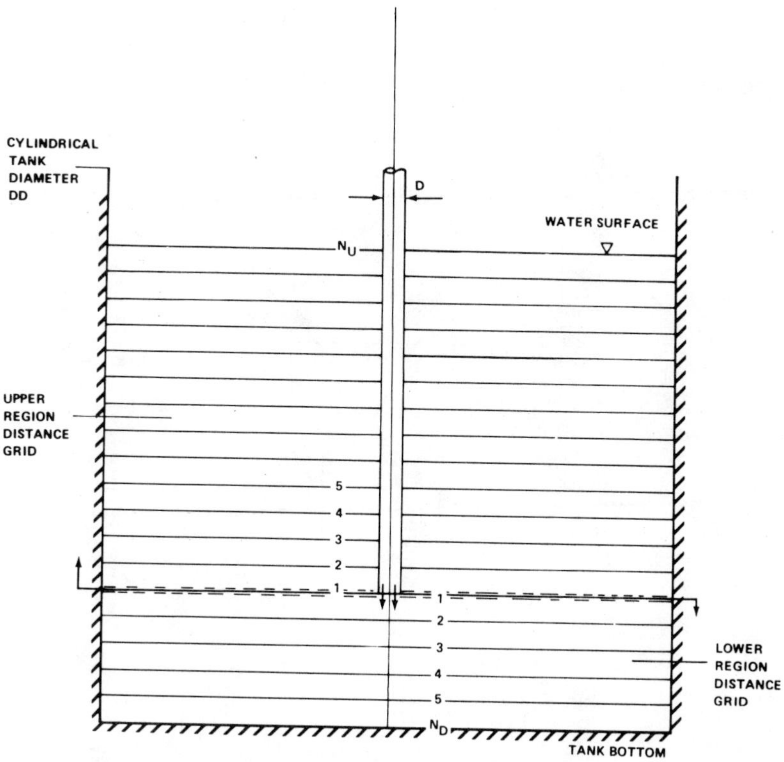

Fig. 4 TWO GRIDS FOR MODELING WATER TANK FLOW USING ONE-DIMENSIONAL APPROXIMATION

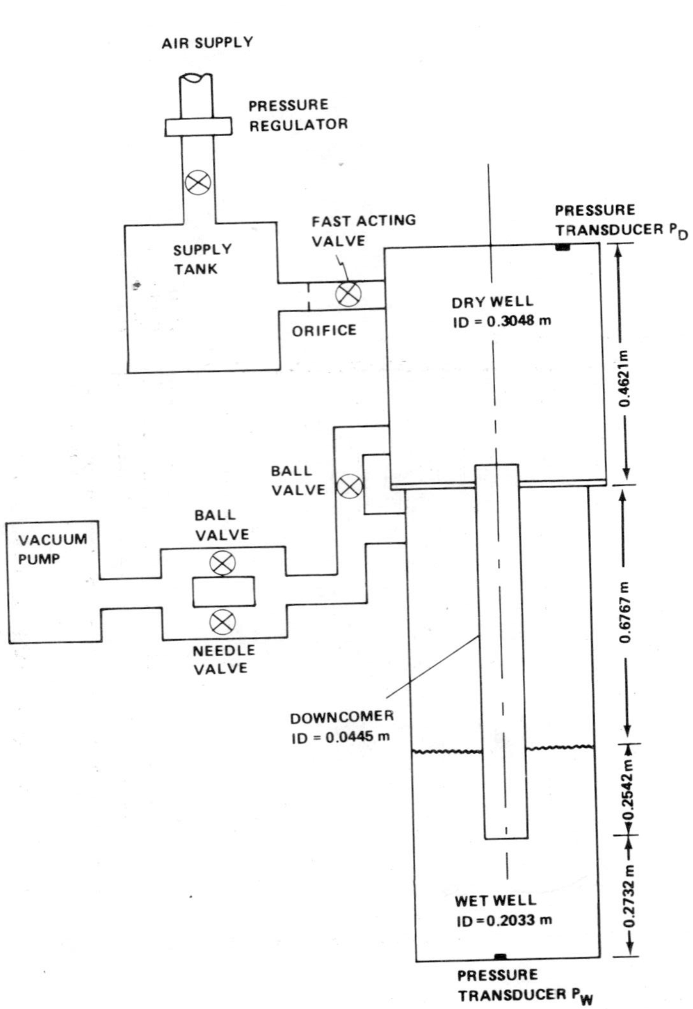

Fig. 5 SCHEMATIC OF THE 1/13.3 SINGLE CELL TEST SYSTEM (REF. 8)

TIME SEC	PRESSURE Nm^{-2}
0.0	163×10^3 *
0.1	141×10^3
0.2	118×10^3
0.3	101×10^3

*ASSUMED CONSTANT THROUGHOUT THE TRANSIENT IN CALCULATIONS

TIME SEC	MASS FLOW** $Kg\ SEC^{-1}$
0.0	0.0272
0.1	0.0232
0.2	0.0202
0.3	0.0175

**REF. 8 AND USED IN THE CALCULATIONS

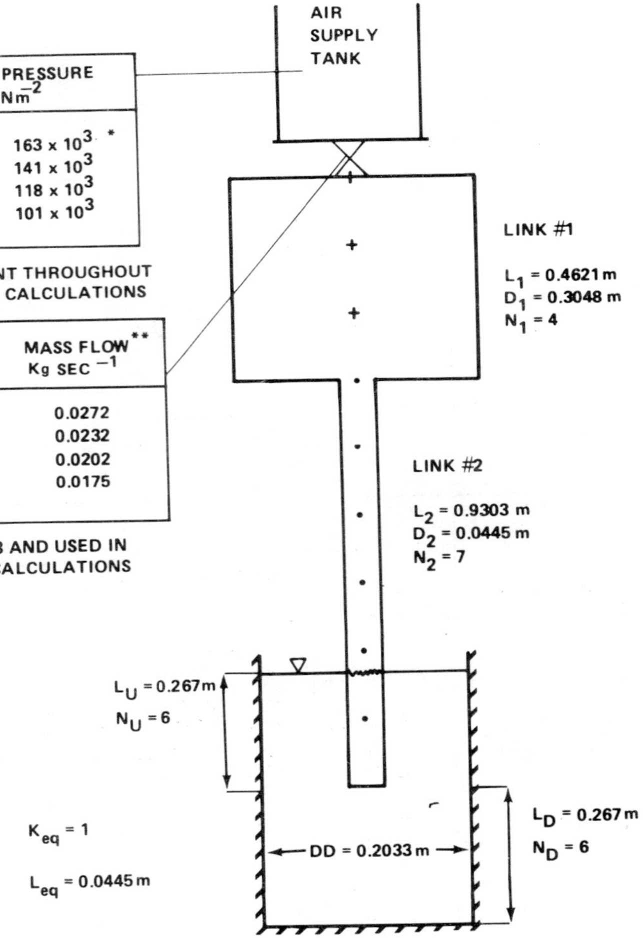

Fig. 6 SCHEME FOR NUMERICAL SIMULATION OF REF. 8 – TEST #387

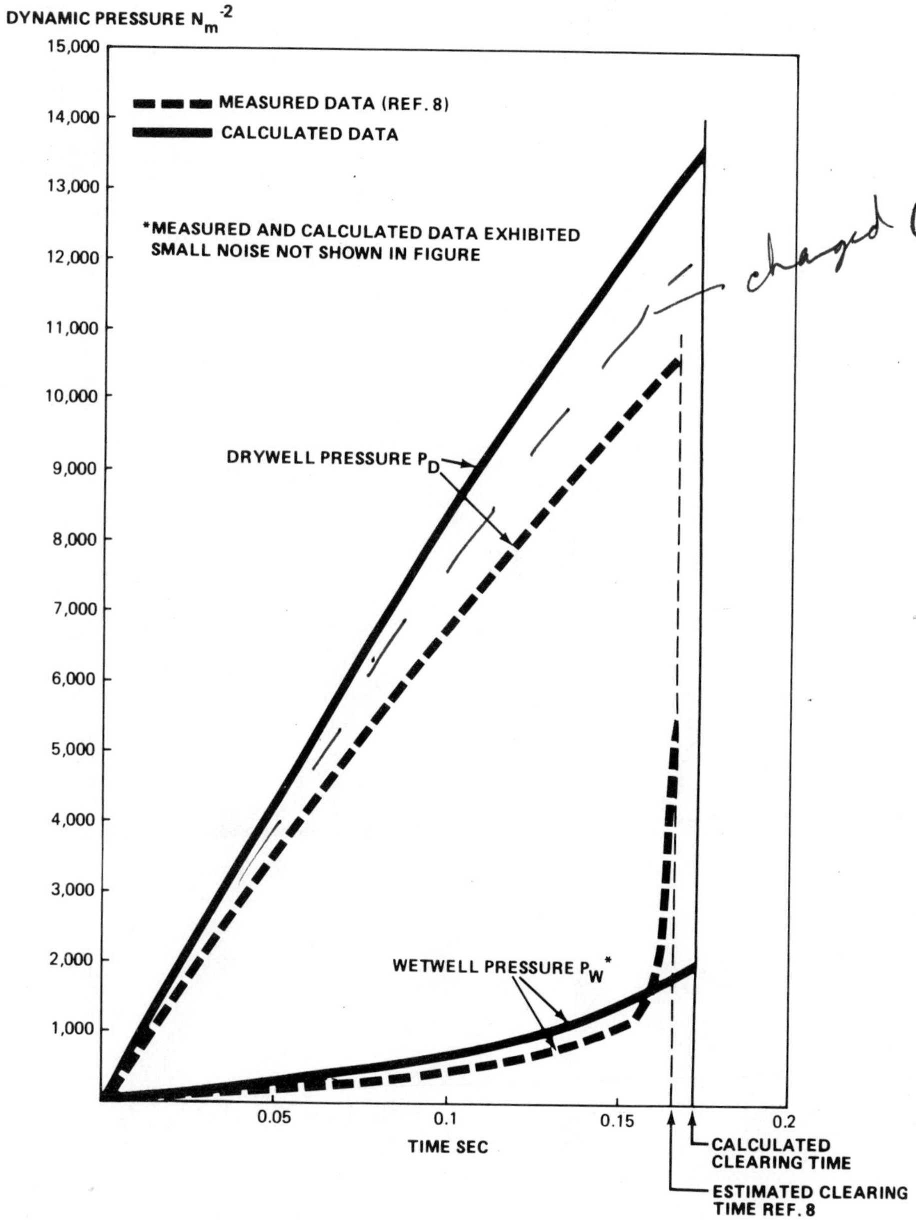

Fig. 7 COMPARISONS BETWEEN MEASURED AND CALCULATED DYNAMIC PRESSURES

Third International Conference on

PRESSURE SURGES

Canterbury, England
March 25-27, 1980

PAPER D

PRESSURE SURGE: DOUBTS IN THE DESIGN PROCESS FOR REAL SYSTEMS

M.U. Papworth

BHRA Fluid Engineering, U.K.

Summary

Pressure surge prediction is not an exact science due to doubts concerning data, methods of system operation, and understanding of problems associated with real systems. The paper outlines some of the problems which exist and suggests that the immediate need is for a combination of education and design data, rather than further improvements in analysis precision.

Held at the University of Kent at Canterbury, U.K.

Conference Organised by BHRA Fluid Engineering, Cranfield, Bedford, England.
In conjunction with The City University, London.

Pressure Surge: doubts in the design process for real systems

In other papers at this conference you will hear many erudite and detailed methods describing more and more complex ways of analysing increasingly specific problems. The question arises of whether researchers are losing sight of the objective of such scrutiny. In the days of graphical analysis, the rigours of the method, for all the drawbacks of difficulty of execution, forced the user to understand the basic phenomena before he or she embarked on the task of predicting events. The advent of fast computational methods has, to a large extent, removed this requirement, enabling novice engineers to 'model' systems without having a fundamental 'feel' for fluid dynamic behaviour under transient conditions. In putting forward this heresy, it must be stressed that there is no intention of advocating the abandonment of the convenience of automatic computing techniques. It is, however, suggested that it is dangerous to assume that automatic analysis of any but the simplest system is desirable, or possible, in the absence of an understanding of the real systems.

The following items illustrate some of the defects present in pressure surge prediction.

1. Failure to consider effects of small system changes

As an example of the deep elephant traps waiting to engulf the unwary, consider the recent trend towards using timed closure valves in the place of non-return valves. In Figure 1 a simple pumped rising main lifts liquid up a hill to a reservoir. For the purpose of illustration, we will assume that the pump has a high inertia, so that when it trips there is no tendency for a cavity to form downstream of the valve. Thus the main problem is to stop any back flow of water through the pump when power is lost. The traditional method is to make the valve a non-return type (this is perfectly feasible for the pipe diameter given). Provided the valve is a high quality device, preferably with a damped rate over its final 20° of closure there should be no problem. However, on occasions non-return valves have been found to stick, permitting reverse flow through the pump. It has been proposed that such valves should be replaced by timed closure butterfly valves. The method suggested is to make the closure rate match the flow reduction, so that the valve shuts just as the flow stops (thus mimicking the action of the non-return valve). For any given flow condition, a simple computer program will give the correct valve timing and it may even work in the field!

There are several reasons for the system parameters to change from those assumed in the computer analysis:

a) the system fouls, affecting the system resistance,
b) the system design flow may be altered to suit local requirements,
c) the valve may be badly maintained, allowing changes to its operating times,
d) the valve timing is usually set up under zero flow: at high flowrates the timing of pneumatically operated valves will be substantially affected.

If one or more of the above occurs, then our benign, pseudo non-return valve may turn into a villain: no longer does its timing match that of the system and it may shut too fast (causing cavity growth), or too slowly (allowing substantial reverse flow before shutting). In either case the potential pressure surges may be substantial. This may appear an obvious and elementary error, but the only true evaluation of design error is experience of actual events, and similar errors do occur.

2. Education of design engineers and plant operators

Another area where the pressure surge experts have failed is that of educating the design engineers into the perils of certain fluid/valve interactions. The water engineers of Victorian times were aware that the fast closure of valves at the ends of long stretches of pipes could cause high pressures and learned their lessons. In spite of this easily understood and remedied problem, systems are still being proposed with valves shutting in times comparable to the system period. As examples, one can cite oil and ammonia unloading lines. Here, for largely economic reasons, loading velocities are high. The product is often toxic or inflammable, or both, and to minimise spillage in the event of overfilling of the ship's tank, an automatic fast closure valve is fitted close to the ship's storage tank (Fig. 2). In normal conditions this valve is not used, but if a filling accident occurs it may operate automatically or, in the event of a fire, it may be triggered manually. There is no doubt that the fast closure of this valve will lead to large pressure rises, and frequently this will exceed the test pressure of either the ship or shore pipework. As a final indication of the dangers of such an event, communication between ship and shore may well be by hand signals or a dubious quality telephone line. In a world where huge oil tankers can collide with the French coast in relatively calm sea conditions, it is surprising that serious loading accidents do not occur more often.

The educational requirement does not stop with the design engineer. With downtime costs of many tens of thousands of pounds per day in large plants, the loss of a cooling water system will be circumvented by the site operator if at all possible. This may result in operational procedures outside the normal plant operation and may not have been considered in any analysis performed at the design stage. Since the timescale for decision is usually hours rather than days, it is extremely unusual for such changes to be considered by pressure surge experts. Cynics have said that if a procedure is possible, then at sometime during the lifetime of the plant this procedure will take place. It is interesting to note that the nuclear engineers adopt this philisophy when considering plant safety and yet unexpected combinations of plant operation still occur.

The answer to this educational deficiency is not easy: the basic information exists in a mass of textbooks, but it is unreasonable to expect design engineers and plant operators to read and understand a mathematical treatise. What is required is a design guide giving simple rules and endeavouring to illuminate the basic fluid dynamics, rather than conceal the problems in mathematical expressions. Such a design guide would need to be supported by a carefully considered series of seminars and courses.

3. System design pressures

Assuming that the appropriate design method has taken place and that the appropriate prediction of pressure/time history has been performed, we arrive at the question of whether the pipework will withstand these pressures. There exist standards for acceptable pressures for most pipe materials, provided that the pressure is applied slowly. However, few, if any, justifiable measurements of the effect of rapid application of pressure on system integrity have been made. In consequence, individual consultants have adopted their own specifications which are not based on technical measurements. To illustrate the problem there follow several of the pressure specifications used in some recent investigations. They are all based on the basic design pressure for steady state from British Standards for the appropriate pipe material.

a) Design pressure

Here it is simply assumed that the material fails due to its tensile strength being exceeded, and hence the form and the duration of the pressure pulse is not considered.

b) Test pressure

The test pressure is usually 150% of the design pressure and frequently it is argued that for a short time it is acceptable to exceed the steady state design pressure and approach the test pressure. Provided the frequency of occurance of transient pressure is low, there is perhaps some justification for this argument, although one must assume that the pipework is well secured.

c) Dynamic pressure doubling

This approach leads from consideration of the dynamics of pressure transients and is very conservative. An equivalent pressure is defined by splitting the pressure time history into a dynamic pressure (DP) and a steady pressure (SP). The equivalent steady state pressure is then defined as:-

Equivalent pressure = SP + 2*DP.

The argument for this approach seems to be based on the effect of release point of a mass/spring system on the maximum tension in the spring. If the mass is put at its final resting position, the maximum spring tension is half that if the mass is released from the zero spring extension position.

In flexible ended systems it is sometimes argued that this approach should be taken.

The common feature of all these approaches is that they are attempts to circumvent a basic lack of information. If standards existed such problems would disappear and the potential material savings would be large. One should note that the steady state standards are largely based on measurements rather than theoretical considerations, and it is probable that any dynamic standards would have to be similarly generated if they are to be convincing to the design engineers.

4. Fluid system interactions

Interaction between system generated pressures and pipe movement is usually either ignored or analysed separately. It is often assumed that the pipework is rigidly fixed by anchor blocks, or, more rarely, that the pipework is not restrained axially. In practice the truth is somewhere between these two extremes and this leads potentially to large bending moments in equipment whose centre of mass is not at the pipe axis. An example frequently found in most piping systems is that of the control valve. Here the centre of mass is well off the pipe centre and consequently pipe movement following a pressure rise will lead to a bending moment in the valve. Thus a valve able to survive a given pressure may still fail due to the jerk applied by the connecting pipework.

It is not obvious how the design engineer overcomes this problem, since the addition of structure response to the basic transient program is unacceptably expensive except in very simple systems.

Probably the best approach is to retain the double analysis concept but change the pipework restraint assumptions between the fluid and structure analysis. Thus in the pressure/time history prediction rigid anchoring would be assumed, but in considering structural response the pipework would be assumed to be only partially restrained.

5. Data errors

For a given data set, the subsequent pressure surge analysis should be precise. Thus comparisons between laboratory models and mathematical models can be very good, since the system data is well controlled. When attempting to predict pressures in a complex system before it is built, this precise knowledge of the system data is not available. One is then left with a need to predict system data before the final equipment is specified and before the operational method is fully specified. There are a number of areas where difficulties arise:-

a) pipe friction and component data is only precise to about ±5%, even for steady state flows,

b) dissolved and free air contents of liquids are functions of ambient conditions and design of sumps,

c) pump characteristic and inertia may only be finally specified late in the design process,

d) large numbers of small design changes occur during the later stages of the design process and these changes may occur after the surge analysis is complete,

e) venting pipes and auxiliary water supply systems are sometimes not specified until shortly before construction is complete,

f) a given system may involve several consultants, none of whome may have overall responsibility for analysing the complete system for pressure surge.

These and other problems raise doubts over the accuracy of transient prediction and lead to a justification of conservative design of systems. The problem can be partially ameliorated by imposing three standard additions to the normal design methods:-

a) surge protection methods adopted to be conservative, or to be flexible enough to be tuned on site to suit the actual severity of the problem,

b) a second surge study to be performed after completion of the design process and before commissioning. This surge study would include system changes accumulated since the initial study,

c) commissioning tests to take place in the presence of a competent and experienced surge expert, with the tests agreed beforehand.

The adoption of these methods will not eliminate problems, but they do reduce the likelihood of unsatisfactory operation of systems.

Conclusions

This paper has stood aside from the detailed arguments of how to predict transient behaviour of fluid systems and examined those doubts and difficulties which cause problems in real systems. It suggests that the rapid advance in solution methods has outstripped the ability to apply these results in real systems. The four main points being made are:

a) there is a lack of knowledge of the mechanism of pressure surge problems by the design engineer,

b) there is a lack of 'feel' for fluid dynamics by mathematically based pressure surge predictors,

c) there is a lack of empirical data for the effect of actual pressure time histories on real systems,

d) there is a lack of consideration for changes to systems during the design process.

It cannot be pretended that there is an easy solution to any of these problems. In the first two there is an obvious requirement for a design guide, but this must be followed up with publicity and seminars on a scale not yet attempted in the field. It is not sufficient to tell design engineers to read pressure surge text books, since these are dominated by the mathematics and are too impoverished in terms of real hardware.

The third item can only be corrected by large scale experimental measurement on a range of pipe materials under realistic pressure/time histories. The cost and timescale of this course of action will be high and long, but it is to be hoped that by the time the next of these conferences takes place some progress will have taken place.

In the meantime, those engineers involved in the design process should concentrate on explaining, as well as predicting pressure surge and adopt some of the conservative measures detailed here.

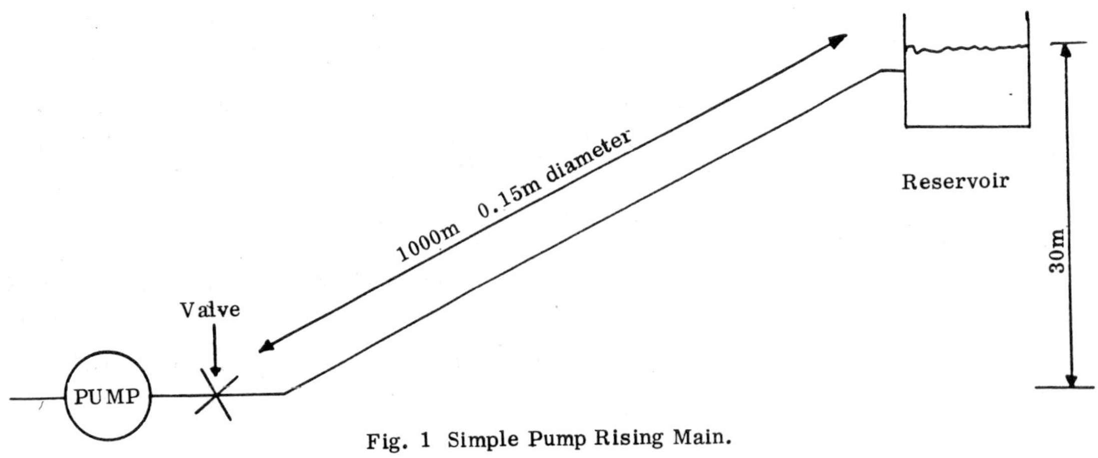

Fig. 1 Simple Pump Rising Main.

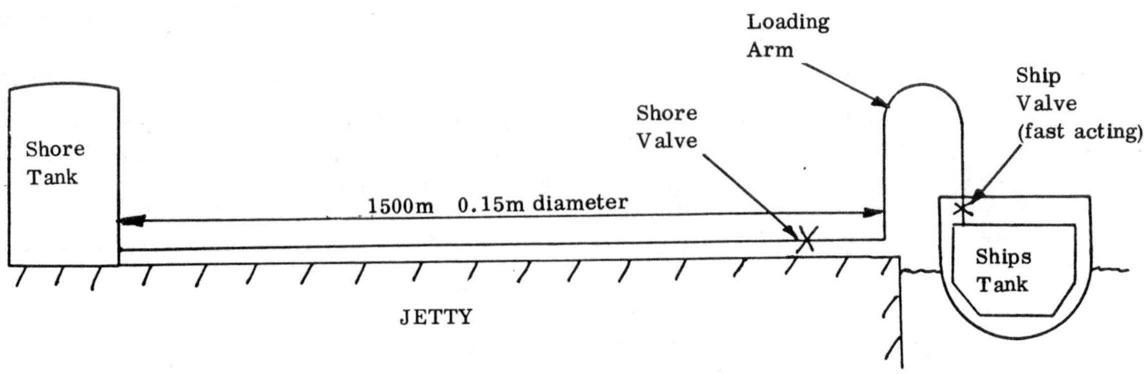

Fig. 2 Product Loading Line.

Third International Conference on
PRESSURE SURGES
Canterbury, England
March 25-27, 1980

PAPER E1

TRANSIENT CAVITATION PHENOMENA IN FLUID-STRUCTURE INTERACTIONS

C.A. Kot, B.J. Hsieh, C.K. Youngdahl and R.A. Valentin

Argonne National Laboratory, U.S.A.

Summary

A generalized column separation model is extended to predict transient cavitation associated with fluid-structure interactions. The essential feature of the combined fluid-structure interaction calculations is the coupling between the fluid transient, which is computed one-dimensionally, and the structural response which can be multi-dimensional. Proper coupling is achieved by defining an average, one-dimensional, structural velocity and by assuming a spatially uniform pressure loading of the structure. This procedure is found to be effective even for very complex finite element structural models for which the required computational time step is orders of magnitude smaller than that for the fluid transient.

Computational experiments show that neglecting cavitation and setting the fluid velocity at all times equal to that of the structural boundary leads to unreal negative pressure predictions and may cause convergence difficulties. In general, the results of such calculations differ drastically from those obtained with the properly coupled column separation model which accounts for the multiple cavity formations and recollapses caused by structural oscillations. While this model is a simplification of the actual phenomena it reproduces the most salient features of fluid-structure interactions, is always rapidly converging and provides reasonable fluid and structural response predictions

Held at the University of Kent at Canterbury, U.K.

Conference Organised by BHRA Fluid Engineering, Cranfield, Bedford, England.
In conjunction with The City University, London.

NOMENCLATURE

b = pipe support parameter (Eq. 7)

c = wave speed in fluid

C_1, C_2 = defined by Eq. (18)

d = disk base diameter

D = pipe diameter

E = elastic modulus of pipe material

f = Darcy-Weisbach pipe friction factor

F = $fu|u|/2D$, friction term

g = acceleration of gravity

G = $g \sin \alpha$, gravitation term

h = disk thickness

H = pipe wall thickness

K = bulk modulus of fluid

p = pressure

R = structural response function (Eq. 16)

S = cavity size

t = time

u, w = fluid velocity

x = axial coordinate along pipe

α = angle of pipe with horizontal (downward)

δt = time step for structure

Δt = time step for fluid

Δx = axial grid spacing

ε = circumferential strain

ρ = fluid density

σ = circumferential stress

$\{d\}$ = nodal displacement vector

$\{\dot{d}\}$ = nodal velocity vector

$\{\ddot{d}\}$ = nodal acceleration vector

$\{F^e\}$ = external nodal force vector

$\{F^i\}$ = internal nodal fare vector

$[M]$ = lumped mass matrix

Subscripts

A, B = intersection points of characteristics with constant time line(see Fig.1)

j, j-1, j+1 = grid points (see Fig. 1)

o = initial- current time line (see Fig. 1)

vap = vaporization

Superscripts

n, n+1 = successive time steps (see Fig. 1)

1. INTRODUCTION

An important aspect of the overall analysis of fluid transients and their consequences in piping systems is the interaction of transient pressure loads with the structural components of the system. In general the response of the structure influences the pressure loading. Therefore, the fluid-structure interaction is a coupled problem in which the behavior of the fluid and the structure must be calculated simultaneously. Whenever fluid transients occur, the possibility of cavitation occurrence cannot be excluded. This holds particularly for systems operating at relatively low pressures. A familiar example of a fluid structure-interaction which may cause cavitation is the reinforcement of rarefaction waves at a closed end or a flow restriction in a pipe (Ref.1). It has also been shown (Ref.2) that plastic pipe deformation may indirectly contribute to the formation of transient cavitation in piping systems. Of interest here, however, is the cavitation which arises directly from the fluid-structure interaction due to the impedance mismatch at the fluid-structure boundary. Essentially the phenomenon can be described as the inability of the fluid to adjust to the motion of the structure without reaching its vapor pressure.

Transient cavitation is a complex phenomenon and both the inception of cavitation and the collapse of the bubbles are nonequilibrium processes. The presence of impurities and in particular dissolved gases further complicates the fluid response. However, when attention is limited to fluids with low vapor pressures and little dissolved gas, a column separation treatment of cavitation has been found adequate (Ref's. 1,3, 4). Here we are concerned with the application of a generalized column separation model to the problem of fluid-structure interaction. Since the structural aspects of the aspects of the interaction may differ drastically from problem to problem the remainder of this paper emphasizes the fluid part of the interaction. The approach is an adaptation of earlier developed methods (Ref's. 1,2,3) and is general, i.e. it can be employed with various structural response calculation procedures. Below we first restate the equations used in the generalized column separation model and then show how the coupling to the structural response is accomplished. This is followed by the outline of the calculational procedure using a specific structural example of a spherical cap located at the end of a pipe. Numerical results which illustrate the importance of treating the fluid cavitation are also presented.

2. FLUID-HAMMER ASSUMPTIONS AND EQUATIONS

The hydraulic transients in the piping system are treated in a standard manner (Ref's. 1,4) by using a one-dimensional fluid-hammer analysis. It is assumed that the axial velocity is the only nonzero velocity component and that the motion in the radial direction is accounted for by the effect of the wall-deformation on the wave speed. In general the pipe wall deformation may have both an elastic and a plastic component (Ref's. 2,5,6) resulting in a variable wave speed in the pipe. For simplicity at the location of a fluid-structure interaction boundary only elastic pipe deformation is considered. As usual changes in fluid density are assumed negligible compared to density itself.

When expressed in terms of dependent variables p-pressure and u-velocity the equations governing fluid transients can be written in characteristic form as follows, (Ref's. 1,3):

Positive characteristic

$$\frac{dx}{dt} = u + c \tag{1}$$

$$\frac{du}{dt} + \frac{1}{\rho c}\frac{dp}{dt} + F(u) - G(\alpha) = 0 \tag{2}$$

Negative characteristic

$$\frac{dx}{dt} = u - c \tag{3}$$

$$\frac{du}{dt} - \frac{1}{\rho c}\frac{dp}{dt} + F(u) - G(\alpha) = 0 \tag{4}$$

Here x and t are the space and time coordinate, respectively; ρ is the density; c is the wave speed; F(u) is the velocity dependent pipe friction effect and G(α) represents the effect of gravity. The wave speed in its most general form is given as (Ref's. 2, 5,6)

$$c^2 = \frac{K/\rho}{1 + \dfrac{KD}{H(\dfrac{d\sigma}{d\varepsilon} - 2\sigma)}} \tag{5}$$

where K is the bulk modulus of the fluid, D is the pipe diameter, H is the pipe wall thickness, ε is the circumferential strain and σ is the stress in the pipe wall. The stress is in equilibrium with the local fluid pressure and is obtained from

$$\sigma = \frac{pD}{2H} \tag{6}$$

At locations where the pipe wall deformation is limited to the elastic regime, equation (5) reduces to the more familiar form (Ref's. 1,4).

$$c^2 = \frac{K/\rho}{1 + \dfrac{KD}{EH} b} \; . \tag{7}$$

where E is the elastic modulus of the pipe material and b is a parameter which depends on the pipe support and has a value of unity if no pipe elongation is permitted (Ref. 4).

3. COLUMN SEPARATION MODEL

The treatment of transient cavitation is based on a generalization of the basic column separation technique. In this approach, cavitation is a local phenomenon which occurs wherever the pressure drops below the vapor pressure. The cavity is assumed to remain at the constant vapor pressure and forms a free boundary for the separated fluid columns. Cavitation is assumed to cease when the cavity vanishes by reimpact of adjacent fluid columns or of a fluid column and a structural boundary. The technique permits the efficient and accurate presentation of pressure wave propagation in cavitated regions, both for linear and nonlinear fluid-hammer formulations (Ref's. 1,3). Large cavitated regions are represented by multiple cavities (at the computational nodes) separated by short fluid slugs. This modeling in effect is equivalent to reducing the wave speed in the cavitated region and leads to a more realistic representation of cavity collapse and in particular of the timing of this event.

The advancement of the solution by the fixed-grid method of characteristic procedure, from time t_o, at which all pressure and velocity values are known, to time $t_o + \Delta t$ at a typical grid point, j, is illustrated in Fig. 1. Points A and B are respectively the intersections of the positive and negative characteristic through grid point j, at time $t = t_o + \Delta t$, with constant time line $t = t_o$. Using simple forward differences, the characteristic equations (1) and (3), and the corresponding compatibility relationships (2) and (4), are expressed in finite difference form. To accommodate the phenomenon of column separation, two values of velocity are introduced at each grid point; u at the left of the point, and w at the right of the point (see Fig. 1). If no cavitation exists, the two velocities at the grid point are equal, i.e., $w_j = u_j$, and one obtains the solution at point j by solving the compatibility relationships (2) and (4) for the pressure and velocity.

$$p_j^{n+1} = \frac{1}{c_A^n + c_B^n}\left\{\left(c_B^n\, p_A^b + c_A^n\, p_B^n\right) + \rho\, c_A^n\, c_B^n \left[\left(u_A^n - u_B^n\right) - \Delta t\left(F_A^n - F_B^n\right)\right]\right\} \tag{8}$$

$$u_j^{n+1} = \frac{1}{c_A^n + c_B^n}\left\{\left(c_A^n\, u_A^n + c_B^n\, u_B^n\right) + \frac{1}{\rho}\left(p_A^n - p_B^u\right) - \Delta t\left[\left(c_A^n\, F_A^n + c_B^n\, F_B^n\right) - G\left(c_A^n + c_B^n\right)\right]\right\} \tag{9}$$

Here the subscripts designate the point location and the superscripts n and n+1 correspond to time t_o and $t_o + \Delta t$ respectively. Values at points A and B are obtained by combining the finite difference forms of Eqs. (1) and (3) with linear interpolation formulae along the constant time line $t = t_o$.

When cavitation occurs at point j, then the velocities are not equal; i.e., $w_j^{n+1} \neq u_j^{n+1}$, and the pressure is given as

$$p_j^{n+1} = p_{vap} \tag{10}$$

where p_{vap} is the vapor pressure of the fluid, at the prevailing temperature. Values at points A and B are obtained by linear interpolation. The velocities are then calculated directly from the finite difference forms of the compatibility equations.

$$u_j^{n+1} = u_A^n - \frac{1}{\rho\, c_A^n}\left(p_j^{n+1} - p_A^n\right) - \left(F_A^n - G\right)\Delta t \tag{11}$$

$$w_j^{n+1} = u_B^n + \frac{1}{\rho\, c_B^n}\left(p_j^{n+1} - p_B^n\right) - \left(F_B^n - G\right)\Delta t \tag{12}$$

The size of the cavity or column separation is then computed from

$$S_j^{n+1} = S_j^n + \frac{1}{2}\left[\left(w_j^{n+1} + w_j^n\right) - \left(u_j^{n+1} + u_j^n\right)\right]\Delta t. \tag{13}$$

The cavity at the node is assumed to have collapsed when $S_j^{n+1} \leq 0$, and the computation reverts then to using Eqs. (8) and (9).

In the general computational procedure, which must be followed at each time step, provision is made for existing cavitation, the inception of cavitation, and for the recollapse of cavities. Since cavitation can occur anywhere in a piping system, including at boundaries and pipe junctions, modified procedures have been developed to treat these cases (Ref. 1). The overall logic of these procedures is the same as outlined above.

4. FLUID TRANSIENTS AT STRUCTURAL BOUNDARIES

Consider the interaction of pressure transients with a structural component located at the end of a pipe. As shown in Fig. 2, the structural boundary may be located at either the last node (Fig. 2(a)) or the first node (Fig. 2(b)) of the pipe. In either case only one characteristic intersects the boundary and therefore only one compatibility relationship is available to solve for the two unknowns (pressure and velocity). The second equation must then arise from the boundary response itself. These relationships are simple when the boundary behavior can be prescribed a priori, e.g. a rigid closed pipe end. However, when the structure and fluid are interactive then their response solutions are coupled.

For a structural boundary at the last pipe node using forward differencing the following finite difference expression is obtained from Eq. (2).

$$u_j^{n+1} - u_A^n + \frac{1}{\rho\, c_A^n}\left(p_j^{n+1} - p_A^n\right) + \left(F_A^n - G\right)\Delta t = 0 \tag{14}$$

Again the values at the intercept, point A in Fig. 2(a), can be obtained by using linear interpolation along the time line $t = t_o$ in combination with the finite difference form of Eq. (1).

$$x_A = x_j - \left(u_A^n + c_A^n\right) \Delta t \tag{15}$$

Since the fluid-transient analysis is one-dimensional the pressure (p_j^{n+1}) acting on the pipe end is assumed to be uniform regardless of the complexity of the structure located there. The response of the structure subject to this uniform pressure load may be expressed as

$$w_j^{n+1} = R\left(p_j^{n+1}\right) \tag{16}$$

Again the notation of two velocities at a grid point (u and w) is being used. $R(p_j^{n+1})$ may be a very nonlinear relationship or may in fact represent the result of elaborate multidimensional structural response calculations. For complex structures the velocity w_j in Eq. (16) is some suitable average (one-dimensional) velocity in the x-direction.

Equations (14) and (16) can be solved simultaneously for the pressure and velocity at the fluid-structure interface. If the junction is not cavitated the fluid is able to follow the structural displacment and the two velocities at the point are equal, i.e. $w_j=u_j$. In this case the fluid and structural response are completely coupled and in general solutions must be obtained by an iterative procedure. Combining Eqs. (14) and (16) one can write the following expression

$$P(p) = R(p) + C_1 p + C_2 = 0 \tag{17}$$

where for simplicity both the subscripts and superscripts on the pressure p_j^{n+1} have been dropped. The values of C_1 and C_2 are constants at each time step and are given as

$$C_1 = \frac{1}{\rho\, c_A^n} \quad ; \quad C_2 = (F_A^n - G)\, \Delta t - u_A^n \tag{18}$$

Any suitable iteration procedure may be used to solve Eq. (17). In the current application a simple resubstitution scheme, in which the value of R(p), i.e. the velocity, from the preceding iteration is used to compute a new value of pressure, is found to be quite adequate and rapidly converging.

The impedance mismatch at the boundary may lead to inertial (velocity) overshoots of the structure to which the fluid cannot adjust before reaching its vapor pressure. In this case the junction becomes cavitated and the column separation technique outlined earlier must be employed. Since the junction pressure is now known ($p_j=p_{vap}$), Eqs. (14) and (16) are solved directly to obtain the two velocities u_j and w_j, respectively, and no iteration is required. The size of the cavity formed between the fluid column and the structural boundary is computed using an expression identical to Eq. (13). Cavitation ceases when this cavity vanishes, and the calculation then reverts to the coupled and iterative solution procedure.

In the above outline it was tacitly assumed that the time steps for the solution of the fluid and structural problems are the same. In general, this is not the case. In fact the structural computations may require, for reasons of numerical stability, a time step as much as two orders of magnitude smaller than that needed for the fluid transient calculations. Since performing the calculations for the entire fluid system using the structural time step is not economical, a modification of the solution procedure becomes necessary. The approach, illustrated in Fig. 3, is to employ the structural time step δt only to compute the interaction at the fluid-structure boundary, while the remainder of the fluid system is calculated using the appropriate fluid time step Δt. Since all variables along the time line $t=t_0$ are known at the node points j and j-1, the values at the intersection points of the characteristics with that line, i.e. at points A_1, $A_2 \ldots A_k \ldots A_N$, can again be obtained by combining linear interpolation formulae with the characteristic equation in finite difference form which for the k-th time interval is given as

$$x_{A_k} = x_j - \left(u^n_{A_k} + c^n_{A_k}\right) k\delta t \tag{19}$$

The solution for each structural time step δt then proceeds in a manner identical to that outlined above for the case of equal time steps. The total number of steps for each fluid time step is simply

$$N = \frac{\Delta t}{\delta t} \ . \tag{20}$$

Since a structure may experience very rapid oscillations the fluid at the boundary may undergo cavitation and subsequent cavity collapse a number of times during the duration of a single fluid time step. In spite of this phenomenon no difficulties have been encountered in the application of the fluid-structure interaction model. For a structural boundary located at the first node of a pipe the solution procedure is analogous, however, as shown in Fig. 2(b) relationships along the negative characteristic must be used.

5. STRUCTURAL ASPECTS

As indicated above the structural response calculations will differ from problem to problem. In most cases the response function R(p) in Eqs. (16) and (17) will be nonlinear and impossible to express in explicit form. However, the relation between the pressure and the average velocity of the structure can always be obtained by solving the structural response problem numerically. The approach is best illustrated by an example such as the spherical cap or disc clamped rigidly at the end of a pipe and subjected to fluid pressure loadings on its convex surface, which is shown schematically in Fig. 4. The equations of motion of this structure can be obtained using the corotational finite element method the details of which are presented elsewhere (Ref's. 7,8). Only the most relevant features are summarized here.

The loading, geometry and response of the structure are assumed to be axisymmetric, therefore it is sufficient to consider only a generator of the shell in a plane. The generator is subdivided into a finite number of segments (elements) with the dividing points called nodes. The displacements inside each element are assumed to be known functions of the nodal displacements. Each node has three generalized displacements referred to the spatial coordinate system. The strain in each element is composed of two components, axial and circumferential strain. The motion of an element will not necessarily cause strain, hence, by introducing a corotational coordinate whose axis passes through the two nodes (Ref's. 7,8), the motion of the shell element can be described by a deformation displacement defined in terms of the corotational coordinates and a rigid-body-motion displacement. The deformation displacement satisfies the Kirchhoff assumptions used for thin shells.

The equation of motion for the discretized disk is, thus obtained as

$$[M]\ \{\ddot{d}\} = \{F^e\} - \{F^i\} \tag{21}$$

where [M] is a lumped diagonal mass matrix, $\{F^e\}$ and $\{F^i\}$ are the generalized external and internal nodal force vectors, and $\{\ddot{d}\}$ is the generalized nodal acceleration vector. This equation is integrated numerically by using the central difference technique where the expressions for the nodal velocity vector, $\{\dot{d}\}$ and the nodal displacement vector, $\{d\}$ are given by

$$\{d(t+\delta t)\} = \{d(t)\} + \delta t\ \{\dot{d}(t)\} + \frac{(\delta t)^2}{2}\ \{\ddot{d}(t)\} \quad , \tag{22}$$

$$\{\dot{d}(t + \delta t)\} = \{\dot{d}(t)\} + \frac{\delta t}{2}\ [\{\ddot{d}(t)\} + \{\ddot{d}(t + \delta t)\}] \quad . \tag{23}$$

Here δt is the time increment for the structural computations.

Once the nodal velocities are obtained by Eqs. (21) to (23), an average velocity can be computed. This velocity is defined as the rate of change of volume due to the disk's motion divided by the disk base area. The exact rate of change of volume is very difficult to obtain. However, when the strains caused by bending are small, good approximations are obtained from the motion of the corotational coordinates or node points. Neglecting the curvatures along the length of each element the area swept out by an element, in the plane of the shell generator, during the time step δt represents a quadrilateral. Using this fact, the rate of change of this area A_i for the i-th element can be readily computed from the known nodal velocities and coordinates. The average velocity of the disk w is then obtained from

$$w = \frac{2\pi \sum \dot{A}_i r_i^*}{\pi d^2/4} \quad . \tag{24}$$

Here the summation (index i) is over all the elements, d is the diameter of the base of the disk and r_i^* is the radial coordinate of the centroid of the quadrilateral area swept out by i-th element during the time step δt. The average velocity defined by Eq. (24) has been found satisfactory in application. It should, however, be remembered that the definition is essentially arbitrary and that other forms of average velocity could also be defined.

6. COMPUTATIONAL RESULTS

To illustrate the computational procedures for fluid-structure interactions outlined above, the response of the spherical cap shown in Fig. 4 is calculated. A prescribed triangular pressure pulse, shown in Fig. 5, is generated at the pressure source which is located at the opposite pipe end. The pipe is filled with water at rest and at an initial pressure of 0.207 MPa (30 psia). The specific problem parameters assumed for the calculations are: pipe diameter - 100 mm, pipe length - 6.1 m, pipe wall thickness - 10 mm, disk thickness - 0.25 mm, disk crown height - 25 mm, and disk radius of curvature 62.5 mm.

To evaluate the importance of cavitation modeling in fluid-structure interactions, two types of computations are performed. The first type of calculation employs the modified column separation model while in the second type cavitation is suppressed completely. In the latter case negative pressures are computed whenever cavitation occurs. Results obtained by assuming a purely elastic, but large displacement, response of the disk are given in Fig's. 6-10. The solid curve in all cases represents the column separation results while the dashed curve is obtained by suppressing cavitation in the calculations. Due to the severity of the loading and the flexibility of the structure, velocity overshoots and consequently cavitation occurs shortly after loading of the structure commences. This is illustrated by the pressure-time histories at the disk given in Fig. 6. As may be expected the collapse of cavitation results in high pressure spikes. These are completely suppressed in the computations which ignore cavitation. At times the latter calculations also predict sustained negative pressures. This phenomenon has no basis in reality. The fluid and disk velocities are shown in Fig's. 7 and 8, respectively. When cavitation occurs the fluid boundary moves at near constant velocity while at the same time the structure, free of the fluid damping, undergoes violent oscillations. Again this phenomenon is suppressed when cavitation is ignored. The displacement-time history of the disk center point, shown in Fig. 9, indicates that the disk buckles rapidly under the applied load. In a time span of approximately 4 ms the center-point displacement is equal to the full initial crown height of the disk and thereafter levels off. The most significant differences in displacement between the two methods of computation are the final level reached by the center point. Large differences are also observed at other locations in the pipe, as illustrated by the pressure-time histories at the pipe midpoint shown in Fig. 10. All similarity between the two curves ceases once cavitation occurs. This is in agreement with results obtained for other piping systems (Ref's. 1, 3), and again demonstrates the importance of properly modeling cavitation.

The influence of cavitation on fluid-structure interactions can also be substantial when the cavitation occurs elsewhere in the fluid system and not at the fluid-structure boundary itself. Again the system shown in Fig. 4 is analyzed and the source pressure history is the one given in Fig. 5. However, the disk response is assumed to be bilinear•elastic-plastic with a yield stress of 138 MPa (20,000 psi). Due to the easier yielding of this disk the pressure load on the structure after an initial spike decays rapidly as shown in Fig. 11. The loading of the structure is thus reduced in comparison with the purely elastic case and no separation between fluid and structure occurs. Initially the computation using column separation and the calculations which suppress cavitation give identical results. However, some regions of the pipe cavitate strongly as indicated for the pipe midpoint in Fig. 12. The pressures at the disk are altered differently, in the two computations, when signals arrive from those regions (Fig. 11). This in turn leads to drastically different disk displacement-time histories as shown in Fig. 13.

7. CONCLUSIONS

A procedure has been developed to compute fluid-structure interactions as part of an overall analysis of fluid transients in piping system. The effects of cavitation on fluid-structure interactions are treated by an extension of a generalized column separation model. The calculational procedure is general and can be employed with a variety of structural response computations including multidimensional finite element codes. The computations are found to be numerically stable and rapidly converging even when there is large disparity in the time steps required for the fluid system and structural response calculations. It is possible to calculate fluid-structure interactions without including the effects of cavitation. However, it is found that the results of such computations differ drastically from those obtained by using the column separation model. This is true both for cases where the cavitation occurs directly at the fluid-structure boundary and where it occurs elsewhere in the fluid system. For structures which are in an unstable configuration and which may undergo buckling it appears, that the structural response prediction is even more sensitive to the fluid modeling assumptions than the fluid response calculations.

The major limitation of the outlined fluid-structure interaction procedure lies in the one-dimensional treatment of fluid response which of necessity leads to the assumption of a spatially uniform load on the structure. To achieve complete coupling between structure and fluid it is therefore necessary to define an average structural velocity. As indicated this definition is by no means unique. In spite of these shortcomings it is believed that the developed procedure treats the most important features of fluid-structure interactions. In particular the separation of the fluid and structure, when the pressure drops to cavitation levels, provides for a more realistic calculation of the structural response and eliminates the unreal effects of fluid damping during periods of cavitation.

8. ACKNOWLEDGMENT

The analysis and results presented here are based on effort supported by the U.S. Department of Energy.

9. REFERENCES

1. Kot, C. A. and Youngdahl, C. K.: "Transient cavitation effects in fluid piping systems." Nuclear Engineering and Design, 45, 1 pp. 93-100. (January, 1978).

2. Youngdahl, C. K., Kot, C. A. and Valentin, R. A.: "Pressure transient analysis in piping systems including the effects of plastic deformation and cavitation." ASME/CSME Pressure Vessel and Piping Conference, Paper No. 78-PVP-56, Montreal, Quebec, Canada. (June 25-30, 1978).

3. Kot, C. A. and Youngdahl, C. K.: "The analysis of fluid transients in piping systems, including the effects of cavitation." Symposium on Fluid Transients and Acoustics in the Power Industry, ASME Winter Annual Meeting, San Francisco, California. (December 10-15, 1978).

4. Streeter, V. L. and Wilie, E. G.: Hydraulic Transients, McGraw-Hill, New York. (1967).

5. Youngdahl, C. K. and Kot, C. A.: "Effect of plastic deformation of piping on fluid-transient propagation." Nuclear Engineering and Design, 35, pp. 315-325. (1975).

6. Youngdahl, C. K. and Kot, C. A.: "Computing the effect of plastic deformation of piping on pressure transient propagation." Computational Methods for Fluid-Structure Interaction Problems, AMD-vol. 26, ASME Winter Annual Meeting, Atlanta, Georgia. (November 27 - December 2, 1977).

7. Belytschko, T. and Hsieh, B. J.: "Nonlinear transient analysis of shells and solids of revolution by convected elements." AIAA Journal, 12, 8 pp. 1031-1035. (August, 1974).

8. Hsieh, B. J.: "Dynamic instability analysis of axisymmetric shells by finite element methods with convected coordinates." Transactions 4th International Conference on Structural Mechanics in Reactor Technology, paper no. M7/2, San Francisco, California. (August 15-19, 1977).

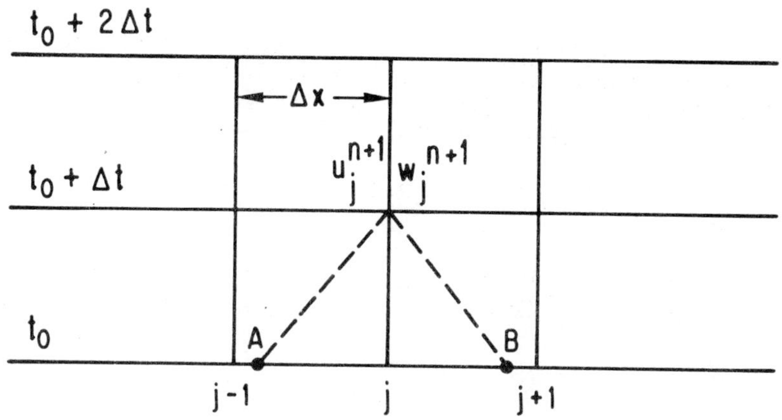

Fig. 1 Finite difference grid for column separation model

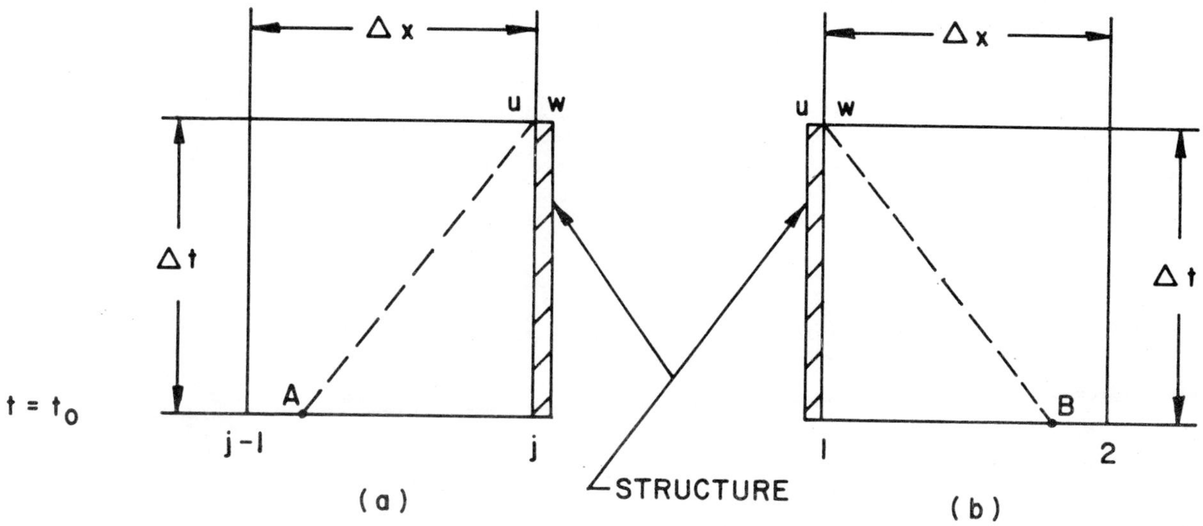

Fig. 2 Finite difference grid at fluid-structure boundary, for equal time steps

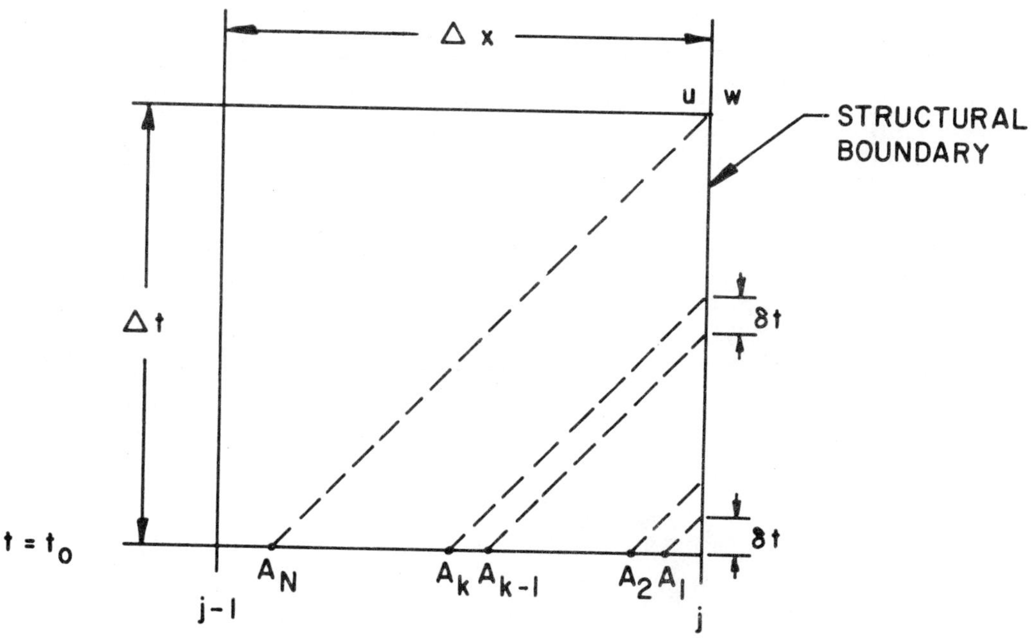

Fig. 3 Numerical treatment of fluid-structure boundary for unequal time steps

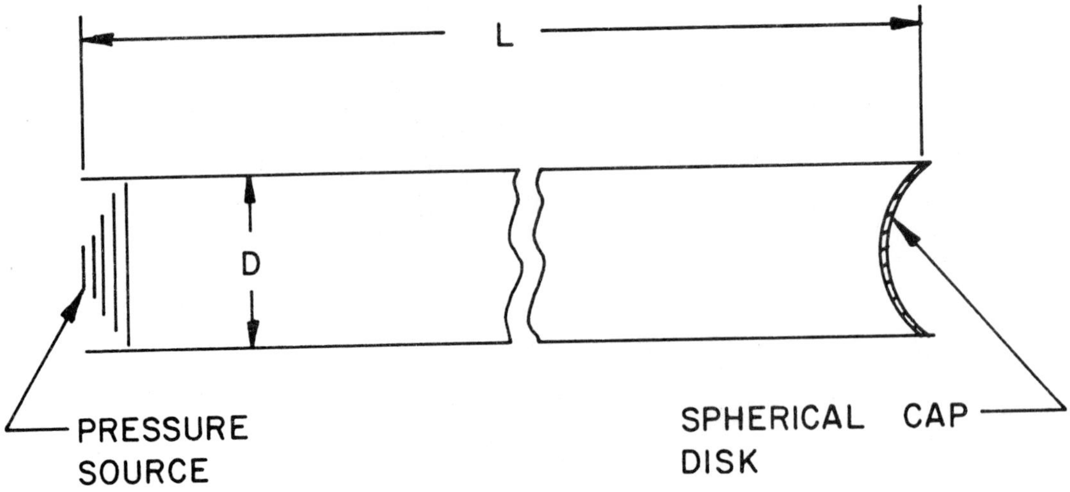

Fig. 4 Schematic of fluid-disk interaction problem

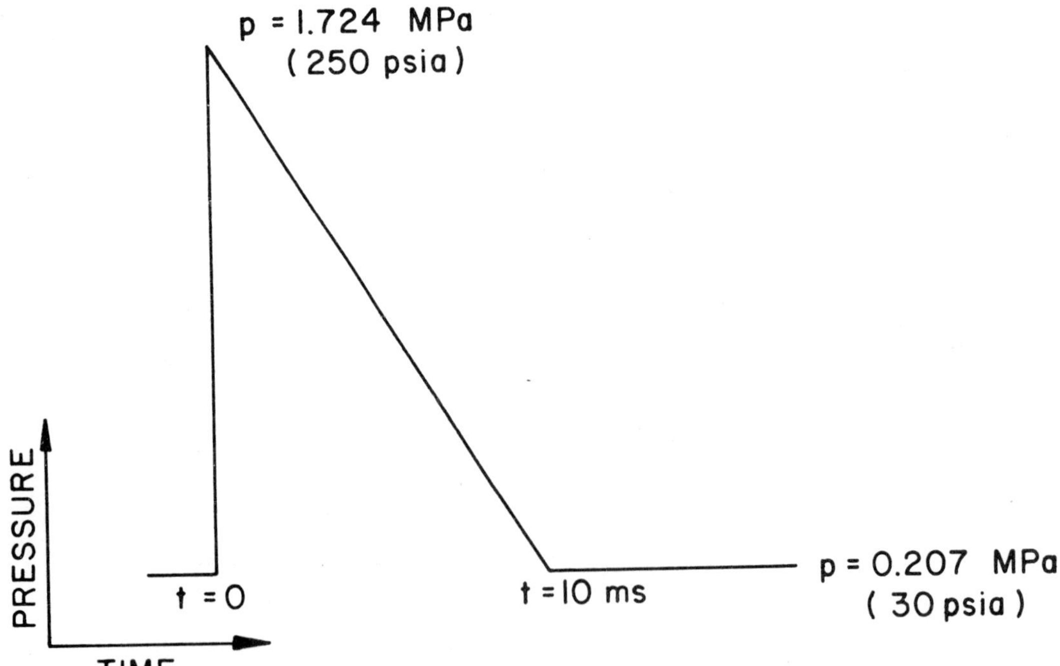

Fig. 5 Pressure-time history at source location

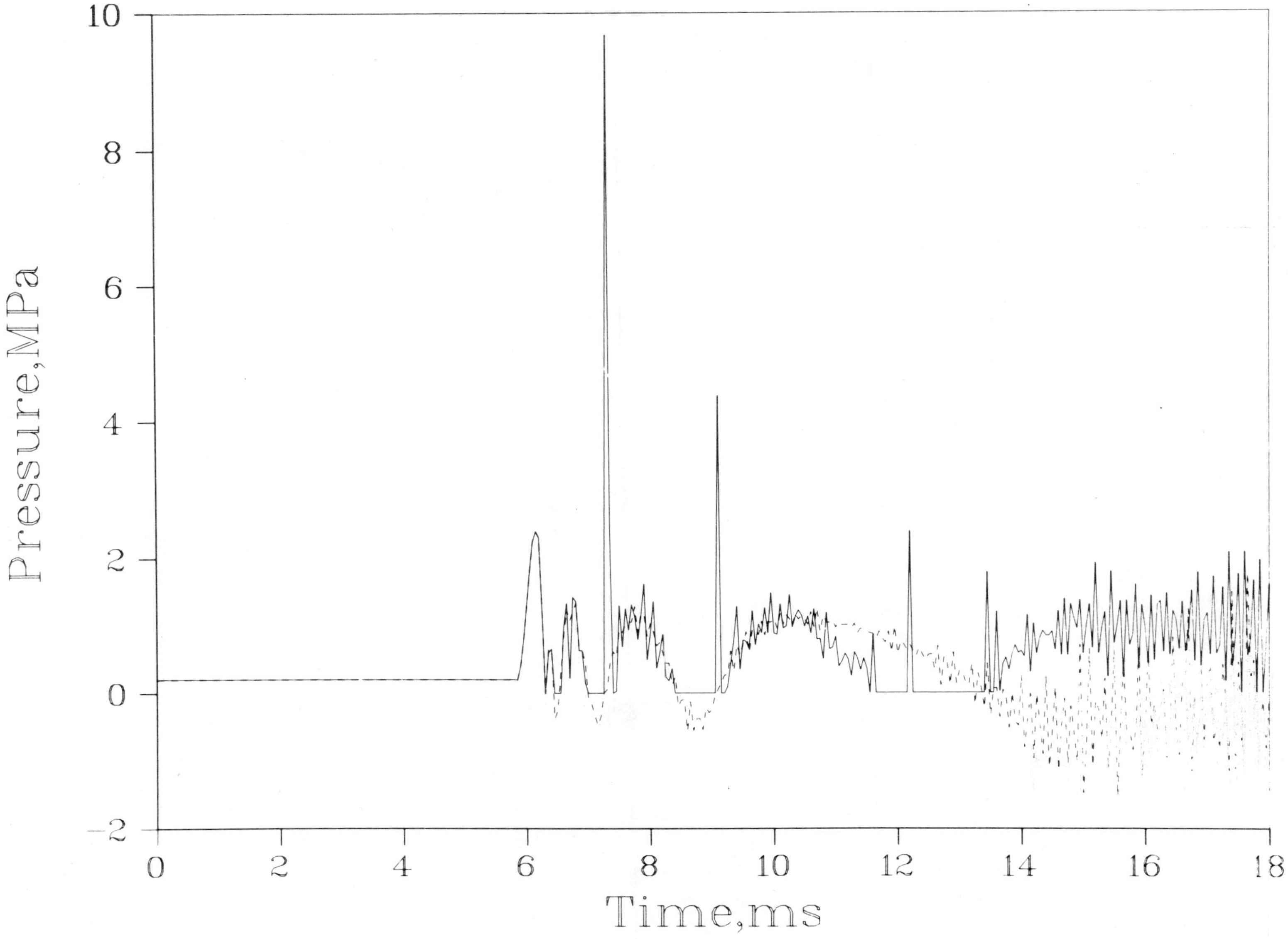

Fig. 6 Pressure variations at fluid-disk boundary, elastic response; (— column separation, - - - no cavitation)

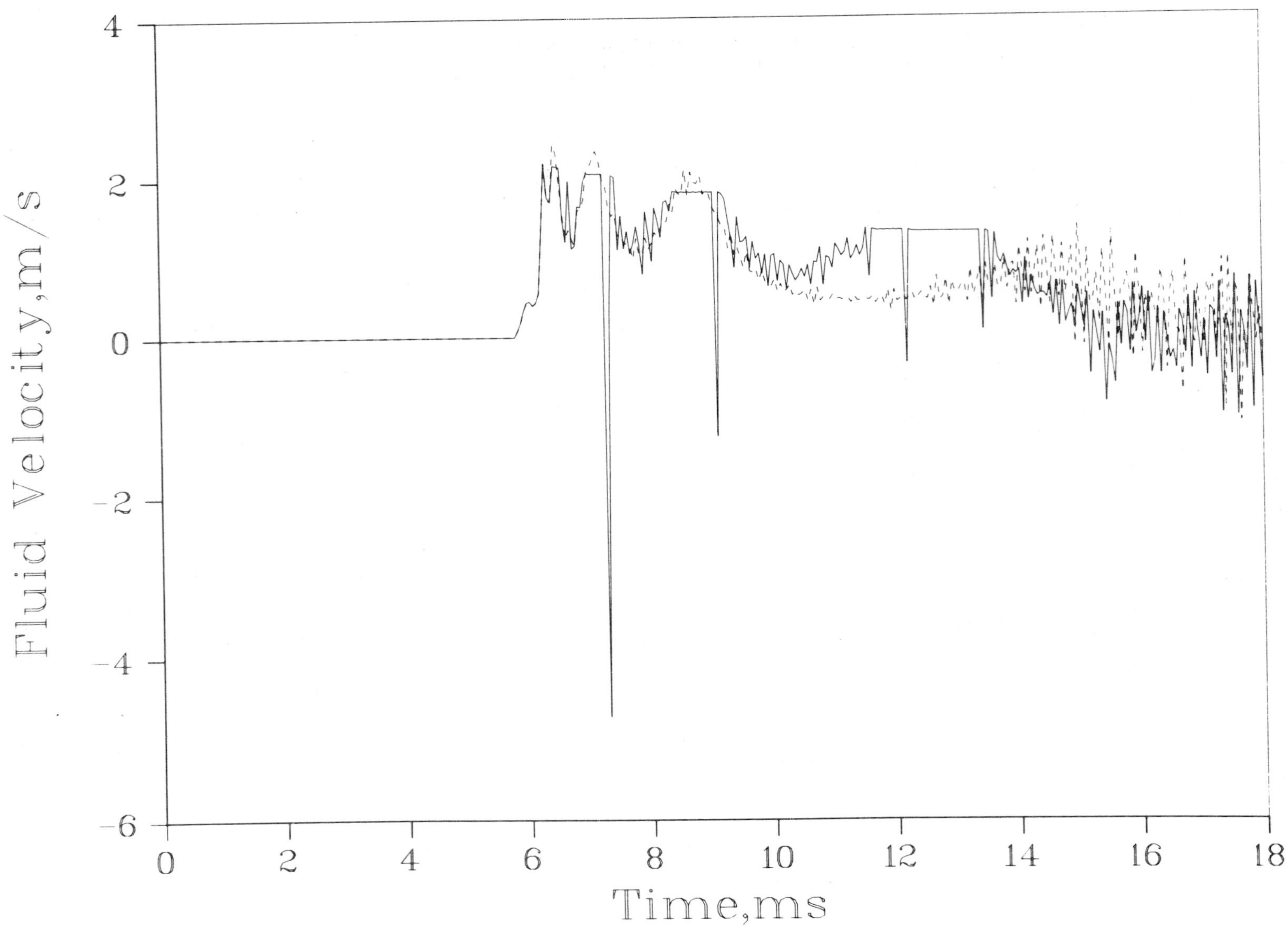

Fig. 7 Fluid velocity variations at fluid-disk boundary, elastic response
(— column separation, - - - no cavitation)

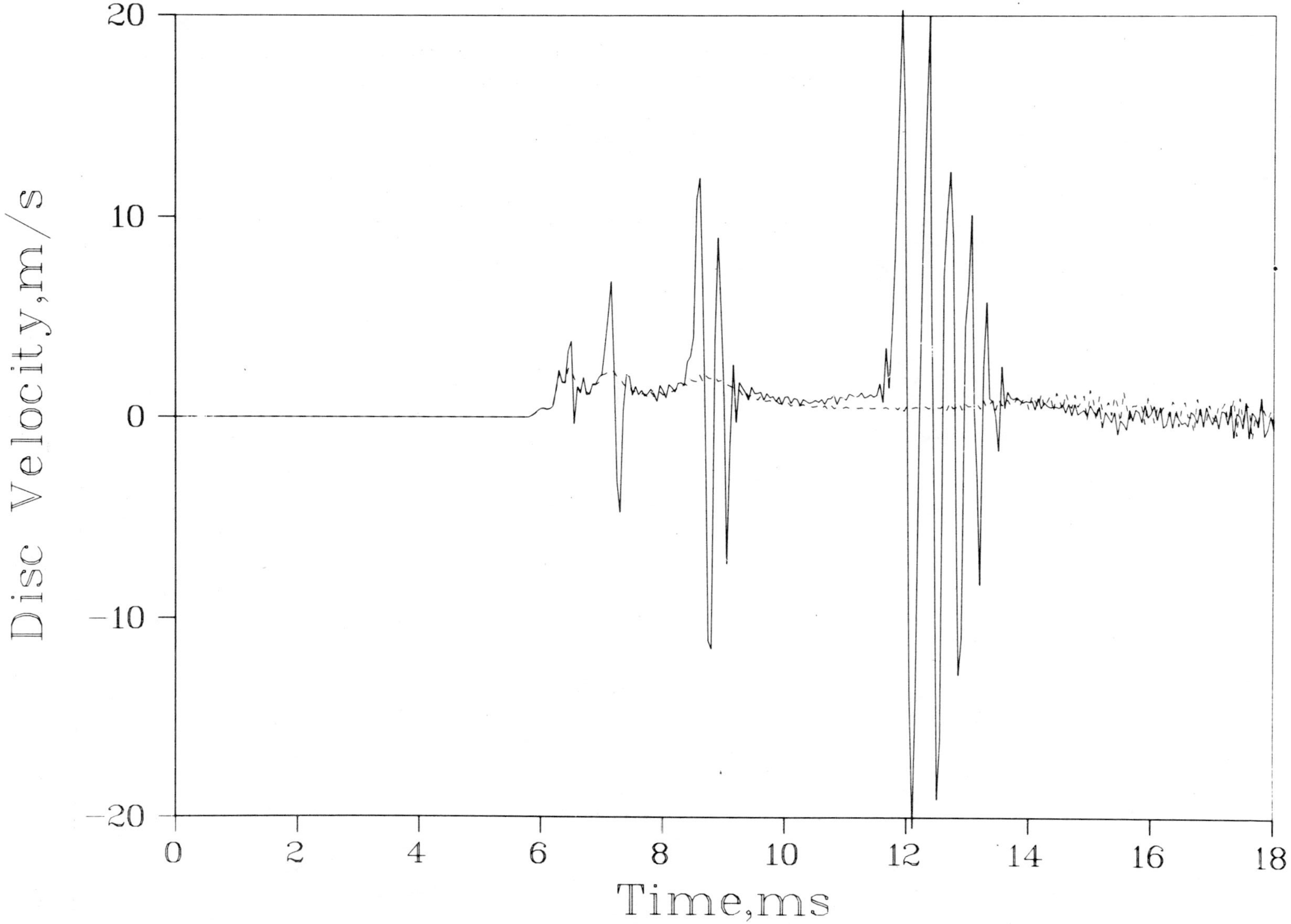

Fig. 8 Disk velocity variations, elastic responses;
(— column separation, - - - no cavitation)

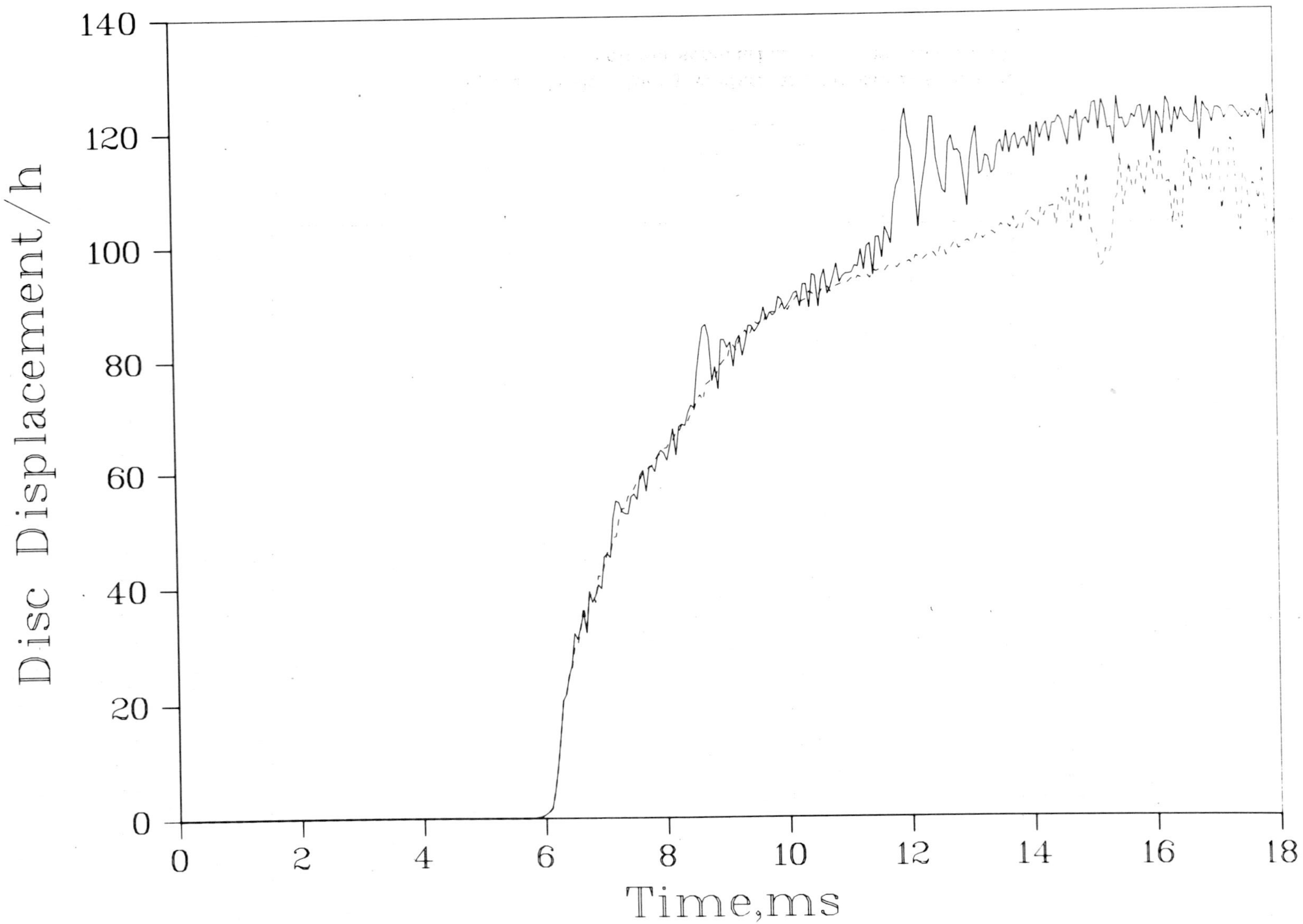

Fig. 9 Disk center point dispacements, elastic response
(— column separation, - - - no cavitation)

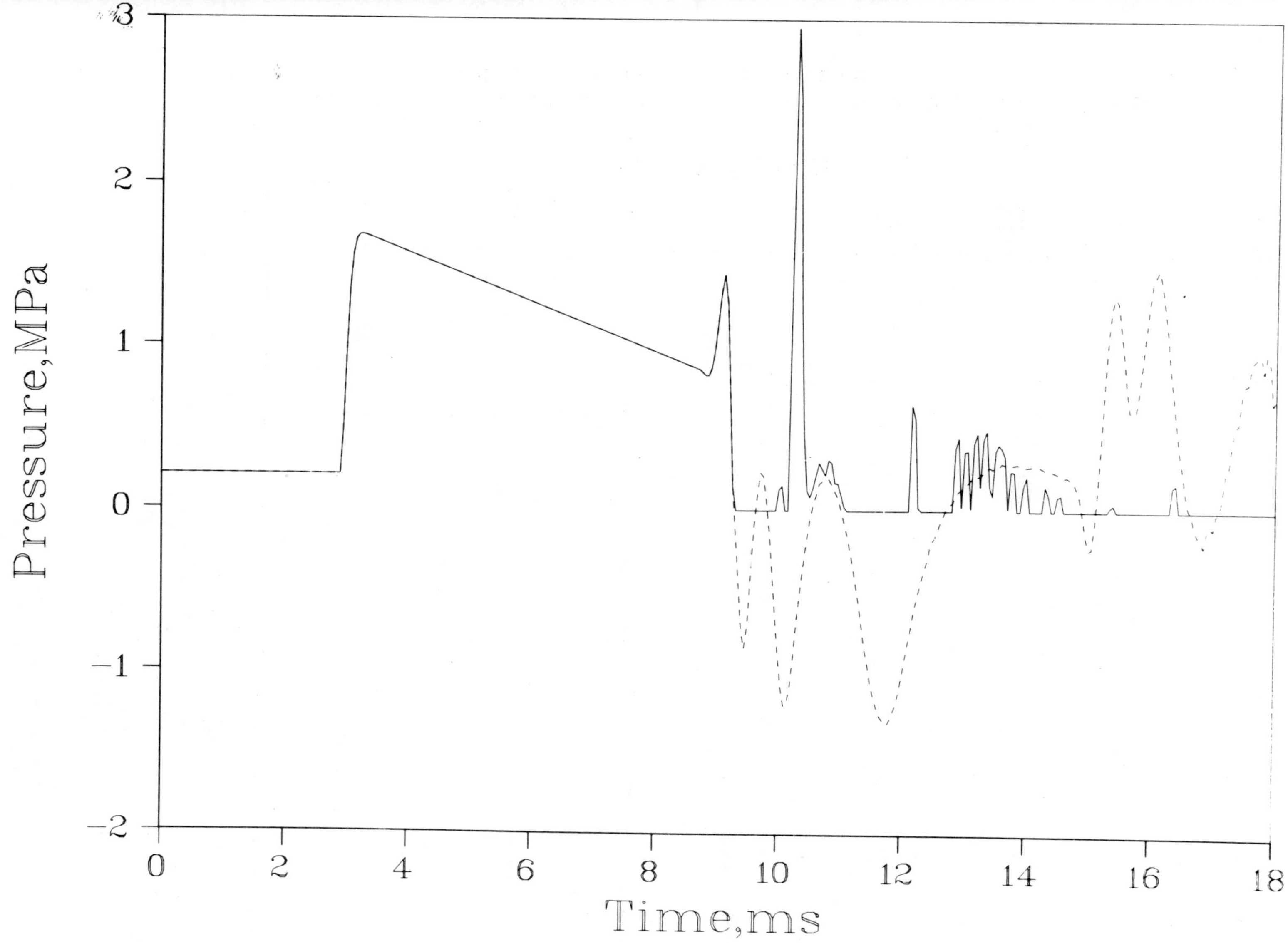

Fig. 10 Pressure variations at pipe midpoint, elastic response
(— column separation, - - - no cavitation)

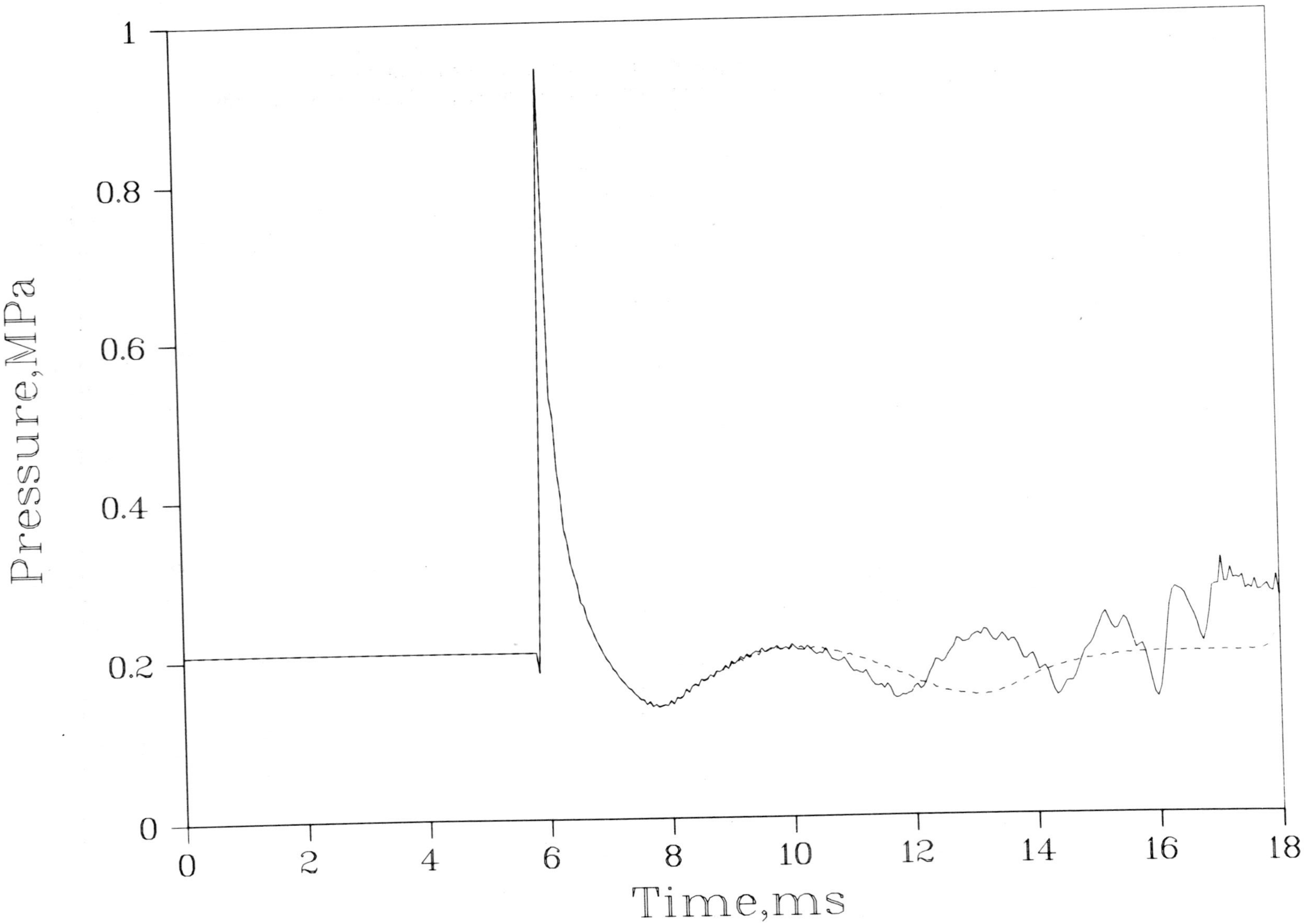

Fig. 11 Pressure variations at fluid-disk boundary, elastic-plastic response
(— column separation, - - - no cavitation)

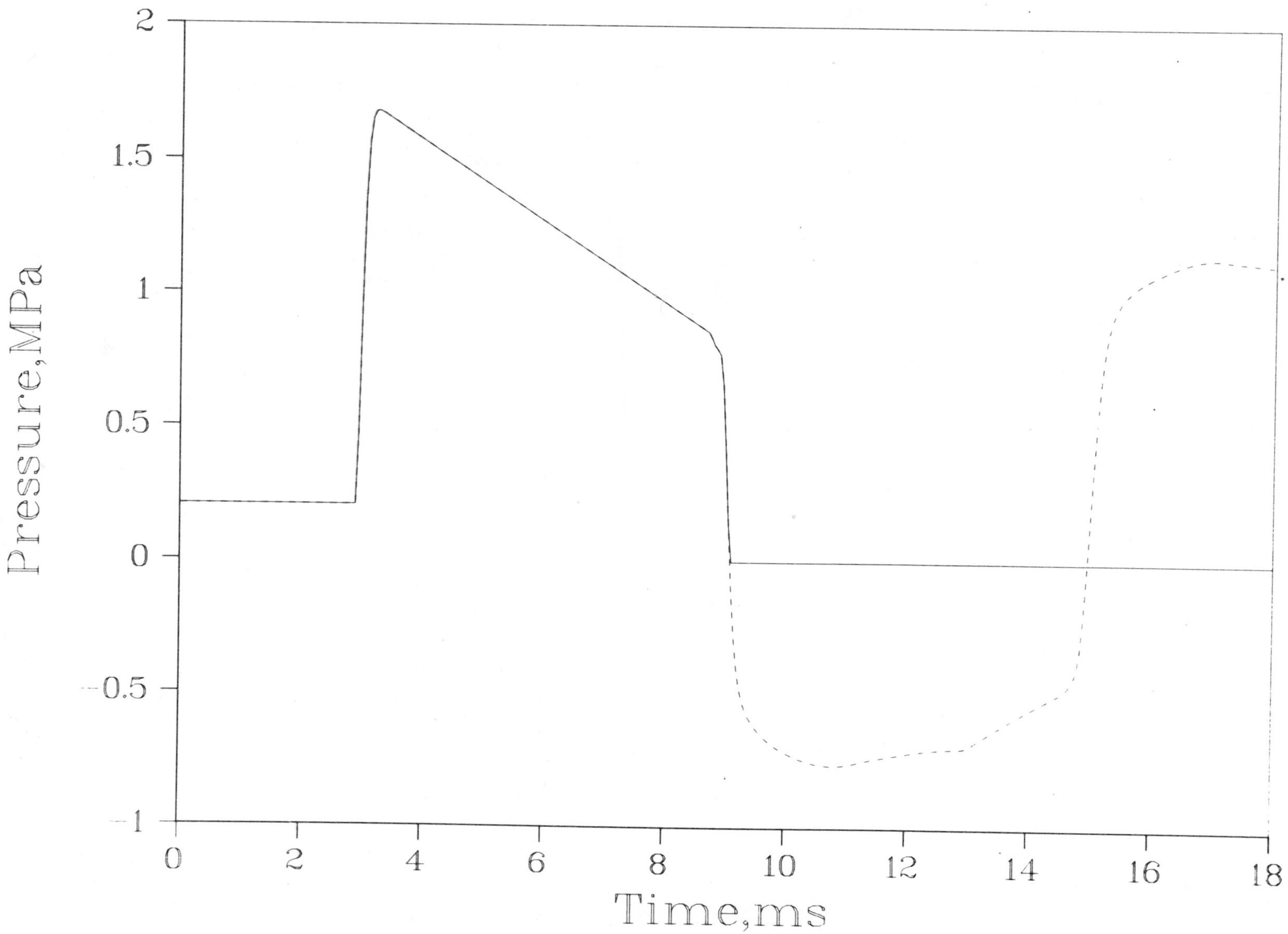

Fig. 12 Pressure variations at pipe midpoint, elastic-plastic response
(— column separation, - - - no cavitation)

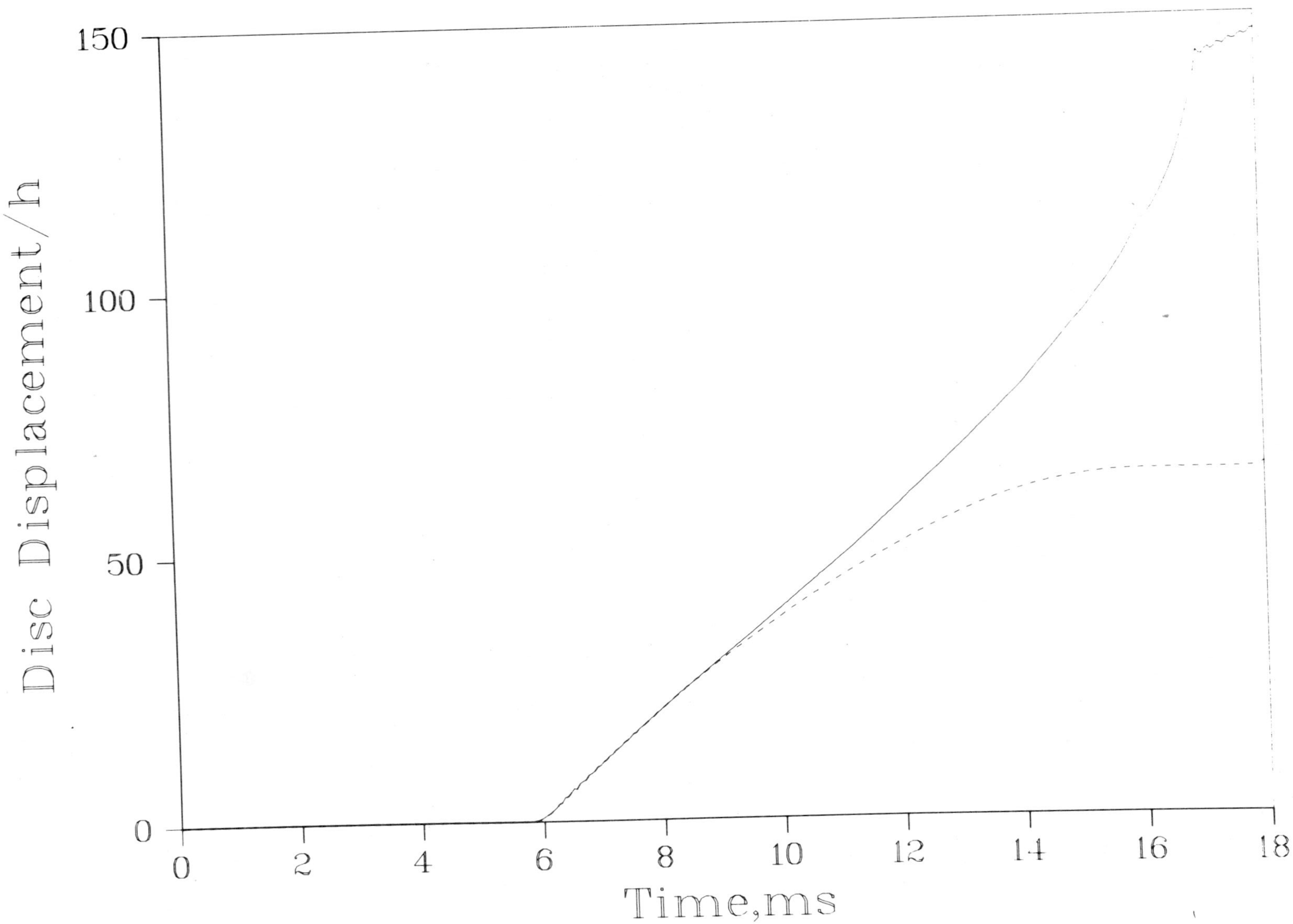

Fig. 13 Disk center point displacements, elastic-plastic response
(— column separation, - - - no cavitation)

Third International Conference on

PRESSURE SURGES

Canterbury, England·
March 25-27, 1980

PAPER E2

DYNAMIC RESPONSE OF PIPEWORK SYSTEMS TO WATER HAMMER

D.H. Wilkinson

C.E.G.B., U.K.

Summary

A theoretical solution for water hammer generation and propagation in a thin-walled straight pipe and comparisons with experiment have been given in a companion paper, Wilkinson & Curtis (1980). The present paper took the two levels of theory given earlier to a more simplified level to give a "basic" theory which still agreed quite well with experiment for straight pipes. The "basic" theory was then extended to cover wave propagation through a "gate" in a straight pipe and through mitre bends and T-junctions. This involved the introduction of shear/bending and torsion waves in addition to the liquid pressure and direct stress waves already included. The predictions for bends were compared with published experimental results. Qualitative agreement was shown but more controlled experiments with measurements of all wave families are needed to check the theory. The elements of a general method for wave propagation in a flexible pipework system were outlined. The simple analytical solutions given allow rapid assessment of the magnitudes of fluid/structure coupling effects in any system.

Held at the University of Kent at Canterbury, U.K.

Conference Organised by BHRA Fluid Engineering, Cranfield, Bedford, England.
In conjunction with The City University, London.

NOMENCLATURE

A	Liquid/Structure Coupling Parameter, eqn. (7)
B	Liquid/Structure Coupling Parameter, eqn. (22)
C	Wave speed (m s^{-1})
C_R	Pressure wave reflection coefficient = reflected Δp/incident Δp
D	Liquid/Structure coupling Parameter, eqn. (22)
E	Young's Modulus (N m^{-2})
F	Shear force (kg)
f	Transverse force factor
G	Modulus of rigidity (N m^{-2})
H	Liquid/Structure coupling parameter, eqn. (27)
I	Second moment of area of pipe cross-section (m^4)
K	Bulk modulus of the liquid (N m^{-2})
k	Shear stiffness constant, eqn. (12)
p	Liquid pressure (N m^{-2})
R	Pipe radius (m)
s	Mean shear stress in pipe wall (N m^{-2})
t	Thickness of pipe wall (m)
V	Liquid velocity (m s^{-1})
V_n	Transverse pipe velocity (m s^{-1})
V_s	Axial pipe velocity (m s^{-1})
x	Axial co-ordinate (m)
Δx	Axial point spacing in numerical solution (m)
y	Transverse pipe deflection (m)
θ	Mitre bend angle
ρ	Density (kg m^{-3})
σ	Direct stress in pipe wall (N m^{-2})
τ	Time (s)
$\Delta\tau$	Time step in numerical solution (s)
ψ	Angle between the normal to the cross-section and the horizontal Fig. 8

NOMENCLATURE - (continued)

SUBSCRIPTS

o	'Original' or 'incident'
1, 2	At the positions indicated in the diagram
b	Bulk liquid or metal
ℓ	Liquid
s	Structure
sh	Shear
t	torsion

1. INTRODUCTION

A companion paper, Wilkinson & Curtis (Ref. 7) has given experimental and theoretical results for water hammer in a straight thin-walled pipe due to impact of a water column with a closed end. The theory was given in two versions

i) an 'exact' theory for elastic or plastic deformation of the pipe wall, allowing large changes of radius and wall thickness and resulting in 21 simultaneous equations, and

ii) a 'simplified' theory, for elastic deformation and small deflection only, requiring solution of a reduced set of 11 equations.

While a computer solution of these sets of eqns. is simple for the straight pipe geometry it is more convenient to have a yet simpler method, referred to as a 'basic' method, both for rapid calculations from an analytical solution and for extension to more complex geometries such as mitre bends, T-junctions, area changes, closed valves and other elements of real pipework systems.

The main part of this paper will be a development of this 'basic' theory of wave propagation through pipework elements with comparisons with the more exact theories for a straight pipe and some very limited comparisons with published experimental work for mitre bends.

2. 'BASIC' THEORY FOR THE STRAIGHT PIPE CASE

Fig. 1 shows the model for conditions before and after impact of a water column of velocity V_o with the closed end of a thin-walled pipe on the 'exact' and 'simplified' theories. Two composite pressure/stress waves at velocities C_s and C_ℓ are generated and the more exact theories mentioned above permit all quantities to be found. However, with a few assumptions, an answer within about 8% of this can be found much more simply. These assumptions are

(a) The weak pressure wave component at velocity C_s and the weak tension wave component at velocity C_ℓ are neglected giving the model shown in Fig. 2.

(b) C_s is given by the bulk wave velocity

$$C_s = C_{sb} = \sqrt{\frac{E}{\rho_s}} \tag{1}$$

(c) C_ℓ is given by the elastic pipe expression,

$$\left.\begin{aligned} C_\ell &= C_{\ell b}\{1 + 2\,\frac{K}{E}\,\frac{R}{t}\}^{-\frac{1}{2}} \\ \text{where } C_{\ell b} &= \sqrt{\frac{K}{\rho_\ell}} \end{aligned}\right\} \tag{2}$$

Applying the momentum eqn. across the waves gives

$$p_1 - p_o = \rho_\ell C_\ell (V_o - V_1) \tag{3}$$

$$\sigma_1 - \sigma_o = \rho_s C_s (V_{s1} - V_{so}) \tag{4}$$

The remaining equations are:

Compatibility of velocities

$$V_{s1} = V_1 \tag{5}$$

Equilibrium of loads on the pipe end, assumed to have negligible mass,

$$p_1 . \pi R^2 - \sigma_1 . 2\pi Rt = 0 \tag{6}$$

For this case $V_{so} = 0$ and $\sigma_o = p_o R/2t$ and solution of the eqns. (3) to (6) for the unknowns $p_1, V_1, \sigma_1, V_{s1}$ gives, taking

$$A = \frac{\rho_s C_s}{\rho_\ell C_\ell} \cdot \frac{2t}{R} \tag{7}$$

$$\left.\begin{aligned} \frac{V_1}{V_o} &= \frac{1}{1 + A} \\ \frac{p_1 - p_o}{\rho_\ell C_\ell V_o} &= \frac{A}{1 + A} \\ \frac{\sigma_1 - \sigma_o}{\rho_\ell C_\ell V_o} &= \frac{A}{1 + A} \cdot \frac{R}{2t} \end{aligned}\right\} \tag{8}$$

A comparison with the more exact solution, Ref. 7, gives, for a thin-walled steel pipe with R/t = 40

$$\begin{aligned} \rho_\ell &= 1{,}000 \text{ kg m}^{-3} \\ \rho_s &= 7{,}893 \text{ kg m}^{-3} \\ E &= 206.8 \text{ GN m}^{-2} \\ K &= 2.196 \text{ GN m}^{-2} \\ C_{sb} &= 5.120 \text{ km s}^{-1} \\ C_{\ell b} &= 1.483 \text{ km s}^{-1} \\ A &= 1.853 \end{aligned}$$

	'Basic'	'Simplified'
C_s/C_{sb}	1.0	1.0223
$C_\ell/C_{\ell b}$	.7353	.7346
V_1/V_o	.3505	.3350
$(p_1 - p_o)/\rho_\ell C_\ell V_o$	.6495	.7063
$(\sigma_1 - \sigma_o)/\rho_\ell C_\ell V_o$	12.99	14.13

The errors are about 8% on stresses and pressures, 4% on liquid velocity and 2% on stress wave speed.

Assuming the wave speeds from eqns. (1) and (2) then many simple water column impact or shock and stress wave propagation estimates can be made using expressions of the type shown in eqns. (3) to (6). An example is a shock wave incident on a 'gate' in a pipe, Fig. 3, in which all pressure and stresses are increments above their original static levels. The equations are

$$\left.\begin{aligned} p_o &= \rho_\ell C_\ell V_o \\ p_1 - p_o &= \rho_\ell C_\ell (V_o - V_1) \\ p_2 &= \rho_\ell C_\ell V_1 \\ \sigma &= \rho_s C_s V_1 \\ \pi R^2 (p_1 - p_2) &= 4\pi R t \sigma \end{aligned}\right\} \tag{9}$$

with solution

$$\left.\begin{aligned} \frac{V_1}{V_o} &= \frac{1}{1 + A} \\ \frac{p_1}{\rho_\ell C_\ell V_o} &= \frac{1 + 2A}{1 + A} \\ \frac{p_2}{\rho_\ell C_\ell V_o} &= \frac{1}{1 + A} \\ \frac{\sigma}{\rho_\ell C_\ell V_o} &= \frac{A}{1 + A} \cdot \frac{R}{2t} \\ &\text{The reflected shock wave is} \\ \frac{p_1 - p_o}{\rho_\ell C_\ell V_o} &= \frac{A}{1 + A} \end{aligned}\right\} \tag{10}$$

Similar equations for an incident tension wave, Fig. 4, give the solution

$$\left.\begin{aligned} \frac{V_1}{V_{so}} &= \frac{A}{1 + A} \\ \frac{p_1}{\rho_\ell C_\ell V_{so}} &= \frac{A}{1 + A} = -\frac{p_2}{\rho_\ell C_\ell V_{so}} \\ \frac{\sigma_1}{\rho_\ell C_\ell V_{so}} &= \frac{A(2 + A)}{1 + A} \cdot \frac{R}{2t} \\ \frac{\sigma_2}{\rho_\ell C_\ell V_{so}} &= \frac{A^2}{1 + A} \cdot \frac{R}{2t} \\ &\text{The reflected tension wave is} \\ \frac{\sigma_1 - \sigma_o}{\rho_\ell C_\ell V_{so}} &= \frac{A}{1 + A} \cdot \frac{R}{2t} \end{aligned}\right\} \tag{11}$$

3. WAVE INTERACTIONS WITH OTHER PIPEWORK ELEMENTS

In a real pipework system tension and pressure waves will encounter bends and T-junctions and the cases of pressure waves traversing these elements will be considered theoretically.

3.1 Single Mitre Bend

For a rigid pipe a pressure wave will go through a bend without reflection or attenuation. For a thin walled flexible pipe the model assumed is shown in Fig. 5, for the situation after the interaction. There are transmitted and reflected pressure waves, tension waves in both pipes and, a new feature, shear waves. Axial movement due to tension waves requires transverse deflection for compatibility and for the moment this is assumed to occur via shear waves alone with properties

$$C_{sh} = \sqrt{\frac{G}{k\rho_{sh}}} \qquad (12)$$

where k = 0.7268 for a pipe

$$\rho_{sh} = \rho_s\left(1 + \frac{\rho_\ell}{\rho_s}\frac{R}{2t}\right) \qquad (13)$$

Then shear stress and normal velocity are related by

$$s = \rho_{sh} C_{sh} V_n \qquad (14)$$

These relations can be derived from the model shown in Fig. 6 using the transverse momentum equation and the stress/strain relation for shear. The validity of using shear waves for transverse deflections will be discussed later and the corrections for bending effects added.

Fig. 5 shows the state after a pressure wave travelling from L to R has interacted with the bend. The eqns. governing the shock wave interaction with the bend are as follows

Upstream liquid momentum

$$p_o - p_1 = \rho_\ell C_\ell (V_1 - V_o)$$

or, since $p_o = \rho_\ell C_\ell V_o$

therefore $$p_1 = \rho_\ell C_\ell (2V_o - V_1) \qquad (15)$$

Downstream liquid momentum

$$p_1 = \rho_\ell C_\ell V_2 \qquad (16)$$

continuity

The partial symmetry of the phenomenon with V_T along the joint line in Fig. 5 implies that both pipes are increasing in length with velocity $V_s = V_T \text{Sin}\ \theta/2$.

Therefore

$$V_2 = V_1 - 2V_s \qquad (17)$$

Displacement compatibility

$$V_s = V_n \text{ Tan } \frac{\theta}{2} \qquad (18)$$

Metal momentum

$$\sigma = \rho_s C_s V_s \tag{19}$$

Shear momentum

$$s_1 = f\rho_{sh} C_{sh} V_n \tag{20}$$

where f is introduced to allow for the reduction in shear force required due to bending of the pipe.

Force balance

Considering forces in the V_T direction acting on the fluid and pipe in the region ABCDA which is taken to be at constant velocity after the waves have passed.

$$p_1 - \sigma \cdot \frac{2t}{R} - s \cdot \frac{2t}{R} \operatorname{Cot} \frac{\theta}{2} = 0 \tag{21}$$

The seven eqns. (15) to (21) may be solved for the unknowns p_1, V_1, V_2, V_s, V_n, σ and s with the following results

$$\left.\begin{aligned} &\text{Putting } B = \frac{f\rho_{sh} C_{sh}}{\rho_\ell C_\ell} \cdot \frac{2t}{R} \\ &\text{and } D = 1 + A + B \operatorname{Cot}^2 \frac{\theta}{2} \end{aligned}\right\} \tag{22}$$

then

$$\left.\begin{aligned} &\frac{p_1}{p_o} = \frac{D - 1}{D} \\ &\frac{V_1}{V_o} = \frac{1 + D}{D} \\ &\frac{V_2}{V_o} = \frac{D - 1}{D} \\ &\frac{V_s}{V_o} = \frac{1}{D} \\ &\frac{V_n}{V_o} = \frac{1}{D} \operatorname{Cot} \frac{\theta}{2} \\ &\frac{\sigma}{p_o} = \frac{A}{D} \frac{R}{2t} \\ &\frac{s}{p_o} = \frac{AB}{D} \left(\frac{R}{2t}\right)^2 \operatorname{Cot} \frac{\theta}{2} \\ &\text{The reflected shock coefficient is} \\ &C_R = \frac{p_1 - p_o}{p_o} = -\frac{1}{D} \end{aligned}\right\} \tag{23}$$

A reflected expansion wave is predicted

The assumption that transverse force and velocity are related by the shear wave eqn. (14) will now be examined in more detail. The full equations for transient transverse beam motion including rotation and shear as well as bending effects are well known and given, for example, by Thomson (Ref. 6) as

$$\frac{\partial^2 y}{\partial \tau^2} = \frac{G}{k\rho_{sh}} \left\{\frac{\partial^2 y}{\partial x^2} - \frac{\partial \psi}{\partial x}\right\} \tag{24}$$

$$\frac{\partial^2 \psi}{\partial \tau^2} = \frac{E}{\rho_s} \frac{\partial^2 \psi}{\partial x^2} + \frac{\pi R^2 G}{k I \rho_s} \left(\frac{\partial y}{\partial x} - \Psi\right) \tag{25}$$

Fig. 8 shows the geometry of a beam element. If the second terms on the RHS in each eqn. are neglected then (24) and (25) become two wave equations, one for shear waves with velocity C_{sh} and one for rotation waves with velocity C_s. However, as these interaction terms are not in general small a simultaneous numerical solution was attempted. The geometry considered was a uniform semi-infinite beam initially at zero velocity with, at time zero, a transverse velocity V_n suddenly produced at the free end, x = o, by an unknown force F. Values of parameters at points on a one-dimensional grid along the beam were found by integrating equations (24) and (25) through time steps $\Delta\tau$. Stability required $\Delta\tau < \Delta x / C_s$ but there were still numerical difficulties of the kind always found in numerical solutions of shock wave propagation. Spurious oscillations in the solution can only be avoided by some form of implicit or explicit numerical smoothing which 'smears out' the shock discontinuity over a minimum of about 4 grid points. There are sound physical reasons for this but it makes interpretation of the results more difficult. A parallel numerical solution of the wave equation forms of (24) and (25), for which analytical solutions were known, was carried out to give an assessment of numerical errors.

With guidance from Roache (Ref. 2) the numerical method of Lax (Ref. 1) was used, with slight modifications.

Figure 9 shows the solution for the shear force and bending moment at the driven end for a 305 mm s^{-1} transverse velocity imposed at the free end of a 508 mm diameter, 6.4 mm thick steel pipe filled with water. The wave equation solution is also given as an indication of the accuracy of the numerical method. The exact solution for the wave equation has F = 185.5 kN and it is seen that the numerical solution gives this quite well with only a small overshoot to 191.3 kN. The full equation solution shows F reaching the same initial value, showing that shear effects dominate in the early stages. However, F then drops off to about 66.7 kN after 1 ms (240 timesteps) and tends to remain fairly constant thereafter, reducing to about half this value after a total of 3 ms. This happens because the beam deflects mainly by bending rather than shear in the later stages and its transverse stiffness is less. However, some force will always be needed to accelerate new beam material sideways as the wave travels along so F is unlikely ever to drop to zero.

Figures 10 and 11 show the form of the disturbance as it progresses along the tube. It is seen that the main effects travel at shear wave speed although small perturbations travel ahead, presumably at tension wave speed. The wave is probably dispersive, becoming less sharp as it travels along but the diffusing effects of the numerical method confuse the results in Figure 11 and the rate at which this occurs cannot be separated out. Even the non-dispersive wave equation has a rapidly increasing rise time. Further development of the numerical method would be needed to clarify the longer term behaviour. Figure 10 shows that a rotation wave travels with the main transverse wave.

Reflection of the combined shear and bending moment wave from a pipe end or junction will not be as simple as for a shear or tension wave alone. The shear/bending wave can in general have two components at right angles to a pipe. Interaction with a bend will, in one case give rise to torsion waves which will travel at a fourth wave speed.

$$C_t = \sqrt{G/\rho_s} \tag{26}$$

It is probable that the proportions of bending to shear will vary depending on their origin and on the type of interaction and the dominant wave speed may also change. If, for example, in Fig. 8 the excitation had been a suddenly imposed rotation speed $\partial\psi/\partial\tau$ at the free end this would probably have propagated at speed C_s with shear phenomena playing only a minor part.

The transient transverse beam solution indicates that forces, deflections and rotations are transmitted approximately at shear wave speed C_{sh} but with the force F to produce a given transverse velocity V_n dropping from its initial value from eqn. (14) to about ¼ of this after 1 or 2 ms, which in rather more dimensionless form is the time for a pressure wave to travel 2½ to 5 diameters. Solution of the bend equations (15) to (21) with a shear stiffness reduced to about ¼ should therefore represent the interaction quite well after the first ms. This can be done approximately by putting f = ¼ in eqns. (20) and (22). Fig. 7 shows the solution for the reflected pressure wave from eqn. (23) as a function of bend angle and for two thickness/diameter ratios. The additional parameters are

$$\text{For } G = 96.5 \text{ GN m}^{-2}$$
$$\rho_{sh} = 27900 \text{ kg m}^{-3}$$
$$B = 2.79$$

Taking a factor of f = ¼ for a 90° bend, C_R of - .28 is predicted for R/t = 40. Shear factors of ½ or ⅛ would give values of - .24 and - .31 so the result is not very sensitive to the assumption made.

3.2 T-Junction

Fig. 12 shows the state after the arrival of a pressure wave through the normal leg of a T-junction, one of many possible interactions between waves and pipe elements. After the interaction there are in general pressure, tension and shear waves in each leg but in this case symmetry ensures zero tension waves in the horizontal legs and zero shear waves in the vertical one. The unknowns are p_1, V_1, V_2, V_n, s, V_s and σ. Writing down the equations in a similar way to the bend case gives the solution

$$\left.\begin{aligned}
H &= A + 2B \\
\frac{p_1}{\rho_\ell C_\ell V_o} &= \frac{2H}{3H + 1} \\
\frac{V_1}{V_o} &= \frac{2(2H + 1)}{(3H + 1)} \\
\frac{V_2}{V_o} &= \frac{2H}{3H + 1} \\
\frac{V_n}{V_o} &= \frac{2}{3H + 1} \\
\frac{\sigma}{\rho_\ell C_\ell V_o} &= A\,\frac{R}{2t} \cdot \frac{2}{3H + 1} \\
\frac{s}{\rho_\ell C_\ell V_o} &= B\,\frac{R}{2t} \cdot \frac{2}{3H + 1}
\end{aligned}\right\} \qquad (27)$$

Reflected shock coefficient is

$$C_R = \frac{p_1 - p_o}{p_o} = - \frac{(H + 1)}{3H + 1}$$

For a pipe strength range from rigid ($H \to \infty$) to zero ($H \to 0$) C_R ranges from -1/3 to -1 and the transmitted wave, p_1/p_o from 2/3 to zero. For the R/t = 40 pipe with shear strength factor of ¼, H = 3.25 and reflected and transmitted waves are - .396 and .605 of the incident wave.

3.3 Other Interactions

In the mathematical model described there are 4 wave types, pressure, tension, torsion and transverse shear/bending with the last counting as two because there are two possible directions for it. This gives 5 possible incident waves. Pipe elements include single mitre or curved bends, T-junctions with two incident wave routes through the straight or normal pipes, 'gates' such as valves, area changes and probably others making at least 6 in all. With 5 waves this gives at least 30 possible interaction cases all of which would need to be solved in producing a general method for predicting water hammer propagation in flexible pipework systems. Only 4 of these interactions are considered above due to shortage of space.

A general numerical method for thin-walled pipework systems has not yet been written partly because of its complexity and partly because the rather approximate treatment of shear and bending waves would limit its accuracy. A rather more general treatment of the very similar quasi-steady acoustic case using transfer matrices has however been given by Wilkinson (Ref. 8). In that paper shear effects are neglected and an exact solution of the transverse beam equation for harmonic excitation allows bending wave effects to be included in the same way as for the other wave types. The transient water hammer case remains less tractable however.

4. COMPARISON WITH EXPERIMENT

The 'basic' theory for straight pipes has been compared with experiment in a companion paper, Ref. 7, and good agreement shown. Further tests on pipes with bends are planned but have yet to be carried out. Meanwhile a few comparisons with published results will be given, but it will be noted that most of these are devoted to discussion of the degree of fixing of the test rigs and to reinterpretation of the results given. This is because of the difficulty of setting up, measuring and understanding this type of test, the main difficulty being that it is not possible to produce just one type of wave, such as a liquid pressure wave, without producing other waves as well. A relatively unwanted tension wave produced with a pressure wave will travel ahead about 4 times as fast and interact with a bend producing an initial pressure rise before the main pressure wave arrives. In fact the tension wave is so fast moving that it may reflect from the far ends of the rig and interact twice more with a bend in this time. The interpretation of the observed overall reflection and transmission of the pressure wave at the bend neglecting the 3 tension wave interactions will therefore give a spurious comparison with the theory above of the pressure wave/bend interaction.

The other point, on the degree of fixing, is that it has been found experimentally in the straight pipe cases to be very difficult to clamp the pipe at a point to give a genuinely rigid boundary condition. Another problem is the generation of a pulse with a short enought rise time. Noting that C_s is about 5 km s^{-1} and that rigs often have pipe legs about 6 m long the tension wave takes about 2.4 ms to return as a reflection. With rise times of about 2 ms or more it becomes difficult to distinguish the initial pulse from its interaction with the system. Rise times of about 0.5 ms are desirable to permit the waves to be distinguished unambiguously.

Swaffield (Refs. 3 & 4) has given the results of a comprehensive investigation into shock wave propagation round bends. The variables included R/t, E/K, bend radius/diameter, bend angle and restraint to movement.

Pipes made of polythene, steel, copper and aluminium were tested and no effects of material thickness/diameter ratio or restraint at the bend were found. An empirical correlation for the reflection and transmission coefficients as a function of bend angle and bend radius/diameter was given. For a 90° single mitre bend C_R was about -17% for all cases. In general the pipe size used for the mitre bend cases was 50 mm diameter with a 4.8 mm wall thickness.

The theory of Section 3.1 gives the following predictions, for f = .25

Material	Steel	Polythene
E (GN m^{-2})	207	0.583
ρ_s (kg m^{-3})	7.90	1.00
C_s (km s^{-1})	5.12	0.764
C_ℓ (km s^{-1})	1.34	0.231
ρ_{sh} (kg m^{-3})	10.6	3.66
C_{sh} (km s^{-1})	3.55	0.290
A	11.33	1.24
B	10.50	1.73
C_R ($\theta = 90^o$)	-.067	-.374

It is clear that the predictions have the right sign but only roughly the right order of magnitude. The widely varying predictions for different materials and the further one that $C_R \simeq o$ if the pipe is prevented from moving at the bend which differ from the near constant experimental results require explanation. For the steel pipes the reason is probably associated with the lack of rigidity in the mounting of the shut-off valve at the end of the pipe which produces the water hammer pulse. Tension waves due to this produce additional pressure changes which may appear to give a larger reflection than actually occurs on the main liquid shock wave.

For the polythene pipe the valve mounting is relatively rigid and the full shock pressure rise is produced at valve closure. However, the radial stretching of the pipe produces an axial strain and hence a tension wave which will interact with the bend to produce a pressure rise. Theory for a tension wave interaction with a bend indicates an initial pressure rise of .09 $\rho_\ell C_\ell V_o$ due to this tension wave interaction. However, since C_s/C_ℓ = 3.3 the reflected tension waves from the bend will be reflected back again from both ends of the rig and will have time to return to produce about a further .08 $\rho_\ell C_\ell V_o$ maximum giving a total pressure rise of about .17 $\rho_\ell C_\ell V_o$ due to tension wave effects. With the 63% of the main shock pressure rise predicted to get round the bend this could give a total of .80 $\rho_\ell C_\ell V_o$ transmitted and .20 $\rho_\ell C_\ell V_o$ reflected theoretically compared to about 0.17 $\rho_\ell C_\ell V_o$ actually measured as reflected, showing fairly good agreement.

The interpretation of experimental results is obviously quite complex because of the many wave families involved particularly since it is difficult if not impossible to set up a single wave type such as a liquid shock wave without associated tension waves. This particular case involved 4 interactions instead of one. A complete comparison with experiment would include measurement of all wave families and a theoretical estimate of all effects to give pressure, strain, etc. as a function of time at a number of points. Since only liquid pressure measurements are available so far, experiments are in hand by the author to measure stress waves as well as shock waves in similar tests.

Wood & Shen (Ref. 9) give measurements of pressure against time for a similar rig to Swaffield's. It is possible qualitatively to identify the main tension wave effects but again a detailed comparison would be beyond the scope of the present

Paper. One difficulty that could perhaps be mentioned with these experiments is that the rise time of the pulse is about 2 ms. Since tension waves travel about 11 m in this time a rig with legs 6 m long rapidly experiences overlap between the wave families making understanding and interpretation difficult. A rise time of 0.5 ms or less is desirable and obtainable by different valve systems.

On the effect of restraints, Swaffield found little or none but decided later, Ref. 5, that the restraints he had used had not been sufficient to reduce the very small displacement that occurred. His results are effectively unrestrained.

Wood & Shen were more successful in clamping their rig and showed that clamping removed the reflection effects, proving that pipe flexibility was the cause. They also showed for T-junctions that flexibility effects were smaller than with bends which, again qualitatively, tends to confirm the analysis above. As a general point the identification of flexibility effects as the cause of shock reflections implies that a very thick and hence rigid pipe will give negligible reflections. This suggests there must be an effect of thickness/diameter and that Swaffield's non-detection of it was due to lack of rigidity in pipe and valve mountings and perhaps an oversimplified interpretation of the multiple wave effects in terms of liquid shock waves alone.

5. CONCLUSIONS

A 'basic' theory for water hammer wave production and propagation in thin-walled pipe systems has been outlined for straight pipes, mitre bends and T-junctions. This involves stress waves in the pipe wall as well as pressure waves in the liquid. Comparison with experiment shows good agreement for straight pipes. Comparisons for bends are difficult because of the incomplete measurements taken and the need to interpret the many wave interactions which will occur. Reflection coefficient are of the correct sign and order of magnitude but further more detailed tests are needed to give adequate comparisons.

6. ACKNOWLEDGEMENT

This work is published with the permission of the Central Electricity Generating Board.

7. REFERENCES

1. Lax, P. D."Weak solutions of non linear hyperbolic equations and their numerical computation". Communications on Pure and Applied Mathematics, 7, 159-193. (1954)

2. Roache, P. J. "Computational Fluid Dynamics". Hermosa. (1972)

3. Swaffield, J. A. "A study of the influence of bends on fluid transients in pipelines". M. Phil. Thesis, City University. (1967)

4. Swaffield, J. A. "The influence of bends on fluid transients propagated in incompressible pipe flow". Proc. Inst. Mech. Engrs. 1968-69, 603. (1968)

5. Swaffield, J. A. Private communication. (1976)

6. Thomson, W. T. Vibration theory and application. Allen & Unwin. (1966)

7. Wilkinson, D. H. & Curtis, E. M. "Water hammer in a thin-walled pipe". 3rd Intnl. pressure surge conference, Canterbury, March 1980. (1980)

8. Wilkinson, D. H. "Acoustic and mechanical vibrations in liquid-filled pipework systems". Paper 8.5 BNES Vibration in Nuclear Plant Conf. Keswick, May 1978. (1978).

9. Wood, D. J. & Shen, P'Eng Chao "Effect of pipeline junctions on water hammer surges". Trans. Engineering J., Proc. Am. Soc. Civil Engrs., August 441. (1971).

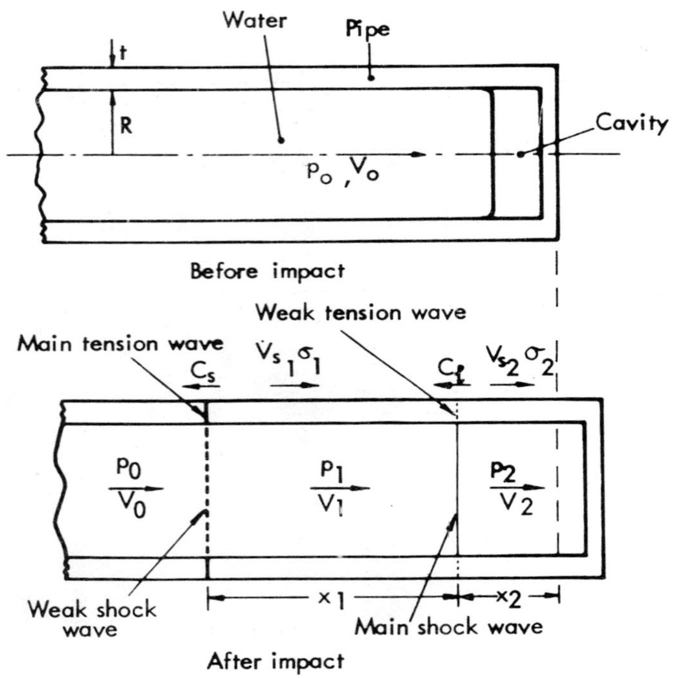

FIGURE 1. LIQUID COLUMN IMPACT WITH THE END OF A STRAIGHT PIPE.

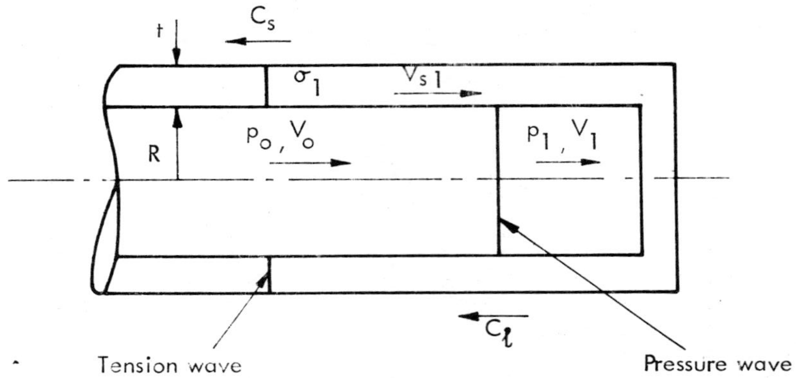

FIGURE 2. 'BASIC' MODEL FOR LIQUID COLUMN IMPACT

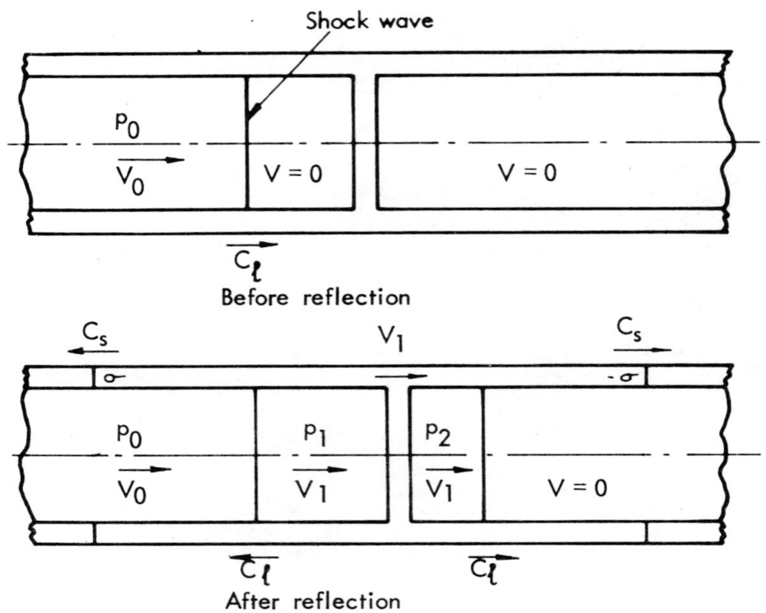

FIGURE 3. INCIDENT SHOCK WAVE CASE.

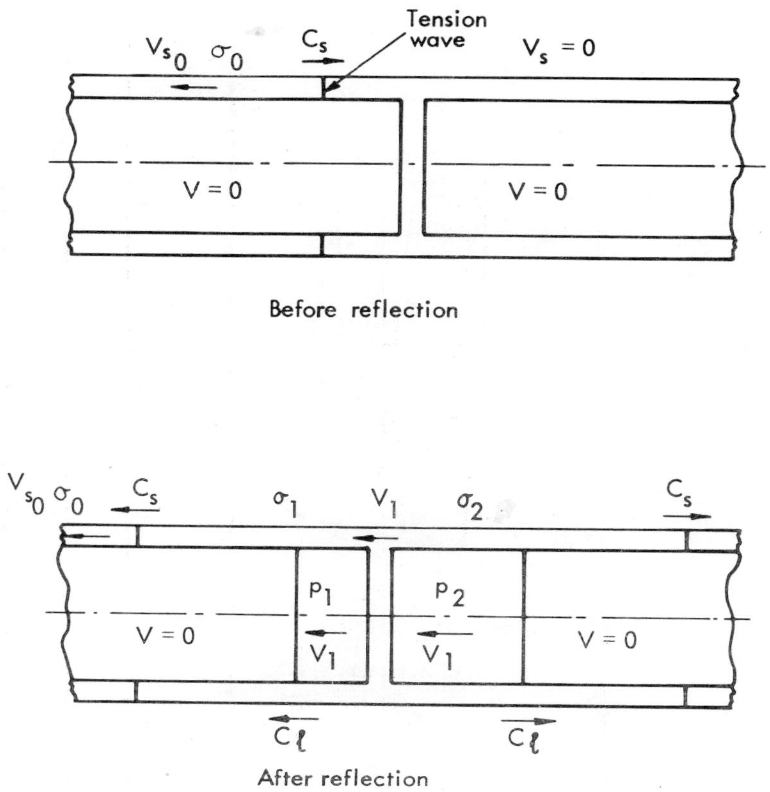

FIGURE 4. INCIDENT TENSION WAVE CASE.

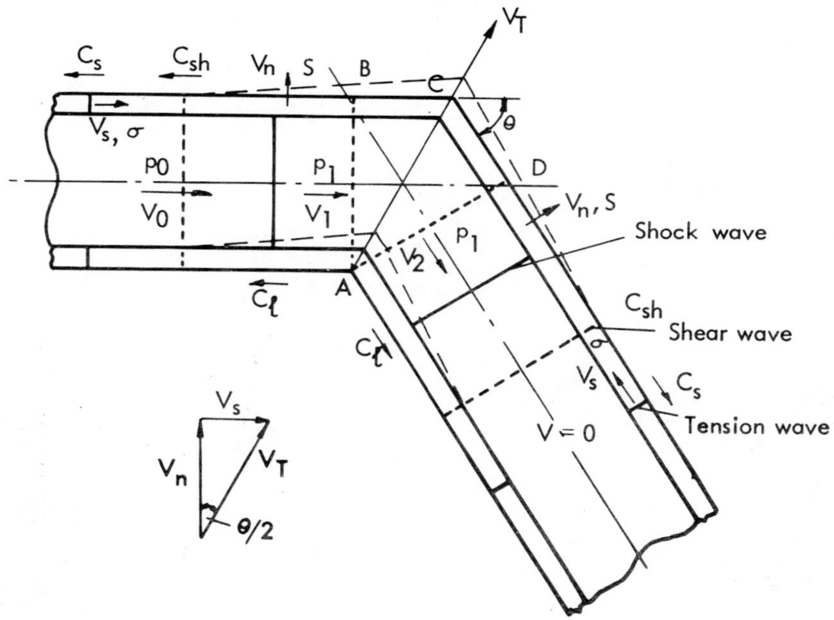

FIGURE 5. SHOCK WAVE INTERACTION WITH A BEND.

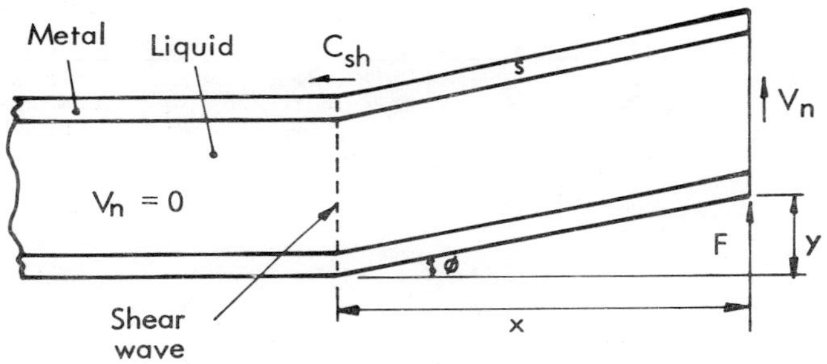

FIGURE 6. SIMPLE CONCEPT OF A SHEAR WAVE IN A LIQUID FILLED PIPE.

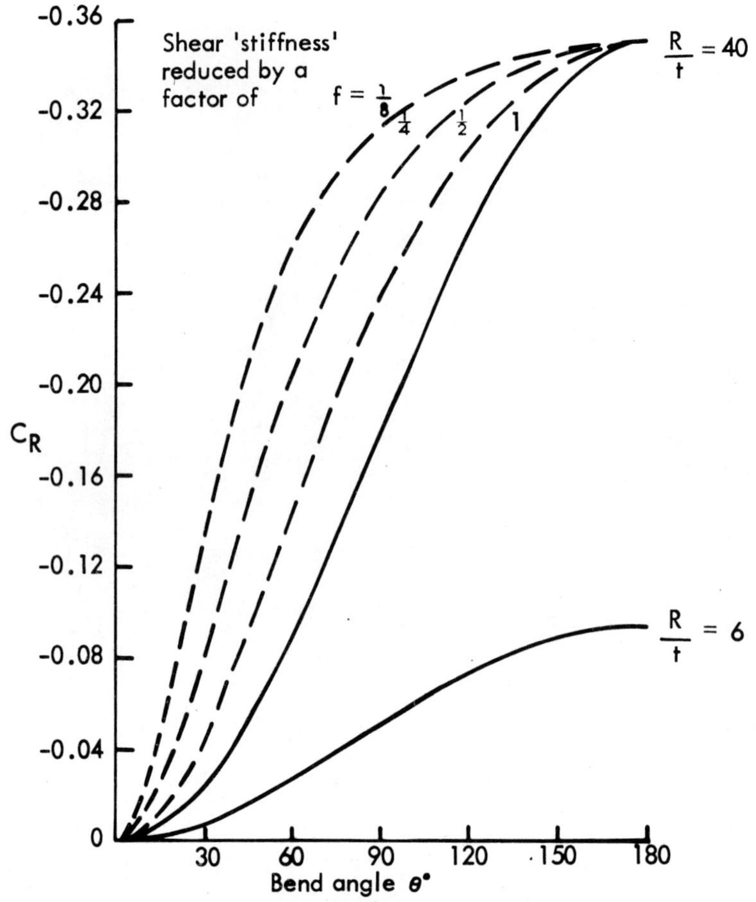

FIGURE 7. PREDICTED REFLECTION COEFFICIENT FOR A SHOCK WAVE AT A PIPE BEND.

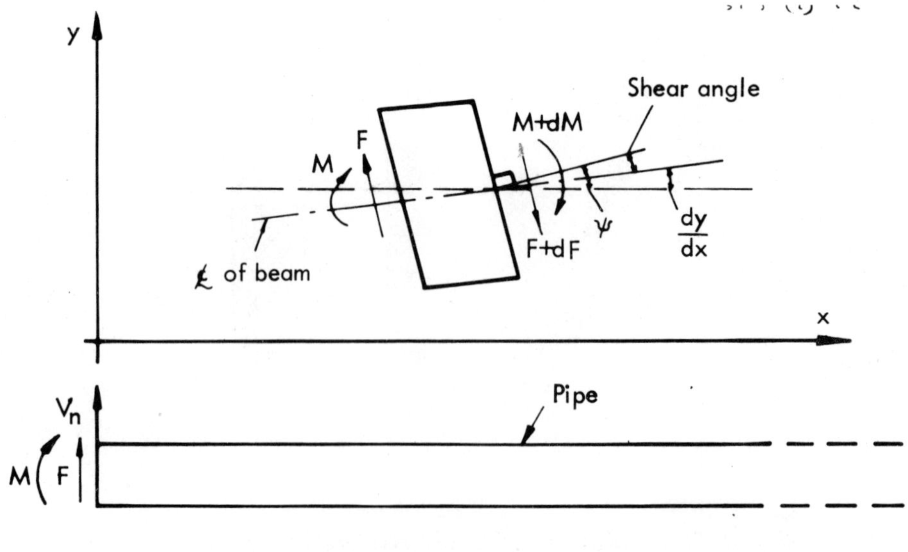

FIGURE 8. TRANSVERSE DEFLECTION OF A BEAM.

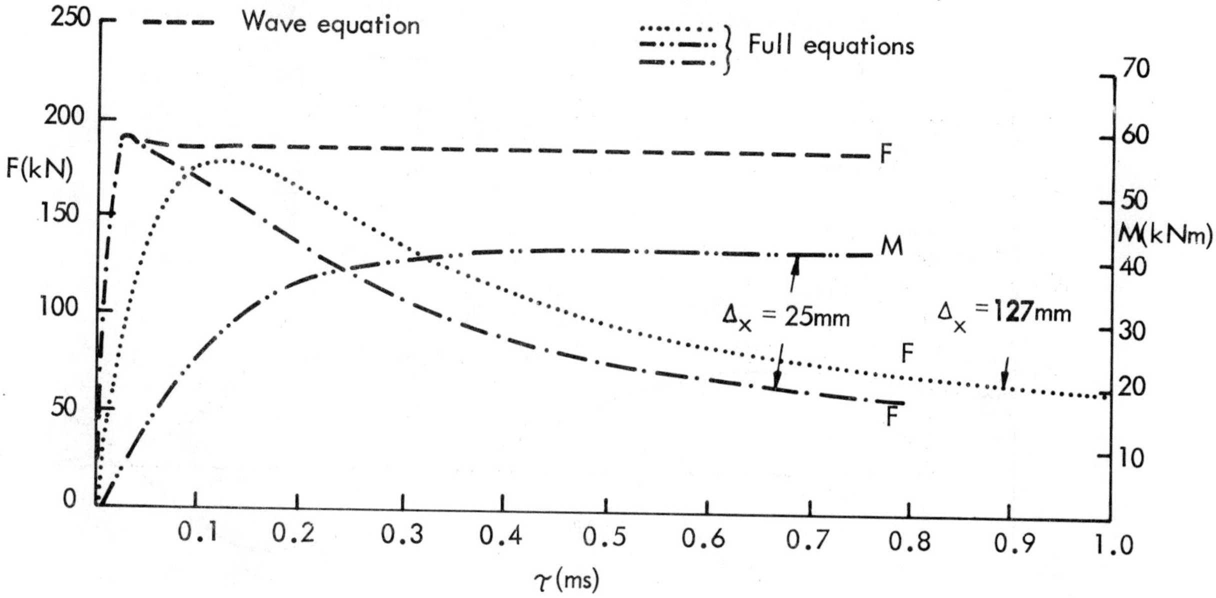

FIGURE 9. BEAM CASE, SHEAR FORCE AND BENDING MOMENT AT DRIVEN END

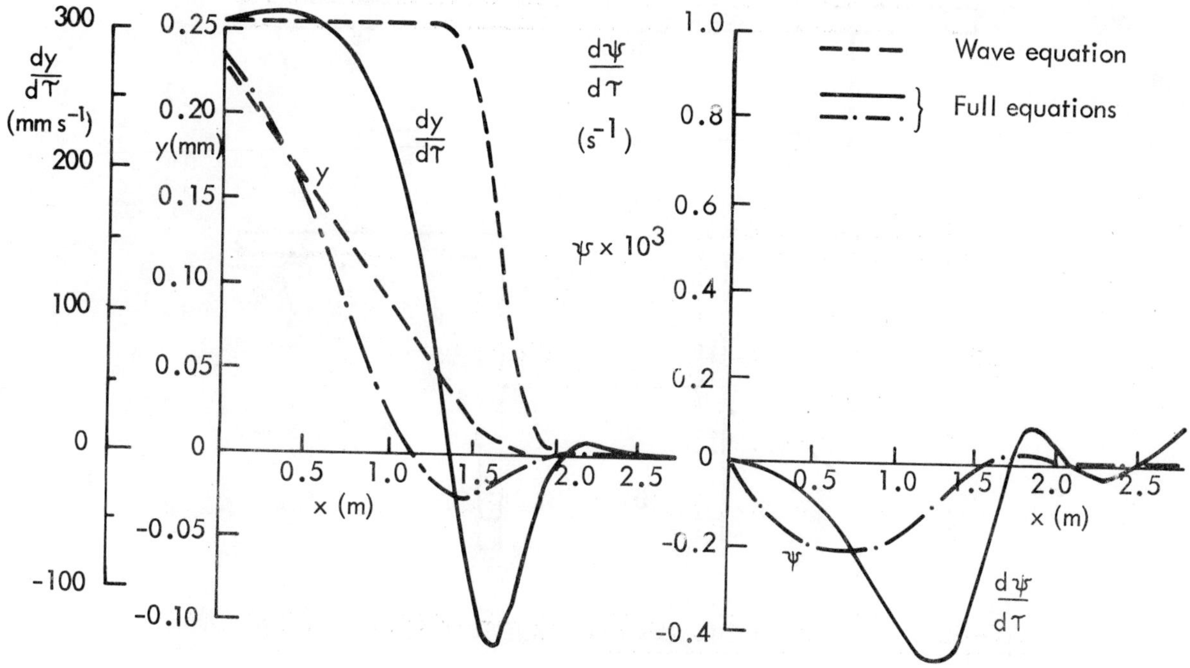

FIGURE 10. BEAM CASE, NUMERICAL SOLUTION AT τ = 0.744ms (180 time steps, Δx = 25mm)

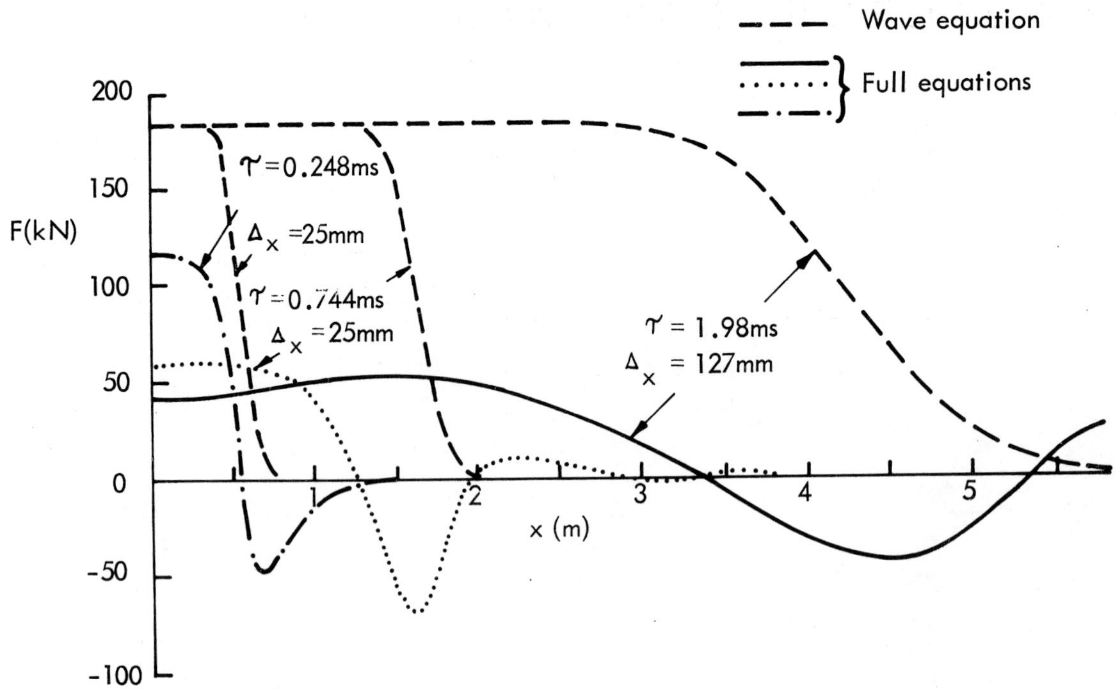

FIGURE 11. BEAM CASE , NUMERICAL SOLUTION FOR SHEAR FORCE DISTRIBUTION

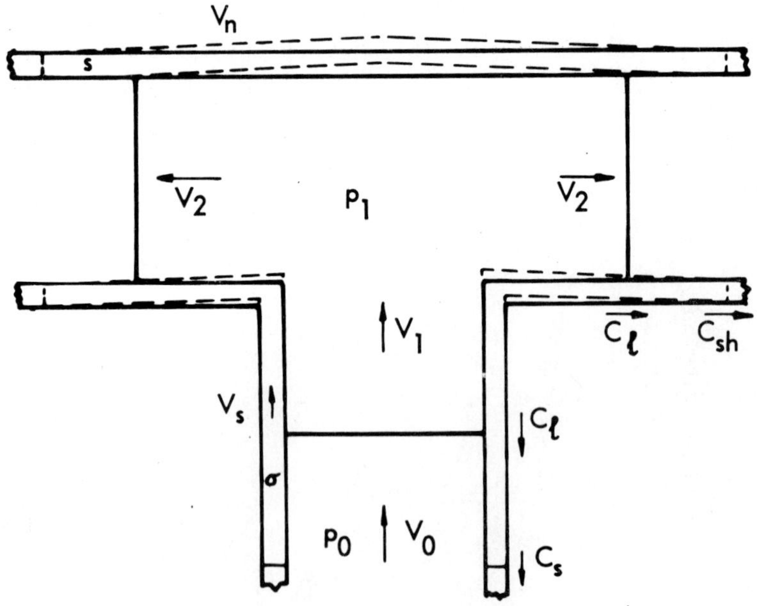

FIGURE 12 . INTERACTION OF AN INCIDENT SHOCK WAVE WITH A T - JUNCTION .

Third International Conference on

PRESSURE SURGES

Canterbury, England
March 25-27, 1980

PAPER E3

A STUDY OF PIPE-LIQUID INTERACTION FOLLOWING PUMP TRIP AND CHECK-VALVE CLOSURE IN A PUMPING STATION

J. Ellis

University of Strathclyde, U.K.

Summary

This paper arose as a consequence of the writer being asked to examine the nature and origins of high-frequency oscillations of liquid pressure in pipelines at a pumping station following pump trip. The study, conducted in two parts, took the form of numerical simulation of events. Firstly, simulation of the non-return valve closure was carried out to determine reasonable estimates of flow conditions within the pumping station at the moment of valve closure. Secondly, the analysis of transient behaviour following closure was considered. The frequency of observed vibrations could not be explained solely by water wave reflections alone. Movement of pipework and fittings was considered as a possible explanation. The liquid-solid behaviour was considered as a two phase system with coupling between the two phases at points such as valves, bends and branches. A numerical method was proposed using two families of characteristics for the liquid phase and two families for the solid phase. Preliminary results are presented using the method. Despite the introduction of a number of simplifying assumptions it is considered that the results of the study justify further development of the technique for analysis of liquid-solid interactions.

Held at the University of Kent at Canterbury, U.K.

Conference Organised by BHRA Fluid Engineering, Cranfield, Bedford, England.
In conjunction with The City University, London.

Nomenclature

a	-	speed of acoustic wave in liquid,
A	-	flow cross-section of pipe,
Ap	-	cross-sectional area of pipe wall,
c	-	speed of acoustic wave in pipe wall,
CA	-	constant in +ve characteristic equation of liquid,
CB	-	constant in -ve characteristic equation of liquid,
D	-	diameter of pipe or valve door,
e	-	thickness of valve door,
E	-	Young's Modulus,
F	-	liquid force and momentum at a cross-section,
g	-	acceleration due to gravity,
h	-	pipe stress head,
H	-	liquid pressure head,
I	-	moment of inertia of valve door about hinge,
K	-	spring stiffness,
l	-	distance of centroid of door from hinge,
M	-	lumped mass at a point,
o	-	suffix denoting conditions at start of a time increment,
Q	-	flow rate
SA	-	constant in +ve characteristic equation of pipe wall,
SB	-	constant in -ve characteristic equation of pipe wall,
t	-	time,
Δt	-	time increment,
v	-	velocity of flow,
x	-	displacement of pipe in X - direction,
x'	-	velocity of pipe in X - direction,
x''	-	acceleration of pipe in X - direction,
y	-	displacement of pipe in Y - direction,
y'	-	velocity of pipe in Y - direction,
y''	-	acceleration of pipe in Y - direction,
θ	-	angle of valve door to the horizontal,
θ'	-	angular velocity of valve door
θ''	-	angular acceleration of valve door,
ρ	-	density of fluid,
ρ_s	-	density of pipe material,
σ	-	pipe stress
S	-	gradient of energy loss due to resistance

Introduction

During tests following initial completion of an effluent pumping station it was noted that the non return valve sited downstream of the pumps closed with a loud bang. Subsequently a high-pressure wave of rapid frequency was recorded on an oscilloscope with a pressure sensor fitted in the pipeline. The maximum pressure head recorded during these high frequency fluctuations was about 160 metres. The maximum surge pressure head allowed by the pipe manufacturers was 40 metres and the design maximum working head was 31 metres. Investigation of the unsteady phenomena was conducted in two parts. Firstly, a study was made of events between the time of pump trip and the time of closure of the non-return valve. This was considered necessary since no information was available on the conditions of flow in the pipework of the station at the instant of valve closure and such data formed an essential part of the information necessary to study the high-frequency oscillations.

Description of the Pumping Station and Pipeline

Each of two parallel Glass Reinforced Polyester (G.R.P.) mains of diameter 0.54 metres and some 4 kilometres in length was designed to carry 0.42 metres ^{3}/second of effluent. The mains were laid over fairly flat ground and terminated in a diffuser on the bed of an estuary. The effluent is supplied to each main via a short length of steel pipeline of diameter 0.35 metres, from one of a pair of pumps mounted in parallel. Since the G.R.P. mains were designed for a minimum pressure head of not less than -0.33 metres, surge suppression equipment in the form of air vessels was provided close to the pumping station. Fig. 1 depicts the arrangement of pipework and surge vessel in the vicinity of the pumping station and fig. 2 shows the layout of the pumping station itself.

Characteristic Equations

A fixed-mesh numerical technique based upon the theory of characteristics was employed for both parts of the study, fig. 3. The equations which were solved within the x-t plane of fig. 3 were;

along the positive path, $dx/dt = v + a$,

$$v_3 + {}^g/a.\ H_3 = v_1 + {}^g/a.\ H_1 - g\int_1^3 S.dt \qquad - \ 1,$$

while along the negative path, $dx/dt = v-a$,

$$v_3 - {}^g/a.\ H_3 = v_2 - {}^g/a.\ H_2 - g\int_2^3 S.dt \qquad - \ 2.$$

Interpolation between adjacent sections is used to dafine conditions at 1 & 2. Denoting the right-hand sides of equations 1 and 2 by CA and CB respectively then,

$$v_3 = (CA + CB)/2 \qquad - \qquad 3$$

and $H_3 = (CA - CB)/(2.{}^g/a) \qquad - \qquad 4.$

Study of the Check-valve closure

The 0.35 metre diameter non-return valves originally fitted to the line consisted of a single door hinged at the top without any recoil device or spring loading, fig. 4. Following pump trip the flow in the line between the sump and the branch to the air vessel will decelerate and the valve will close under the out of balance action of gravity, buoyancy and hydrodynamic forces. Accordingly the valve closure rate is determined partially by the flow conditions themselves. The rate of closure could be calculated given the flow, head loss relationship for the valve as a function of gate opening and the corresponding pressure distribution over each face of the valve as it closes. With such information then the out of balance of gravity, buoyancy and pressure moments about the hinge may be equated to the inertial torque. At the time of carrying out the study no data on pressure distribution or head-discharge relationship were available. In order that some estimate could be made of valve closure time the following assumptions were made,

(i) a uniform pressure was assumed over each face of the valve at any time and

(ii) the head drop through the valve was equal to, $\{\text{flow through the valve}/(A.(1 - \sin\theta))\}^2/(2.g)$

Then with reference to the definition sketch of fig. 5, and taking moments about the fulcrum, (clockwise + ve),

$$\text{gravity and buoyancy moments} = \frac{-\pi}{4} D.^2 e.1(\rho_s - \rho).g. \cos\theta ,$$

$$\text{moment due to differential pressure} = \frac{\pi}{4} D.^2 1.\rho.g.(Hu/_s - Hd/_s),$$

$$\text{inertial torque} = I.\theta''$$

Summing these moments then,

$$I.\theta'' + \frac{\pi}{4} D.^2 1.g.(e(\rho_s - \rho).-\cos\theta + \rho(Hu/s - Hd/s)) = 0 \qquad - \quad 5.$$

Discharge in the pipeline at the valve is comprised of two components,

(i) flow past the valve door

(ii) flow as a result of changing valve door angle.

If the angular velocity of the gate is θ' then the volume flux induced by gate movement is equal to $\frac{\pi}{4} D.^2\theta'.1.$

$$\text{Therefore } v \times A = \frac{\pi}{4} D.^2\theta'.1 + (Hu/_s - H\,d/s).(1 - \sin\theta).\frac{\pi}{4}D^2.\sqrt{(2g)} \qquad - \quad 6.$$

Introducing the characteristic equations upstream and downstream of the valve then,

$$H\,u/_s = (CA - v)/\frac{g}{a} \qquad - \quad 7$$

$$\text{and} -H\,d/_s = (CB - v)/\frac{g}{a} \qquad - \quad 8$$

The set of equations 5 to 8 was solved for the unknown valve position θ, v, $H\,u/_s$ and $H\,d/_s$.

The arrangement of sections for the numerical model is shown in fig. 6. The time increment used was 0.004 seconds. Computations at the pump were carried out using a technique similar to that described by Streeter and Wylie (Ref.1). Air vessel calculations assumed isentropic behaviour of the gas volume. The boundary condition at the node representing the beginning of the 4 kilometre long main was that no reflection returned from the main during the time period of interest.

Results of Check Valve Closure

Typical results of computations showed a considerable reversed flow at the time of closure. Time histories of velocity at the valve are shown in fig. 7 for different weights of gate. It was noted from these results that the addition of mass to the valve in the form of weights did provide an additional closing moment due to gravity but the moment of inertia was also increased resulting in little change of significance in the time of closure. Bearing in mind the limitations of the simulation because of the assumptions involved it is evident that a considerable reversed flow developed prior to the time of closure.

From this initial study a reversed velocity of about 1.0 metre/second was taken to exist in the pipeline at the moment of valve closure. Also from fig. 7 it is evident that the velocity of the valve door was close to the velocity of flow in the pipe immediately prior to closure.

Study of Behaviour following Closure of the Check Valve

The high frequency pressure fluctuations which were recorded in the pipeline downstream of the non-return valve subsequent to its closure were at a frequency much higher than would be anticipated from hydraulic pressure wave effects alone. The dimensions of pipework, fig. 2, suggests that water waves would have frequencies much less than the 200 Hz. recorded. Hence some longitudinal movement of the pipeline itself was considered as a possibility for inducing the observed frequencies.

Recent years have seen an increased amount of attention devoted to study of the interactions, under unsteady conditions, between fluids flowing in closed conduits and the conduits themselves. Calculations designed to predict the behaviour of pressure transients have generally considered the liquid and conduit as a composite system and an acoustic wave velocity is introduced to the calculations which comprises the elastic properties of both liquid and pipeline media. More detailed mathematical analysis of the behaviour of such a system by Thorley (Ref.2) predicted the existence of a wave travelling along the pipeline and a wave travelling through the liquid at speeds close to the individual acoustic speeds as evaluated independently from properties of pipeline and fluid respectively, and the occurrence of these separate waves was also confirmed by experiment. Further, this study showed the significance of the Poisson effect in that changes in axial strain in the pipe wall produced changes in the cross-sectional area of the pipeline and corresponding changes within the liquid. Thus a wave travelling axially through the pipe wall continually creates flow and pressure changes within the liquid giving the appearance of a wave effect in the liquid travelling at a rate close to that of the wave in the pipe wall. The study by Williams (Ref. 3) confirmed the importance of this form of coupling between the two phases in producing dispersion and damping of the wave motion particularly in the case of softer pipe materials.

A second form of coupling between the liquid and its surroundings can arise where there is a component of boundary pressure directed along the axis of the line. Such components are present at valves, bends, tees, changes of cross-section and the like. Where rigid anchorages are provided at such locations then any change in pressure within the liquid will be transmitted to the anchorage, but if some movement can take place at the anchorage then pressures within the liquid may require to be partially resisted by axial stresses in the conduit walls. This response of the pipeline and its supports has been shown to give rise to vibrations. Such vibrations have been examined both theoretically and in the laboratory by Wood (Ref. 4), Jones and Wood (Ref. 5) and Wood (Ref. 6) for a wide range of structural parameters and it has been demonstrated, by representing a pipe and its fittings as a spring-mass system, that the effects of movement in the line due to this form of coupling between fluid and pipeline can greatly alter the pressure surge from that which would be predicted neglecting such movement.

The present paper considers the second form of coupling between liquid and pipeline, but acknowledges that both forms will always be present in some degree. Interestingly in the results presented by Wood (Ref. 4) the agreement achieved between experiment and calculation suggests that the dissipative effect due to coupling through the Poisson effect may not be very important for strong pipeline materials during the first few cycles of vibration.

In these studies representing a pipeline as a single degree of freedom spring-mass system it was concluded that vibrations in more complex prototype installations would be difficult to analyse. The writer has therefore taken the opportunity to employ an alternative method which should be capable of extension to more complex systems. A set of equations may be written for the pipeline which are analogous to the characteristic equations introduced for the liquid behaviour. Normally employed for such solid body calculations as pile driving (Ref. 7), forging operations (Ref. 8) or other structural calculations, these may be written in terms of axial stress and displacement,

$$x' \pm \sigma/(c\,\rho_s) \quad = \quad \text{Constant} \qquad - \qquad 9$$

where $c = \sqrt{(E/\rho_s)}$.

While this value of 'c' is strictly accurate only for a Poisson's ratio of zero it is a close approximation in other circumstances, provided that the wavelength involved is large compared with lateral dimensions. In terms of an equivalent pressure head of liquid then,

$$x' \pm \rho.g.h/(c\rho_s) \quad = \quad \text{Constant}$$

writing $c' = c\rho_s/\rho$

$$\text{then } x' + {}^{g}/_{c'}.\,h = SA \qquad - \qquad 10$$

along the path $^{dx}/dt = x' + c$

$$\text{and } x' - g/c'.\,h = SB \qquad - \qquad 11$$

along the path $dx/dt = x' - c$.

Because of the similarity of equations 1 and 2, and 10 and 11, the same fixed mesh technique was employed for both phases of the system. The time increment is the same for both liquid and pipewall calculations but the spacing of sections representing the pipe wall required to be $^{c}/a$ x the spacing of the liquid phase sections, giving a computational arrangement as shown in fig. 8.

The following assumptions were made in the analysis, (i) minor losses at bends and branches were ignored. While acknowledging the desirability of including the results of studies such as those of Contractor (Ref. 9) and Swaffield (Ref. 10) on the losses and partial reflections at bends and fittings, especially where a number of such occur in series, it was felt that a preliminary analysis such as this, designed to establish the feasibility of the method, could reasonably be conducted neglecting these effects,

(ii) changes of liquid momentum at points of coupling were ignored. The total force and momentum of the liquid at a cross-section may be written,

$$F = \rho.g.\,H.A + \rho.Q.v. \qquad - \qquad 12$$

$$\text{or} \quad \frac{F}{\rho g A} = H + v^2/g.$$

Since the pressure head changes being considered are of the order a.v/g then,

$$\frac{F}{\rho g A} \simeq \frac{av}{g} + \frac{v^2}{g} \simeq \frac{av}{g} \qquad - \qquad 13$$

$\therefore$ it seems reasonable to neglect the momentum component in a first analysis,

(iii) at changes of direction within the pipeline there is a possibility of bending stresses being set up depending upon the method of pipe jointing. This effect has been ignored, that is, a hinge has been assumed to operate at the change of direction.

(iv) friction effects were neglected as was the head loss through the pump in this part of the study.

The following points of coupling between liquid and solid phases were considered,

(a) the check valve,

With reference to the definition sketch of fig. 9, characteristic equations for the liquid,

$$v + g/a.\,H_{u/s} = CA \qquad - \qquad 14$$

$$v - g/a.\ H\ d/s = CB \qquad - \qquad 15$$

characteristic equations for the pipe wall,

$$x' + g/c'.\ h\ u/s = SA \qquad - \qquad 16$$

$$x' - g/c'.\ h\ d/s = SB \qquad - \qquad 17$$

assuming fluid remains in contact with the valve face at all times then,

$$v = x' \qquad - \qquad 18$$

equating the forces acting,

$$\rho.g.A(H\ u/s - H\ d/s) + \rho A_P g(h\ u/s - h\ d/s) - M.x'' - K.x = 0 \qquad - \qquad 19$$

Equations 14 to 19 were solved simultaneously for each time increment. Two modified solutions were also included in the analysis, these catered for the possibility of the valve re-opening and the occurrence of a cavity at vapour pressure on the low pressure side of the valve following closure.

(b) the branch.

Referring to the definition sketch of fig. 10, then the following equations were solved,

characteristic equations for the liquid,

$$v_1 + g/a.H = C1, \qquad - \qquad 20$$

$$v_2 + g/A.H = C2, \qquad - \qquad 21$$

and $$v_3 - g/a.H = C3, \qquad - \qquad 22$$

characteristic equation for the pipewall,

$$x' + g/c'.h = S1 \qquad - \qquad 23$$

conservation of mass at the junction,

$$v_1 + v_2 - v_3 = x' \qquad - \qquad 24$$

equating the forces operating in the direction 1, then,

$$\rho.g.(A.H + A_p.h) - M.x'' - k.x = 0 \qquad - \qquad 25.$$

Equations 20 to 25 were solved simultaneously for each time movement.

(c) 90° bend.

From the definition sketch of fig. 11, then, characteristic equations for the liquid,

$$v_1 + g/a.\ H = CA \qquad - \qquad 26$$

and $$v_2 - g/a.\ H = CB \qquad - \qquad 27$$

characteristic equation for the pipe/wall,

$$x' + g/c'.h_1 = SA \qquad - \qquad 28$$

and $$y' - g/c'.h_2 = SB \qquad - \qquad 29$$

force balance in X - direction, gives,

$$\rho.g(A.H + A_p.h_1) - M_x x'' - K_x.x = 0 \qquad - \qquad 30$$

and balancing forces in the Y - direction gives,

$$\rho . g . (A.H + A_p h_2) - M_y . y'' - K_y . y = 0 \qquad - \qquad 31$$

conservation of volume gives,

$$v_1 - v_2 = x' - y' \qquad - \qquad 32.$$

Equations 26 to 32 were solved simultaneously at the bend for each increment of time.

Model Arrangement

The arrangement of sections for this model was as shown in fig. 12. Fifty-one sections were used to define liquid behaviour and thirteen sections to define solid phase conditions. Points of coupling between solid and liquid phases were at the two check valves, the 90^0 bend and the 'tee' branch. At liquid section no. 51 the boundary condition was that no reflection occurred at this point, and at liquid sections no.'s 1 and 30 a constant sump level was assumed. For the solid phase the boundary condition at sections no. 1 and 9 a mass and spring constant were applied. The time increment used in this model was 0.00028 seconds and liquid sections were spaced 0.30 metres apart while solid phase sections were 1.5 metres apart. Wave velocities chosen for the liquid and solid phases were respectively 1050 metres/second and 5250.0 metres/second. Other data used in preliminary tests using the model were; pipe wall cross-sectional area = 0.014 metres^2,
velocity of flow in line = -1.0 metres/second and
initial head in the system = sump level = 0.0 metres.

Preliminary Results using Model

At the present time only some preliminary results have been obtained from the model. These initial calculations have been undertaken using nominal valves of such parameters as wavespeeds and pipewall area, and purely arbitrary valves of spring stiffnesses. The influence of parameters related to structural behaviour, such as the masses of valves, pumps and other fittings has not been explored to any extent at this stage.

The results displayed in fig. 13 indicates the effect of the bend rigidly fixed for closure of the valve adjoining pump no. 1. Fig. 14 shows results following closure of the same valve but with the 90^0 bend unrestrained. In fig. 15, for a test on closure of the check valve at pump no. 2, the bend is restrained by springs of stiffness = 10^4 kN/metre^2 in both directions. Also considering this valve, fig. 16 displays results for the bend rigidly fixed but with the "tee" branch restrained by a spring of equal stiffness.

In general the results obtained from these initial calculations show an amount of damping greater than that observed. Comparing results at the 90^0 bend from tests on the valve at pump 1, then it is seen from figs. 13 and 14 that greater amplitude of vibration occurs when the bend is unrestrained. When the bend is free of restraint then following closure of the non-return valve, a tensile wave will be set up in the pipe wall between the valve and the bend. On arrival at the bend this wave will tend to pull the bend in the direction of the valve. This tendency will be resisted by a liquid pressure force within the conduit at the bend and a compression wave is established within the liquid at this point through this coupling action. This compression wave will also influence the 'Y' direction, fig. 10 and set up a tensile wave effect in the pipe wall between the bend and the "tee" branch.

An interesting aspect of these initial computations is that while only the trip of one pump is involved, the liquid pressure wave effects propagating into the branch leading to the second pump sets up vibrations in that line and it appears both from the observed results and from the computations that this effect is of importance in prolonging the time of vibrations. In some instances the computed peak values at the valve adjoining the stationary pump are higher than those at the valve which initiated these transient effects.

Conclusions

The present study has shown that calculations involving the interaction of a number of factors controlling vibration can be carried out by considering the pipe-liquid combination as a two-component system and employing the appropriate characteristic equations for each phase. The method should be capable of further refinement to include the influence of the other form of coupling mentioned earlier.

Many aspects of the behaviour of the pipeline system remain to be explored. In particular information relating to the response of different types of support is lacking. Also the possibility of differential movement of one length of pipe relative to another requires investigation. For example, at the joint between the 90° bend and an adjacent pipe length there may be some differential movement. Further study will be concentrated on examining the nature of the interaction between adjacent elements of the pipeline themselves.

REFERENCES

1) STREETER, V.L., and WYLIE, E.B.: "Hydraulic Transients", McGraw Hill Book Co., New York 1967.

2) THORLEY, A.R.D.: "Pressure Transients in Hydraulic Pipelines", Journal of Basic Engineering, Trans. A.S.M.E. pp 453-461, (September 1969).

3) WILLIAMS, D.J.: "Waterhammer in non-rigid Pipes, Precursor Waves and Mechanical Damping". Journal of Mechanical Engineering Science. Institution of Mech. Engrs. Vol. 19 No. 6 pp 237-242 (1977).

4) WOOD, D.J.: "A Study of the Response of Coupled Liquid Flow -- Structural Systems Subjected to Periodic Disturbances". Journal of Basic Engineering, Trans. A.S.M.E., pp 532-540 (Dec. 1968).

5) JONES, S.E., and WOOD, D.J.: "The Effect of Axial Boundary Motion on Pressure Surge Calculations". Journal of Basic Engineering, Trans. A.S.M.E. pp 441-446, (June 1972).

6) WOOD, D.J.: "Influence of Line Motion on Waterhammer Surge". Journal of the Hydraulics Division, Proc. A.S.C.E. HY3 pp 941-959, (May 1969).

7) CLOUGH, R.W. and PENZIEN, J.: "Dynamics of Structures". McGraw-Hill Book Company New York, Chapter 22 (1975).

8) COLE, B.N.: "Impact, Stress Waves and Rebound". International Journal of Mechanical Engineering Education, Vol. 7, No. 2, pp 91-97 (April 1979).

9) CONTRACTOR, D.N.: "The Reflection of Waterhammer Pressure Waves from Minor Losses". Journal of Basic Engineering, Trans. A.S.M.E. pp 445-452, (June 1965).

10) SWAFFIELD, J.A.: "The Influence of Bends on Fluid Transients propagated in Incompressible Pipe Flow". Proceedings, Institution of Mechanical Engineers, 183 pt. 1 No. 29 pp 603-614 (1968-69).

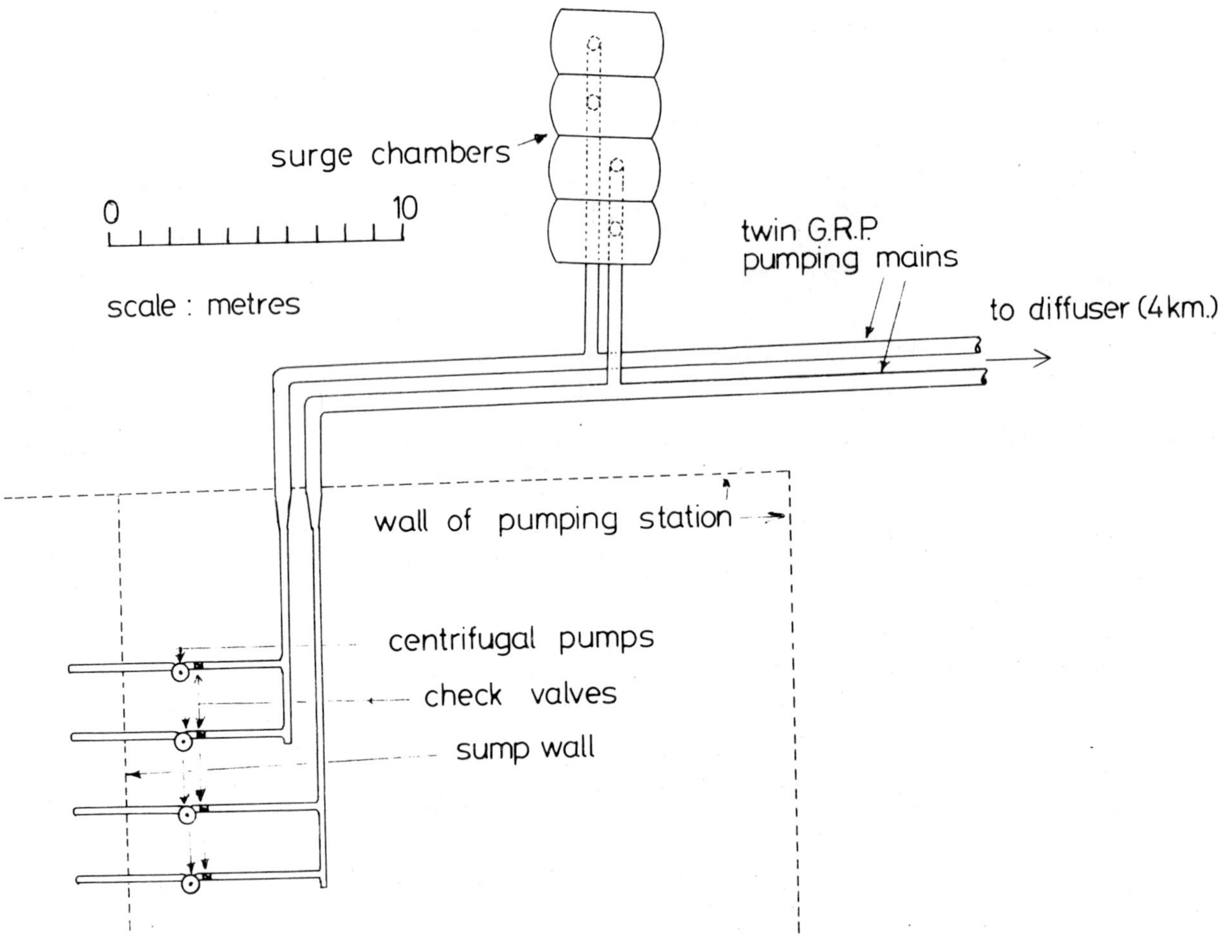

Fig.1 layout of pipelines near pumping station

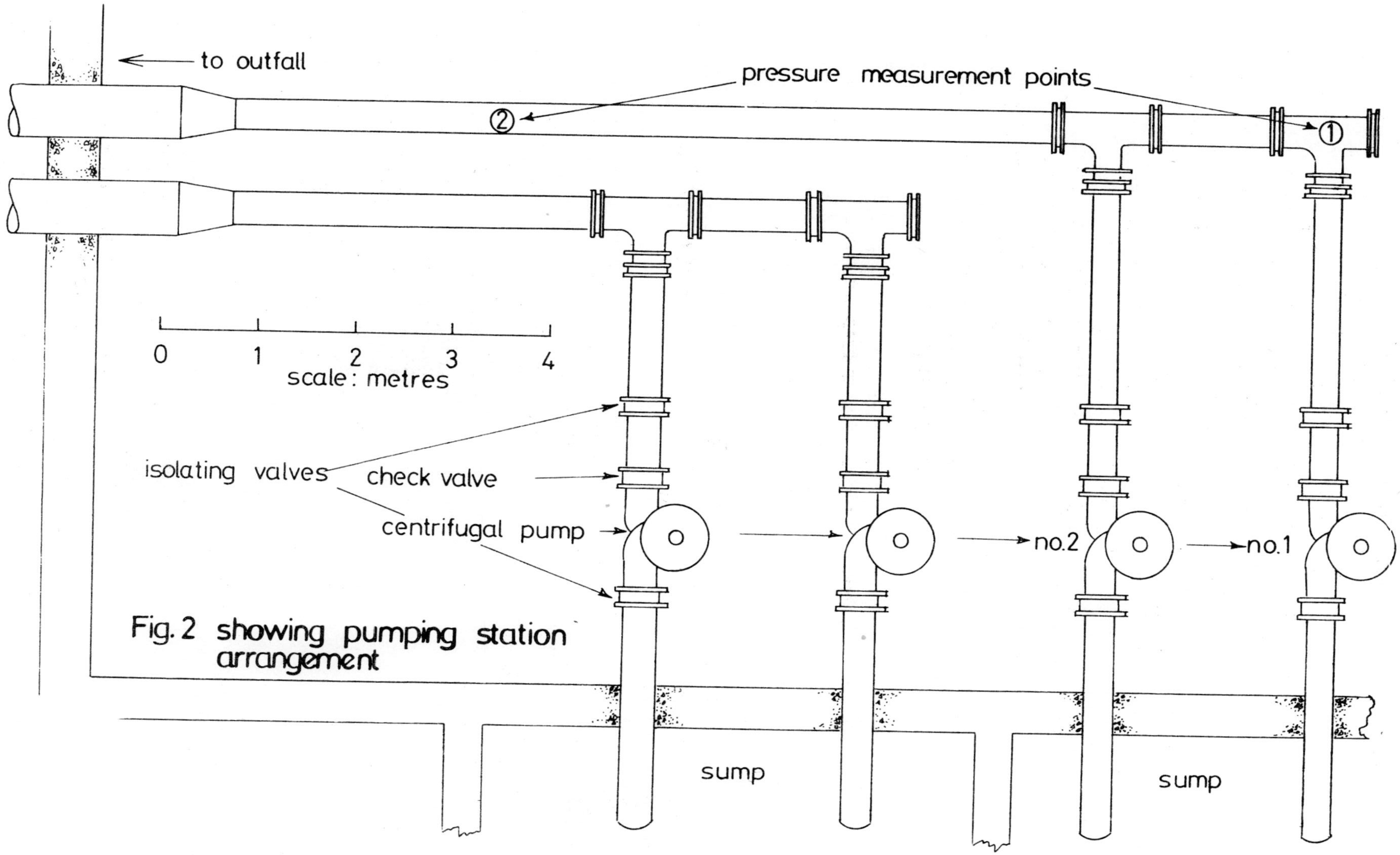

Fig. 2 showing pumping station arrangement

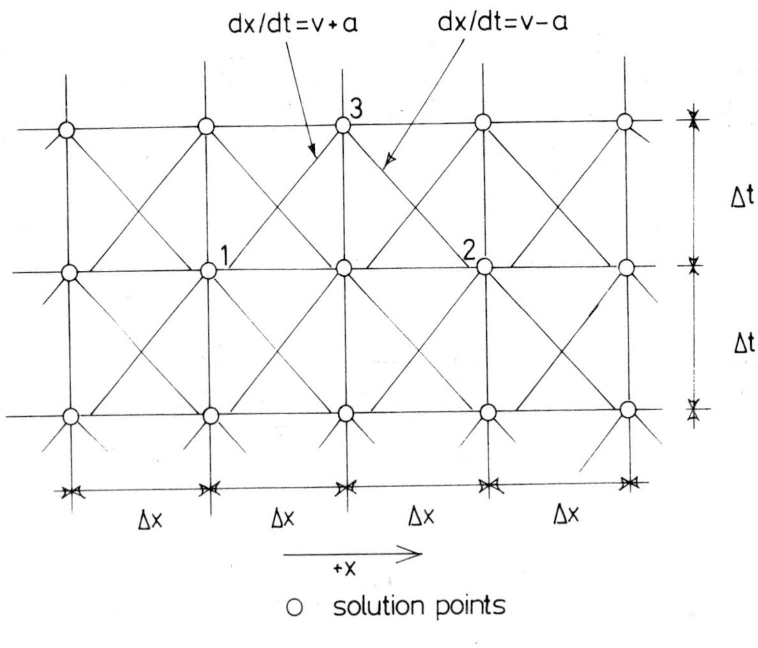

Fig.3 showing numerical integration scheme.

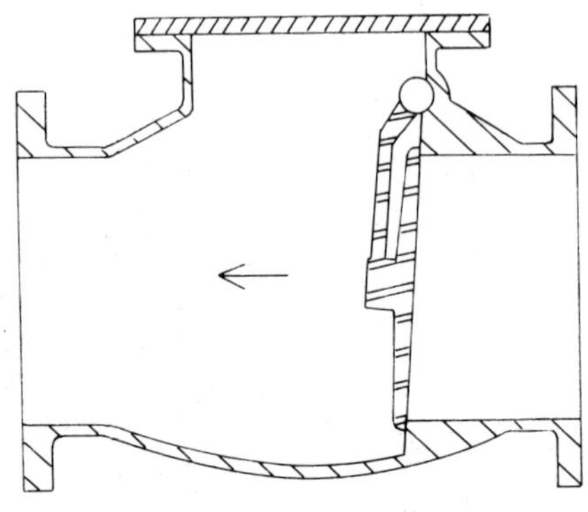

Fig.4 sketch depicting longitudinal section of check valve

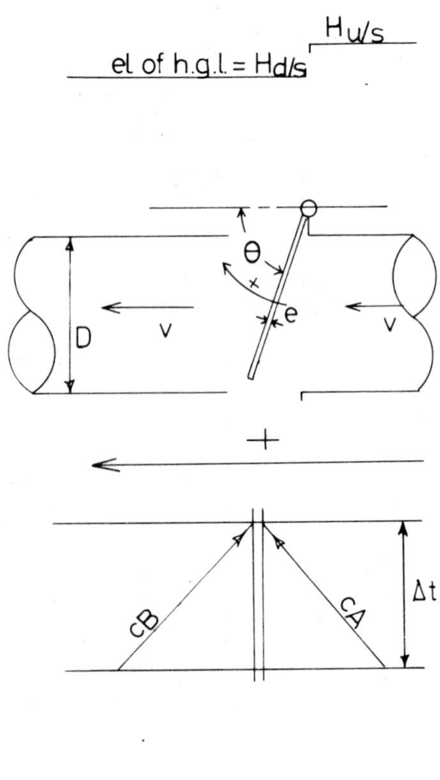

Fig. 5 definition sketch for check valve closure.

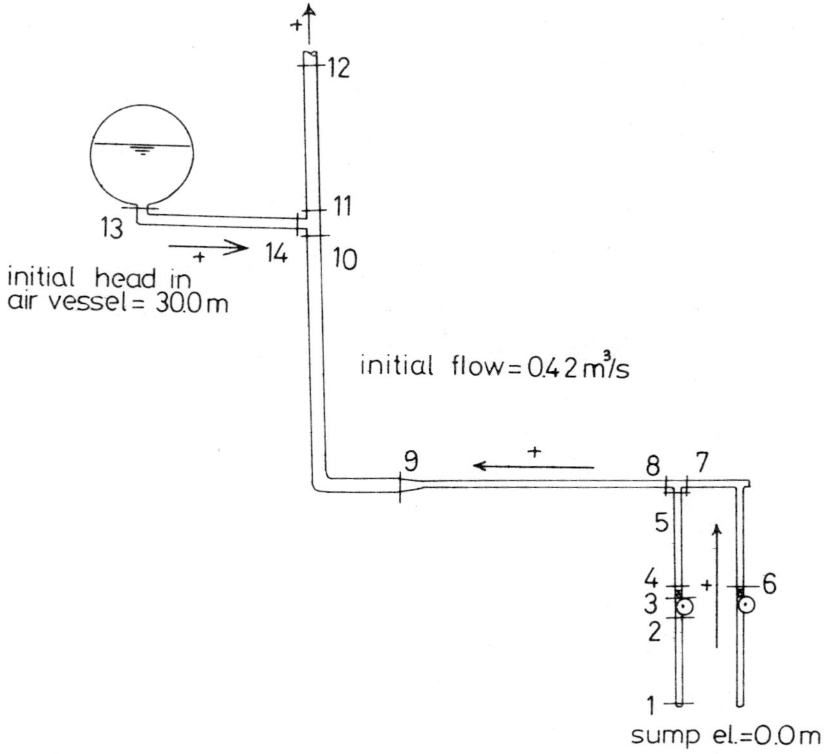

Fig. 6 arrangement of model sections for study of check valve closure at pump no. 2.

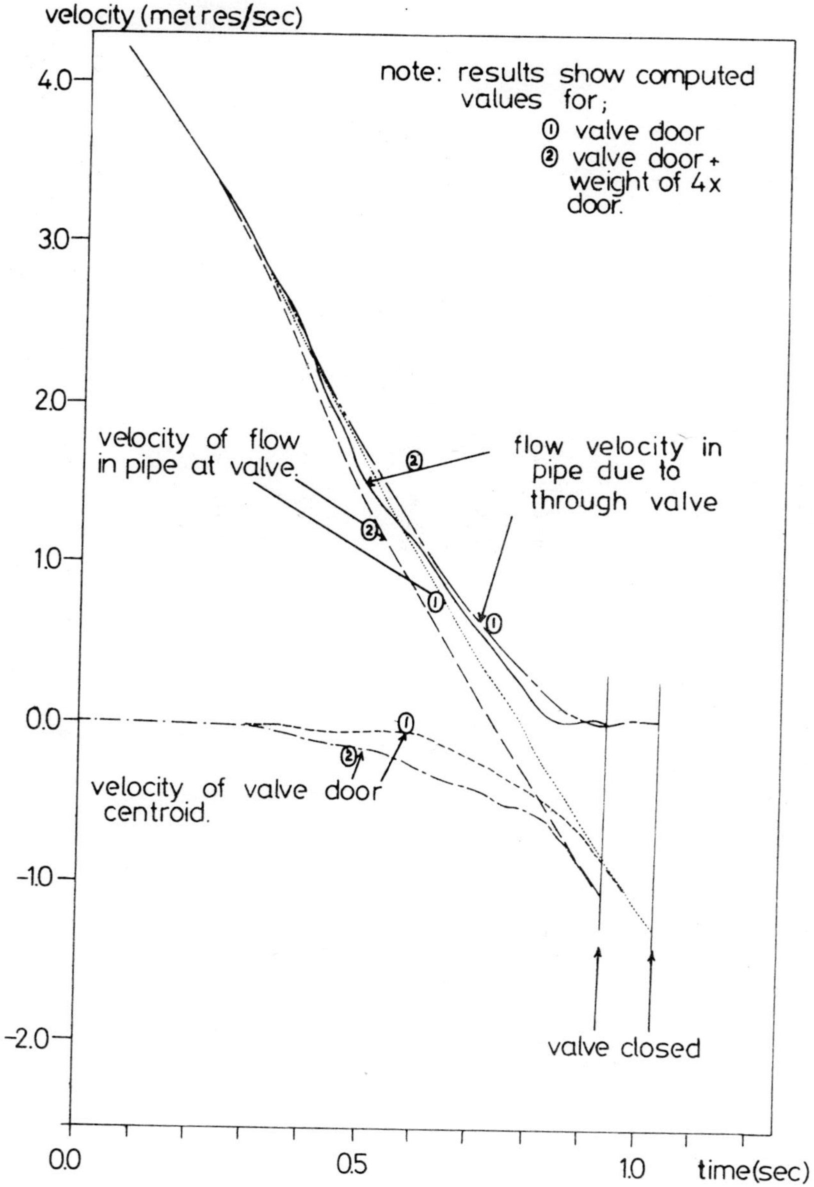

Fig.7 showing computed results of valve closure.

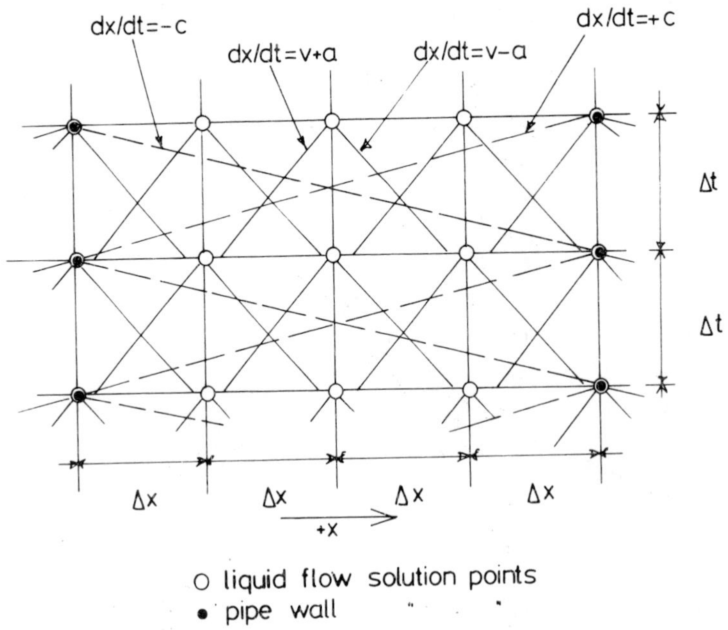

Fig. 8 showing two-phase integration scheme.

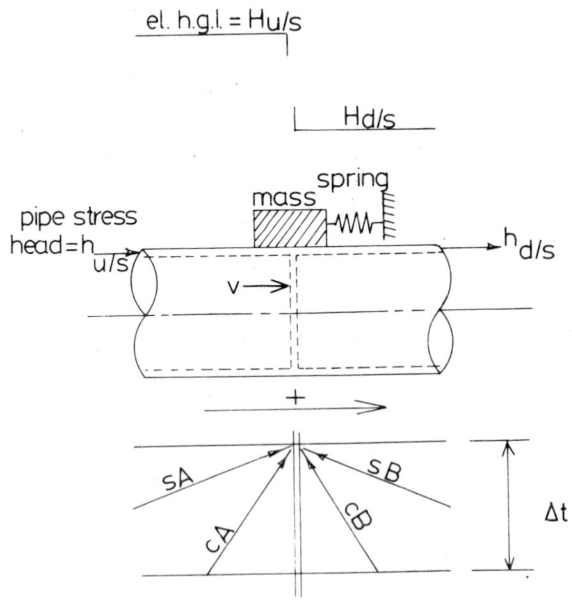

Fig.9 sketch for check valve analysis.

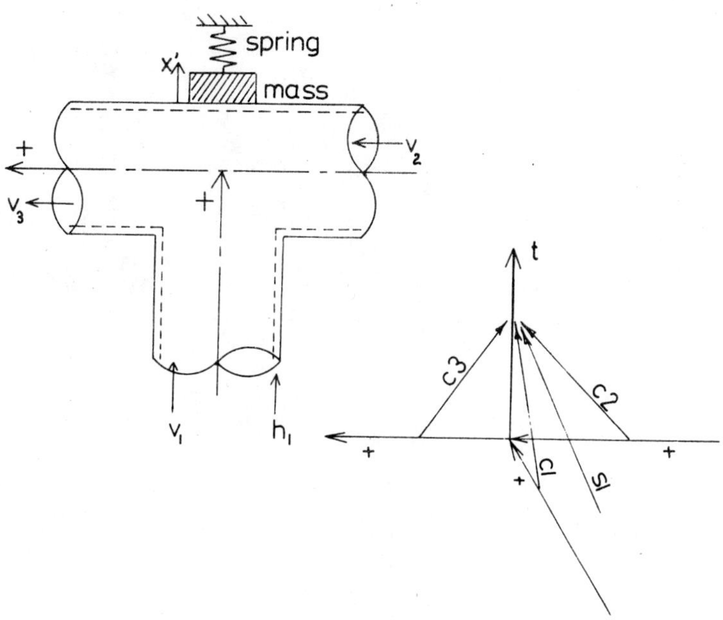

Fig. 10 sketch for analysis of "tee" branch.

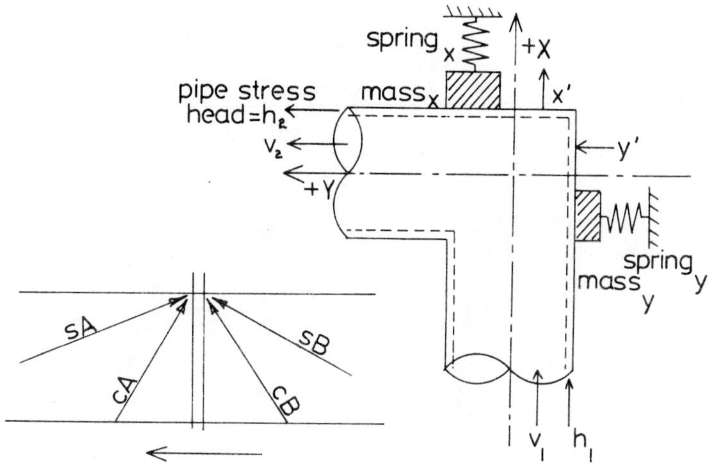

Fig. 11 sketch for analysis of 90° bend.

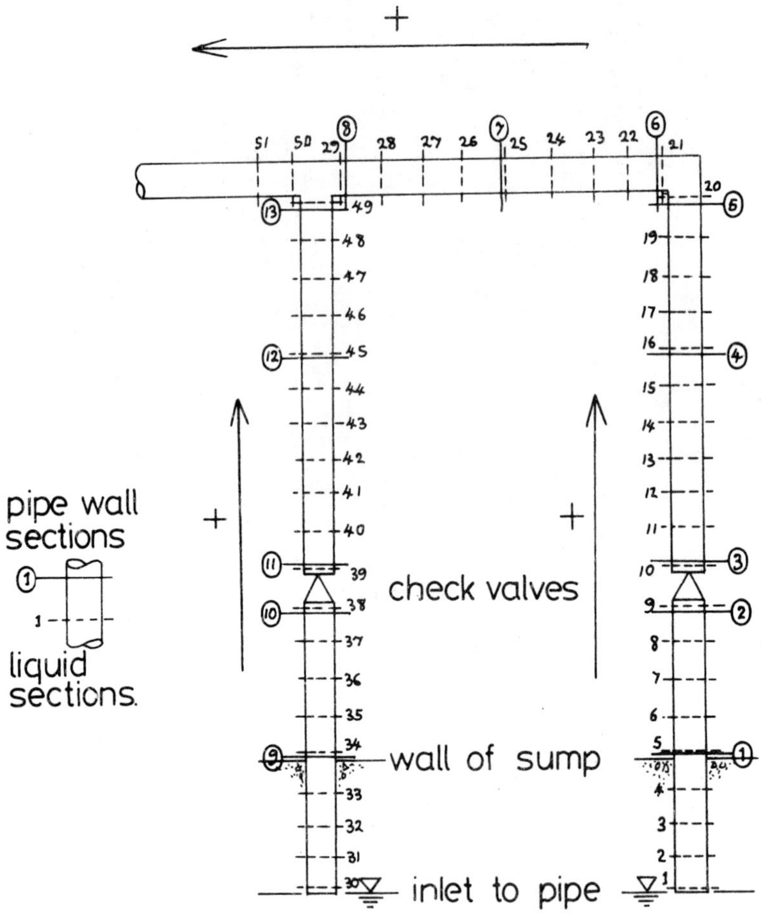

Fig.12 section arrangement for study of high frequency oscillations.

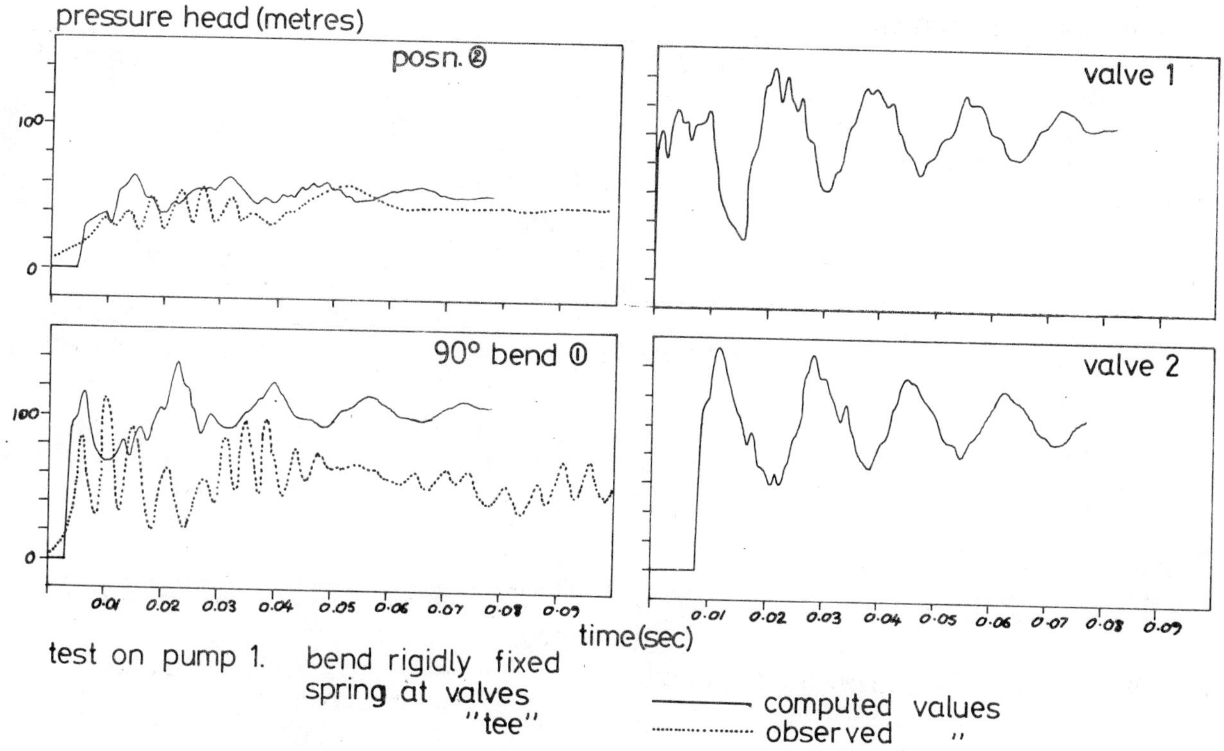

Fig. 13 Comparisons of computed values, pump 1

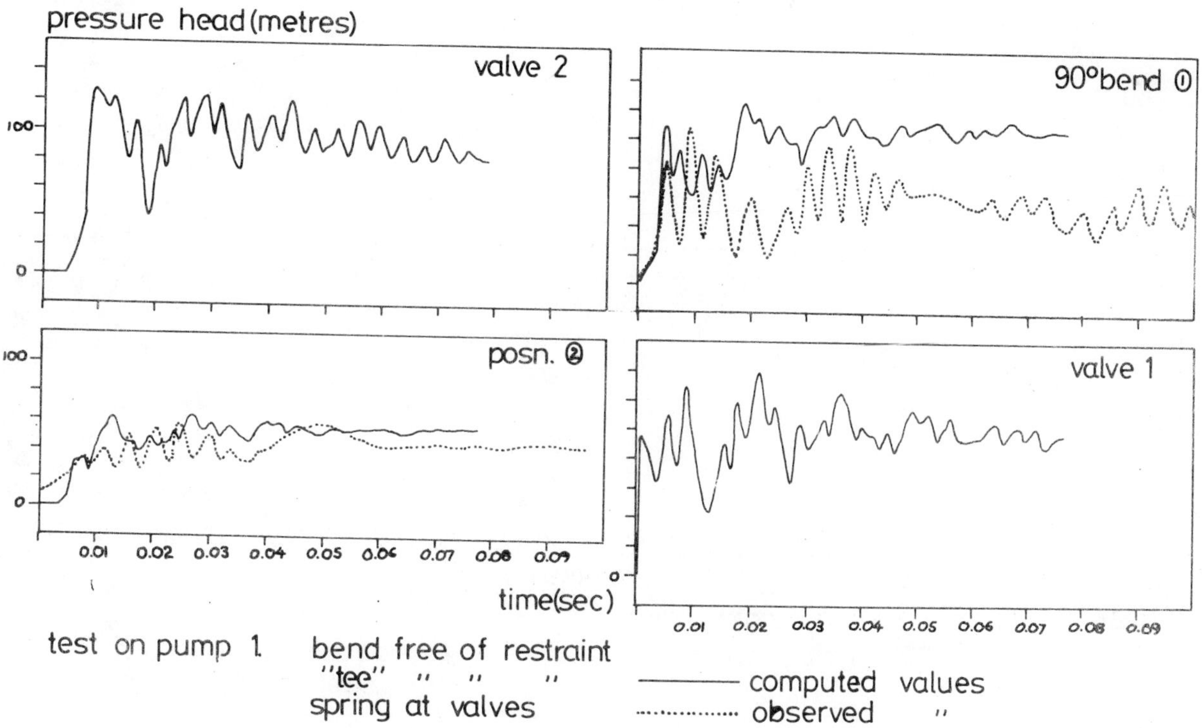

Fig. 14 comparisons of computed values, pump 1

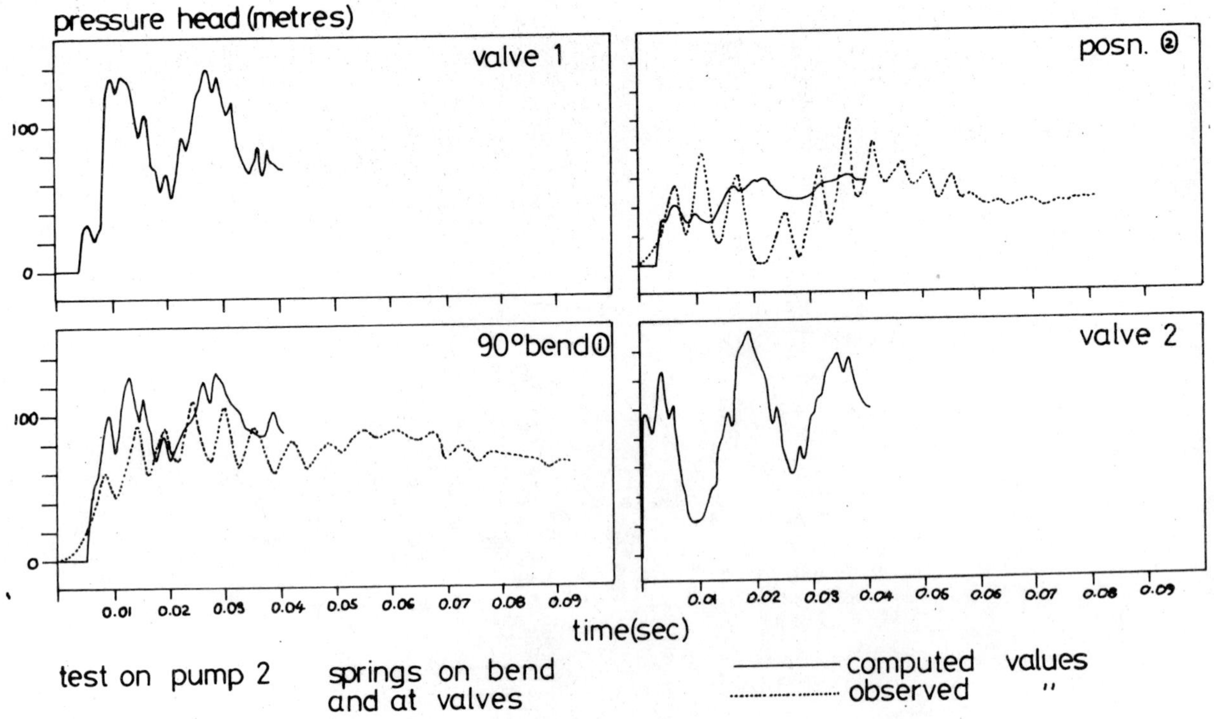

Fig. 15 Comparisons of computed values, pump 2

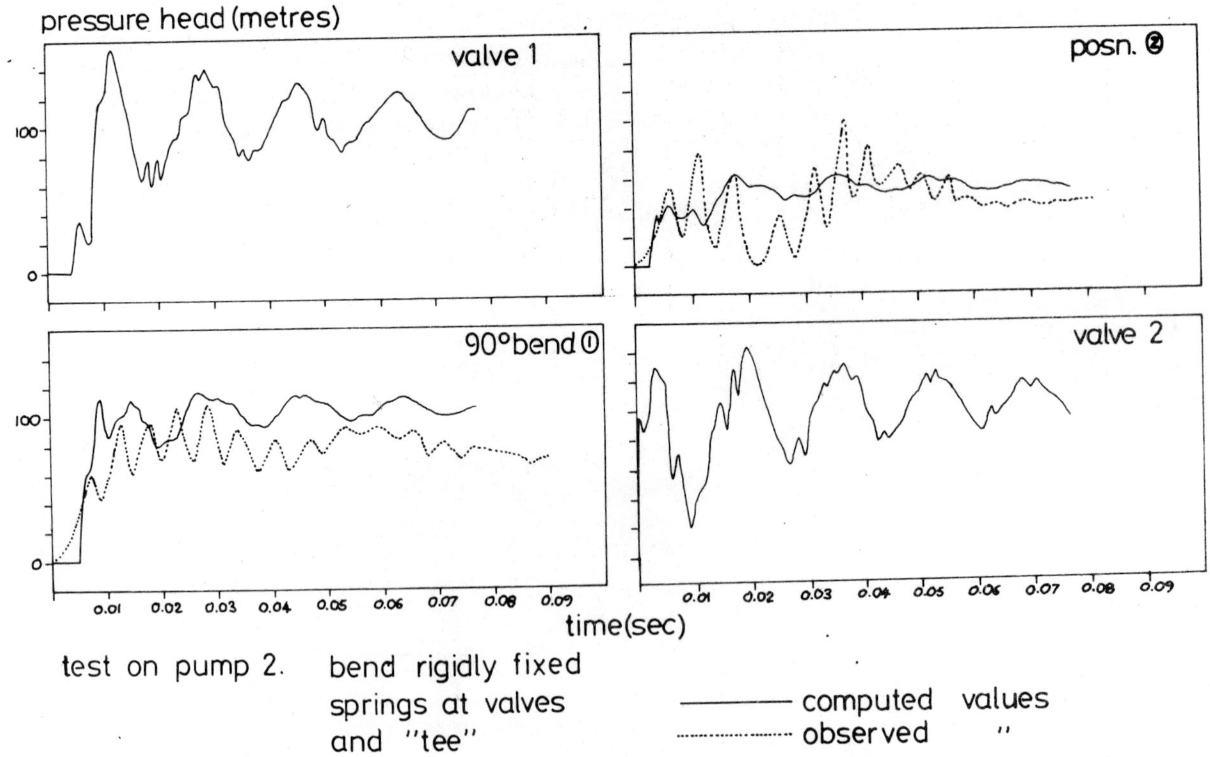

Fig. 16 Comparisons of computed values, pump 2

Third International Conference on
PRESSURE SURGES
Canterbury, England
March 25-27, 1980

PAPER E4

WATER HAMMER IN A THIN-WALLED PIPE

D.H. Wilkinson and E.M. Curtis
C.E.G.B., U.K.

Summary

Water-hammer in a thin-walled pipe differs from that in a rigid pipe because of the stretching of the pipe walls which may be in the elastic or plastic ranges. Conventional analysis allows for elastic radial strain of the pipe wall, the main effect of which is to reduce the speed of pressure waves in the liquid. A new analysis is given here which includes both axial and radial deformation of both elastic and plastic types. While general equations are given they are applied to the particular case of water column impact with the closed end of a pipe. Two composite pressure/stress waves were assumed, with large variations in pipe radius, wall thickness, liquid density, etc. permitted. Twenty-one non-linear equations were formulated to describe the phenomena and solved numerically over a range of water column impact velocities. As water velocity increased, the deformation state after the two waves went from elastic/elastic through elastic/plastic to plastic/plastic. Large variations in wave speeds occurred in the plastic cases with the final prediction of an impact velocity to rupture the pipe of about 7 times that to cause yield.

An experimental investigation was carried out for the elastic/elastic case and very good agreement with theory shown for wave speeds and transient pressures.

Held at the University of Kent at Canterbury, U.K.

Conference Organised by BHRA Fluid Engineering, Cranfield, Bedford, England.
In conjunction with The City University, London.

NOMENCLATURE

A	Pipe (flow) area
C_ℓ	Liquid wave velocity
C_s	Metal wave velocity
$C_{\ell b}$	Speed of sound in bulk liquid ($= \sqrt{K/\rho}$)
C_{sb}	Speed of sound in metal ($= \sqrt{E/\rho_s}$)
E	Young's Modulus of pipe metal
K	Bulk Modulus of liquid
p	Pressure in the liquid
Δp	Pressure change across liquid wave
R	Pipe (internal) radius
t	Pipe wall thickness
V	Velocity of liquid
V_s	Velocity of pipe metal
ε	Strain
$\bar{\varepsilon}$	'Effective' strain
ν	Poisson's ratio
ρ	Liquid density
ρ_s	Metal density
σ	Stress
$\bar{\sigma}$	'Effective' stress (equations 20 and 21)
$\Delta\sigma$	Stress change across tension wave in pipe metal

Subscripts

a	Ambient
o'	Unstressed condition
o	Initial condition prior to water-hammer
1	Region between C_ℓ and C_s waves (See Fig. 1)
2	Region adjacent to end of pipe (See Fig. 1)
Z	Axial
θ	Tangential
r	Radial
u	Ultimate
y	Yield

1. INTRODUCTION

This paper describes a theory which has been developed for the prediction of water-hammer phenomena in thin walled pipes, together with the experimental verification of part of this theory for a particular case with elastic deformation. The work is part of a programme being undertaken to investigate the response of thin walled pipework to water-hammer loading, with particular application to power station systems.

A simple straight tube geometry is considered, and is analysed in terms of the propagation of axial stress waves in the tube wall and pressure waves in the liquid. The method neglects radial inertia, assumes non-dispersive waves, and is rather simpler than the solutions developed by Skalak (Ref. 4) and Thorley (Ref. 5) but has been formulated particularly to deal with cases in which both elastic and plastic deformation of the tube wall occur. The theory gives a complete prediction for the plastic case up to the rupture condition.

Experiments were conducted on thick walled and thin walled tubes to develop experimental techniques and to obtain verification of the theory for the elastic deformation case.

A companion paper by Wilkinson (Ref. 6) applies a simplified version of the theory to more complex geometries, mitre bends, T-junctions, etc. Future experiments will include both complex geometries and plastic deformation.

2. COMPLETE THEORY

Figs. 1a and 1b show the case considered; impact of a water column with the closed end of a thin walled pipe. After impact it is assumed that two composite liquid pressure and metal stress waves travel upstream at velocities C_ℓ and C_s. All changes in properties are taken to occur across these waves which are essentially discontinuities. Changes in pipe radius and wall thickness and in liquid density are included as plastic deformation may be large. The deformation condition after each wave has passed may be either elastic or plastic and these states remain to be determined as part of the calculation. Radial velocities after the waves have passed are zero, and radial inertia effects, and all the complex deformation processes occurring in the neighbourhood of the actual waves, are not dealt with explicitly, only their overall effects being considered. This is the standard procedure with shock waves in gas dynamics where the shock structure is not normally considered. The analogy is not exact however, and in the plastic cases it is possible that significant effects have not been included. Although Fig. 1 depicts R_1/R_o as greater than unity, the numerical solutions give R_1 as slightly less than R_o.

Since plastic effects are non-linear it is necessary to work in absolute pressures and stresses rather than in terms of changes in them, and the initial state of deformation due to internal pressure p_o and ambient pressure p_a must be included to give the subscript o values. The superscript ' values are for the completely unstressed pipe at zero pressure everywhere.

The quantities specified as known are as follows

R'_o/t'_o, p_o, p_a, V_o, ρ'_o, E, K, ρ_s, σ_y, σ_u, ε_u & ν

The 21 unknowns to be found are then

R_1, R_2, t_1, t_2, p_1, p_2, V_1, V_2, V_{s1}, C_ℓ, C_s, σ_{Z1}, $\sigma_{\theta 1}$, σ_{r1}, σ_{Z2}, $\sigma_{\theta 2}$, σ_{r2}, $\bar{\sigma}_2$, $\bar{\sigma}_1$, ρ_1 & ρ_2

A further unknown is V_{s2}, but as this equals V_2 it is not worth including with another equation. Some 'static' unknowns, independent of V_o, are R_o/t_o, ρ_o, σ_{Zo}, $\sigma_{\theta o}$ and σ_{ro} which are found by a preliminary calculation.

2.1 Liquid Continuity

For 'C_s' wave

$$\rho_1 A_1 V_1 - \rho_o A_o V_o + C_s(\rho_1 A_1 - \rho_o A_o) = 0 \qquad (1)$$

For 'C_ℓ' wave

$$\rho_2 A_2 V_2 - \rho_1 A_1 V_1 + C_\ell(\rho_2 A_2 - \rho_1 A_1) = 0 \qquad (2)$$

2.2 Metal 'Continuity'

In terms of axial velocities and displacements

$$\frac{V_{s1}}{C_s} - \frac{1}{E_1}\{\sigma_{Z1} - \nu_1(\sigma_{\theta 1} + \sigma_{r1})\} + \frac{1}{E}\{\sigma_{Zo} - \nu(\sigma_{\theta o} + \sigma_{ro})\} = 0 \qquad (3)$$

$$\frac{V_{s2} - V_{s1}}{C_\ell} - \frac{1}{E_2}\{\sigma_{Z2} - \nu_2(\sigma_{\theta 2} + \sigma_{r2})\} + \frac{1}{E_1}\{\sigma_{Z1} - \nu_1(\sigma_{\theta 1} + \sigma_{r1})\} = 0 \qquad (4)$$

2.3 Liquid Momentum

$$(p_o - p_1)\frac{(A_o + A_1)}{2} - \rho_1 V_1^2 A_1 + \rho_o V_o^2 A_o - C_s(\rho_1 A_1 V_1 - \rho_o A_o V_o) = 0 \qquad (5)$$

$$(p_1 - p_2)\frac{(A_1 + A_2)}{2} - \rho_2 V_2^2 A_2 + \rho_1 V_1^2 A_1 - C_\ell(\rho_2 A_2 V_2 - \rho_1 A_1 V_1) = 0 \qquad (6)$$

The mean pressure $\frac{p_1 + p_2}{2}$ acting on the area difference $A_1 - A_o$, has been included as an axial force on the liquid in the derivation of eqn. (5) and similarly $(p_1 + p_2)/2$ in eqn. (6). The pressure difference forces, such as $\frac{(p_1 + p_o)}{2} - p_a$ acting on $(A_1 - A_o)$ are included in the next two eqns.

2.4 Metal Momentum

$$2\pi(R_1 + \frac{t_1}{2})t_1\{\sigma_{Z1} - \rho_s V_{s1}(C_s + V_{s1})\}$$

$$- 2\pi(R_o + \frac{t_o}{2})t_o\sigma_{Zo} - (\frac{p_1 + p_o}{2} - p_a)(A_1 - A_o) = 0 \qquad (7)$$

$$2\pi(R_2 + \frac{t_2}{2})t_2\{\sigma_{Z2} - \rho_s V_{s2}(C_\ell + V_{s2})\}$$

$$- 2\pi(R_1 + \frac{t_1}{2})t_1\{\sigma_{Z1} - \rho_s V_{s1}(C_\ell + V_{s1})\} - (\frac{p_2 + p_1}{2} - p_a)(A_2 - A_1) = 0 \qquad (8)$$

2.5 Stresses

$$\sigma_{\theta 1} = (p_1 - p_a)\frac{R_1}{t_1} \qquad (9)$$

$$\sigma_{r1} = -(\frac{p_1 + p_a}{2}) \qquad (10)$$

$$\sigma_{\theta 2} = (p_2 - p_a)\frac{R_2}{t_2} \qquad (11)$$

$$\sigma_{r2} = -(\frac{p_2 + p_a}{2}) \qquad (12)$$

2.6 Strains

Using terms such as $\ell n \frac{R_1}{R'_o}$ instead of $\frac{R_1 - R'_o}{R'_o}$

$$\varepsilon_{\theta 1} = \ell n \frac{R_1}{R'_o} = \frac{1}{2} \ell n \frac{A_1}{A'_o} = \frac{1}{E_1} \{\sigma_{\theta 1} - \nu_1(\sigma_{Z1} + \sigma_{r1})\} \qquad (13)$$

$$\varepsilon_{r1} = \ell n \frac{t_1}{t'_o} = \frac{1}{E_1} \{\sigma_{r1} - \nu_1(\sigma_{Z1} + \sigma_{\theta 1})\} \qquad (14)$$

$$\varepsilon_{\theta 2} = \ell n \frac{R_2}{R'_o} = \frac{1}{2} \ell n \frac{A_2}{A'_o} = \frac{1}{E_2} \{\sigma_{\theta 2} - \nu_2(\sigma_{Z2} + \sigma_{r2})\} \qquad (15)$$

$$\varepsilon_{r2} = \ell n \frac{t_2}{t'_o} = \frac{1}{E_2} \{\sigma_{r2} - \nu_2(\sigma_{Z2} + \sigma_{\theta 2})\} \qquad (16)$$

2.7 Liquid Density

$$\rho_1 = \rho'_o + (p_1 - p_a)\frac{\rho'_o}{K} \qquad (17)$$

$$\rho_2 = \rho'_o + (p_2 - p_a)\frac{\rho'_o}{K} \qquad (18)$$

This implies that ρ'_o is the density at pressure p_a. Variations in metal density ρ_s are neglected because they are small in the elastic range and zero in the plastic range due to putting $\nu = 0.5$.

2.8 Compatibility of Velocities

$$V_2 = V_{s2}$$

2.10 End Load

The mass of the disc at the closed end of the pipe is taken to be low enough that the time taken to accelerate it is small and the spread of compression and expansion waves is neglected.

For equilibrium of forces:

$$(p_2 - p_a)A_2 - 2\pi(R_2 + \frac{t_2}{2})t_2\sigma_{Z2} = 0 \qquad (19)$$

2.11 Stress Strain Curve

A bi-linear stress strain curve as shown in Fig. 1 is used. This is for mathematical simplicity but it is possible to use any general curve instead for the particular material of the pipe.

For plastic deformation the Levy-Mises 'effective' stresses $\bar{\sigma}_1$ and $\bar{\sigma}_2$ are defined, from Dieter (Ref. 3) as

$$\bar{\sigma}_1 = \frac{1}{\sqrt{2}}\{(\sigma_{Z1} - \sigma_{\theta 1})^2 + (\sigma_{\theta 1} - \sigma_{r1})^2 + (\sigma_{r1} - \sigma_{Z1})^2\}^{\frac{1}{2}} \qquad (20)$$

$$\bar{\sigma}_2 = \frac{1}{\sqrt{2}}\{(\sigma_{Z2} - \sigma_{\theta 2})^2 + (\sigma_{\theta 2} - \sigma_{r2})^2 + (\sigma_{r2} - \sigma_{Z2})^2\}^{\frac{1}{2}} \qquad (21)$$

If 1 or 2 are elastic regions

$$E_1, E_2 = E$$

$$\nu_1, \nu_2 = \nu$$

If 1 or 2 are plastic regions then it is assumed that the stresses are reached in a 'proportional' way to avoid the need for numerical integration.

The slope of the plastic σ~ε curve is b, given by

$$b = \frac{\sigma_u - \sigma_y}{\varepsilon_u - \varepsilon_y}$$

where

$$\varepsilon_y = \frac{\sigma_y}{E}$$

Note that the stress σ_u is a true value based on the calculated cross sectional area $2\pi(R_1 + t_1/2)t_1$ rather than on the original area $2\pi(R_o' + t_o'/2)t_o'$.

In general then $\bar{\sigma} = a + b\bar{\varepsilon}$ where

$$a = \frac{\sigma_y\varepsilon_u - \sigma_u\varepsilon_y}{\varepsilon_u - \varepsilon_y}$$

From the 'effective' stress $\bar{\sigma}$ from (20) or (21) the stress strain curve gives 'effective' strain as

$$\bar{\varepsilon} = \frac{\bar{\sigma} - a}{b}$$

$$\text{or } \frac{\bar{\varepsilon}}{\bar{\sigma}} = \frac{\bar{\sigma} - a}{\bar{\sigma}b}$$

However, any other relation between $\bar{\varepsilon}$ and $\bar{\sigma}$ could be used. Then component stresses and strains are related by expressions of the form

$$\varepsilon_Z = \frac{\bar{\varepsilon}}{\bar{\sigma}}\{\sigma_Z - 0.5(\sigma_\theta + \sigma_r)\}$$

This is the form used in eqns. (3), (4), (13), (14), (15), & (16) where it is understood that, for i = 1 or 2

	elastic	plastic
ν_i	ν	.5
$\frac{1}{E_i}$	$\frac{1}{E}$	$\frac{\bar{\sigma} - a}{\bar{\sigma}b}$

The 21 eqns. (1) to (21) permit solution for the 21 unknowns. The actual number of eqns. and unknowns used is slightly arbitrary as some variables such as V_{s2}, $\bar{\varepsilon}_1$, $\bar{\varepsilon}_2$, $\bar{\varepsilon}_{Z1}$, $\bar{\varepsilon}_{Z2}$ etc. only appear linearly and are eliminated in formulating the present 21 eqns. but could appear explicitly in a larger set. Alternatively, some of the present ones could be eliminated, such as ρ_1 and ρ_2, the actual number used being a matter of choice and convenience provided the theoretical method is kept the same.

2.12 Initial Conditions

The calculation of the initial statically pressurised state involves the following relations

$$\left.\begin{aligned}
\sigma_{Zo} &= (p_o - p_a)\frac{R_o}{2t_o} \\
\sigma_{ro} &= \frac{-(p_o + p_a)}{2} \\
\sigma_{\theta o} &= (p_o - p_a)\frac{R_o}{t_o} \\
\ln\frac{R_o}{R'_o} &= \frac{1}{E}\{\sigma_{\theta o} - \nu(\sigma_{Zo} + \sigma_{ro})\} \\
\ln\frac{t_o}{t'_o} &= \frac{1}{E}\{\sigma_{ro} - \nu(\sigma_{\theta o} + \sigma_{Zo})\}
\end{aligned}\right\} \qquad (22)$$

Starting with $R_o/R'_o = 1$, $t_o/t'_o = 1$ these equations are solved iteratively for the 5 variables σ_{Zo}, σ_{ro}, $\sigma_{\theta o}$, R_o/R'_o, t_o/t'_o in a preliminary calculation.

2.13 Method of Solution

The equations are generally non-linear, involving products of up to 5 unknowns in some terms. A numerical solution was therefore necessary and an IBM FORTRAN library subroutine was available for this. Starting with estimated initial values this solved the equation using Newton's method and 'steepest descent' techniques.

It was necessary to non-dimensionalise and scale the equations and unknowns so that all variables were of order 1 and all residuals of similar size. Limitations of space preclude a complete exposition of the equations actually used but the unknowns were non-dimensionalised as follows

i	Non-dimensional variables
1	A_1/A'_o
2	A_2/A'_o
3	t_1/t'_o
4	t_2/t'_o
5	$p_1/\bar{p}$
6	$p_2/\bar{p}$
7	V_1/V_o
8	V_2/V_o
9	V_{s1}/V_o
10	$C_\ell/C_{\ell b}$
11	C_s/C_{sb}
12	σ_{Z1}/σ_y
13	$\sigma_{\theta 1}/\sigma_y$
14	σ_{r1}/σ_y
15	σ_{Z2}/σ_y
16	$\sigma_{\theta 2}/\sigma_y$
17	σ_{r2}/σ_y
18	$\bar{\sigma}_2/\sigma_y$
19	$\bar{\sigma}_1/\sigma_y$
20	ρ_1/ρ'_o
21	ρ_2/ρ'_o

where $C_{\ell b} = \sqrt{\frac{K}{\rho'_o}}$

$C_{sb} = \sqrt{\frac{E}{\rho_s}}$

$\bar{p} = p_o + \rho'_o C_{\ell b} V_o$

The starting values had to be near enough to the solution or convergence was not obtained. It is not clear how the uniqueness of the solutions could be proved as there might be perhaps up to 5 solutions with one equation being 5th order. The procedure was to work up from low V_o with initial values from simpler elastic-elastic solutions. Eventually it was found that $\bar{\sigma}_2$ reached σ_y whilst $\bar{\sigma}_1$ was still below σ_y. For further increase in V_o an elastic-plastic solution was then taken with region 1 elastic and 2 plastic. As V_o was increased further $\bar{\sigma}_1$ also reached σ_y, and a plastic-plastic solution was then taken. Different starting values were needed in each deformation regime, but these were based on results at lower V_o.

3. THEORETICAL RESULTS

Solutions over the full V_o range have been calculated for the geometry of the experimental thin walled pipe described in Section 5, and the main results are plotted in Figs. 2 and 3.

3.1 Elastic-Elastic

Up to about $V_o = 9.2\ \mathrm{m\ s^{-1}}$ variables are either substantially constant or vary linearly with V_o. The quantities that can be compared most directly with other analyses are C_ℓ and C_s. The simplest versions of these are given by the elastic pipe formula and C_{sb}.

$$\left.\begin{aligned} C_\ell &\simeq C_{\ell b}\{1 + 2\frac{K}{E}\frac{R}{t}\}^{-\frac{1}{2}} \\ C_{sb} &= \sqrt{\frac{E}{\rho}} \end{aligned}\right\} \qquad (23)$$

The next simplest values come from solution of eqn. (24) given later

	'Exact'	Eqn.(24)	Eqn.(23)	
C_ℓ	1.002	1.009	1.010	$\mathrm{Km\ s^{-1}}$
C_s	4.814	4.817	4.712	

It is seen that C_ℓ is in good agreement with the elastic pipe formula while C_s is slightly higher than C_{sb}. Otherwise $V_2 \simeq 0.36\ V_o$ and compared to a non-extending pipe solution, p_2 is reduced to about $0.69\ \rho_o'\ C_\ell\ V_o$; p_1 is always small; σ_{Z1} is about $p_2R_o'/2t_o'$ and σ_{Z2} is slightly less than σ_{Z1}, so that the weak tension wave component at speed C_ℓ is slightly negative, a compression.

3.2 Elastic-Plastic

Above $V_o = 9.2\ \mathrm{m\ s^{-1}}$ the deformation in region 1 is still elastic but in region 2 is plastic. C_ℓ drops off almost inversely with V_o so that p_2 tends to level out and only rises by about 50% between $V_o = 9.2\ \mathrm{m\ s^{-1}}$ and $60\ \mathrm{m\ s^{-1}}$; p_1 is still small and R_1/R_o' and t_1/t_o' remain almost exactly at 1.00. However, R_2 increases and t_2 reduces almost inversely.

The curves above and below $V_o = 9.2\ \mathrm{m\ s^{-1}}$ do not meet exactly because, although $\bar{\sigma}/\bar{\varepsilon}$ from the stress/strain curve is continuous ν_2 is not, jumping from 0.28 to 0.5. At about $V_o = 66\ \mathrm{m\ s^{-1}}$ $\bar{\sigma}_1$ reaches σ_y indicating that region 1 is plastic above this velocity.

3.3 Plastic-Plastic

The boundary between elastic-plastic and plastic-plastic regimes is not clear cut because of the sudden change in ν from 0.28 to 0.5 assumed. This results in discontinuities in most variables with $\bar{\sigma}_1$ dropping from σ_y to just below it while $\bar{\sigma}_2$ jumps up to σ_u indicating rupture at about $68\ \mathrm{m\ s^{-1}}$. The plastic-plastic region therefore has a rather doubtful existence even theoretically. A different assumption of a continuous variation of ν from 0.28 to 0.5 through a transition region would probably give a tidier result. Several points with stresses above σ_u have been shown to indicate the form of the variation in this region. The main changes are that C_s starts to drop off rapidly after the initial discontinuity, much as C_ℓ does when region 2 goes plastic, and p_1, $\sigma_{\theta 1}$ and V_2 start to increase.

4. EXPERIMENTS

The experimental rig is essentially a U-tube, partly full of water, with the top of one limb closed, and the top of the other limb open to atmosphere. To initiate water-hammer the air above the free surface in the closed limb is evacuated and water is sucked up on this side. Water-hammer occurs when the rising water column impacts the closed end of the tube. The arrangement is illustrated in Fig. 4. For the experiment described here the test section was a stainless steel (AISI 304) thin walled tube, 6.1 m long, 25.3 mm diameter with a wall thickness of 0.276 mm, giving a radius to thickness ratio (R/t) of 45.24. At its lower end the test section was securely attached to a relatively massive steel plate forming the top of the reservoir vessel. This provided a 'built in' end condition at this end. At the closed top end of the test section there was no restraint against axial motion although at three positions along its length lateral restraint was provided. The vacuum connection to the top end of the test section was fitted with a flap valve made from 0.1 mm brass shim. This valve remained open during the evacuation of the space above the rising water column but was closed by the impact of the water column when this reached the top of the test section thus sealing off the vacuum line and preventing the high pressure water in the test section from leaking out into the vacuum line during the transient. The valve was essential in order to achieve the full water-hammer pressure rise; without the valve the flow of water out of the test section during the transient would roughly halve the pressure rise. The other limb of the U-tube was a 5 cm bore acrylic tube, open to atmospheric at the top and connected to the reservoir through two one inch ball valves.

Instrumentation was fitted to measure water pressures in the test section, water velocity at impact, and liquid wave velocity. Two miniature semiconductor pressure transducers (Entran type EPK-125) were flush mounted in the test section wall, one 0.13 m below the top and one 1.20 m above the bottom (Fig. 5). The velocity of the rising water column just prior to impact was measured using conductivity devices located near the top of the test section. These consisted of pairs of parallel wires, 6 mm apart, stretched across chords of the test section. The wires were secured with epoxy adhesive and contact with the tube wall was prevented by insulating sleeves. The drilling of the tube for these wires was done using a jig and slip gauges to ensure accurate spacing along the tube. Each pair of wires was connected to a 1.5 volt battery in series with a 10 k resistor which gave approximately 0.5 volts across the resistor when the wires became immersed.

The signals from pressure transducers and conductivity wires were measured on double-beam digital storage oscilloscopes.

Measurement of transit time between conductivity wires gave the velocity of the rising water column; a small correction was made to allow for the deceleration of the column as it approached the closed top end of the test section. The velocity of the liquid pressure waves was derived from the transit time for a wave travelling between the two pressure transducers.

The oscilloscopes were calibrated using frequency and voltage sources and errors due to parallax in the photographic recording of traces were also allowed for. With these precautions voltage and time intervals could be measured with errors of about 0.5%.

As estimate of the remaining sources of error in the measurements has been made and is of interest in deciding whether differences between experimental and theoretical results are significant. The probable errors in each quantity are:-

Pressure rise ± 2%
Water velocity ± 1.25%
Sonic velocity ± 1%

The comparison with theory generally involves comparing a calculated value of the quantity $\Delta p/\rho_\ell C_\ell V_o$ with the measured value. The error in the measured value with the errors quoted above is about ± 3%. However, the theoretical calculation involves using material properties (Young's Modulus and Poisson's Ratio) for which there seems uncertainty in the literature as to the correct values for a work-hardened stainless steel (Woolman and Mottram Ref.7) Assuming a likely error of ± 5% in Young's Modulus gives a further error of ± 1% in the calculation of $\Delta p/\rho_\ell C_\ell V_o$ for this test section tubing. Taking this into account differences between measurement and theory will be significant if the discrepancy exceeds about ± 4%.

Despite the simplicity of the apparatus there were difficulties initially in achieving satisfactory water-hammer pulses. (At first a thick walled test section was employed ($R/t = 4.6$) in which pipe flexibility effects were small and a square pulse was to be expected). Although the ragged pulses initially produced were thought to be affected by air coming out of solution they were not improved by prolonged evacuation, nor by repeated testing, and the technique eventually used to provide good pulses in a repeatable manner was to minimise the time at low pressure. The procedure was to open valves V1, V3 and V4 (Fig. 4) and to set V5 to control the velocity. On opening V2 the pressure above the pressure above the free surface starts to fall and the water moves up the test section, the pressure falling continuously as it does so. At impact the pressure indicated on the gauge P1 (Fig. 4) was about 30 torr. This operating technique gave consistently satisfactory results.

A typical pressure trace record is shown in Fig. 6. The upper trace is for the transducer near to the top of the test section and the lower trace is for the lower transducer. The pipe flexibility effects are quite large; for a rigid pipe the pulses would be flat topped and of the same amplitude at both locations. The pressure pulse terminates when, after about 12 ms, the pressure at the top of the tube falls to a low value and the flap valve opens. At the lower transducer a small negative pressure is recorded at the end of the pulse. When the flap valve opens a void occurs at the top of the tube and is analogous to column separation in a closed system. This process produces three further water-hammers over the next 500 ms, of successively smaller amplitudes as flow losses dissipate the energy. A liquid pressure wave from the top transducer takes about 5 ms to reach the lower transducer and the pressure rise seen at the lower transducer after only 1 ms is due to the arrival of the tension wave and associated pressure wave originating from column impact at the top of the tube and travelling at the tubewall sonic velocity (approximately 4700 m s^{-1}).

5. COMPARISON OF EXPERIMENT AND THEORY

For making a comparison between theory and experiment some simplifying assumptions were made to the theory. These were applicable to the experimental situation and assumed that the tube material remains in the elastic region, implying that the strains are small, that the tube thickness is small compared with the radius, and that the velocities of the liquid and metal are small compared with the wavespeeds. With these assumptions the original set of equations is reduced to a smaller number of simpler equations which are as follows:-

(Note $A_1 - A_o' = 2\pi R^2 dR/R = 2\pi R^2(\varepsilon_{\theta 1} - \varepsilon_{\theta o}) = 2\pi R^2\{\sigma_{\theta 1} - \sigma_{\theta o} - \nu(\sigma_{Z1} + \sigma_{r1} - \sigma_{Zo} - \sigma_{ro})\}$)

<u>Liquid continuity</u>

$$(V_o - V_1)/C_s = (2/E)\{(p_1 - p_o)R/t - \nu(\sigma_{Z1} - \sigma_{Zo})\} + (p_1 - p_o)/K \quad (1a)$$

$$(V_1 - V_2)/C_\ell = (2/E)\{(p_2 - p_o)R/t - \nu(\sigma_{Z2} - \sigma_{Z1})\} + (p_2 - p_1)/K \quad (2a)$$

Metal Continuity

$$(V_{s1} - V_{so})/C_s = (1/E)\{(\sigma_{Z1} - \nu\sigma_{\theta 1}) - (\sigma_{Zo} - \nu\sigma_{\theta o})\} \tag{3a}$$

$$(V_{s2} - V_{s1})/C_s = (1/E)\{(\sigma_{Z2} - \nu\sigma_{\theta 2}) - (\sigma_{Z1} - \nu\sigma_{\theta 1}) \tag{4a}$$

Liquid Momentum

$$p_o - p_1 = C_s\rho\,(V_1 - V_o) \tag{5a}$$

$$p_1 - p_2 = C_\ell\rho(V_2 - V_1) \tag{6a}$$

Metal Momentum

$$\sigma_{Z1} - \sigma_{Zo} = \rho_s C_s (V_{s1} - V_{so}) \tag{7a}$$

$$\sigma_{Z2} - \sigma_{Z1} = \rho_s C_\ell (V_{s2} - V_{s1}) \tag{8a}$$

Stresses

$$\sigma_{\theta 1} = (p_1 - p_a)R/t \tag{9a}$$

$$\sigma_{\theta 2} = (p_2 - p_a)R/t \tag{10a}$$

End Load

$$p_2 - p_a = (2t/R)\sigma_{Z2} \tag{19a}$$

Compatibility

$$V_2 = V_{s2}$$

Solution of equations (1a), (3a), (5a), (7a) and (9a) gives a quadratic equation in $C_s{}^2$:-

$$\left\{1 + (1 - \nu^2)\,\frac{\rho_\ell C_{\ell b}{}^2}{\rho_s C_{sb}{}^2}\,\frac{2R}{t}\right\}\left(\frac{C_s}{C_{sb}}\right)^4 - \left\{1 + \left(\frac{C_{\ell b}}{C_{sb}}\right)^2 \left(1 + \rho_\ell/\rho_s \cdot \frac{2R}{t}\right)\right\}\left(\frac{C_s}{C_{sb}}\right)^2 + \left(\frac{C_{\ell b}}{C_{sb}}\right)^2 = 0 \tag{24}$$

The same equation, but in $C_\ell{}^2$, is derived from (2a), (4a), (6a), (8a) and (10a) and its two roots give the wavespeeds C_ℓ and C_s for the two waves.

When the wavespeeds have thus been calculated the equations are solved for the other unknowns.

In order to calculate the transient pressures which follow the column impact it is necessary to consider the reflection of waves at the ends of the test section. There are two cases to be considered

(a) Reflection of a wave at a closed free end
(b) Reflection of a wave at an open fixed end

The incident wave in either case may be a strong liquid wave, plus the associated weak tension wave, both travelling at velocity C_ℓ, or a strong tension wave plus the accompanying weak liquid wave, both travelling at velocity C_s. For both of these possible incident waves there are, after reflection, two reflected waves, one travelling at velocity C_ℓ, the other at velocity C_s and each having components in both liquid and metal. Fig. 7 illustrates the two basic cases of wave reflection at a closed free end and wave reflection at an open fixed end which are the basic interactions needed for a theoretical analysis of the experiment. For reflection of a wave at a closed free end, the equations derived previously for the column impact case apply with the velocity upstream of the incident wave substituted for the velocity of the water column. For waves reflecting at the bottom (open) end the velocity compatibility and end load conditions are replaced by new boundary conditions of constant (reservoir) pressure and zero tube wall velocity at the end of the tube, which is rigidly clamped to the reservoir.

The equations have been solved for the reflection of unit strength liquid and tube metal waves at both ends of the test section and for the column impact. The following numerical values of tube and liquid parameters, applying to the experiment, were used

Radius/thickness ratio of tube wall	R/t =	45.24
Metal/liquid density ratio	ρ_s/ρ_ℓ =	7.92
Poisson's Ratio	ν =	0.28
Liquid acoustic velocity	$C_{\ell b} = \sqrt{K/\rho_\ell}$ =	1465 $m\ s^{-1}$
Metal acoustic velocity	$C_{sb} = \sqrt{E/\rho_s}$ =	4712 $m\ s^{-1}$
	Tube length =	6.10 m

The solution gives the wavespeeds as:-

Liquid wave velocity C_ℓ = 1009 $m\ s^{-1}$

Metal wave velocity C_s = 4817 $m\ s^{-1}$

It also gives values for the pressures, stresses, and velocities existing after reflection. In the application of the theory to derive the pressure/time history after column impact for comparison with the measurements, it is convenient to deal in terms of wavestrengths and wave positions, and it is not necessary to solve for the liquid and tube metal velocities. Table 1 gives the strengths of the reflected waves for the required interactions. New waves are generated each time a reflection occurs and their strengths are calculated from Table 1, knowing the strength of the incident wave. As the transient proceeds the number of waves increases as each reflection generates two reflected waves. When the wave velocities C_ℓ and C_s have been calculated the wave positions can be plotted as in Fig. 8 and the data in Table 1 used to derive the strengths of the two reflected waves (one at speed C_ℓ, one at speed C_s) resulting from each end reflection. The wave strengths are given in Table 2; the ratio of the amplitude of the strong wave to that of its associated weak component is constant, although dependent on whether a wave at velocity C_ℓ or at velocity C_s is involved. This ratio is obtained, for a wave travelling at velocity C_s, from equations (3a), (7a) and (9a) as

$$\Delta\sigma/\Delta p = (\nu R/t) \div (1 - C_{sb}^2/C_s^2) \tag{25}$$

and similarly for a wave travelling at velocity C_ℓ the ratio is

$$\Delta\sigma/\Delta p = (\nu R/t) \div (1 - C_{sb}^2/C_\ell^2) \tag{26}$$

The numerical values in the present case are

$$\Delta\sigma/\Delta p = +\ 294.2 \text{ from eqn. (25) (wavespeed} = C_s)$$

and

$$\Delta\sigma/\Delta p = -\ 0.61 \text{ from eqn. (26) (wavespeed} = C_\ell)$$

In calculating the strengths of the waves produced by reflection it is convenient to work initially in terms of the main components only ($\Delta\sigma$ for waves with velocity C_s and Δp for waves with velocity C_ℓ) and to use the values derived above to obtain the weak components.

For making a comparison with experimental measurements the pressure variation at the transducer locations is required for the period of the transient. This is readily obtained by identifying on Fig. 8 the waves passing the particular location, noting their arrival times and, from Table 2, their strengths. The corresponding pressure at the point being considered at a given time is then obtained by summation of the strengths of all the waves which have passed that point up to the given time. This procedure has been carried out to obtain the pressure variation with time at the upper and lower transducer locations and the results are shown in Fig. 9. These plots include both a prediction from the theory just described and an alternative theoretical prediction from a much simplified theory in which the weak wave components are neglected (Wilkinson Ref. 6). Agreement with the more exact theory is good apart from the effects of dispersion. The very simple theory is rather less successful in predicting the shape of the transient.

6. CONCLUSIONS

A theory has been derived and verified by experiment which gives a good prediction of the pipe flexibility effects in water-hammer in a straight pipe for elastic deformation of the pipe wall. The theory has also been used to predict behaviour at higher water impact velocities with plastic deformation in one or both regions.

7. ACKNOWLEDGEMENT

This work is published with the permission of the Central Electricity Generating Board.

8. REFERENCES

1. Curtis, E. M. "Experiments on Water-Hammer in Straight Tubes". Internal CEGB Report MM/MECH/TG 154. (1979).

2. Curtis, E. M. "A Comparison of Experimental and Theoretical Results for Water-Hammer in a thin walled tube". Internal CEGB Report MM/MECH/TG 155. (1979).

3. Dieter, G. E. "Mechanical Metallurgy", pp. 91-92, 2nd Ed. McGraw-Hill. (1976).

4. Skalak, R. "Extension of the Theory of Water-Hammer". Trans ASME. (1956).

5. Thorley, A. R. D. "Pressure Transients in Hydraulic Pipelines". ASME Trans 91 Set D. (1969).

6. Wilkinson, D. H. "Water-hammer in Thin Walled Pipe Systems" 3rd Int. Conf. on Pressure Surges, Canterbury, (March 1980).

7. Woolman, J. & Mottram, R. A. "The Mechanical and Physical Properties of EN Steels" Pergamon (1964).

Interaction	Incident Wave Strength		Reflected Wave Strengths	
	Δp	$\Delta\sigma$	Δp	$\Delta\sigma$
Liquid column impact with closed end			(+ 0.053) + 0.623	(+ 15.665) - 0.384
Tension wave reflection at open end	(+ 0.003)	(+ 1.000)	(+ 0.003) - 0.007	(+ 0.980) + 0.004
Tension wave reflection at closed end	(+ 0.003)	(+ 1.000)	(- 0.0005) + 0.034	(- 0.149) - 0.021
Liquid wave reflection at open end	+ 1.000	- 0.610	(- 0.020) - 0.980	(- 5.764) + 0.598
Liquid wave reflection at closed end	+ 1.000	- 0.610	(+ 0.102) + 0.187	(+ 29.870) - 0.114

Table 1 Reflected Wave Strengths for Unit Incident Waves

NOTES: (1) All wave strengths in the Table have been non-dimensionalised by dividing by $\rho C_\ell V_o$.

(2) Figures enclosed in brackets ( ) are for waves travelling at velocity C_s and figures without brackets are for waves travelling at velocity C_ℓ.

Wave No.	$\frac{\Delta p}{\rho C_\ell V_o}$	$\frac{\Delta \sigma}{\rho C_\ell V_o}$	Speed	Direction*	Start Time ms	End Time ms
1	+ .053	+ 15.665	C_s	d	0	1.27
2	+ .623	− .380	C_ℓ	d	0	6.04
3	+ .052	+ 15.358	C_s	u	1.27	2.54
4	− .105	+ .065	C_ℓ	u	1.27	7.31
5	− .008	− 2.291	C_s	d	2.54	3.81
6	+ .521	− .321	C_ℓ	d	2.54	8.58
7	− .008	− 2.246	C_s	u	3.81	5.08
8	+ .015	− .009	C_ℓ	u	3.81	9.85
9	+ .001	+ .335	C_s	d	5.08	6.35
10	− .076	+ .046	C_ℓ	d	5.08	11.12
11	− .012	− 3.589	C_s	u	6.04	7.31
12	− .610	+ .372	C_ℓ	u	6.04	12.08
13	+ .001	+ .328	C_s	u	6.35	7.62
14	− .002	+ .001	C_ℓ	u	6.35	12.39
15	− .009	− 2.613	C_s	d	7.31	8.58
16	− .141	+ .086	C_ℓ	d	7.31	13.35
17	− .000	− .049	C_s	d	7.62	8.89
18	+ .011	− .007	C_ℓ	d	7.62	13.66
19	− .019	− 5.563	C_s	u	8.58	9.85
20	− .492	+ .300	C_ℓ	u	8.58	14.62
21	− .000	− .048	C_s	u	8.89	10.16
22	+ .000	.000	C_ℓ	u	8.89	14.93
23	+ .004	+ 1.290	C_s	d	9.85	11.12
24	− .186	+ .113	C_ℓ	d	9.85	15.89
25	+ .000	+ .007	C_s	d	10.16	11.43
26	− .002	+ .001	C_ℓ	d	10.16	16.20
27	+ .006	+ 1.703	C_s	u	11.12	12.39
28	+ .066	− .040	C_ℓ	u	11.12	17.16
29	.000	+ .007	C_s	u	11.43	12.70
30	.000	.000	C_ℓ	u	11.43	17.47
31	− .062	− 18.221	C_s	d	12.08	13.35
32	− .114	+ .070	C_ℓ	d	12.08	18.12
33	− .001	− .323	C_s	d	12.39	13.66
34	+ .057	− .035	C_ℓ	d	12.39	18.43
35	+ .000	− .001	C_s	d	12.70	13.97
36	+ .000	.000	C_ℓ	d	12.70	18.74
37	− .058	− 17.049	C_s	u	13.35	14.62
38	+ .261	− .159	C_ℓ	u	13.35	19.39
39	− .001	− .380	C_s	u	13.66	14.93
40	− .009	+ .005	C_ℓ	u	13.66	19.70
41	.000	− .001	C_s	u	13.97	15.24
42	.000	.000	C_ℓ	u	13.97	20.01
43	− .041	− 12.167	C_s	d	14.62	15.89
44	− .670	+ .409	C_ℓ	d	14.62	20.66
45	.000	+ .066	C_s	d	14.93	16.20
46	− .013	+ .008	C_ℓ	d	14.93	20.97
47	.000	.000	C_s	d	15.24	16.51
48	.000	.000	C_ℓ	d	15.24	21.28
49	− .037	− 10.858	C_s	u	15.89	17.16
50	+ .190	− .116	C_ℓ	u	15.89	21.93

Table 2 Wave Strengths

*NOTE: 'u' refers to upward moving waves, i.e. wave travelling from the open, fixed, end, towards the closed, free, end.
'd' refers to downward moving waves.

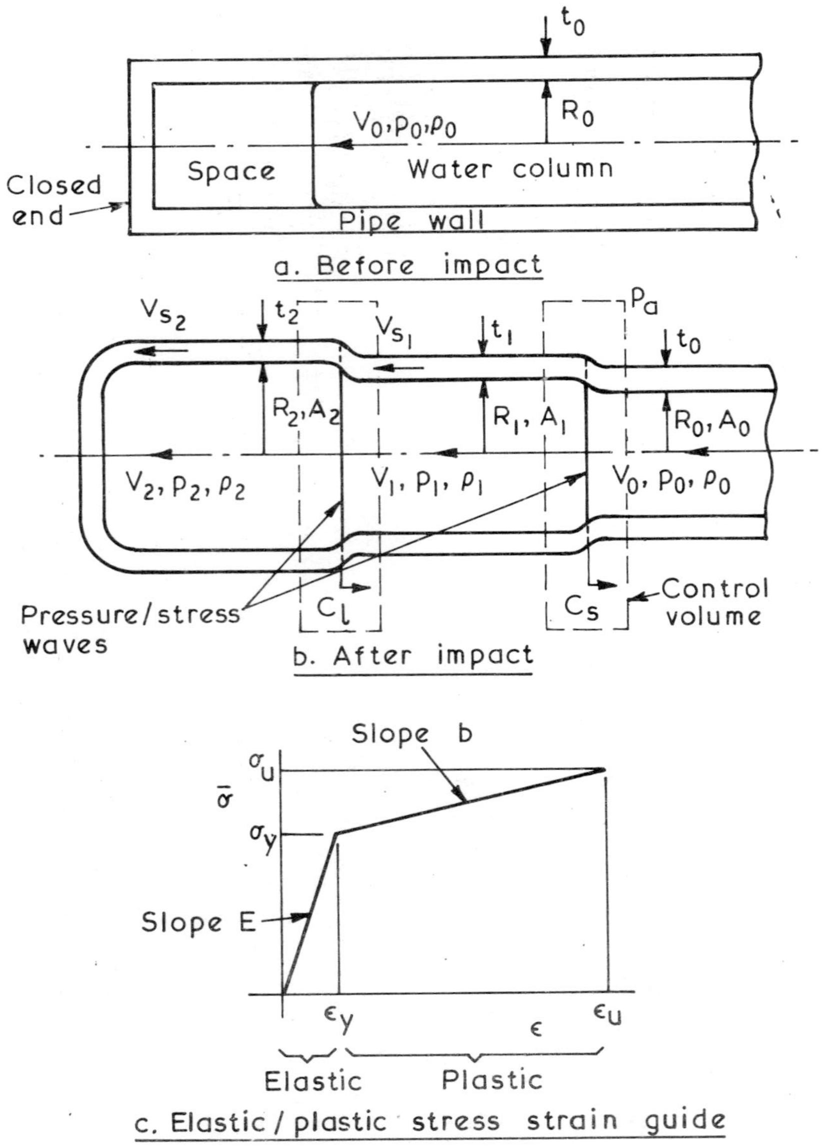

Fig. 1 Water column impact at a closed end

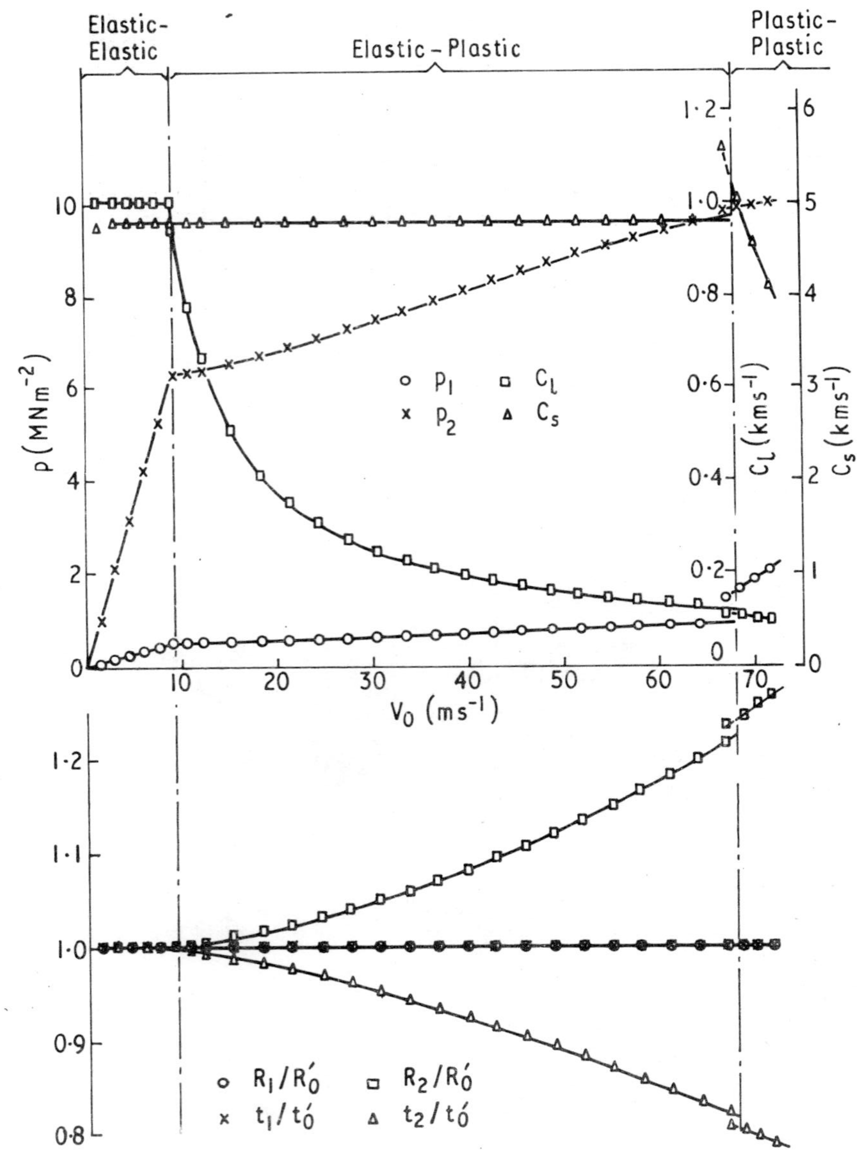

Fig. 2 Theoretical results for P, C_ℓ, C_s, R and t

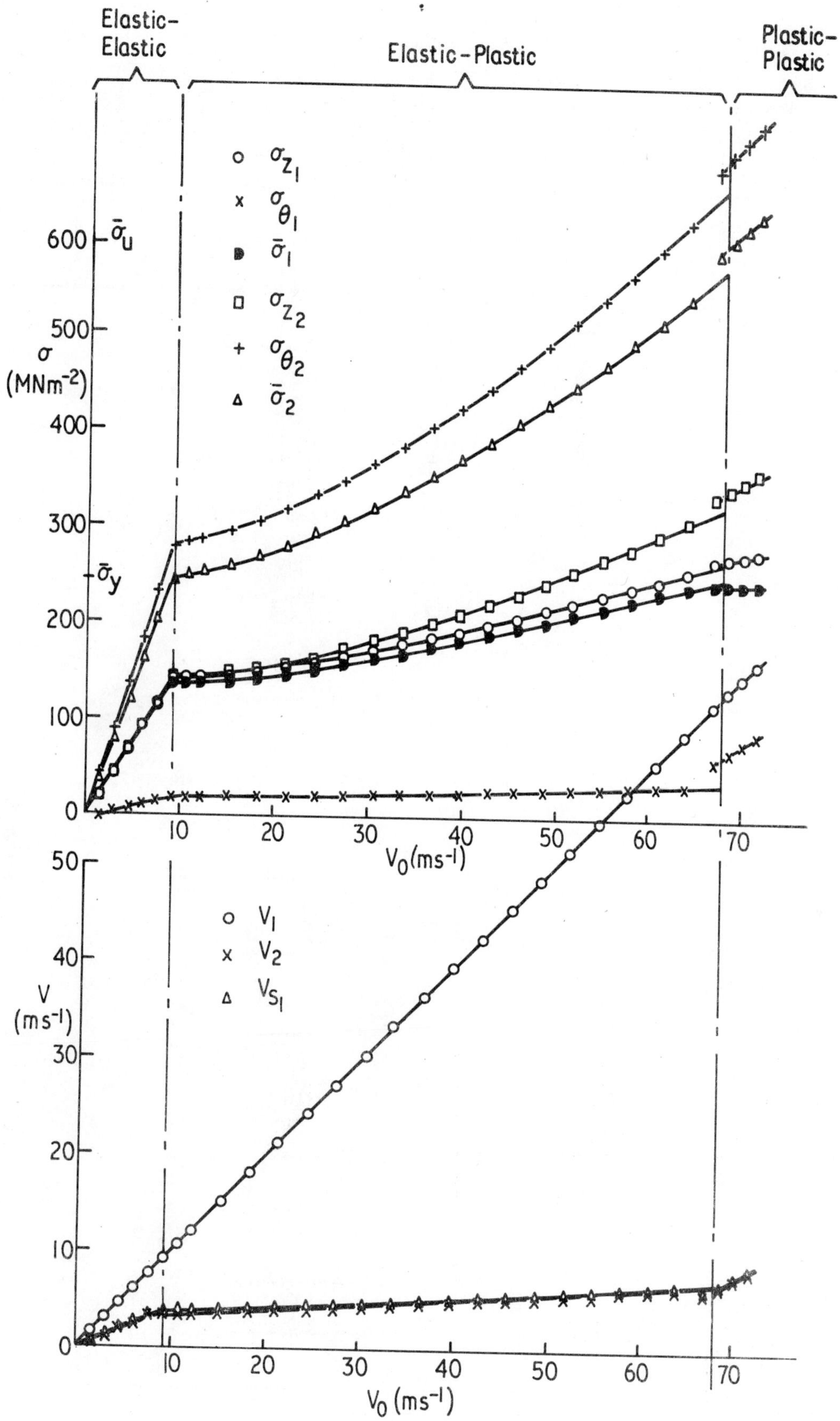

Fig. 3 Theoretical results for σ and V

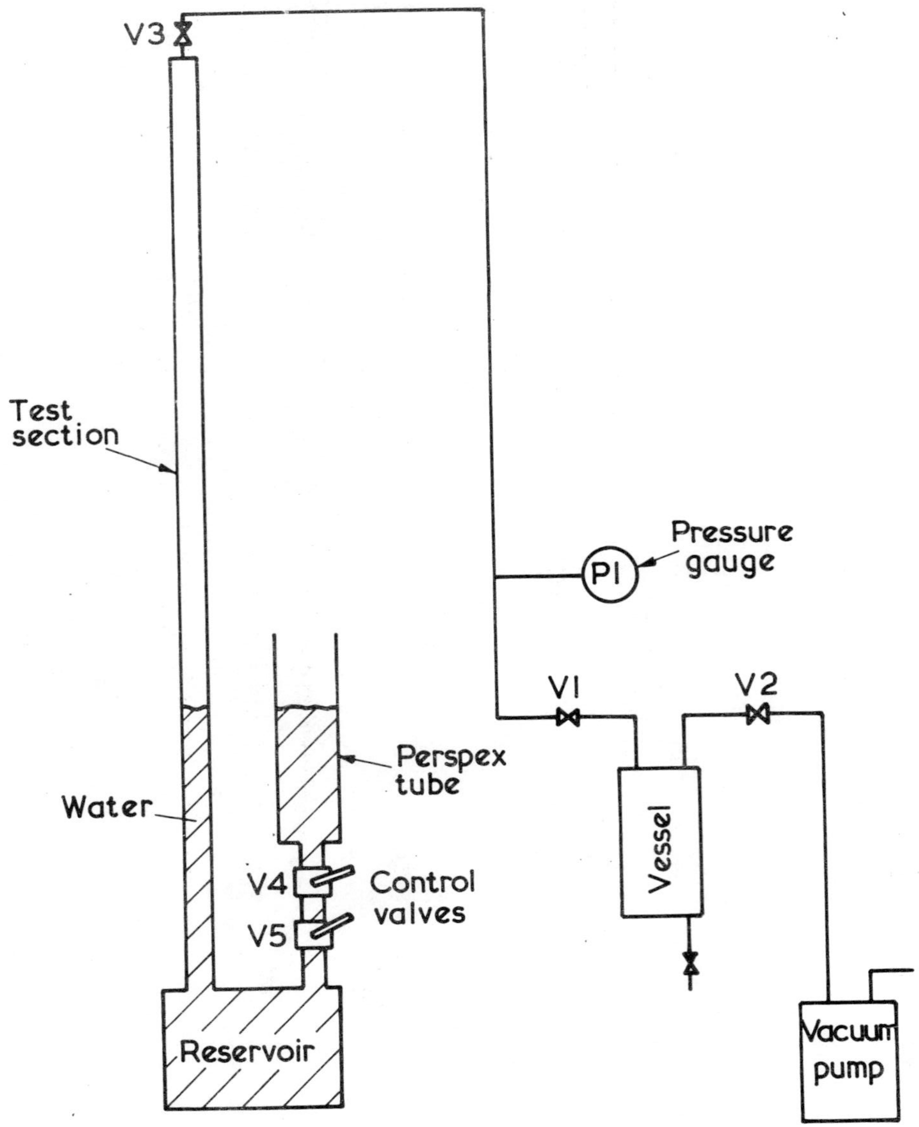

Fig. 4 Water hammer rig

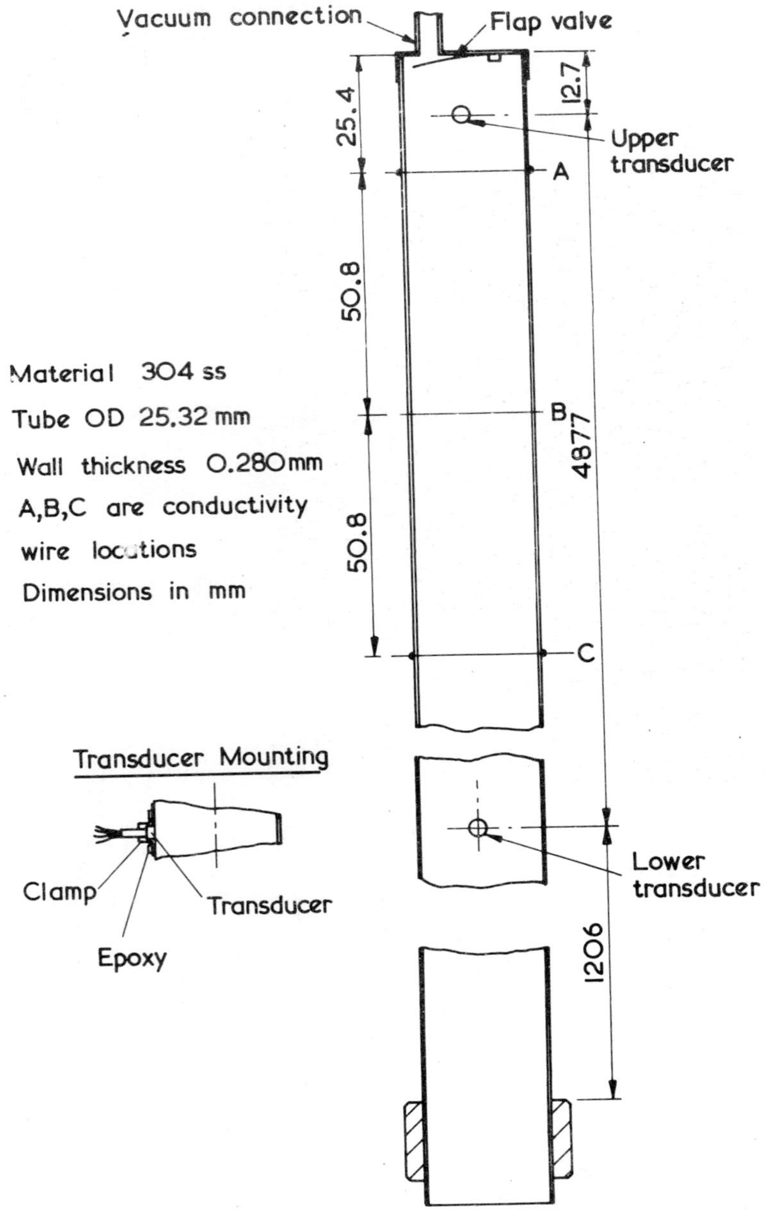

Fig. 5 Test section details

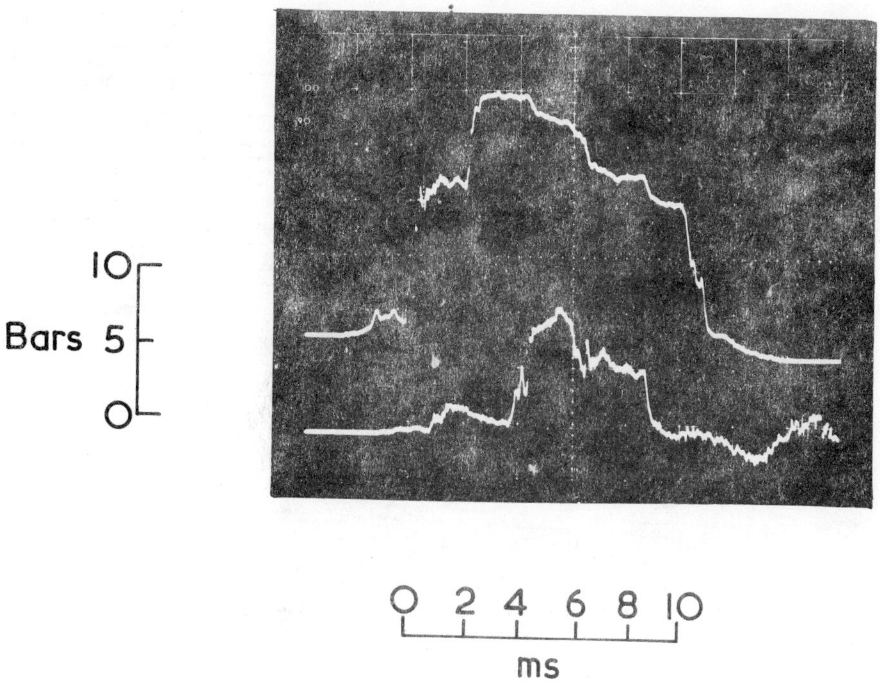

Fig. 6 Typical pressure traces

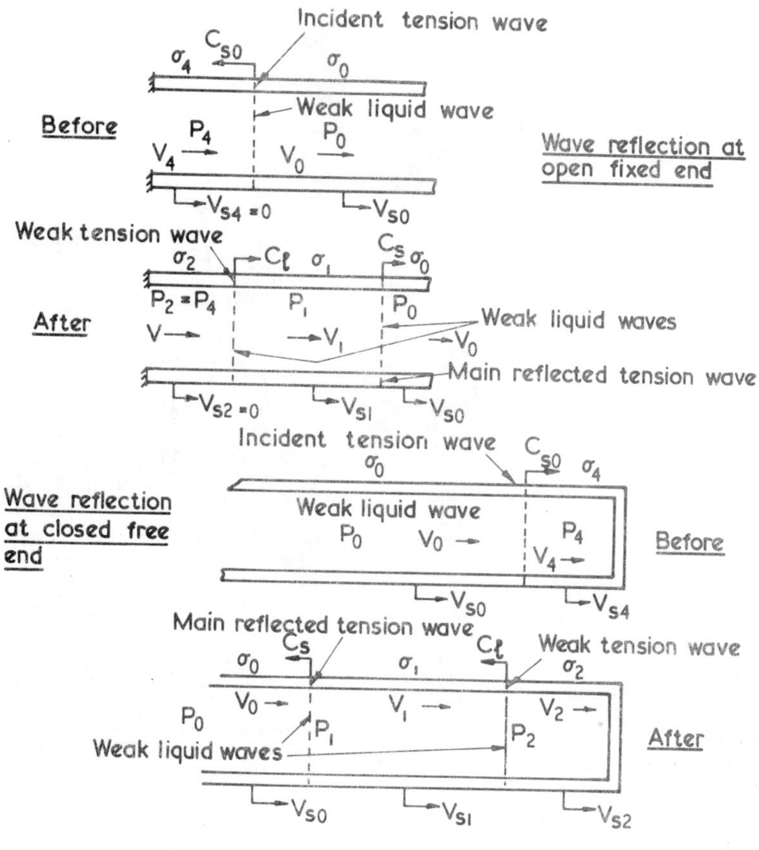

Fig. 7 Wave reflections

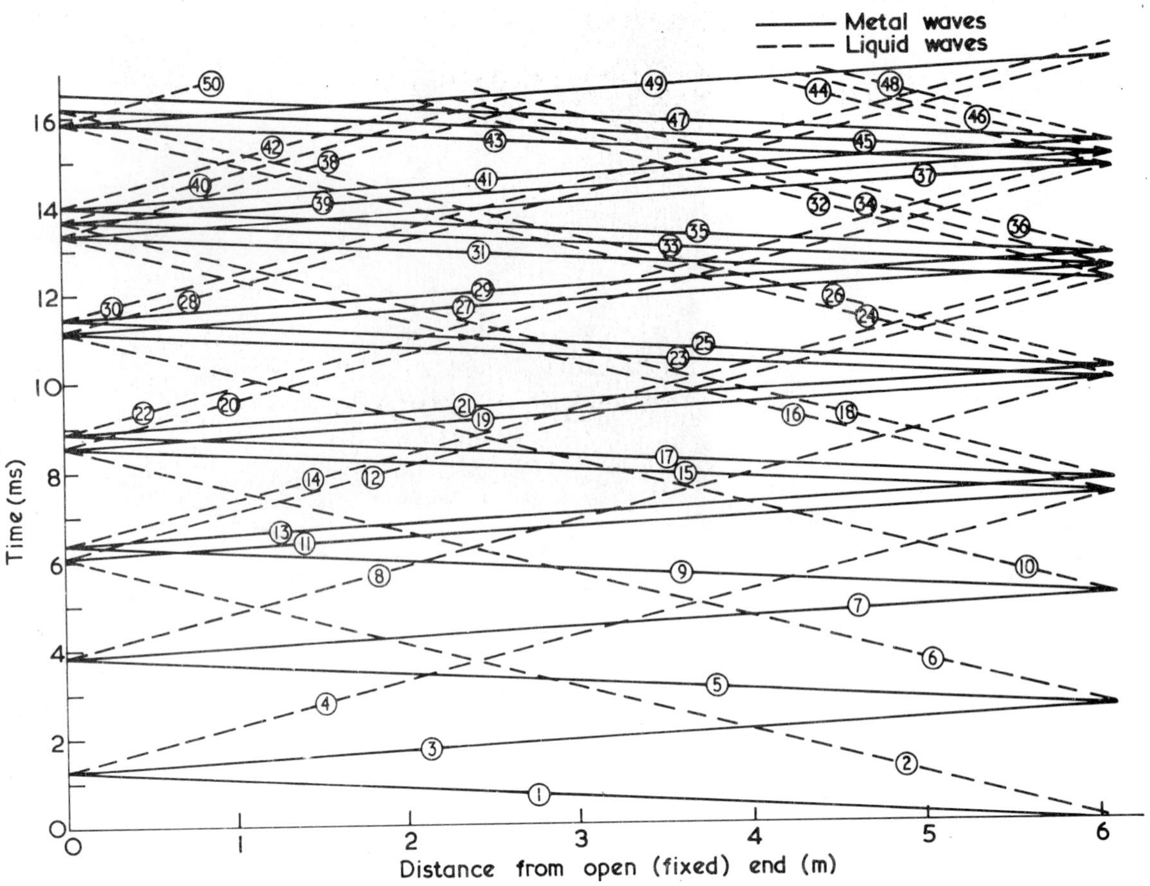

Fig. 8 Calculated wave positions

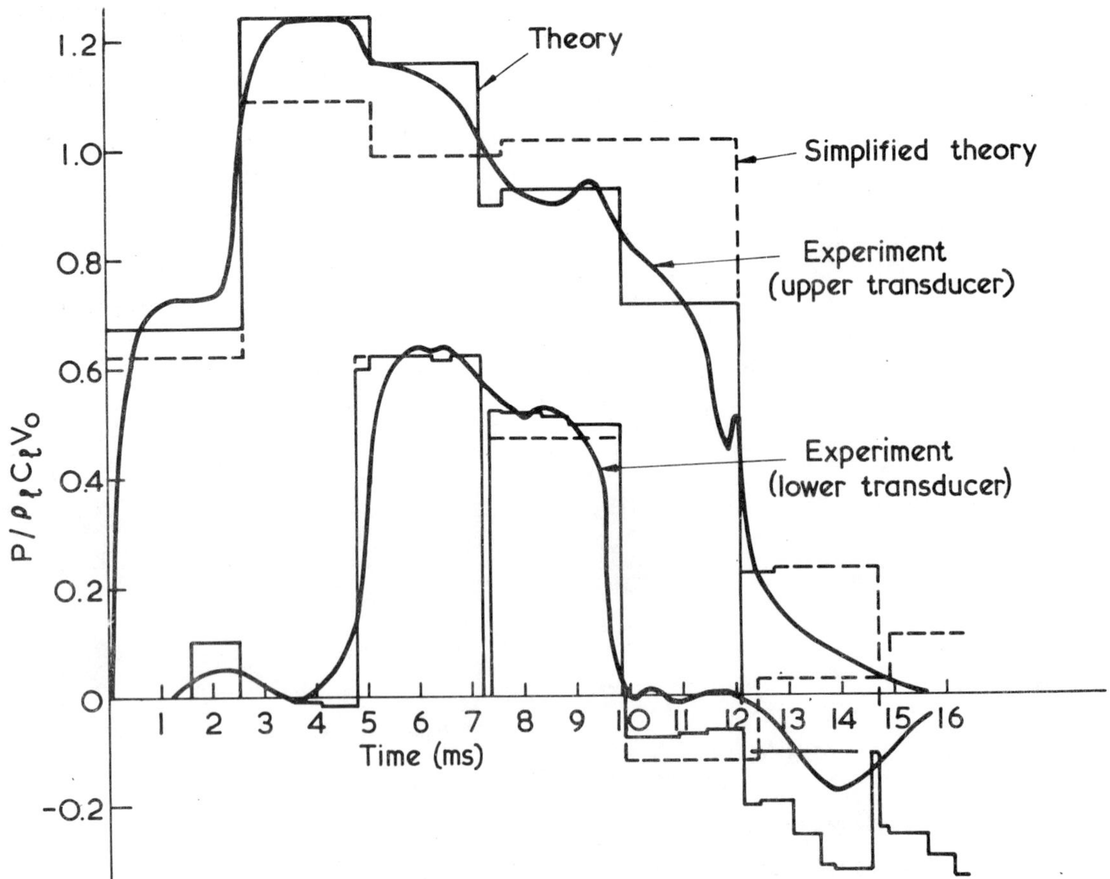

Fig. 9 Comparison of theoretical and experimental pressures

Third International Conference on

PRESSURE SURGES

Canterbury, England
March 25-27, 1980

PAPER E5

THE EFFECT OF COMPRESSIBLE PIPE-LINERS ON PRESSURE TRANSIENT GENERATION

A.K. Grundy

Shell Internationale Petroleum Maatschappij B.V., The Netherlands

and

J.A. Fox

University of Leeds, U.K.

Summary

The design of a surge attenuator is presented which is based on the 'reduction of wave-speed' concept. The transient wavespeed along the length of the attenuator is reduced by increasing the compressibility of the enclosed fluid. This has been achieved by introducing a compressible material into direct contact with the fluid, without interrupting the flow.

An experimental investigation performed to assess the feasibility of such a device as a method of surge protection is described.

Two mathematical models are developed which describe the effect of the attenuator on transient flow within a pipeline. The first method of analysis is based on the method of characteristics and traces the propagation of pressure waves through the attenuator. The second, a simpler technique, considers the attenuator as a boundary condition combining the storage effects of the device at a single grid point, and ignoring internal transients. The results of the two theoretical analyses are compared with data obtained from the experimental tests. Agreement is generally good.

Finally, the possible applications of such a device are discussed.

Held at the University of Kent at Canterbury, U.K.

Conference Organised by BHRA Fluid Engineering, Cranfield, Bedford, England.
In conjunction with The City University, London.

NOMENCLATURE

a_F	area of flow
a_M	effective cross-sectional area of compressible material
a_P	area of pipe
a_T	area of attenuator
c	wavespeed
c_1	pipe constraint constant
d	diameter
e	pipe wall thickness
E	Young Modulus of elasticity
f	Darcy f, (friction factor)
F_i	function $= \frac{2.f_i.v_i\|v_i\|}{d}$
g	acceleration due to gravity
h	absolute head
K_F	bulk modulus of fluid
K'	effective bulk modulus of fluid
ℓ	length of attenuator
P	Pressure
R_M	material parameter (analogous to bulk modulus, equation 1)
t	time
v	mean flow velocity
w	weight density of fluid
x	distance along pipeline
α	ratio of area of compressible pipe-liner to total area of cross-section
ρ	mass density

Subscripts

0	original

INTRODUCTION

There are many methods of surge control available for the protection of pipeline systems, but for particular applications these may be impractical, or prohibitively expensive. In this paper the design of a new type of surge attenuator is presented which offers an alternative method of surge suppression.

The design philosophy of the attenuator is based on the 'reduction of wavespeed' concept. Since the head rise for a given change in velocity is directly proportional to the transient wavespeed, as defined by the Joukowsky equation

$$\Delta h = c.\Delta v/g \qquad (1)$$

the reduction of wavespeed may be beneficial in the control of transients. Primarily, the wavespeed depends upon pipe-wall properties and geometry, liquid compressibility, and gas content in free bubble form. One method of wavespeed reduction is the injecting of air into the liquid flow at low pressure points along a pipeline. The air can be beneficial not only in reducing maximum pressure magnitudes, but also in reducing the severity of cavitation, and the transient oscillations associated with resonance conditions. Although this technique can be used effectively, it must be noted that in certain instances entraining air into a system can in fact greatly worsen the circumstance.

This concept has been extended: rather than bleed air into a pipeline to increase the compressibility of the fluid, one may hold the air along the length of the pipe within a small flexible hose that is sealed at closely spaced sections so that the air is trapped. This is analogous to air in free bubble form; the bulk modulus of the system is reduced, and thus the transient wavespeed is lowered. This device was developed by Remenieras (ref.1) and was patented in France in 1951. Unfortunately this device has severe limitations, primarily caused by the positioning of the flexible hose inside the pipe. The hose restricts the free passage of any entrained solids within the flow and encourages the build up of deposits. This, in turn, causes the performance of the pipeline to deteriorate and can lead to problems with internal corrosion. In addition, routine maintenance of such a pipeline is hampered. The passing of scrapers (pigs) through the line is made impossible, and the replacement of a damaged hose would be troublesome, if not totally impractical. Later, Waller (ref.2) proposed the use of rubber hose and hollow rubber spheres trapped within the liquid to provide extra capacitance in the system, but similarly the application of these methods are limited.

The surge attenuator presented in this paper overcomes the disadvantages associated with the devices discussed above. The schematic diagram, Fig. 1, shows the basic design of the attenuator. The interspace between the two cylinders (the outer casing and the perforated core) is occupied by a compressible material, hence, the compressibility of the fluid in the device is increased, reducing the local wavespeed without interrupting the flow. The performance and feasibility of such a device have been investigated experimentally using a prototype attenuator, and two analytic techniques have been developed which model the transient behaviour of the attenuator: the first, based on the method of characteristics; and the second, a simpler model, which treats the attenuator as a single boundary condition.

The Experimental Investigation

The attenuator built for test purposes was 2.75 m in length with an outer casing of Class D P.V.C. pipe, external diameter 12 inch and wall thickness 0.625 inch, enclosing a perforated core of 4 inch diameter. An exploded view, Fig. 2, shows the end plate assembly; the outer casing located into a recess in the end plate where a rubber gasket coated with silicon grease formed a watertight seal. The mild steel end plates were secured using six high tensile steel tie rods. Three tie rods, set 120° apart, were permanently fixed into each end plate, the free ends passed through locating holes in the opposite plate and were bolted and tensioned. The perforated core was formed from a sheet of mild steel, 3 mm thickness with 6 mm bore perforations, and was galvanised. Two P.V.C. locating rings fitted onto each end of the core restricted any lateral movement, and enabled the steel flanges of the pipeline, which bolted directly into each end plate, to hold the core in position. The compressible pipe-liner comprised twelve inflatable neoprene tubes arranged longitudinally as shown in Fig. 3. Each tube was 2.75 m long (the length of the attenuator), diameter 2 inch and wall thickness 1/16 inch. The tube dimensions were chosen to ensure high flexibility, and the material - neoprene - because it would not deteriorate while immersed in water. The tubes were sealed at one end and connected via aluminium adaptors, screwed into the end plate, to a compressed air supply manifold at the other. Thus, they could be inflated to any desired pressure.

The attenuator was installed and tested in a pipeline test loop, shown in a schematic diagram Fig. 4. The pipeline was constructed of P.V.C. Class D pipe, external diameter 4 inch, the total length being 110 m, forty times the length of the attenuator. Flow through the system was provided by a centrifugal pump with a 'no flow head' of 20 metres of water and a maximum flow velocity of 3.5 m/sec. The initial transient causing unsteady flow in the pipeline was generated by a 4 inch diameter butterfly valve situated downstream. A pneumatic actuator operated the valve and could produce closure times of 0.1 sec. Pressure measurement was by means of differential pressure gauges, with output to an ultra-violet light recorder.

A series of test runs were carried out with the air pressure in the neoprene tubes at 5, 10 and 15 p.s.i. For each run, the system was allowed to reach steady state conditions for that particular flow rate, before closure of the butterfly valve was initiated.

THE METHODS OF ANALYSIS

The mathematical modelling of the attenuator is considered in two discrete sections:-

The 'transient' model

This model traces the propagation of pressure waves through the attenuator and requires the development of a stress-strain relationship for the compressible pipe-liner, and the redevelopment of the wavespeed and characteristic equations to include variations in the effective cross-sectional area of flow caused by the passage of a pressure wave; (i.e. compression and expansion of the pipe-liner).

The wavespeed equation:

When a pressure change δP is applied to the fluid in the attenuator a number of volume changes occur. Considering a unit length, the outer pipe will extend diametrically (ref.3)

$$\delta a_P = a_P \cdot \frac{dc_1}{Ee} \cdot \delta P \qquad (2)$$

where a_P is the cross-sectional area of the attenuator, E the Young's Modulus of the pipe-wall material, e the wall thickness, and c_1 a constant dependent on the pipe constraint.

The fluid volume increase per unit length is given by

$$\delta a_F = -\frac{a_F}{K_F} \cdot \delta P \qquad (3)$$

where a_F is the cross-sectional area of flow, and K_F the bulk modulus.

The volume change per unit length of compressible pipe-liner can be expressed as

$$\delta a_M = -\frac{a_M}{R_M} \cdot \delta P \qquad (4)$$

where a_M is the effective cross-sectional area of the compressible material, and R_M is a material parameter analogous to bulk modulus. In the case of neoprene tubes, R_M relates volume change with pressure variation in terms of the bulk modulus of the trapped air and unit hoop strain in the tube wall. The development of this stress: strain relationship was based on the basic relationships for pulsatile flow through highly distensible vessels (ref.4).

The net volume change per unit length is therefore

$$\delta a_T = -\left[\frac{a_F}{K_F} + \frac{a_M}{R_M} + a_P \cdot \frac{dc_1}{Ee}\right] \cdot \delta P \tag{5}$$

If α is defined as the ratio of the effective cross-sectional area of the compressible material to the total area in a given cross-section, a_T, then the wavespeed can be defined by

$$c = \left[\frac{K'}{\rho}\right]^{\frac{1}{2}} = \frac{1}{\sqrt{\rho\left[\frac{(1-\alpha)}{K_F} + \frac{\alpha}{R_M} + \frac{dc_1}{Ee}\right]}} \tag{6}$$

where K' is the effective bulk modulus of flow through the attenuator, and ρ the density of the fluid.

The continuity equation:

Referring to Fig. 5, considering a horizontal pipe, and lumping the compressible material, the net mass inflow and mass storage for continuity can be equated (ignoring second order terms) as follows

$$\frac{\partial v}{\partial x} + \frac{v}{a_F} \cdot \frac{\partial a_F}{\partial x} + \frac{v}{\rho} \cdot \frac{\partial \rho}{\partial x} + \frac{a_T}{a_F} \cdot \frac{\partial P}{\partial t} \cdot \frac{1}{K'} = 0 \tag{7}$$

Now $$\frac{v}{\rho} \cdot \frac{\partial \rho}{\partial x} = \frac{v}{\rho} \cdot \frac{\rho_o \frac{\partial P}{\partial x}}{K_F}$$

also $\frac{\rho}{\rho_o} = 1 + \frac{P}{K_F}$ and as K_F is extremely large in comparison to P

(in the practical circumstance), ρ/ρ_o can be approximated to unity. Using the relationships from the last section, equations 2 and 4, the change in the area of flow can be defined as the sum of the attenuator distension and the compression of the pipe-liner, hence

$$\frac{v}{a_F} \cdot \frac{\partial a_F}{\partial x} = \frac{v}{a_F}\left[\frac{a_M}{R_M} + a_P \cdot \frac{dc_1}{Ee}\right]\frac{\partial P}{\partial x}$$

∴ using these relationships and transforming into total area, equation 7 becomes

$$\frac{\partial v}{\partial x} + \frac{v}{(1-\alpha)}\left[\frac{\alpha}{R_M} + \frac{dc_1}{Ee}\right]\frac{\partial P}{\partial x} + \frac{v}{K_F} \cdot \frac{\partial P}{\partial x} + \frac{1}{(1-\alpha)} \cdot \frac{\partial P}{\partial t} \cdot \frac{1}{K'} = 0 \tag{8}$$

rearranging, substituting equation 6 and converting into head

$$\frac{\partial h}{\partial t} + v\frac{\partial h}{\partial x} + (1-\alpha)\frac{c^2}{g} \cdot \frac{\partial v}{\partial x} = 0 \tag{9}$$

This is the continuity equation.

The dynamic equation:

The dynamic equation of unsteady flow in the attenuator

$$\frac{\partial h}{\partial x} + \frac{1}{g} \cdot \frac{\partial v}{\partial t} + \frac{v}{g} \cdot \frac{\partial v}{\partial x} + \frac{2fv|v|}{gd} = 0 \tag{10}$$

is also the standard dynamic equation for a horizontal pipe. The derivation of equation 10 is virtually identical to that of the standard equation (ref.5) if the frictional resistance between the flow and the pipe-liner is considered to be negligible.

The characteristic equations:

Using a standard method of transforming equations 9 and 10 (ref.6), the following characteristic equations were obtained

$$\frac{dx}{dt} = v \pm c\sqrt{1-\alpha} \tag{11}$$

and

$$\frac{\pm g}{c\sqrt{1-\alpha}} \cdot \frac{dh}{dt} + \frac{dv}{dt} + \frac{2fv|v|}{d} = 0 \tag{12}$$

These four equations were then cast into finite-difference form and solved numerically in a manner similar to that of a 'standard' method of characteristics analysis (refs. 5 and 7). A regular rectangular grid was employed for the numerical scheme, in which the Δx length representing the attenuator was subdivided into a number of $\Delta\Delta x$ lengths over which equations 11 and 12 were valid, see Fig. 6. The upstream and downstream boundaries of the attenuator UB and DB marked the limits between which C_A^+ and C_A^- could travel. If a characteristic crossed either UB or DB then the boundary condition was solved or a time wise interpolation was performed at that boundary.

Initially, at the beginning of a Δt time interval, all information is known. The initial values α and R_M can be used to calculate the wavespeed c, and equations 11 define points R and S. The integration of equations 12 in their finite difference forms can now be performed; the boundaries being dealt with first and then the internal grid points. Using the new heads, the values of α and R_M can be redefined at each grid point, then reassigned as initial values and the operation repeated.

The interpolation scheme employed was proposed by Vardy (8). This allowed flexibility of the Courant-Lewy (ref.9) stability criterion. The advantage of the scheme is that it minimizes the numerical attenuation and dispersion associated with interpolation in a region of varying wavespeeds.

The 'boundary condition' model

If the ratio of the length of the attenuator to the total length of the pipeline is small, the attenuator model can be simplified by concentrating the capacitance of the device at a grid point and treating it as a boundary condition, similar in many ways to the treatment of an airvessel (5).

Referring to Fig. 7, it is assumed that both the effective cross-sectional area of the pipe-liner, and the pressure, are constant throughout the length of the attenuator at any instant in time. The model is defined by the following four finite-difference equations: the two characteristic equations for the pipeline, the forward characteristic

$$h_P = h_R - \frac{c_0}{g}(v_{P1} - v_R + F_R \cdot \Delta t) \tag{13}$$

and the backward characteristic;

$$h_P = h_S + \frac{c_0}{g}(v_{P3} - v_S + F_S \cdot \Delta t) \tag{14}$$

the continuity equation

$$a_P v_{P1} = a_P v_{P3} + (a_F' - a_F)\ell/\Delta t \tag{15}$$

where the prime, a_F' refers to the value at P rather than the initial value, a_F, at 0;

and the stress:strain relationship for the compressible pipe-liner

$$a'_M = a_M \left[1-(h_P - h_O)w/R_M\right] \tag{16}$$

The solution of these equations is an iterative process. Initially a value of h_P is estimated (using the result from the previous time-interval as a guide), and substituted into equations 13 and 14 giving v_{P1} and v_{P3}. With these values, equation 15 can be solved yielding a'_E, hence, a'_M, and by back substitution into equation 16 a new value of h_P is obtained, h_{PNEW}. If $| h_{PNEW} - h_P |$ is within the chosen limits of accuracy, then the boundary condition has been solved, otherwise, h_{PNEW} is assigned to h_P and the computation repeated.

RESULTS OF THE EXPERIMENTAL TESTS AND THEORETICAL

Analyses

Examples of the results obtained are presented in Figs. 8 to 13, each graph comparing the recorded pressure histories at transducer 'a' with a computer simulation of the surge pressures generated in the unprotected line, and results of a theoretical analysis of the line with the attenuator installed; Figs. 8 to 10 for the 'transient' model, and Figs. 11 to 13 for the 'boundary condition' model.

The effectiveness of the attenuator in suppressing surge pressures was most impressive, the maximum surge pressure being restricted to within 30% of the value in the unprotected pipeline for the range of flows investigated. An effect of varying the air pressure in the tubes is immediately apparent on examining the practical traces; the higher the air pressure the faster the initial pressure rise. This was due to the relative compressibility of air at different pressures. However, another less obvious feature of the practical results was that, as the air pressure increased the attenuator became more effective. Over the range of flows investigated, the pressure rises for the 15 p.s.i. tests were smaller than for the 10 p.s.i., and they in turn were smaller than for the 5 p.s.i. tests. Yet obviously the effectiveness of the attenuator cannot continue to improve indefinitely as the air pressure is increased, hence, there exists an optimum air pressure for a specified range of flows at which the device operates with maximum efficiency.

The correlation between the experimental results and the theoretical data produced using the 'transient' model are good. The graphs show that the discrepancies between the two traces are small in comparison with the relative magnitudes of surge pressures in the unprotected pipeline. Referring to the 'boundary condition' model, Figs. 11 to 13, the correlation between results was also good. A property of these traces unlike those produced using the 'transient' model was the lack of pressure fluctuations on the initial surge. This serves to illustrate the fundamental differences between the two analytic techniques; the first, tracing the propagation of transients through the device and corresponding changes in the cross-sectional area of flow; the second, simply considering the storage effects of the device. The pressure fluctuations seen on the theoretical traces Figs. 8 to 10 were produced by the numerical scheme. They developed due to the interaction between step-waves generated by the theoretical treatment of valve closure, and reflections of these waves from the upstream boundary of the attenuator.

CONCLUSIONS

The performance of the attenuator in the experimental tests showed that the device can be used as an effective method of surge protection.

An attractive feature of the attenuator design presented in this paper, is that the compressibility of the pipe-liner can be altered by changing the air pressure in the neoprene-tubes. Thus, the performance of the attenuator can be 'tuned' to suit the transient behaviour of a particular pipeline system, and modified in situ to deal with varying operational requirements. Also, a puncture in one of the tubes would not necessarily disable the device. By adjustment of the air pressure, the attenuator would be able to operate satisfactorily for periods with a number of the tubes deflated.

The results of the theoretical analyses show that the 'transient flow' and 'boundary condition' models accurately simulated the unsteady conditions in the test pipeline with the attenuator installed. It is contended that the transient behaviour of any pipe network incorporating one or more devices of this type, can be predicted by the techniques developed with sufficient accuracy for engineering design requirements. However, there are important criteria to be born in mind when choosing between the two methods of analysis:-

(i) the more rigorously formulated characteristics analysis considers the propagation of transients through the attenuator, and thus computes the modified 'pipe period' of the system and the interaction between reflected pressure waves. This model is suitable for use in the transient analysis of any pipe configuration. The numerical scheme is not particularly difficult to program, no more so than the 'standard' method of characteristics (ref.5), although the computation time can be increased significantly if a large number of additional grid points are specified within the attenuator.

(ii) in contrast the 'boundary condition' model is a much simpler technique which requires little extra programming and virtually no additional computing time. However, the analysis takes no account of the time it takes a wave to pass through the device, and hence, the 'pipe period' of the system is incorrectly modelled; the larger the proportion of pipeline occupied by the attenuator the greater the error. This is not a serious criticism of the technique but rather an indication of its limitations. In the vast majority of engineering applications the device would necessarily be restricted to a small fraction of the pipeline length simply because of economic and practical considerations. Therefore, the 'boundary condition' model is considered suitable for the engineering design analysis of most pipe networks. However, it is foreseen that this method of analysis may not be adequate for all situations.

The main applications of the attenuator are thought to be in the protection of 'short' pipelines. For the attenuator to be effective, it is necessary that pressure waves transmitted from the device are reflected at the upstream boundary of the pipeline, and return before the total storage capacity of the attenuator is exhausted. The more fluid that is stored in the attenuator, the less compressible the pipe-liner becomes. Eventually, if the flow into the attenuator is not checked, the pipe-liner will cease to be effective allowing large surge pressures to develop which may damage the pipeline. Therefore, if the dimensions of the device are to be kept within acceptable limits, its most attractive application must be the protection of 'short pipelines'.

A possible application of the attenuator is the protection of offshore tanker loading systems. The floating hose strings used for connecting the pipeline to the tanker are at risk from unsteady flow pressures owing to the relatively high flow velocities maintained during loading. The attenuator has the advantage that it could be incorporated in a floating hose, close to the source of transient generation.

Also, the use of this type of device in the Medical field has been considered. Here the attenuator, with a pipe-liner of compressible material, could be used to replace a section of schlerotic aorta in a person suffering from a heart complaint such as angina pectoris or a heart condition resulting from a coronary infarction. This replacement, close to the heart, would drastically reduce stress within the body's arterial system, decreasing the possibility of a subsequent heart attack and its consequences.

ACKNOWLEDGEMENTS

The authors are grateful to the Science Research Council for their financial support during this work.

REFERENCES

1. Remenieras, G — Dispositif simple pour reduire la celerite des ondes elastiques dans les conduites en charge. La Houille Blanche, no. Special A, pp. 172-196. (1952)

2. Waller, E.J. — Pressure surge control in pipeline systems. Oklahoma State University, Pub. 102, Stillwater, Okla. (1958)

3. Streeter, V.L., Wylie, E.B. — Hydraulic Transients McGraw-Hill. (1967)

4. Streeter, V.L., Keitzer, W.F., Bohr, D.F. — Pulsatile pressure and flow through distensible vessels. Circ. Res. v.13, pp.3-20. (1963)

5. Fox, J.A. — Hydraulic Analysis of Unsteady Flow in Pipe Networks. Macmillan. (1977)

6. Forsythe, G.E., Wasow, W.R. — Finite-Difference Methods for Partial Differential Equations. John Wiley and Sons. (1960)

7. Lister, M. — The numerical solution of hyperbolic partial differential equations by the method of characteristics. In: A. Ralston and H.S. Wilf (eds.), Mathematical Methods for Digital Computers. John Wiley and Sons. New York. (1960)

8. Vardy, A.E. — On the use of the method of characteristics for the solution of unsteady flows in networks. Second Inter. Conf. on Pressure Surges, Sept. (1976)

9. Courant, R., Friedricks, K., Lewy, H. — On the partial differential equations of mathematical physics. New York University Institute of Mathematics, translated by P. Fox. (1956)

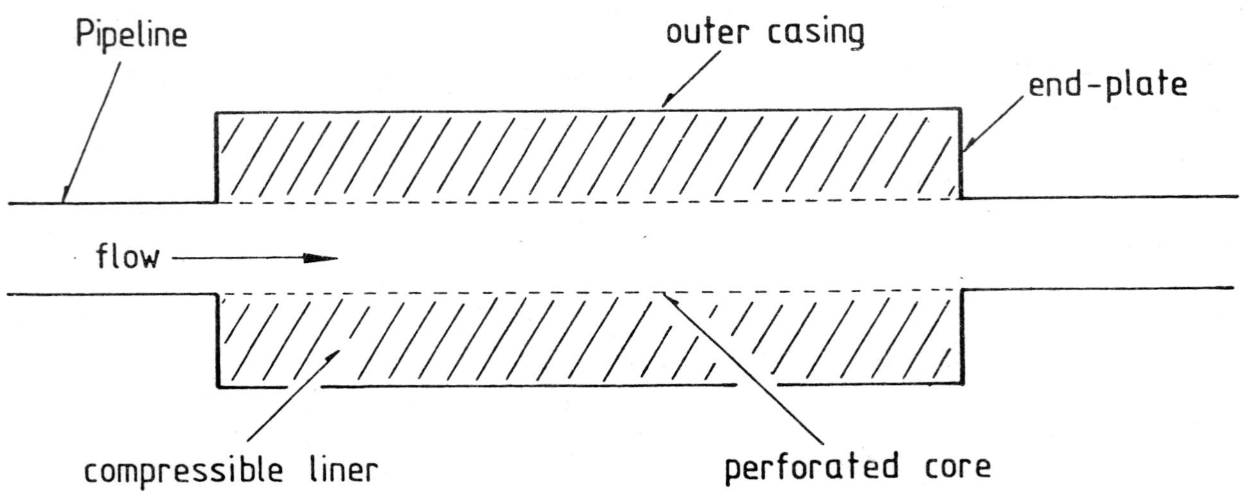

Fig. 1 Schematic diagram of the Attenuator

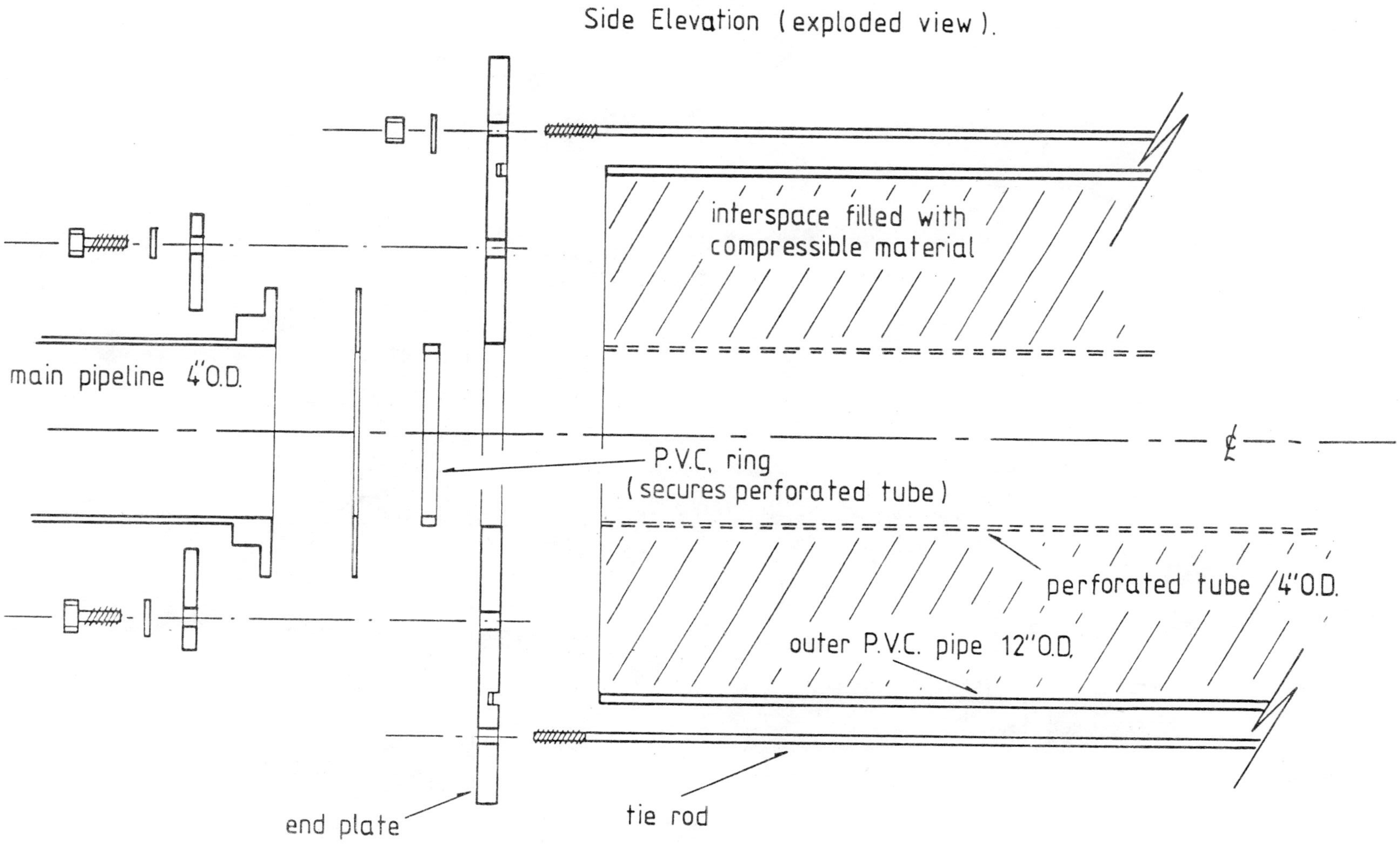

Fig. 2 The Attenuator (basic design)

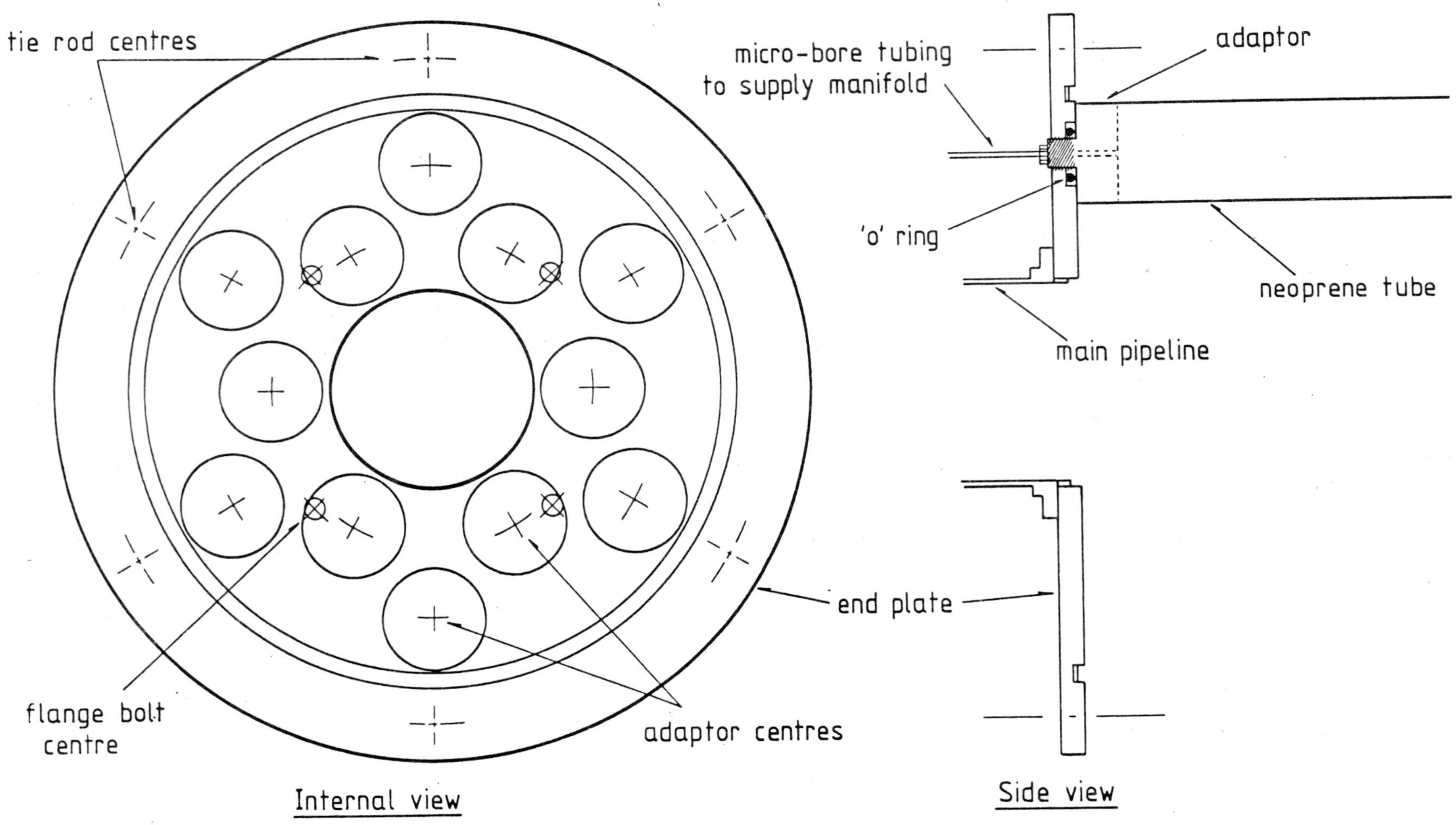

Fig. 3 Attenuator end plate (showing tube attachment)

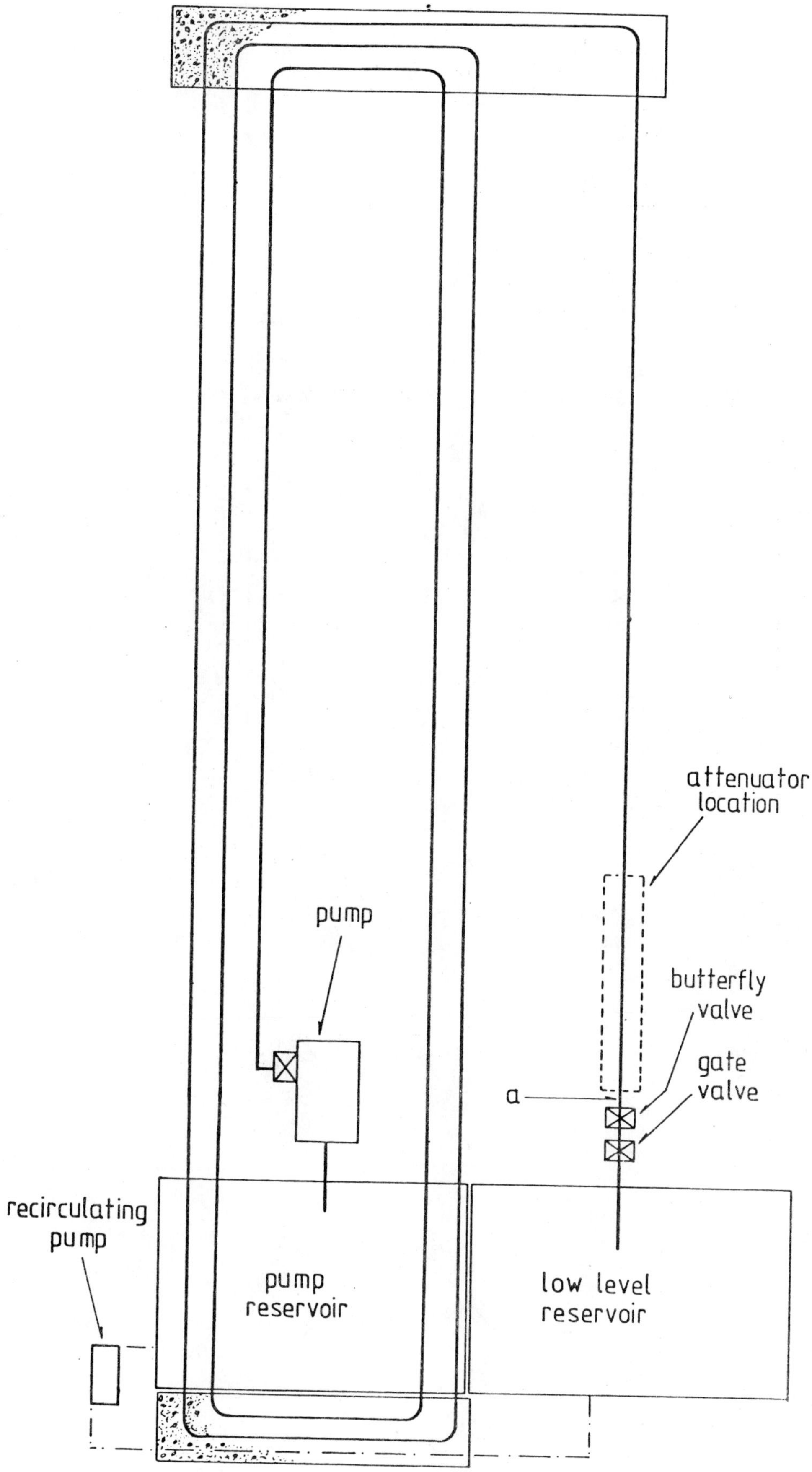

Fig. 4 The test pipeline (plan)

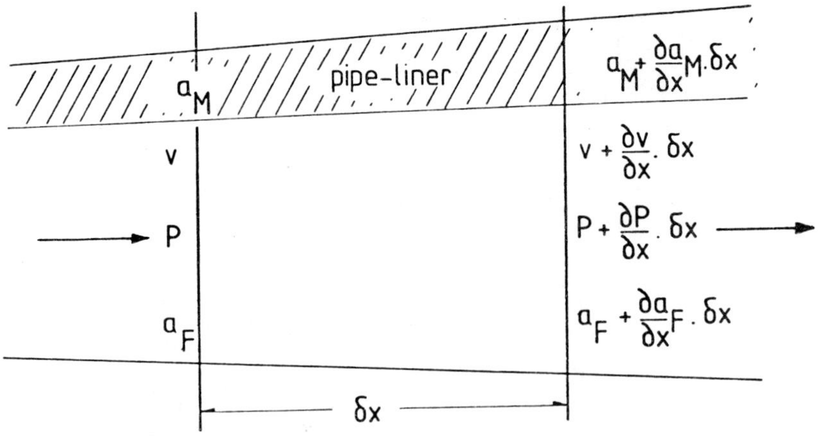

Fig. 5 Elemental length of the Attenuator ('Transient' model)

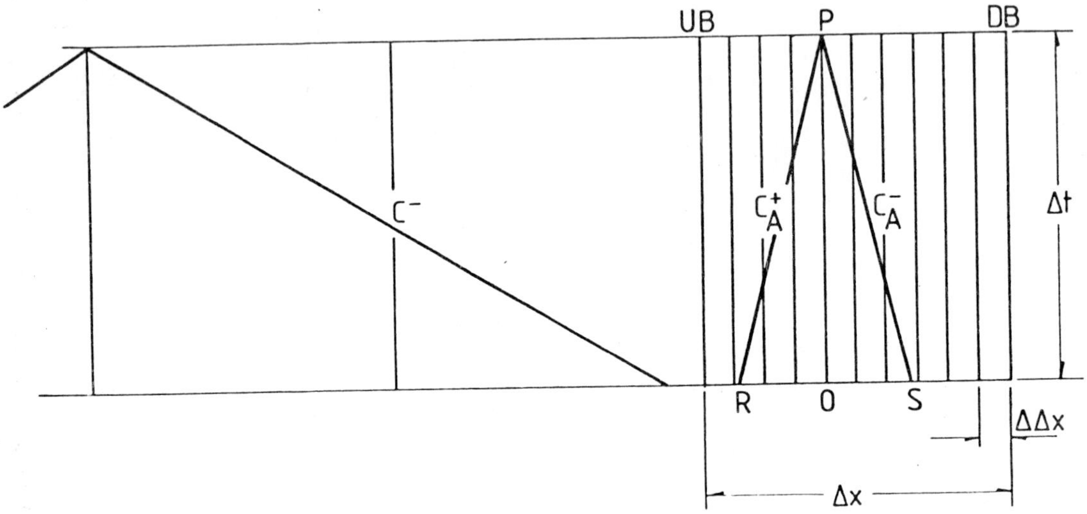

Fig. 6 Computation grid for the Attenuator and adjoining pipeline

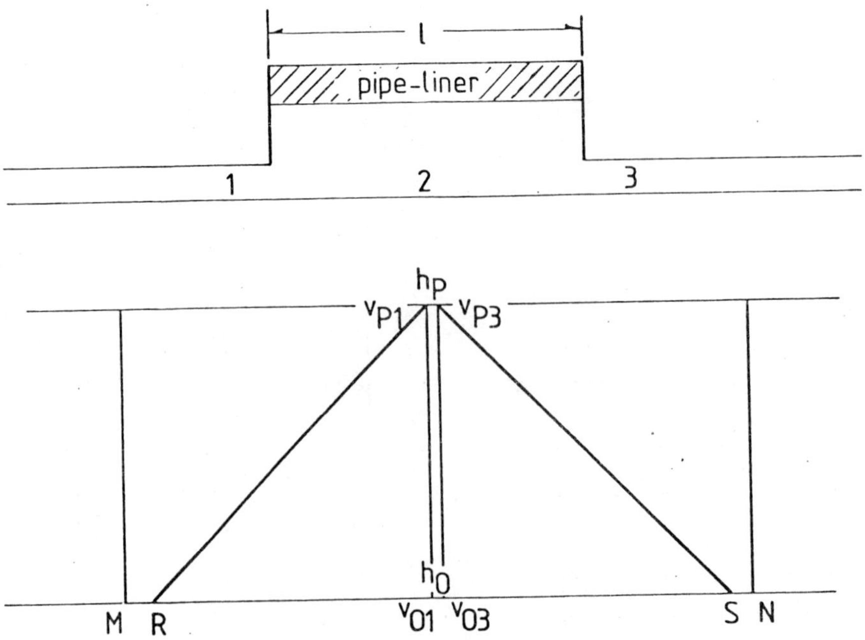

Fig. 7 The Attenuator ('Boundary condition' model)

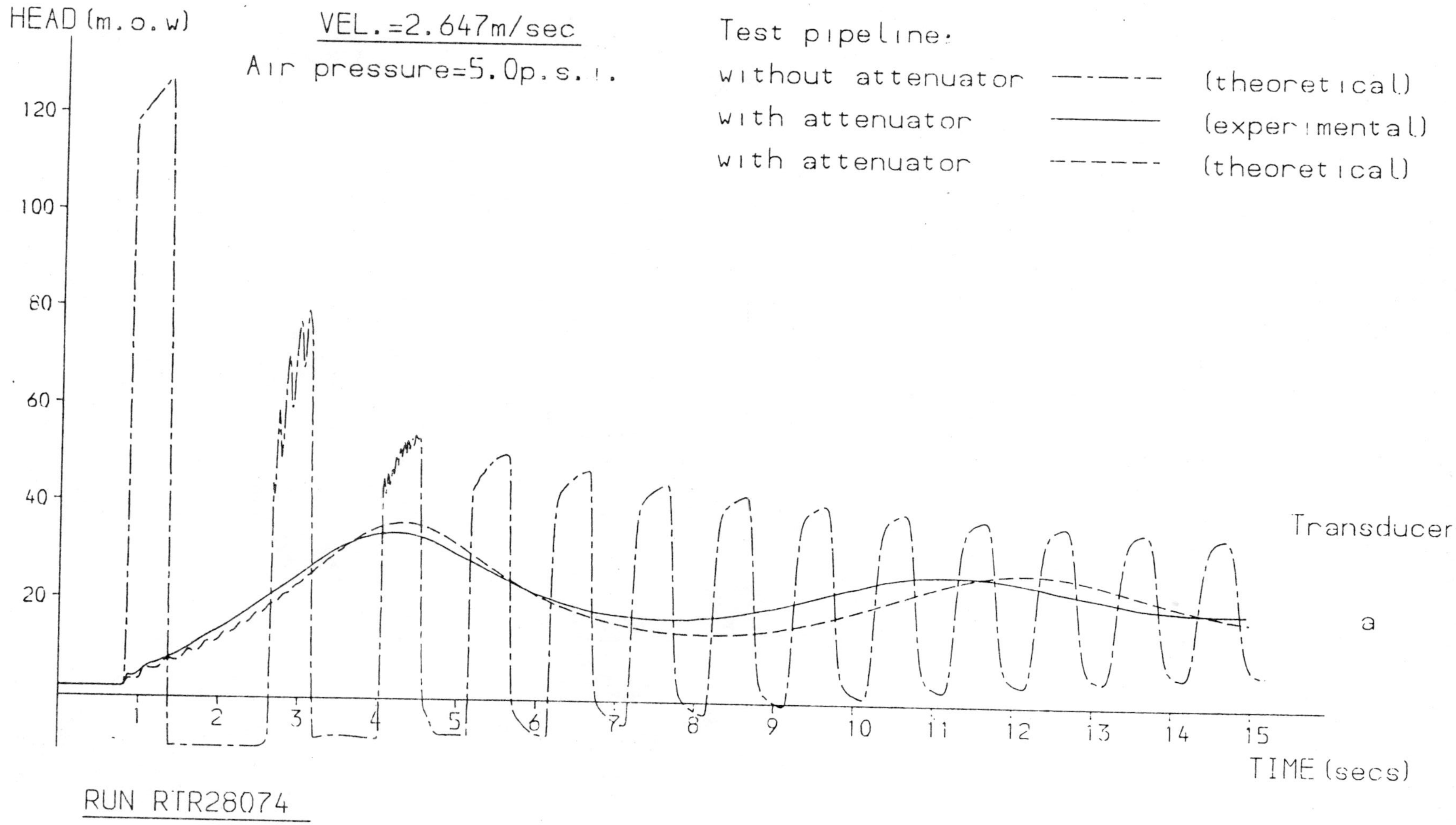

Fig. 8 Graph comparing experimental and theoretical results (Attenuator analysis – 'Transient' model)

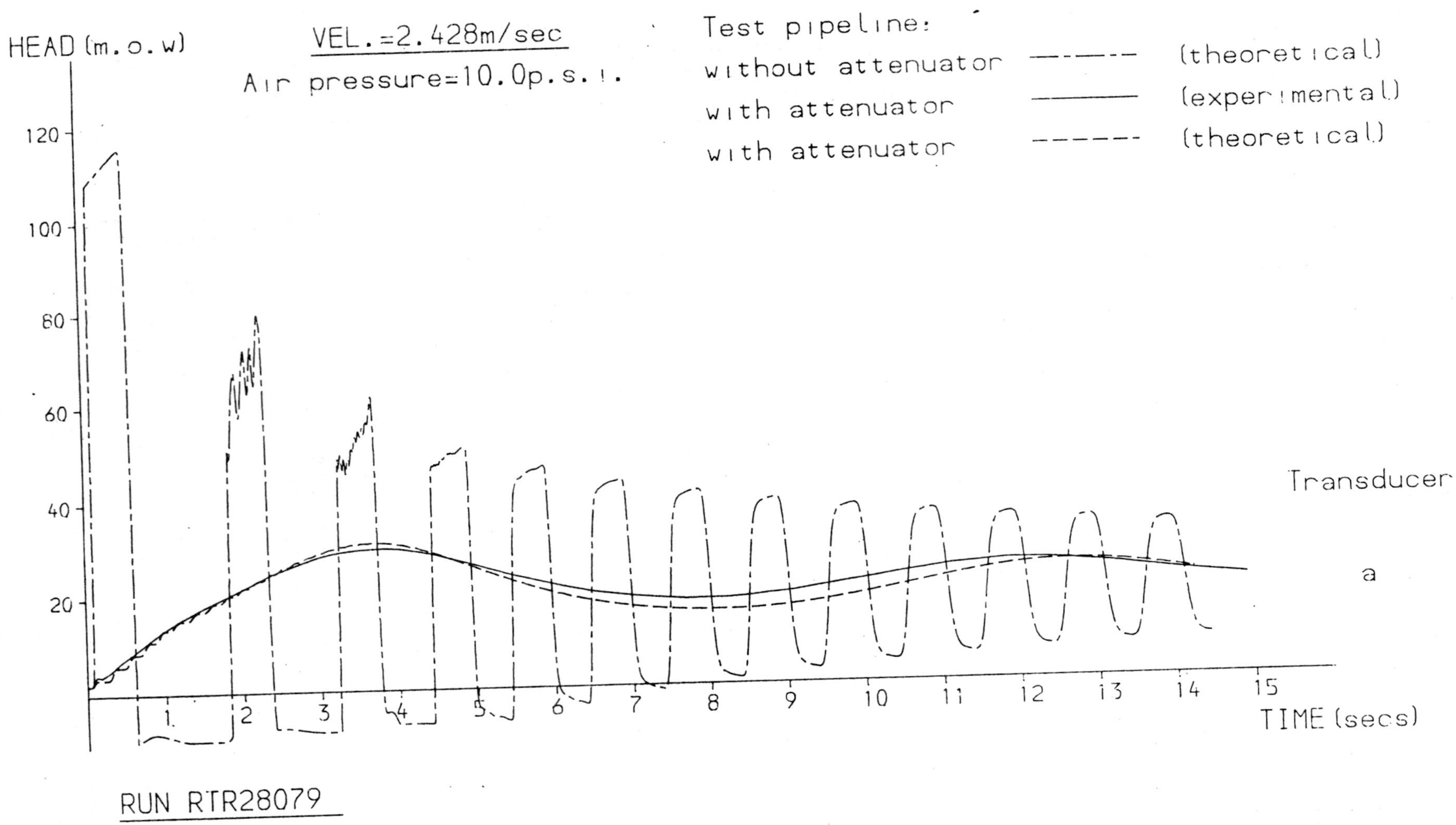

Fig. 9 Graph comparing experimental and theoretical results (Attenuator analysis – 'Transient' model)

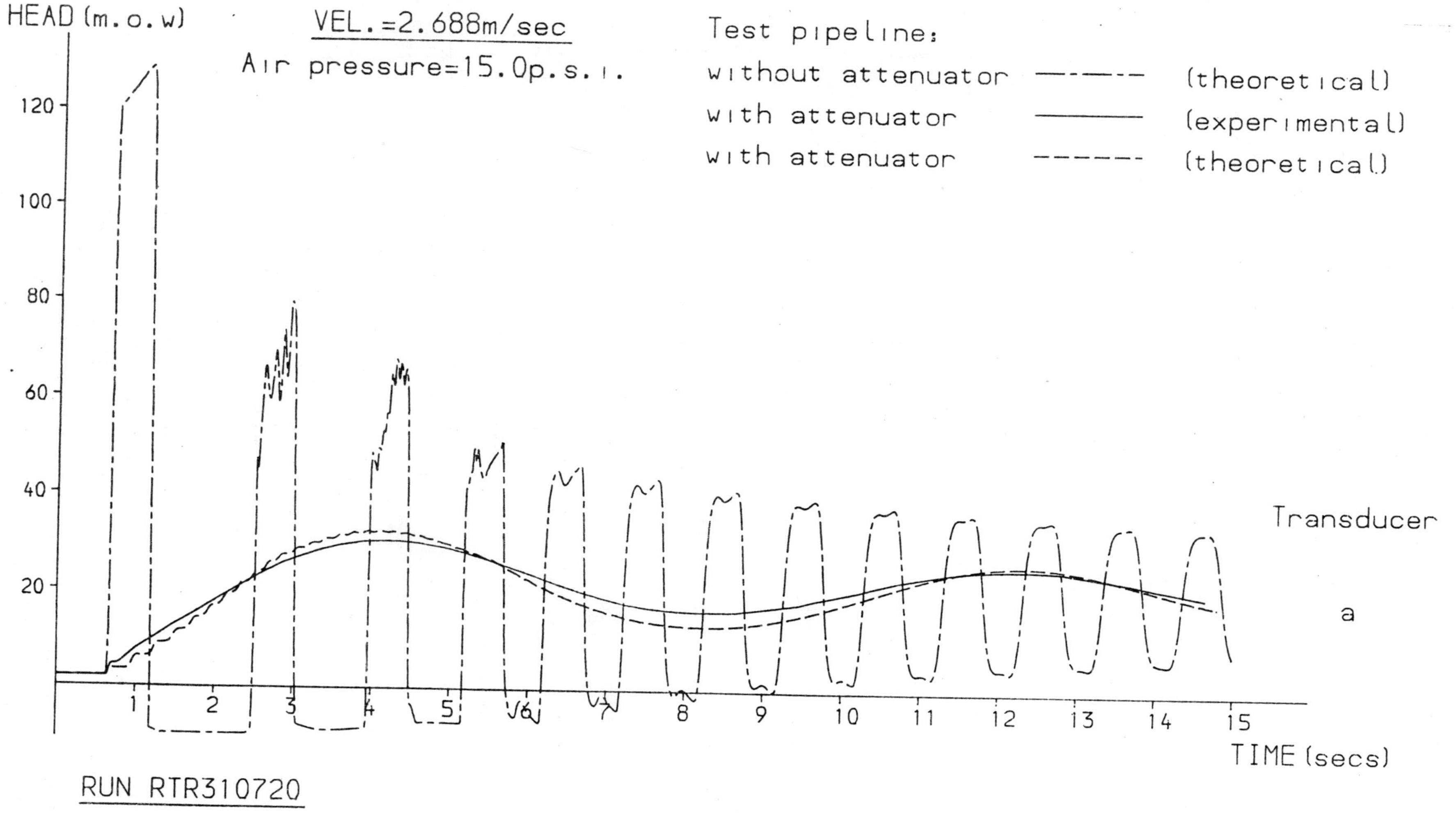

Fig. 10 Graph comparing experimental and theeoretical results (Attenuator analysis — 'Transient' model)

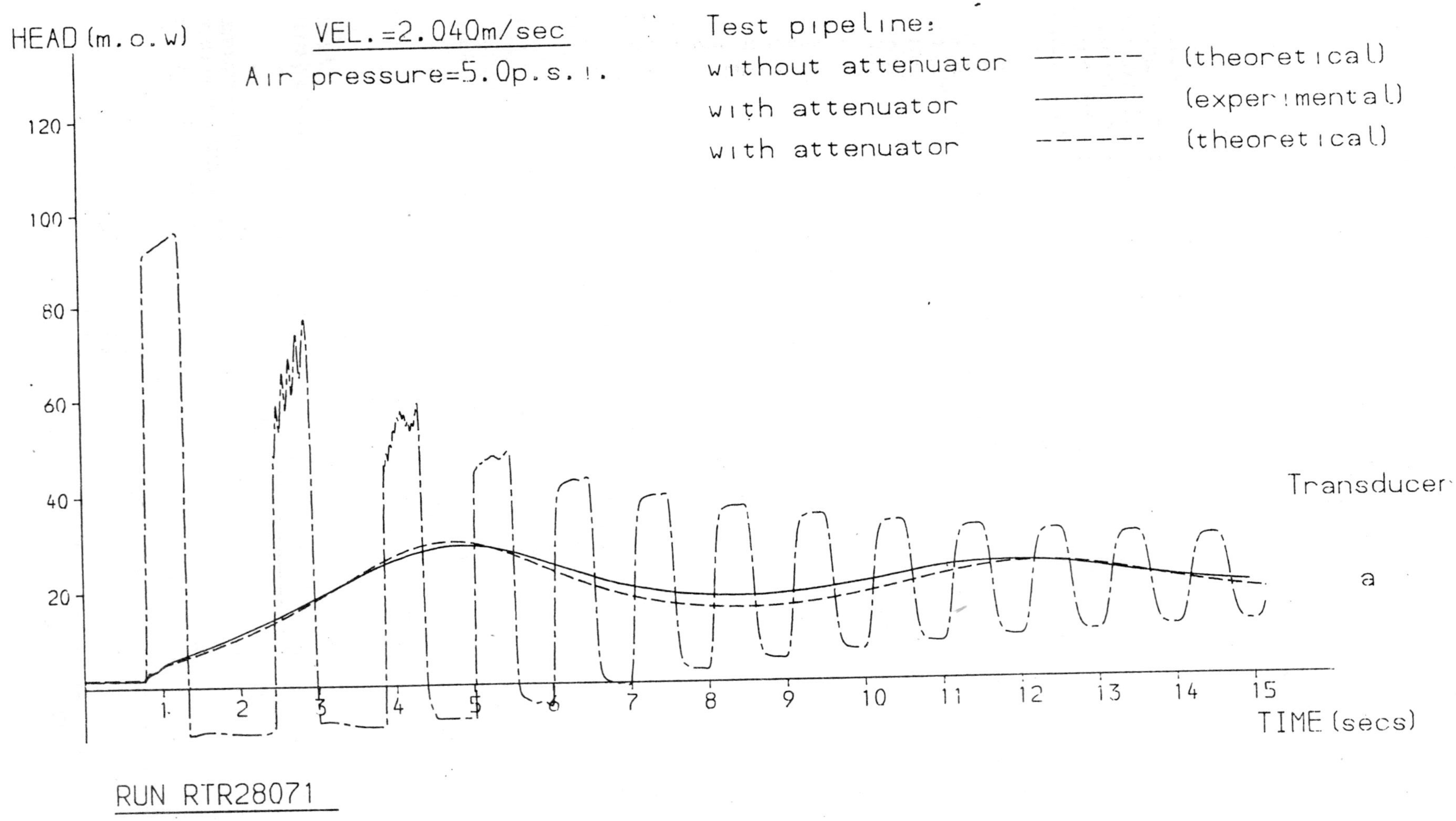

Fig. 11 Graph comparing experimental and theoretical results (Attenuator analysis — 'Boundary condition' model)

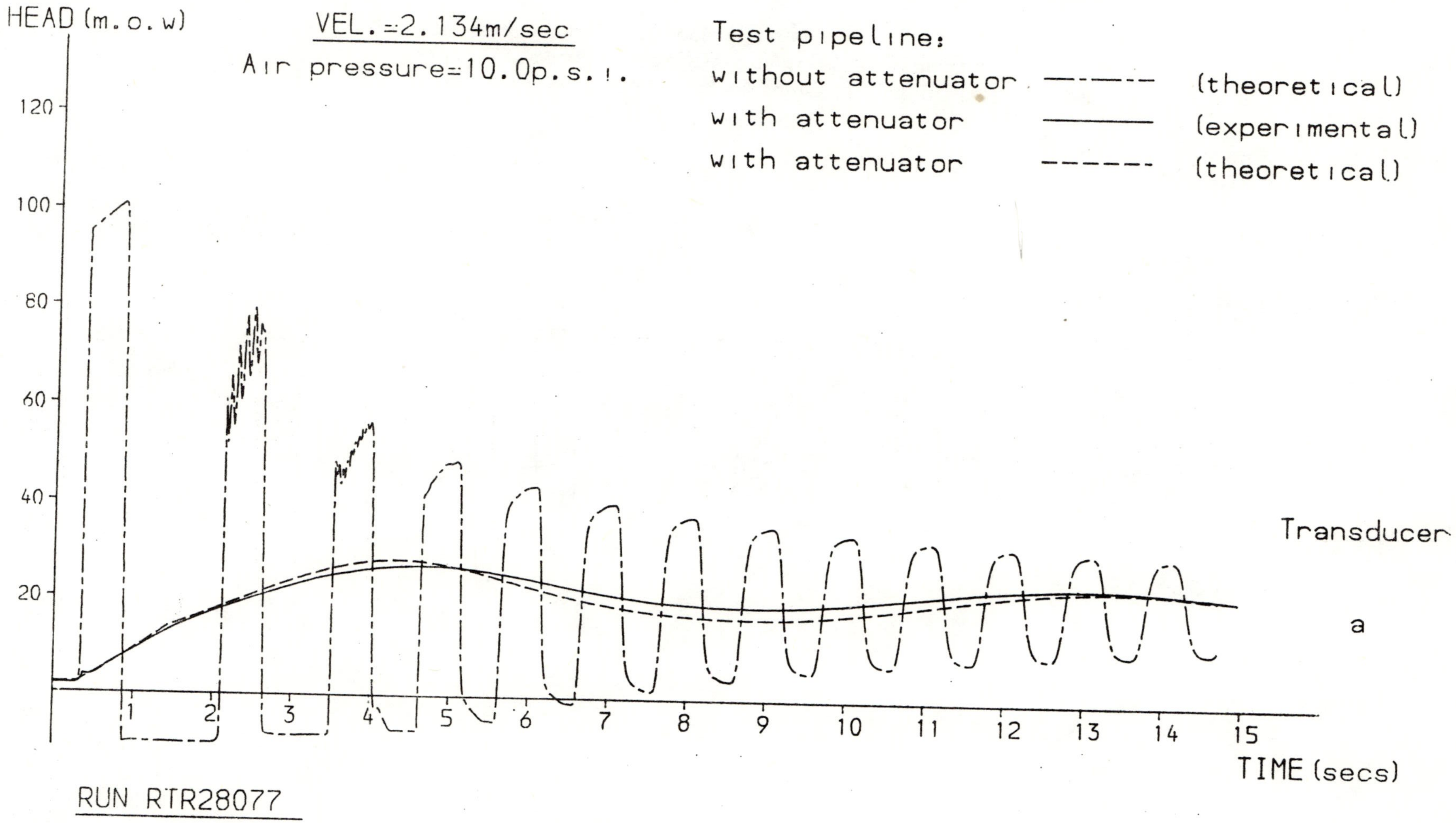

Fig. 12 Graph comparing experimental and theoretical results (Attenuator analysis — 'Boundary condtion' model)

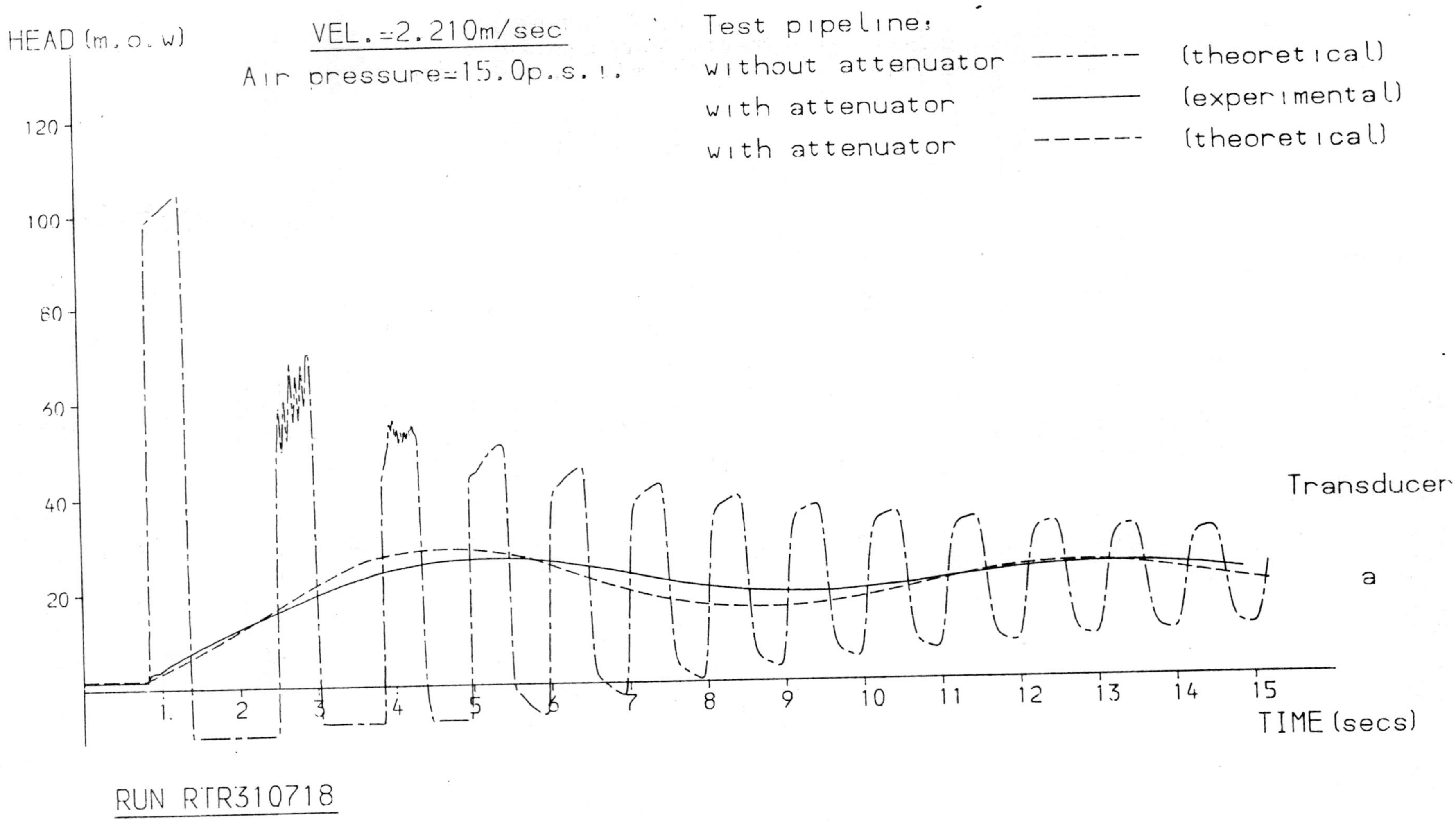

Fig. 13 Graph comparing experimental and theoretical results (Attenuator analysis — 'Boundary condition' model)

Third International Conference on
PRESSURE SURGES
Canterbury, England
March 25-27, 1980

PAPER F1

RECENT DEVELOPMENTS OF AN EXISTING TRANSIENT-SIMULATION PROGRAM FOR COOLING WATER CIRCUITS

M. Fanelli and G Fenini

ENEL-DSR-CRIS, Italy.

Summary

This work is a first approach to extend the application rules of a computation program which describes transients in cooling water systems.

The program has been developed since the end of the 60's and increased in scope and computational power later. It can simulate transients originated by the trip-off of the pump (or pumps) with or without operation of air intake valves both at the condenser and at a downstream syphon. Cavitation phenomena can be represented.

The newly added part concerns the starting up of pump with the circuit and the condenser full of water or partially empty.

In this case manoeuvres are possible concerning the rev-up of the pump, butterfly valve opening rate, and on request rate and duration of opening for the aeration valve on the syphon.

Some comparison tests have been performed taking as reference measurements taken on existing Italian plants.

Each simulation output can be obtained both as numerical print-outs and as automatically done graphical results.

Held at the University of Kent at Canterbury, U.K.

Conference Organised by BHRA Fluid Engineering, Cranfield, Bedford, England.
In conjunction with The City University, London.

The Centre for Hydraulic and Structural Research (CRIS), belonging to the Research and Development Department (DSR) of the Italian National Power Board (ENEL), has been studying since the end of the Sixties the problems connected with hydraulic transients in the condenser cooling water circuits of thermal power plants.

As well known, such circuits are often exposed to situations of water hammer overpressures in the events of pump trip-off or pump trip-on. The magnitude of these pressure peaks can be very high and may even exceed the maximum allowed by the condenser manufactures. In this event the condenser and the generating unit itself would be out of work for weeks.

The designers of cooling water circuits carefully try to avoid those situations, and are always fond of information about the true safety limits of their plants. They ask for modelling techniques able to clear up any doubts even at the cost of modifying the circuit lay-out and its controls.

Numerical models seem to be able to give satisfactory answers easier than physical models : in fact they can describe different kinds of circuits and allow several modifications both in geometry description and in simulating many kinds of manoeuvres as often as requested.

However, also important physical models of cooling water circuits have been built and operated by CRIS (Ref. 3), but the present report is not concerned with this part of research.

The activity of writing down numerical models is still going on and many problems to improve their reliability are not yet resolved. The complexity of basic physical phenomena to simulate (expecially when water column separation and/or reattachment occurs) is quite high and there are of course some arguments which representation is more or less adequate or which phenomena can nearly be neglected. (Ref. 4 and 5).

Although incomplete, numerical models can help the designer when for their plants they are concerned more about safety conditions than with debugging or optimizing all parts of the layout (Ref. 6).

The numerical model we are talking about (TRAID computer program) was developed mainly looking at this purpose: it is not a sophisticated model, but has become, year by year, more powerful and able to treat, with sufficient degree of reliability, different kinds of situations (Ref. 7 and 8), the last one concerning starting-up manoeuvres.

STRUCTURE OF THE "TRAID" SIMULATION PROGRAM

a) Physical assumptions

- Five basic physical configurations (or layouts) are considered for trip-off situations, three of which can also be used for the trip-on (or start-up) situations (Fig. 1).
- Celerity is assumed constant during transient time but with independent values in upstream and downstream conduits.
- Steady-state pump characteristic curves are used.
- Unsteady-motion hydraulic friction losses are assumed to follow the same law as steady-state friction losses (always the same function of the discharge rate).

- Valve is of the butterfly type: opening and closing is assumed to follow a linear or bilinear law of variation (angle v.s. time). Its head losses are assumed to follow the Love-Sarpkaya formula (Ref. 9).
- Cavitation occurs as soon as water absolute pressure falls down to vapour saturation pressure: if there is no air introduction, the cavity pressure is held fixed at vapour saturation pressure corresponding to the water temperature.
- A cavity in the condenser may also be found at the beginning of start-up procedures, in that case the air inside the condenser and the pipes may have different values in pressure, related to the levels of water remaining in the circuit (for example as the result of a previous trip-off transient).
- Air introduction can be operated at a pre-set time with a linear or parabolic law for opening and closing of air valves. Air introduction can be effected either at the top of the condenser downstream water-box or at another high point along the downstream (delivery) pipes.
- Air motion in the pneumatic circuit is markedly subsonic to allow a single dynamic equation for the whole air column to be used, taking into account inertia and head losses.
- Air evolution inside pneumatic circuit and condenser cavity follows a polytropic law. Air entrainment in the outcoming flow is taken in account.
- Inlet and outlet constancy of levels is assumed, outlet negative waves due to a long outfall free surface canal can be simulated.

b) Mathematical formulation and method of solution

The equation representing the different parts of the system are well-known and have been extensively reported on elsewhere (Ref. 1 and 2). They will not repeated there, due to lack of space, but only briefly mentioned.

The method of characteristics is used to represent wave propagation in the upstream and downstream conduits considering each one as a single uniform pipe element with its own properties. Continuous head losses are concentrated at the two ends of each conduit.

Pump characteristics are represented as non-dimensional discretized curves (both for $h = \pm 1$ and $m = \pm 1$) in the plane q, n of reduced discharge and reduced speed (Fig. 2).

Butterfly valve head loss curve vs. degrees of opening/closing angle is also discretized by points (Fig. 3).

Integration of the dynamic equations has required a predictor--corrector finite differences scheme with under - or over-relaxation procedures to get optimum convergence.

A common time interval has been adopted with time-interpolations for the two conduits (upstream and downstream the condenser) to keep free of necessity to divide them into many small portions.

The two air pipes leading to air valves on top of the condenser are represented as a single column having linearly-variable physical properties; its motion follows a polytropic law with a prefixed exponent intermediate between 1.0 and 1.4 (from isothermal to adiabatic conditions). The air column equilibrium is expressed by a global dynamic equation suitable for mass oscillations modelling instead of that for wave propagation.

In the present version of TRAID, the condenser filling and emptying operations are calculated by formulas taking into account both the surface of water inside the condenser and "weir effect" of the tubes connecting the condenser with the downstream pipes .

PROGRAM ORGANIZATION

TRAID program is divided into five main parts (Fig. 4).

1) Acknowledgement of basic data such as discretized characteristic curves of the pump and butterfly valve head-loss curve, geometrical and physical constants describing the circuit, definition of operating manoeuvres. This part also provides computations of steady-state work conditions before the transient phenomenon. Time is set to zero.

2) Increasing of time (actual transient time) and computation of numerical values about the first part of the circuit: head losses across the butterfly valve, discharges, piezometric heights and pump relation speed (checking on the pump characteristic curves). This part of the circuit takes into account, by means of wave propagation equations, pressure surges and flows proceeding from the condenser.

3) Computation of numerical values about the second part of the circuit: discharges and piezometric heights through the condenser and at the outlet using two different sets of equations, according to circumstances :

   a) when there is no water-column separation: wave propagation equations, as for upstream zone (butterfly valve), also for downstream zone (outlet syphon);

   b) when water-column separation occurs, whereas wave propagation equations are still used for upstream conduit, mass oscillation equations, with or without air injection, are used for the downstream conduit.

4) Printing and storing all numerical values, as far as they are requested for further calculations, and looping back to point 2 until the given time-limit is reached.

5) Optional manipulation of the most important values and generation of a sequential stream of data necessary to drive automatically a plotter.

TRAID runs on an Honeywell 66/60 computer, needs up to 22 Kwords of core and an additional direct access file up to 180 thousand characters. A transient time history, divided into 1000 instants (time steps), takes 10 or 20 seconds of CPU (time depends mainly on pneumatic equations) and 7 seconds of I/O. As input data TRAID asks for three different groups of informations :

1) steady-state pump characteristic curves, discretized with 360 points in the reduced plane q, n (Fig. 2);
2) butterfly valve head-losses curve, discretized with 40 points (Fig. 3) ;
3) geometrical and physical data of the circuit (diameters, lengths, celerities, elevations, roughness coefficients, etc. see Fig. 5) and other general informations such as: river level, times of operating manoeuvres , kind of basic configurations and options.

All input data are described in 63 punched cards.

TRAID output consists of the echo list of input data and, for each instant, a list of values about flow rates, piezometric heads, pressures, speeds, volumes and elevations computed at few typical points of the circuit.

Another optional output that may be either additional or substitutional of the numeric output, is a sequential file ready to be plotted. It contains data for plotting seven typical values, as piezometric heads upstream and downstream the butterfly valve (points A, B Fig. 5), piezometric heads upstream and downstream of the condenser (points C, D Fig. 5), piezometric head at outlet (point E Fig. 5), pump rotation speed and pressure of the admitted air, if any. The plot is completed with the draw ing of graded axes and a summary of operating manoeuvres (as the behaviour of the opening/closing of butterfly valve).

COMPUTATIONS VS. SURVEYS - COMPARISONS

The main validation we have always sought for each TRAID implementation, has been comparison with results got during field tests.

Field tests have been performed, by different survey organizations, on circuits n. 3 and n. 4 of the "La Casella" power station, with inlet and outlet on river Po.

Each test report contains graphic informations (oscillograms) about:

- Pump rotation speed
- Absolute pressures at characteristic points of the circuit
- Differential pressures in pump spiral case, inlet and outlet condenser elbows
- Angular displacement of butterfly valves downstream of pumps.
- Displacement of moving parts of vacuum breaking valves
- Angular displacement of a by-pass butterfly valve interconnecting circuits 3 and 4 .

The survey transient tests are 43, for the two circuits both in the interconnected and in the isolated mode. These tests cover a large range of situations, thus making precious data available for program validation. These data, collected in internal ENEL reports, have been made available upon request to other interested organizations.

Testing the last version of program TRAID we used among others the survey oscillogram n. 13 which shows (Fig. 6) a start-up transient with the river level very low and the conduits and the condenser partially filled of water before start--up.

A) Main test values

This test refers to circuit n. 4 in the isolated mode (without interconnection with circuit n. 3); its layout is showed in fig. 5. Nominal operating conditions are :

- Discharge: 8.2 m^3/s
- Generating unit rated electric power : 320 MW
- Conduits diameters: 2.0 m upstream, 1.6 m downstream the condenser
- Wave celerity (propagation speed): 880 m/s for both conduits
- Conduits lengths : 418 m upstream, 232 m downstream the condenser
- Condenser free horizontal surface: 24 m^2 (constant in whole height)
- Pumps: RIVA, centrifugal mixed-flow, vertical axis nominal revolution speed 238 turns minute, power 910 KW, PD^2 (reduced to pump shaft) 21,130 $kgfm^2$.
- Butterfly valve, downstream of the pump with 1.2 m diameter, 80 degrees of rotation to full opening.

B) Conditions at the beginning of test

- River level: 49.64 m (elevation above mean sea level; notice that the top of condenser is at 61.2 m a. s.l.)
- Butterfly valve closed
- Pump rotation: 30 turns a minute
- Upstream conduit full of water
- Water level in the condenser: 2.0 m below the top, that means a half metre below the bottom sill ("weir" crest) of the two tubes outcoming downstream
- Water level in the two vertical tubes immediately downstream the condenser: 8 m above the river level (that is the level of the syphon at outlet); this level is 3.56 m below the top of condenser
- Downstream conduit full of water
- Air pressure in the condenser: 10.33 m of water column (atmospheric)
- Air valve on the syphon at outlet: opened
- Pneumatic equilibrium between air pressure and levels.

C) Manoeuvres during test

- Pump speed reaches the nominal rotation of 238 turns in 3.5 seconds
- Butterfly valve opens following a bilinear law in 14 seconds (after 6 seconds the rotation of valve is 48 degrees instead of 46: valve full open corresponding to 0 degrees)
- Air valve on the syphon is always opened.

D) Comments upon simulation results

Starting conditions of piezometric heads are quite the same as requested and operating manoeuvres follow the same pattern.
After 5, 6 seconds from starting-up time piezometric heads upstream and downstream the condenser become different. The whole discharge is filling the condenser (before this time internal air pressure attempts to empty the condenser through the opening butterfly-valve) until the "weir" sill of outcoming tubes is reached.

Then both piezometric heads and air pressure grow continuosly with different speed and piezometric heads overcome air pressure.
The crossing point is at 10 seconds.
From now on the air pressure is kept fixed at 12.88 m of water column by the automatic opening of a disc valve on the top of the condenser, while piezometric heads go on increasing.

During this time the greatest part of discharge falls downstream the condenser and only a little portion is available for increasing water level in the condenser.
A meaningful increase is taking place meanwhile in the two vertical tubes downstream, causing the mo vement of water at outlet to initiate a mass oscillation.

From 25 seconds on, air pressure begins to decrease as a result of mass oscillations. Air is carried off continuously by discharge flow.
After 80 seconds from the beginning, the water discharge rate becomes about 60% of the nominal one.

This situation goes on as far as air is present in the condenser but unfortunately the survey record stops here.

As can be seen comparing the two diagrams (oscillograms n. 13 and Fig. 7), the behaviour of main quantities exposed is reproduced with good degree of accuracy by the numerical simulation.

Of course there are several parameters of the real circuit which are not exactly known and which have been more or less adjusted in the computation to obtain a good fit: very important is the law describing levels of water outcoming the condenser and the law about the carrying out of air. Both, as a few other assumptions, need to be improved in the near future.

However, the agreement being quite good for main quantities, it is deemed that this can be considered a reasonably reliable model allowing designers to try different kinds of situations (even those for which there are no survey reports to compare).

Figure 8 shows the same transient extended in time. The scale of time is reduced and the sketch of butterfly valve manoeuvre is missing to allow an intelligible representation. At 310 seconds the valve on the syphon begins to close (closure lasts 50 seconds) and at the end of the 400 seconds we have the normal steady-state situation in the circuit.

Figure 9 shows a situation quite similar to the Fig. 7 : the water level in the condenser is 0, 50 m higher and equals the elevation of the "weir" sill of the two tubes outcoming the condenser.

The piezometric heads at the beginning are of course 0, 50 m higher but what is interesting is the time of top air pressure: it is 5 seconds shorter than in Figure 7 , which allows to close earlier the syphon valve with the same behaviour of transient.

Finally, Figure 10 illustrates a situation where both conduits and condenser are supposed full of water (fully primed circuit) . Though high wave oscillations, are present, we can close the syphon valve much earlier than before (30 second from start-up) and with a short time of manoeuvre (only 10 seconds) making the circuit operating at steady-state only 65 seconds after beginning of the transient.

CONCLUSIONS

This new range of applications makes TRAID more useful for designers than before. Its basic philosophy aiming to ensure the safety of hydraulic systems, its low cost, easy input and various output options, sufficient reliability, have been already appreciated by designers.

We know that it is only a design tool having still several limitations and there is a long way to make it better or to transform it into a research tool. This stage has been only a little step.

BIBLIOGRAPHIC LIST

(1) G. Evangelisti
"Water hammer analysis by the method of characteristics"
L'Energia Elettrica n° 10, 11, 12 - 1969

(2) M. Fanelli
"Il regime transitorio nei circuiti di raffreddamento dei condensatori per le grandi unità termoelettriche - Studi di carattere analitico"
L'Energia Elettrica n° 12 , Dic. 1971 (in Italian)

(3) J. Bonnin, M.Fanelli, C.Jolas
"Etude des régimes transitoires dans les circuits hydrauliques de réfrigeration CEA-EdF : Cicles de conférences sur le source froides des centrales électriques". III Meeting on transients in cooling water circuits. Royaumont, 6-10 Oct. 1975 (in French)

(4) M.Fanelli, M.Reali
"A theoretical determination of the celerity of water hammer waves in a two-phase fluid mixture"
L'Energia Elettrica n° 4, April 1975

(5) E.Raiteri, F. Siccardi
"Transients in conduits conveying a two-phase bubbly flow: experimental measurements of celerity"
L'Energia Elettrica n° 5, June 1975

(6) R.P.Rossini
"I fenomeni di distacco di vena e la progettazione degli impianti idraulici delle centrali termiche e nucleari"
III Meeting on transients in cooling water circuits
Royaumont 24-26 June 1976 (in Italian)

(7) M.Fanelli, S.Fedato, G.Fenini
"Numerical prediction of transients in cooling water circuit for thermal power stations"
Joint Symposium on design and operation of fluid machinery
Fort Collins, June 1978

(8) D.Bonolis, S.Fedato, R.P.Rossini, G.Fenini
"Hydraulic transients in three new thermal plants of ENEL"
IV Meeting on transients in cooling water circuits, Cagliari Sept.1979

(9) G. Pezzoli
"Determinazione teorica delle perdite di carico negli organi di regolazione: valvole a saracinesca e e farfalla", X Meeting on Hydraulics:
Cagliari Feb. 1967.

Additional definitions of symbols

$PD^2 = 4\,g\,J$ where : $g = 9.81\ m/s^2$ = acceleration of gravity.
$J = 538,5\ kgfm\ s^2$ = polar moment of inertia of the rotating masses.
$PD^2 = 21,130\ kgfm^2$

$h = \frac{H}{H_o}$ (H = head across the pump in m
H_o = reference value for H)

$m = \frac{T}{T_o}$ (T = Torque exerted by the water on the pump runner in kg m
T_o = reference value for T).

$q = \frac{Q}{Q_o}$ (Q = discharge across the pump in m^3/s
Q_o = reference value for Q)

$n = \frac{N}{N_o}$ (N = revolution speed in rev/min
N_o = reference value for N).

NTIP = 1 LAYOUT OF CIRCUIT WITH TWO PUMPS AND TWO INTER-CONNECTED PIPES.

TRIP-OFF OF SINGLE PUMP.

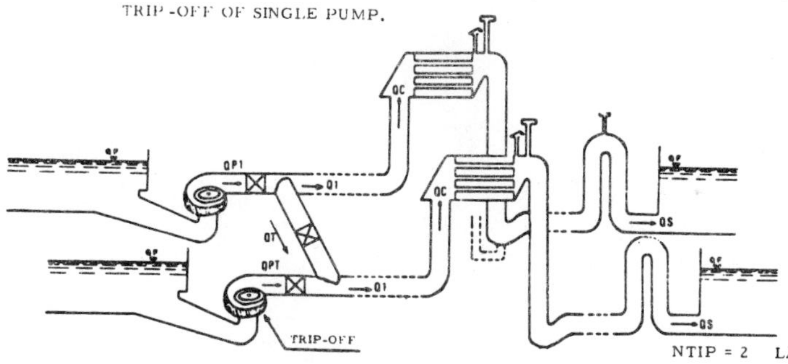

QP1 Q1 + QT. QT Q1 + QPT. QP1 Q1 + QPT + Q1

THE CODE'S INTERNAL CONSTANTS ARE NOT MODIFIED.

NTIP = 2 LAYOUT OF CIRCUIT WITH SINGLE PUMP AND TWO PIPES.

PUMP TRIP-OFF.
PUMP TRIP-ON.

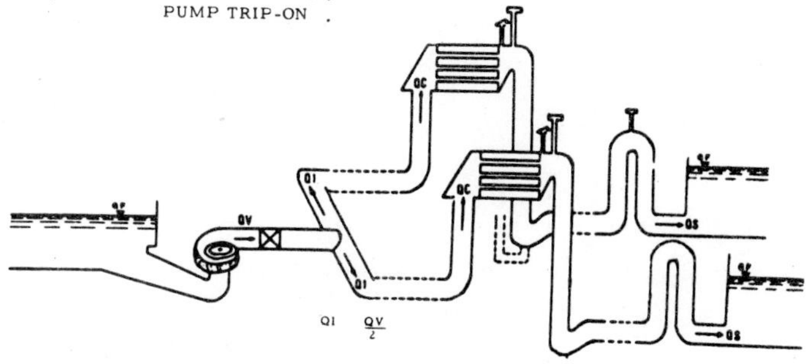

$Q1 \quad \frac{QV}{2}$

THE CODE'S INTERNAL CONSTANTS ARE AUTOMATICALLY MODIFIED:

$Q_0 \quad Q_0$ orig. $\cdot \frac{1}{2}$. $C_{valv} \quad C_{valv}$ orig. $\cdot$ 4

THE VALVE DISCHARGE IS DOUBLED BEFORE PRINTING (THE CODE CONSIDERS TWO SEPARATE CIRCUITS WITH TWO PUMPS, EACH FOR HALF THE TOTAL DISCHARGE, TRIPPING OFF SIMULTANEOUSLY).

NTIP = 3 LAYOUT OF CIRCUIT WITH ONE PIPE TWO INTER-CONNECTED PUMPS.

BOTH PUMPS TRIP-OFF.
BOTH PUMP TRIP-ON.

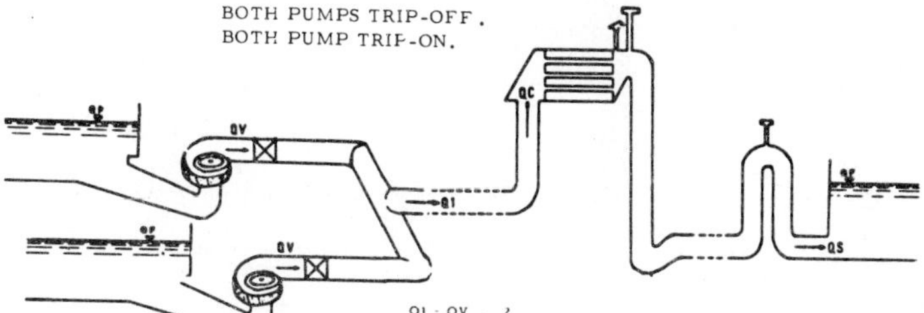

Q1 = QV . 2

THE CODE'S INTERNAL CONSTANTS ARE AUTOMATICALLY MODIFIED:

$Q_0 = Q_0$ orig. . 2 ; $C_{valv} = C_{valv}$ orig. $\frac{1}{4}$

THE VALVE DISCHARGE IS HALVED BEFORE PRINTING.

NTIP = 4 LAYOUT OF CIRCUIT WITH ONE PIPE AND TWO INTER-CONNECTED PUMPS.

TRIP-OFF OF SINGLE PUMP.

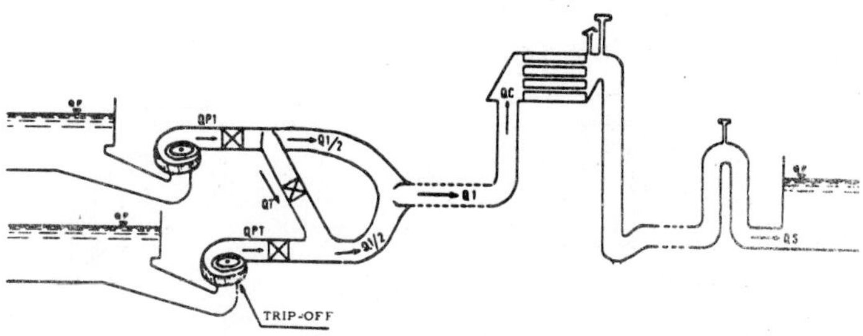

QP1=Q1/2+QT ; QT=Q1/2-QPT ; QP1=Q1-QPT

THE CODE'S INTERNAL CONSTANTS ARE AUTOMATICALLY MODIFIED

Diameters	=	Diam. orig. $\sqrt{1/2}$
Hydr. Radius	=	Hydr. radius orig. $\sqrt{1/2}$
Sections	=	Sections orig. $\sqrt{1/2}$
Gaukler-Strickler	=	Gaukler-Strickler orig. $\sqrt[3]{2}$
C_{cond}	=	$C_{cond,}$ orig. $\cdot$ 4

THERE IS ONLY ONE CONDENSER WITH LOSSES PERTAINING TO TOTAL DISCHARGE. DISCHARGES AT CONDENSER AND OUTLET ARE DOUBLED BEFORE PRINTING: THE CODE SIMULATES TWO CIRCUITS IN PARALLEL, EACH FOR ALL THE TOTAL DISCHARGE.

NTIP = 5 LAYOUT OF CIRCUIT WITH ONE PIPE AND SINGLE PUMP.

PUMP TRIP-OFF.
PUMP TRIP-ON.

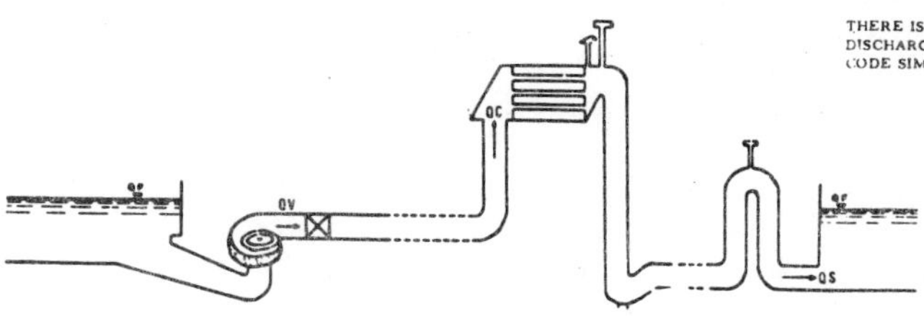

THE CODE'S INTERNAL CONSTANTS ARE NOT MODIFIED.

Fig. 1

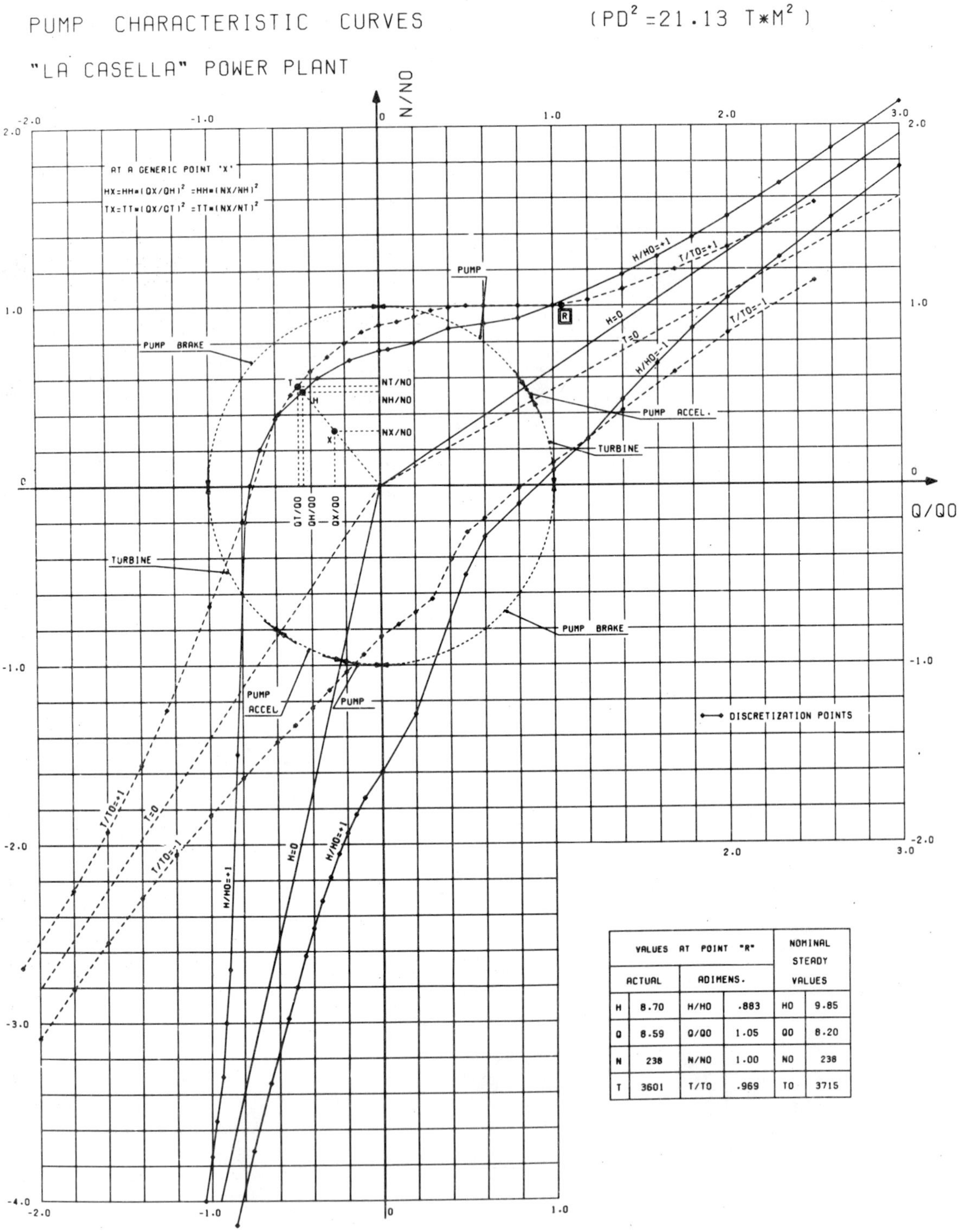

VALUES AT POINT "R"				NOMINAL STEADY VALUES	
ACTUAL		ADIMENS.			
H	8.70	H/HO	.883	HO	9.85
Q	8.59	Q/QO	1.05	QO	8.20
N	238	N/NO	1.00	NO	238
T	3601	T/TO	.969	TO	3715

Fig. 2

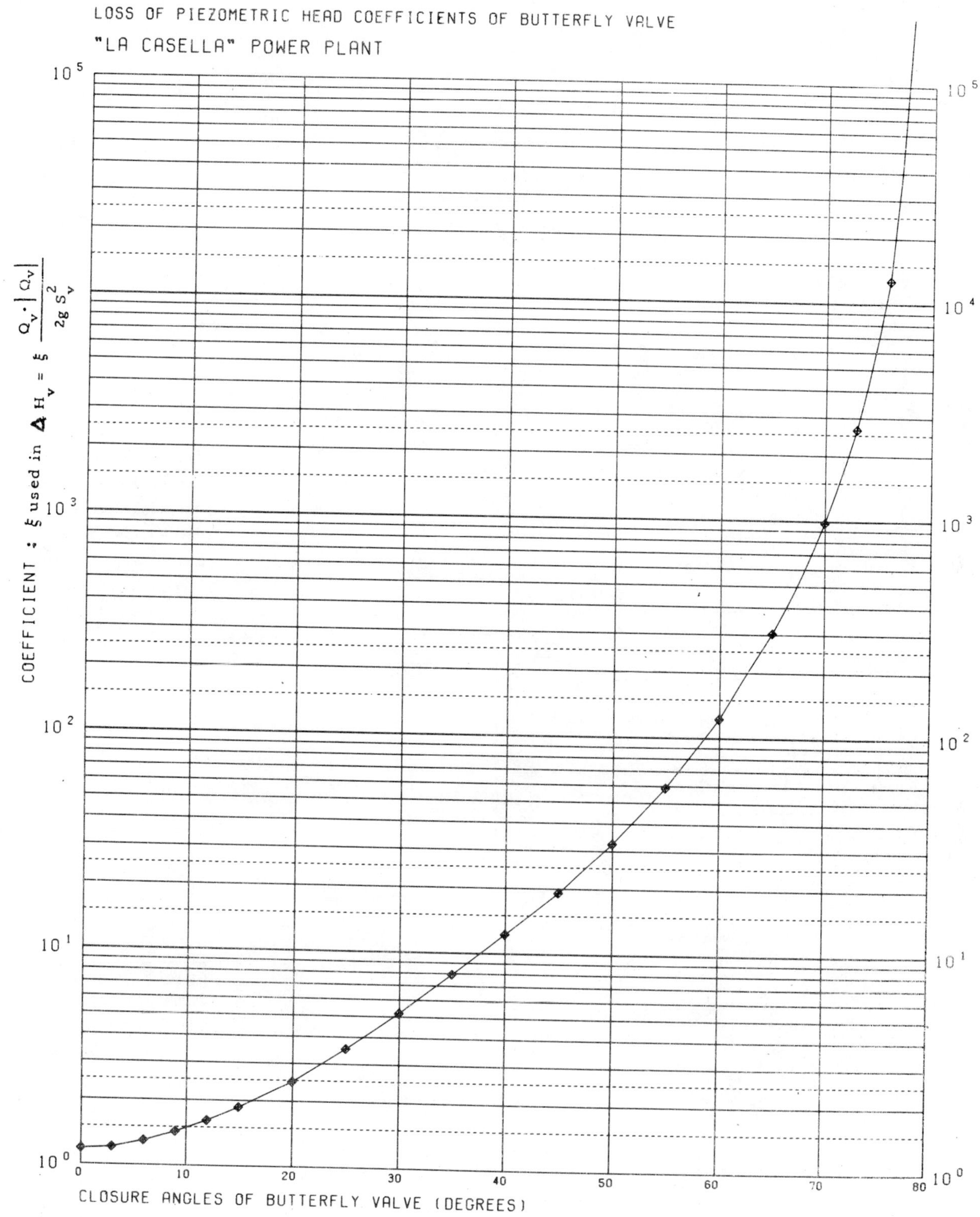

LOSS OF PIEZOMETRIC HEAD COEFFICIENTS OF BUTTERFLY VALVE
"LA CASELLA" POWER PLANT
COEFFICIENT : ξ used in ΔH_v = ξ Q_v·|Q_v| / 2g S_v^2
10^5
10^4
10^3
10^2
10^1
10^0
0
10
20
30
40
50
60
70
80
CLOSURE ANGLES OF BUTTERFLY VALVE (DEGREES)

Fig. 3

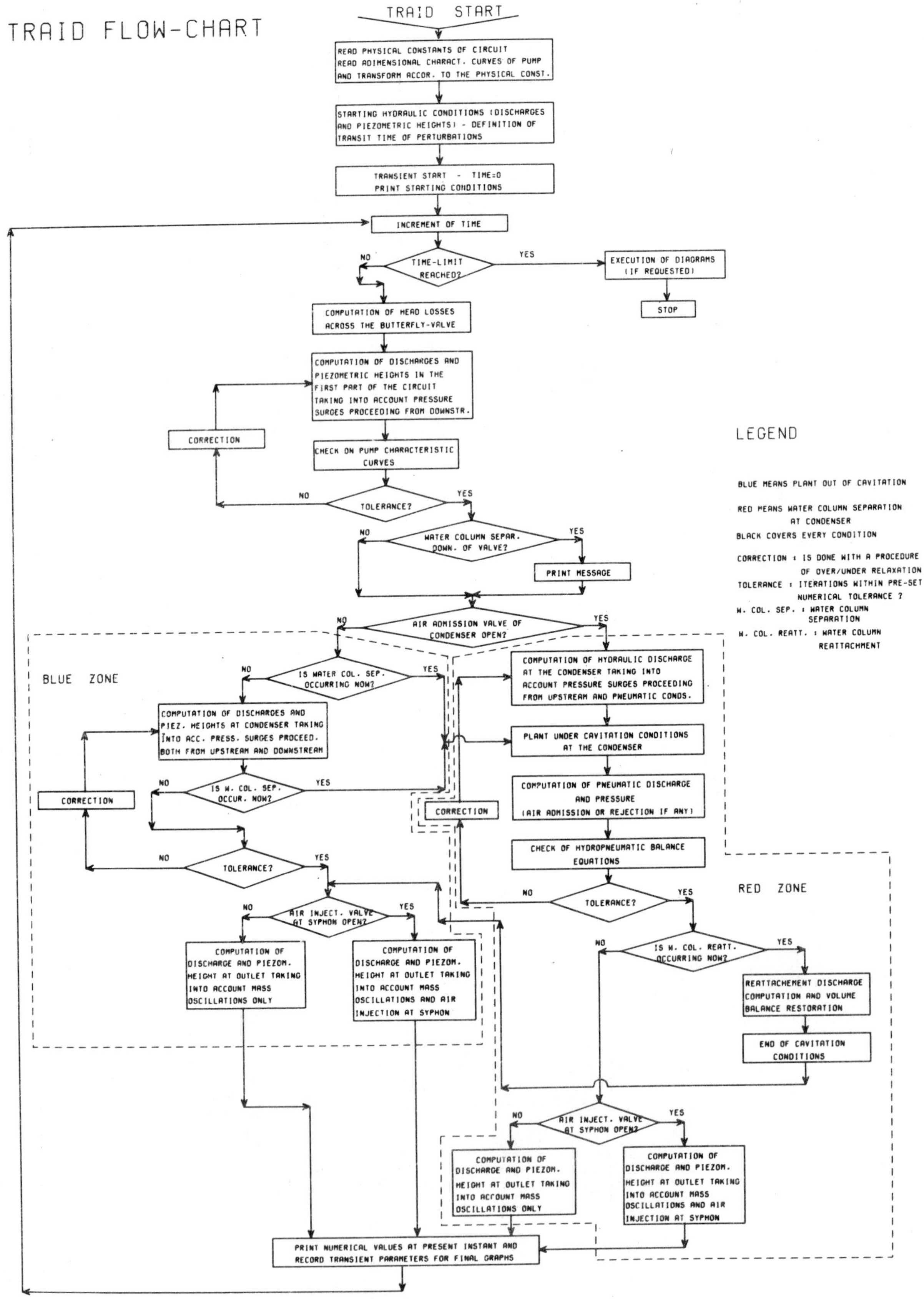

TRAID FLOW-CHART
TRAID START
READ PHYSICAL CONSTANTS OF CIRCUIT READ ADIMENSIONAL CHARACT. CURVES OF PUMP AND TRANSFORM ACCOR. TO THE PHYSICAL CONST.
STARTING HYDRAULIC CONDITIONS (DISCHARGES AND PIEZOMETRIC HEIGHTS) - DEFINITION OF TRANSIT TIME OF PERTURBATIONS
TRANSIENT START - TIME=0 PRINT STARTING CONDITIONS
INCREMENT OF TIME
TIME-LIMIT REACHED?
NO
YES
EXECUTION OF DIAGRAMS (IF REQUESTED)
STOP
COMPUTATION OF HEAD LOSSES ACROSS THE BUTTERFLY-VALVE
COMPUTATION OF DISCHARGES AND PIEZOMETRIC HEIGHTS IN THE FIRST PART OF THE CIRCUIT TAKING INTO ACCOUNT PRESSURE SURGES PROCEEDING FROM DOWNSTR.
CORRECTION
CHECK ON PUMP CHARACTERISTIC CURVES
TOLERANCE?
WATER COLUMN SEPAR. DOWN. OF VALVE?
PRINT MESSAGE
AIR ADMISSION VALVE OF CONDENSER OPEN?
BLUE ZONE
IS WATER COL. SEP. OCCURRING NOW?
COMPUTATION OF DISCHARGES AND PIEZ. HEIGHTS AT CONDENSER TAKING INTO ACC. PRESS. SURGES PROCEED. BOTH FROM UPSTREAM AND DOWNSTREAM
IS W. COL. SEP. OCCUR. NOW?
TOLERANCE?
AIR INJECT. VALVE AT SYPHON OPEN?
COMPUTATION OF DISCHARGE AND PIEZOM. HEIGHT AT OUTLET TAKING INTO ACCOUNT MASS OSCILLATIONS ONLY
COMPUTATION OF DISCHARGE AND PIEZOM. HEIGHT AT OUTLET TAKING INTO ACCOUNT MASS OSCILLATIONS AND AIR INJECTION AT SYPHON
COMPUTATION OF HYDRAULIC DISCHARGE AT THE CONDENSER TAKING INTO ACCOUNT PRESSURE SURGES PROCEEDING FROM UPSTREAM AND PNEUMATIC CONDS.
PLANT UNDER CAVITATION CONDITIONS AT THE CONDENSER
COMPUTATION OF PNEUMATIC DISCHARGE AND PRESSURE (AIR ADMISSION OR REJECTION IF ANY)
CHECK OF HYDROPNEUMATIC BALANCE EQUATIONS
RED ZONE
IS W. COL. REATT. OCCURRING NOW?
REATTACHEMENT DISCHARGE COMPUTATION AND VOLUME BALANCE RESTORATION
END OF CAVITATION CONDITIONS
PRINT NUMERICAL VALUES AT PRESENT INSTANT AND RECORD TRANSIENT PARAMETERS FOR FINAL GRAPHS
LEGEND
BLUE MEANS PLANT OUT OF CAVITATION
RED MEANS WATER COLUMN SEPARATION AT CONDENSER
BLACK COVERS EVERY CONDITION
CORRECTION : IS DONE WITH A PROCEDURE OF OVER/UNDER RELAXATION
TOLERANCE : ITERATIONS WITHIN PRE-SET NUMERICAL TOLERANCE ?
W. COL. SEP. : WATER COLUMN SEPARATION
W. COL. REATT. : WATER COLUMN REATTACHMENT

Fig. 4

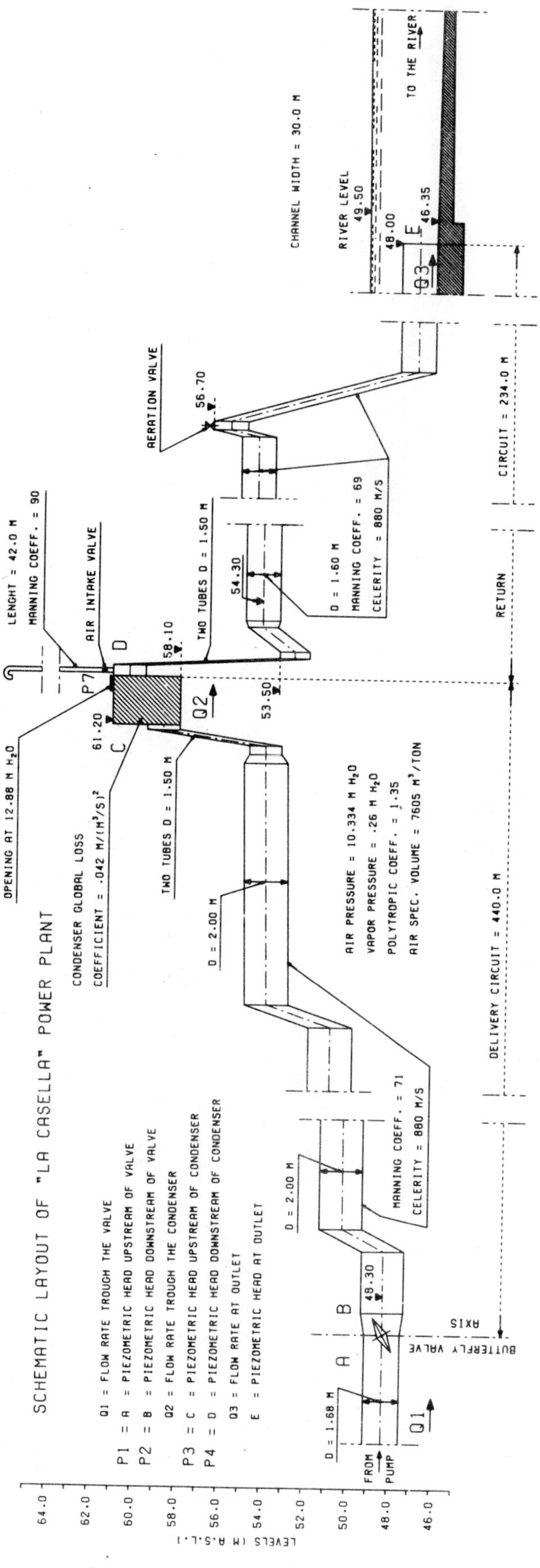
SCHEMATIC LAYOUT OF "LA CASELLA" POWER PLANT
Q1 = FLOW RATE TROUGH THE VALVE
P1 = A = PIEZOMETRIC HEAD UPSTREAM OF VALVE
P2 = B = PIEZOMETRIC HEAD DOWNSTREAM OF VALVE
Q2 = FLOW RATE TROUGH THE CONDENSER
P3 = C = PIEZOMETRIC HEAD UPSTREAM OF CONDENSER
P4 = D = PIEZOMETRIC HEAD DOWNSTREAM OF CONDENSER
Q3 = FLOW RATE AT OUTLET
E = PIEZOMETRIC HEAD AT OUTLET
LEVELS (M A.S.L.)
64.0
62.0
60.0
58.0
56.0
54.0
52.0
50.0
48.0
46.0
FROM PUMP
D = 1.68 M
Q1
A
B
48.30
BUTTERFLY VALVE AXIS
D = 2.00 M
MANNING COEFF. = 71
CELERITY = 880 M/S
D = 2.00 M
DISC VALVE
OPENING AT 12.88 M H2O
CONDENSER GLOBAL LOSS
COEFFICIENT = .042 M/(M3/S)2
TWO TUBES D = 1.50 M
61.20
C
Q2
P7
D
53.50
58.10
AIR PRESSURE = 10.334 M H2O
VAPOR PRESSURE = .26 M H2O
POLYTROPIC COEFF. = 1.35
AIR SPEC. VOLUME = 7605 M3/TON
DELIVERY CIRCUIT = 440.0 M
AIR INLET
TWO TUBES D = 0.20 M
LENGHT = 42.0 M
MANNING COEFF. = 90
AIR INTAKE VALVE
TWO TUBES D = 1.50 M
54.30
D = 1.60 M
MANNING COEFF. = 69
CELERITY = 880 M/S
RETURN
CIRCUIT = 234.0 M
AERATION VALVE
56.70
CHANNEL WIDTH = 30.0 M
RIVER LEVEL
49.50
48.00
E
46.35
Q3
TO THE RIVER

Fig. 5

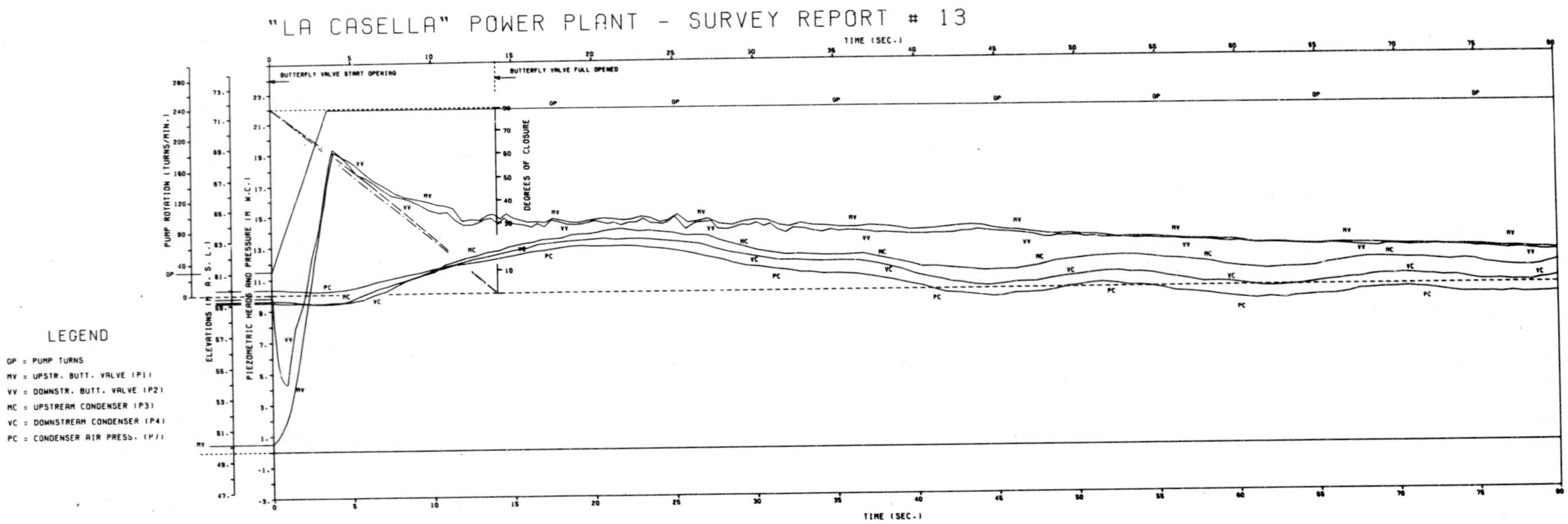
"LA CASELLA" POWER PLANT - SURVEY REPORT # 13
TIME (SEC.)
BUTTERFLY VALVE START OPENING
BUTTERFLY VALVE FULL OPENED
DEGREES OF CLOSURE
PUMP ROTATION (TURNS/MIN.)
ELEVATIONS (M. A. S. L.)
PIEZOMETRIC HEADS AND PRESSURE (M W.C.)
LEGEND
GP = PUMP TURNS
MV = UPSTR. BUTT. VALVE (P1)
VV = DOWNSTR. BUTT. VALVE (P2)
MC = UPSTREAM CONDENSER (P3)
VC = DOWNSTREAM CONDENSER (P4)
PC = CONDENSER AIR PRESS. (P7)

Fig. 6

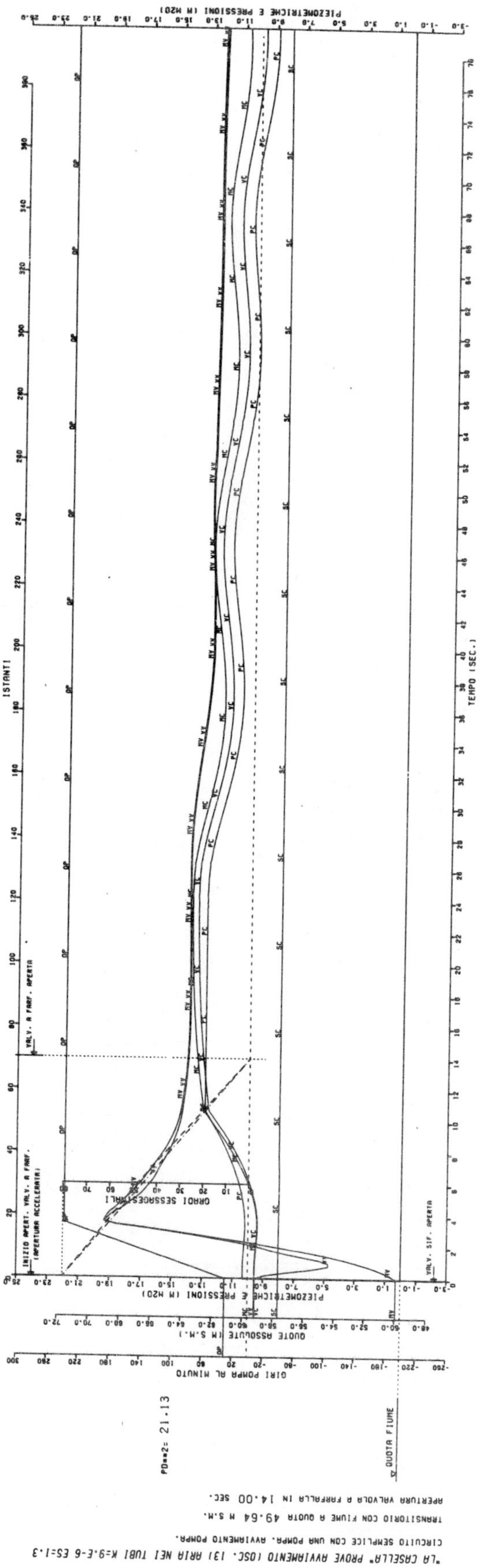
"LA CASELLA" PROVE AVVIAMENTO (OSC. 13) ARIA NEI TUBI K=9.E-6 ES=1.3
CIRCUITO SEMPLICE CON UNA POMPA. AVVIAMENTO POMPA.
TRANSITORIO CON FIUME A QUOTA 49.64 M S.M.
APERTURA VALVOLA A FARFALLA IN 14.00 SEC.
PD**2= 21.13
QUOTA FIUME
GIRI POMPA AL MINUTO
QUOTE ASSOLUTE (M S.M.)
PIEZOMETRICHE E PRESSIONI (M H2O)
GRADI SESSAGESIMALI
ISTANTI
TEMPO (SEC.)
VALV. SIF. APERTA
INIZIO APERT. VALV. A FARF. (APERTURA ACCELERATA)
VALV. A FARF. APERTA

Fig. 7

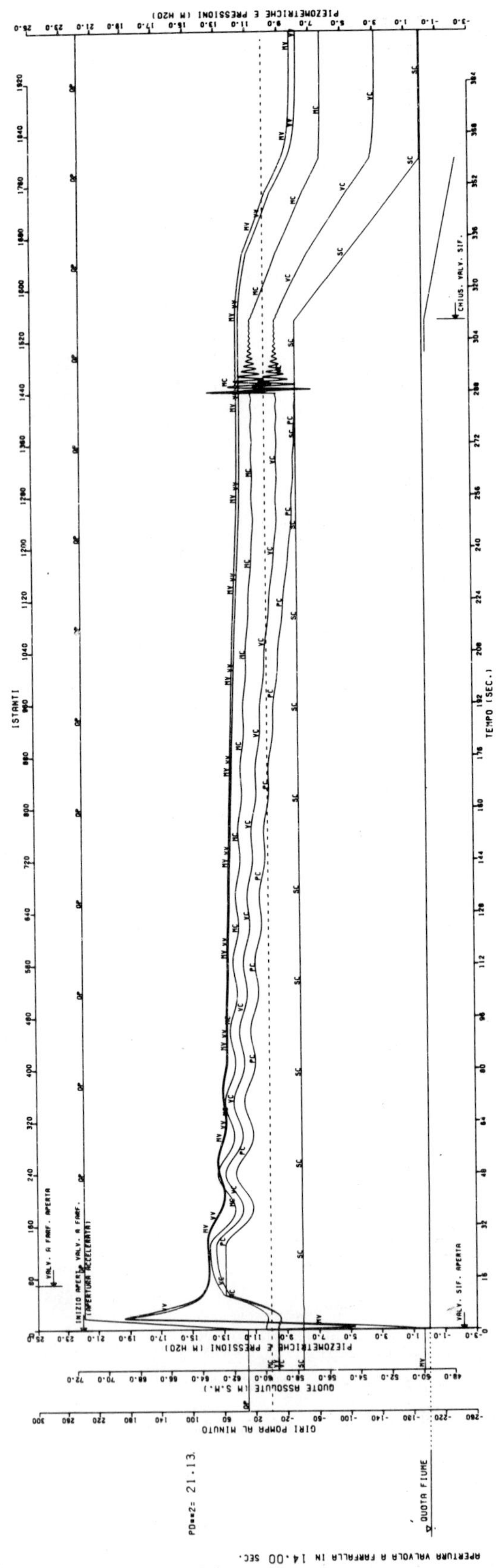

Fig. 8

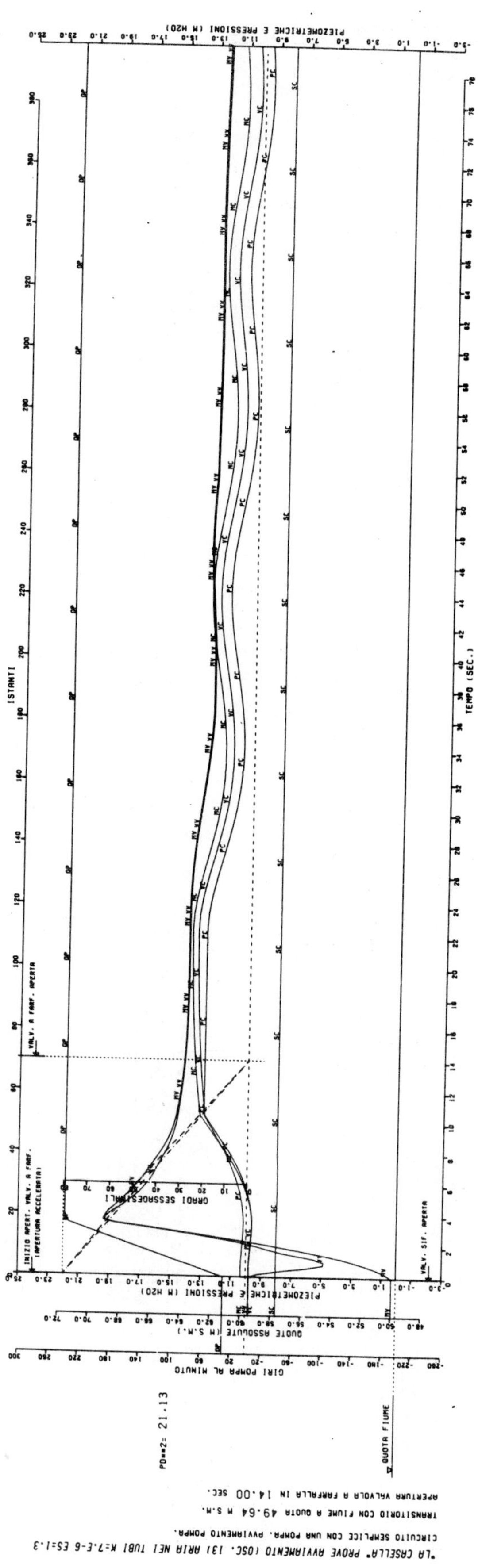
"LA CASELLA" PROVE AVVIAMENTO (OSC. 13) ARIA NEI TUBI K=7.E-6 ES=1.3
CIRCUITO SEMPLICE CON UNA POMPA. AVVIAMENTO POMPA.
TRANSITORIO CON FIUME A QUOTA 49.64 M S.M.
APERTURA VALVOLA A FARFALLA IN 14.00 SEC.
PD**2= 21.13
QUOTA FIUME
GIRI POMPA AL MINUTO
QUOTE ASSOLUTE (M S.M.)
PIEZOMETRICHE E PRESSIONI (M H2O)
GRADI SESSAGESIMALI
ISTANTI
TEMPO (SEC.)
VALV. SIF. APERTA
VALV. A FARF. APERTA

Fig. 9

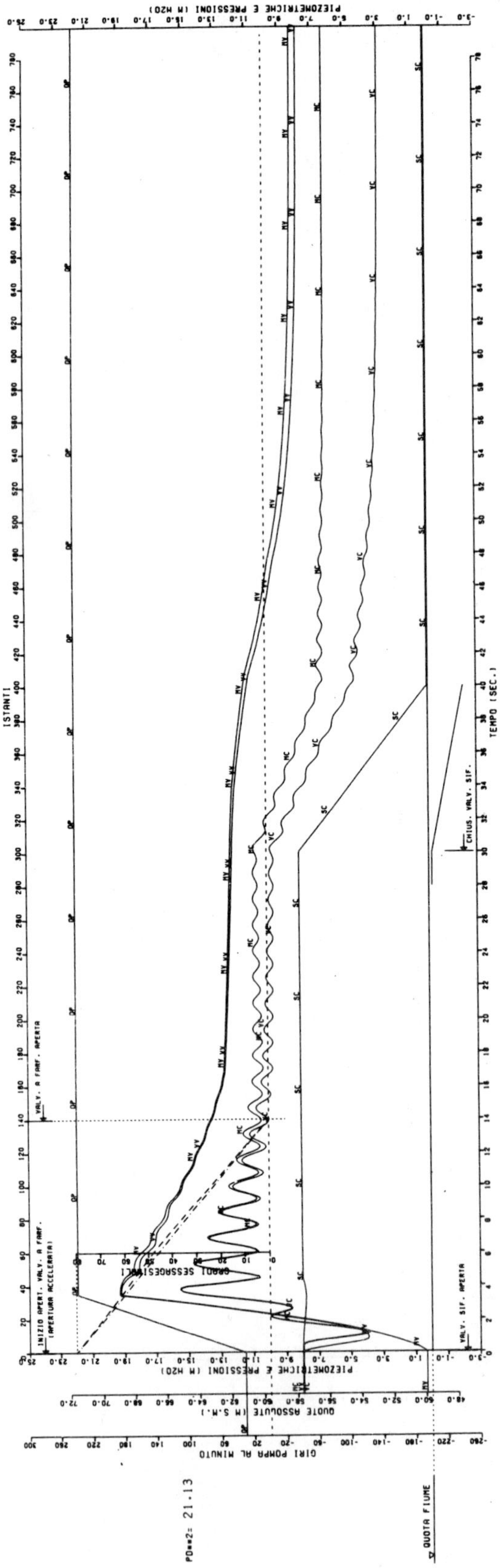
"LA CASELLA" PROVE AVVIAMENTO (OSC. 13) CIRCUITO PIENO
CIRCUITO SEMPLICE CON UNA POMPA. AVVIAMENTO POMPA.
TRANSITORIO CON FIUME A QUOTA 49.64 M S.M.
APERTURA VALVOLA A FARFALLA IN 14.00 SEC.
PD**2= 21.13
QUOTA FIUME
GIRI POMPA AL MINUTO
QUOTE ASSOLUTE (M S.M.)
PIEZOMETRICHE E PRESSIONI (M H2O)
ISTANTI
TEMPO (SEC.)
GRADI SESSAGESIMALI

Fig. 10

Third International Conference on

PRESSURE SURGES

Canterbury, England
March 25-27, 1980

PAPER F2

LEVEL SURGES IN THE CIRCULATING WATER DOWNSHAFT OF LARGE GENERATING STATIONS

B.T. Goldring and W.T. Mawer

C.E.G.B., U.K.

and

N. Thomas

Cambridge University, U.K.

Summary

Several power stations in this country have experienced level surges caused by air entrained in the condenser circulating water flow at outfall downshafts. These surges produce rapidly fluctuating pressures on the concrete lining of a downshaft and this phenomenon might cause damage to the structure.

Considerable care was taken in designing the outfall downshaft for the new Thameside Station to ensure that level surges would be minimised. Constraints on the site positioning and layout made it necessary to develop the downshaft design entirely by hydraulic modelling, but the inability to accurately model air entraining processes made it difficult to assess the severity of air entrainment in the prototype. The hydraulic model testing is described with emphasis on minimising air entrainment by the plunging weir flows.

A fundamental study of air entrainment in weir type flows was carried out, and whilst this was not able to influence the hydraulic modelling work, the results of this study are given because of their potential application on future projects.

Held at the University of Kent at Canterbury, U.K.

Conference Organised by BHRA Fluid Engineering, Cranfield, Bedford, England.
In conjunction with The City University, London.

1. INTRODUCTION

Power Station condenser cooling systems comprise three basic methods of water circulation with the cooling water taken from (i) a river, estuary or the sea with all the heated water returned to the cooling sink, (ii) a closed recirculation design employing cooling towers or a spray distribution network, and (iii) a mixture of the two previous systems. The phenomenon which is described in the paper is primarily associated with (i) which is defined as a once-through system.

Fig. 1 shows the main network for a once-through cooling water system indicating the position of the syphon seal weir and downshaft. It will be seen that a weir/downshaft combination is sited at the outlet from the condenser and the downshaft normally serves as a method of returning the heated water from the site to the cooling medium or an offshore discharge structure via a deep tunnel. The water level surge which is the subject of this paper occurs in the system between the weir/downshaft and the discharge point.

2. COOLING WATER SYSTEM DESIGN

Normally all once-through cooling water system arrangements adopt the principle of syphonic recovery with a safeguard to prevent the minimum system pressure reaching a 2 phase water/water vapour state. Fig.2 shows the hydraulic gradient for a once-through system and illustrates the region of minimum pressure at the condenser outlet together with the seal weir which ensures a single phase water flow. It is CEGB policy to adopt a minimum syphon head of minus 9.65metres and the factors which affect the calculation of the seal position in Fig.2 are:-

(i) maximum condenser outlet temperature;

(ii) maximum main condenser and auxiliary cooling system flow;

(iii) lowest system operational head loss; for a station with three large turbo generators (units) and four cooling water pumps the losses are evaluated ranging from three turbines and two cooling water pumps in service to a single unit with an emergency operation of one pump;

(iv) margin for system fouling throughout the station life;

(v) The component head losses are calculated in accordance with BHRA Report RR1064.

It will be seen from the hydraulic gradient that a variation of water level occurs at the downshaft corresponding to the tidal state of the cooling medium. This results in a jet plunging on to the free water surface when the weir is not flooded, with subsequent air entrainment in the water column. Entrained air has the effect of altering the density of the water column in the downshaft with a severity depending on the shaft geometry, bubble rise velocity and the water velocity in the downshaft, and the length/inclination of the tunnel leading to an offshore discharge point. The level surges arise by air escaping upwards from the shaft after concentrating in large pockets at the first bend.

3. DOWNSHAFT DESIGN FOR A 2000MW PROJECT

Fig.3/4 shows a syphon seal and downshaft design for a 2000MW project. The scheme was designed for a total cooling water flow of 58.4m^3/s and high level outlet culverts from the condensers. The seal weir position ensured that the system pressure did not fall below minus 9.65metres. The height of the high level culverts was determined by the local site subsoil conditions.

The difference in height between the seal weir and the low level free water surface in the downshaft is larger than most previous designs and this resulted in a large fall of the plunging jet with an anticipated high degree of air entrainment.

Means of reducing the jet plunge were examined and it was subsequently decided to instal a secondary weir in the river outfall which would regulate the seal pit level. The pump duty had already been specified and it was necessary to site the offshore weir crest at a position which would not incur additional pumping head. Fig. 3 shows the arrangement and the design ensures that under maximum flow with four pumps there is no jet plunge. At lower flows with one or two turbines in operation the maximum free fall of the jet will be 2.7metres.

4. DOWNSHAFT OPERATIONAL PROBLEMS

Level instability in downshafts have been reported within the U.K. on projects for the South of Scotland Electricity Board and the CEGB (Ref 1 & 2) with maximum surges of 9m. Hydraulic model tests for the SSEB project, which entailed a jet plunge of 7m affirmed that the interchange of entrained air between the downshaft and tunnel was responsible for this instability and the model work examined various methods of venting the air from the first bend in the downshaft to a point above the free water surface. A subsequent venting system gave some reduction in the surge height.

Site tests were carried out for a CEGB project to ascertain whether the surges resulted in high pressures on the downshaft and tunnel base. The results did not yield conclusive data and it would have required a very extensive instrumentation project to determine the pressure adjacent to collapsing air voids. This feature has been discussed with civil engineers who have stated that they are reluctant to guarantee the civil works integrity for this type of oscillation which would occur continuously over a period of 30-40 years.

5. HYDRAULIC MODELLING

5.1 General Principles

A number of design changes were made during the modelling needing changes in the latter arrangement. However this description deals with the final design of an 11m diameter shaft supplied by three culverts as shown in Figs. 3 & 4.

In view of the level surges experienced at other stations, it was necessary to design the outfall downshaft so as to minimise the entrainment of air into the water flow. This posed the first problem, because air entrainment cannot be accurately simulated in a scale model. However, the flow patterns which cause air entrainment can be reproduced in a scale model, though air entrainment rates are underestimated, and situations where air entrainment is likely to occur can also be identified.

Water flowrates were adjusted to give prototype Froude numbers so that water surface effects would be reproduced. This results in low Reynolds numbers, but viscous effects can be reduced by using smooth perspex and varnished wood surfaces in all flow areas which is standard modelling practice.

Two scale models of the outfall downshaft were used. The first was a small (1/40 scale) model for the purpose of an early model start and an investigation of bulk flow patterns, and identification of problem areas within the downshaft. A larger (1/13.3 scale) model was built so that the details of the solutions to the flow problems could be resolved and to provide more accurate estimates of residual air entrainment.

5.2 Small Scale Model Results

The changing flow patterns in the downshaft will be described by starting with the lowest tidal level and progressing up the tidal range.

At low water level the flow from the priming weir is in the form of a submerged jet, guided downwards parallel to the weir nose, and gradually diverging as it approaches the farther wall of the downshaft. With the water level in the downshaft much below the level on the priming weir, this plunging jet entrains air into the water flow. Very little of this air is released at the free surface. Some of the air bubbles penetrate through the jet and form a large recirculation area below the overhanging weir noses, whilst the remainder is carried down and around the bottom bend as shown in Fig.5.

When the water level in the downshaft approaches the water level over the priming weir, the jet detaches from the nose of the weir and flows in a horizontal direction with an undulating surface wave towards the farther wall of the downshaft. There is considerable agitation of the water surface at the downshaft wall opposite the inlets with consequential air entrainment.

With asymmetric inflow (e.g. from culvert numbers 1 and 2) there is a massive rotation of the body of water in the downshaft with an air core which persists through the conical transition into the vertical shaft. This rotation within the downshaft exists up to the highest water level, although the accompanying surface agitation decreases with increasing water depth.

These different forms of behaviour all contributed to air entrainment into the downshaft, thus increasing the potential for level surges, so various ways of eliminating them were investigated.

The massive rotation in the downshaft was stopped by extending the inter-culvert walls to span the downshaft, supporting them on pillars carried down to the top of the conical transition (shown in Fig.6). The scour down the rear wall of the downshaft was reduced by fitting a cascade of vanes across the downshaft to turn the water flow downwards (also shown in Fig.6).

The low-tide air entrainment caused by the plunging jets could not be completely stopped. It was, however, considerably reduced by designing a discharge weir in the outfall structure so that the maximum fall to the free surface in the downshaft was reduced from 5.4m to 2.7m (prototype values). Any further reduction would have incurred unacceptable head losses elsewhere in the system.

At this stage of the investigation, work was transferred to the larger model.

5.3 Large Scale Model Results

The results from the optimum vane positions show that, in the downshaft opposite the two outer weirs, the velocity distribution below the cross-walls is much more uniform with the vanes in use (Fig.7). Air entrainment by surface flows was also reduced with this vane arrangement.

The behaviour of air entrained by the plunging weir flows at low tide levels was examined. There were two separate aspects; the air carried down and around the lower bend, and the air hung-up in the large recirculation area below the weirs (Fig.5). The latter was of concern because the oscillations on the South of Scotland Electricity Board project mentioned earlier had been associated with this feature.

On starting the flow at low tide, some of the air entrained by the plunging weir flows was carried down and around the bottom bend, whilst air hang-up increased progressively until there was a large cloud of air bubbles recirculating in the downshaft. The apparent density of this recirculating mixture stabilised after a short time; from then on the amount of air entrained by the plunging weir flows (less the small amount of air released at the free surface) was carried down past the recirculation area and around the bottom bend, with little interchange occurring.

There were, at thisjuncture, two important unknowns; the amount of air entrained by the plunging weir flows, and the amount of air held in suspension in the recirculation area. Attempts to measure the latter included differential head measurements at the conical transition which were reasonably successful, and acoustic measurements which were unsuccessful. The former readings were used to calibrate visual estimates of bubble load, and the results for nominal flow in all three culverts are shown in Fig.8. It can be seen that air hang-up was negligible down to a water level of +1.6m which is of special interest because the lowest operating water level in the downshaft under these flow conditions is +1.9m. At this water level the air volume recirculating in the downshaft is only approximately 0.1% of the water volume in the downshaft. However, scale models are known to underestimate air entrainment, so this value is likely to be exceeded in the prototype.

Because air entrainment at low water could not be eliminated, two measures to intercept and remove the hung-up air were tried (Fig.9). The first was an attempt to trap the recirculating air under overhangs below the weirs and vent it to atmosphere, but the flow patterns in the downshaft kept the air away from this region so that very little was collected. The second idea was to catch the air bubbles in the stagnation area below an inverted v-shaped beam, but this too resulted in very little air removal.

The other unknown, the amount of air carried down and around the bottom bend, could not be easily measured due to the configuration of the model. Also, neither the shaft nor the bend below the conical transition were transparent, so even a visual estimate was not possible.

5.4 Effect of Water Flow and Tide Level

One of the worst situations occurs with the combination of low tide level and the minimum throughput as defined in Section 2. This combination creates the maximum jet plunge and air entrainment and the low net downward water velocity in the downshaft under these conditions results in much of the entrained air remaining in the downshaft - the "hang up" condition. At higher flowrates, most of the entrained air is carried down into the tunnel, where a proportion of it agglomerates on the tunnel soffit and forms a void immediately downstream of the shaft/tunnel bend.

5.5 Susceptibility to Level Surges

A void can only form at the lower bend if air is being entrained into the downshaft water flow. This occurred at low tide levels for all pump and unit combinations except 4 pumps and 3 units, and is likely to be worst for a 2 pump and 2 unit operation. Above a water level in the downshaft of approximately 1.7m, air entrainment will be slight and void formation is unlikely.

Some voiding tests were conducted on a circular bend of radius ratio 0.75, of a design not unlike the lower bend below the downshaft. Preliminary results are shown in Fig.10, with water velocities scaled up to prototype values. At the nominal water velocity of 4m/s, the results show that a sudden reduction of water velocity to below 3.0m/s would cause a void to vent, and it is expected to rise all the way to the free surface. However, tests to be conducted shortly on a more representative bend may modify this value.

A different situation exists under steady operation at water velocities below 3m/s. A void starts to form, but vents before it can reach a large size. As it starts to rise up the shaft it breaks-up into smaller voids and air bubbles, and is then carried back down and around the bend by the water flow. Thus, the air problem at low water flows consists of large bubbles of air recirculating in the area of the bend, and the problems associated with violent venting will not be present.

6. FUNDAMENTAL STUDIES (Ref 3-10)

6.1 Objectives

Two targets were proposed for the research programme at Cambridge:

(i) To deduce the fundamental scaling relations between the entrained air and the plunging water flow.

(ii) To identify the fluid mechanisms responsible for the entrainment and transport of the bubbles.

To expedite these studies, an experimental plunging wall-jet system was constructed in Cambridge. As a result of observations of the flow structure in this system, a third target was adopted towards the end of the programme:

(iii) To investigate simple methods of controlling air entrainment.

The main findings on each of these topics are summarised below.

6.2 Fundamental scaling relations

6.2.1 Rationale & Background

Individual contributions to the total air entrainment arise from the plunge and from the jet impingement. These contributions possess different scalings and must be assessed separately. As a first approximation they may be regarded as independent.

In the system under consideration a significant fraction of the entrained air is detrained immediately downstream of the impingement region. It is the net entrainment, i.e. the total entrainment less the detrainment, that is of practical interest. The ultimate aim is to provide theoretical scalings for this quantity. The present effort, apparently the first attempt at a rational theoretical treatment, was mounted before the experimental plunging jet system was available. It was directed primarily at providing a sensible physical basis for the diversity of published results.

There is a considerable literature on hydraulic studies directed at bulk data correlations for the air entrainment; some examples for plunging jets, hydraulic jumps and syphon flows are shown in Table 1. These data will be compared with the theoretical scalings.

6.2.2 Theoretical Scalings

The basic physical premise in the present treatment is that air entrainment occurs via a surface layer of foam; air is entrained into the foam and bubbles are subsequently entrained from the foam by the turbulent water flow. The foam is maintained by vigorous surface activity such as breaking waves. This premise, later confirmed by experimental observations, provides a convenient reference for the entrainment calculations - the air concentration in a foam is typically 50% by volume.

Given this premise it is a straightforward matter to set up the theoretical problem using the classical approaches of turbulent entrainment and diffusion. The details are reported elsewhere (Ref. 9). The basic scalings for the volume flux ratio air to water (Φ) are as follows:

(a) hydraulic jump

$$\Phi \sim F_r - 1, \quad F_r \lesssim 10$$

$$\Phi \sim F_r^2, \quad F_r \gtrsim 10$$

(b) plunging jet

$$\Phi \sim (l/d)^{\frac{1}{2}} ;$$

l=length of plunge from entrainment inception

d=jet depth

In addition a first approximation was deduced for

(c) bubble detrainment

$$\Phi\,(B) = \frac{(1-B)^2}{(1-\frac{1}{3}B)}\;\Phi\,(B=o), \quad B = \frac{v_b \cos\phi}{v}$$

$v_b \cos\phi$ = effective bubble escape speed
v = turbulence intensity

Note that the scalings (a) & (b) for hydraulic jumps and plunging jets accord with the trends of the experimental data summarised in Table 1. A quantitative comparison is critically discussed below.

6.2.3 Quantitative estimates

(a) Hydraulic jumps

The scale factors in relations (a) depend to a first approximation only on fundamental physical properties like the coefficients of turbulent entrainment and diffusion in one-phase flows, and on the air concentration in the foam. The major uncertainty arises from the length of the entrainment region, which clearly also has to be prescribed. In the present study it was either assumed equal to the change in flow depth for small Froude number jumps, or inferred from a momentum balance for the extensive region of foam exhibited by high Froude jumps (see Ref. 9). Values assigned on this basis give numerical estimates in reasonable quantitative agreement with the data. Comparisons appear on Fig.11.

In the case of small Froude number jumps, it is likely that the detrainment characteristics largely determine the effective length of the entrainment region. The assumption that the latter is equal to the change in flow depth is probably the weakest feature of the quantitative estimates - the good agreement with data can be attributed to this judicious choice. More likely, the entrainment length is of the order of the length of the jump (in which case Φ would be increased by a factor of about three), but at the same time detrainment (neglected in the quantitative estimate) is probably substantial, contributing a reducing factor of similar magnitude. Further considerations are obviously required.

More confidence is attached to the large Froude number estimate, mainly because the higher flow speeds mean there is less dependence on the uncertain detrainment characteristics.

(b) Plunging jets

As with the hydraulic jumps, the scale factor in relation (b) for plunging jets depends only on fundamental physical properties. For a quantitative estimate it is necessary to specify in addition the entrainment inception point. The present study adopted, without much justification, a critical Reynolds number hypothesis, i.e. entrainment starts when the plunging jet Reynolds number (based on the length of plunge) exceeds a critical value - here taken to be 2.10^5. This value was deduced from photographs of high speed orifice jets (Ref. 11). As Fig.12 shows, this hypothesis seems to be adequate for correlating Ervine's slot jet data.

Clearly the inception conditions require further study. It may be that a critical surface speed, or in the case of wall-jets, the location of free surface penetration by the wall boundary layer, is more relevant than the plunge Reynolds number. For free jets, the problem is virtually intractable in view of their sensitivity to initial conditions, angle of plunge and curvature.

It should be pointed out that hydraulic models operating under Froude similarity fail to reproduce plunging jet entrainment, as studies of spillway flows have clearly demonstrated (Ref. 6).

(c) Buoyant detrainment

For a test of the bubble detrainment scaling relation (c) it is necessary to prescribe the effective bubble escape speed $v_b \cos \phi$. This was done by invoking the several experimental observations of a critical flow speed q_w^* for entrainment inception - as Table 1 shows, these values are around 1m/s, independent of flow angle. Invoking this empirical inception condition, a bubble escape speed of 5cm/s was assigned assuming, reasonably, a turbulence intensity of about 5%. With this experimental closure, relation (c) agrees very well with Casteleyn's data correlation, as Fig. 13 shows.

Two fundamental questions arise from this analysis:

- why is the effective bubble escape speed (5cm/s) so very much less than the bubble rise speed in still water (about 20cm/s)?

- why is the escape speed apparently independent of flow angle?

Answers to both questions will be provided by the flow structure observations described as follows.

6.3 Fluid mechanism study

6.3.1 Rationale

The experimental problem as posed has too many degrees of freedom for a basic study - significant parameters include speed and angle of the plunging jet, length of plunge, wall-jet boundary curvature, chamber boundary geometry and chamber downflow speed. In the interests of precision it should be recorded that the present study was of short wall jets plunging down a plane wall into a large chamber with a negligible chamber downflow speed.

It should also be noted that the primary motivation was a basic understanding of the fluid mechanisms responsible for entrainment and transport, and not an extensive but rig-specific data correlation. To this end, one parameter in particular was exploited - the angle of plunge. This was felt intuitively to hold the key to a general insight into the problem, particularly in view of Goldring's hydraulic model tests, which had revealed a strong dependence of flow pattern and structure on the downshaft water level (equivalent to the plunge angle for flow over a curved weir).

Some results of this study are described below.

6.3.2 Experimental system

The experimental system was designed primarily to achieve good all round visualisation of the flow structure in the impingement and submerged jet regions. A perspex tank 15cm wide, 2m long and 1m deep was used as the receiving chamber. A rotable slot-nozzle (15cm wide, 2cm deep) was employed to supply a two-dimensional plunging wall-jet at any desired orientation from horizontal to vertical and at speeds up to 4m/s. The water was recirculated to the nozzle via a 5cm pump and a stilling reservoir of $2m^3$ capacity. Water level control in the receiving chamber was achieved by adjustable drain valves.

6.3.3 Flow patterns and structure

A broad classification of the steady flow patterns is sketched in Fig.14. Flow types include "quiescent" and "wavy", obtained only for small jet speeds and small angles of plunge, and "jumping" and "penetrating" obtained for larger jet speeds at small and large angles of plunge respectively. Transitional flows intermediate to these types are on the whole unsteady. The sketches also illustrate the submerged jet mixing regions responsible for the air entrainment into jumping and penetrating flows, as hypothesised in the theoretical analysis.

The key feature, of course, is the appearance of large vortex structures within the jet mixing region. As shown by Plate 1 air is entrained from a surface layer of foam and is transported in clusters of bubbles within energetic discrete vortices. The vortices maintain the surface layer of foam by inducing vigorous surface flows and breaking waves.

At large angles of plunge the submerged jet and the mixing layer with its bubbly vortices penetrate to great depths, with the result that bubbles are dispersed throughout the receiving chamber (Fig.14 & Plate 2). If the angle of plunge is reduced sufficiently, the mixing layer suddenly reattaches to the free surface and the bubbly vortices are confined to a surface jet flow. In this case no bubbles are found in the interior of the chamber (Fig.14 & Plate 3) - assuming of course the chamber is sufficiently large - in practice complications arise when the surface jet encounters the chamber boundaries. This problem will not be discussed here.

6.3.4 Entrainment & transport mechanisms

The above observations provide the promised answers to the questions posed by the theoretical analysis of buoyant detrainment (Section 6.2.3(c)). The mixing-region vortices are clearly the dominant mechanisms responsible for entrainment and transport, so inception must be determined mainly by the vortex strength. This scales on the jet speed and not on the jet angle. In addition, the radial pressure gradient in each vortex tends to trap the bubbles, resulting in a reduction in the effective escape speed compared with the bubble rise speed in still water; this effect is also independent of jet angle.

It is possible to go further and offer an order of magnitude estimate of the critical speed for entrainment inception. The experimental studies indicate that the nascent vortices in the mixing layer travel about 5 to 10cm and grow to diameters of between 1 and 2cm before they are shed and travel on as discrete entities. For substantial entrainment to occur, the nascent vortices must be capable of withdrawing bubbles from the surface foam and retaining them by the radial pressure gradient mechanism. This can only happen if effective g in the shed vortices exceeds gravitational g by a considerable margin. If a factor of 5 to 10 is adopted for this margin, it follows that substantial entrainment may be expected only when the jet speed exceeds $\frac{1}{2}$ to 1m/s. This compares favourably with the experimental values quoted in Table 1.

Further studies of these mechanisms are being pursued at Cambridge. They clearly hold the key to a more profound understanding of the basic mechanisms.

6.4 Air entrainment control

With the knowledge that discrete vortices are mainly responsible for the entrainment and transport of air, it was thought likely that some measure of control might be offered by disrupting the organised flows in the vortex formation region.

With this in mind some preliminary tests were conducted using square mesh screens mounted normal to the backing plate of the submerged wall jet. The upstream pressure field set up by the screens causes the mixing region of the submerged jet to be deflected upwards towards the free surface as sketched in Fig.15, which should be compared with Fig.14; for a sufficient screen blockage factor the flow pattern appears rather like a hydraulic jump, with the mixing region reattaching to the free surface (Plate 4, which should be compared with Plates 2 & 3). An essential difference from the hydraulic jump flows obtained at small impingement angles in the absence of screens is that the downstream jet no longer exhibits large vortical structures but consists only of small-scale fully three-dimensional turbulence. Most of the air is detrained from the reattaching mixing region upstream of the screen, but significantly, the bubbles travelling through the screen are no longer trapped in coherent vortices and are also efficiently detrained from the flow.

To check that this result is not simply a surface tension scale effect ("bubble blocking" by a fine mesh screen), additional tests were conducted using arrays of bars mounted normal to the backing plate of the submerged wall jet. The previous observations were reproduced for three different bar sizes used in arrays with nominally similar overall blockage factors. These results, recorded on movie film, show that the screen efficiency scales primarily on the blockage factor. This indicates a predominantly inertial scaling for the screen effect which, together with the obviously inertial scaling of the mixing region structure, augurs well for future tests in large scale flows.

7. CONCLUSIONS

The downshaft level surging has caused much anxiety about the ability of outfall structures to withstand this phenomenon for the life of a power station. Current field data on localised pressure changes caused by the collapse or movement of air pockets is not adequate for a civil engineering design appraisal and it might be necessary to mount a large scale instrumentation survey on at least one project to gain more insight into the structure-air pocket-level surge damage relationship. In the meantime a research programme comprising hydraulic modelling and fundamental studies has examined means of minimising surges.

For the new CEGB Thameside project the surge potential was reduced during the initial design stage by the addition of a secondary weir in the offshore outfall structure. This curtailed the height of the jet plunge from the syphon seal weir to the downshaft free water surface at low river levels.

The project orientated hydraulic model investigation defined a recirculation pattern in the downshaft which was a major influence on the air entrainment process. With this understanding of the flow patterns it was possible to devise a baffle vane system which considerably reduced the recirculating and downwards induced air bubbles.

The 1:13.3 model was used to obtain approximate measurements of the recirculating air content as a function of the downshaft water level. At the lower water condition the air/water volume ratio entering the downshaft bend has been deduced as considerably less than 0.1%. Previous model work has suggested that this is the critical area for inducing level surges and that a low air/water ratio of this order will result in a satisfactory design. Work continues on the model to examine transient void venting caused by changes in the downshaft water velocity.

The fundamental theoretical work has shown the relationship between air entrainment and Froude index and it was possible to establish a first order approximation of the major parameters which determined bubble detrainment.

The supporting experimental work has demonstrated that vortices in the mixing region are the most significant factor in bubble entrainment. The vortex pressure gradient appears to be responsible for the different bubble escape velocities in a downshaft system and still water.

The outcome of the fundamental work has been to recommend a screen array which results in a reattachment of the critical mixing region to the surface and modifies the vortex structure into a small scale turbulence. These factors produce a high degree of bubble detrainment.

Further work is taking place both on the project orientated model and the fundamental studies and these will be reported at the discussion. It is planned to carry out a comprehensive test on the Thameside site installation to compare the model and fundamental findings with prototype data.

8. REFERENCES

1. CEGB Sutton Office to W.T. Mawer Feb. 1977
   Private Communication

2. Miller D.S. Oscillation in a circulating water outfall
   system at Hunterston B Power Station. BHRA June 1973

3. Casteleyn J.A. & Kolkman, P.A. Manuscript.
   Delft Hydraulic Lab. 1977

4. Ervine, R.A. Water Power, Dec , 27-10 1976

5. Kalinske, A.A. & Robertson, J.M. Trans. ASCE 108 1435. 1943

6. Keller, R.J. & Wood, I.R. J. Hydr. Div. ASCE HY4, 553. 1974

7. Rajaratnam, N. Adv. Hydrosciences, 4 197. 1967

8. Renner, J. BHRA Sympos. 'Siphons', A4 49. 1975

9. Thomas, N.H. Unpublished M.S., D.A.M.T.P., Cambridge Univ. 1978

10. Wisner, P. IAHR. 11th Cong., Leningrad. Paper 1.15. 1965

11. Van de Sande, E. & Smith, J. M., Chem. Eng. Sci., 28,1161. 1973

9. ACKNOWLEDGEMENTS

The Authors acknowledge the assistance and advice of K.P. Gibbs and Partners, Dr. D.J.F. Ewing of CERL, Leatherhead and Mr. D.C. Cheesely of DAMTP, Cambridge University.

This paper is published by permission of the Central Electricity Generating Board.

System and Investigator	Correlation (Bulk air/water flux ratio)	Symbols
Plunging free (slot) jet (Ervine 1976)	$.13(h/d)^{.446}(1-q_w^*/q_w)$ $q_w^* = 1.1 m/s$	h = plunge height d = jet depth q_w = jet speed q_w^* = inception speed
Forced hydraulic jump (Renner 1975)	$(.002 \pm .0005)F_R^2$ $(0 < \varphi < \pi/4)$ $(3 \lesssim F_R \lesssim 20)$	$F_R = q_w/\sqrt{gd}$ g = grav. acceleration ϕ = flow incidence
Hydraulic jump (Rajaratnam 1967)	$.018(F_R-1)^{1.245}$, $\phi = 0$ $(2 \lesssim F_R \lesssim 8)$	
Duct (Wisner 1965)	$.014(F_R-1)^{1.4}$, $\phi = 0$	
(Kalinske 1943)	$.0066(F_R-1)^{1.4}$, $\phi = 0$	
Siphon (Casteleyn 1977)	$.001(q_w-q_w^*)^3/q_w d$ (S.I. units) $(q_w^*) = 0.8 m/s$, $\phi = 45°$	

Table 1 Bulk data correlations for air/water flux ratio

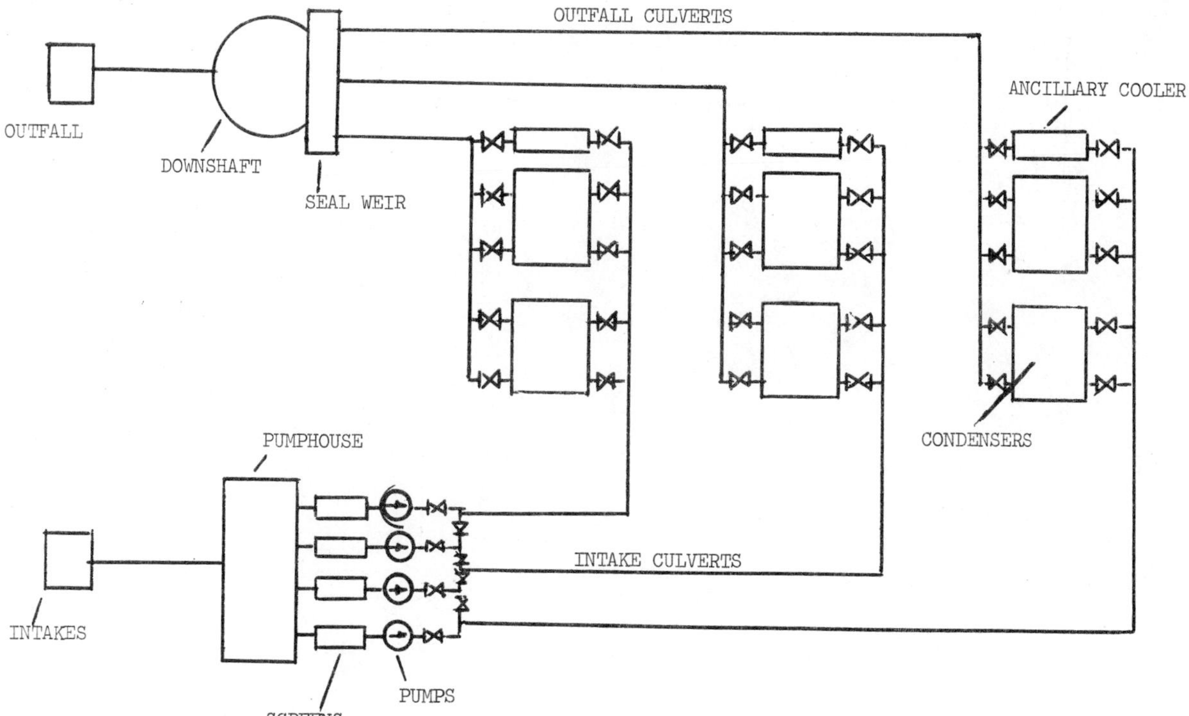

Fig. 1 Diagram of cooling water system

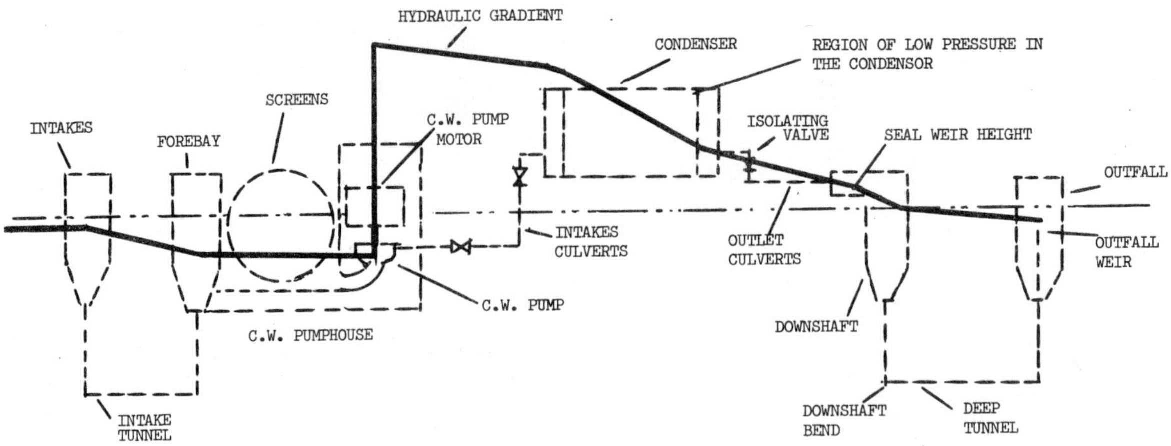

Fig. 2 C.W. hydraulic gradient

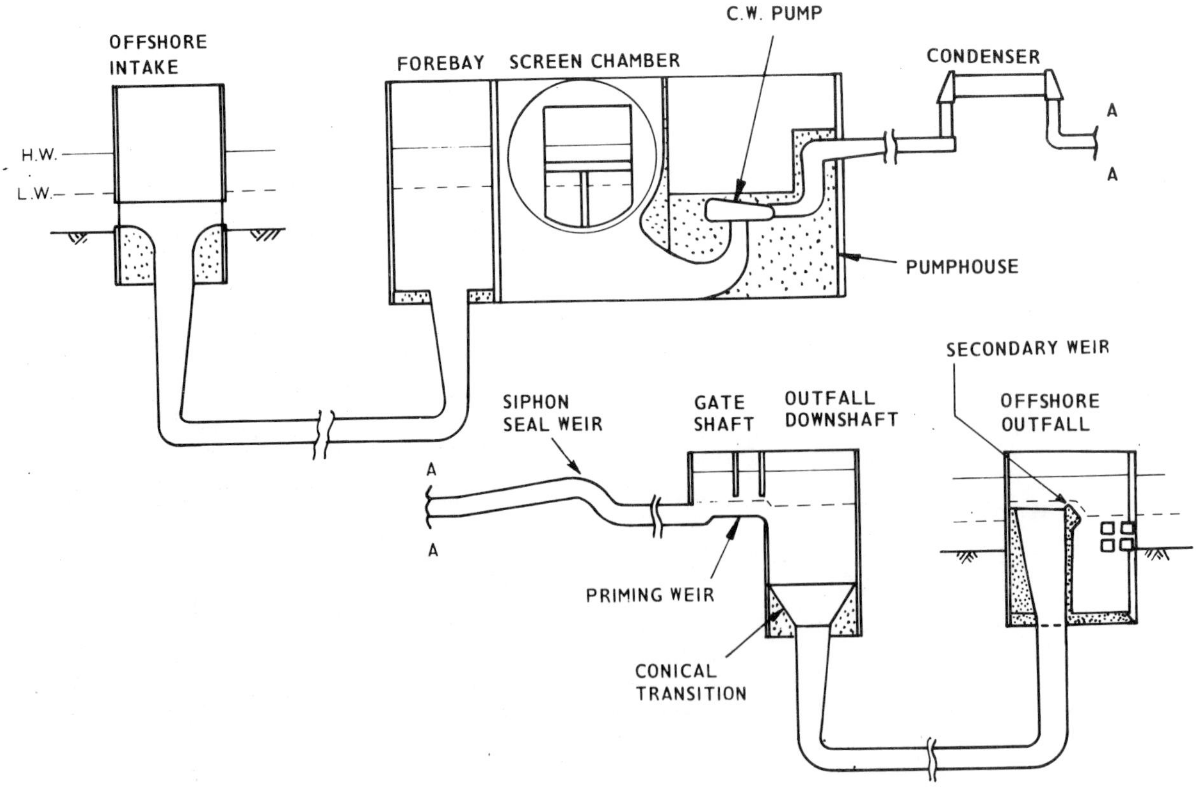

Fig. 3 Elevation sketch of cooling water system

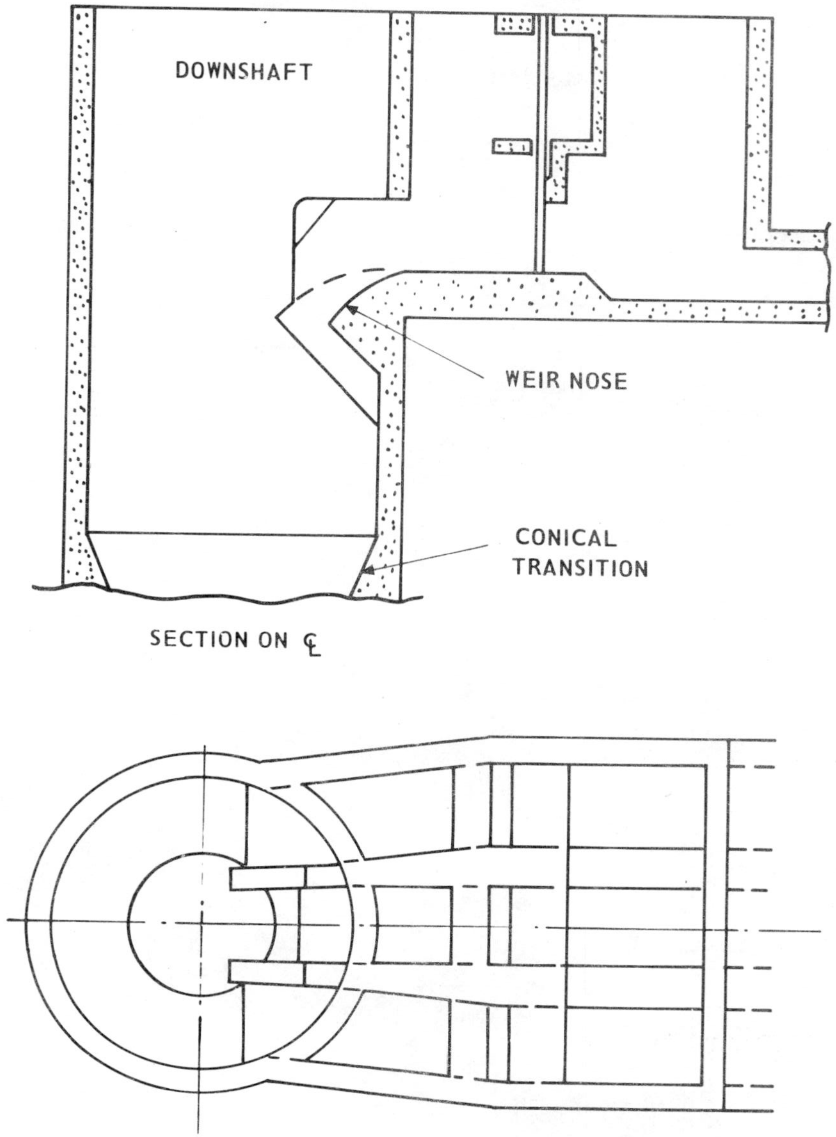

Fig. 4 11m diameter outfall downshaft

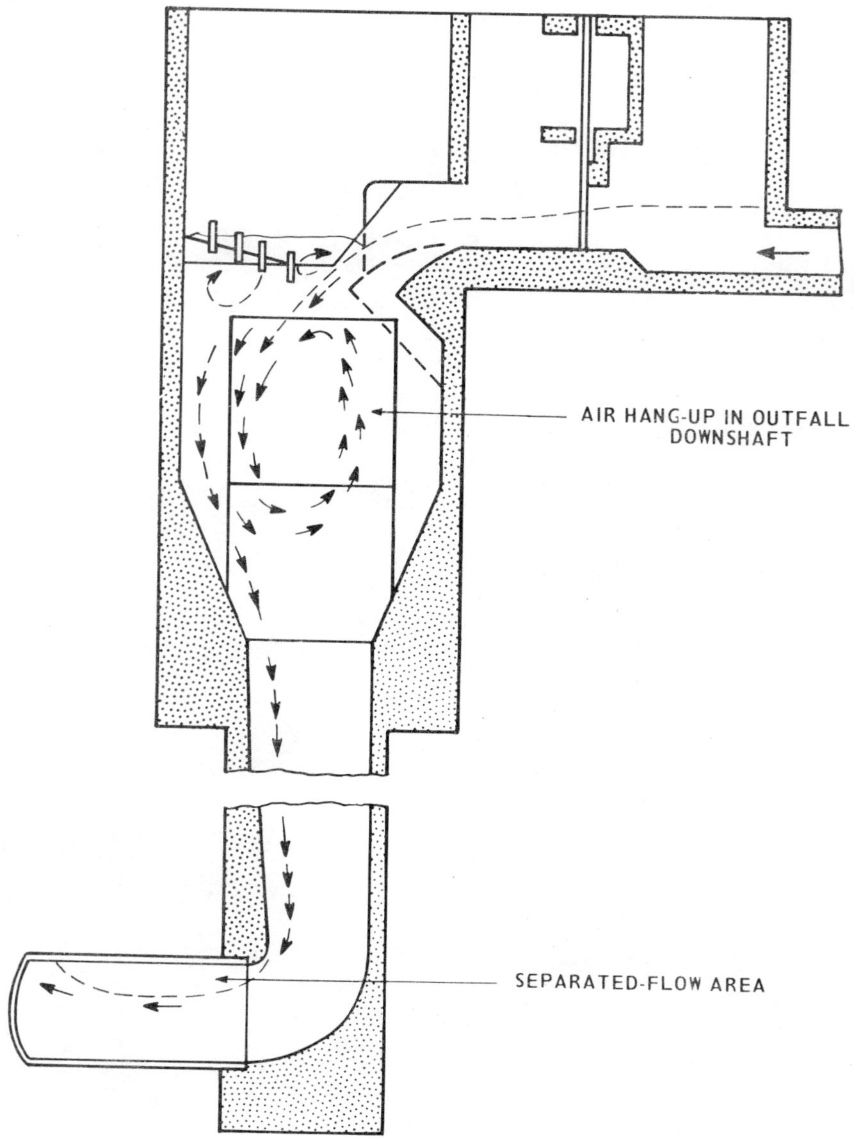

Fig. 5 Air recirculation patterns for single culvert flow (No. 3) at E.L.W.

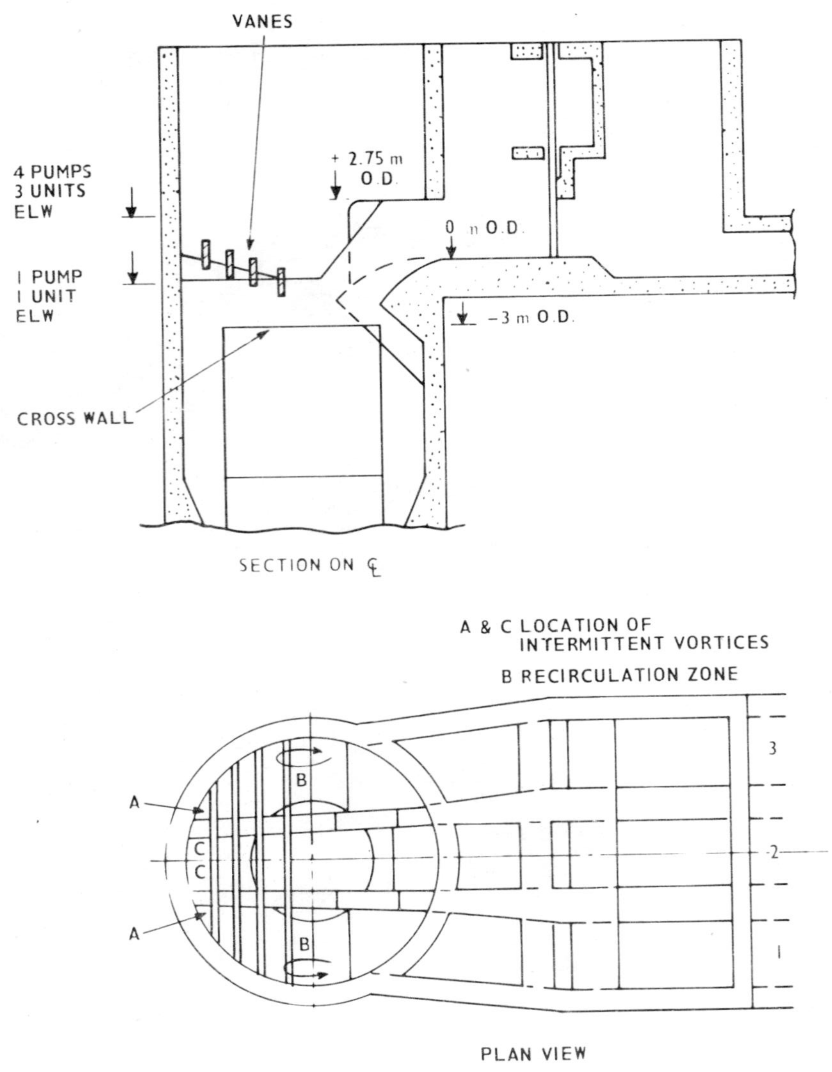

Fig. 6 11 m diameter outfall downshaft showing levels and air entrainment zones

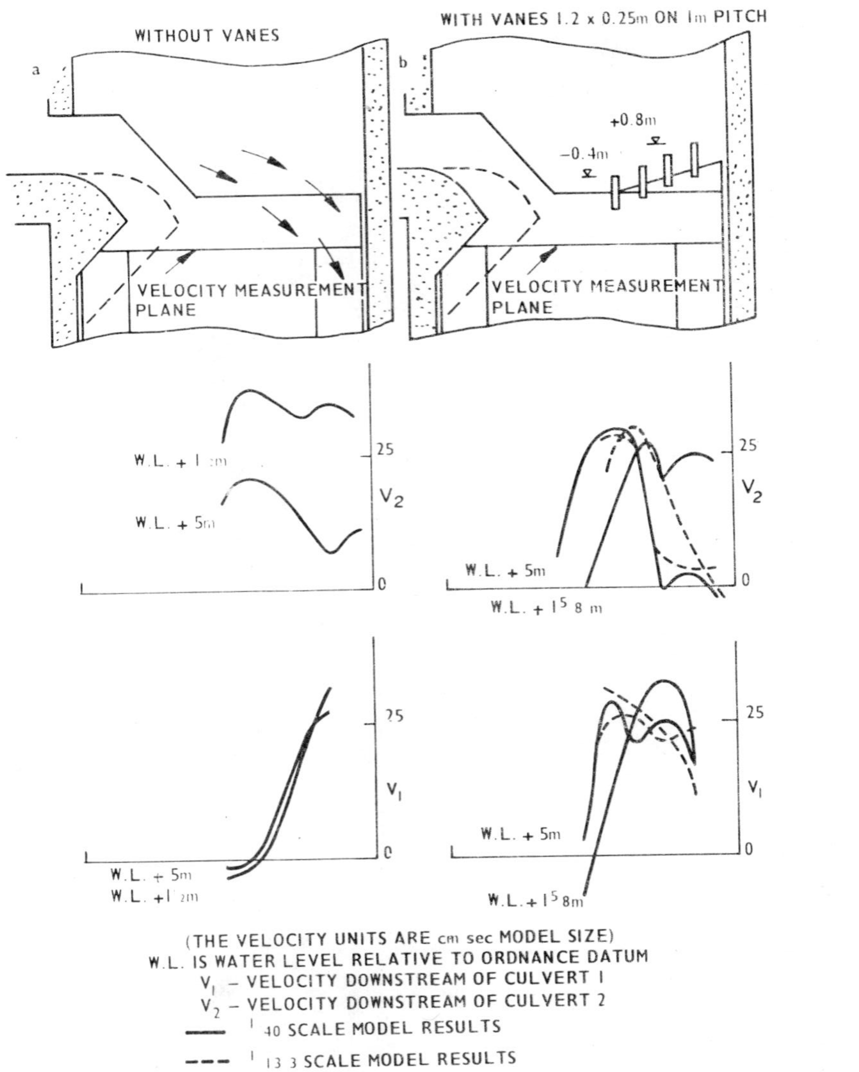

Fig. 7 Distribution of vertical component of water velocity below cross-walls in outfall models

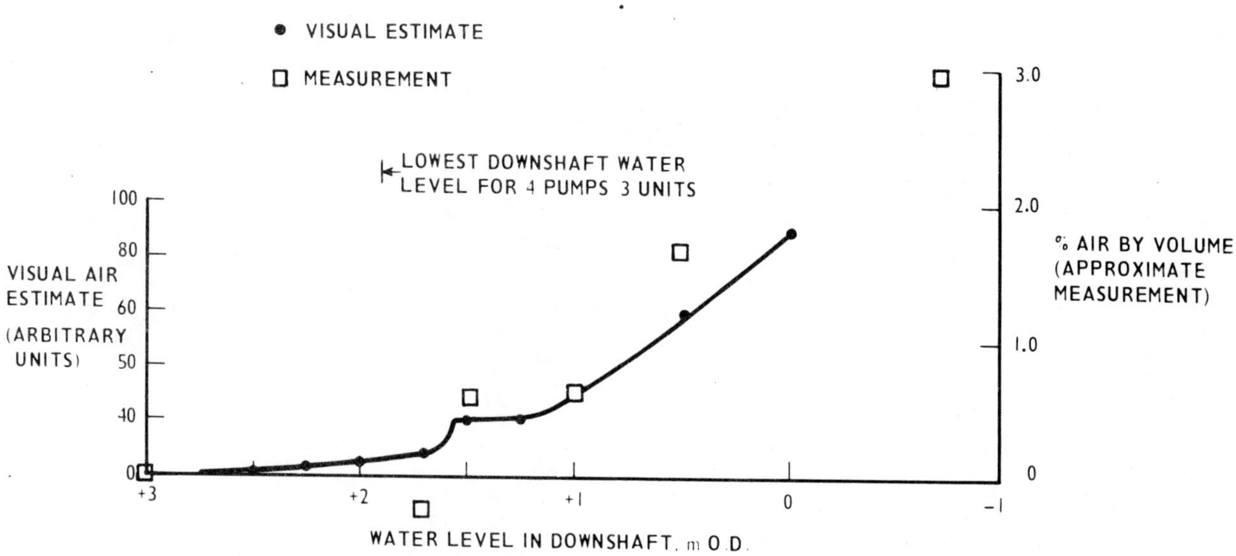

Fig. 8 Visual estimates of air hang-up compared with differential head measurements (nominal flow in 3 culverts)

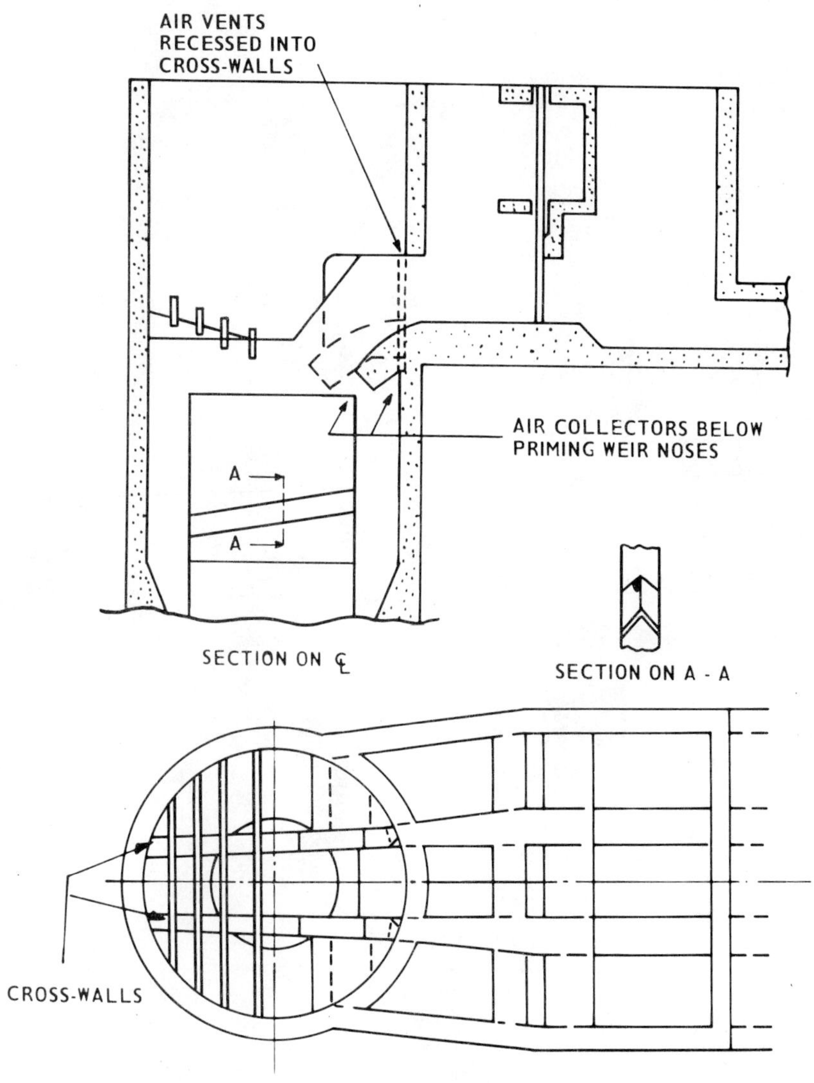

Fig. 9 Air collectors in model of outfall downshaft

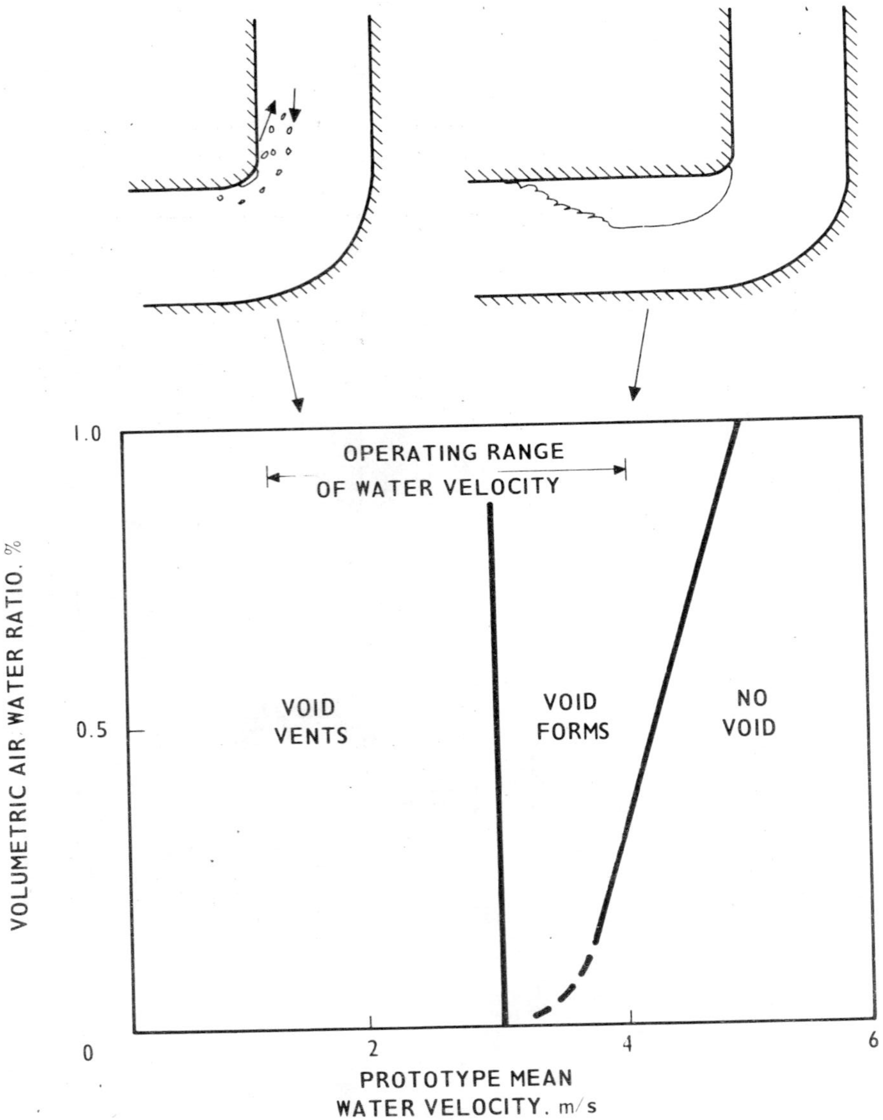

Fig. 10 Void formation at circular bend of radius ratio 0.75

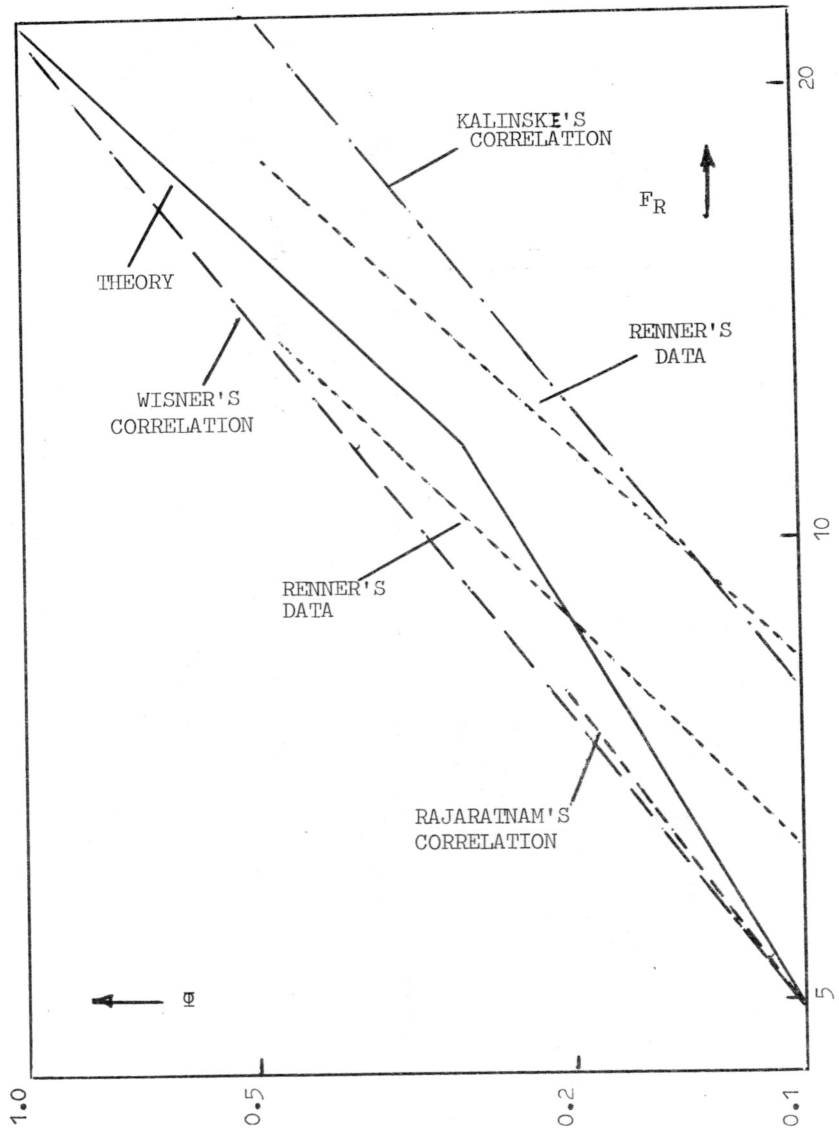

Fig. 11 Theoretical scalings for air-entrainment hydraulic jumps

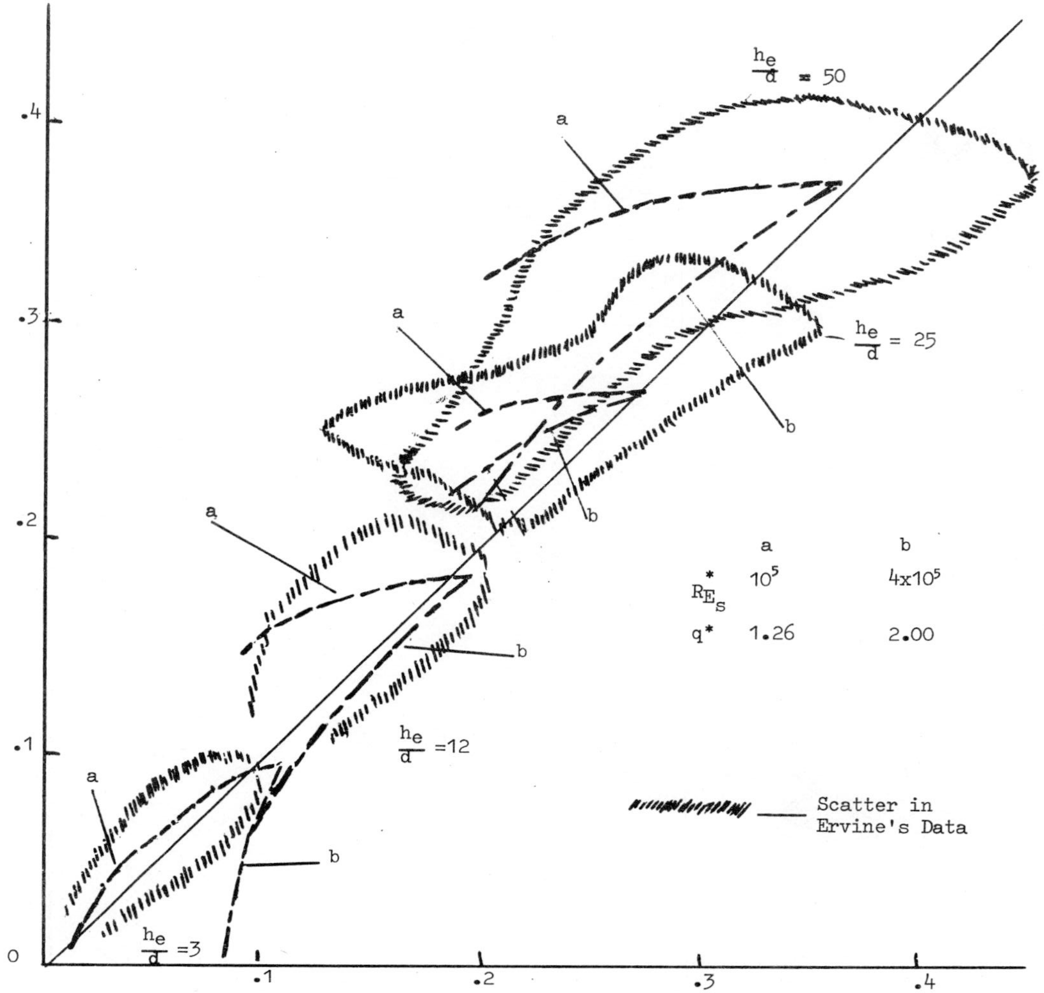

Fig. 12 Comparison of 'free' entrainment model with Ervine's correlation for supported flows

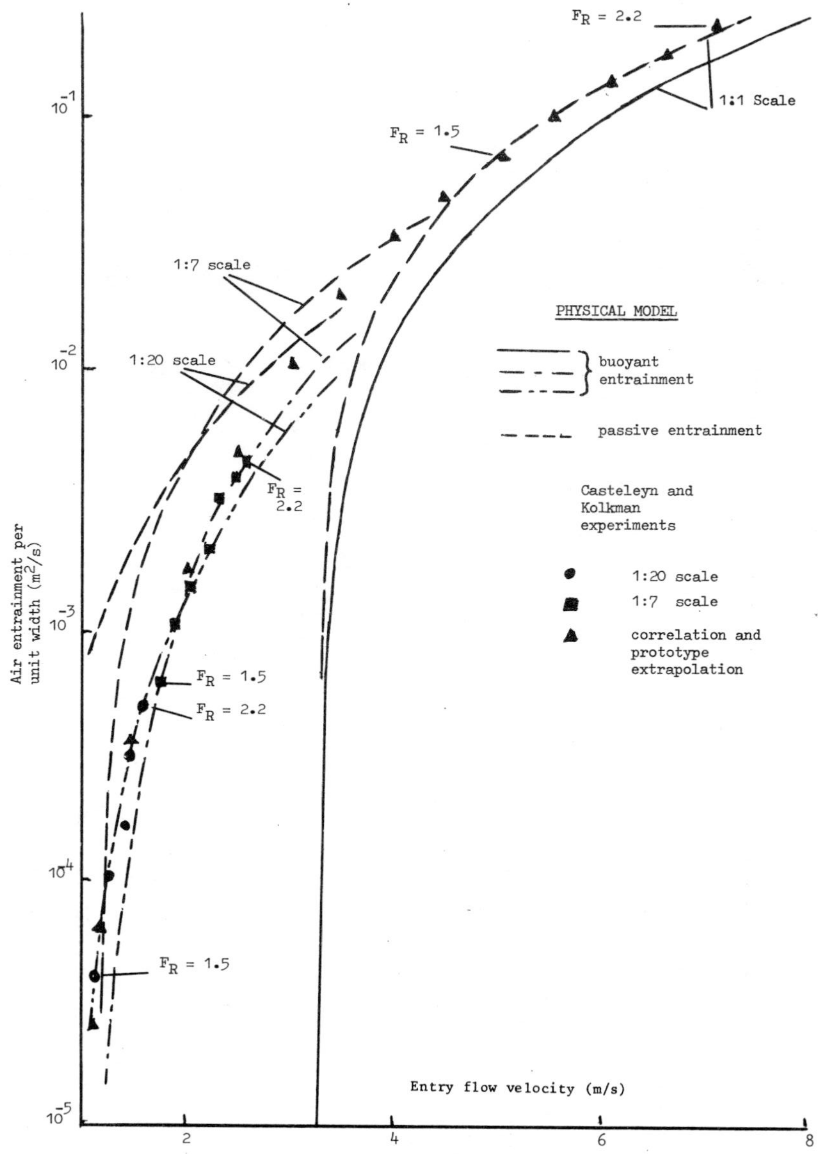

Fig. 13 Buoyant entrainment model compared with Casteleyn and Kolkman results

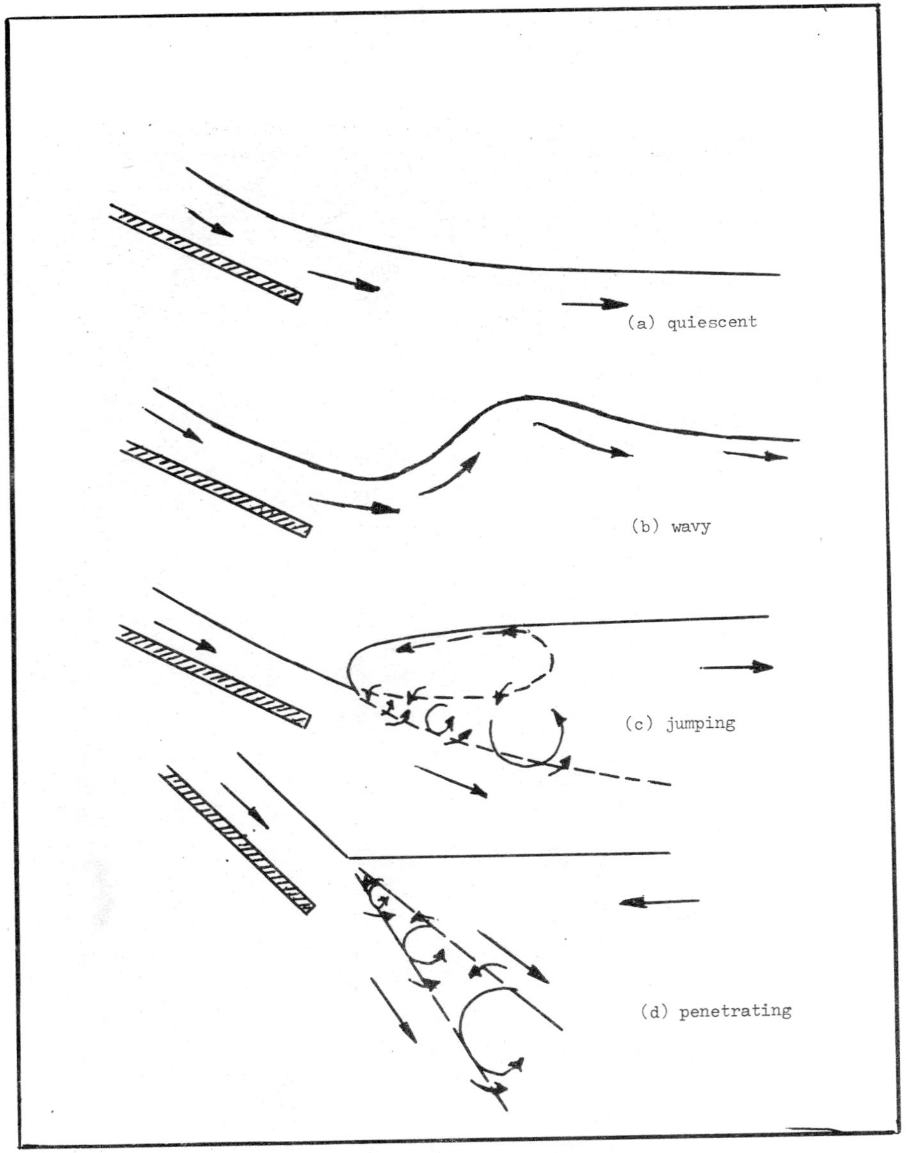

Fig. 14 Classification according to configuration of the surface flows

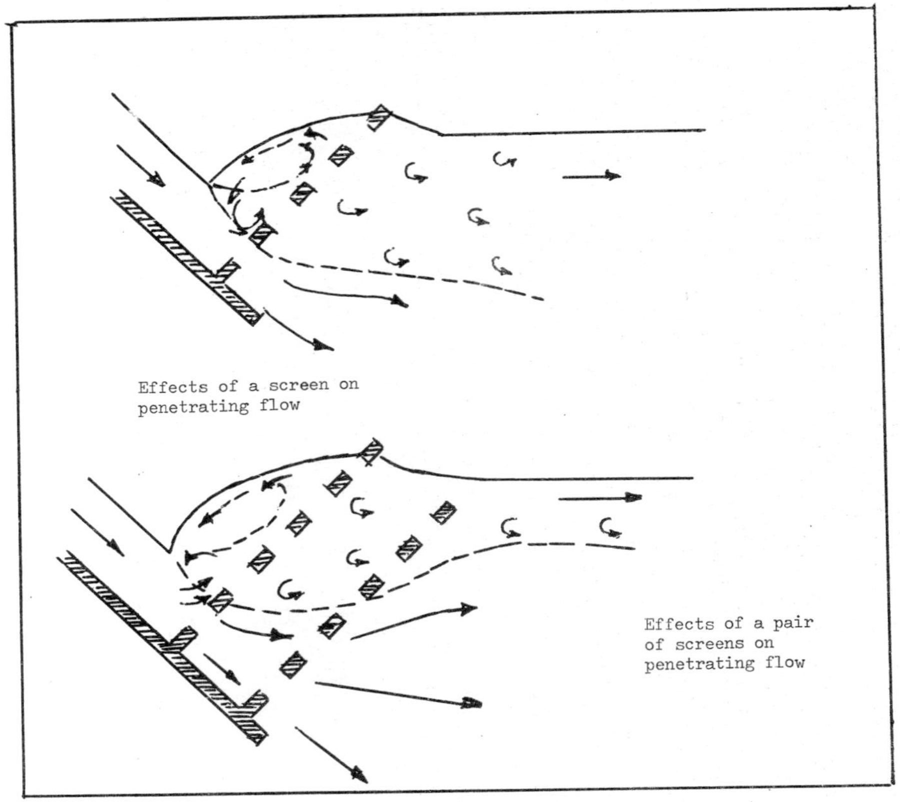

Fig. 15 Effects of screens on penetrating flow

PLATE 3 JUMPING FLOW AT 0° INCIDENCE.
BUBBLES CONFINED TO THE SURFACE JET

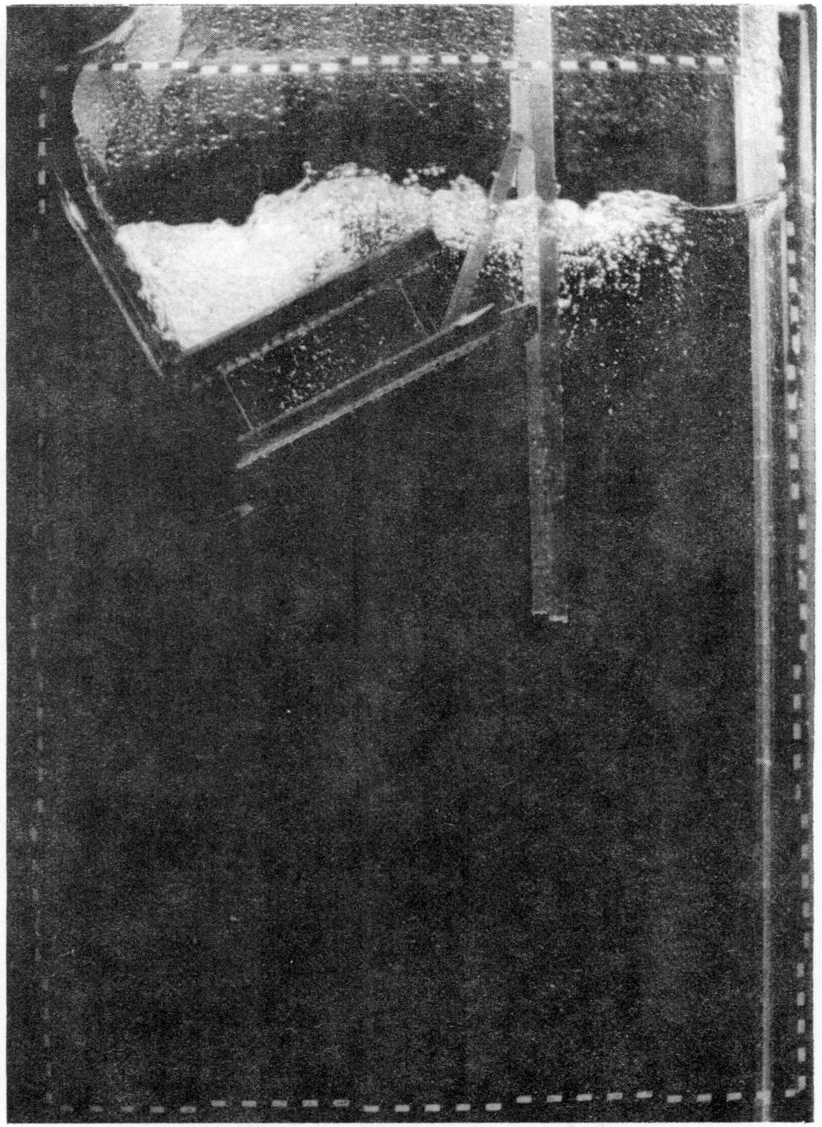

PLATE 4 AIR ENTRAINMENT CONTROL USING SCREENS.
MIXING LAYER REATTACHES TO THE FREE SURFACE

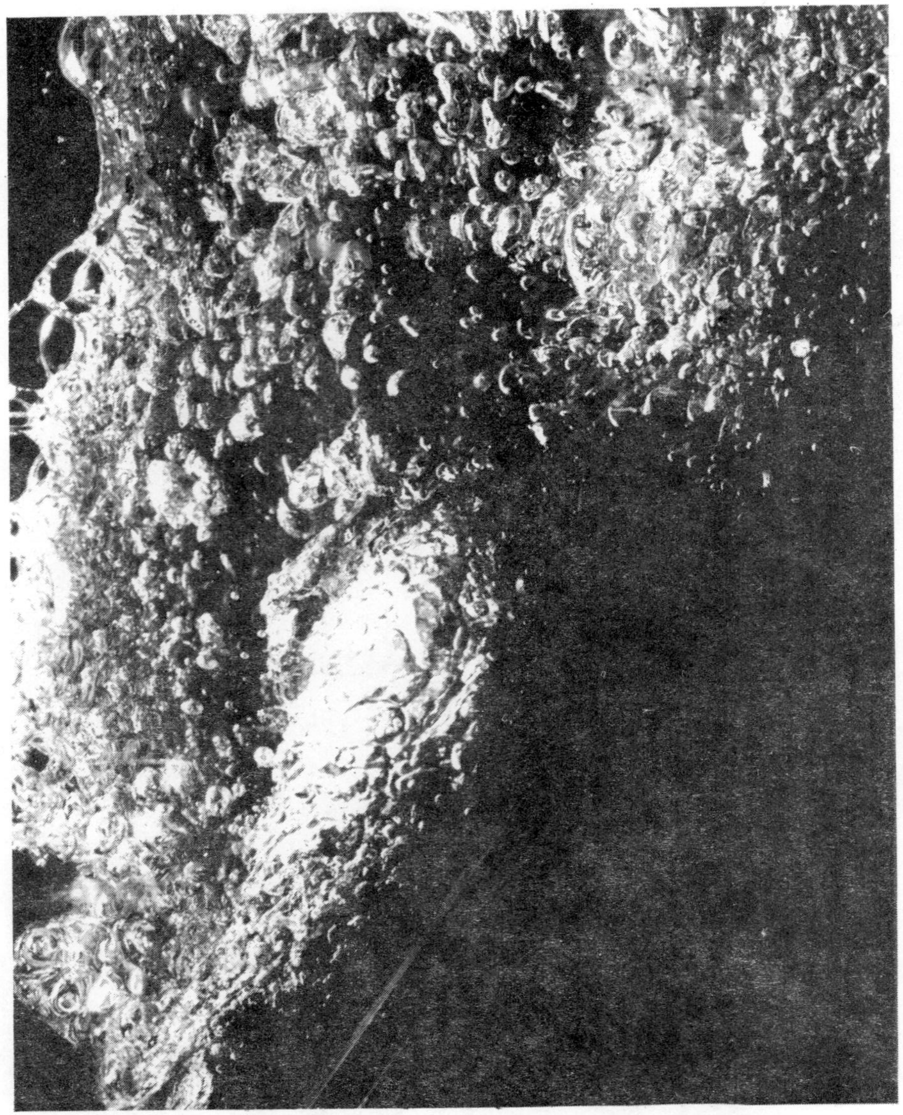

PLATE 1. VORTEX FORMATION IN THE SUBMERGED JET MIXING REGION. ENTRAINMENT OF BUBBLES FROM THE SURFACE FOAM.

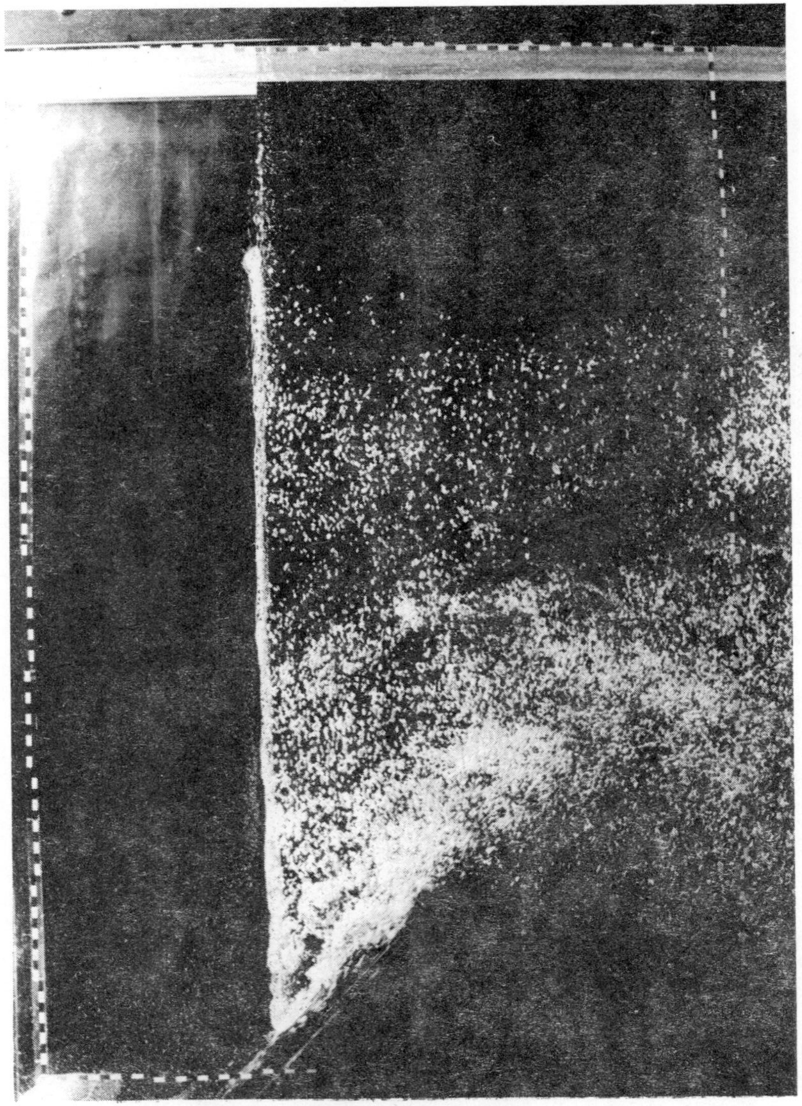

PLATE 2. PENETRATING FLOW AT AN INCIDENCE OF 45°. BUBBLES DISPERSED THROUGHOUT THE CHAMBER BY SUBMERGED JET.

Third International Conference on

PRESSURE SURGES

Canterbury, England
March 25-27, 1980

PAPER G1

A NONLINEAR MATHEMATICAL MODEL FOR ANALYSIS OF TRANSIENTS CAUSED BY A GOVERNED FRANCIS TURBINE

M.H. Chaudhry

Old Dominion University, U.S.A.

Summary

A mathematical model is presented to analyze transient conditions in a hydro-electric power plant caused by generator-load changes. Transient conditions include opening or closing of wicket gates under governor control, speed of the turbogenerator set and the penstock pressures and flows. The model is suitable for both small and large load changes since all major nonlinearities of the system are included.

The method of characteristics is used to solve the dynamic and continuity equations describing the unsteady flow in closed conduits. The turbine characteristics over the complete range and nonlinearities and saturation of various governor components are included. Both temporary-droop and proportional-integral-derivative governors are considered.

To verify the model, the computed wicket-gate closure, turbine speed, and penstock pressures are compared with prototype test data. There is good agreement between the computed and measured results.

Held at the University of Kent at Canterbury, U.K.

Conference Organised by BHRA Fluid Engineering, Cranfield, Bedford, England.
In conjunction with The City University, London.

NOMENCLATURE

a = waterhammer wave velocity
A = cross-sectional area of pipe
C_a, C_n, C_p = constants in Eqs. 2 to 4
D = pipe diameter
d = runner diameter
e_t = output of dashpot
f = Darcy-Weisbach friction factor
g = acceleration due to gravity
H = piezometric head at the beginning of time step
H_n = net head
H_P = piezometric head at the end of time step
H_{tail} = tailwater level above datum
I = moment of inertia of rotational parts
k_d = distributing-valve gain
k_{da} = gain of derivative element
k_i = gain of integral element
k_p = gain of the proportional element
k_s = gate-servomotor gain
n = N/N_R
N = rotational speed
N_R = rated speed of turbine
P_g = power of generator
P_t = turbine power
p = unit power
q = unit flow
Q = flow at the beginning of time step
Q_P = flow at the end of time step
s = Laplace operator
T_a = actuator time constant
T_c = effective wicket-gate closing time
T_d = distributing-valve time constant
T_{da} = time constant of derivative element
T_i = time constant of integral element
T_o = effective wicket-gate opening time
T_r = dashpot time constant
v_a = actuator output
v_d = output of distributing valve
v_{da} = output of derivative element
v_{ia} = output of integral element
v_{pa} = output of proportional element

Greek letters

δ = temporary speed droop
η_g = generator efficiency
η_t = turbine efficiency
σ = permanent speed droop
τ = wicket-gate opening
ϕ = unit speed
ω = rotational speed in radians/sec.

INTRODUCTION

The transients caused by load changes on a turbogenerator set are analyzed to determine: (1) the maximum and minimum pressures along the upstream and downstream conduits; (2) the maximum turbine speed following full-load rejection; and (3) the loading and unloading characteristics and governing stability of the system. A mathematical model is required for such analyses.

Investigation of transients caused by governed hydraulic turbines began with the advent of the turbines and control governors; early investigators being Maxwell (Ref. 1), Stodola (Ref. 2), Frizell (Ref. 3), Rateau (Ref. 4), Allievi (Ref. 5,6), Thoma (Ref. 7), and Tolle (Ref. 8). Frizell (Ref. 3) appears to be the first to investigate the effect of waterhammer on the speed regulation of turbines. Arithmetic integration was used by Kerr (Ref. 9) to compute the pressure drop in penstocks due to opening of turbine gates and by Strowger and Kerr (Ref. 10) to determine the speed changes of hydraulic turbines for sudden load changes. Governing stability using linear mathematical models was studied by several investigators (Ref. 11-18). Swiecicki (Ref. 19) investigated the regulation characteristics of a hydraulic turbine by using a step-by-step arithmetical integration. A computer program was developed in Ref. 20 in which the dynamic and continuity equations were solved by an implicit finite-difference method, the turbine characteristics were included in a simplified form and the governor was represented by a linear third-order ordinary differential equation having constant coefficients, i.e. the saturation of various elements was not included. Newey (Ref. 21) presented the details of a computer program for studying the speed regulation of Bay D'Espoir Hydroelectric generating station. Wozniak and Fett (Ref. 22) developed a computer program for a closed-loop simulation of the transient behavior of a hydroelectric power plant. In this program, the governor nonlinearities were included, the turbine characteristics were approximated by polynomials, and the conduits were represented by MacLaurin-series expansion of the closed-form solution of the linerized partial differential equations. In another computer program, Blair and Wozniak (Ref. 23) used the slopes of the turbine characteristics at various points to represent the turbine. Streeter and Wylie (Ref. 24) presented a mathematical model in which the method of characteristics was used to simulate the conduit, turbine characteristics were included and the governor was represented by a second-order, ordinary differential equation. Chaudhry presented a mathematical model (Ref. 25) in which turbine characteristics were globally represented, method of characteristics was used to solve the continuity and dynamic equations representing transient flow in closed conduits and all major nonlinearities of the governor were included. This model was for a temporary-droop governor only and required two iteration loops to solve the equations.

In this paper, a computational procedure is presented in which the number of iteration loops is reduced from two loops to one. In addition, the simulation of a PID governor is also included. Computed results are compared with the prototype data to verify the model.

COMPONENTS OF MATHEMATICAL MODEL

The following components have to be mathematically simulated in the model of a hydroelectric power plant:

1. Upstream and downstream conduits and associated boundaries
2. Turbine
3. Governor

A description of each follows:

1. CONDUIT SYSTEM

The conduit system and appropriate boundaries in a hydroelectric power plant are comprised of the following (see Fig. 1): (a) upstream reservoir; (b) upstream pressurized conduit with or without surge tanks; (c) turbine spiral case; (d) draft tube; (e) downstream conduit with or without surge tanks; and (f) downstream reservoir. If the downstream water passages consist of a free-flow tunnel, a canal, a river, or a channel, then the draft tube as well as the downstream water passages may be neglected to simplify the analysis.

The method of characteristics have been widely used for computing the transient-state conditions in closed conduits. Since the details of this method are easily available in the literature (see Ref. 24-27), only equations required for developing the boundary conditions for a turbine are included herein:

For an intermediate section, i, of a pipeline (Ref. 25):

$$Q_{P_i} = 0.5\ (C_p + C_n) \tag{1}$$

in which

$$C_p = Q_{i-1} + C_a\ H_i - \frac{f\Delta t}{2DA}\ Q_{i-1}|Q_{i-1}|; \tag{2}$$

$$C_n = Q_{i+1} - C_a\ H_{i+1} - \frac{f\Delta t}{2DA}\ Q_{i+1}\ |Q_{i+1}| \tag{3}$$

$$C_a = \frac{gA}{a} \tag{4}$$

Q = transient-state flow; g = acceleration due to gravity; A = cross-sectional area of the pipe; a = waterhammer-wave velocity; H = piezometric head above the datum; f = Darcy-Weisbach friction factor; D = pipe diamter; Δt = computational time step; subscript P indicates conditions at the end of time step and the subscripts i, i+1 and i-1 denote the pipe sections. For a downstream boundary

$$Q_{P_{n+1}} = C_p - C_a\ H_{P_{n+1}} \tag{5}$$

in which the subscript (n+1) denotes the last section on the pipe. Eq. 5 ties the downstream boundary to the rest of the system. In this equation, $Q_{P_{n+1}}$ and $H_{P_{n+1}}$ are two unknowns and we need another equation for a unique solution. This equation is provided by the relationship between the net head and the turbine flow.

The boundary conditions such as upstream reservoir, various types of surge tanks and junction of two or more pipes are available in the literature (Ref. 24-26) and will not be discussed here; only the boundary conditions for a turbine will be developed herein.

2. FRANCIS TURBINE

As discussed above, the relationship between flow, Q, and net head, H_n, has to be specified to develop the boundary conditions. The flow through a Francis turbine is a function of the net head, wicket-gate opening, τ, rotational speed, N, and efficiency, η, of the turbine. The relationship between these five variables, namely Q, H_n, τ, N, and η are usually represented by a family of curves which are called turbine characteristics. These curves are obtained from tests conducted on the model of the turbine runner and usually turbine power, instead of the turbine efficiency, is used as the fifth variable. Fig. 2 shows the typical characteristics for a high-head Francis turbine. In this figure

$$\text{Unit speed, } \phi = \frac{d\ N}{84.45\sqrt{H_n}} \tag{6}$$

$$\text{Unit flow, } q = \frac{Q}{d^2\ \sqrt{H_n}} \tag{7}$$

$$\text{Unit power, } p = \frac{P_t}{d^2 H_n^{1.5}} \qquad (8)$$

in which d = throat diameter of the runner, in m; N = rotational speed, in rpm; P_t = turbine power, in kW; Q = turbine flow, in m^3/s and H_n = net head, in m.

Following load reduction on a turbogenerator set, its speed increases (speed will decrease following a load increase) due to the unbalanced torque, T_u, according to the following equation:

$$T_u = I \frac{d\omega}{dt} \qquad (9)$$

in which I = moment of inertia of the rotating parts of the turbogenerator set, in kg m^2, and ω = rotational speed of the unit, in radians/s. Eq. 9 may be written as:

$$P_t - \frac{P_g}{\eta_g} = I\omega \frac{d\omega}{dt} \qquad (10)$$

or

$$\frac{dN}{dt} = \frac{3600}{I(2\pi)^2 N}\left(P_t - \frac{P_g}{\eta_g}\right) \qquad (11)$$

or

$$\frac{dn}{dt} = \frac{91.19}{I\, N_R^2 n}\left(P_t - \frac{P_g}{\eta_g}\right)$$

in which the subscripts t and g denote quantities for the turbine and for the generator; $n = N/N_R$; and N_R = synchronous speed.

3. HYDRAULIC-TURBINE GOVERNORS

Governors are used to keep the speed of the turbogenerator set constant by increasing or decreasing the turbine flow. They may be classified using their corrective action as the criterion. For example, the corrective action in a tempoary -droop governor depends upon speed deviation from the synchronous speed and its integral; and it depends on the combination of proportional, integral and derivative of the speed deviation in a proportional integral-derivative (PID) governor.

Figs. 3 and 4 show the block diagram for a temporary-droop and for a PID governor. In these figures, s = Laplace operator which is equivalent to d/dt if the initial conditions are zero; other notations are as follows: T_a = actuator time constant; T_r = dashpot time constant; δ = temporary speed droop; σ = permanent speed droop; T_d = distributing valve time constant; k_d = distributing valve gain; k_s = gate-servomotor gain; n_{ref} = 1.0; k_p = gain of the proportional element; k_{da} = gain of the derivative element, and k_a = gain of the integral element. The transfer function for each component is listed in the block representing it while the inputs and outputs are marked on the incoming and outgoing arrows from each block.

Saturation limits of various components are listed in Figs. 3 and 4. The typical values of T_a, σ, T_d, k_d, k_s, T_{da} and $e_{t_{max}}$ are 0.05 to 0.1s, 0.03 to 0.05, 0.05 to 0.1s, 10 to 15, 0.2, 0.01s, and 0.2 to 0.5 respectively. The optimum values of T_r, δ, k_i, k_{da}, and k_p are determined for a particular power plant by conducting simulation studies; emperical relationships reported in Refs. 12-14 and 18 may be used to estimate their values as a first approximation.

The differential equations for various components may be written from the transfer functions listed in Figs. 3 and 4.

(a) PID Governor

(i) Proportional

$$v_{pa} = (n_{ref} - n - \sigma\, v_a)\, k_p \qquad (13)$$

(ii) Derivative

$$\frac{dv_{da}}{dt} = \frac{k_{da}}{T_{da}} \left(- \frac{dn}{dt} - \sigma \frac{dv_a}{dt} \right) - \frac{v_{da}}{T_{da}} \tag{14}$$

(iii) Integral

$$\frac{dv_{ia}}{dt} = \frac{1}{T_i} (n_{ref} - n - \sigma v_a) \qquad 0 \leqslant v_{ia} \leqslant 1. \tag{15}$$

(iv) Actuator

$$\frac{dv_a}{dt} = \frac{1}{T_a} (v_{pa} + v_{da} + v_{ia} + v_a) \tag{16}$$

Substituting for v_{pa} from Eq. 13, this equation becomes

$$\frac{dv_a}{dt} = \frac{1}{T_a} [k_p (n_{ref} - n - \sigma v_a) + v_{da} + v_{ia} + v_a] \tag{17}$$

$$0 \leqslant v_a \leqslant 1$$

(v) Distributing Valve

$$T_d \frac{dv_d}{dt} + v_d - k_d v_i = 0 \qquad v_{d_{min}} \leqslant v_d \leqslant v_{d_{max}} \tag{18}$$

in which

$$v_{d_{max}} = \frac{1}{k_s T_o} \tag{19}$$

$$v_{d_{min}} = \frac{-1}{k_s T_c} \tag{20}$$

and T_o and T_c are the effective wicket-gate opening and closing times. These are defined as twice the time taken by the wicket gates to open or close between 25 and 75 percent openings.

(iv) Gate Servomotor

$$\frac{d\tau}{dt} - k_s v_d = 0 \qquad 0 \leqslant \tau \leqslant 1 \tag{21}$$

(b) Temporary-droop Governor

Proceeding similarly as above, the following equations may be written for a temporary-droop governor:

$$\frac{dv_a}{dt} = \frac{1}{T_a} (n_{ref} - n - e_t - \sigma v_a) \qquad 0 \leqslant v \leqslant 1 \tag{22}$$

$$\frac{de_t}{dt} = \frac{1}{T_r} \left(\delta T \frac{dv_a}{dt} - e_t\right) \qquad -e_{t_{max}} \leqslant e_t \leqslant e_{t_{max}} \tag{23}$$

$$\frac{dv_d}{dt} = \frac{1}{T_d} [k_d (v_a - \tau) - v_d] \qquad v_{d_{min}} \leqslant v_d \leqslant v_{d_{max}} \tag{24}$$

$$\frac{d}{dt} = k_s v_d \qquad 0 \leqslant \tau \leqslant 1 \tag{25}$$

BOUNDARY CONDITIONS

To develop the boundary conditions for the turbine end, one has to solve the positive characteristic equation (Eq. 5) with the relationship between the flow, Q_P, and net head, H_n, imposed by the boundary. The flow through a Francis Turbine depends upon the head, H_n, gate opening, τ_P, and the speed of the unit N_P. Since the wicket gates are opened or closed under governor control depending upon the turbine speed which is also unknown, equations between these variables have to be solved simultaneously.

Referring to Fig. 5

$$H_P = H_n + H_{tail} - \frac{Q_P^2}{2gA^2} \tag{26}$$

in which H_{tail} = tailwater level above datum.

There are four unknowns: H_P, Q_P, τ_P and N_P. Because the tansient-state speed and gate opening vary gradually, their values may be estimated as a first approximation by parabolic extrapolation. To determine the range of ϕ for which turbine characteristics have to be used during the time step, H_n is also extrapolated. Let the values of these variables estimated by extrapolation be denoted by subscript e; the corresponding value of ϕ be ϕ_e; and the characteristic for gate opening τ_e be approximated by straight line EF (see Fig. 6). The values at E are interpolated from the known values at grid points A and B, and the values at F are interpolated from the known values at points C and D.

The equation of the straight line may be written as

$$q = a_o + a_1 \phi \tag{27}$$

in which a_o and a_1 are determined from the known coordinates of E and F. Substituting for q and ϕ from Eqs. 6 and 7 into Eq. 27 and simplifying

$$a_2\sqrt{H_n} = Q_P - a_3 \tag{28}$$

in which $a_2 = a_o D^2$ and $a_3 = N_e D^3 a_1/84.45$. Combining Eq. 5 with Eq. 26, eliminating H_n from the resulting equation and Eq. 28, one obtains

$$a_4 Q_P^2 + a_5 Q_P + a_6 = 0 \tag{29}$$

in which $a_4 = C_a/(2gA^2) - C_a/a_2^2$; $a_5 = (2a_3 C_a/a_2^2) - 1$; and $a_6 = C_p - C_a H_{tail} - C_a a_3^2/a_2^2$. Solution of Eq. 29 yields

$$Q_P = \frac{-a_5 - \sqrt{a_5^2 - 4a_4 a_6}}{2 a_4} \tag{30}$$

Note that the positive sign with the radical term is neglected. Now H_P is determined from Eq. 5 and H_n from Eq. 28.

For the computed value of H_n and for the estimated values of N_e, the value of ϕ is determined and a parabola ABC (Fig. 7) for the turbine characteristics for τ_e is fitted, i.e.,

$$p = c_0 + c_1 + c_2 \phi^2 \tag{31}$$

in which the coefficients c_0, c_1, and c_2 are determined from the known coordinates of points A, B and C. Substituting Eqs. 6 and 8 into Eq. 31 and simplifying

$$P_t = a_7 + a_8 n + a_9 n^2 \tag{32}$$

in which $a_7 = c_o d^2 H_n^{1.5}$; $a_8 = c_1 d N_R H_n$; and $a_9 = c_2 d N_R^2 \sqrt{H_n}$. Hence Eq. 12 can be written as

$$\frac{dn}{dt} = \frac{91.19}{I\ N_R^2 n} \qquad a_7 + a_8 n + a_9 n^2 - \frac{P_g}{\eta_g} \tag{33}$$

The governing equations for a PID governor, (Eqs. 14, 15, 17, 18, 21 and 33) are a set of nonlinear ordinary differential equations in six unknowns, namely v_{pa}, v_{da}, v_{ia}, v_a, v_d, τ, and n. Similarly, Eqs. 22 to 25 and Eq. 33 are governing equations for a temporary-droop governor in five unknowns, namely v_a, e_t, v_d, τ, and n. These equations may be solved by the fourth-order Runge Kutta method (Ref. 28).

Note that even though the computed values of n and H_n may be slightly different from their estimated values, no iterations are required since the latter were used only to determine the grid points on the turbine characteristics on either side of ϕ_e. However, if $|\tau_p - \tau_e| > 0.005$, then the above procedure should be repeated to refine the solution.

VERIFICATION

To verify the mathematical model, the computed results were compared with the prototype test results recorded during load-rejection tests at G.M. Shrum Generating Station (Ref. 20) and at Mica Powerplant (Ref. 29, 30). The former power plant has a temporary-droop governor while the later has a PID governor.

The computed and measured results are shown in Figs. 8 and 9. It is clear from these figures that there is good agreement between the computed and measured results for G.M. Shrum Generating station. For the Mica power plant, however, there is satisfactory agreement till the occurance of an unusual pressure spike in the recorded results. The cause of this spike is unknown and is presently under investigation.

SUMMARY

A nonlinear mathematical model for the analysis of transients produced by a governed Francis turbine is presented. The transient-state conditions in the conduits are computed by using the method of characteristics; the governor nonlinearities are included and the turbine characteristics are globally represented. The model is verified by comparing the computed results with those measured on two power plants.

ACKNOWLEDGEMENTS

The block diagram of Figs. 3 and 4 were provided by R. M. Rockwell and J. H. Gurney. Permission by B. C. Hydro and Power Authority, Vancouver, Canada for the inclusion of the prototype test results is appreciated.

REFERENCES

1. Maxwell, J. C., "On Governors," Philosophical Magazine, 35, no. 4, p. 385 (May 1968)

2. Stodola, A. "Ueber die regulierung von turbinen," Schweiz. Bauzeit, Oct. 1893, May 1894)

3. Frizell, J. P., "Pressure Resulting from Changes of Velocity of Water in Pipes," Trans., Amer. Soc. Civil Engrs., 39, pp. 1-18 (June 1898)

4. Rateau, A., "Traite des Turbomachines," Chez Dunod, Paris (1900)

5. Allievi, L., "Theorie generale du mouvement varie de l'eau dans les tuyaux du conduite," Revue de Mecanique (Jan., Mar. 1904)

6. Allievi, L., "Teoria del colpo d'ariete," Reale Academic dei Lincei, Anno CCCIX (1912)

7. Thoma, D., "Zur theorie des wasserschlosses bei selbsttatig geregelten turbinenanlagen," Munchen, Verlag R. Oldenbourg, (1910)

8. Tolle, M., "Regelung der Kraftmaschinen," Springer, Berlin, 1921

9. Kerr, S. L., "Fall in Pressure in Hydraulic Turbine Penstock Due to Acceleration of Flow," Hyd Power Commission Report, National Electric Light Association (1924)

10. Strowger, E. B., and Kerr, S. L., "Speed Changes of Hydraulic Turbines for Sudden Changes of Load," Trans., Amer. Soc. of Mech. Engrs., 48 (1926)

11. Stein, T., "The Influence of Self-Regulation and of the Damping Period on the WR^2 Value of Hydroelectric Power Plant," The Engineers' Digest, May-June 1948 (Translated from Schweizerische Bauzeitung, Sept.-Oct. 1947)

12. Paynter, H. M., "A Palimpsest on the Electronic Analog Act," A Philbrick Researches Inc., Boston, Mass., 1955

13. Hovey, H. M., "Optimum Adjustment of Governors in Hydro Generating Stations," Engineering Jour., Engineering Inst. of Canada, pp. 64-71, (Nov. 1960)

14. Chaudhry, M. H., "Governing Stability of a Hydroelectric Power Plant," Water Power, London, pp. 131-136 (April 1970)

15. Chaudhry, M. H., and Ruus, E., "Analysis of Governing Stability of Hydroelectric Power Plants," Trans., Engineering Inst. of Canada, 13, pp. I-V., (June 1970)

16. Schleif, F. R., "Governing Characteristics for Large Hydraulic Turbines," Report REC-ERC- 71-14, United States Bureau of Reclamation, Denver, Colorado, Feb. 1971

17. Thorne, D. H., and Hill, E. F., "Extension of Stability Boundaries of a Hydraulic Turbine Generating Unit," Trans. on Power Apparatus and Systems, Amer. Inst. of Elect. & Electronic Engrs., PAS 94, no. 4, pp. 1401-1409, (July/August 1975)

18. Hagihara, S., Yokota, H., Goda, U., and Isobe, K., "Stability of a Hydraulic Turbine Generating Unit Controlled by PID Governor," presented at Winter Meeting, Amer. Inst. of Elect. & Electron. Engrs., (Feb. 1979)

19. Swiecicki, I., "Regulation of a Hydraulic Turbine Calculated by Step-by-Step Method," Jour. Basic Engineering, Amer. Soc. of Mech. Engrs., pp. 445-455 (Sept. 1961)

20. Perkins, F. E., Tedrow, A. C., Eagleson, P. W., and Ippen, A. T., "Hydropower Plant Transients," Dept. of Civil Engineering, Hydrodynamics Lab Report No. 71, Part II to IV, Massachusetts Inst. of Tech., (Sept. 1964)

21. Newey, R., "Speed Regulation Study for Bay D' Espoir Hydroelectric Generating Station," Paper No. 67-WA/FE-17, presented at Winter Annual Meeting, Amer. Soc. of Mech. Engrs., (Dec. 1967)

22. Wozniak, L., and Fett, G. H., "Conduit Representation in Closed Loop Simulation of Hydroelectric Systems," Jour. Basic Engineering, Amer. Soc. of Mech. Engrs., pp. 599-604 (Sept. 1972) (See also discussion by Chaudhry, M. H., pp. 604-605)

23. Blair, P., and Wozniak, L., "Nonlinear Simulation of Hydraulic Turbine Governor," Water Power and Dam Construction, London, England, (Sept. 1976)

24. Streeter, V. L., and Wylie, E. B. Hydraulic Transients. McGraw-Hill Book Co., New York, N. Y., (1967)

25. Chaudhry, M. H., Applied Hydraulic Transients, Van Nostrand Reinhold Co., New York, N. Y. (1978)

26. Wylie, E. B., and Streeter, V. L. Fluid Transients, McGraw-Hill Book Co., New York, N. Y. (1978)

27. Chaudhry, M. H., and Portfors, E. A., "A Mathematical Model for Analyzing Hydraulic Transients in a Hydroelectric Power Plant," Proc., First Canadian Hydraulic Conference, published by the University of Alberta, Edmonton, Canada, pp. 298-314 (May 1973)

28. McCracken, D. D., and Dorn, W.S., Numerical Methods and Fortran Programming, John Wiley & Sons, Inc., New York, (1964)

29. Chaudhry, M. H., and Gurney, J. H., "Mica Project: Memorandum on Effect of Increased Wicket Gate Closing Time on Governing Characteristics of Units 1 and 2," Development Department Report No. DD115, B. C. Hydro and Power Authority, Vancouver, Canada (March 1977)

30. Chaudhry, M. H., "Governing Stability of Hydraulic Turbines," Chapt. 6, in Transient Flow and Hydromachinery, edited by Shen, H. T., Water Resources Publications, (in press, 1979)

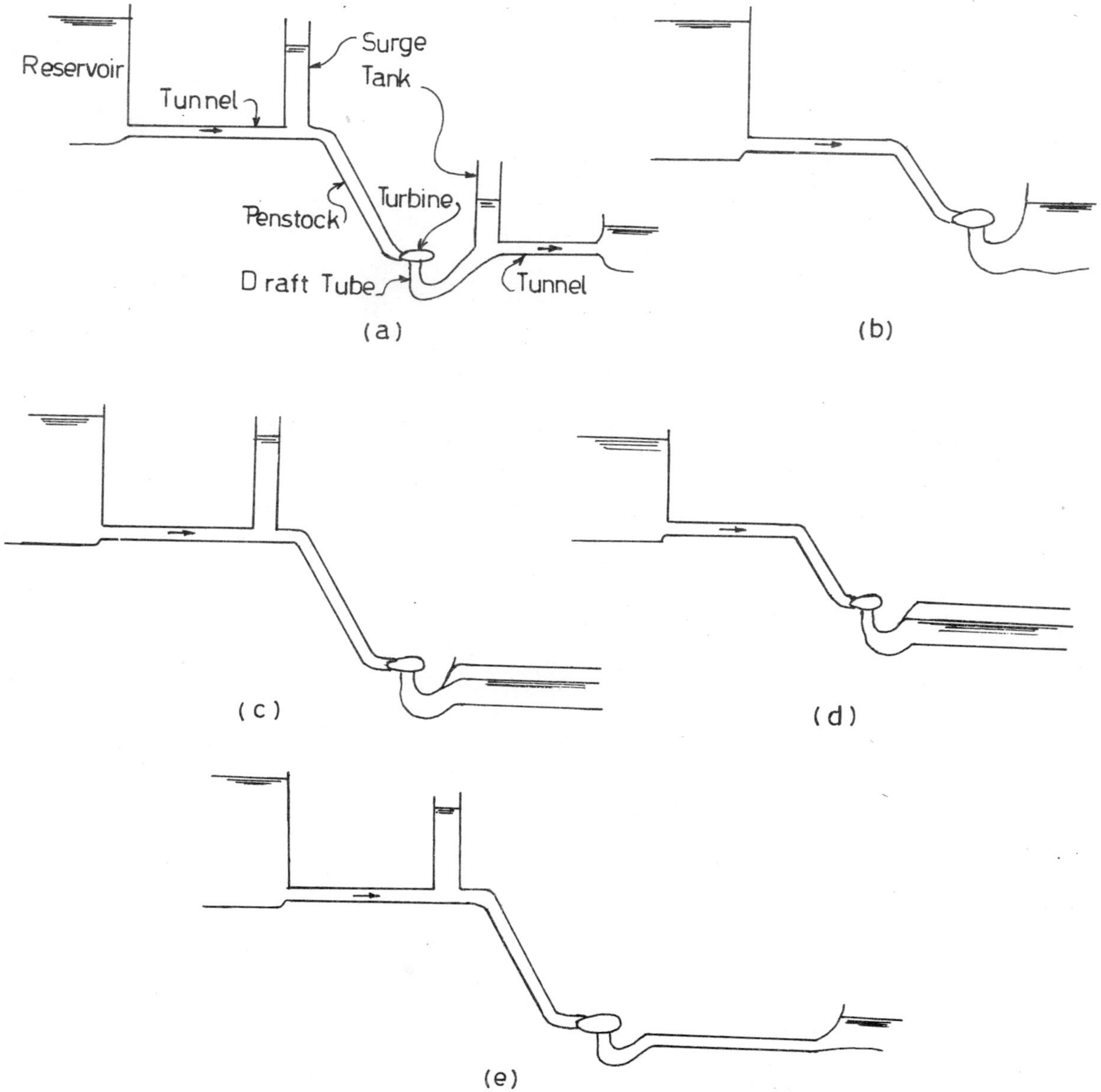

Fig. 1 Typical Water Passages of a Hydroelectric Power Plant

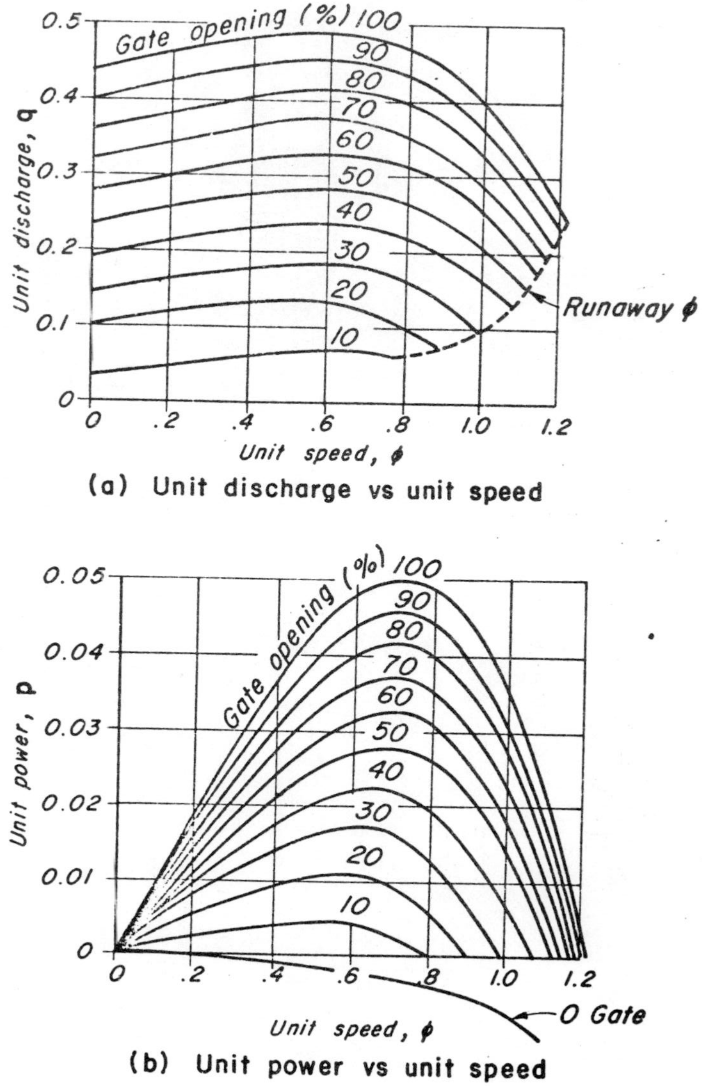

Fig. 2 Turbine Characteristics

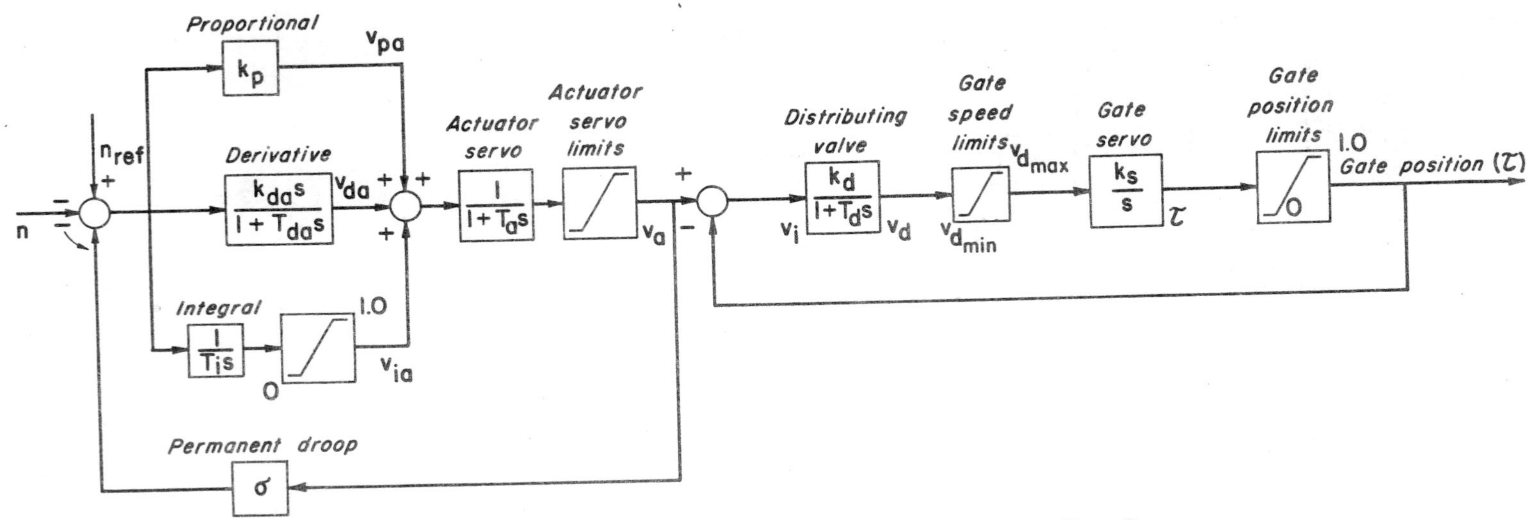

Fig. 3 Block Diagram for a Proportional-Integral-Derivative Governor

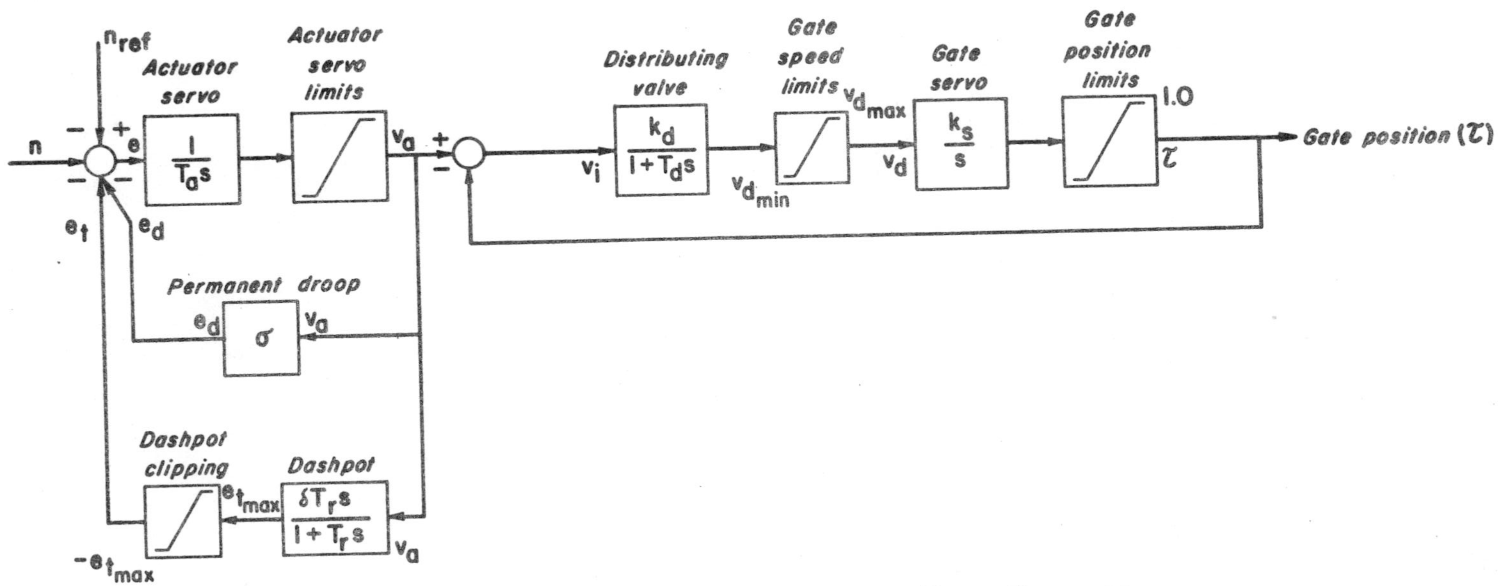

Fig. 4 Block Diagram for a Temporary-Droop Governor

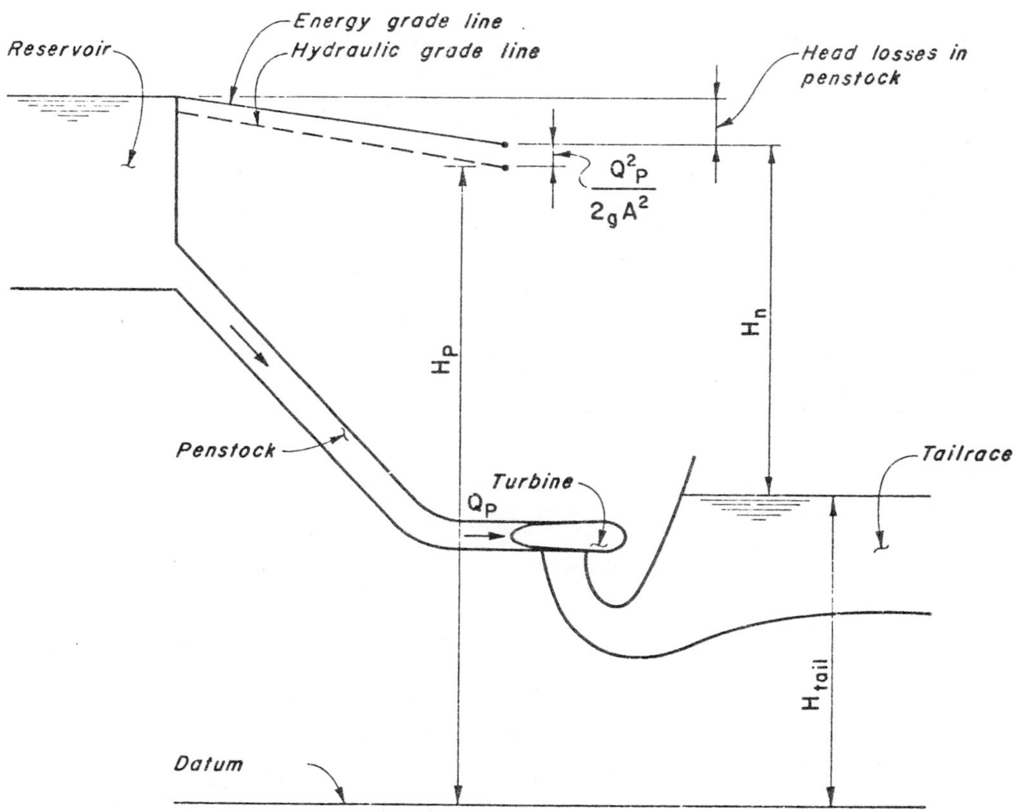

Fig. 5 Definition Sketch

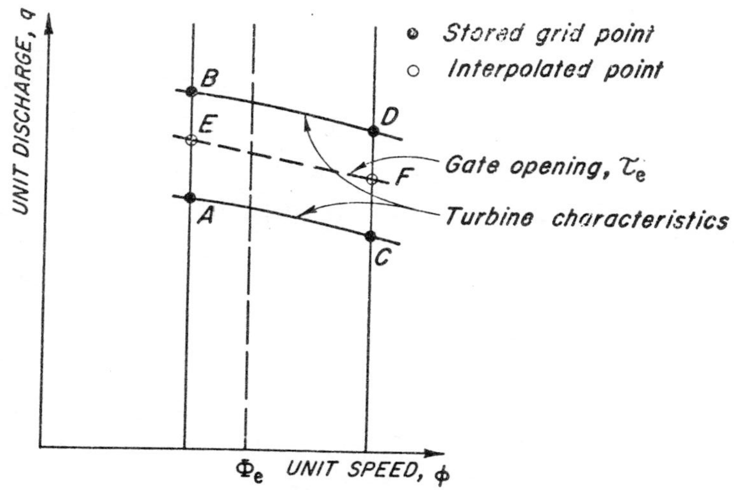

Fig. 6 Approximation of Characteristic for Unit Discharge

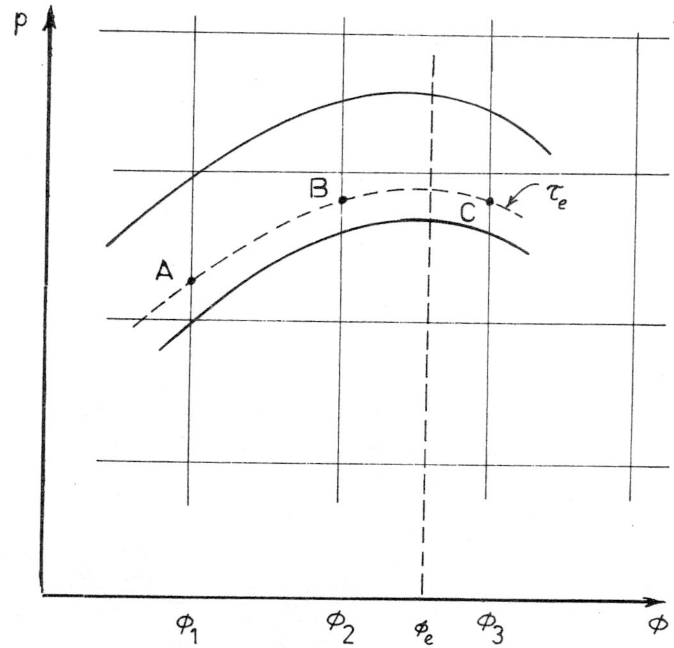

Fig. 7 Approximation of Characteristic for Unit Power

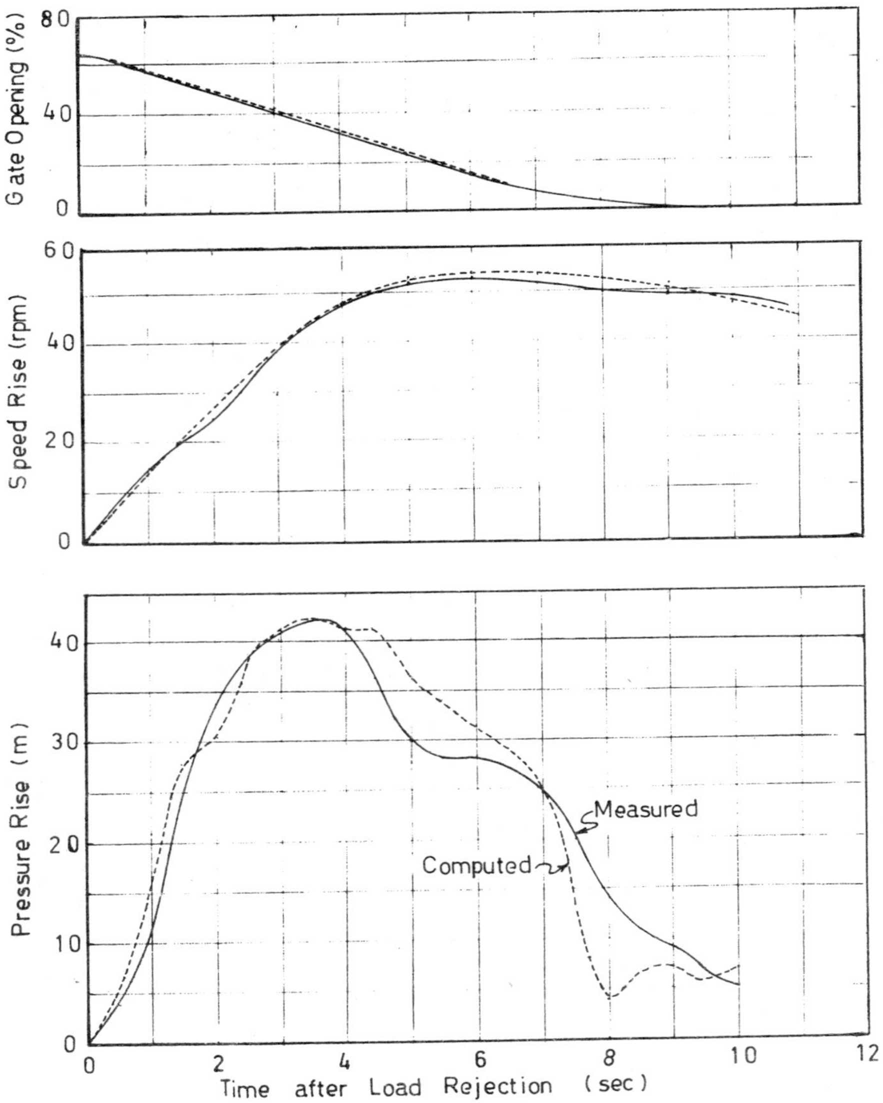

Fig. 8 G. M. Shrum Generating Station: Comparison of Computed and Measured Results

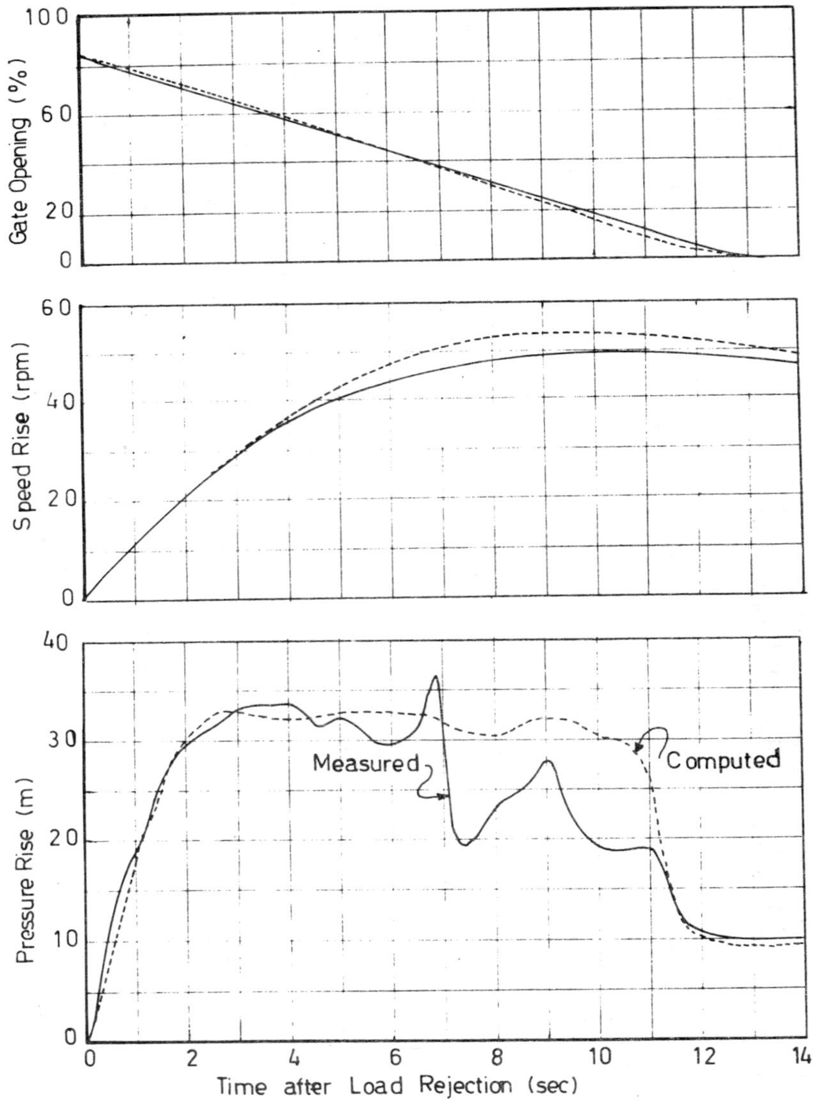

Fig. 9 Mica Power Plant: Comparison of Computed and Measured Results

Third International Conference on

PRESSURE SURGES

Canterbury, England
March 25-27, 1980

PAPER G2

CHIBRO KHODRI TANDEM CONTROL

J. Khosa,

Bharat Heavy Electricals Ltd., India

Summary

Chibro and Khodri power stations form the stage II of the Yamuna Hydro-Electric scheme in India. The water conductor system has over 12 kms long conduit, three surge tanks, an expansion chamber connected to the collection gallery at Chibro, and the two hydro stations.

The control adopted for the satisfactory operation of the two power houses is described taking into account the constraints imposed by the low availability of water in the collection gallery, high friction losses in the head race tunnels and the net heads at Khodri being much lower than the net head at Chibro.

The aim of the control scheme is to monitor the level in the collection gallery, located downstream of Chibro. Suitable corrective signals will be given to vary the guide vane opening of Chibro machines for flow matching between the two power stations. The signals will take the form of pulses, the frequency being determined by the level.

Additional controls include shedding of load at Khodri at dangerously low level and tripping of all Khodri machines at the trip level.

The operation of the two power stations following load acceptance and rejection is detailed out to show the role of the control adopted. Control of Khodri gate (thereby load) from Chibro and other indications and controls are also described.

Held at the University of Kent at Canterbury, U.K.

Conference Organised by BHRA Fluid Engineering, Cranfield, Bedford, England.
In conjunction with The City University, London.

1. INTRODUCTION

Chibro and Khodri Hydro Stations have to be operated in tandem in view of the constraints imposed by the low availability of water in the Collection Gallery downstream of Chibro, high friction losses in the headrace tunnels and the net head at Khodri being much lower than the net head at Chibro. This paper describes the controls that have to be adopted for the operation of the two power houses in tandem.

2. YAMUNA HYDRO SCHEME STAGE II

Chibro and Khodri power houses form the stage II of the Yamuna Hydro-electric scheme in India. See Figure No.1 for the layout of this scheme.
A 50 m high concrete diversion dam has been built across river TONS at Ichari. The river water is diverted through a 7 m diameter, 6200 m long concrete lined headrace tunnel which is connected to a 21 m diameter Surge Tank. See figure 2 for details.
Four penstocks lead water from this surge tank to the powerhouse consisting of 4 units each of 60 MW rating with 10 percent continuous overload capacity. A collection gallery (see figure 3 for details) downstream of the power house with an expansion chamber for all practical purposes is the tailrace surge tank.
Water from the collection gallery will be led into another head race tunnel 7.5 m diameter, 5941 m long, connected to a twin surge tank (see figure 4 for details) upstream of Khodri power house equipped with four units of 30 MW each. Two units are connected to one penstock through suitable bifurcations.
Thus the Chibro-Khodri water conductor system has over 12 km long conduit with 3 surge tanks and 2 power houses. A total gross head of over 190 m is utilised for generation. (Ref.1).
All the four machines at Chibro are in operation for the last four years. An outlet weir, directly connected with the collection gallery, has been provided to bypass Chibro discharge into the river. Bulk head gates between the collection gallery and the expansion chamber have been lowered to prevent entry of water into the Khodri head race tunnel, under construction at present.

3. IMPORTANT CHARACTERISTICS OF TUNNELS, SURGE TANKS AND COLLECTION GALLERY

The maximum velocity of flow allowed through the tunnels is approximately 6m/s resulting in head losses to the extent of 27 percent of the static head against the recommended figure of 2 to 6 percent. Since the head loss is proportional to square of the velocity, it means that the net head drops quickly and any advantage of higher flows is lost due to near saturation in power output. The operation of the two power houses, as such, is ideal upto a flow of about 180 m^3/s.
The total fluctuation of water level in the surge tanks is of the order of 10 percent of net head for a simple load change of 20 percent. This again is a high value.
The Chibro surge tank is of restricted orifice type with one expansion chamber to limit upsurges. Khodri surge tank is connected to another overflowing type surge tank. (Ref.2)
The total volume of the collection gallery and the expansion chamber had to be kept to a minimum in view of economic reasons. The whole water conductor system was proved on a distorted hydraulic model built to a length scale of 1/120, a diameter scale of 1/23 and a surge scale of 1/40. The model tests were carried out by Irrigation Research Institute, Roorkee, India (Ref.5). The significant observations made from the model studies is the formation of unsteady flow with likelihood of air entry into Khodri head race tunnel at low levels of collection gallery. This aspect naturally has to be taken into consideration for reliable operation of Khodri power station, by preventing low level in the collection gallery. It was observed that the model results tallied very closely with those obtained through computer studies. The permissible low levels in the surge tanks were also defined on the basis of model studies.
The outlet weir elevation can be adjusted by log gates to suit operating requirements.

4. TURBINES

Both the power houses are equipped with Francis Turbines with Butterfly Valves. The significant feature of these turbines is that the rated discharge of the machines is

almost equal. This means that the operation of the two power houses with their m/c's operating at rated head is smooth. However the losses in the two tunnels being almost equal, the net head drops considerably at Khodri with the result that corrective action has to be carried out for equalising the flows.
The important Turbine parameters are given below:-

	CHIBRO	KHODRI
Rated Head , m	110	57.9
Maximum Head, m	123	70.5
Minimum Head, m	100	49.7
Rated Discharge , m^3/s	62.5	61.25
Normal Speed , RPM	250	200
Rated Output , MW	62.6	32.5

5. GOVERNORS

All the Turbines at Chibro and Khodri are controlled by Electro-Hydraulic Governors using Temporary Speed Droop characteristics for stability and Permanent Speed Droop based on frequency vs. guide vane relationship.
Some of the important features of the Governors are given below:-

i. Frequency Dead Band. This device will enable the two hydro stations to operate independent of the grid within the dead band limits, adjustable between 0 to ± 0.5 Hz frequency.

ii. Loading and unloading of the machines by Gate setting device i.e the signal from this manually operated device will be compared with the gate (guide vanes). This setting will not be affected by any changes made in the setting of permanent speed droop.

iii. Joint Control. All the machines at Chibro and Khodri can be operated under their respective joint control units. Single motorised joint gate setting potentiometers are provided for loading and unloading of station output. Joint permanent droop helps in altering the regulation characteristics of the station. Follow up circuits for smooth changeover from one mode to the other have been provided.

The response of the hydro-sets for changes in gate (thereby load) is dependent upon the governor parameters. The factors influencing the setting of these parameters are given below:-

	CHIBRO	KHODRI
Water time constant, Tw	1.13 s.	1.37 s.
Machine time constant, Ta	7.50 s.	7.50 s.
Ratio Tw/Ta	0.15	0.176
Maximum speed rise	42.00 %	42.00 %
Maximum pressure rise	40.00 %	24.00 %
Closing time of guide vanes	5.00 s.	5.60 s.

The governor parameters--Temporary speed droop and its decay time constant will be set to ensure similar response from the governed hydro-sets. Chibro machines have shown good response to small changes in command signal (gate setting device). Large changes namely acceptance of load from 0 to 60 MW is possible within 30 to 40 seconds.
The motorised joint gate setting or individual gate setting devices can accept signals from tandem control equipment for correcting the flows through the machines.

6. SUPERVISORY CONTROLS

The automatic and/or manual control of the generating sets will be carried out from the Contrl Room at the underground Chibro power house. The emphasis is more on monitoring the hydraulic system, therefore the following supervisory controls have been recommended for location at Chibro Control Room:-

i. Ichari Reservoir — a. Level indication and recording.

	b. Intake trash rack choking alarm.
ii. Chibro Surge Tank	a. Level indication and recording. b. Low level alarm. c. Low level contact for control sequences.
iii. Collection Gallery	a. Level indication and recording. b. Float switches for control purposes. c. Low level alarm. d. High level alarm.
iv. Outlet Weir	a. River Tons level indication and recording. b. Outlet Weir level indication and recording. c. High level alarm.
v. Khodri Surge Tank	a. Level indication and recording. b. Low level alarm. c. Low level contact for control sequences.
vi. Khodri Tailrace	a. Level indication and recording.

All the transducers will be located locally and the indicators,recorders and control devices will be located on Tandem Control Panels with provision for Hydraulic Mimic diagram.Thus the information available to the operators at Chibro will be complete in respect of the whole scheme as far as the hydraulic system including the river is concerned.

7. SELF REGULATION

As explained above the feature of permanent droop related to guide vane position is best suited to the control of the flow between the two hydro stations. Since governors in both stations see the same frequency,then provided the same droop settings are used,guide vane positions in both stations will normally remain in step or vary in a particular ratio in case the droops are set to specified ratio.
Calculations were made to determine what head changes would take place in the system if the Chibro and Khodri stations summated guide vane settings were mismatched.This showed that the system possessed a certain amount of self regulation.It will be able to accomadate 1.25 percent mismatch of summated guide vane settings under any operating conditions for one metre change of level at the collection gallery.

8. STABILITY

Any control mechanism will have the basic function of maintaining the mean level of water in the collection gallery and the Khodri surge tank as nearly constant as possible.That is,the quantity of water downstream of Chibro and Upstream of Khodri is to be maintained constant.Under steady state conditions it would be sufficient to measure the level in the collection gallery to determine this quantity approximately. However,in the presence of load transients there will be swings between the Khodri surge tank and collection gallery and hence the level in the collection gallery represents the quantity of water between Chibro and Khodri with an undesirable oscillation superimposed.It was feared that if one attempted to maintain equal flow through the two stations monitoring collection gallery level then false signals would be given owing to the surge tank swing and that an unstable oscillation of increasing magnitude would result.
Similarly the flow through a turbine at any given gate position is proportional to the square root of the net head.The net head is in turn affected by surge tank levels. Hence measurement of flow at the respective stations in an attempt to control levels could also lead to unstable conditions.
So it is obvious that monitoring of collection gallery level is the best method of controlling flow and loads at Chibro and Khodri.The storage capacity,above trip level, of the collection gallery for four machine operation at Khodri is for a couple of minutes only.

9. PRELIMANARY STUDIES

Premiminary studies were conducted to define the controls to be adopted for operation of Chibro and Khodri in tandem. These studies are:-

i. Instantaneous closure of 100 percent load at Chibro.
This showed the surge tank design to be adequate.

ii. Simultaneous closure of Chibro and Khodri from 100 percent load with the collection gallery at both normal and minimum operating levels.
Minimum drawdown in collection gallery was verified.

iii. Behaviour of Chibro and Khodri when subjected to small load changes.
Each station was considered individually to determine simply variation of parameters.

10. PROPOSED TANDEM CONTROL

The surge tank and collection gallery level fluctuations as a result of load changes reveal that the best method of equalising the flows through the machines of the two power houses is by correcting the Chibro guide vane opening based on the level in the collection gallery. Thus a level control equipment has been proposed whose function is to give timed pulses to the individual unit or joint control gate setting potentiometers depending on the state of level. A directional indicator will continuously monitor whether the level is rising or falling and will permit control action to suit.

10.1. The pulses to Chibro governors will be given as per the following scheme:-

Collection Gallery Level	State of Level	Corrective action of Chibro Guide vanes.
1.0 metre high	Rising	1 pulse to LOWER every 75 seconds.
1.0 ,, ,,	Falling	1 pulse to LOWER every 150 seconds.
0.5 ,, ,,	Rising	1 ,, ,, ,, ,, ,, ,,
0.5 ,, ,,	Falling	1 ,, ,, ,, ,, ,, ,,
0.5 metre low	Falling	1 ,, ,, RAISE ,, 150 ,,
0.5 ,, ,,	Rising	1 ,, ,, ,, ,, ,, ,,
1.0 ,, ,,	Falling	1 ,, ,, ,, ,, 75 ,,
1.0 ,, ,,	Rising	1 ,, ,, ,, ,, 150 ,,
1.5 ,, ,,	Falling	1 ,, ,, ,, ,, 37.5 ,,
1.5 ,, ,,	Rising	1 ,, ,, ,, ,, 150 ,,

Note: 'Lower' means to reduce the guide vane opening
'Raise' means to open the guide vane.
The magnitude of each pulse is fixed to give the same amount of change in the guide vane opening.

The three pulsing schemes(Slow, Medium and Fast) enable the level in the collection gallery to come within the normal working zone of +0.5 metre to -0.5 metre. Any change of level within this band will not make appreciable effect due to the self-regulation characteristics.

10.2. The proposed working level for collection gallery is 518.5 m with spilling over the outlet weir at 519.5 m and level below 514.5 m is considered dangerous as such the following protective controls have been recommended:-

i. Alarm at 519.5 m to indicate spilling over the outlet weir.

ii. Alarm at 516.0 m and control action to Khodri machines to unload by 25 percent.

iii. Alarm at 514.5 m and tripping of all Khodri machines to prevent air entrainment into Khodri head race tunnel.

10.3. The following controls and indications of Khodri will be located at Chibro control room:-

i. Raise/Lower of Khodri joint gate setting device, which is motorised for remote control.

ii. Emergency tripping of Khodri machines.

iii. Guide vane position indication of all machines.

iv. Guide vane limit indication of all machines.

v. Raise/Lower of guide vane limit of all machines.

vi. Summated load of Khodri.

10.4. The machines at Khodri will be started locally, synchronised and switched over to joint control. Khodri machines under joint control will now be remotely controlled from Chibro. All signals to and from Chibro will be transmitted by telemetry system. Double channels have been specified for alarm and trip control actions of Khodri for reliable operation of the two power stations.

10.5. The operator can always override the tandem control action by altering the loads at the two power stations to suit hour to hour load demand and to make any coarse corrective action leaving fine control to the level controller.

11. DETAILED STUDIES

Using the pulsing and control scheme as detailed above, the following specific studies were carried out to ascertain the reliable operation of the power stations (Ref. 3):-

11.1 MAXIMUM LOAD ACCEPTANCE

Initial conditions:-	Level in Ichari Reservoir	641.0 m
	Level in Collection Gallery	518.5 m
	Level in Khodri Tailrace	456.3 m

Initially all machines at Chibro and Khodri are loaded partially such that the flow through Chibro and Khodri is the same i.e 108 cu.m/sec. This is the condition at approx half of maximum station loads. Other stations on the grid are assumed to be working under constant heads. Now all machines at Chibro, Khodri and other stations in the grid are loaded instantaneously to give power twice its original value, to approach full load. The level control is assumed to have been in operation and for convenience load acceptance is considered to occur at zero time. 8 pulses were required in this case to create a sufficient difference in Chibro and Khodri gates in order to overcome the effects of tunnel friction at high flows, and equalise the flow through the stations after the transient. As expected the level in the collection gallery rises and spillage takes place for some period (See Figure 5).
The pulses were discontinued once the level had started to fall. However the actual level controller has been designed so that the slow (150 sec) pulsing will be continued even though the level is falling. This will cause the level in the collection gallery to return within the working band (519.0 to 518.0 m) more quickly. This effect is shown by dotted lines.
Each pulse magnitue has been taken as one percent change in guide vane opening.

11.2. LOAD REJECTION

Initial conditions of levels are similar to those for load acceptance. Initially all machines at Chibro and Khodri are loaded in such a way that the flows through the two stations is the same i.e 210 m^3/s. The difference in gates of Khodri and Chibro of 9.2 percent is due to tunnel friction. Other stations in the grid are assumed to be

working under constant heads. Now the total load of the grid is reduced in such a manner as to cause all machines in the grid to unload instantaneously to half the original value.
The load rejection has been studied for the following variants:-

i. Level control not working.
ii. 1% pulsing.
iii. 2% pulsing.
iv. 4% pulsing.

It can be seen that the proposed pulse magnitude of 1% fails to avoid the tripping of Khodri. But on the other hand increasing the pulse magnitude to 4% is no solution. This is because the larger pulse, while still failing to avoid the trip, introduced larger magnitude of swings at Chibro surge tank.
These cases highlighted the need of avoiding tripping Khodri at the trip level by sending a corrective signal at Alarm level to unload Khodri by 25 % retaining 1 % corrective pulses. See Figure 6 for details of the level fluctuations.

11.3 EFFECT OF UNLOADING A UNIT AT CHIBRO FOLLOWED BY UNLOADING ONE UNIT AT KHODRI AT LOW LEVEL ALARM IN THE COLLECTION GALLERY.

Initial conditions are same as in the earlier cases.
Now a unit is tripped or inadvertantly unloaded at Chibro without compensating action at Khodri. It is worthwhile to note that the heavy mismatch of flows between Khodri tunnel and Chibro turbine is far greater than the mismatch in the case of load rejection. See figure 7.
In order to avoid the trip and the fall in level as shown by dotted line, a unit at Khodri is tripped at alarm level. The collection gallery level recovers to the normal working range. Surge tank and flow swings die out quickly.

11.4. LOAD REJECTION BY CHIBRO AND KHODRI IN ISOLATION

Chibro and Khodri under steady state conditions for a flow of 235 m^3/s gives total output of 302 MW. All the machines at Chibro and Khodri are unloaded to total output of approx. 152 MW. The collection gallery level drops to 514.75 after the 25 % Khodri gate closure has operated at 75th second. Without this facility the level would have dropped to 512.59 m.
Collection gallery outlet weir spills for 1275 seconds while Khodri surge tank spill lasts for 140 secs. and 125 secs. in two spells. See figure 8 for these spillages.

11.5. LOAD REJECTION BY GRID INCLUDING CHIBRO AND KHODRI

The initial conditions of levels etc. are the same as in the earlier studies.
The initial output from Chibro and Khodri works out to 206 and 96 MW respectively, a total of 302 MW.
Following two cases, and as shown in figure 9, were studied:-

11.5.1. The instantaneous drop in output from Chibro and Khodri is more than 180 MW due to low net heads available as a result of high friction losses.
25 % gate closure at Khodri takes place at alarm level of 516.0 m. The collection gallery level falls to 513.6 m which is 0.9 m below the Khodri trip level. With further reduction of load from Khodri at alarm level, the total power obtained from Chibro and Khodri is far less than 150 MW after the transients are over.

11.5.2. The instantaneous change in Chibro and Khodri output at the begining of transients is 146 MW and output is 120 MW at 2500th second. The minimum collection gallery levels attained with and without 25 % Khodri gate closure are 514.91 m and 514.39 m respectively.

12. INFLUENCE OF GRID ON CHIBRO -KHODRI SYSTEM

The studies under the foregoing paras reveal that the grid has a marked effect on Chibro-Khodri system because the power unloaded by Khodri at alarm level is shared by all ststions in the grid including Chibro and Khodri machines. Thus the instantaneous change in gate at Khodri will be less than 25 % depending on the size of the grid. Simultaneously the Chibro gates will also open to absorb proportional power. The larger the grid the less will be the increase in Chibro gate opening and the change in Khodri will approach 25 %. Increase in Chibro gates will help in filling the collection gallery and try to prevent complete tripping of Khodri at trip level.
If the Chibro-Khodri system works in isolation and looses 150 MW from 300 to 150 MW then the increase in Chibro gate will be approx. 8 % at the instance of 25 % gate closure at Khodri at alarm level. This results in an increased discharge of Chibro turbines into the collection gallery of approx. 23 m^3/s compared to 11 m^3/s if a sizable grid is taken into account. For an infinite grid this value will be zero. These facts show that the isolated condition is the best to prevent tripping of Khodri for 150 MW load rejection.
The frequency dead band provided in the Governors can be effectively used to prevent any change in the two power house outputs due to small changes in frequency. A higher value of permanent speed droops in relation to the settings at other power stations will also help in reliable operation of Chibro and Khodri. Isolated operation is not possible. The total grid considered in these studies is 800 MW, however the grid size will be more than 2000 MW by the time the Chibro-Khodri tandem contrl is put into operation. A sudden change of load, to the extent of unloading of Chibro and Khodri by 150 MW, cannot be expected unless there is a large system swing.

13. INFLUENCE OF GATE MISMATCH AND PERMANENT SPEED DROOP ON CHIBRO-KHODRI SYSTEM

As explained earlier, the tunnel frictions in Chibro and Khodri tunnels are almost equal thus resulting in disproportionate net heads at the two stations. The level controls automatically maintain the working level in the collection gallery by giving necessary pulses to Chibro governors, to vary the gates, so that at steady conditions the flow through the two stations is kept equal. The gate mismatch between Chibro and Khodri for equal flows when 1,2,3 or 4 machines are working is shown in Figure 10. Necessary pulses are given to Chibro machines to allow the new gate relationship at reduced flow following a load rejection. Similarly for load acceptance the level control will give pulses to close gates for a gate relationship at higher flows where Khodri gates are more than the Chibro gates. The amount of mismatch for equal flows through the two power houses also depends on the water levels at Ichari Reservoir and Khodri tailrace. (Ref.4).
The mismatch required for equal flows can be achieved by setting the permanent speed droop at Khodri lower than that set at Chibro. This feature will enable a larger change in gate movement due to frequency changes in the grid. Computer studies reveal that the gate change between Chibro and Khodri in the ratio of 0.8:1.0 will considerably help in avoiding tripping of Khodri for large load rejections.

14. OTHER CONTROLS

14.1 CHOKING OF TRASHRACK

Quite often the Trashrack at the intake of Chibro head race tunnel chokes due to the presence of driftwood, leaves etc., resulting in quicker drop of level in the Chibro surge tank. Abnormal difference in the upstream and downstream levels is detected to give alarm at Chibro control room for desired manual action.

14.2 FLOODING OF RIVER TONS

The Tons river level during rainy season can rise beyond the danger mark and flooding of Chibro power house may result due to water trying to enter into the cavern through the draft-tube gate slots. The following controls have been recommended:-

i. Controlled generation with level in the river above 525.0 m.
ii. 25 % unloading of Chibro generation at river level of 530.0 m.
iii. Tripping of Chibro and Khodri power houses at river level of 534.0 m after a set time.

15. CONCLUSION

A hydro-electric scheme involving three surge tanks,two power houses,head race tunnel of over 12 km. unfavourable conditions of friction and equalising of large quantities of flow has to be unique.The controls specified,as a result of these studies,will ensure reliable operation of both the power houses.
Chibro power house is in operation since April 1975.The friction loss in the head-race tunnel is as per the calculated values.Load acceptance and rejection tests were carried out to prove that the actual results are favourable.
Khodri head race tunnel is under construction . A 1000 m. length of the tunnel is being modified due to the presence of weak rock.The modified tunnel imposes extra frictional losses,however the tandem control scheme can take care of this extra loss. The operation of Chibro in the last four years has revealed that flows of 175 m^3/s should not be exceeded from the point of view of tunnel velocities and presence of silt in water during rainy season.This condition will limit the maximum power from the two power houses to 250 MW instead of 300 MW.Chibro and Khodri will meet block load requirements and generation can be optimised by operating the two power houses at lesser loads for longer periods,knowing that the total availability of water over a period of a day or week is limited due to low flows in the river.

16. ACKNOWLEDGEMENT

This paper is being published with the kind permission of Bharat Heavy Electricals Limited,Bhopal.

17.REFERENCES.

i.Central Design Directorate Irrigation Dept. U.P.,India. "Notes on the revised project data of Yamuna Hydel Scheme ,Stage II".

ii.Sharma,H.R " Some engineering aspects of the Yamuna Stage II hydraulic design " Water Power,Aug. 1973.

iii.Pilkington,J Khosa,J.L and McDonald,A "Chibro-Khodri Tandem Control " BHEL Reports.

iv.Khosa,J.L " On the influence of gate mismatch and permanent speed droop on Chibro-Khodri tandem control " BHEL Report.

v.Garg,S.p et.el. " Hydraulic model of Chibro-Khodri system " Proc.,40th Annual Research session, Shillong India,June 1970.Central Board of Irr. and Pow.

APPENDIX NO. I

LEVEL INDICATION AND RECORDING EQUIPMENT

1. Ichari Reservoir.

Two gauge wells,one located upstream of the intake trashrack and the other located downstream of the sedimentation chamber,have been provided for level measurement. Float operated level measurement of Ichari reservoir is a mechanical/electrical device which converts the movement of the float operated pulley into electrical signal compatible with the indicator and recorder.
Pressure transducer,to be lowered in the gauge well,will be used for monitoring the choking of trashrack.The transmitter will be locally mounted.

2. Chibro Surge Tank

A 75 mm bore pipe connected to the Chibro surge tank at an elevation of 590 m has been terminated in the Anchor gallery.The sensing instrument of the force balance type ,will be located on the upstream wall of the anchor gallery.This transducer will be connected electrically to the indicator and recorder mounted in the control room at Chibro.A direct access from the anchor gallery to the control room is available for laying of the cables.The system is designed to indicate and record level fluctuations between 590 m to 665 m.

3. Collection Gallery

A 150 mm bore pipe runs all along the draft tube of Chibro 4th unit from collection gallery to the valve pit.This pipe is connected to a vertical Stand Pipe 225 mm bore through a diaphragm and an isolating valve.A suitable tapping will be provided in this pipe for connection to the level sensing equipment of the force balance type. The level indicator and recorder system is designed for level variation of 510 m to 525 m.

4.Outlet Weir

A gauge well has been provided for monitoring the level at the outlet weir.
The level sensing equipment will be of the float operated type,electrically connected to the indicator and recorder mounted in the Chibro control room.
Similar equipment has been specified for level measurement of river Tons.

5. Khodri Surge Tank

A 75 mm bore pipe is taken out at elevation 480 m from the surge tank and routed to the power house and terminated near the upstream wall of Khodri unit no. 1.
The level sensing equipment,of the force balance type,will be located near unit no.1. This equipment is similar to that provided for Chibro surge tank and collection gallery level measurement.
Level variation of 480 m to 530 m will be recorded and indicated.

6. Khodri Tailrace

A float operated level measurement system has been specified for Khodri tailrace.

APPENDIX NO. II

LEVEL CONTROLLER

The 225 mm bore Stand Pipe, located in the Butterfly Valve pit of Chibro 4th unit, will be used for detecting the level conditions in the collection gallery. Float switches will be located on the stand pipe at various levels and suitably connected to the pipe.
A float operated pulley connected to two limit switches through a spring loaded friction clutch, will detect whether the level is rising ix or falling. One limit switch will detect rise in level and the other a drop in the level. These limit switches will operate within a small angular movement of the float operated pulley.
Other devices provided on the stand pipe are:-

i. Local level indication.
ii. Detection of failure of float operated pulley. Alarm contact available.
iii. Flushing arrangement for cleaning the stand pipe and the connecting pipe to the collection gallery.

Duplicate float switches will be provided for the alarm and trip levels.

Automatic operation of the isolating valve is envisaged to provent flooding of the power house in the event of any failure.

The float level switches etc. will be electrically connected to the Tandem control circuits. The timers for the pulsing scheme are of the electronic type.
Suitable control and relaying system has been specified for achieving the objectives of the tandem control operation.

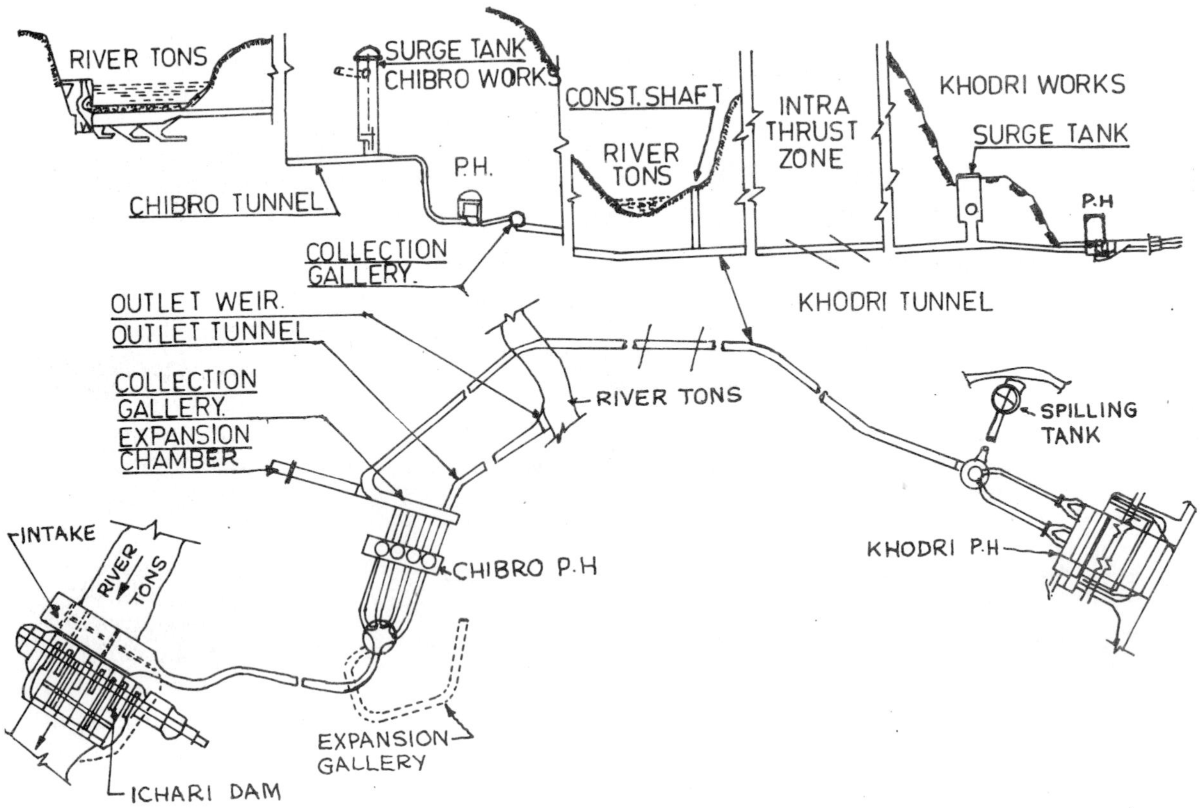

Fig. 1 Layout of Yamuna Hydel Scheme

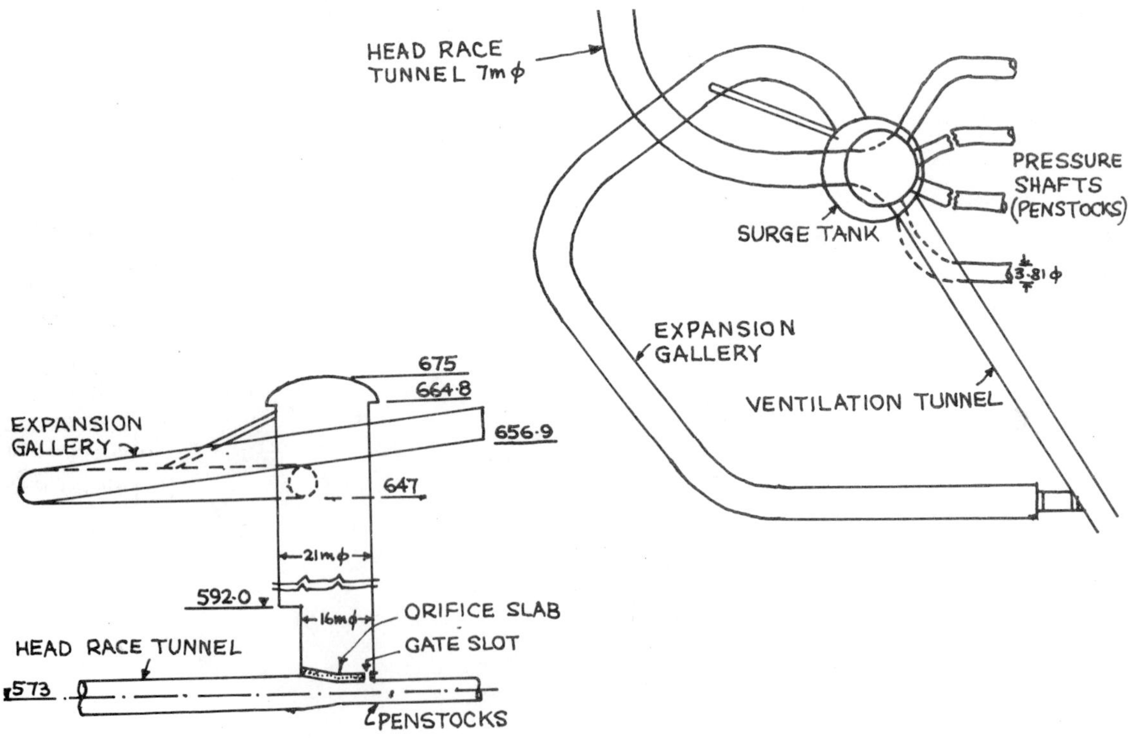

Fig. 2 Chibro Surge Tank and Expansion Gallery

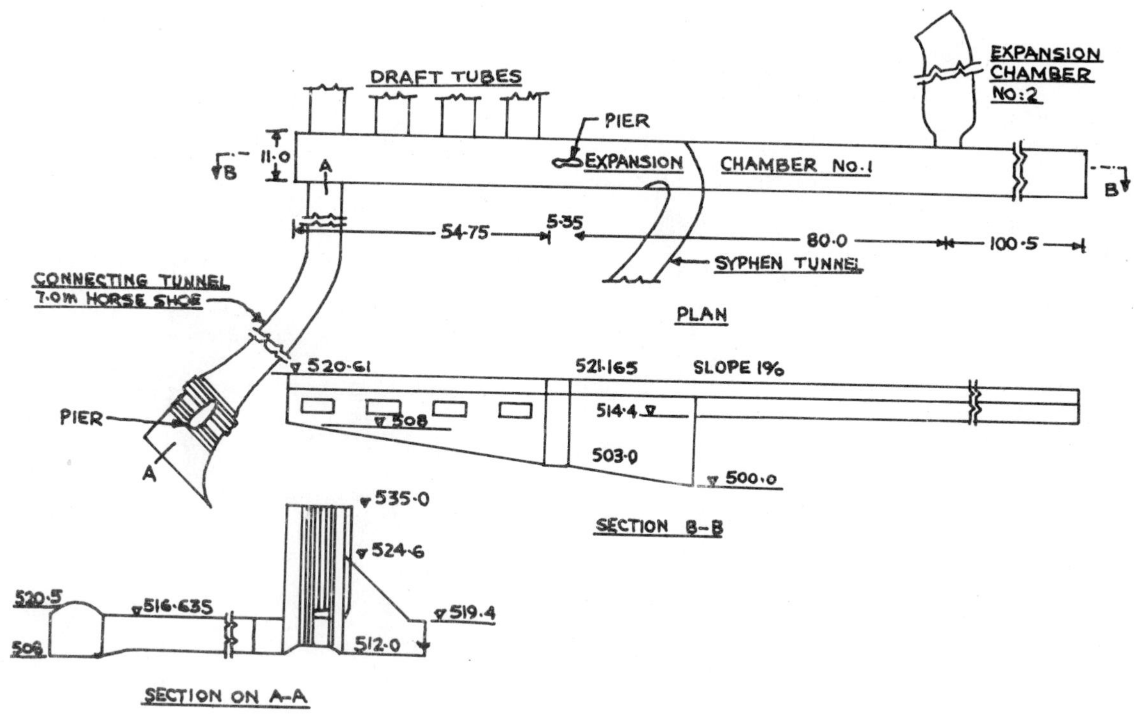

Fig. 3 Collection Gallery with Expansion Chamber and spilling arrangement

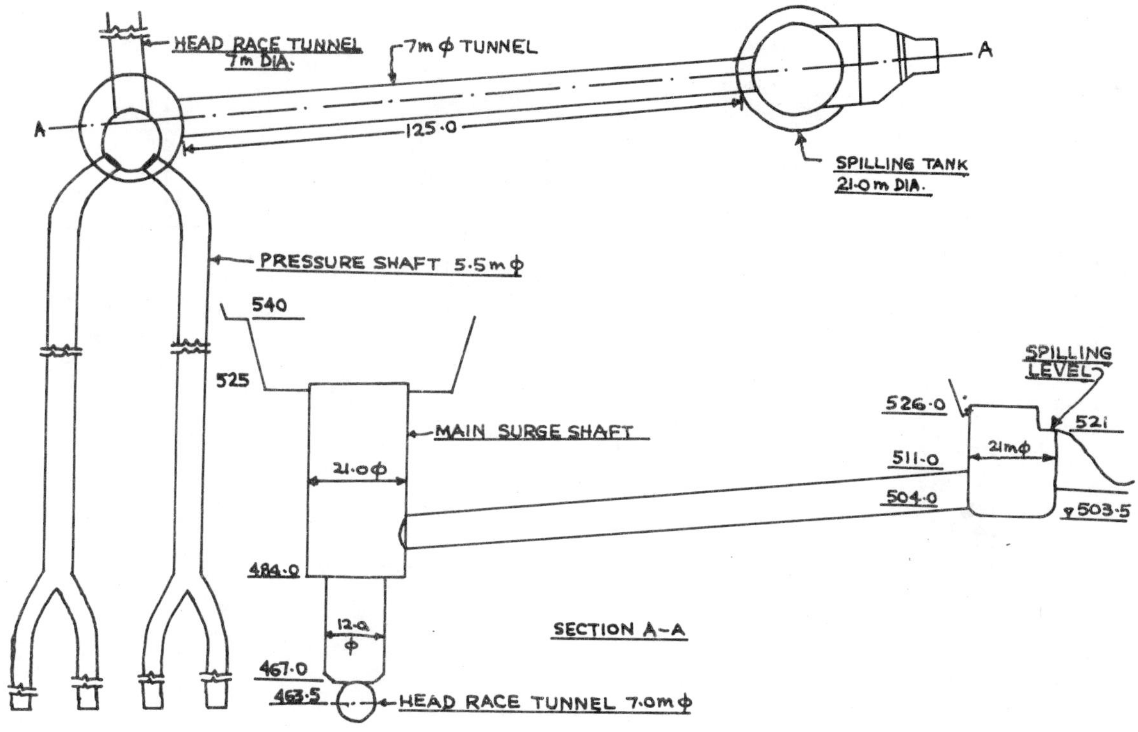

Fig. 4 Khodri Surge Tank and spilling arrangement

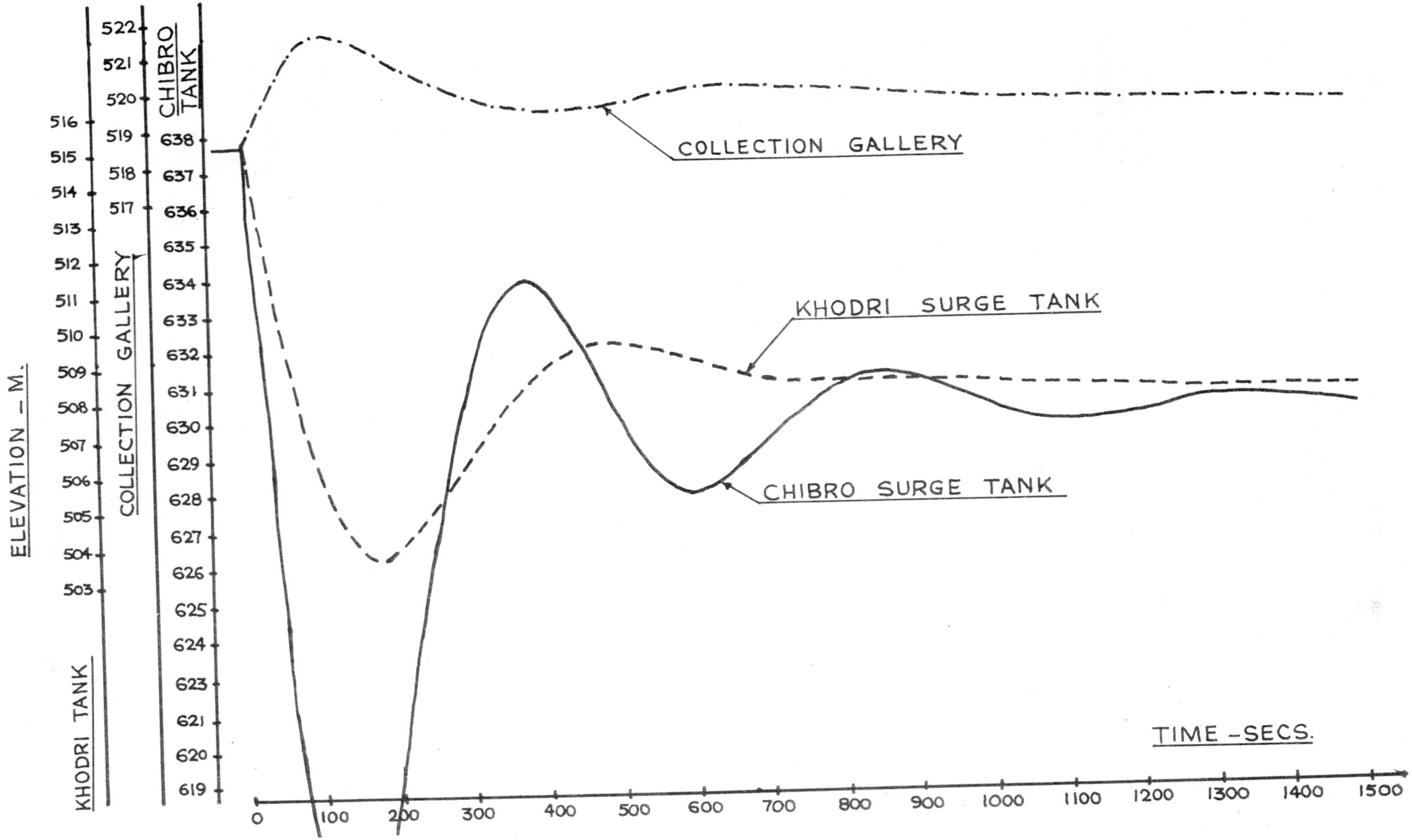

Fig. 5 Maximum Load Acceptance by Chibro, Khodri and other power stations

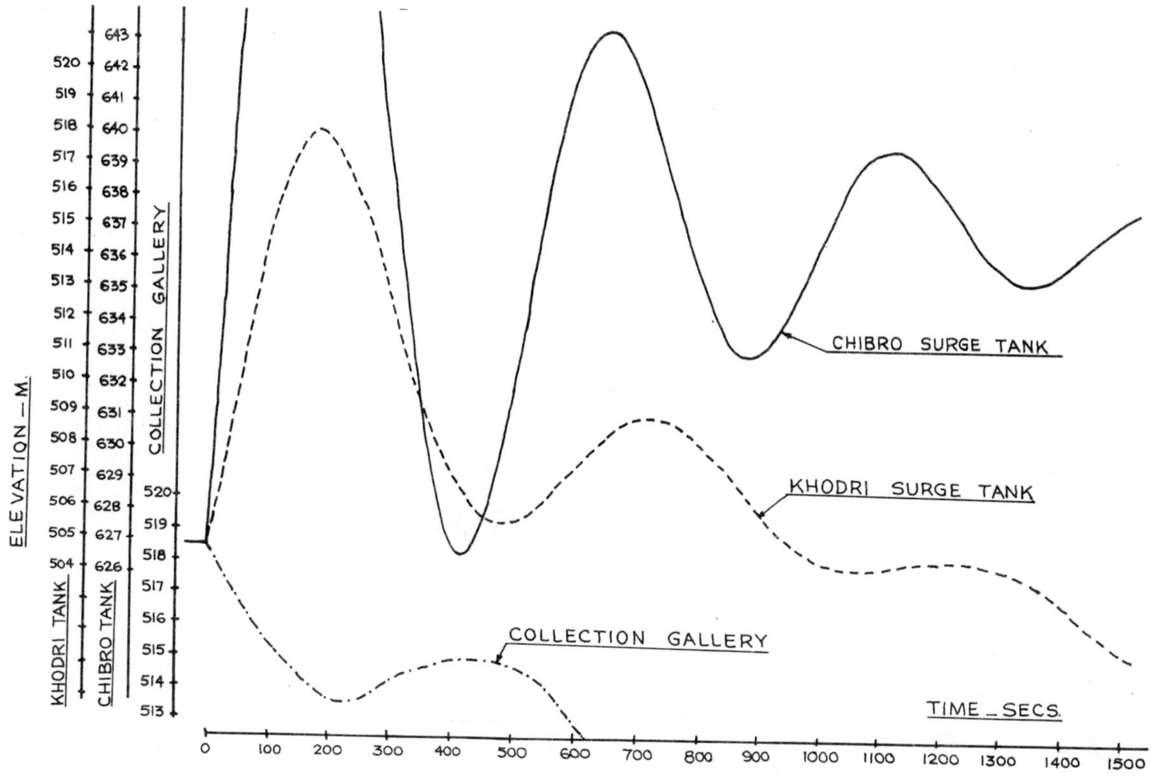

Fig. 6A Half Load Rejection by Chibro, Khodri and other stations. Level control not working

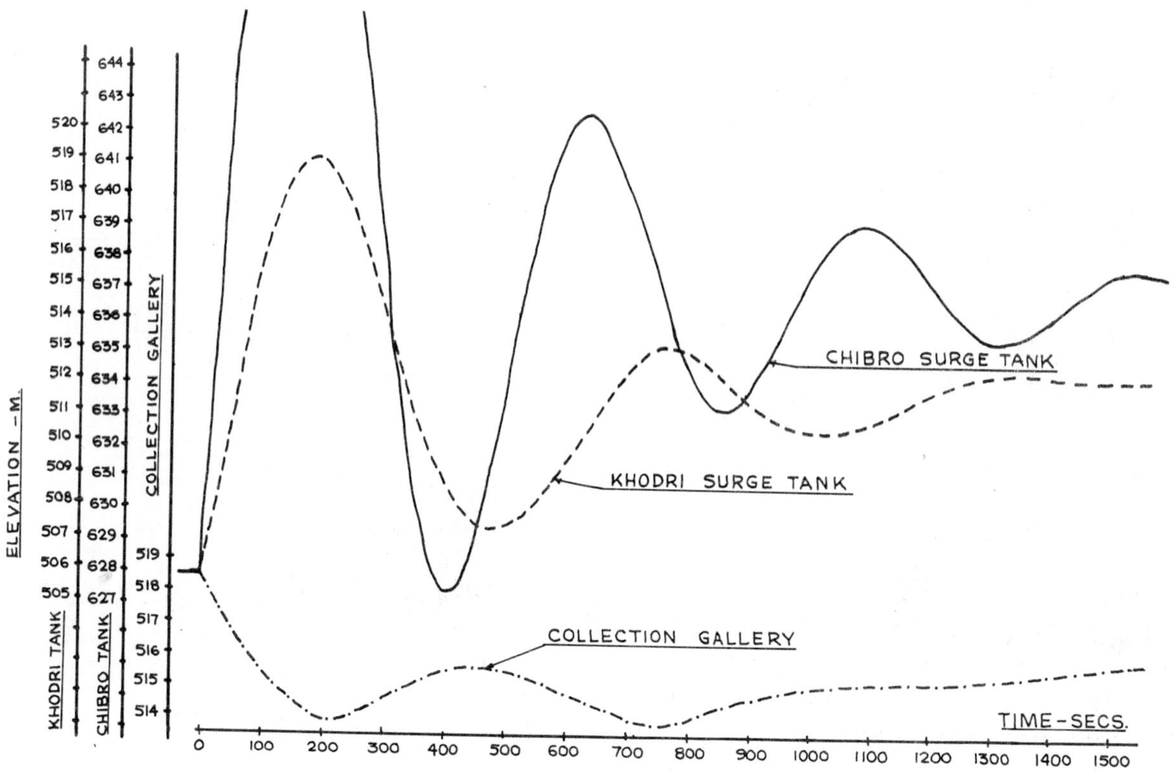

Fig. 6B Half load rejection by Chibro, Khodri and other stations

1 % pulsing

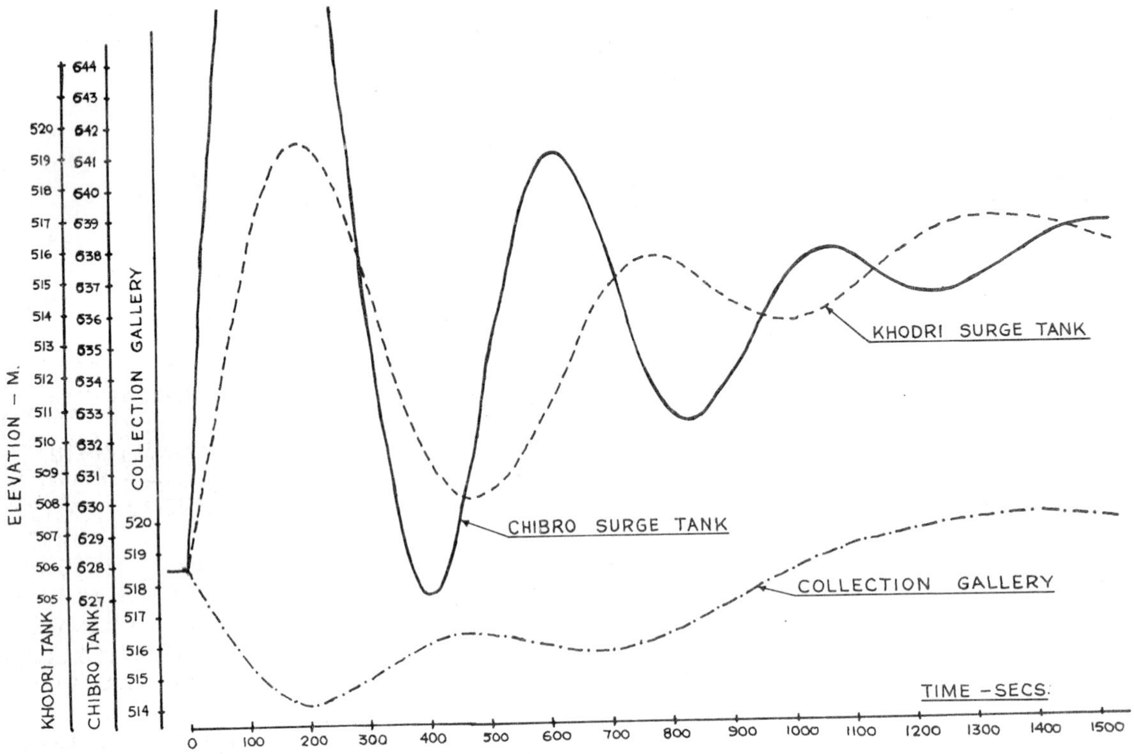

Fig. 6C Half load rejection by Chibro, Khodri and other stations 2 % pulsing

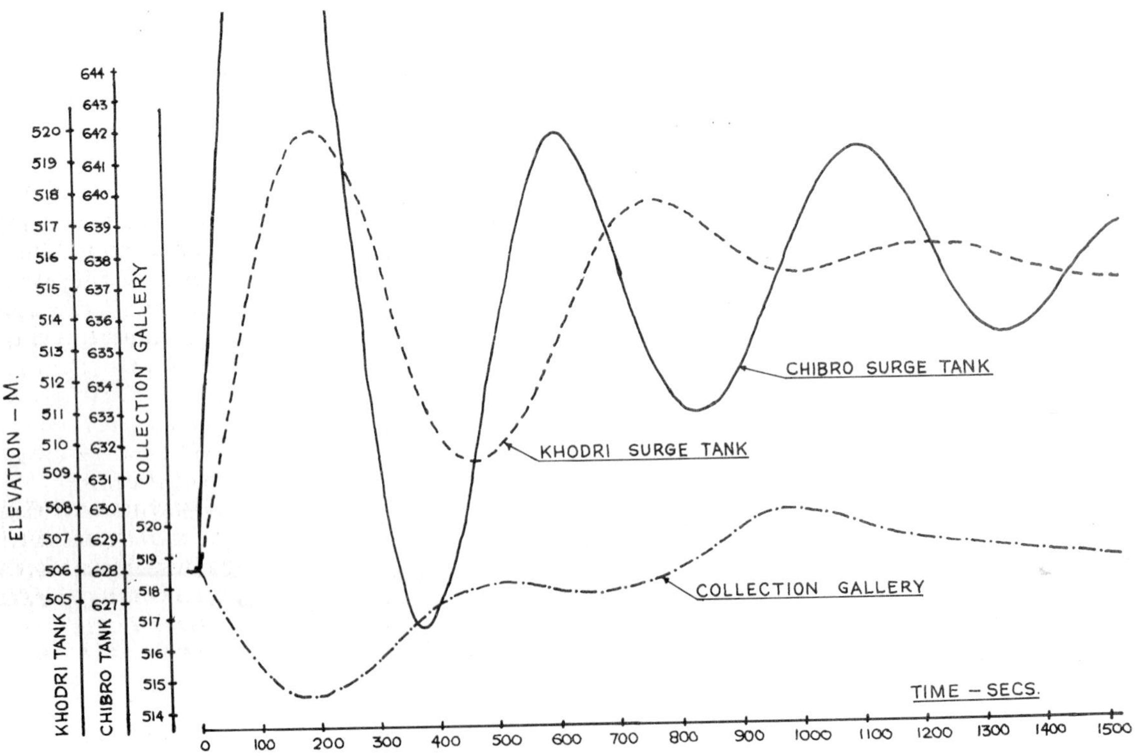

Fig. 6D Half load rejection by Chibro, Khodri and other stations 4 % pulsing

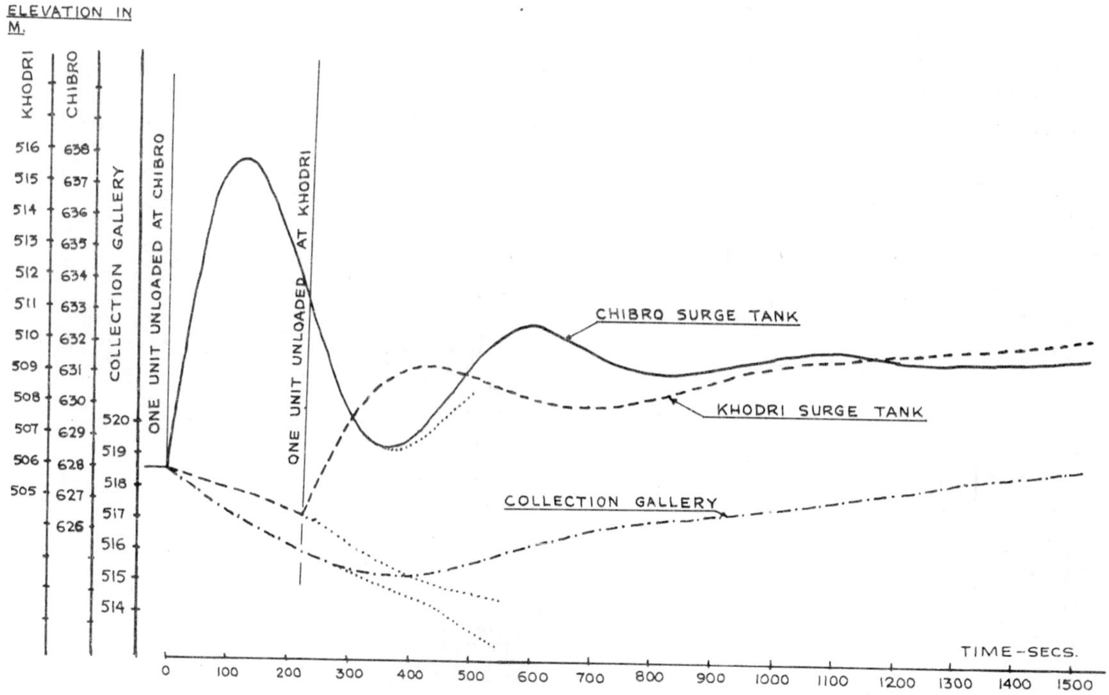

Fig. 7 Effect of unloading one unit at Chibro followed by unloading one unit at Khodri at alarm level in Collection Gallery

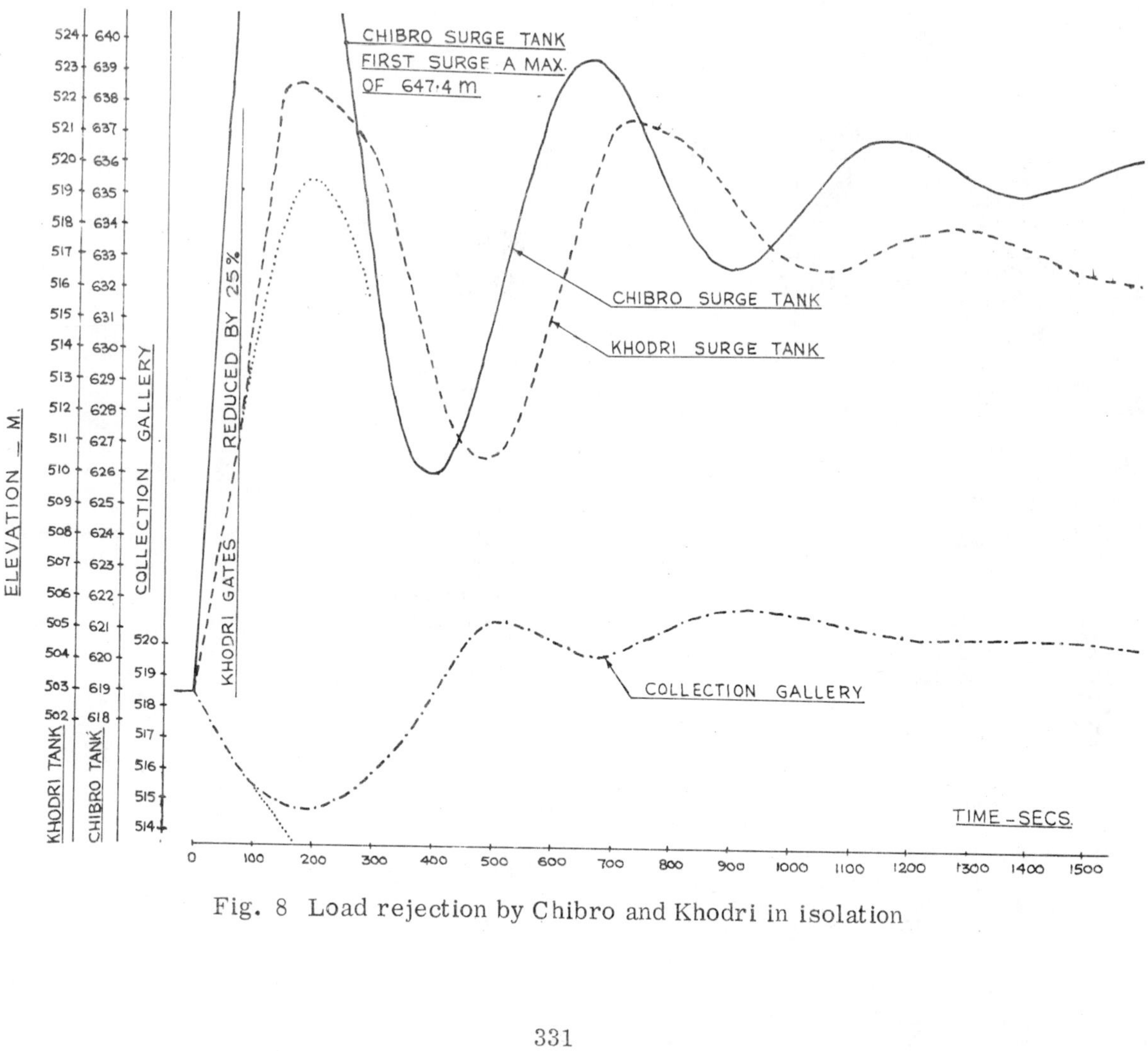

Fig. 8 Load rejection by Chibro and Khodri in isolation

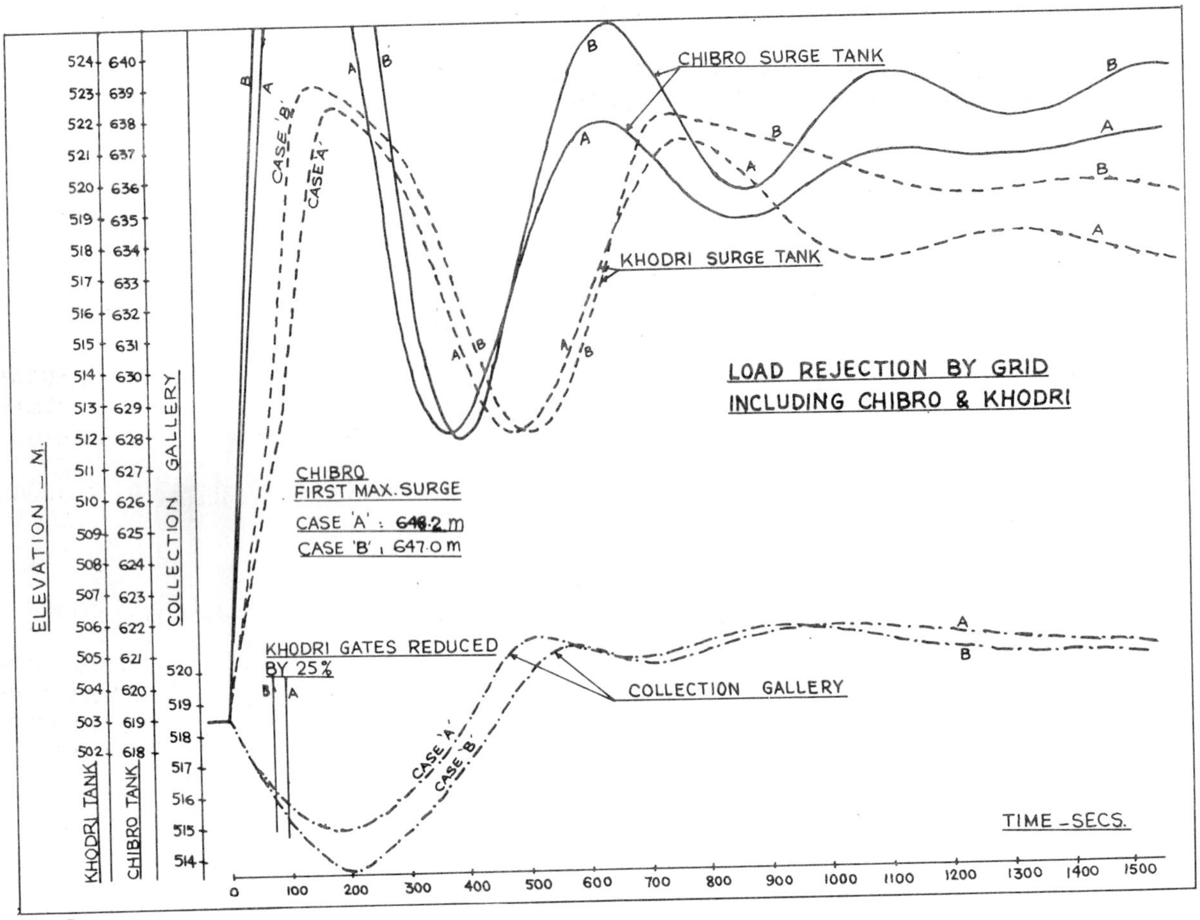

FIG. 9

Fig. 9 Load rejection by Grid including Chibro and Khodri

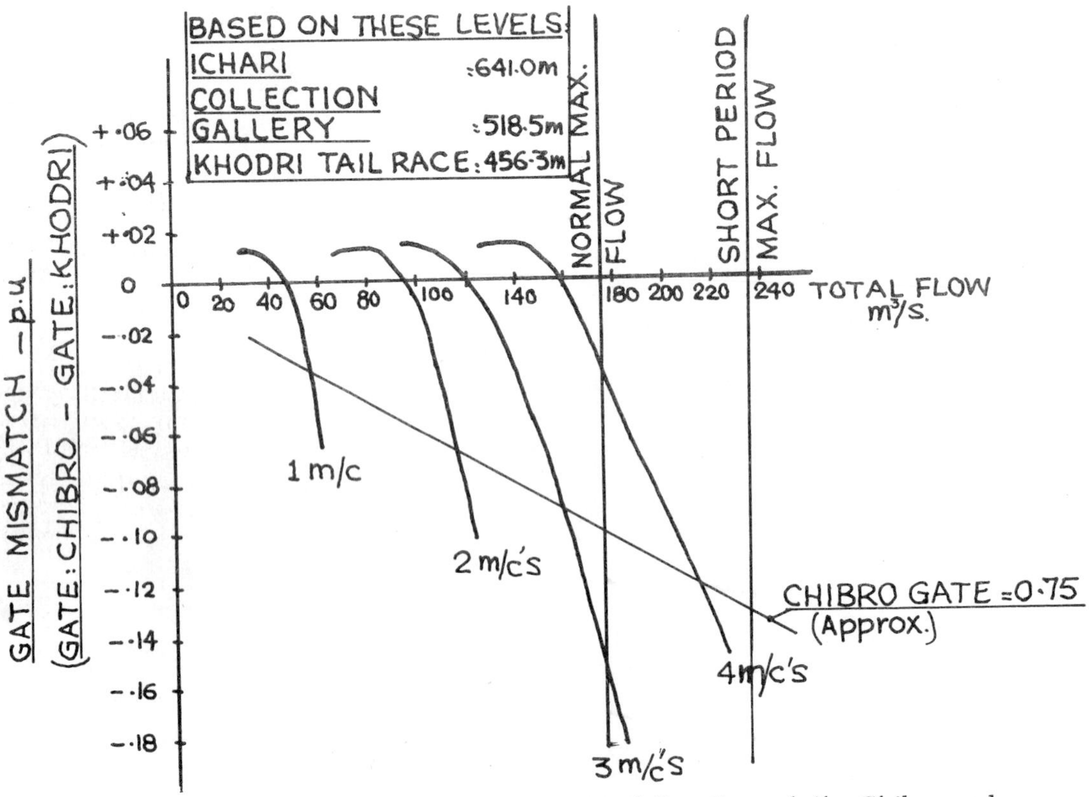

Fig. 10 Effect of tunnel friction for equal flow through the Chibro and Khodri turbines

Third International Conference on

PRESSURE SURGES

Canterbury, England
March 25-27, 1980

MASS OSCILLATION PREDICTION - A COMPARATIVE STUDY OF MASS SURGE AND WATERHAMMER METHODS

J.A. Forrest and J.F. Robbie

University of Aberdeen, U.K.

Summary

The transient response of hydro power schemes has been traditionally split into independent studies of mass oscillation and waterhammer.

In recent times it has been suggested that the application of computer packages developed for waterhammer analysis, should be extended to include the prediction of mass oscillation behaviour.

Whereas this may be attractive in principle, it is only through the application of both methods to the prediction of mass oscillation in a representative selection of hydro power installations that a satisfactory conclusion can be reached.

This paper reports on the effectiveness, accuracy, economy and relative merits of each approach in analysing the response of simple, throttled, galleried, differential and enclosed surge chambers.

Held at the University of Kent at Canterbury, U.K.

Conference Organised by BHRA Fluid Engineering, Cranfield, Bedford, England.
In conjunction with The City University, London.

Nomenclature

A	=	area of conduit, and also penstock
A_m	=	area of main tank for differential chamber
A_p	=	area of port for differential chamber
A_r	=	area of riser for differential chamber
A_o	=	area of throttle at base of chamber
A_s	=	constant horizontal cross section of chamber
A_v	=	total area available for heat flow from air chamber
a_n	=	constants in mass surge analysis
CH1, CH2	=	characteristics in waterhammer analysis
g	=	acceleration due to gravity
H_c	=	absolute head in air chamber
K	=	head loss coefficients of the form $\frac{KQ\|Q\|}{gA^2}$
K_f	=	coefficient of conduit friction loss
K_t	=	coefficient of conduit shape loss
K_o	=	coefficient of throttling loss
K_p	=	coefficient of port head loss for differential chamber
K_{pi}, K_{po}	=	port coefficients; + ve flow, - ve flow
k_i, k_o	=	$\frac{gAp^2}{K_{pi}}$, $\frac{gAp^2}{K_{po}}$, port constants
L	=	length of conduit, and also penstock
* Q	=	conduit discharge
* Q_{mc}	=	penstock discharge to machine
*† Q_m	=	port discharge to main tank
*† Q_r	=	riser discharge
*† Q_s	=	total discharge to surge chamber
Q_*	=	flow reference value for dimensionless reduction
QH	=	heat flow out of air chamber
* T	=	time
T_*	=	time reference value for dimensionless reduction
T_i	=	absolute temperature in air chamber
T_{ex}	=	absolute external ambient temperature
V	=	volume of air chamber
Z	=	equivalent surge level for air chamber
Z_*	=	surge level reference value for dimensionless reduction
* Z_c	=	top of riser level above reservoir datum
*† Z_m	=	main tank water level above reservoir datum
*† Z_r	=	riser water level above reservoir datum
Δt	=	time interval in waterhammer analysis
ΔT	=	difference in temperature across air chamber walls
γ	=	ratio of specific heats for air (= 1.4)
ρ	=	density of water
ϕ	=	reference time in waterhammer analysis

* lower case donates dimensionless form

† subscript p denotes updated values in waterhammer analysis at time $T = \phi$

1.0 Introduction

The prediction of the transient behaviour of hydro electric power schemes traditionally has been separated into consideration of mass surge and waterhammer phenomena.

The mass surge approach has normally been used for the design of surge chambers, when present, and the prediction of surging in the associated tunnels. Complementary studies using elastic wave theory have been generally applied to the prediction of pressure surges along the conduit, to the checking of the effectiveness of the surge chambers (when present) and to more detailed studies on the behaviour of the turbo-machines in operation.

From time to time however, variour authors [1,2,3] have suggested that waterhammer analysis should be used in the prediction of the mass oscillation behaviour. Whilst this may appear attractive in principle, Robbie and Anderson [4] have suggested that there is likely to be little technical advantage and indeed economic disadvantage in so doing relative to the numerically efficient incompressible methods.

This paper presents calculations for a representative selection of hydro power schemes and surge chamber types using both methods in an attempt to resolve the practical issues for the benefit of designers. Simple, galleried, differential and enclosed air chambers are considered and the relative merits of each approach compared on the basis of accuracy, economy and numerical efficiency.

In relation to plant operation, simple load changes are considered in the main. Only one example of the more complex sequential load cases, described by Robbie and Robson [5] is considered on account of the great disparities in calculation time for waterhammer apparent in the simple cases.

The method of characteristics has been adopted for the solution of the waterhammer equations and the equally well tried numerical solutions discussed by Bullough and Robbie [6] are used for mass surge calculations. Since formulation has generally followed standard practice, detailed descriptions have been restricted in the paper to certain innovations in relation to differential and enclosed air chambers.

2.0 Mass Surge Equations

The formulation of mass surge equations for simple, throttled and galleried surge chambers has been presented by many previous authors [4,5,6].

The following additional formulation however, has been undertaken for the differential surge chamber and the enclosed air chamber.

2.1 Differential Surge Chamber

The typical arrangement of a scheme with a single differential surge chamber between the upper reservoir and the machine is shown in Fig. 1, for which the relevant mass surge equations are:

$$Q_s = Q - Q_{mc} \tag{1}$$

$$Q_s = Q_r + Q_m \tag{2}$$

$$Q_m = \pm \sqrt{\frac{gA_p^2}{K_p} \left| Z_r - Z_m \right|} \tag{3}$$

$$\frac{dZ_r}{dT} = Q_r/A_r \tag{4}$$

$$\frac{dZ_m}{dT} = Q_m/A_m \tag{5}$$

$$\frac{L}{gA}\frac{dQ}{dT} = - Z_r - \frac{K_o}{gA_o^{\,2}} Q_s|Q_s| - \left\{ \frac{K_f}{gA^2} + \frac{K_t}{gA^2} \right\} Q|Q| \tag{6}$$

These equations can be reduced to dimensionless form in a variety of ways, as discussed by Anderson [7]. By introducing the dimensionless variables -

$$\left. \begin{array}{l} z_m = Z_m/Z_*,\ z_r = Z_r/Z_*,\ z_c = Z_c/Z_* \quad t = T/T_*, \\ q = Q/Q_*,\ q_p = Q_p/Q_*,\ q_r = Q_r/Q_*,\ q_{mc} = Q_{mc}/Q_*,\ q_s = Q_s/Q \end{array} \right\} \tag{7}$$

and reference values of flow and time defined by -

$$Q_* = A\sqrt{gZ_*} \qquad T_* = L/\sqrt{gZ_*} \tag{8}$$

Eqs. (1) to (6) then reduce to the following dimensionless form:

$$q_s = q - q_{mc} \tag{9}$$

$$q_s = q_r + q_m \tag{10}$$

$$q_m = \pm\sqrt{|z_r - z_m|/a_5} \tag{11}$$

$$\frac{dz_r}{dt} = a_1\, q_r/A_r \tag{12}$$

$$\frac{dz_m}{dt} = a_1\, q_m/A_m \tag{13}$$

$$\frac{dq}{dt} = - z_r - a_2\, q_s|q_s| - a_3\, q|q| - a_4\, q|q| \tag{14}$$

The constants a_n are defined by -

$$a_1 = \frac{LA}{Z_*},\ a_2 = K_o\left(\frac{A}{A_o}\right)^2,\ a_3 = K_f,\ a_4 = K_t,\ a_5 = k_p\left(\frac{A}{A_p}\right)^2 \tag{15}$$

Flow through the ports depends on the differential head between riser and main tank and is given by Eq. (11). Care must be taken to allocate the appropriate sign during computation and this can be readily achieved through independent logic statements.

In the special case where the riser is spilling Eqs. (12) and (13) are replaced by the following equations:

$$z_r = z_c + 0.5\, q_r^{\,2} \tag{16}$$

$$\frac{dz_m}{dt} = (q_r + q_m)/A_m \tag{17}$$

In this case q_m is determined by Eq. (11) and q_r by satisfying continuity in Eq. (10), knowing the value for q_s.

The allowance made for the raised water level of the riser, by including the kinetic head of water flowing in the riser, represents an innovation in the formulation of Eq. (16).

2.2 Enclosed Air Chamber

The thermodynamic behaviour of the air in the enclosed air chamber has been modelled using the rational heat transfer equation (RHT) as developed and described by Graze [8,9,10,11] in preference to the more generally used polytropic equation [13] The RHT approach avoids the necessity to assume a value for the polytropic exponent and also offers a more realistic approach to heat losses which may prove to be a significant factor in attenuation [10].

The arrangement for a scheme with a single enclosed air chamber between the upper reservoir and the machine is shown in Fig. 2. The RHT equation is represented by Eq. (18), the perfect gas laws by Eq. (19) and heat flows, assuming SI units and free convection [11] by Eq. (20).

$$\frac{dH_c}{dT} = -\gamma \frac{H_c}{V}\frac{dV}{dT} - (\gamma - 1)/\rho g V \frac{dQH}{dT} \qquad (18)$$

$$\left(\frac{H_c V}{T_i}\right)_{T=0} = \left(\frac{H_c V}{T_i}\right)_{T=\phi} \qquad (19)$$

$$\frac{dQH}{dT} = 1.4\, A_v |\Delta T|^{\frac{1}{3}}\, \Delta T \qquad (20)$$

In Eq. (20) the area A_v is the total area over which heat flow may occur, and varies with the volume of air in the chamber.

The continuity and dynamic relationships are expressed by the following equations:

$$Q_s = Q - Q_{mc} \qquad (21)$$

$$\frac{dV}{dT} = -Q_s \qquad (22)$$

$$\frac{dZ}{dT} = \frac{dH_c}{dT} + Q_s/A_s \qquad (23)$$

In this study a variable Z, representing the equivalent level of a conventional surge chamber, has been introduced to facilitate comparisons. It moderates the conduit flow through the dynamic relationship represented by Eq. (6). In the computations, these equations have been reduced to the dimensionless form in a standard manner.

The variable H_c is the absolute head of the air in the chamber, measured in piezometric terms in equivalent metres of water. In the initialization of the variables for the numerical solution, Z must be determined in pressure head terms relative to atmosphere.

3.0 Waterhammer by Method of Characteristics

The method of characteristics has been described extensively by Streeter and Wylie [14] and others, and the formulation of the boundary conditions for simple and differential surge chambers by Martin [2,3]. Amplification of the method of characteristics will be restricted in this paper, to the following features associated with the differential and enclosed air chambers.

3.1 Differential Surge Chamber

The formulation for a spilling differential surge chamber can be improved by allowing for the raised water level in the riser as described in section 2.1.

Martin's [2] formulation for a differential surge chamber, is repeated here, for convenience as Eqs. 24-29. The spilling differential surge chamber is illustrated in Fig. 3.

$$Q_{sp} = Q_{mp} + Q_{rp} \tag{24}$$

$$Q_{sp} = CH1 - CH2\ Z_{rp} \tag{25}$$

For flow into main tank

$$Q_{mp} = \sqrt{\frac{gA_p^2}{K_{pi}}(Z_r - Z_{mp})} \tag{26}$$

For flow out of main tank

$$Q_{mp} = -\sqrt{\frac{gA_p^2}{K_{po}}(Z_{mp} - Z_r)} \tag{27}$$

$$Z_{mp} = Z_m + Q_{mp}\ \Delta t/A_m \tag{28}$$

$$Z_{rp} = Z_r + Q_{rp}\ \Delta t/A_r \tag{29}$$

Improved formulation can be achieved through manipulation of the equations as follows, to yield Z_{rp} and Z_{mp} directly and thereafter using these values to obtain improved updated values of the discharges Q_{mp} and Q_{rp} from Eqs. (26) or (27) and (24) as appropriate.

For flow into the main chamber the simultaneous solution of Eqs. (26) and (28) yields the following quadratic equation:

$$Z_{mp}^2 + Z_{mp}\left(k_i\left(\frac{\Delta t}{A_m}\right)^2 - 2\ Z_m\right) + Z_m^2 - k_i\left(\frac{\Delta t}{A_m}\right)^2 Z_r = 0 \tag{30}$$

As $k_i > 0$ $Z_{mp} > 0$ and $Z_r > Z_{mp} > Z_m$ the solution requires the positive sign of the radical term to be taken - giving the following equation for Z_{mp} :

$$Z_{mp} = Z_m - \frac{k_i}{2}\left(\frac{\Delta t}{A_m}\right)^2 + \frac{1}{2}\sqrt{k_i\left(\frac{\Delta t}{A_m}\right)^2\left(k_i\left(\frac{\Delta t}{A_m}\right)^2 - 4(Z_m - Z_r)\right)} \tag{31}$$

Similarly Eq. (32) can be derived for Z_{rp} by the simultaneous solution of Eqs. (24), (25), (26) and (29).

$$Z_{rp} = \frac{Z_r + \frac{\Delta t}{A_r} CH1 + \frac{k_i}{2} \frac{\Delta t}{A_m A_r} - \frac{k_i}{2} \frac{\Delta t}{A_m} \sqrt{\left(\frac{\Delta t}{A_m}\right)^2 + \frac{4}{k_i} (Z_r - Z_m)}}{1 + \frac{\Delta t}{A_r} CH2} \tag{32}$$

For flow out of the main chamber the corresponding equations are:

$$Z_{mp} = Z_m + \frac{k_o}{2}\left(\frac{\Delta t}{A_m}\right)^2 - \frac{1}{2}\sqrt{k_o \left(\frac{\Delta t}{A_m}\right)^2 \left(k_o \left(\frac{\Delta t}{A_m}\right)^2 + 4(Z_m - Z_r)\right)} \tag{33}$$

$$Z_{rp} \frac{Z_r + \frac{\Delta t}{A_m} CH1 - \frac{k_o}{2}\left(\frac{\Delta t}{A_m} - \sqrt{\left(\frac{\Delta t}{A_m}\right)^2 - \frac{4}{k_o} (Z_r - Z_m)}\right)}{1 + \frac{\Delta t}{A_m} CH2} \tag{34}$$

The following equations represent a more rational approach to the case of riser spilling by providing directly updated values of the new levels. If the riser is spilling (i.e. $Z_{rp} > Z_c$), then due allowance can be made for the kinetic head in the riser by using the riser discharge from the previous time interval.

$$Z_{rp} = Z_c + \frac{Q_r^2}{2gA_r^2} \tag{35}$$

and Z_{mp} will be given by:

$$Z_{mp} = Z_m + (CH1 - CH2\ Z_{rp}) \frac{\Delta t}{A_m} \tag{36}$$

3.2 Enclosed Air Chamber

The equation representing the thermodynamic behaviour of the enclosed air chamber is given by the combination of Eqs. (18), (19), (20) and (22):

$$\frac{dH_c}{dT} = - \gamma \frac{H_c Q_s}{V} - 1.4(\gamma - 1)/\rho g V \left[\left|\left(\frac{\left(H_c V\right)_{T=\phi}}{\left(\frac{H_c V}{T_i}\right)_{T=0}}\right. - T_{ex}\right| \left(\frac{\left(H_c V\right)^{\frac{1}{3}}_{T=\phi}}{\left(\frac{H_c V}{T_i}\right)_{T=0}} - T_{ex}\right)\left(2A_s + V\sqrt{\frac{4\pi}{A_s}}\right)\right]$$

..... (37)

which is solved in conjunction Eqs. (22) and (23) and the waterhammer relationship:

$$Q_s = CH1 - CH2\ Z \tag{38}$$

These equations are of ordinary differential form amenable to solution by numerical techniques.

The same precautions to determine Z in appropriate head terms are necessary as described in section 2.2.

4. Comparison of Mass Surge and Waterhammer Predictions

4.1 General Conditions and basis of comparison

A representative selection of types of upstream surge chambers and machine load changes have been investigated and wherever possible, schemes with available field measurements have been analysed.

The essential physical and hydraulic details of each scheme used in the study, are listed in Table 1 together with the sources of data. Further information on each scheme may therefore be obtained from the sources listed.

Accuracy and numerical efficiency (in terms of calculation time) are taken as the two most significant criteria for practical comparison. The predicted oscillations calculated by each method are shown superimposed, one upon the other, in Figs. 4 to 12 over a period of 1000 seconds (real time).

Computational details of these studies are given in Table 2.

The calculation time recorded in Table 2, has been taken as the time to execute the relevant numerical procedures in each case - excluding time overheads for compilation, input, storage and output of data and results. In this way distortions attributable to particular computer packages have been eliminated as far as possible.

Surge chamber throttling losses, tee junction losses and tunnel kinetic heads have been ignored in this instance, but entrance losses have been allowed for, especially when estimating frictional losses from field data.

Linear changes of flow with time have been assumed in all cases and calculations have been executed in double precision (72 bit word size) using a Honeywell Level 66 Computer.

4.2 Aspects of Waterhammer analysis

In relation to waterhammer, computation time is directly related to the number of conduit intervals chosen for the analysis. The ultimate controlling factors governing the permissible coarseness of these intervals are the relative lengths of the penstock and conduits, the degree of interpolation to be tolerated and the sensitivity of the scheme to adjustment of wave celerity.

Since this study was principally concerned with the prediction of mass oscillation effects, the coarseset representation possible of the penstock generally has been adopted (i.e. 1 or 2 reaches).

4.3 Aspects of Mass oscillation analysis

For consistency, a single numerical integration technique was used in evaluating the mass oscillations. The Fehlberg 6th Order Runge Kutta was selected from the many whose accuracy and behaviour in the solution of equations of mass oscillation have been thoroughly investigated by Bullough [6].

The time step of 2 seconds used throughout could be regarded by many as being excessively small in relation to wavelength, particularly since Bullough's work [6] indicates that step sizes up to ten times this value would give adequate accuracy. This has been verified in relation to Lietzach II using a step of 20 seconds. The corresponding decrease in computer execution time should be borne in mind when the following detailed comparisons between methods are assessed.

4.4 Simple Surge Chamber

4.4.1 Lietzach II Scheme

The Lietzach II surge chamber has been widely reported in the literature [2,3,15,16]

The results for surge chamber level oscillations, shown in Fig. 4, are for a waterhammer analysis with a total of 24 reaches - 10 in the penstock and 14 in the low pressure conduit. The results from both methods agree so closely, as can be seen from Fig. 4, that they cannot be differentiated and both give a fair representation of the actual surge. A comparison of the calculation details in Table 2 however shows that the mass surge analysis can be completed in approximately $^1/_{54}$ of the time required for the waterhammer solution. It is true that waterhammer computation time can be reduced by a factor of 10 by using only 1 interval in the penstock. However in the Lietzach case, the lengths of the low pressure conduit and penstock are of the same order and consequently if the technique of celerity adjustment is used to avoid interpolations (as it is in this study) the system reduces to an unrealistic total of only 2 reaches. Although giving savings in computer time, accuracy is sacrificed on maximum and minimum values as shown in Fig. 5.

4.4.2 Tanzmuhle Pumped Storage Scheme

Field measurements for both turbine and pump load rejections are available in the literature [16] for the Tanzmuhle Pumped Storage scheme. Once again the close agreement of both methods is demonstrated in Figs. 6 and 7 for turbine load rejection and pump load rejection, respectively.

For this scheme the mass surge method required approximately $^1/_{42}$ of the waterhammer computation time with the minimum possible number of reaches.

4.5 Galleried Surge Chamber

4.5.1 Dinorwic Pumped Storage Scheme

The Dinorwic surge chamber has a circular upper headpond effectively forming an upper expansion gallery as described by Mawer et al [17]. The surge chamber oscillations for a pump load accept case are shown in Fig. 8. Close aggrement between the two methods in again demonstrated and the mass surge calculations are completed in approximately $^1/_{28}$ of the time for a waterhammer analysis when a total of 20 reaches is used.

Contemporary pumped storage schemes, like Dinorwic, are being designed not only for very rapid but also complex sequences of load changes since the extremes in surge chamber levels are very seldomly realised by a simple load change. In general it is more likely that one of a number of critical sequences of machine operation will be applicable as described by Robbie and Robson [5]. One such sequential load case taken from their paper and listed below, is intended to produce the minimum surge level in the upstream surge chamber. This case, presented in Fig. 9, is based on calculations by the mass surge method only for the reasons discussed presently.

The sequence of events involved is as follows.

1. Steady pumping with all machines but one, and upper reservoir empty.
2. Last pump started at time 0 seconds.
3. All pumping load dropped at time 130 seconds.
4. Machines allowed to reverse and then synchronised on "no load" turbining.
5. Full generating load accepted at time 380 seconds.

The computer time required for this case of sequential loading at fixed time intervals by mass surge was 21.0 seconds which is much less than that generally required for a waterhammer analysis of a simple load case. There can therefore be little justification for conducting a waterhammer analyses for sequential load cases since each load change would require commensurate increases in computer time for no greater accuracy. In design studies for actual schemes, the disparity in efficiency will be even greater because then it is necessary to establish the critical points in time at which load changes, if initiated, produce the worst subsequent surge. The establishment of such points manually can necessitate an extended series of computer runs and to do so automatically certainly requires very many more.

4.6 Differential Surge Chamber

4.6.1 Apalachia

The Apalachia surge chamber, highlights an important feature which could, in certain circumstances, justify the specific use of the waterhammer approach in surge chamber design. The riser of the Apalachia surge chamber experiences higher frequency oscillations as noted in the paper by Martin [2], from which the details of the scheme, summarized in Table 1, are abstracted. The results presented in Fig. 10, show that the water level oscillations are virtually coincident for both methods except that the waterhammer method reveals the higher frequency oscillations in the riser. This is shown more clearly in the enlargement on Fig. 10.

These secondary oscillations represent a reduced effectiveness of this chamber in suppressing the propogation of waterhammer pressure waves into the low pressure conduit, and are caused by pressure waves reflected from the upper reservoir. Although in this instance of little practical concern, it is clearly advisable that the waterhammer method should be used to check the possible existence of such features. In this case the mass surge approach requires $^{1}/_{33}$ of the computer time of the waterhammer analysis. The results in Fig. 10 demonstrate that the mass surge represents the mean of these secondary oscillations.

4.6.2 Santeetlah

Secondary oscillations are again demonstrated in the results for the Santeetlah spilling differential surge chamber as can be seen in Fig. 11. The effect of the spilling riser is also clearly demonstrated. Martin and Jackson [3] give details of the Santeetlah scheme summarized here in Table 1. Once again the results are virtually coincident but the relative amounts of computer time for this case are even more disproportionate to those of some of the other schemes discussed so far. The mass surge calculations can be completed in approximately $^{1}/_{90}$ of the waterhammer time.

This is due to the low pressure tunnel being very long in relation to the penstock and consequently requiring more reaches for reasonable representation. In such cases implicit methods may be preferable in order to reduce computer time as discussed by Martin [2].

It may be noted that the spilling action of the riser at Santeetlah serves to virtually remove the effect of the secondary oscillations.

4.7 Enclosed Air Chambers

4.7.1 Driva Hydro Electric Scheme

This type of chamber represents one of the latest innovations in surge chamber design for Hydro Power schemes and in appropriate site conditions could be adopted more generally in the future. As is well known, it was first pioneered at the Driva installation in Norway approximately 50 km South of Trondheim.

A comprehensive description of the scheme has been presented by Rathe [18] from which some of the data in Table 1 has been established, although it was necessary to make some reasonable assumptions in relation to friction losses and pressure wave velocities.

Whereas, in all previous cases examined in this study, agreement between methods has been virtually precise, air chamber predictions, presented in Fig. 12 reveal unacceptably wide disparities in amplitude and period, although the general trend of the oscillations is similar.

Since field observations were not available for Driva, the accuracy of the methods could not be established directly. Recourse was made therefore, to an analysis of the Villingen pumping scheme, as presented by Graze, Forrest et al [12], for which field results are available.

The conclusions which may be drawn from the results of this scheme, presented in Figs. 13,14 and 15 are that the mass surge approach yields predictions which are much closer to the observed response. It is evident that further careful consideration, which is outwith the scope of this paper, is required to be given to the waterhammer approach for air chamber installations.

The long calculation time shown against the waterhammer solution in Table 2, is due, in part, to the high order, variable step, numerical method attributable to Gear and described by Craigie [19] used in the analysis in an attempt to narrow the disparity in results.
For the Driva scheme the mass surge method required approximately $^1/_{100}$ of the waterhammer computation time.

5.0 CONCLUSIONS

5.1 The feasibility of calculating surge chamber oscillations by means of the elastic waterhammer equations has been further demonstrated for an extended number of surge chamber forms and load change cases.

5.2 Under certain circumstances, it may be necessary to check the effectiveness of the surge chamber in suppressing the transmission of waterhammer waves into the low pressure conduit, but in general mass surging in conduit systems can be accurately and much more efficiently predicted using incompressible methods.

5.3 Even using excessively short time intervals in the mass oscillation calculations and excessively coarse reaches in the waterhammer solutions, computer times for waterhammer analyses in cases of simple load changes, for simple, galleried and differential chambers varied between 30 and 90 times that required for mass surge with no increase in accuracy. Such disparity renders the use of the waterhammer approach in serious design studies of sequential machine load changes economically unattractive.

5.4 The difficulties encountered, and as yet unresolved, in obtaining acceptable waterhammer simulation of oscillations in air chamber installations are believed to lie in the parallel solution of the waterhammer and thermodynamic equations.

5.5 With the exception of air chamber installations, the results obtained in this study validate both mass surge and waterhammer approaches to surge chamber design.

The choice of method, in a given investigation, in practice may be heavily influenced by the type of computational facilities available, but it is hoped that this paper will prove to be of real value to the practical designer in assessing the technical and economic considerations.

6. Acknowledgements

The Authors wish to thank Professor T.M. Charlton formerly of the University of Aberdeen, for his encouragement. Mr. Forrest was studying as a Commonwealth Scholar in the Department of Engineering, University of Aberdeen, when this work was undertaken. Computer Services were provided by the University Computing Centre, University of Aberdeen.

7.0 REFERENCES

1. Jeager, C., "General Theory of Waterhammer", (Theorie Generale du Coup de Belier), Dunod, Paris, 1933 (In French).

2. Martin, C.S., "Method of Characteristics applied to calculation of Surge Chamber Oscillations", Paper El, Proceeding First International Conference on Pressure Surges, September 6-8, 1972, BHRA Fluid Engineering, London, (1972).

3. Martin, C.S., and Jackson, H.C., "Combined Surge Tank and Waterhammer Analysis by digital Computer", Water Power, 24, 4, April, 1972.

4. Anderson, A., and Robbie, J.F., "Some novel experiences in Computer aided design of Dinorwic Pumped Storage Surge System", Paper B2, Proceedings Second International Conferences on Pressure Surges, September 22-24, 1976, BHRA Fluid Engineering, Cranfield, UK (1977).

5. Robbie, J.F., and Robson, F.M., "Computer Aided design of surge chambers in Pumped Storage Conduit Systems", Paper E2, Proceeding First International Conference on Pressure Surges, September 6-8, 1972, BHRA Fluid Engineering, London (1972).

6. Bullough, J.B.B., and Robbie, J.F., "Theoretical and experimental analysis of the error in the numerical solution of the equations of mass oscillation in Pumped Storage Schemes", Paper H3, Proceedings Second International Conference on Pressure Surges, September 22-24, 1976, BHRA Fluid Engineering, Cranfield, UK, (1977).

7. Anderson, A., "Non-dimensional Equations for Computations in Hydraulic Transients", The Institution of Civil Engineers, Paper No. 7839, 1975.

8. Graze, H.R., "A Rational approach to the Thermodynamic behaviour of Air Chambers", Thesis (Ph.D.), University of Melbourne, 1967.

9. Graze, H.R., "A Rational Thermodynamic Equation for Air Chamber Design", Third Australasian Conference on Hydraulics and Fluid Mechanics, Sydney, November 25-29, 1968, Paper No. 2607, p 57.

10. Graze, H.R., "New Air Chamber Characteristics", Fourth Australasian Conference on Hydraulics and Fluid Mechanics, Melbourne, November 29 - December 3, 1971, p 259.

11. Graze, H.R., "The Importance of Temperature in Air Chamber Operations", Paper F2, Proceeding First International Conference on Pressure Surges, September 6-8, 1972, BHRA Fluid Engineering, London (1972).

12. Graze, H.R., Forrest, J.A., et al "Analysis of field measurements of air chamber Installations", Paper K2, Proceedings Second International Conference on Pressure Surges, September 22-24, 1976, BHRA Fluid Engineering, Cranfield, UK, (1977).

13. Svee, R., "Surge Chamber with an enclosed compressed air Cushion", Paper G2, First International Conference on Pressure Surges, September 6-8, 1972, BHRA Fluid Engineering, London (1972).

14. Streeter, V.L., and Wylie, E.B., "Hydraulic Transients", McGraw-Hill, 1967.

15. Lottes, G., "Pumped Storage Planning in Germany - Part II", Water Power, 15, 1, p 21-26, January 1963.

16. Naber, G., "Mass Oscillations in Surge Tanks", Water Power, 16, 6, p 264-270, June 1964.

17. Mawer, W.T., Buchanan, R.W., and Queen, B.B., "Dinorwic Pumped Storage Project - Pressure Surge Investigations", Paper B3, Proceedings Second International Conference on Pressure Surges, September 22-24, 1976, BHRA Fluid Engineering, Cranfield, UK (1977).

18. Rathe, L., "An Innovation in Surge Chamber Design", Water Power and Dam Construction, p 244, June/July, 1975.

19. Craigie, J.A.I., " A variable order multi-step method for stiff systems of ordinary differential equations", University of Manchester, Numerical Analysis Report 11, 1975.

TABLE 1. Physical data for schemes compared

Scheme	Conduit L m	Conduit A m^2	Conduit K_f +flow / -flow	Conduit K_t +flow / -flow	Conduit Celerity m/s	Penstock L m	Penstock A m^2	Penstock K_f +flow / -flow	Penstock K_t +flow / -flow	Penstock Celerity m/s	Surge Chamber A_s m^2	A_r m^2	A_m m^2	A_p m^2	K_o +flow / -flow	K_{pi}	K_{po}	Z_c m O.D.	$(H_c)_{T=0}$ m	$(V)_{T=0}$ m^3	Flow Q m^3/s +gen. -pump	Time of linear valve movement s	Upper reservoir level m O.D.
Simple Surge Chamber																							
Lietzach (Ref. 2,3,15,16)	1560.0	12.57	3.003 / 3.003	0.50 / 0.00	1310.0	800.0	12.57	1.540 / 1.540	0.00 / 0.00	920.0	314.16	-	-	-	0.00 / 0.00	-	-	-	-	-	+18.20	5.0	653.17
Tanzmuhle (Ref. 16)	2665.0	9.62	5.901 / 6.701	0.50 / 0.00	1300.0	105.0	9.62	0.232 / 0.264	0.00 / 0.00	1100.0	95.03	-	-	-	0.00 / 0.00	-	-	-	-	-	+27.95 / -17.14	26.0 / 45.0	457.35 / 457.28
Gallery Surge Chamber																							
Dinorwic (Ref. 4,17)	1518.5	86.60	2.272 / 1.943	0.50 / 0.00	1310.0	1119.2	67.50	0.608 / 0.548	0.00 / 0.00	920.0	706.9	-	-	-	0.00 / 0.00	-	-	-	-	-	-329.00	10.0	633.00
Differential Surge Chamber																							
Apalachia (Ref. 2)	13167.0	24.72	22.532 / 22.532	0.50 / 0.00	1305.0	223.0	9.46	0.460 / 0.460	0.00 / 0.00	1100.0	-	18.70	299.37	3.70	0.00 / 0.00	0.913	0.604	408.14	-	-	89.48	5.0	389.63
Santeetlah (Ref. 3)	7653.5	8.83	15.601 / 15.601	0.50 / 0.00	1200.0	98.15	4.67	0.284 / 0.284	0.00 / 0.00	1000.0	-	3.58	73.49	1.23	0.00 / 0.00	1.183	0.617	564.95	-	-	29.90	7.0	553.70
Enclosed Air Chamber																							
Driva (Ref. 18)	18805.0	20.50	46.00 / 46.00	0.50 / 0.00	1310.0	1140.0	21.98	2.694 / 2.694	0.00 / 0.00	920.0	780.0	-	-	-	0.00 / 0.00	-	-	-	416.74	5000.0	30.0	2.0	660.55
Villengen (Ref. 12)	11581.0	0.041	679.958 / 679.958	0.50 / 0.00	1250.0	1000.0	0.041	0.000 / 0.000	0.00 / 0.00	1250.0	1.082	-	-	-	0.00 / 0.00	-	-	-	221.12	0.986	-0.0413	0.00	888.62

TABLE 2. Comparisons between mass surge and waterhammer calculations

Type of Chamber	Figure No.	Waterhammer Calculations: Number of reaches penstock	Number of reaches conduit	Step Length (Secs)	Number of Calculations	Calculation time (Secs)	Mass Surge Calculations: Step Length (Secs)	Number of Calculations	Calculation time (Secs)	Ratio of times of Calculation
Simple Surge Chamber										
Lietzach II - turbine load reject	4 5	10 1	14 1	0.087 0.87	287356 3448	146.9 2.0	2.0 10.0 20.0 2.0	500 100 50 500	2.74 0.54 0.27 2.8	1/54
Tanzmuhle - turbine load reject	6	1	21	0.096	239583	115.6				1/42
Tanzmuhle - pump load reject	7	1	21	0.096	239583	115.9	2.0	500	2.7	1/42
Gallery Surge Chamber										
Dinorwic - pump load accept	8	10	10	0.122	172131	84.1	2.0	500	3.0	1/28
Dinorwic - sequential load	9	-	-	-	-	-	2.0	500	21.0	-
Differential Surge Chamber										
Apalachia - turbine load reject	10	1	50	0.203	256157	123.6	2.0	500	3.7	1/33
Santeetlah - turbine load reject	11	1	65	0.098	683673	322.6	2.0	500	3.6	1/90
Enclosed Air Chamber										
Driva - turbine load reject	12	1	12	1.239	Varies for high order multi-step Gear method	532.0	2.0	500	5.3	1/100

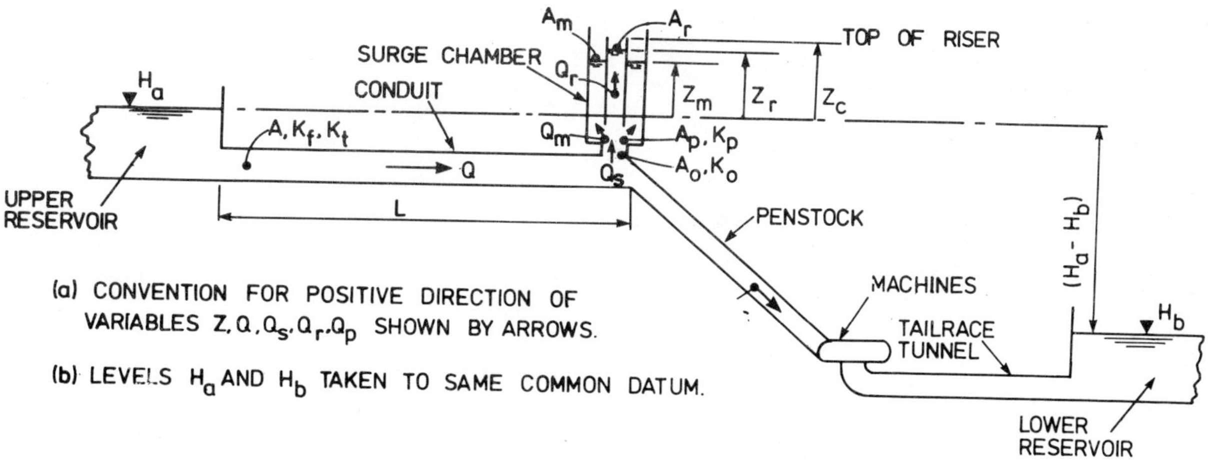

Fig. 1 Scheme arrangement for single upstream differential surge chamber

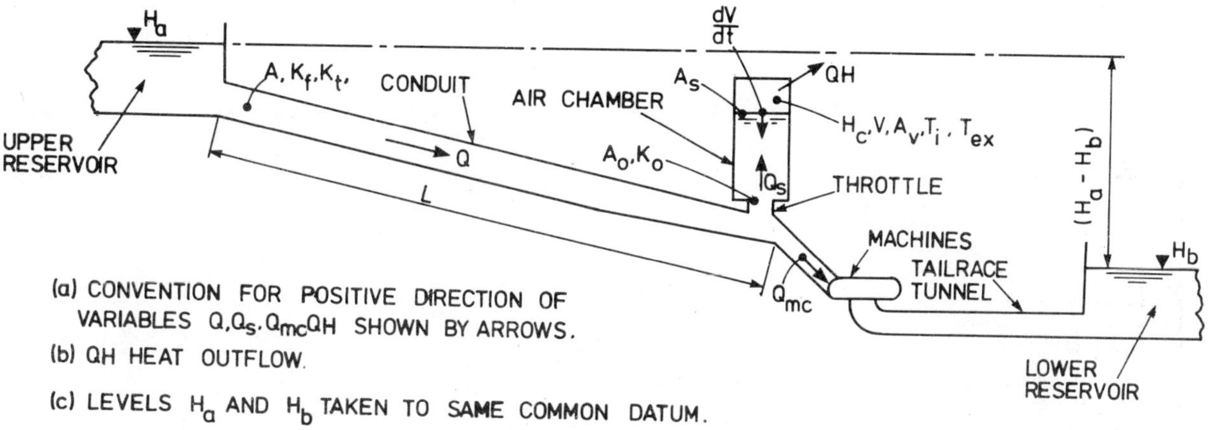

Fig. 2 Scheme arrangement for single upstream enclosed air chamber

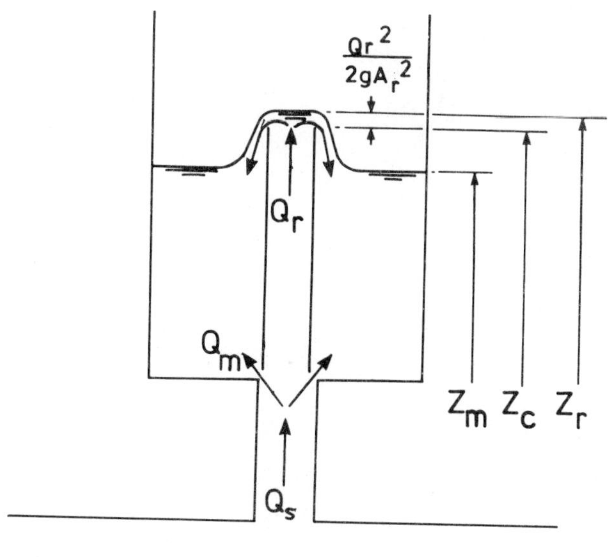

Fig. 3 Notation for spilling differential surge chamber

CALCULATION OF SURGE CHAMBER OSCILLATIONS - MASS SURGE AND WATERHAMMER
LIETZACH SIMPLE SURGE CHAMBER
DIAMETER OF CHAMBER 20.0 METRES LENGTH OF CONDUIT 1560.0 METRES

MASS SURGE BY DASHED LINE WATERHAMMER BY SOLID LINE
FIELD MEASUREMENT BY NABER SHOWN BY DOT

MAXIMUM VALUE= 656.3 MINIMUM VALUE= 650.5
LOAD CHANGE INDICATED BY ▽

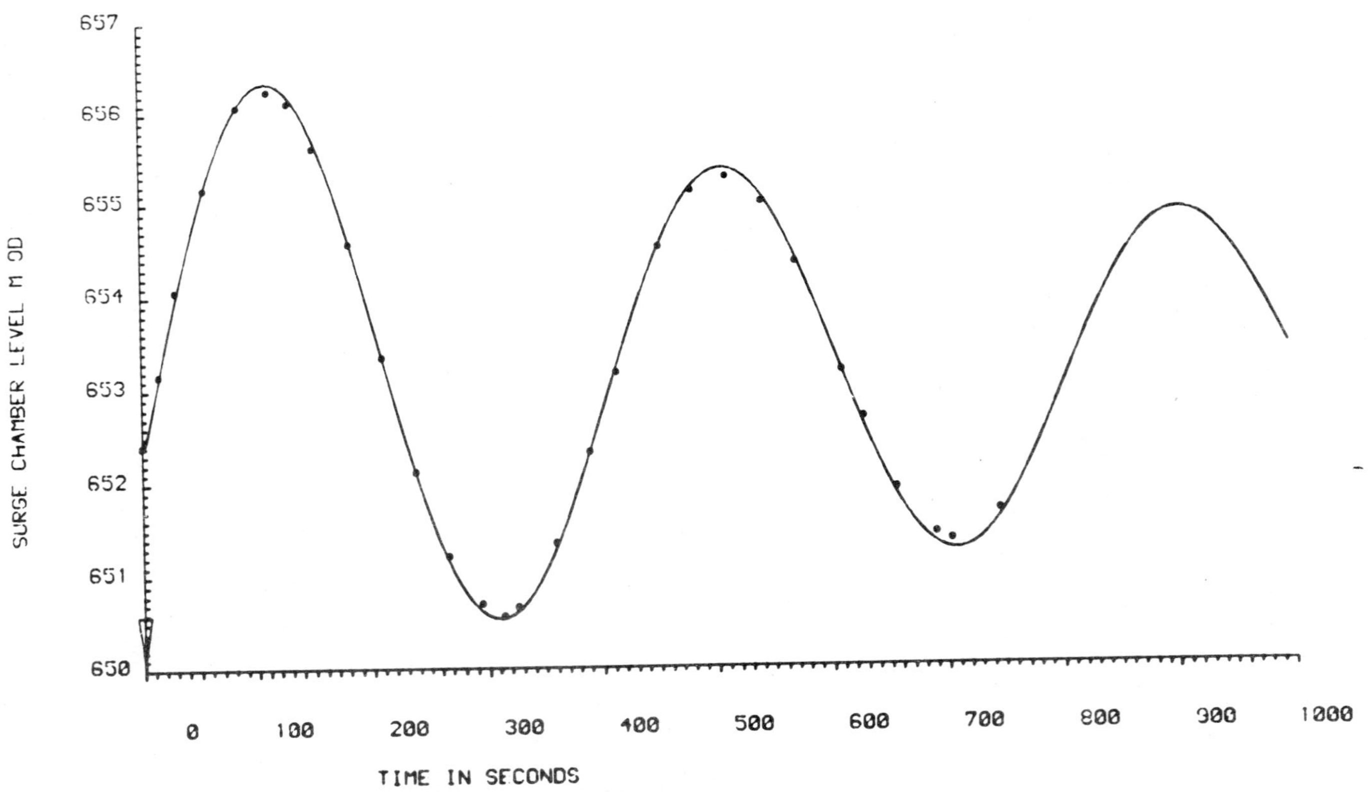

Fig. 4 Lietzach II P.S. Scheme - waterhammer with 10 reaches in penstock

CALCULATION OF SURGE CHAMBER OSCILLATIONS - MASS SURGE AND WATERHAMMER
LIETZACH SIMPLE SURGE CHAMBER
DIAMETER OF CHAMBER 20.0 METRES LENGTH OF CONDUIT 1560.0 METRES

MASS SURGE BY DASHED LINE WATERHAMMER BY SOLID LINE
FIELD MEASUREMENT BY NABER SHOWN BY DOT

MAXIMUM VALUE= 656.3 MINIMUM VALUE= 650.5
LOAD CHANGE INDICATED BY ▽

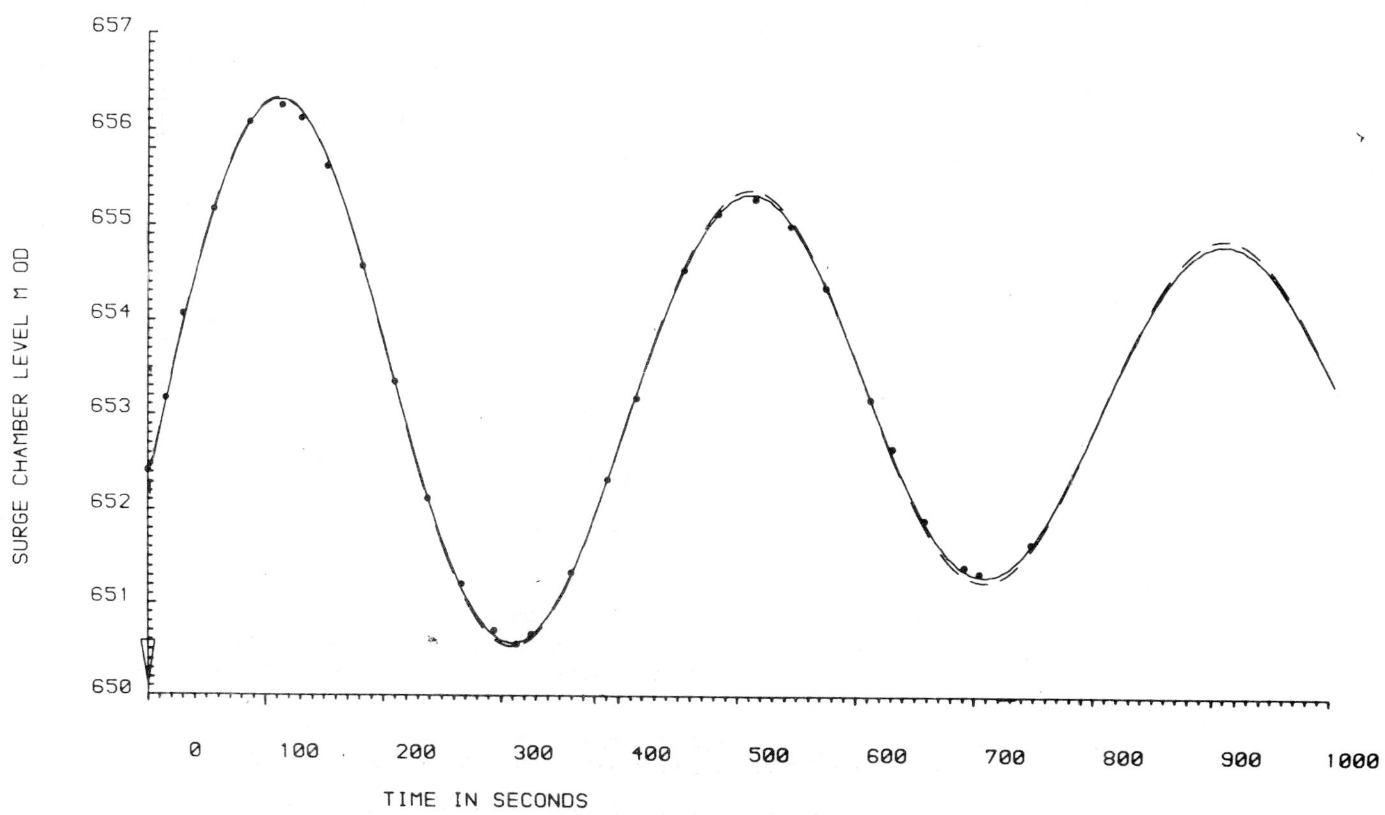

Fig. 5 Lietzach II P.S. Scheme - waterhammer with 1 reach in penstock

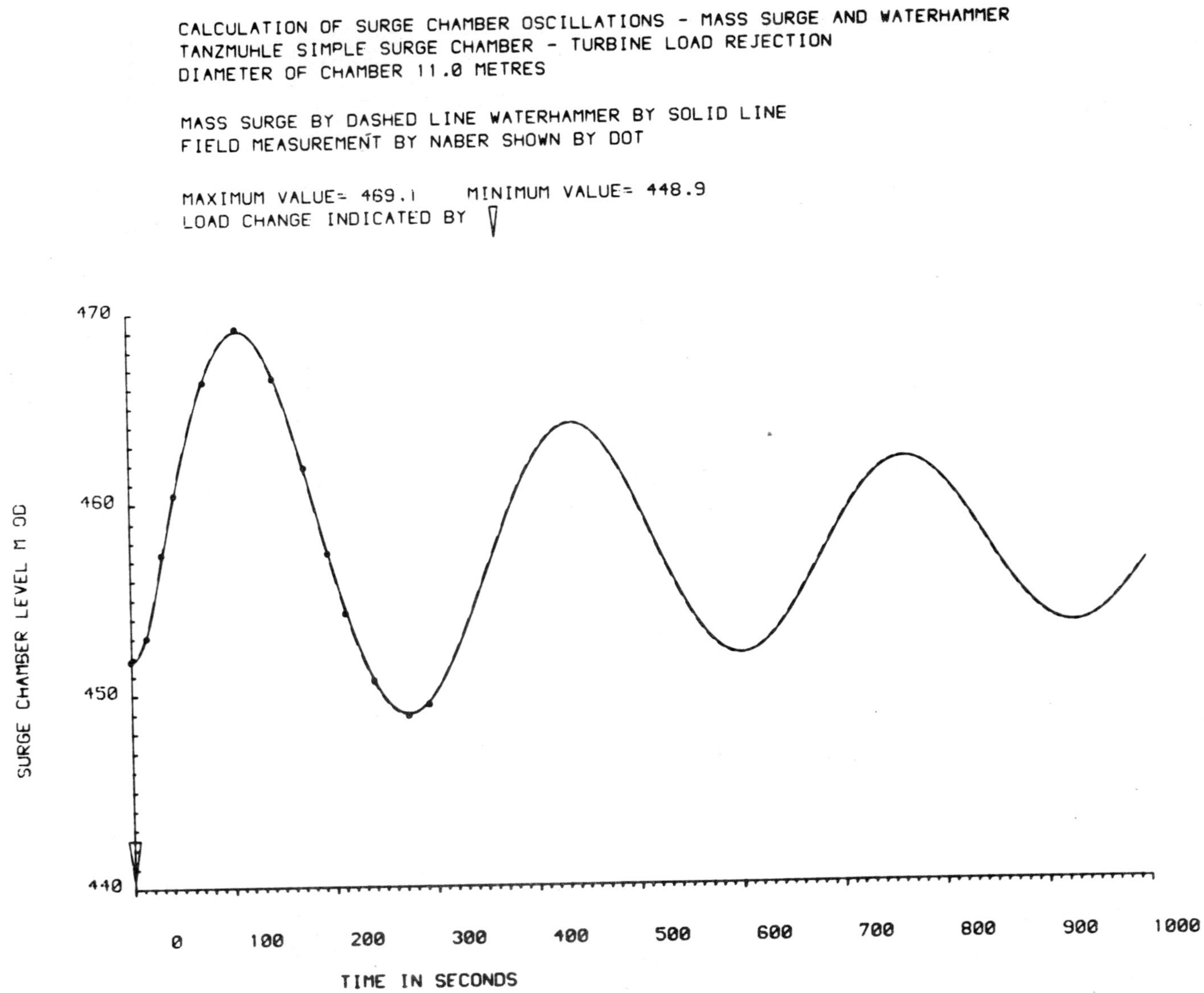

Fig. 6 Tanzmuhle P.S. Scheme - turbine load rejection

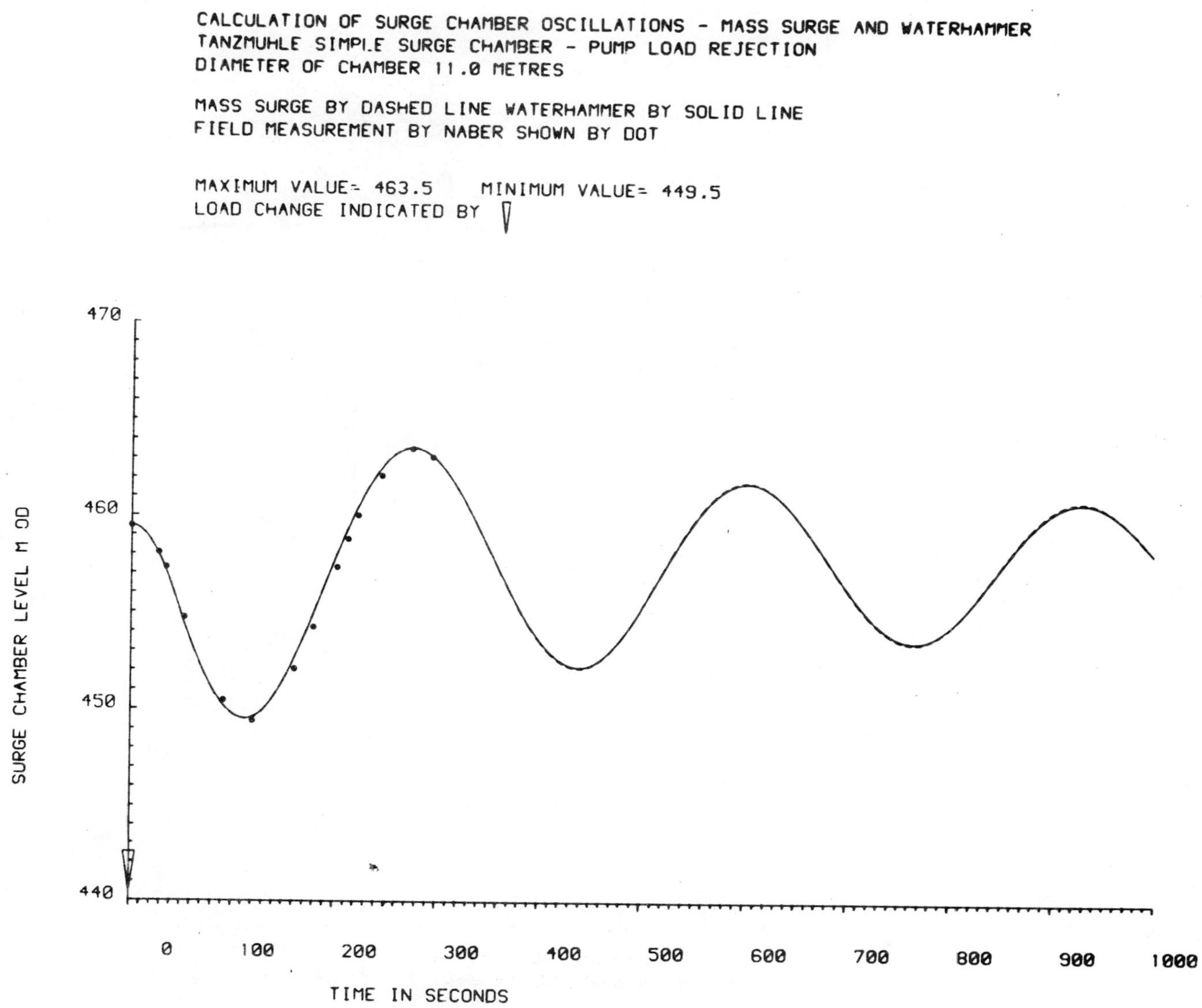

Fig. 7 Tanzmuhle P.S. Scheme - pump load rejection

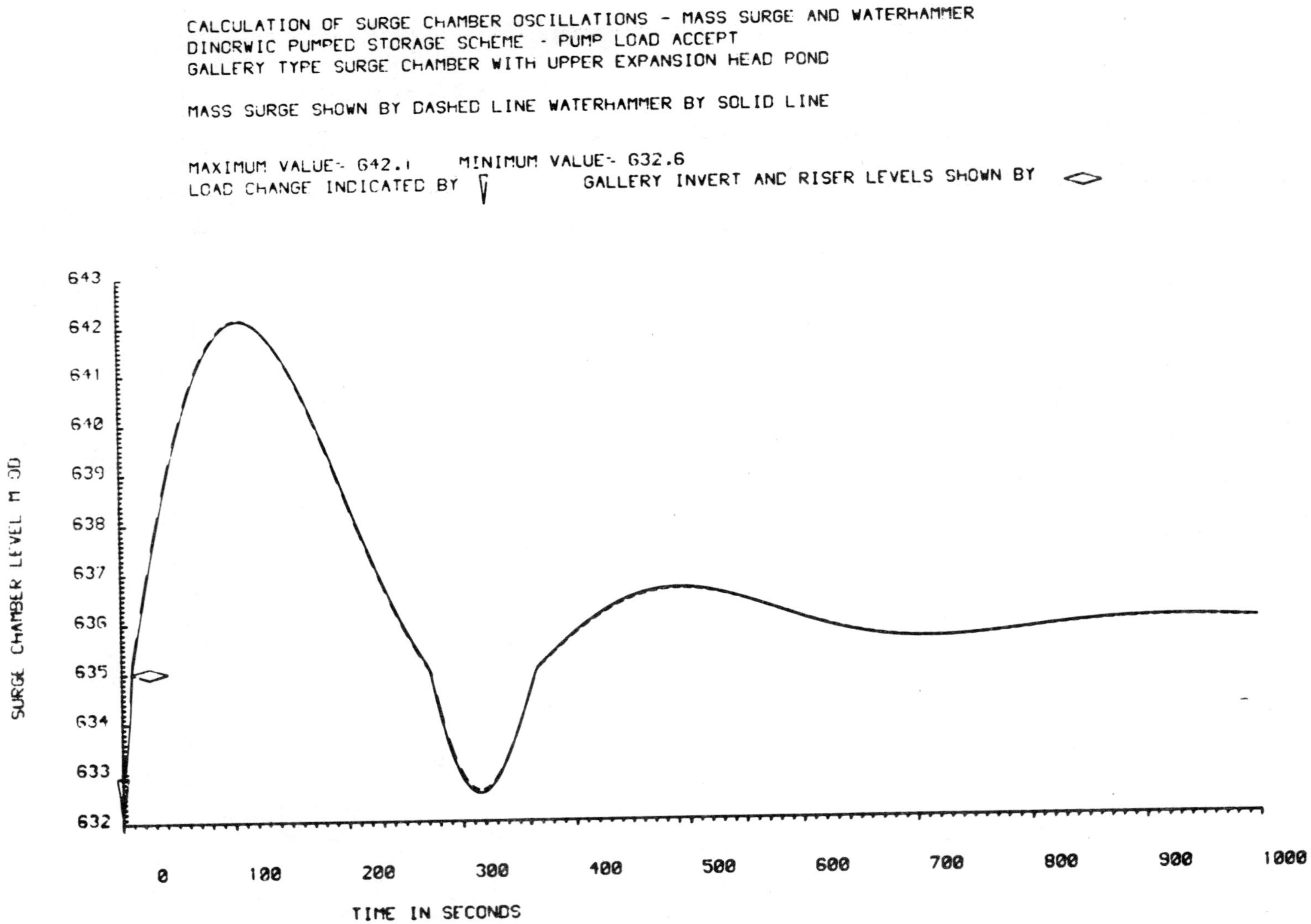

Fig. 8 Dinorwic P.S. Scheme – pump load accept

CALCULATION OF SURGE CHAMBER OSCILLATIONS - MASS SURGE ONLY
DINORWIC PUMPED STORAGE SCHEME - SEQUENTIAL LOAD OPERATING SEQUENCE B
GALLERY TYPE SURGE CHAMBER WITH UPPER EXPANSION HEAD POND
ALL PUMP LOAD ACCEPTED/REJECTED/NO LOAD FLOW/FULL TURBINE LOAD ACCEPTED
DISCHARGE 0-(-334)-41.4-414 CU.M/S   TIMES 0/130/380 SECS.
LOAD CHANGE INTERVALS 10.0 SECS.
UPPER RES 600.0M OD . LOWER RES 106.0M OD

MASS SURGE BY DASHED LINE
MAXIMUM VALUE- 615.3   MINIMUM VALUE- 574.0
LOAD CHANGE INDICATED BY ▽   GALLERY INVERT AND RISER LEVELS SHOWN BY ◇

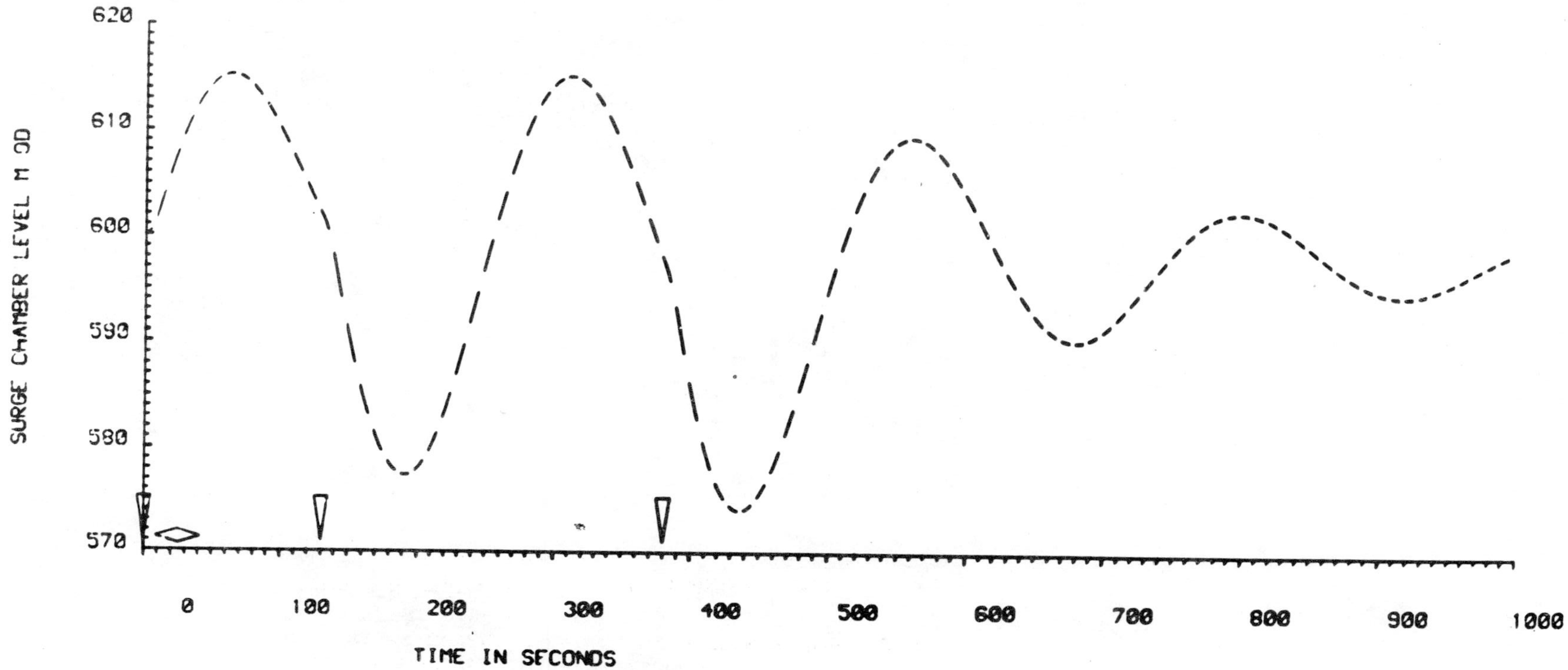

Fig. 9 Dinorwic P.S. Scheme - sequential load changes on the machines

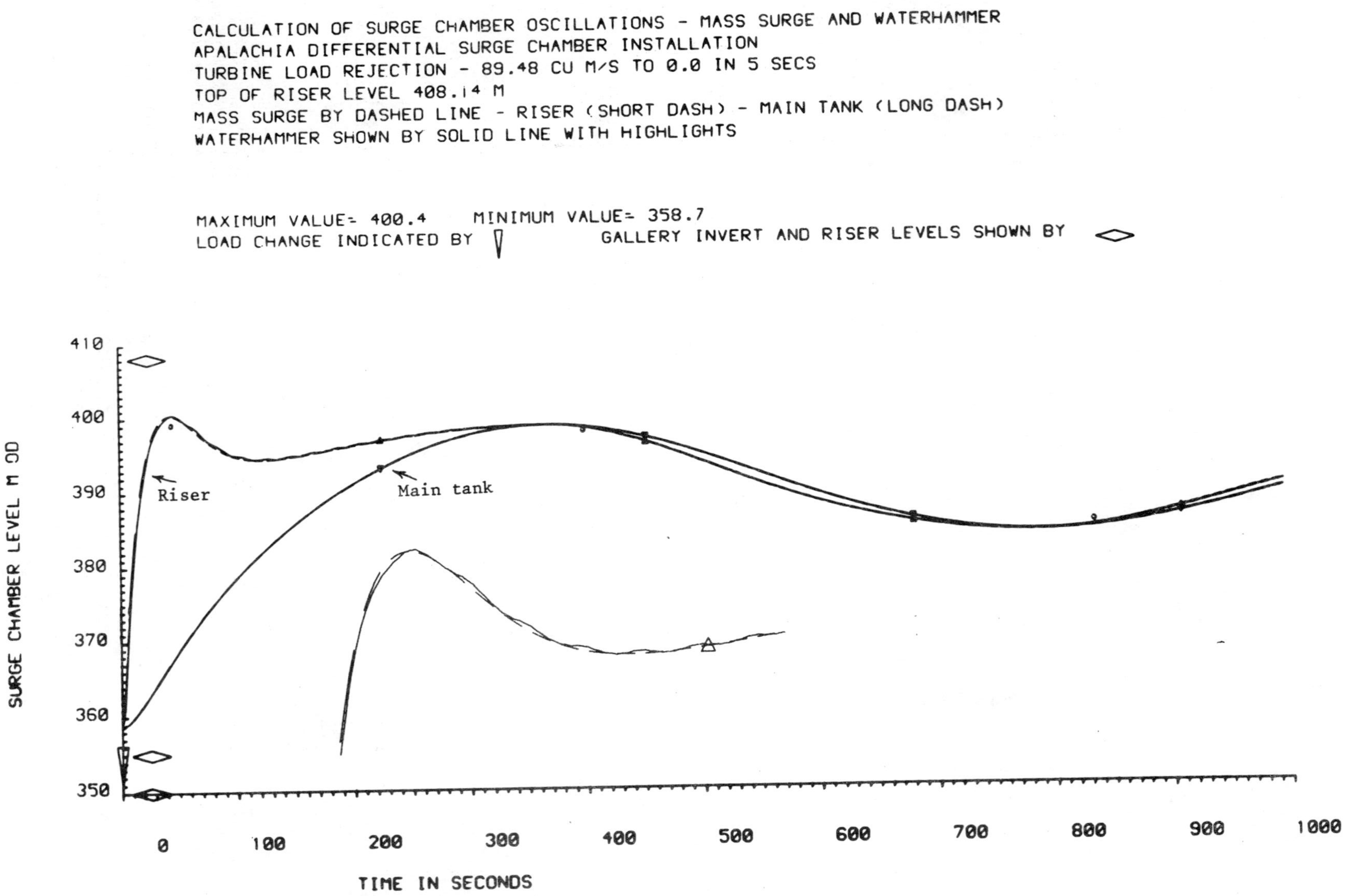

Fig. 10 Apalachia Hydro Scheme - Turbine load rejection

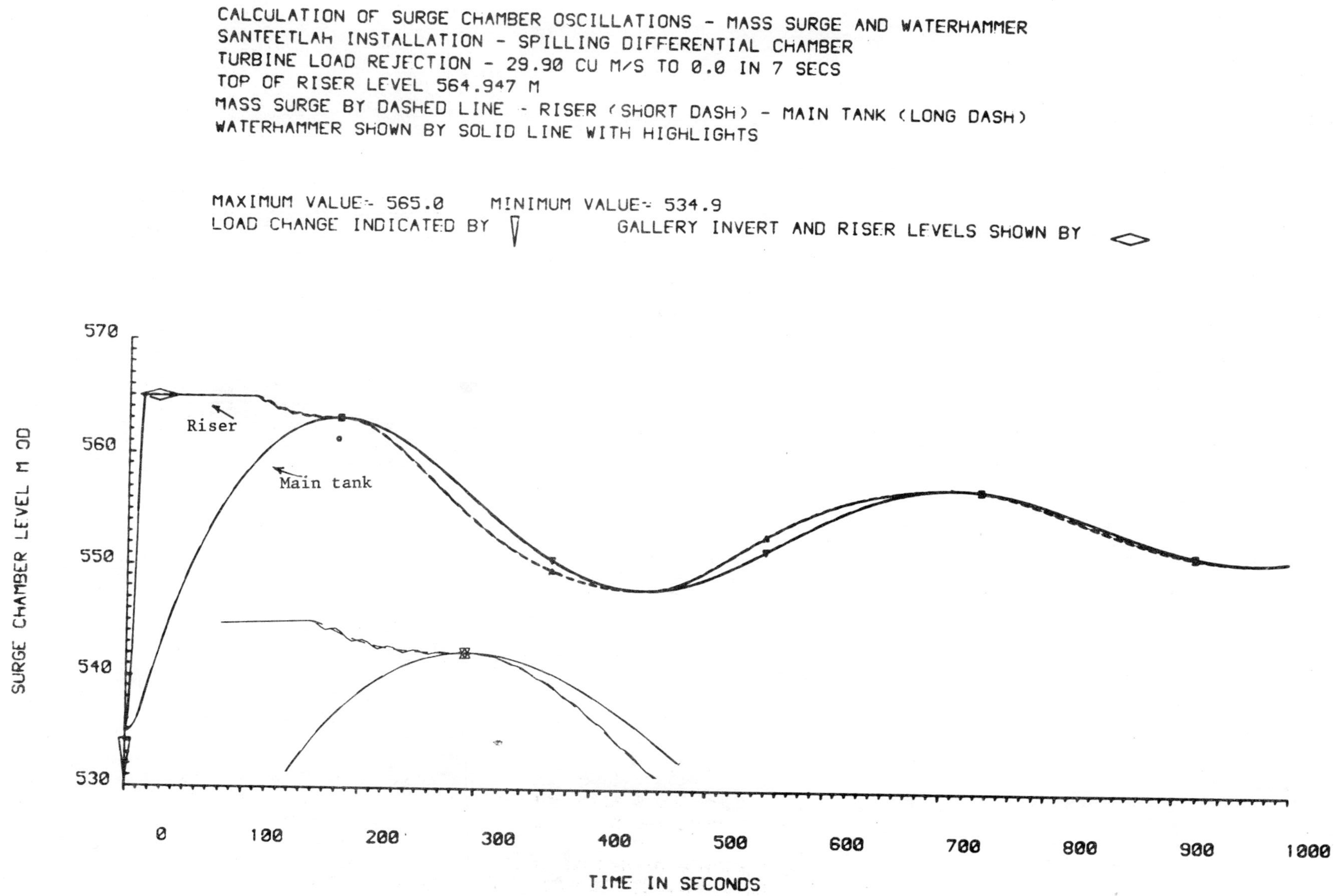

Fig. 11 Santeetlah Hydro Scheme – turbine load rejection

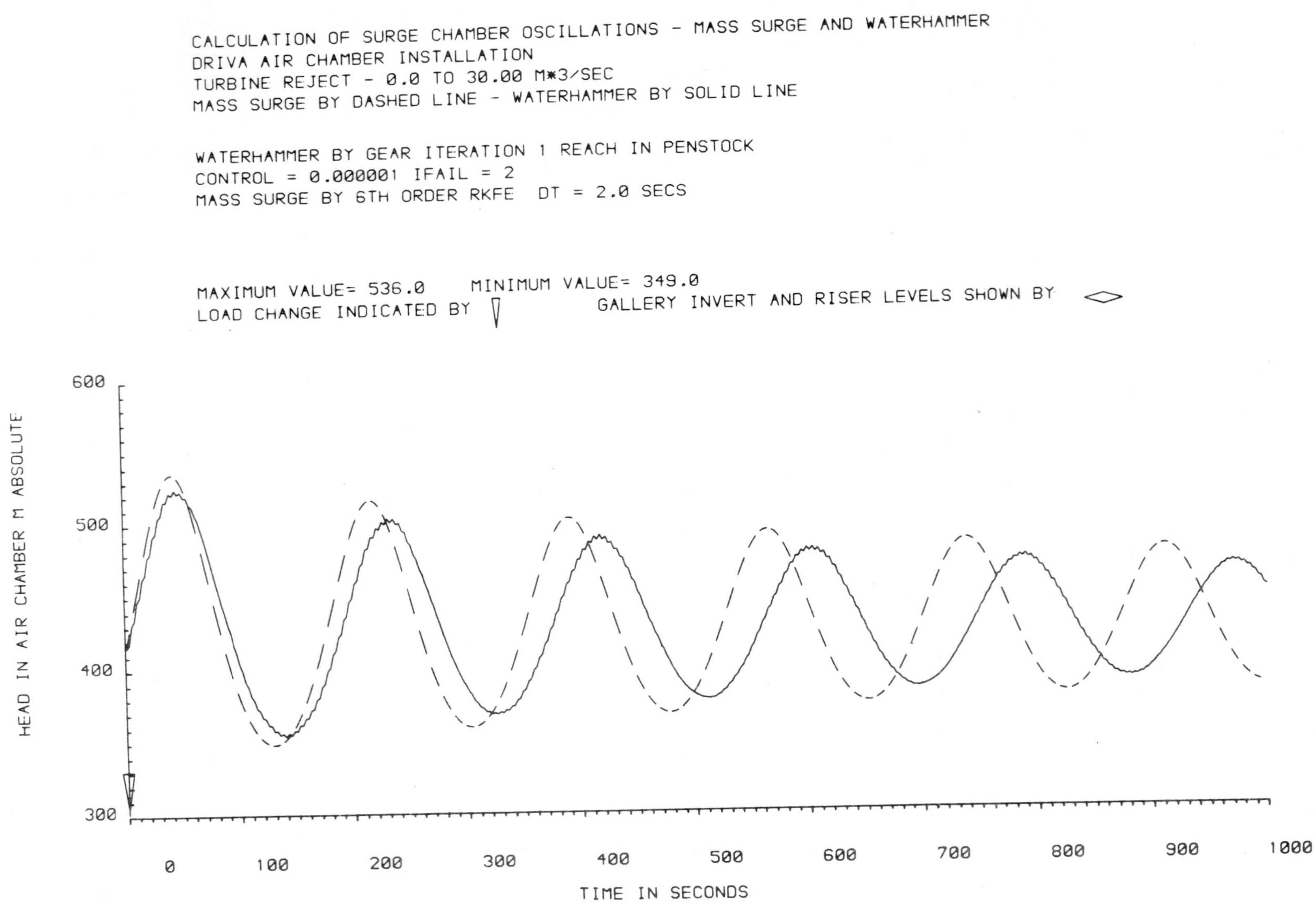

Fig. 12 Driva Hydro Scheme - effective level in air chamber

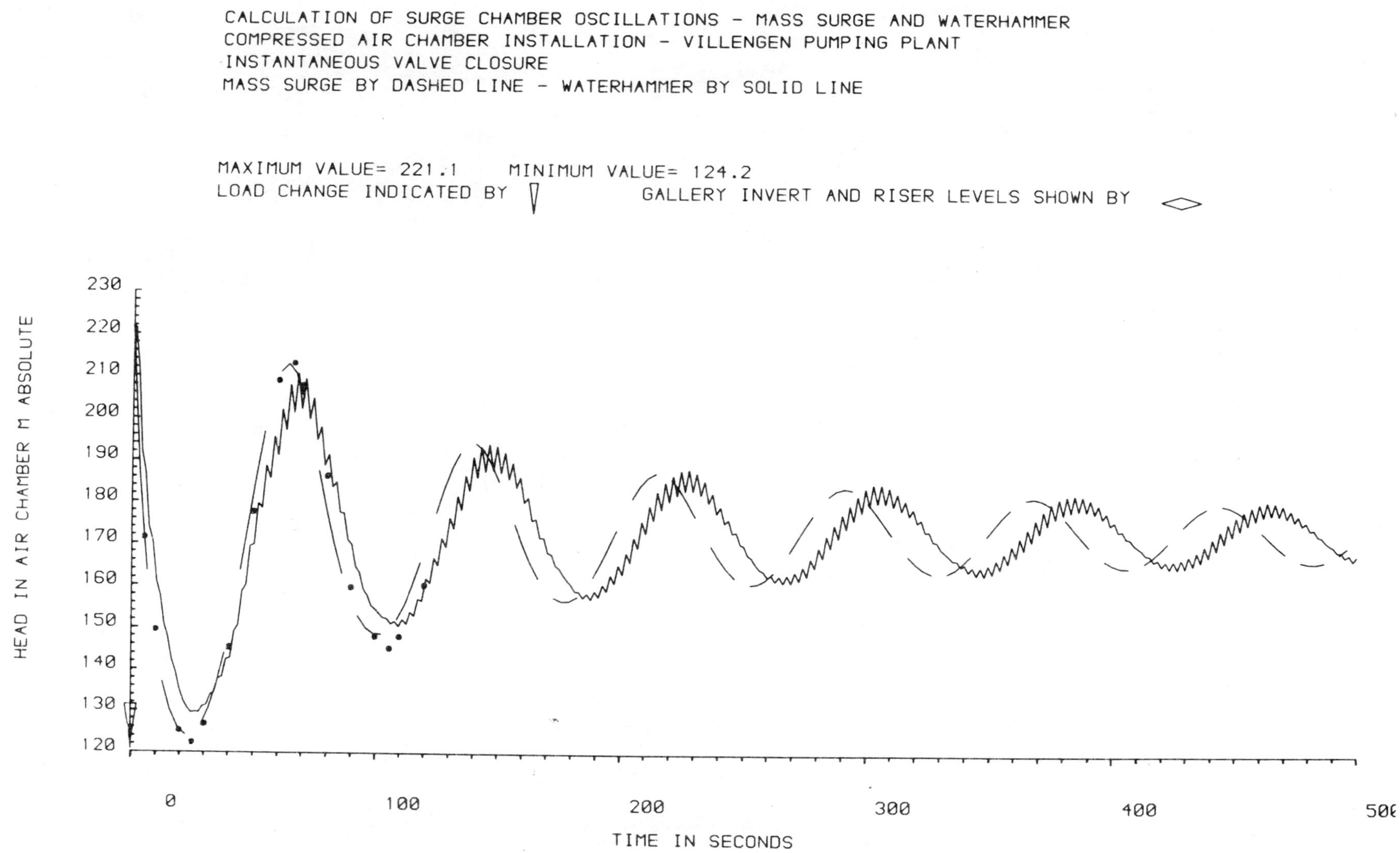

Fig. 13 Villingen Pumping Scheme - absolute head in chamber

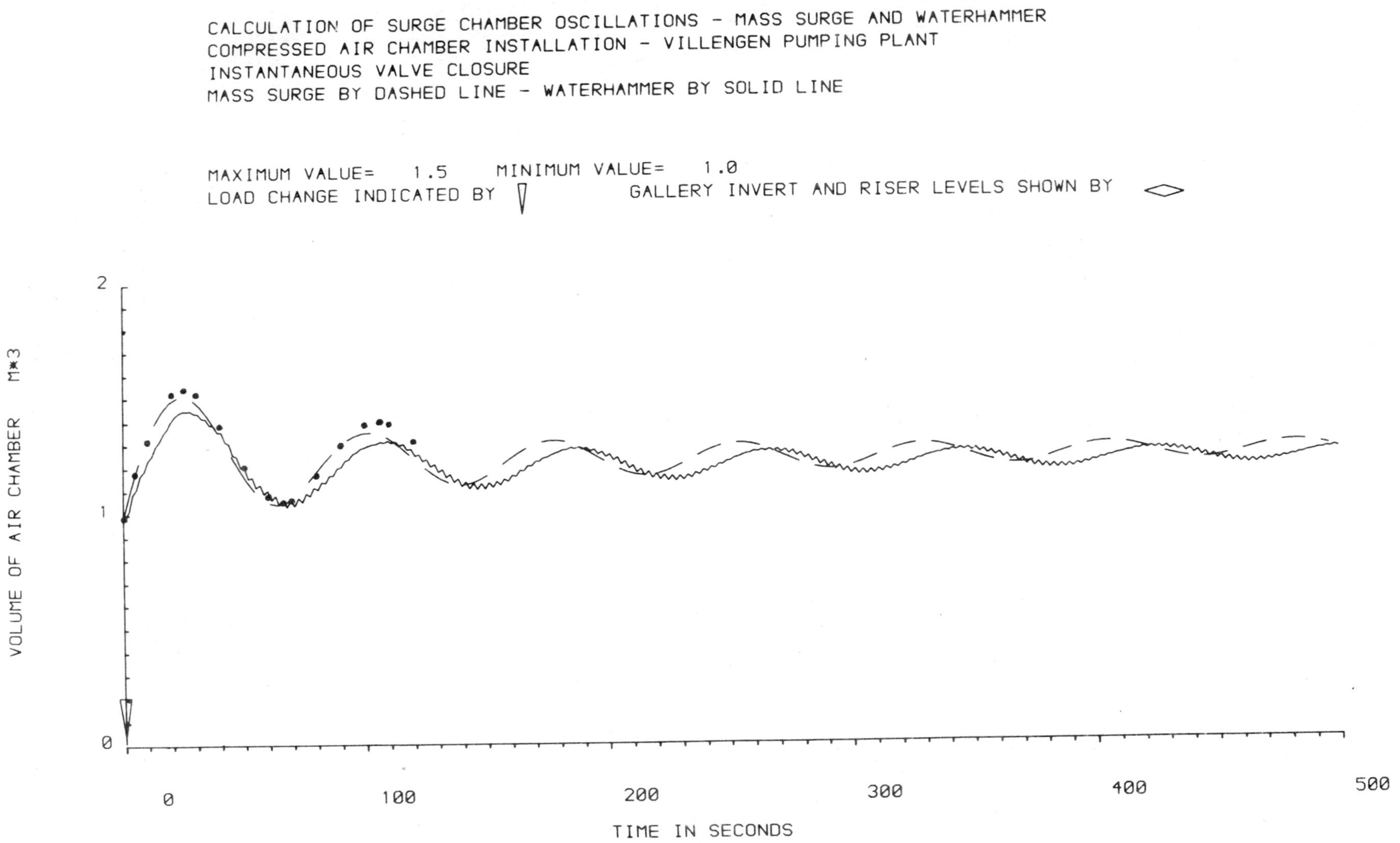

Fig. 14 Villingen Pumping Scheme - volume of air in air chamber

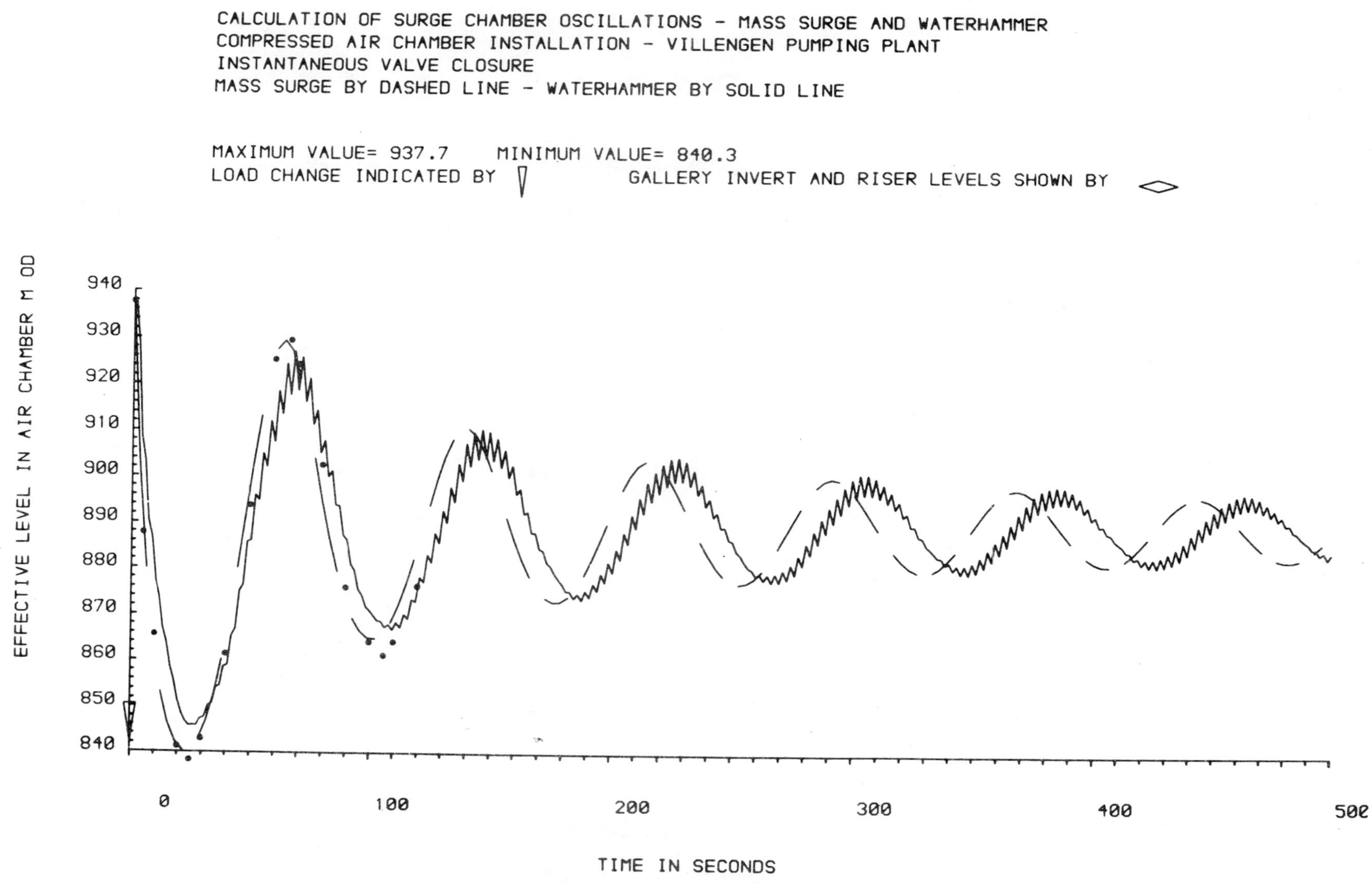

Fig. 15 Villingen Pumping Scheme - effective level in air chamber

Third International Conference on
PRESSURE SURGES
Canterbury, England
March 25-27, 1980

PAPER G4

EFFECT OF EQUATION FORMULATION ON THE PREDICTION OF MASS OSCILLATIONS IN CLOSED CONDUITS AND SURGE TANKS

A. Anderson

University of Newcastle upon Tyne, U.K.

and

J.F. Robbie

University of Aberdeen, U.K.

Summary

The 1972 and 1976 Pressure Surge Conferences confirm Jaeger's view that "the range of engineering problems connected with hydraulic transients has been considerably widened in the last 20 years" and that "new methods and accurate analyses of pressure waves are of paramount importance to modern engineering". In addition to the traditional hydroelectric applications, surge shaft mass oscillation studies increasingly are being undertaken for pumped-storage plant with more onerous operational requirements and for cooling water and other systems with more complex multiple shaft layouts. "Secondary effects" that were previously thought to be unimportant may now assume greater significance. Certain innovations in mass oscillation analyses relating to specific pumped-storage schemes were reported at the previous Conferences but there is a more general need to review mass oscillation formulations in order to resolve ambiguities arising in some earlier models. A formulation procedure is developed which facilitates accurate representation of all terms in the equations of complex multi-shaft systems. Particular attention is drawn to junction losses. As other studies have demonstrated their influence on surge predictions the emphasis is instead on elucidating their physical role and how it may be reflected in inconsistencies in simplified formulations.

Held at the University of Kent at Canterbury, U.K.

Conference Organised by BHRA Fluid Engineering, Cranfield, Bedford, England.
In conjunction with The City University, London.

NOMENCLATURE

A	Mean cross sectional area along low pressure conduit (length = L)
A_c	Cross sectional area in low pressure conduit immediately adjacent to surge shaft junction
A_i	Flow cross sectional area at specific section i (i = a, b, etc)
A_o	Area of throttle at base of surge shaft
A_s	Horizontal free surface area in surge shaft at level y_f or Z
$\overline{A}_s$	Mean cross sectional area up surge shaft (height = $y_f - y_d$)
$B(T)$	Bernoulli function (constant at any instant along a streamline)
c	Nondimensional factor for unsteady loss coefficients (Table 3)
D	Pipe diameter
$f_{u/s}$	Darcy friction factor in unsteady/steady flow
F_{ij}	Friction head loss for flow from section i to j (ij = ab, bc, etc)
g	Gravitational acceleration
H_{ij}	Sum of local head losses for flow from section i to j (ij = ab, bc, etc)
J_{ij}	Junction loss for flow from branch i to j (ij = ab, etc and particularly cs - low pressure conduit to surge shaft, cp - low pressure conduit to penstock, sp - surge shaft to penstock)
K	Head loss coefficient for low pressure conduit (area A)
K_o	Throttle head loss coefficient (area A_o)
L	Length of low pressure conduit (area A)
$\underline{n}$	Unit vector tangential to streamline
P	Total pressure across any conduit section (Eq 2)
P	Period of oscillation of surge tank (Table 4)
p_{at}	Atmospheric pressure
p_i	Fluid pressure across specific section i (i = a, b, etc)
Q	Volumetric flow rate along low pressure conduit
$Q_{a/b/c}$	Volumetric flow rate in branch a/b/c at junction (Fig 1)
Q_i	Initial steady state value of Q and Q_m
Q_m	Machine flow (volumetric flow rate in penstock)
Q_s	Volumetric flow rate into surge chamber
T	Time (independent variable)
V	Mean flow velocity across pipe section of diameter D
$\underline{v}$	Velocity tangential to streamline
y_i	Level above some common datum at section i (i = a, b, etc)
Z	Surge shaft surface level measured above reservoir level (Fig 2)
Z_i	Initial steady state value of Z corresponding to flow Q_i
ΔT	Time step in numerical solution (Table 4)
ρ	Density of fluid

1. INTRODUCTION

Examination of many standard texts (Ref 7, 19, 22, 31, 33, 45, 47, 51) reveals that the formulation of incompressible fluid, "rigid column", mass oscillation equations for surge tank design has remained virtually unchanged since the publication of Calame and Gaden's monograph (Ref 14) over 50 years ago.

Some recent surge shaft design studies (Ref 43, 48), however, have exposed two areas in which traditional approaches to equation formulation have limitations :

(i) The incorporation of new head loss data at the surge shaft to conduit junction (eg Ref 5, 6, 10, 28, 35, 36, 40, 41) reopens the debate over the inclusion or omission of a tunnel velocity head term in mass oscillation equations, previously tentatively discussed by Jaeger (Ref 31 pp 6 - 7, 88 - 89)

(ii) The analysis of complex multiple shaft layouts (eg Ref 1, 4) requires a general derivation of mass oscillation equations that will include all terms including those which have previously tended to be "lost" in simplified derivations but which may assume significance in contemporary design studies.

Through re-examining mass oscillation equations in order to overcome these limitations an approach has been developed which can be applied to systems of any degree of complexity but which, in any given case, allows appropriate simplification to be achieved through conscious planned deletion of terms rather than by default. Thus it becomes possible to take full advantage of advances in the use of numerical procedures (Ref 11 - 13) and the continually expanding base of hydraulic loss information to achieve more reliable predictions.

2. DERIVATION OF MATHEMATICAL MODELS FOR MASS OSCILLATIONS

2.1 Approaches to formulating mass oscillation equations

The two usual approaches to deriving mass oscillation equations are :

(a) As eg in Jaeger (Ref 31), the hydrodynamic Bernoulli equation :

$$(1/g) \int (\partial \underline{v}/\partial T) \cdot \underline{n} dL + \underline{v} \cdot \underline{v}/2g + \int dp/\rho g + (y - F) = B(T)/g \quad (1)$$

may be integrated along streamlines drawn through the conduit system, applying appropriate integration limit conditions.

(b) Following Calame and Gaden (Ref 14), a momentum equation :

$$\rho \int A(\partial \underline{v}/\partial T)\ dL + \int d(PA\underline{n}) = 0 \quad (2)$$

can be applied across appropriate sections of the conduit system, again applying appropriate limiting values.

The validity of such approaches is not in question but they can involve practical difficulties in application, particularly in visualising the correct forms of the limiting conditions and streamline paths, eg at junctions. These have tended to be less critical for simple layouts examined in the past but they do become compounded and important for the more complex systems found in contemporary studies.

Fortunately these difficulties can be minimised if the mathematical model is built up out of standard "components", using one-dimensional, cross-section averaged equations for each element, linked together to represent a system of any complexity. This approach, which has been partially developed before (Ref 7 Section 16), makes it possible to systematically formulate each element in a conduit system without having to approximate the details of any local streamline distributions, etc.

In any approach, however, care must be taken to ensure that each element is described in full, without simplification at the initial stage. This is especially important at junctions where the treatment is often deficient because it is not always realised that "head losses" associated with division or combination of flows do not always behave in an analogous manner to friction and other local shape losses.

2.2 The role and physical significance of junction losses

Where division or combination of flow occurs at a junction of more than two conduits (eg Fig 1) it is necessary to provide not only a continuity equation :

$$Q_a = Q_b + Q_c \tag{3}$$

but also to incorporate the juction head losses J_{ab}, J_{ac} and J_{bc} (Ref 22, 23) defined as (Fig 1a) :

$$J_{ab} = \{(p_a/\rho g) + y_a + (Q_a^2/2gA_a^2)\} - \{(p_b/\rho g) + y_b + (Q_b^2/2gA_b^2)\} \tag{4}$$

$$J_{ac} = \{(p_a/\rho g) + y_a + (Q_a^2/2gA_a^2)\} - \{(p_c/\rho g) + y_c + (Q_c^2/2gA_c^2)\} \tag{5}$$

$$J_{bc} = \{(p_b/\rho g) + y_b + (Q_b^2/2gA_b^2)\} - \{(p_c/\rho g) + y_c + (Q_c^2/2gA_c^2)\} \tag{6}$$

Published data on junctions show that associated head losses differ from normal shape and friction losses in the following respects :

(a) Junction "losses" may not simply represent loss of energy but may include recoverable velocity head components. Manifestations of this phenomenon include apparently negative junction losses (see Ref 1 for a literature survey) and losses occurring across branches which have no flow between them.

(b) The nature of junction losses alters with the flow configuration at the junction (orientation and magnitude of flow streams).

These points can be illustrated with respect to the two limiting regimes for the case of dividing flow at a typical "tee" junction (Fig 1b) :

(i) $Q_b = 0$:

In this case the lateral branch could be idealised as a simple piezometric tapping on to the undivided flow through the main. If this is so then :

$$\{(p_b/\rho g) + y_b\} = \{(p_a/\rho g) + y_a\} = \{(p_c/\rho g) + y_c\} \tag{7}$$

which substituted into Eq 4 - 6 yields idealised values for the junction losses :

$$J_{ab} = - J_{bc} = + (Q_a^2/2gA_a^2), \; J_{ac} = 0 \tag{8A}$$

Actual experimental results for a right-angled junction of pipes of equal size (Ref 22, 23) yield confirmatory values :

$$J_{ab} \simeq - J_{bc} \simeq + 0.9\,(Q_a^2/2gA_a^2), \; J_{ac} \simeq 0 \tag{8B}$$

with the small deviations from the idealised Eq 8A being caused by turbulent flow separation at the corners of the pipe intersection (Ref 18, 22, 23). Eq 8 indicates that even with no flow from the main to the branch an apparent "loss" of approximately one velocity head exists. This, however, is in turn "recoverable" on returning to the main downstream from the branch. These illustrate "losses" which do not represent degradation of energy in the fashion normally associated with other head losses.

(ii) $Q_c = 0$:

With all the flow turning through a right angle into the lateral branch then pipe 'c' could be idealised as a pitot total head tapping into the main flow of pipe 'a' and the flow course could be idealised as a mitre elbow bend. Thus, if pipe 'c' does "measure" the stagnation head of pipe 'a' :

$$\{(p_a/\rho g) + y_a + (Q_a^2/2gA_a^2)\} = \{(p_c/\rho g) + y_c\}$$

$$= \{(p_b/\rho g) + y_b + (Q_b^2/2gA_b^2)\} + \text{mitre elbow loss} \tag{9}$$

which substituted into Eq 4 - 6 yields the idealised values for the junction losses :

$$J_{ac} = 0,\ J_{ab} = -\ J_{bc} = \text{mitre elbow loss} \simeq +\ (Q_a^2/2gA_a^2) \tag{10A}$$

It might be suspected that this would be a less satisfactory idealisation than case (i) above and actual experimental results confirm this, showing a greater divergence from the ideal than before :

$$J_{ac} \simeq +\ 0.25\ (Q_a^2/2gA_a^2),\ J_{ab} \simeq +\ (Q_a^2/2gA_a^2),\ J_{bc} \simeq -\ 0.75\ (Q_a^2/2gA_a^2) \tag{10B}$$

Nevertheless Eq 10 indicates a more apparently normal result for J_{ab} than Eq 8, leading to degradation of flow energy around the bend.

Measured data which illustrate these two roles played by local losses at junctions are widely reported in the literature and the work of Gardel (Ref 22, 23) is recommended for its comprehensive coverage of the possible flow regimes at junctions. Junction loss coefficients derived from Gardel's work for a pumped-storage scheme surge shaft are presented in Fig 3. Not only do these validate Eq 7 - 10 in specific instances and demonstrate the existence of "negative losses" but they indicate the effect of flow regime on the losses. Similar effects have been recorded for surge tanks at Raccoon Mountain (Ref 28), Foyers (Ref 10), Dinorwic (Ref 6), etc.

2.3 Example - the single simple surge tank layout

The proposed method is well illustrated by example and, although it is most effective with complex layouts, the simple case of a single surge tank (Fig 2) is chosen because it can be used to reveal the source of the uncertainty over the velocity head mentioned in the Introduction above.

The basis of the procedure, illustrated in Fig 2 and Tables 1 and 2, consists in essence of five steps :

(i) Any conduit/shaft system can be made up from a number of components as in Fig 2. Each component can be mathematically modelled by equations as in Table 1, taking care not to oversimplify the representation at this stage, especially at the junction (Eq 3 - 6).

(ii) The Bernoulli equations for the constituent parts in Table 1 can be linked together by addition. Any convenient grouping can be adopted. That illustrated in Table 2 merely follows tradition for this conduit layout.

(iii) In addition to the equations obtained by this procedure continuity equations at junctions and surfaces free to move are selected from Table 1 to complete the set.

(iv) It may be possible to clarify the equation set, eg following conventional presentation for the example refer the surge shaft level (Z) to the constant upper reservoir level (y_a) as datum :

$$Z = \left[p_{at}(y_f)/\rho g + y_f\right] - \left[p_{at}(y_a)/\rho g + y_a\right] \simeq (y_f - y_a) \quad (11)$$

$$dZ/dT \simeq (dy_f/dT - dy_a/dT) = dy_f/dT \quad (12)$$

(v) A principal advantage of the method, as illustrated in Table 2, is that all the non-cancelling terms are systematically incorporated with the correct sign, provided that each of the individual components is initially correctly and adequately described. The equations may then subsequently be scrutinised for simplifying assumptions appropriate to the problem under investigation, eg : where the surge shaft has a very much smaller volume than the low pressure conduit coupled with a larger cross section (ie $LA >> (y_a - y_d)\bar{A}_s$ and $\bar{A}_s > A$) it is common to ignore the shaft friction losses (F_{df}) and inertia term (in dQ_s/dT), though these terms may be significant (eg Ref 4, 8, 21, 44). Simplifying at this final stage rather than at step (i) helps to ensure that potentially significant "secondary effects" are ignored by intention rather than by default.

2.4 Notes on head losses in unsteady flow

(i) It is usual to combine turbulent losses together in terms of a loss coefficient and a single velocity head, eg :

low pressure conduit : $(H_{ab} + F_{bc} + H_{bc}) = (K/2gA^2)Q|Q|$ (13)

surge tank throttle : $H_{df} = (K_o/2gA_o^{\ 2})Q_s|Q_s|$ (14)

(ii) Using the form $Q|Q|$ rather than Q^2 for the velocity head ensures that the change in sign of losses with flow direction (which indicates irreversibility) is automatically implemented. The difference between dissipative ($Q|Q|$) and recoverable (Q^2) effects should be distinguished.

(iii) Bearing this in mind, and because as already noted junction losses (J_{cs}, J_{cp}, J_{sp}) depend on the flow regime at the junction (Ref 22, 23), they are best kept separate.

(iv) In prototype mass oscillation studies the loss coefficients in unsteady flow are generally taken as those applying at equivalent but steady flow rates. Sometimes (Ref 39, 50) unsteady loss coefficient formulas (see Table 3) are suggested for surge tank studies but the literature reveals that these have been obtained for pipe sizes and acceleration (or frequency) values which are not appropriate at full scale, though they may be for small models (Ref 1).

(v) Loss coefficients at equivalent steady flows may vary with flow direction, particularly for shape losses (eg expansion/contraction), and also with Reynolds' Number, though this latter effect is generally not significant.

3. DISCUSSION OF MASS OSCILLATION FORMULATIONS OBTAINED

3.1 The velocity head term in the mass oscillation equations

From Sections 2.3 and 2.4 above the mass oscillation equations for the single shaft scheme of Fig 2 are, in summary :

$$Q_s = Q - Q_m \quad (15)$$

$$dZ/dT = Q_s/A_s(Z) \quad (16)$$

$$(L/gA)dQ/dT = -Z - \left[(Q_s^{\ 2}/2gA_s^{\ 2}) + J_{cs}\right] - (K/2gA^2)Q|Q| - (K_o/2gA_o^{\ 2})Q_s|Q_s| \quad (17)$$

Corresponding to the unsteady conduit dynamic equation, Eq 17, there is a steady state ($d/dT = 0$, $Q_s = 0$, $Q = Q_m = Q_i$) equation :

$$Z_i = -(K/2gA^2)Q_i|Q_i| - J_{cs}(Q_i) \tag{18}$$

The above conduit dynamic equation, Eq 17/18, is as given by Gardel (Ref 22 pp 82 - 83). A number of recent texts come close to this formulation but either omit the surge shaft velocity head ($Q_s^2/2gA_s^2$) while including the junction loss J_{cs} (eg Ref 8, 26, 29, 38, 46) or vice versa (eg Ref 7 pp 82 - 84, 17 p 388, 44). Traditionally, however, the conduit dynamic equation for Fig 2 is usually given in one of the following three forms in place of Eq 17/18 :

(a) The velocity head term is taken as the velocity head ($Q^2/2gA_c^2$) in the conduit at the base of the surge shaft :

$$(L/gA)dQ/dT = -Z - [Q^2/2gA_c^2] - (K/2gA^2)Q|Q| - (K_o/2gA_o^2)Q_s|Q_s| \tag{19A}$$

$$\text{steady :} \quad Z_i = -(K/2gA^2)Q_i|Q_i| - (Q_i^2/2gA_c^2) \tag{20A}$$

(b) The velocity head term is omitted altogether :

$$(L/gA)dQ/dT = -Z - (K/2gA^2)Q|Q| - (K_o/2gA_o^2)Q_s|Q_s| \tag{19B}$$

$$\text{steady :} \quad Z_i = -(K/2gA^2)Q_i|Q_i| \tag{20B}$$

(c) The velocity head term as in (a) is included in the dissipative conduit loss term :

$$(L/gA)dQ/dT = -Z - Z_i(Q/Q_i)|Q/Q_i| - (K_o/2gA_o^2)Q_s|Q_s| \tag{19c}$$

steady : Z_i as Eq 20A.

Finally there are texts (eg Ref 32 pp 162 - 167) where the unsteady equation can be in form (b) without the conduit velocity head, Eq 19B, but the steady equation is in form (a) with the velocity head, Eq 20A. The resulting discontinuity between steady and unsteady states is probably concealed in practice by the change imposed on Q_m to create the transient.

Attempts have been made to rationalise this situation by distinguishing between surge shafts which are assumed to act as either piezometers, or as reservoirs (eg Ref 7 pp 82 - 83, 51 p 44). Following this principle consider each of the three forms of Eq 19/20 in turn :

Form (a) with Eq 19A/20A :

For a surge shaft which in steady conditions could be idealised as a piezometric tapping on the conduit (Fig 4a) Eq 20A approximates the correct initial level by Eq 7 - 8, but for unsteady conditions with flow in the shaft ($Q_s \neq 0$) it is apparent from the difference between the two roles of junction losses discussed before in Section 2.2 that the idealisation breaks down and the shaft cannot continue to act as a piezometer. Thus Eq 19A should not continue to contain the recoverable conduit velocity head but should take account of the change in junction behaviour from piezometric to non-piezometric as Eq 17 does.

Form (b) with Eq 19B/20B

For a surge tank acting as a reservoir at the end of the conduit (Fig 4b) or idealised as a pitot-tapping on the conduit (Fig 4c) then in steady conditions Eq 20B approximates the correct initial level by Eq 9 - 10. For unsteady conditions with flows in the shaft ($Q_s \neq 0$) it would certainly cease to act as a pitot-tapping, though probably continuing to act as a reservoir.

Form (c) with Eq 19C/20A

This can be considered as either :

(i) a special case of form (b) in which the reservoir intake loss (H_{ab}) has been removed from the loss coefficient K in Eq 13 and arbitrarily expressed as a separate velocity head, or

(ii) an incorrect application of form (a) in that a recoverable velocity head has been made dissipative by expressing it in the irreversible form $Q|Q|$ rather than the reversible Q^2.

In summary, therefore, it is apparent that :

(i) Junction losses can play an important role in surge shaft response on account of their idiosyncratic and changing behaviour. Although adequate simulation may require interpolation between tabulated discrete values junction losses are essential in ensuring consistency in formulation.

(ii) Appropriate formulation for different types of element together with appropriate junction losses generalise the mass oscillation dynamic equations (eg Eq 17) for a range of shaft configurations more accurately than is possible with idealised and simplified models of junction behaviour.

(iii) In particular, the application of junction losses reveals the cause of the confusion over the velocity head term which arises from the above points.

It follows, therefore, that Eq 17 with both junction loss J_{cs} and surge shaft velocity head ($Q_s^2/2gA_s^2$) is the preferred form of the conduit dynamic equation.

3.2 Thoma's surge shaft stability theory

Thoma's classic study of surge shaft stability was concerned with incipient instability for small deviations from the steady state (Ref 14, 31). Though it has long been realised that this limits the accuracy and thus applicability of the result it has proved to be a useful guide in surge shaft design.

Thoma's study was developed using Eq 19A (form (a) above including the conduit velocity head at the base of the shaft) instead of the now preferred Eq 17. Fortunately this does not invalidate the classic analysis because :

(a) Steady state conditions derived from Eq 19A/20A are very nearly the same as the more correct Eq 17/18 because as it happens the junction loss J_{cs} for a simple shaft is approximately equal to the conduit velocity head ($Q^2/2gA_c^2$) as in Eq 7 - 8. Since the stability analysis is valid only for small deviations about the steady state then it seems reasonable to assume that this value for J_{cs} remains valid in the dynamic equation, Eq 17, under these circumstances (Ref 3).

(b) Terms of order greater than one in the small variations can be ignored so that the additional surge shaft velocity head in Eq 17 does not appear in the linearised equations (Ref 3).

Thus the linearised equation for small amplitude deviations from the steady state is effectively the same whether derived from the traditional Eq 19A or from the more correct Eq 17 so that the Thoma result remains unaltered.

3.3 The "venturi contraction" surge tank

A form of surge tank with a venturi contraction occurring in the conduit immediately beneath it (Fig 5) has been described (eg Ref 30, 31 p 48, 45 p 125). Where Eq 19A or C are assumed to represent the conduit dynamic equation it is tempting to conclude that increasing the conduit velocity head ($Q^2/2gA_c^2$) by reducing A_c at a venturi contraction would decrease surge amplitudes and contribute to shaft stability. Jaeger (Ref 30) seems to confirm this when he reports that laboratory models of such devices have demonstrated additional damping of surges. Examination of the preferred Eq 17, however, suggests that additional damping is due to an increase in the junction loss J_{cs} which is proportional to the velocity head at the base of the shaft during normal turbine operation. In other words the venturi contraction acts as an additional dissipator similar in action to a throttle in the shaft.

At steady low levels in experimental models of such surge shafts sudden irregular jumps in shaft level have been recorded. It has been tentatively suggested that these might be attributed to changes in the energy level "measured" by the shaft, ie the system fluctuates between Fig 4a (Eq 19A or C) and Fig 4c (Eq 19B) with changes from "the piezometric level to the energy level" (Ref 30, 31 pp 88 - 89), but the mechanism for this cannot be explained by either formulation of Eq 17 or 19.

A similar phenomen has been demonstrated, however, in a study (Ref 18) of piezometric measurement tappings. In that case it was demonstrated experimentally that turbulent flow fluctuations at the tapping/pipe interface were responsible and that the larger the size of tapping the greater the effect. At a surge shaft the shaft/conduit interface is equivalent to a very coarse "tapping" and the effect would be accentuated if a venturi contraction were present. It is anticipated that the effect would be diminished in full size prototypes where the relative magnitude of turbulent fluctuations could be less and the inertia of the water mass in the shaft proportionately greater.

4. APPLICATION OF JUNCTION LOSSES IN SURGE STUDIES

The influence of junction losses may assume greater importance in multiple shaft schemes than in the basic case of Fig 2. Attention has been drawn (Ref 4) to a novel type of surge shaft instability which occurs in two shaft arrangements and a study (Ref 3) of the incipient instability for very small disturbances has been carried out using the Thoma assumptions as in Section 3.2 above, but the prediction of stability is strongly influenced by the inclusion of junction losses as is shown by Fig 6 (Ref 52).

Even for the single shaft case (Fig 2), however, previous studies illustrate behaviour reflecting the above discussion of the properties of junctions. For example a comparative error study (Ref 29) has indicated that :

(a) junction losses do not always have a negligible influence on mass oscillations in surge tanks or on the transmission of waterhammer pressures, and

(b) deviations in solutions due to neglecting junction losses do not necessarily err on the safe side.

Another study (Ref 28) has suggested that surging behaviour may be influenced at the design stage by altering the form of the surge shaft junction and that in certain cases the ultimate surge amplitudes depend to a great extent on the head loss characteristic of the surge tank/conduit junction.

Junction behaviour may have a more marked influence on other elements in the hydraulic system than on surge shaft levels. Experience with pumped-storage schemes which have a rapid load-on requirement (eg Ref 43) shows that oversimplification at surge shaft throttle and junction may introduce significant errors into prediction of turbine loading performance by neglecting the rapid variations in loss coefficients as the flow regime at the junction alters.

More generally, using the traditional forms of the equations, Eq 19A or B, Escande (Ref 20) and others (eg Ref 9) have shown that, while instantaneous closure gives the worst upsurge for unthrottled ($K_o = 0$) surge tanks it does not do so for throttled surge tanks ($K_o \neq 0$), eg Table 4°. Since real surge shafts will always have some loss associated with them due to the junction effect then the concept of the instantaneous closure cannot always be relied on to give safe results.

The presence and influence of losses at a junction is illustrated by the example of Table 6 which presents results from a study (Ref 34) in which observed surges in a hydraulic model of the Foyers pumped-storage scheme were compared with predicted surges using :

(i) a simplified representation of the junction using a conventional throttle loss effect $(K_o/2gA_o^2)Q_s|Q_s|$ as in Eq 19, and

(ii) tee-junction losses as in Eq 17 based on Gardel's work (Ref 23) and presented in Fig 3.

Use of appropriate junction losses as in Fig 3 may yield not only improved accuracies of surge extremes, as shown in Table 6, but also better representation of the oscillation curve between extremes. Such improvement can assume additional significance where it is required to study sequences of superimposed load changes, in eg contemporary pumped-storage schemes (Ref 43, 48).

Notwithstanding the above examples, however, it also remains true that in the past good results have been obtained using the traditional forms of the equations which ignore junction losses, Eq 19A or B, typically within 5% of field observations. Analysis of a study (Ref 2) of 9 different schemes, plus those in Tables 4 - 6, suggest that among the reasons for this are :

(i) When field results are quoted, as for Tables 4 and 5, it is often difficult to obtain also details of the surge shaft/conduit junction for comparison purposes, but it might be expected that Eq 19B should give good results where the surge shaft is of the "reservoir" type (Fig 4b). It is not always easy, however, to predict when the traditional idealisations will give good results.

(ii) For simple single shaft layouts (Fig 2) it is often noted that predictions for load rejections are more accurate than for load acceptances. This is especially true for relatively rapid closures because, with zero flow in one branch (the penstock) the junction losses become purely dissipative with a constant loss coefficient (as case (ii) in Section 2.2 above), ie again the traditional equations might be expected to give good results.

5. CONCLUSIONS

Although this is not the first study to suggest that junction losses are a most important "secondary effect" which should be included in the prediction of surge shaft level oscillations, surprisingly few comprehensive junction loss studies have been undertaken for surge tanks. It is hoped that this paper will encourage the collection and publication of such data.

Rather than cover the ground of previous comparisons of solutions with and without junction effects an attempt has been made to explain the special nature of junction losses and how their neglect has contributed in the past to certain puzzles (eg the nature of the velocity head term) or misunderstandings (eg the venturi contraction surge tank) in mass oscillation theory.

The inclusion of junction losses and other terms in mass oscillation equations can be facilitated by the formulation process illustrated in the paper. The method is simple and flexible, enabling the assumptions made to be readily identified (eg Table 1) and reducing the possibility of errors in including or omitting terms or in the signs of terms (eg Table 2).

Acknowledgements

This paper is partly based on studies (Ref 1) at the University of Aberdeen supported by the Science Research Council. Considerable assistance in its preparation was received from the Partners of Merz and McLellan, Consulting Engineers, Newcastle upon Tyne, UK.

REFERENCES

1. Anderson, A: "On the formulation of problems in hydraulic transients". PhD Thesis. University of Aberdeen. (1976).

2. Anderson, A: "Surge tank solutions using programmable pocket calculators". Water Power and Dam Construction, 30, 7, pp 50 -59. (July 1978).

3. Anderson, A: "A novel type of surge shaft instability". Proceedings, Institution of Civil Engineers, 67, 2, pp 695 - 706. (Sept 1979).

4. Anderson, A and Robbie, JF: "Some novel experiences in the computer aided design of Dinorwic pumped-storage surge system". Proc. 2nd International Conference on Pressure Surges, London. Paper B2. Organised by Brit. Hydromech. Res. Assoc., Cranfield. (22 - 24 Sept 1976).

5. Arato, EG: "Tedzani hydro-electric scheme. Part 2. Hydraulic model study of the surge chamber". RR 1133. Brit. Hydromech. Res. Assoc., Cranfield. (May 1972).

6. Arato, EG and Wightman, MJ: "Hydraulic model studies of Dinorwic surge chamber". RR 1406. Brit. Hydromech. Res. Assoc., Cranfield. (April 1977).

7. Aronovich, GV, Kartvelishvili, NA and Lyubimtsev, YK: "Water hammer and surge tanks". Trans. D. Ledermann. Israel Program for Scientific Translations, Jerusalem. (1970).

8. Barbarossa, NL: "Hydraulic analysis of surge tanks by digital computer". Journal of the Hydraulics Division, Proc. ASCE, 85, HY4, pp 39 - 78. (April 1959).

9. Barr, JN: "A comparative study of surge chamber design using an interactive digital computer facility". BSc(Eng) Honours Dissertation (Mechanical). University of Aberdeen. (1975).

10. Boldy, AP: "Foyers pumped-storage scheme". RR 1099. Brit. Hydromech. Res. Assoc., Cranfield. (May 1971).

11. Bullough, JBB: "The numerical solution of the equations of mass ocillation in a pumped-storage scheme". PhD Thesis. University of Aberdeen. (May 1975).

12. Bullough, JBB and Robbie, JF: "The accuracy of certain numerical procedures when applied to the solution of ordinary differential equations of the type used in the digital computer prediction of mass oscillation in closed conduits". Proc. International Conference on Pressure Surges, Canterbury. Paper A6. Organised by Brit. Hydromech. Res. Assoc., Cranfield. (6 - 8 Sept 1972).

13. Bullough, JBB and Robbie, JF: "Theoretical and numerical analysis of the error in the numerical solution of the equations of mass oscillation in pumped-storage schemes". Proc. 2nd International Conference on Pressure Surges, London. Paper H3. Organised by Brit. Hydromech. Res. Assoc., Cranfield. (22 - 24 Sept 1976).

14. Calame, J and Gaden, D: "Theory of surge chambers". (Théorie des chambres d`équilibre). La Concorde, Lausanne and Gauthier-Villars, Paris. (1926). (In French). Trans. into English by O. Elsden and FWA Whitaker, Sir Alexander Gibb and Partners, London. (1950).

15. Carstens, MR and Roller, JE: "Boundary shear stresses in unsteady turbulent pipe flow". Journal of the Hydraulics Division, Proc. ASCE, 85, HY2, pp 67 - 81. (Feb 1959).

16. Daily, JW, Hankey, WL, Olive, RW and Jordaan, JM: "Resistance coefficients for accelerated and decelerated flows through smooth tubes and orifices". Transactions, American Society of Mechanical Engineers, 78, 5, pp 1071 - 1077. (1956).

17. Daugherty, RL and Franzini, JB: "Fluid mechanics with engineering applications". 6th edn, McGraw Hill, Tokyo. (1965).

18. Emmett, WW and Wallace, JR: "Errors in piezometric measurement". Journal of the Hydraulics Division, Proc. ASCE, 90, HY6, pp 45 - 61. (Nov 1964).

19. Escande, L: "New methods for surge chamber calculations". (Méthodes nouvelles pour le calcul des chambres d`équilibre). Dunod, Paris. (1950). (In French).

20. Escande, L: "Superposed oscillations in ordinary constant section or throttled surge tanks". (Oscillations superposées dans les cheminées d`équilibre à section constante ordinaires ou à étranglement). La Houille Blanche, Numéro Spéciale A, pp 283 - 292. (May/June 1955). (In French).

21. Fanelli, M: "Prototype tests on surge tank as compared with numerical analysis". Proc. International Symposium on Waterhammer in Pumped-Storage Projects, Chicago. pp 144 - 153. American Society of Mechanical Engineers, New York. (Nov 1965).

22. Gardel, A: "Surge chambers". (Chambres d`équilibre). F. Rouge et Cie, Lausanne. (1956). (In French).

23. Gardel, A and Rechsteiner GF: "Head losses in tee junctions of circular section conduits". (Les pertes de charge dans les branchements en té des conduits de section circulaire). Bulletin Technique de la Suisse Romande, 25, pp 363 - 391. (12 Dec 1970). (In French).

24. Gibson, AH and Cowen, W: "A comparison of the results of observations on surge tank installations and on their scale models; with an investigation of the dead-beat surge tank, and of surge tanks of non-uniform cross section". Minutes of Proceedings, Institution of Civil Engineers, 235, 1, pp 327 - 351, 398 - 432. (1932 - 33).

25. Gvazava, GN: "Application of the electronic analog computer for calculation of unsteady flow in conduits and surge tanks of hydroelectric plants". Proc. 9th Convention International Association for Hydraulic Research, Dubrovnik. pp 732 - 739. (Sept 1961).

26. Haindl, K: "Water hammer protection of low-head conduits and networks by air chambers with natural air content". Proc. International Conference on Pressure Surges, Canterbury. Paper B7. Organised by Brit. Hydromech. Res. Assoc., Cranfield. (6 - 8 Sept 1972).

27. Hirose, K and Oka, T: "The friction factors of unsteady pipe flows". Memoirs of the School of Engineering Okayama University, 4, 1, pp 9 - 12. (Sept 1969).

28. Hsu, S-T and Elder, RA: "Raccoon Mountain pumped-storage plant hydraulic transient studies". Transactions 6th Symposium Section for Hydraulic Machinery Equipment and Cavitation, Rome. Part 1, Paper G1. International

Association for Hydraulic Research, Rome. (1972).

29. Hussain Khan, M: "Surge chambers". (Chambres d'équilibre). SPES/Dunod, Lausanne. (1966). (In French).

30. Jaeger, C: "Present trends in surge tank design". Proceedings Institution of Mechanical Engineers, 168, 2, pp 91 -124. (1954).

31. Jaeger, C: "Fluid transients in hydro-electric engineering practice". Blackie, Glasgow and London. (1977).

32. James, ML, Smith, GM and Wolford, JC: "Analog and digital computer methods in engineering analysis". International Textbook Co, Scanton Penn. (1964).

33. Kleinschroth, A: "Hydromechanics of unsteady flow of free water surfaces and in pressure conduits". (Hydromechanik instationärer Bewegungsvorgänge bei freiem Wasserspiegel und in Druckleitungen). Mitteilungen, Institut für Hydraulik und Gewässerkunde, Technische Universitat München, 13. (1973). (In German).

34. Leask, DS: "A hydraulic and computational study of a typical pumped-storage scheme". BSc(Eng) Honours Dissertation. University of Aberdeen. (1978).

35. Li, Y-T: "Orifice head loss in the T-section of a throttled surge tank". Water Power, 24, 9, pp 326 - 334. (Sept 1972).

36. Li, Y-T: "Head losses in T-junctions and manifolds". Water Power, 25, 7, pp 266 - 272. (July 1973).

37. Liiv, UR: "Of losses for the unsteady motion of incompressible fluid in a rigid pressure conduit. Of hydraulic laws for decelerated motion of fluid in rigid cylindrical containers. Of hydraulic laws for accelerated motion of fluid in rigid cylindrical conduits". Trudy Tallinskogo Politekhnicheskogo Instituta, Series A, 223, pp 21 - 50. (1965). (In Russian).

38. Lin, C-N and Gladwell, JS: "Non-dimensional surge tank analysis". International Journal for Numerical Methods in Engineering, 4, 1, pp 23 - 37. (Jan - Feb 1972).

39. Macagno, EO and Macagno, M: "Effects of acceleration dependent resistance on oscillatory turbulent internal flows". Proc. 17th Congress International Association for Hydraulic Research, Baden-Baden. Vol 2, Paper A119, pp 351 - 358. Institut Wasserbau, Universität Karlsruhe. (1977).

40. Maione, U: "Head losses in the contracted section of a surge tank". (Perdite di carico nella strozzature di un pozzo piezometrico). L'Energia Elettrica, 38, 4, pp 330 - 338. (1961). (In Italian).

41. Maione, U: "Head losses due to sharp angled contractions in surge tanks - experimental research". (Perdite di carico delle strozzature a spigolo vivo dei pozzi piezometrici - ricerca sperimentale). L'Energia Elettrica, 45, 4, pp 237 - 253. (April 1968). (In Italian).

42. Martin, CS and Jackson, HC: "Combined surge tank and water-hammer analysis by digital computer". Water Power, 24, 4, pp 123 - 129. (April 1972).

43. Mawer, WT, Buchanan, RW and Queen, BB: "Dinorwic pumped-storage project - pressure surge investigations". Proc. 2nd International Conference on Pressure Surges, London. Paper B3. Organised by Brit. Hydromech. Res. Assoc., Cranfield. (22 - 24 Sept 1976).

44. Milano, V, Sassoli, F and Todini, E: "Simulation by means of the 'Continuous Simulation Modeling Program' of a mathematical model of transients in an elevated station equipped with an air chamber". (Simulazione mediante il

'Continuous Simulation Modeling Program' di un modello matematico di transitorio in un impianto elevatorio munito di camera d'aria). L'Energia Elettrica, 51, 6, pp 325 - 331. (1974). (In Italian).

45. Pickford, J: "Analysis of surge". Macmillan, London. (1969).

46. Popescu, M: "Mathematical model to study the stability of a system of two surge chambers". (Modèle mathématique pour l'étude de la stabilité d'un systeme de deux chambres d'équilibre). Comptes Rendus de l'Academie des Sciences, Paris, 273, A, pp 129 - 132. (July 1971). (In French).

47. Popescu, M: "Present problems in the field of surge tank hydraulics". (Probleme actuale in domeniul hidraulicii castelelor de echilibru). Hydraulic Researches 21, Institutul de Studii si Cercetări Hidrotehnice, Bucharest. (1969). (In Rumanian).

48. Robbie, JF and Robson, FM: "Computer aided design of surge chambers in pumped-storage conduit systems". Proc. International Conference on Pressure Surges, Canterbury. Paper E2. Organised by Brit. Hydromech. Res. Assoc., Cranfield. (6 - 8 Sept 1972).

49. Safwat, HH and van der Polder, J: "Friction - frequency dependence for oscillatory flows in circular pipe". Journal of the Hydraulics Division, Proc. ASCE, 99, HY11, pp 1933 - 1945. (Nov 1973).

50. Sarpkaya, T: "The effect of unsteadiness on hydrodynamic resistances and on mass oscillations in surge tanks". Transactions Engineering Institute of Canada, 3, 3, pp 96 - 99. (Nov 1959).

51. Stucky, A: "Surge chambers". (Chambres d'équilibre). École Polytechnique de l'Université de Lausanne. (1952). (In French).

52. Svehla, KM: "Hydraulic stability of a pumped-storage scheme with double surge shafts". BSc(Eng) Honours Dissertation. University of Aberdeen. (1979).

TABLE 1. BASIC COMPONENTS AND EQUATIONS FOR SINGLE TANK SCHEME (FIG 2)

COMPONENT	EQUATION FOR COMPONENT
1. Large upper	$p_a = p_{at}(y_a)$
reservoir	$dy_a/dT \simeq 0$ ie $y_a \simeq$ constant
2. Intake to conduit	$-\{(p_a/\rho g) + y_a\} + \{(p_b/\rho g) + y_b + (Q^2/2gA_b^{\,2})\} + H_{ab} = 0$
3. Low pressure conduit	$(L/gA)\,dQ/dT - \{(p_b/\rho g) + y_b + (Q^2/2gA_b^{\,2})\} + \{(p_c/\rho g) + y_c + (Q^2/2gA_c^{\,2})\} + F_{bc} + H_{bc} = 0$
4. Junction at the	$Q_s = Q - Q_m$
base of the	$-\{(p_c/\rho g) + y_c + (Q^2/2gA_c^{\,2})\} + \{(p_d/\rho g) + y_d + (Q_s^{\,2}/2gA_d^{\,2})\} + J_{cs} = 0$
surge shaft	$-\{(p_c/\rho g) + y_c + (Q^2/2gA_c^{\,2})\} + \{(p_e/\rho g) + y_e + (Q_m^{\,2}/2{}_gA_e^{\,2})\} + J_{cp} = 0$
(connecting 3 to 5 and 6)	$-\{(p_d/\rho g) + y_d + (Q_s^{\,2}/2gA_d^{\,2})\} + \{(p_e/\rho g) + y_e + (Q_m^{\,2}/2gA_e^{\,2})\} + J_{sp} = 0$
5. Riser leading	$[(y_f - y_d)/g\bar{A}_s]\,dQ_s/dT - \{(p_d/\rho g) + y_d + (Q_s^{\,2}/2gA_d^{\,2})\} + \{(p_f/\rho g) + y_f + (Q_s^{\,2}/2gA_s^{\,2})\} + F_{df} + H_{df} = 0$
to free surface	$p_f = p_{at}(y_f)$
surge chamber	$dy_f/dT = Q_s/A_s(y_f)$
6. High pressure penstock	etc

TABLE 2. COMBINATION OF BERNOULLI EQUATIONS FROM TABLE 1 (UPPER RESERVOIR TO SURGE SHAFT SURFACE)
NB: LHS for cancelling terms, RHS for non-cancelling terms

$$
\begin{aligned}
-\{(p_b/\rho g) + y_b + (Q^2/2gA_b^{\,2})\} &= \{p_{at}(y_a)/\rho g + y_a\} + H_{ab} \\
-\{(p_c/\rho g) + y_c + (Q^2/2gA_c^{\,2})\} + \{(p_b/\rho g) + y_b + (Q^2/2gA_b^{\,2})\} &= (L/gA)\,dQ/dT + F_{bc} + H_{bc} \\
+\ (p_c/\rho g) + y_c + (Q^2/2gA_c^{\,2})\} - \{(p_d/\rho g) + y_d + (Q_s^{\,2}/2gA_d^{\,2})\} &= +J_{cs} \\
+\{(p_d/\rho g) + y_d + (Q_s^{\,2}/2gA_d^{\,2})\} &= +\{p_{at}(y_f)/\rho g + y_f + (Q_s^{\,2}/2gA_s^{\,2})\} \\
&\quad +[(y_f - y_d)/g\bar{A}_s]\,dQ_s/dT + F_{df} + H_{df}
\end{aligned}
$$

$$
\begin{aligned}
0 = {} & (L/gA)\,dQ/dT + \{[p_{at}(y_f)/\rho g + y_f] - [p_{at}(y_a)/\rho g + y_a]\} + (Q_s^{\,2}/2gA_s^{\,2}) + (H_{ab} + F_{bc} + H_{bc}) \\
& + J_{cs} + (F_{df} + H_{df}) + [(y_f - y_d)/g\bar{A}_s]\,dQ_s/dT
\end{aligned}
$$

TABLE 3. FORMULAS FOR UNSTEADY TURBULENT FRICTION LOSS COEFFICIENTS FROM THE LITERATURE

$$f_u = f_s + c\ [(DdV/dT)/V^2]$$

REF.	VALUES OF FACTOR c	
	ACCELERATING FLOW (dV/dT)>0	DECELERATING FLOW (dV/dT)<0
16 (quoted 50)	0.02 to 0.03	1.24
15	0.31 to 0.449	
37	1.28	
27 (quoted 39)	0.46	0.274
49	-0.088 to +0.044 (averages)	

TABLE 4. LOAD REJECTION FOR MINGECHAUR HEP (AZERBAIJAN SSR)

Source	Closure time (Period P≃72.5s)	1st Upsurge (m)	1st Downsurge (m)
Field results (Ref 25)	not given	+7.20	-4.05
Analog computer (Ref 25)	P/∞ = instant	+7.37	-3.84
Using programmable pocket calculator solution (Ref 2) with time step $\Delta T \simeq P/24 = 3s$ and linear reduction in machine flow Q_m	P/∞ = instant	+7.32	-4.00
	P/48	+7.55	-4.06
	P/24	+7.66	-4.09
	P/12	+7.73	-4.10
	P/8	+7.85	-4.14
	P/6	+7.92	-4.16
	P/4	+7.95 (10% error)	-4.18 (3% error)
	P/3	+7.81	-4.14
	P/2	+6.90	-3.85
Data (from Ref 25):	L = 284.2 m Z_i = -3.41 m	$A = A_o = 22.6\ m^2$ $Q_i = 128\ m^3/s$	$A_s = 103.9\ m^2$ $K_o/K = 3.78$

TABLE 5. LOAD REJECTION FOR TALLULAH FALLS, GEORGIA

Source	First five surge extremes (ft) NB: 1 ft = 0.3048 m				
Field results (Ref 24)	+18.6	-13.7	+12.5	-10.0	+9.3
Model study (Ref 24)	+18.6	-13.8	+12.6	-10.1	+9.0
Waterhammer calculation (Ref 42)	+20.1	-16.2	+13.5	-11.7	+10.2
Pocket calculator (Ref 2) solution	+20.1	-16.2	+13.5	-11.7	+10.3

Data (from Ref 24, 42): L = 6666 ft $A = 151\ ft^2$ $A_s = 2130\ ft^2$

$Z_i = -5$ ft $Q_i = 925\ ft^3/s$ $K_o = 0$

Closure assumed linear over 32.4s

TABLE 6. INFLUENCE OF JUNCTION LOSSES ON SURGE LEVEL PREDICTION FOR A PUMPED-STORAGE SCHEME (Ref 34)

	Observed model surge extremes		Prediction using form (i) - simplified throttle loss			Prediction using form (ii) - Gardel junction losses		
	level (mm)	time (s)	level (mm)	time (s)	error (%)	level (mm)	time (s)	error (%)
Instantaneous Turbine Accept	0.0	0.0	0.0	0.0	-	0.0	0.0	
	-83.0	5.5	-80.0	6.0	+3.6	-81.4	6.0	+1.9
	-28.0	13.75	-23.0	16.0	+6.0	-27.8	16.5	+0.2
	-38.0	23.0	-34.9	26.0	+3.7	-36.5	26.5	+1.8
	-35.0	33.0	-31.7	35.5	+4.0	-34.6	36.5	+0.5
Instantaneous Pump Accept	0.0	0.0	0.0	0.0	-	0.0	0.0	-
	+75.0	5.5	+67.9	6.0	-9.5	+74.1	6.0	-1.2
	+18.0	14.0	+16.0	16.0	-2.7	+18.5	16.0	+0.7
	+32.0	23.0	+28.6	25.5	-4.5	+31.4	25.5	-0.8
	+29.0	32.0	+24.8	35.0	-5.6	+27.6	35.0	-1.8

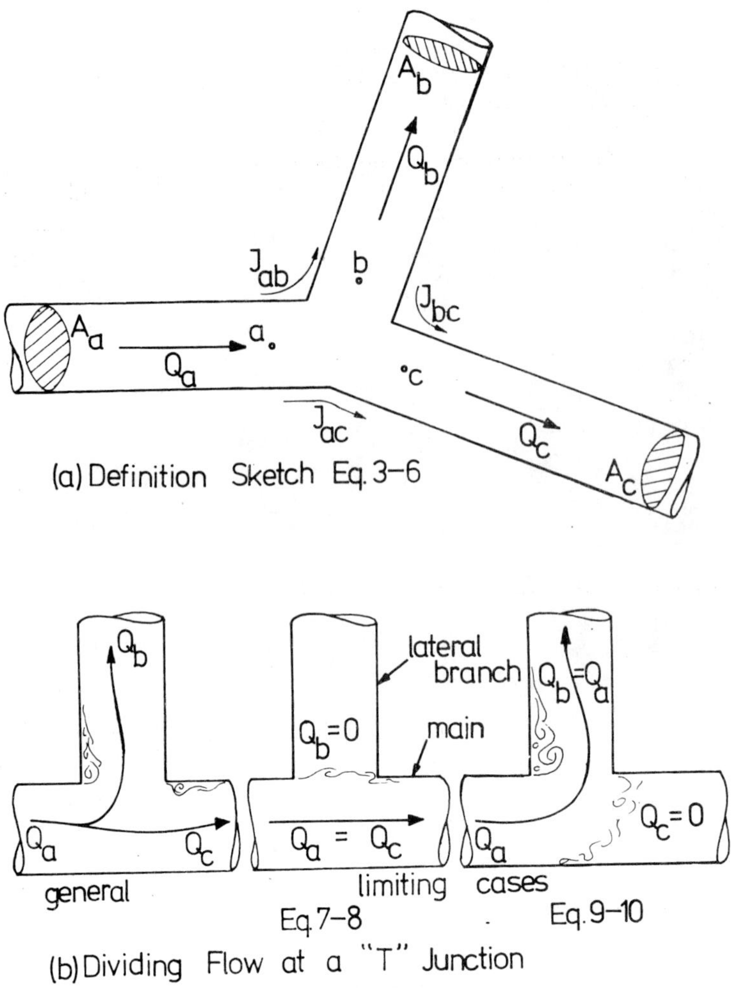

Fig. 1 Three branch junction

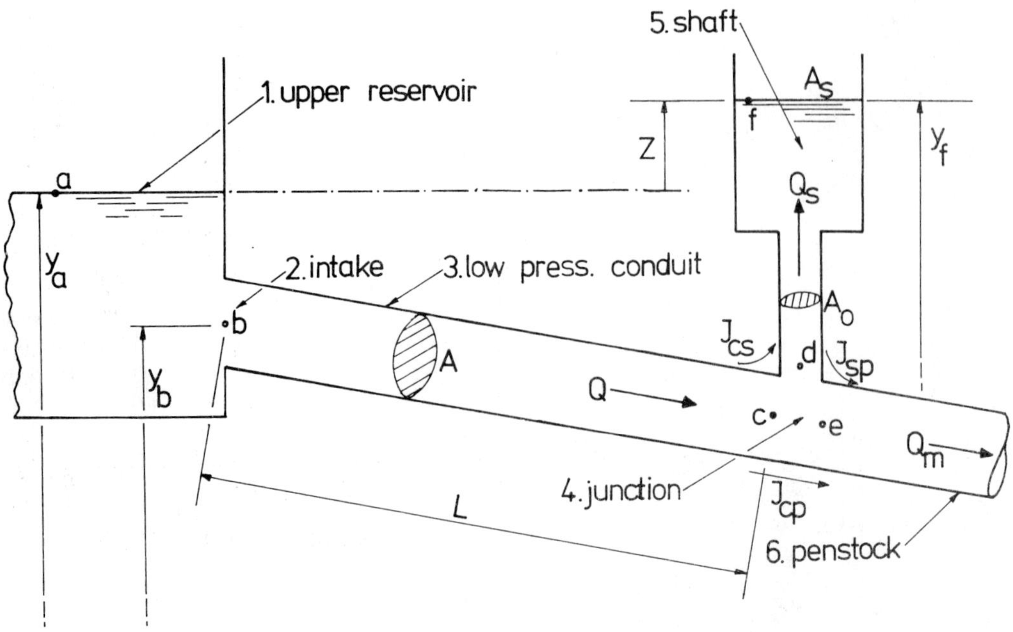

Fig. 2 Traditional single surge tank layout

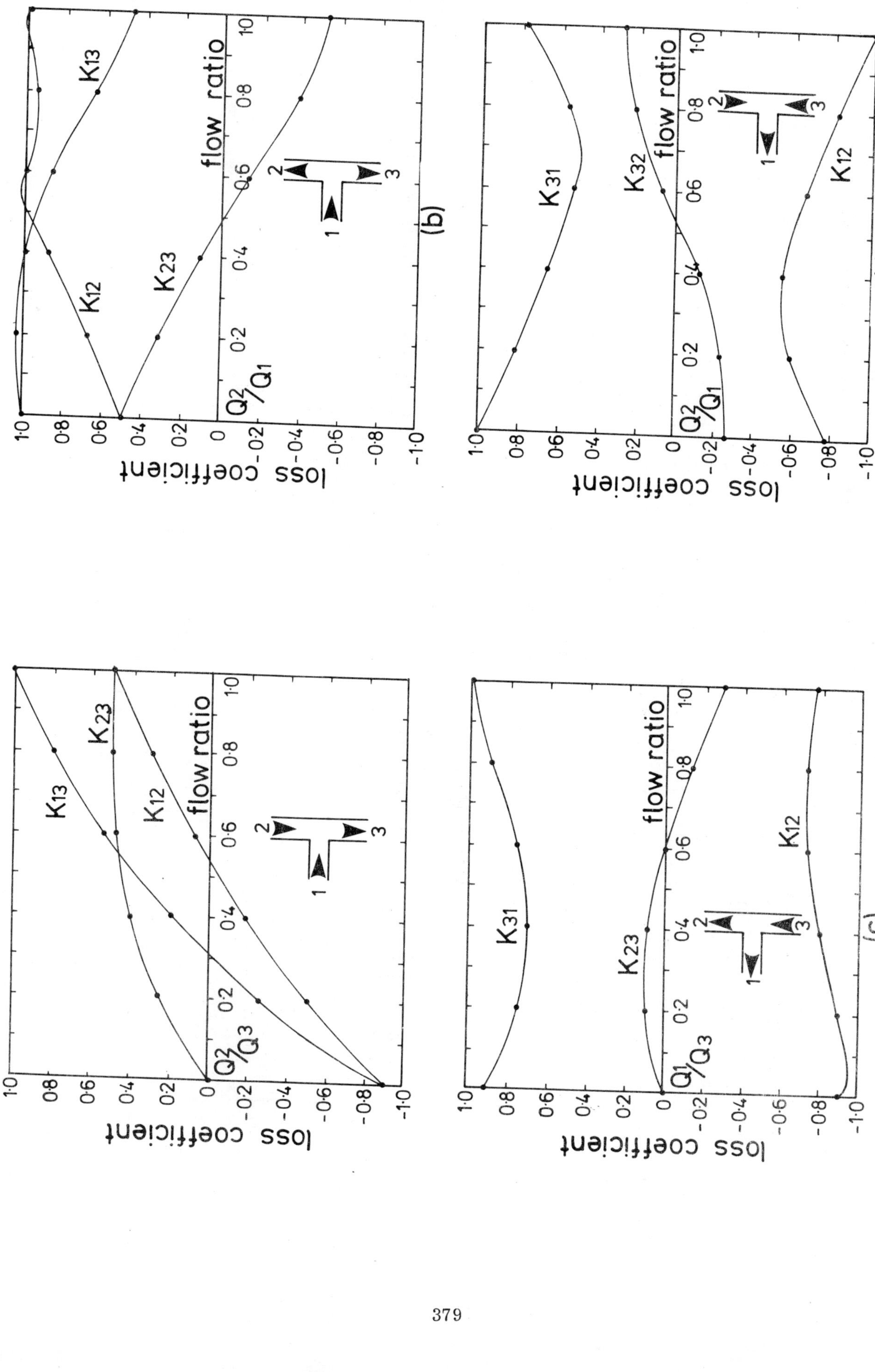

Fig. 3 Losses for typical surge tank (Ref 34)

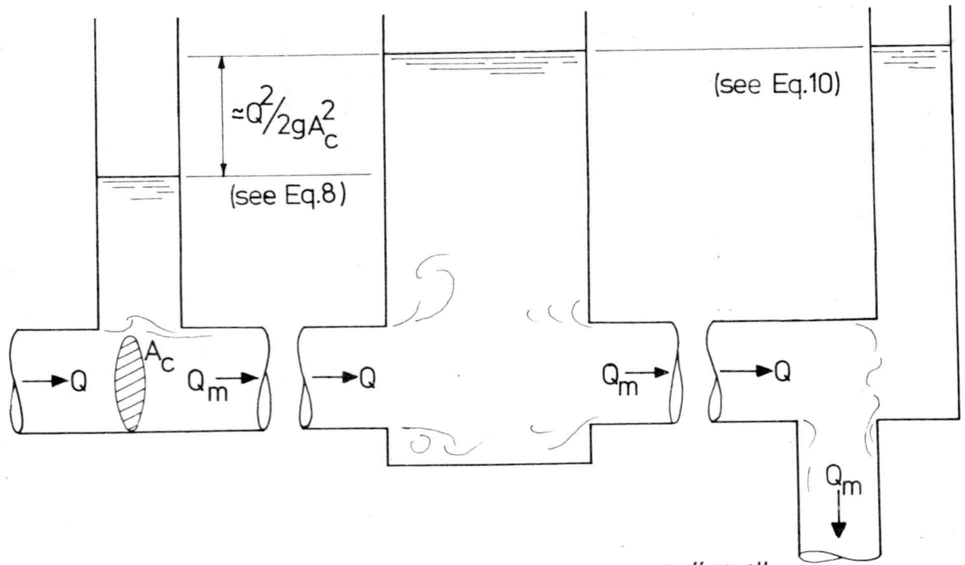

(a) "piezometric" (b) reservoir (c) "pitot"

Fig. 4 Types of surge shaft/conduit junction

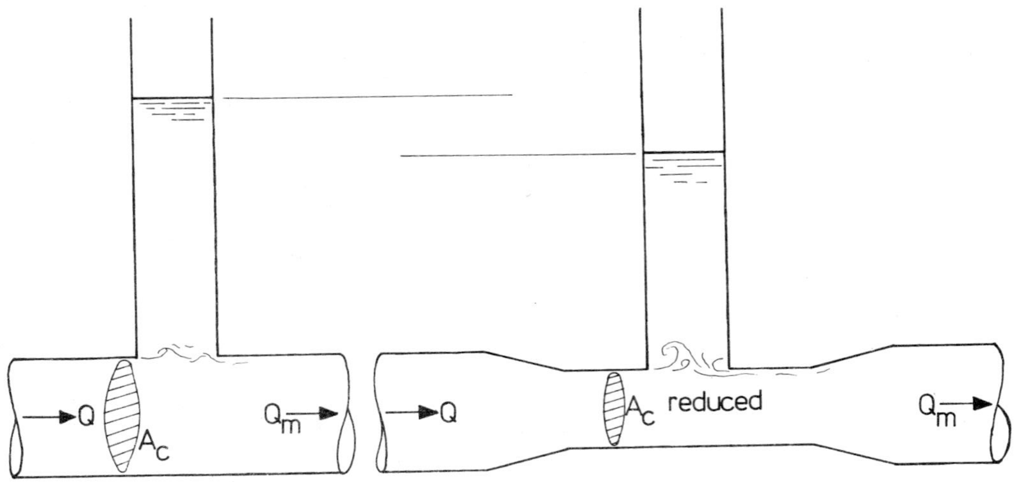

(a) conventional (b) with "venturi contraction"

Fig. 5 Venturi contraction surge chamber

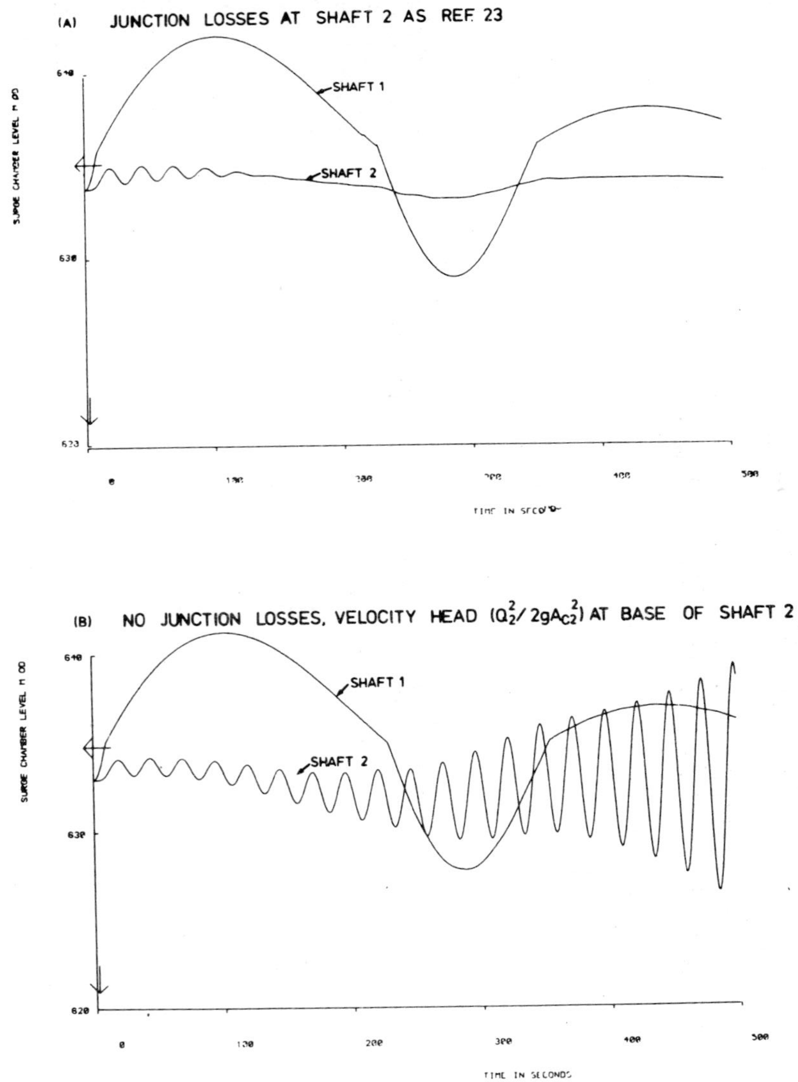

Fig. 6 Effect of junction losses on prediction of shaft stability (Ref 52)

Third International Conference on
PRESSURE SURGES
Canterbury, England
March 25-27, 1980

FIRE MAIN FAILURES DUE TO RAPID PRIMING OF DRY LINES

P. Hope, and
M.U. Papworth

BHRA Fluid Engineering. U.K.

Summary

Automatic fire protection systems can be a very effective means of saving lives and property in the event of fire. Such systems can, however, have pressure surge problems. It is common practice for the exposed pipework leading to fire fighting devices to be kept empty when not in use, to prevent damage by freezing during winter months, for example. In the event of a water demand, empty pipework is rapidly primed. Large pressure transients can be generated when the priming flow is totally or partially obstructed by the fire fighting device. Three examples of systems having this type of problem are described, and proposed solutions given. Weaknesses in the currently available means of assessing the damaging effects of predicted transient pressures on system pipework are highlighted.

Held at the University of Kent at Canterbury, U.K.

Conference Organised by BHRA Fluid Engineering, Cranfield, Bedford, England.
In conjunction with The City University, London.

Introduction

Each year the cost of fire damage, quite apart from suffering caused by the loss of human lives, is considerable. It is generally acknowledged that these losses could be substantially reduced if sufficiently early warnings could be obtained of the initiation of fires. Frequently in practice no human observers are present when a fire starts, and even when there are, the fire brigade is not always called, or it is not called soon enough.

Automatic fire protection systems which can detect the onset of fire at a very early stage, and which can perform the dual functions of extinguishing action and alarm sounding can be very effective. Their effectiveness is reflected in reduced insurance premiums paid by organisations utilising them. Such systems can, however, present certain pressure surge problems, and this paper is concerned with one of these problems.

In such fire main installations it is common practice for pipework downstream of the deluge valves supplying sprinklers and monitors, for example, to be kept empty when the main is not being used. This is generally done for safety reasons; typically to prevent damage due to freezing during winter months, or to reduce corrosion damage. Unfortunately, this safety precaution can can sometimes be the cause of system failure in the event of a sudden water demand. This paper outlines three fire fighting systems that have been analysed at BHRA recently for their transient behaviour during normal operating conditions, two of which had failed during commissioning tests.

The basic cause of failure typically occurs following the rapid opening of deluge valves. The pressure upstream of deluge valves is normally kept at a high pre-set level, and so the priming flow is rapidly accelerated into the empty pipework. At the point of filling the pipework, the high velocity fluid is suddenly obstructed, for example at a monitor nozzle, and this causes a sharp pressure transient which can seriously damage the system. Problems due to transient pressures can also occur when the system is shut down following a water demand, but these problems are more easily remedied because the shut down procedure can be executed more gently than the system start up.

The paper discusses the remedial measures proposed for each of the three described systems, to ensure that failure due to pressure transients would be unlikely to occur in the event of a fire, whilst maintaining maximum speed of operation.

1. System A: Hose failure due to a pressure transient caused by rapid priming against a blank end with trapped air

Fig 1 shows a schematic diagram of system A. The installation comprises a ring main which is supplied with water from an underground pumphouse. The ring main feeds hydrant spurs, a hose reel system and a sprinkler system. Because the sprinkler system has to be maintained at 70 m head of water, the entire system as shown in fig 1, must therefore be at this pressure in the no flow condition, and is maintained at this level by an electrically driven jockey pump. When there is a water demand, for example during a fire drill, up to three diesel driven pumps can start in order to supply the required amount of water. The pumps are triggered by a sequence of pressure sensing switches located at the pump house. During a normal fire drill procedure a fire tender is coupled to one of the hydrant points using a flexible canvas hose and the hydrant valve is then rapidly opened, with the tender valve closed, to prime the hose. The tender pump is then started and this can supply several high pressure hoses. It was found that as water was rapidly accelerated into the canvas hose and then abruptly halted at the tender valve, a very large pressure rise was generated. The impact was cushioned to some extent by trapped air but neverthless the pressure transient was sufficiently large to burst the canvas hose on a number of occasions. Fig 2 shows a pressure:time curve recorded at the hydrant point during such a fire drill. Large pressure rises were also generated upon rapid simultaneous closure of the high pressure hose nozzles when these were previously providing full flow. However, these pressures could be contained satisfactorily by the high pressure hoses.

A remedy was proposed for the system and a computer analysis was required to verify the proposed modifications. The proposed solution was to insert a pressure reducing station immediately upstream of the ring main in such a way that the sprinkler system could remain at 70 m head whilst the ring main was maintained at about 17m. An air vessel was to be located upstream of the pressure reducing station, with a capacity large enough to permit priming of two canvas hoses and tender pumps without triggering startup of any of the main pumps. The effect of omitting the air vessel was also to be investigated in the analysis. Fig 1 shows the proposed system modifications in addition to the original system layout.

Ref 2 describes the computer analysis in detail but some of the results are given here. Fig 3 shows the predicted pressure:time curve corresponding to the operating conditions of fig 2. There were a number of critical parameters in the modelled system; for example the amount of trapped air in the hose, the assumed rate of opening of the hydrant valve and the hydrant valve characteristics. In principle fig 2 could be used to "calibrate" these sensitive data parameters, but the predicted curve of fig 3 was obtained using data obtained by site observations without the need for further adjustment. Fig 4 shows a computer pressure:time curve for the same operating conditions but with the modified system. A further critical data parameter in this representation was the assumed response rate for the pressure reducing valves. As this rate is adjustable on the real valves the choice of data was to some extent arbitrary. By performing several computer runs it was found that in fact the peak collapse pressure reached an asymptotic level of about 78 m as the response rate increased. It was found that if the air vessel was omitted the resulting collapse pressure was about 10% higher than in fig 4 for equivalent operating conditions. Also large reflected pressure peaks were superimposed on line pressures upstream of the pressure reducing station. The analysis therefore showed that the proposed system, including the air vessel, would satisfactorily reduce the priming collapse pressures, and these were incorporated in the changes in the real system. A commissioning test was carried out upon completion of the modifications, and fig 5 shows a measured pressure:time curve at the same hydrant as in fig 2 (ref 3). It can be seen that the predicted curve of fig 4 gives a very pessimistic estimate of the collapse pressure. This was due to a combination of the following reasons:

(i) the response rate of the pressure reducing valves (PRV) was slower than expected

(ii) the hydrant valve was opened more slowly and more erratically than expected, probably due to a worn key fitting on the hydrant, and

(iii) attenuation of the pressure transient by the hose wall was neglected in the theoretical analysis.

The primary problem was therefore solved satisfactorily by the proposed modifications, although the reliability and therefore the effectiveness of the new system is critically dependent upon the performance of both PRV's. If one PRV were to stick in a substantially open position, then the ring main would be pressurised to 70 m and thus in effect be the same as the original system. Alternately, if one PRV were to stick in a closed position and remain unnoticed whilst the other PRV in parallel operated normally, then if the second PRV stuck in a closed position at some later date the ring main would be incapable of supplying water.

2 System B: Rapid priming against a partial flow obstruction

Fig 6 shows system B in schematic form. The pumphouse supplies water to a branched firemain which feeds three "deluge houses", and various hydrants which are not shown in the figure. The deluge houses each comprise a sprinkler/sprayer network which is isolated from the main by deluge valves , see fig 7. Thus the deluge lines are empty, whilst the main is kept at 70 m head in the no flow condition by a jockey pump. A water demand from any part of the system would cause an electrically driven pump to start up, followed by a diesel pump if necessary. The pumps are triggered by a sequence of pressure sensing switches, as in system A. Pressures in excess of 70 m at the pumphouse are prevented by a relief valve.

A computer analysis was required to investigate operating conditions during which potentially large pressure transients could be generated. Three such conditions were anticipated: pump failure, pump startup against a closed system, and priming of the empty deluge lines. The analysis showed that with this particular system the pressure transients generated by pump startup or failure were within the design limits of the system pipework. However, priming of the empty deluge lines could cause excessive pressure rises.

In the event of a fire it is likely that only one deluge valve will open initially, thus one empty line will be primed by flow from a pump which is capable of supplying perhaps five similar lines. Very high priming velocities may therefore occur, and these could be even higher than anticipated because pipe wall friction and other losses are initially lower than the steady state turbulent losses typically assumed in numerical analyses. The worst priming condition envisaged was that of deluge house number 1 in figs 6 and 7. This deluge system is the one nearest to the pumphouse and is therefore subject to the most rapid reaction from the pumps in the event of a water demand. Also the inertia of the water in the pipework between the pumphouse and this system is lower than for other parts of the main, and the water would therefore be more rapidly accelerated into a dry line.

The type of priming collapse occurring in this system is slightly different from that occurring in system A. In this case an accelerating column of water enters a sprinkler/sprayer unit, which can be conservatively regarded as a "large discrete loss" or a single partial obstruction to the flow. If, however, the sprinkler/sprayer unit involves lossy connecting pipework, then the priming velocity will be progressively reduced as the flow fills the pipework. In system A the column of water was completely arrested at a blank end, but cushioned by a trapped pocket of air. In both cases the actual magnitude of the transmitted pressure wave will be lower than the value predicted for an all rigid pipework system by an amount dependent upon the elasticity of the obstruction/termination (the pipe work itself will acquire some momentum due to the impact). For the proportion of elastic energy transmitted to the fluid following such a collapse event, it is necessary to investigate the magnitude of transmitted pressure waves as they propagate through the pipework system. Different classes of pipework are often used in such systems and account must be taken of this in addition to the area ratio effect of pipework size on pressure waves.

The computer analysis indicated that priming collapse events could cause pressure transients which were approximately equivalent to twice the permitted design limits for the system pipework. It was therefore recommended that the primary velocities are limited by the inclusion of some form of flow inhibitor such as an orifice plate, or plates. The optimum sizing and location of such devices would depend upon the detailed design of the system and the proposed operating conditions. A compromise is inevitably necessary between the required reduction in primary velocity and the desirable steady flowrate under full demand conditions.

The possibility of using air vessels to overcome the problem was discarded in favour of the orifice plate approach on the grounds of cost.

3 System C: Structural failure due to dynamic resonse of pipework to rapid priming against a partial flow obstruction

Fig 8 shows system C in schematic form. The system comrpises a reservoir which supplies water to a manifold which feeds various sprinklers and monitors. The reservoir level is maintained by pumping, and a PRV between the reservoir and the manifold maintains a set pressure upstream of the manifold. The lines to each of the fire fighting devices from the manifold are kept empty as in systems A and B. Flow to each of these lines is controlled by a motorised valve.

The commissioning tests for this system revealed problems caused by pressure transients on both system startup and shutdown. It was found that on priming of each of the monitors, large collapse pressures were measured and fracture of the monitor base castings occurred. System shutdown was also found to generate significant pressure rises and pipework movement. It was therefore necessary to carry out remedial measures in order to permit startup and shutdown of the monitors without causing excessive pressure rises, but without significally reducing the fire fighting effectiveness of the system.

A computer analysis confirmed that a priming collapse event in one of the monitors could generate a pressure peak which would exceed the design limit for the system pipework by a factor of at least three, if all the kinetic energy of the priming flow was converted to elastic energy in the fluid.

However, it seems probable that in the site test the collapse event imparted considerable momentum to the monitor structure, causing large dynamic bending stresses at the base castings, and thus failure.

A solution to the problem required a compromise in order to achieve a reduction in the priming velocity whilst minimizing the priming time and flow obstruction once priming was complete. If the motorised valves were opened sufficiently slowly to limit the priming velocity, then at a constant rate of opening the total opening time to achieve design flow would be excessive. A two rate opening valve was one possible solution, and another was to separate the priming and running requirements by installing a bypass line across each motorised valve connected to a monitor. Each bypass could be operated by a solenoid valve to prime the monitor and the desired rate of priming achieved by appropriate sizing of the bypass line. The motorised valve could then be actuated to open by a pressure sensing switch when priming was complete. This pressure switch would have to be located downstream of the motorised valves. It would be necessary to ensure that the pressure switch was not actuated by any sharp transient pressures caused by the solenoid operated valve in the pass line.

Conclusions

It has been shown that the fire fighting effectiveness of certain automatic protection systems could be completely lost in the event of a real fire, due to structural failure of the system pipework on rapid priming of empty lines. Three examples of this type of problem have been outlined, together with proposed solutions.

The problem can arise because it is common practice to leave the exposed pipework leading to fire fighting appliances empty when not in use. Typically, therefore, in the event of a water demand, an empty section of pipework has to be rapidly filled by a pump which may be capable of supplying many such lines. If this pipework is tortuous and lossy, then the priming velocity will be limited. However, the pipework is frequently designed to be of minimum loss in order to maximise the flow throughput in the event of a fire. The accelerating column of fluid may therefore acquire considerable velocity before being obstructed; either totally at a blank end, or partially by a monitor nozzle. In the case of total arrest of the fluid, the impact will usually be cushioned by trapped air.

In all cases a proportion of the fluid momentum change will be imparted to the associated pipework by an amount dependent upon the elasticity of the structure. Most of the rest of the dissipated flow kinetic energy is converted into elastic energy within the fluid. The pipework is thus subjected to transient stresses as a result of its structural response to the fluid impact, and also as a result of the propagating pressure waves within it. Either of these stress wave mechanisms could cause structural failure.

The general solution to the problem is to reduce the priming velocity to an acceptable level by throttling the pump delivery.

For applications where the maximum flowrate capability of the system is critical, the priming and running functions are best separated. This can be achieved, for example, by using a double closure rate deluge valve, or by using a throttled bypass line for priming followed by normal deluge valve operation upon completion of priming.

The usefulness of computer modelling techniques in the design process has been illustrated, and some of their limitations highlighted. It would appear that a better understanding of the fluid-structural interactions occurring during a collapse event would reduce the conservatism inherent in current methods of assessment of the damaging effects of such an event. Given that reasonably accurate predictions of transient loading can be obtained, there is a need for satisfactory criteria relating this loading to the potential damage caused by it.

References

1. Pucill, P M — "Automatic Fire Protection", Proceedings of the Third National Fire Protection Conference, organised by the Fire Protection Association, London (1968)

2. Hope, P — "Computer Analysis of Hydraulic Transients in a Fire Main - The Air Freight Terminal at RAF Mildenhall". BHRA Report TRC 1089 (October 1976)

3. Priddin, K and Hope, P — "Further Site Tests at RAF Mildenhall". BHRA Report TRC 1189 (September 1979)

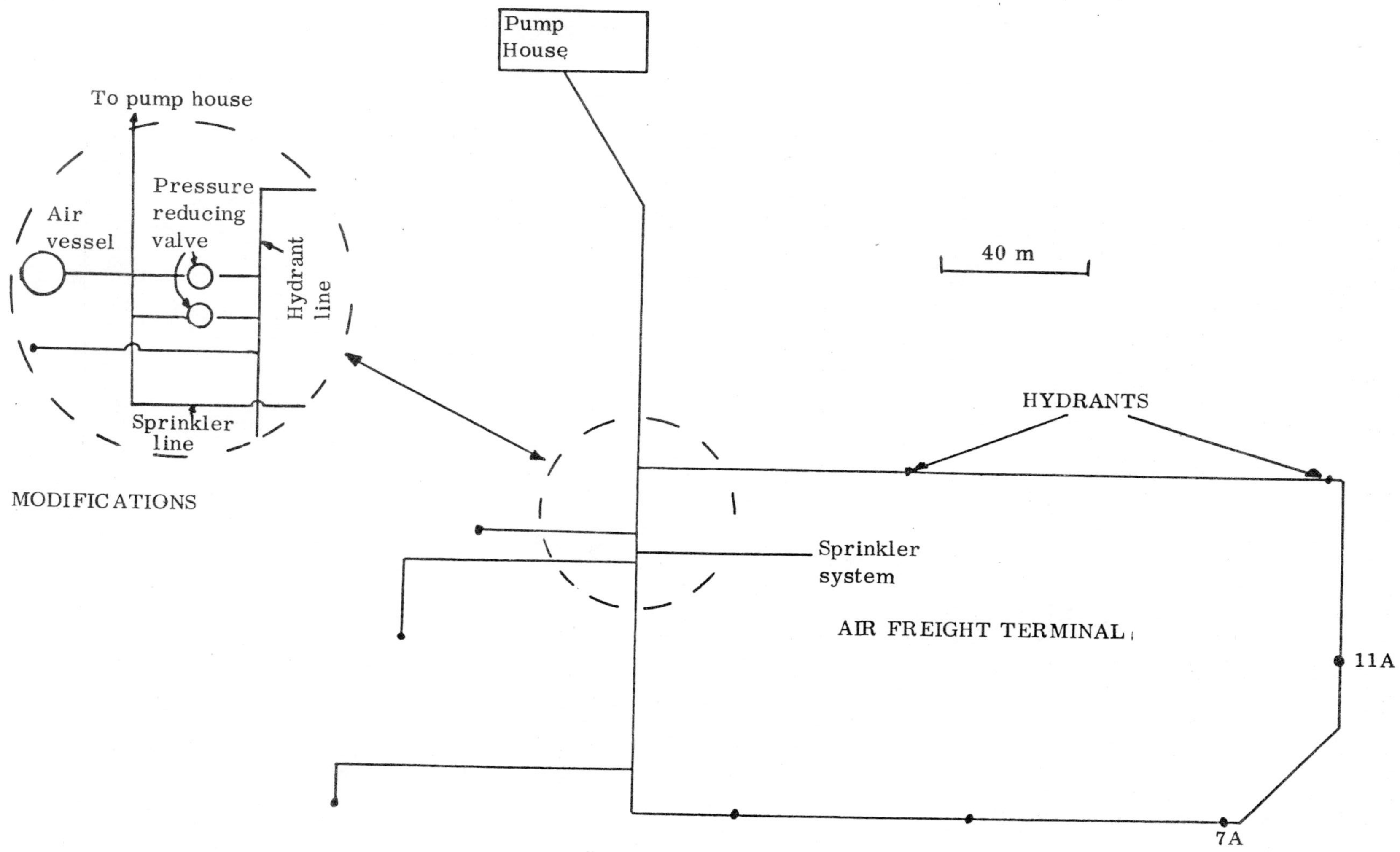

Fig. 1 Schematic of original and modified system

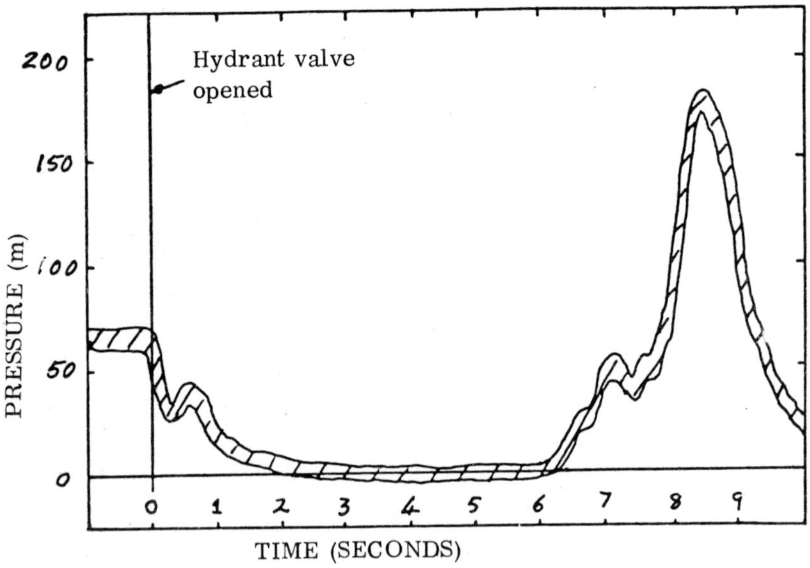

Fig. 2 Pressure history recorded at hydrant 7a

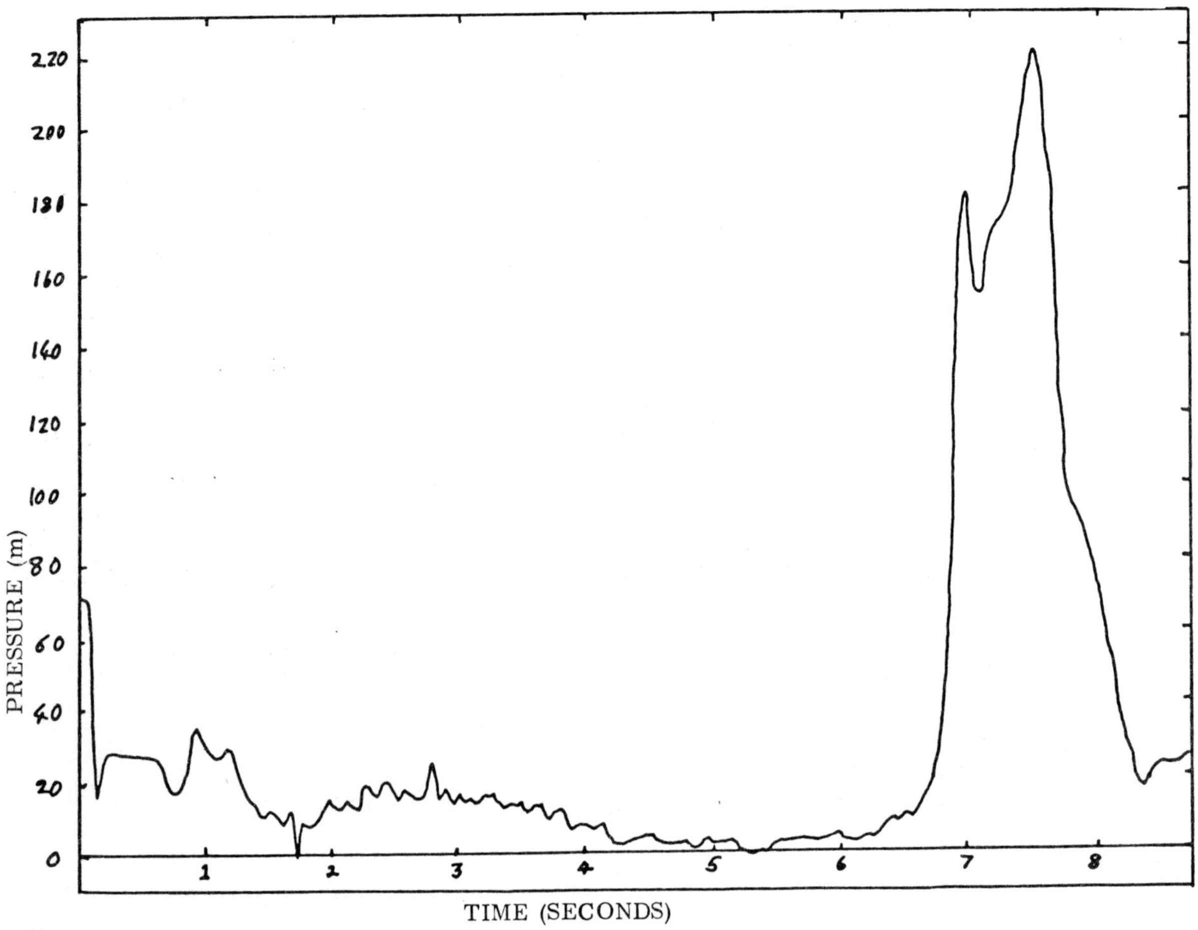

Fig. 3 Predicted pressure history at hydrant 7a

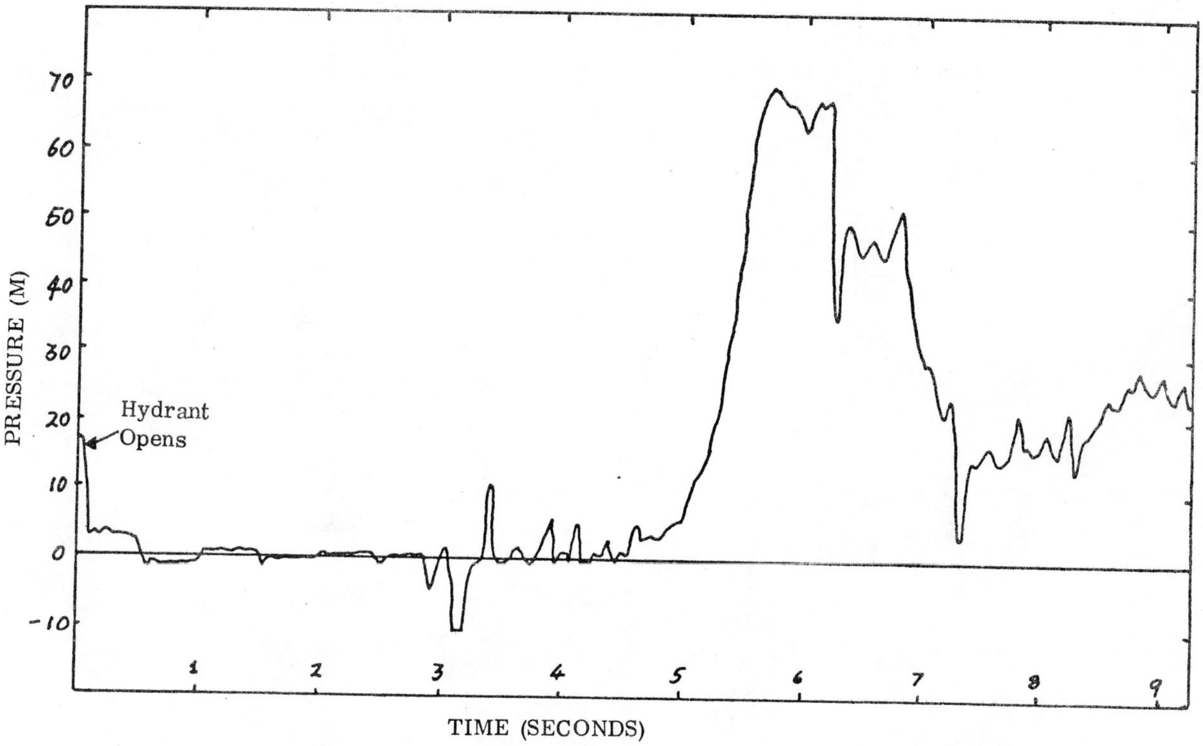

Fig. 4 Predicted pressure at hydrant 7a for modified system with fast pressure reducing valve response

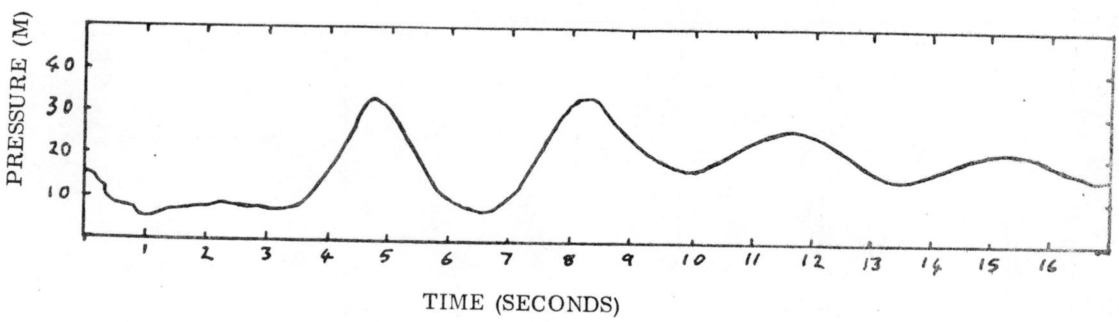

Fig. 5 Measured pressure at hydrant 7a for modified system

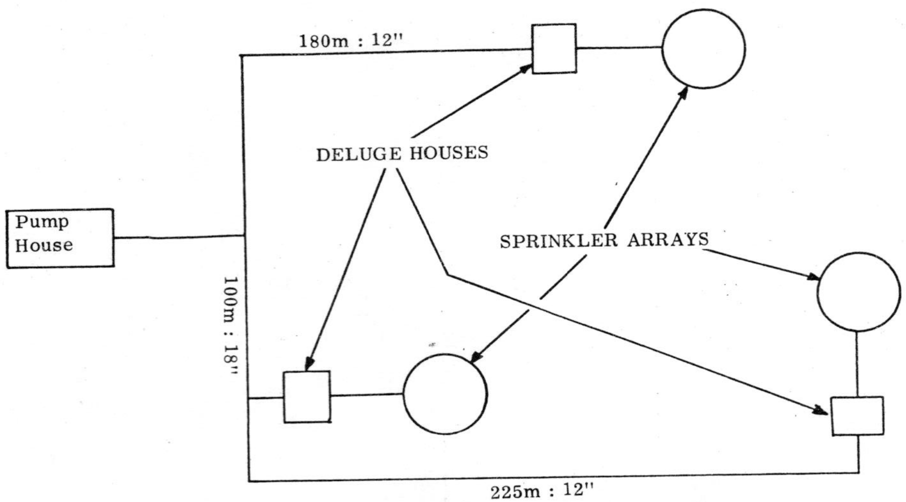

Fig. 6 Schematic of fire protection system

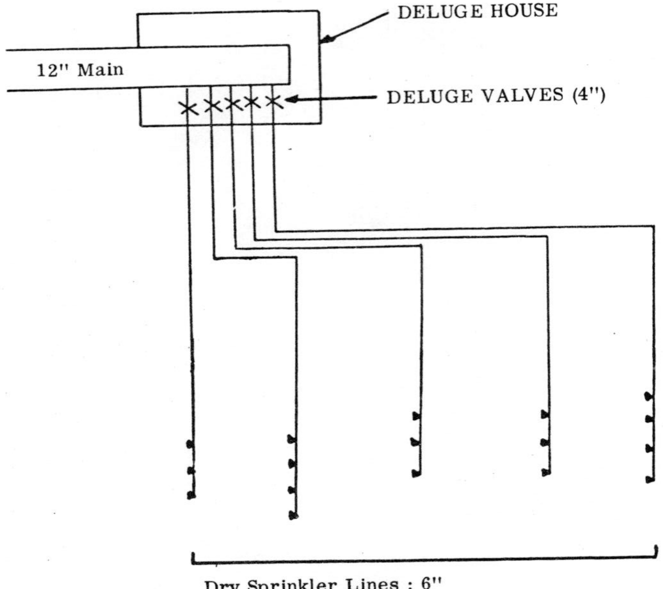

Fig. 7 Typical deluge house and sprinkler line arrangement

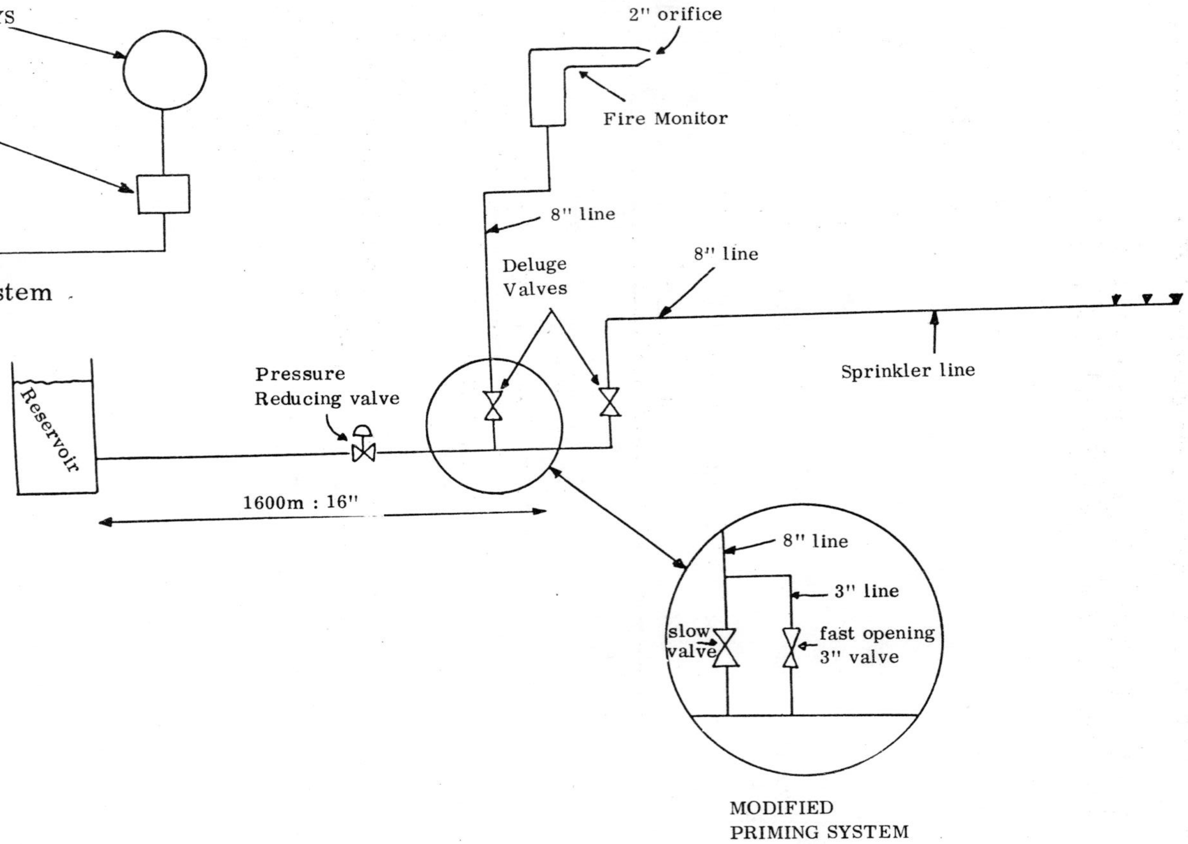

Fig. 8 Schematic of fire monitor system

Third International Conference on

PRESSURE SURGES

Canterbury, England
March 25-27, 1980

PAPER H2

KAPLAN TURBINE ACCIDENTS AND REVERSE WATER HAMMER

S. Pejovic, L. Krsmanovic and A. Gajic
Faculty of Mechanical Engineering, Belgrade, Yugoslavia.

and

D. Obradovic
"Energoprojekt", Belgrade, Yugoslavia.

Summary

The paper describes a Kaplan turbine accident, caused by a reverse water hammer. Some interesting results of field and model tests are presented along with a procedure to analyze danger of accidents due to the reverse water hammer from the draft tube onto the runner.

Held at the University of Kent at Canterbury, U.K.

Conference Organised by BHRA Fluid Engineering, Cranfield, Bedford, England.
In conjunction with The City University, London.

Nomenclature

A_R - runner sectional area
A, a_o - guide vane opening
c - flow velocity
c_m - mean axial velocity of water in the runner
c_{rez} - limited velocity
D - runner diameter
d_g - diameter of runner hub
Fa - axial hydraulic thrust
G - weight of the rotor
g - acceleration due to gravity
H - head
H_{at} - atmospheric pressure
H_{rez} - limited pressure
H_{vp} - saturated vapor pressure
H_s - suction head
ΔH_s - pressure rise behind the runner due to the water hammer
H_k - mean pressure on turbine cover
ΔH_k - mean pressure drop through the runner
h_1, h_2 - pressure on turbine head cover
h_{sp} - pressure in spiral casing
K_ϕ - constant in flow coefficient equation
K_ψ - constant in pressure coefficient equation
K_σ - constant in axial thrust coefficient equation
n - speed of revolutions
P - power
Q - discharge
$r_g=d_g/D$ - dimensionless diameter of runner hub
β - runner blade inclination
$\phi=\frac{K_\phi Q}{D^3 n}$ - flow coefficient
$\Psi=\frac{K_\psi H}{(Dn)^2}$ - pressure coefficient
$\sigma_F=\frac{K_\sigma F a}{\rho D^4 n^2}$ - axial thrust coefficient
ρ - density
η_s - draft tube efficiency
mWC - meters of water column

Introduction

For the past few years, some Kaplan turbine quick closure accidents in Yugoslav hydroelectric power stations have occured. Reverse water hammer was the cause of great damage to the plant units.

The Faculty of Mechanical Engineering, University of Belgrade, and ZEP - Association of Power Industry, Belgrade, collaborate on a study of reverse water hammer in hydraulic turbines, the first stage of which has been completed. This paper presents some of the results in a condensed form.

1. Description of an accident

A generating unit, power P=22,4 MW, head H=19,3 m, discharge Q=136 m^3/s, speed n=150 rpm, was operating during the night with a small power output of 8 MW. Due to the governor failure (the feedback tape was torn off), the guide vanes opened very quickly and the power output increased. The instruments registred power oscillations between 20 and 25 MW, and roaring sounds were heard from the turbine room.

The unit was closed by pressing the turbine emergency shutoff button. The generator was disconnected from the electric-power-supply network. The unit overspeed control device went into operation. When the machine came to a stop, a banging noice was heard from the turbine and the water was leaking out through the turbine head cover. A serious accident took place.

One of the runner blades was broken in the root (Fig.1.1). The fracture line went obliquely from the hub towards the inlet edge. A smaller part of the blade root and inlet section stayed with the runner hub, while the larger part was broken off. The fracture (Fig.1.2) showed no cracks or defects in casting, and no signs on the material fatique. It was clear that the breaking had started on the lower part of the blade.

Damages, caused by collision with the torn-off runner blade, were discovered at the inlet edges of all other blades (Fig.1.3). Damages were also found in the lower guide-vane ring, upper guide-vane ring, runner head, runner pit liner cone and inner head cover. The damages on the guide vanes were in the form of scratches. It can be concluded from the position and shape of score marks that the guide vanes were either very little opened or completely closed the moment the broken blade struck at them.

2. Some investigations

After the accident described in section 1, a series of tests were made with another identical unit. An interesting strip chart, recorded during the turbine's emergency shutoff, is shown in Fig.2.1. It is important to notice that the pressure on the turbine head cover (h_1 and h_2) reaches an absolute vacuum at the end of the servomotor stroke. It means that an air-vapor cavity is formed above the runner, resulting in water column separation. Pressure on the turbine head cover begins to rise only twenty seconds after the beginning of the transient process.

To make a more complete analysis of the reverse water hammer phenomenon, the axial thrust was recorded on a model turbine under various steady operating conditions. Fig.2.2 shows the results in a dimensionless form, where $\phi=K_\phi Q/(D^3 n)$ denotes flow coefficient, $\Psi=K_\psi H/(nD)^2$ - pressure coefficient, $\sigma_F=K_\sigma F_a/(\rho D^4 n^2)$ - axial thrust coefficient, Q - discharge, H - head, F_a - axial hydraulic thrust, D - runner diameter, n - speed of rotation, ρ- density.

It is a frequent case in practice that the axial force raises the runner, and the diagrams drawn in Figs. 2.3, 2.4 and 2.5 show that at a rated speed and a runner blade angle of inclination on 23^0 the runner will not be raised (Fig.2.3). With an increase in runner speed the axial force is increased and raises the unit's runner at small openings of the guide vanes (hatched zone in Figs. 2.4 and 2.5). The largest hydraulic force of runner lift is with closed guide vanes and depends on the runner blade inclination angle (Fig.2.6a). This diagram was used to draw Fig.2.6b, which shows the allowable zone of operation and the zone in which the axial force raises the runner. As will be seen later on, the runner lift zone is at the same time the zone of danger of accident due to reverse water hammer. It can be seen that at completely open runner blades ($\beta=27{,}7^0$) when the speed is increased by 12% only, the runner will be raised.

3. Prevention of reverse water hammer

To prevent the reverse water hammer phenomenon it is necessary to prevent dewatering of the runner during the guide vane closure. As it is known and can be seen from the strip chart in Fig.2.1, it is not possible to prevent the phenomenon of absolute vacuum under the turbine cover (in some cases, even if the turbine closure time is very long). Therefore, an adopted criterion of the danger of reverse water hammer states that the maximum pressure (usually found near the guide vanes) on the turbine head cover must, through the transient process, be higher than the pressure of saturated water vapor increased by a reserve of pressure that takes into account the approximate nature of computing procedures. This condition, however, proves to be too severe, and an additional condition is introduced for the case when the maximum pressure drops below a permissible value. The condition in this case is that the mean velocity of water outflow under the runner should be lower than a certain limiting small value.

If no measured data on the pressure distribution at the turbine head cover is available, the mean pressure drop ΔH_k through the runner can be calculated from the axial hydraulic thrust F_a:

$$\Delta H_k = \frac{F_a}{A_R(1-r_g^2)} = \frac{\sigma_F D^4 n^2}{K_o \frac{D^2\Pi}{4}(1-r_g^2)}$$

where $A_R = D^2\Pi/4$ - is runner sectional area, $r_g = d_g/D$ - is dimensionless diameter of runner hub, and the mean value of pressure on the turbine cover will be |Ref.3|:

$$H_K = H_{at} - H_s + \Delta H_s - \eta_s \frac{c_m^2}{2g} + \Delta H_K$$

where H_{at} - is atmospheric pressure, H_s - suction head, ΔH_s - pressure rise behind the runner due to the water hammer ($\Delta H_s < 0$ at turbine closure), η_s - draft tube efficiency, and

$$c_m = \frac{Q}{\frac{D^2\Pi}{4}(1-r_g^2)}$$

denotes the mean value of the axial component of flow velocity at the runner outlet. The condition which prevents the runner dewarting is:

$$H > H_{vp} + H_{rez} = 2 \div 3 \text{ mWC}$$

where H_{vp} - saturated vapor pressure, and H_{rez} - is introduced on account of pressure pulsations. The pressure pulsations depend on the guide vane opening |Ref.4| and the cavitation number value |Ref.5|, but cannot be calculated accurately enough.

In case the above condition is not satisfied, the following additional criterion is introduced:

$$c_m < c_{rez} = 0{,}5 \text{ m/s} ,$$

which means that when the water reverses from the draft tube and strikes against the runner blades, provided the water reverses at the same speed with which it leaves the runner, the impact force will be a small one and will not cause any damage. The values of H_{rez} and c_{rez} must still be corrected on the basis of practical experience gained in turbine operation.

An additional condition is of equal importance. This is the time during which the runner remains waterless, because both the size of cavity and the speed with which the water returning from the draft tube strikes at the runner depend on this time. This condition is specially suitable for analyzing oscillograms recorded when examining transient processes in water power plants. If the duration of absolute vacuum is shorter than one second, it is considered that there is no danger of accident due to a water hammer. These short pressure drops are caused by pressure fluctations and oscillations in the generating unit.

Results of such a calculation are presented in Fig.3.1. The lines of guide vane opening A, head H, discharge Q, and speed n, are obtained by calculating the unit quick closure. The line F_a shows the hydraulic thrust variation as obtained by calculation from model test data. The mean pressure on the turbine cover H_k can be obtained from the pressure change in the draft tube and the hydraulic thrust variation. According to the adopted criterion, water column separation occurs in the eight second. The water outflow speed is c_m=1,6 m/s >c_{rez}=0,5 m/s, and there is danger of an accident.

In the case of two-speed turbine closure (Fig.3.2) with the total closure time 14,5 sec., the pressure H_k drops below its limiting value,

$$H_k = -0,9 \text{ mWC} < H_{vp} + H_{rez} = 2 \div 3 \text{ mWC} ,$$

but

$$c_m = 0,3 \text{ m/s} < c_{rez} = 0,5 \text{ m/s}$$

consequently, there is no danger of accident.

In the transient process, given in Fig.3.3, there is no danger of a reverse water hammer and a unit accident since the lowest pressure on the turbine head cover is H_k=3 mWC $\geq H_{vp}+H_{rez}$=2÷3 mWC

In the further treatment of this problem, the limiting values of the maximum pressure on the turbine head cover and the allowable values of water outflow speed under the runner should be thoroughly tested. Likewise, an analysis must be made of the influence of quantities that are not modelled, such as: air supply to the turbine, change in the cavitation coefficient, and other quantities.

Conclusions

The reverse water hammer may be prevented by adjusting the time of turbine closure. A sufficiently long time is the best protection. Turbines with two closing speeds, the one at the end of closure being lower (slower closure), are very suitable since they permit a sufficient decrease in the unit´s rotational speed before the guide vanes are closed.

When control conditions require a minimum closure time, comprehensive and expensive investigations must be conducted. On a model, under steady operating conditions, the pressure distribution on the turbine head cover and/or the hydraulic thrust should be measured.

If the governor fails to operate, both the guide vanes and the runner blades must be closed slowly. If, at a high speed of the runner and a large inclination angle of runner blades, the guide vanes close completely, the axial thrust will raise the unit rotor and there is serious danger of accident due to the reverse water hammer.

References

|1| PEJOVIĆ,S.at all., "A Study of Reverse Water Hammer" (Studija povratnog hidrauličkog udara), Fac.of Mech. Engg., Belgrade (in collaboration with the ZEP - Power Industry Association, Belgrade) (1978) (In Serbo-Croatian)

|2| KRIVTCHENKO, G.I., "Hydrodynamic Transient in Hydroelectronic Plants" (Гидромеханические переходные процессы в гидроэнергетических установках) Энергия, Москва (1975) (In Russian)

|3| PEJOVIĆ,S., KRSMANOVIĆ, Lj., "Control Problems in Hydraulic Generating Units" - Prof.Dr. G.I. Krivtchenko´s lectures (Problemi regulisanja hidroenergetskih agregata - predavanja prof.dr G.I.Krivčenka), Fac.of Mech.Engg., Belgrade - (1978) (In Serbo-Croatian)

|4| Koichi NAKANISHI and Tsuneo UEDA, "Air Supply into Draft Tube of Francis Turbine", Fuji Electric Review, Vol.10, No.3, p.p.81÷91 (1971)

|5| MOLLENKOF, G., RAABE, J., "Measurements of Fluctuations of Velocity and Pressure in the Draft Tube on a Francis Turbine", IAHR Symposium, Stockholm, Paper B1, p.p.1÷12 (Part 1) (1970)

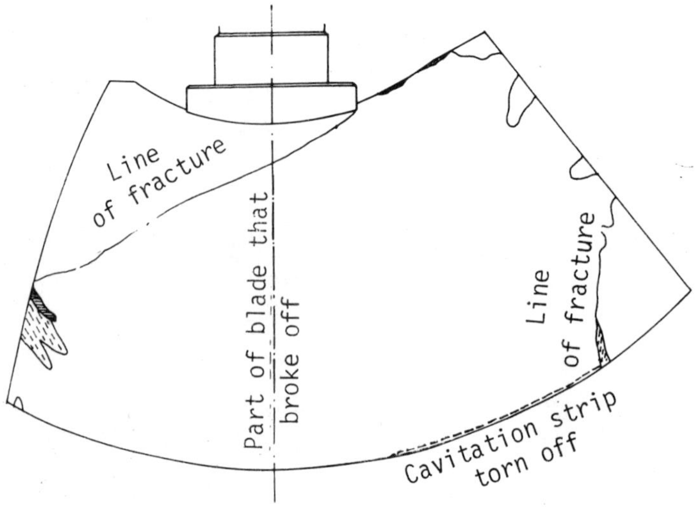

Fig.1.1 - Runner blade (No.2) broken in the accident

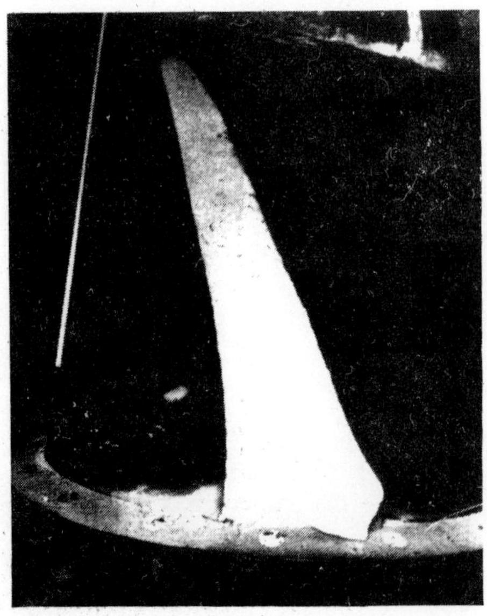

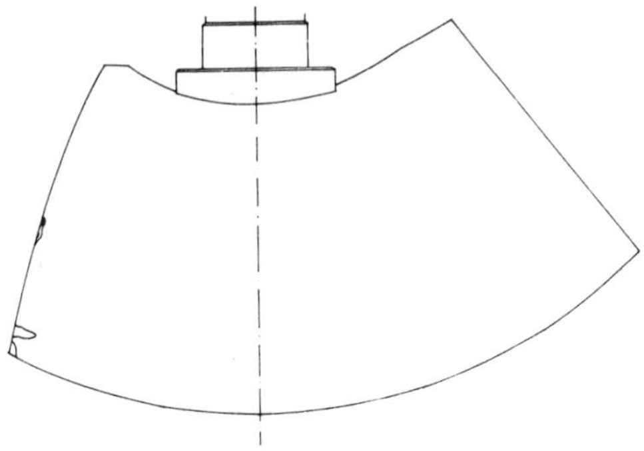

Fig.1.2 - Appearance of fracture on the part of runner that remained on the runner

Fig.1.3 - Runner blade (No.4) damaged in collision with the torn-off part of blade No.2

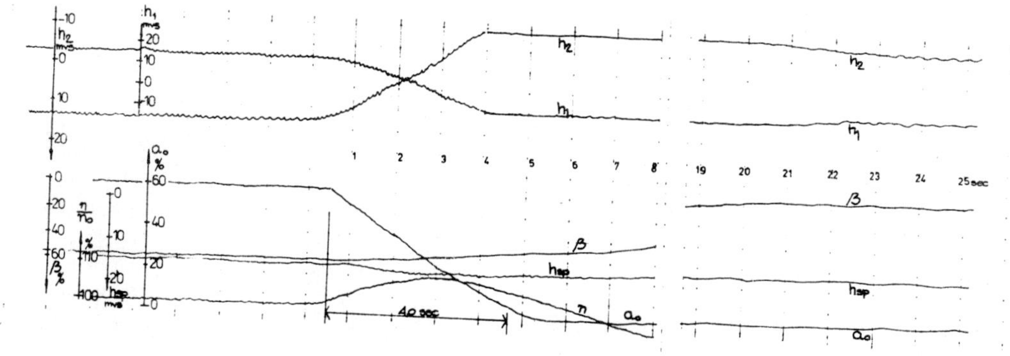

Fig.2.1 - Quick closure from power output of 16 MW. Turbine closure time - 4 sec; a_0(%) - guide-vane opening; β(°) - runner blade inclination; n/n_0(%) - rotational speed; h_{sp}(mVS) - pressure in spiral casing; h_1 and h_2 - pressure on turbine head cover.

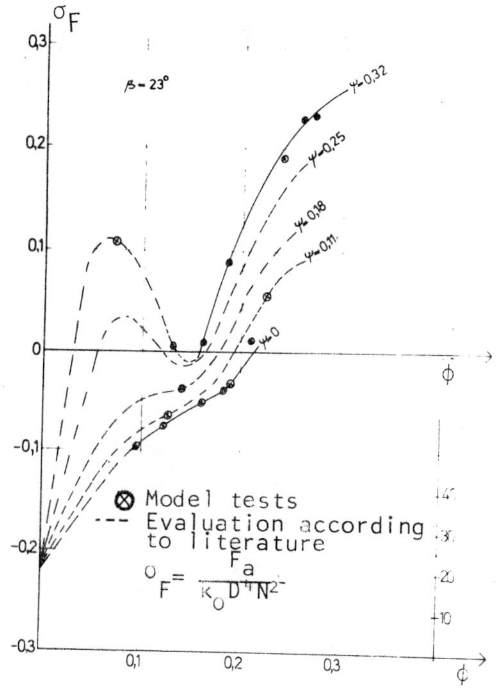

Fig.2.2 - Hydraulic thrust characteristic for runner blade inclination β=23°

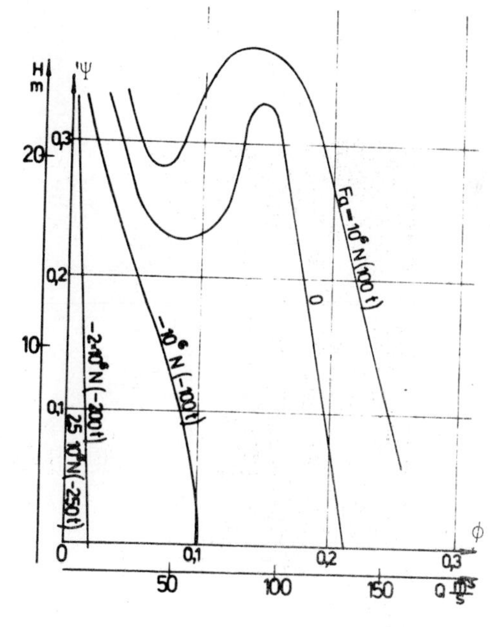

Fig.2.3 - Axial thrust for rated speed ($n=n_0$) and runner blade inclination β=23°

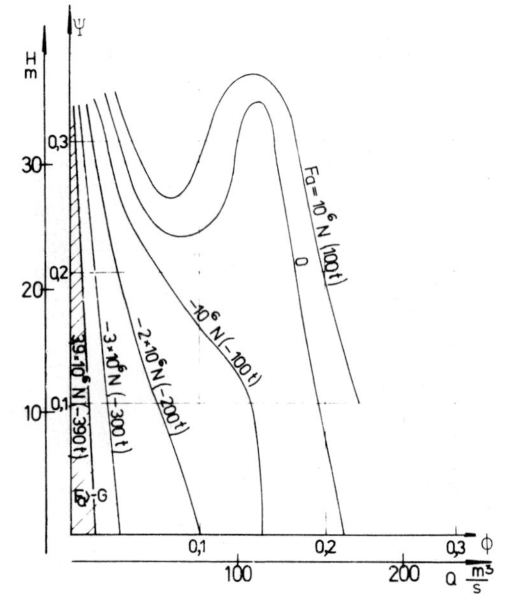

Fig.2.4 - Axial thrust at 1,25 rated speed ($n=1,25\ n_0$) and runner blade inclination $\beta=23^o$. Hattched zone where the rotor is raised.

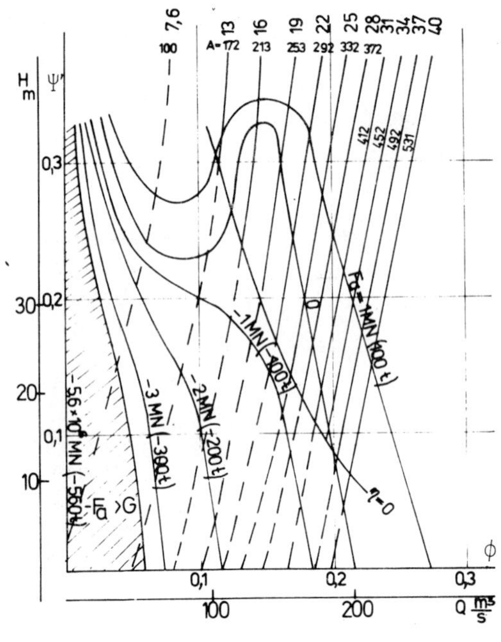

Fig.2.5 - Axial thrust at 1,5 rated speed ($n=1,5\ n_0$) and runner blade inclination $\beta=23^o$

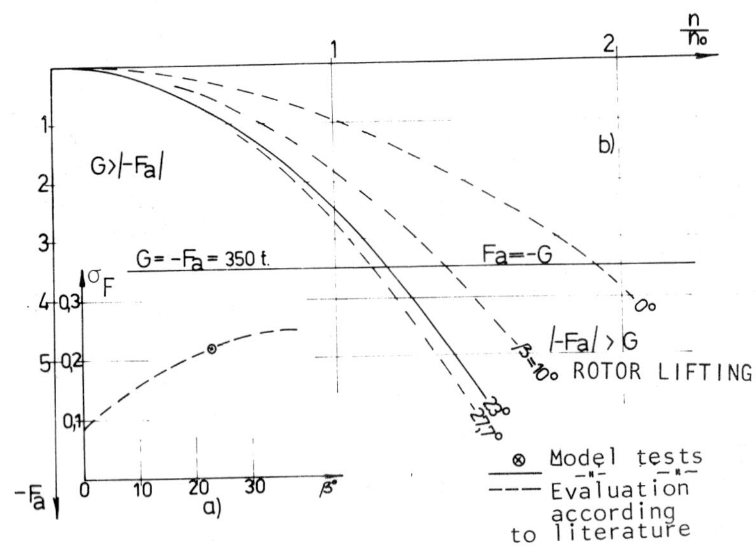

Fig.2.6 - a) Axial thrust coefficient σ_F as a function of runner blade inclination β;
b) Axial thrust as a function of runner blade inclination and speed of rotation zone $|-Fa|>G$ is not permited for unit operation.

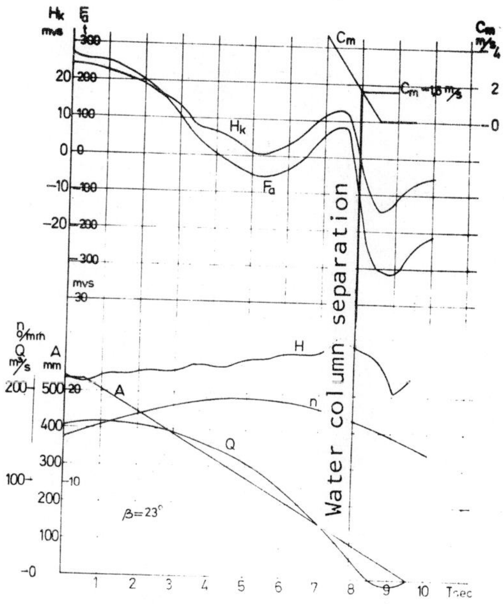

Fig.3.1

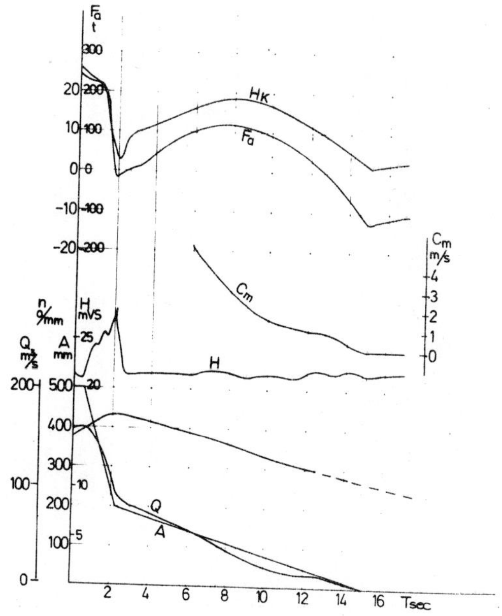

Fig.3.2

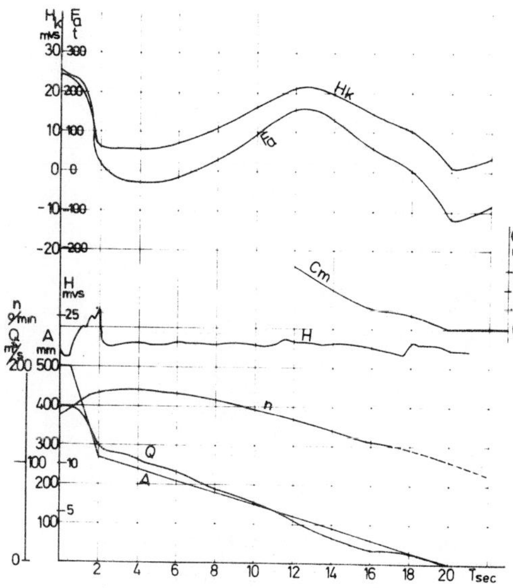

Fig.3.3

Calculation of unit quck closure:
A - guide-vane opening; H - head;
Q - discharge; n - speed;
H_k- mean pressure on turbine cover;
F_a- hydraulic thrust; c_m - mean meridional velocity of water in the runner

Third International Conference on

PRESSURE SURGES

Canterbury, England
March 25-27, 1980

PAPEF H3

CAVITATION AND CORROSION IN A PUMPING MAIN

W.L. Glass

Stewart Sviridov & Oliver, S. Africa.

Summary

The paper raises the question of whether cavitation during pressure surges can be expected to cause damage such as that experienced in pumps and turbines.

A description is given of a steel pumping main which was severely damaged internally, apparently as a result of pressure surge cavitation. A contributory cause is considered to be the release of dissolved gases during cavitation which made possible and encouraged the growth of iron corroding bacteria.

The dam from which the water was drawn receives coal mine drainage which is considered to be largely responsible for the dissolved gases and the bacteria. In such cases cavitation should clearly be prevented by adopting such measures as surge tanks or air vessels.

Held at the University of Kent at Canterbury, U.K.

Conference Organised by BHRA Fluid Engineering, Cranfield, Bedford, England.
In conjunction with The City University, London.

1. INTRODUCTION

As a result of the higher velocities now being adopted in pumps, turbines and pipelines, cavitation appears to be increasingly encountered and frequently causes severe damage. Such damage, however, appears to occur only where the cavitation is sustained for relatively long periods such as in pump impellers, venturi meters or needle valves. In these cases it has been well established that the damage does not occur at the low pressure areas where cavitation is induced but at adjacent areas where the vapour bubbles are imploded. The implosion apparently results in extremely high-velocity miniscule jets which can disintegrate metal surfaces to a marked degree in a few weeks.

Although cavitation is also known to occur as a result of pressure surges it does not generally appear to cause such disintegration of the pipe surfaces. It is noteworthy that at the symposium on Cavitation in Edinburgh in 1974 over 40 papers were presented (mostly relating to cavitation damage) not one of which dealt with pressure surge cavitation (Ref. 1).

It has often been stated that pressure surge cavitation can cause large vapour pockets which on collapsing result in damaging high pressures. The damage envisaged in this case is, however, not disintegration of the surface of the walls, but bursting of the pipe. Such pressures are normally calculable with a fair degree of accuracy. At the 1976 International Conference on Pressure Surges, for example, a description is given of a large pipeline in Holland where it was decided not to instal any surge tanks or other devices to prevent cavitation; instead the pipeline was designed to withstand the estimated extra pressures caused by cavity collapse (Ref. 2). Many pipelines in South Africa have also been designed on this basis in the past.

2. WITBANK PUMPING MAIN

As in other parts of the world, pumping mains in South Africa are generally equipped with electrically driven pumps often supplied by means of long power lines. Power failures due to lightning are common and the accompanying low pressure surges in many cases result in vaporisation and cavitation at high points on the main. No cavitation damage, however, had been encountered on some pipelines which had been in service for nearly 50 years.

It was accordingly somewhat alarming when, in 1974, it was learnt that two bitumen lined steel pumping mains had been severely damaged by what appeared to be cavitation erosion. One supplies water to the town of Witbank and the other to a power station nearby; both obtain water from entirely different catchment areas.

The power station pipeline is some 30 km long and is owned and operated by the Electricity Supply Commission (ESCOM) who have carried out some independent investigations. This paper is mainly concerned with the Witbank pumping main.

A longitudinal section of the pumping main is shown in Figure 1 and further particulars of the pipeline are as follows:

(a) Type of pipe : spirally welded mild steel.

(b) Type of joints : butt-welded

(c) Internal lining : 3 mm thick bitumen.

(d) Thickness of steel : 6.4 mm and 8 mm.

(e) Finished internal diameter : 584 mm.

(f) Length : 7.8 km.

(g) Rate of pumping : 400 to 450 ℓ/s.

(h) Velocity in pipe : 1.5 to 1.7 m/s.

3. DESCRIPTION OF DAMAGE

Fig. 2 is a detail of the longitudinal section of the pipe in the vicinity of the damaged length and gives the variation in the average width and depth of the damage; these are based on X-ray photographs taken in 1974 and ultrasonic measurements of the steel thickness carried out in 1979. It will be noted that the pipe has been damaged over an area varying in width from 40 to 200 mm for a distance of some 360 m. From piping which has been replaced it is evident that the bitumen lining has been removed in all the damaged areas.

Near the middle of the damaged length a groove about 10 mm wide has been formed for a distance of about 130 m. This groove is illustrated in Fig. 3 which is taken from a sample cut from the pipeline; a photograph of the sample is shown in Fig. 4. In places the groove has penetrated right through the pipe but generally the depth (averaged over 100 mm) varies from 5% to 50% of the 6.4 mm steel plate.

The groove is located along the inside crown of the pipe and is remarkably straight throughout its length as shown in Fig. 3 and 4. Another striking feature is the fact that the damaged area on one side of the groove is a mirror image of the other side.

4. POSSIBLE CAUSES OF DAMAGE

Initially it was considered that the most likely cause of the damage was cavitation erosion coupled with cavitation corrosion similar to that which occurs in pumps. This was supported by a computer analysis of the transient pressures on pump shut-down which showed that vapour pockets could be expected at the failed section. It was also noted that on tripping a pump a loud 'pistol shot' noise occurred at the damaged section.

In further support of this theory was the unusual situation that, for various reasons, the pumps could not be shut down gradually but had to be tripped soon after they began to be throttled; this resulted in severe pressure surges and cavitation at the damaged section each time the pump was shut down. In the first few years the pipeline was intermittently stopped and started owing to the relatively low demand and the need to keep reservoirs full. There were also many power failures due to lightning.

One difficulty with this theory was that similar damage, even if on a much smaller scale, does not appear to have been reported on other pumping mains, some of which have similar profiles and flows. Another was that microscopic examination of the metal revealed no evidence of distortion of the metal crystals such as might be expected from severe cavitation erosion (Ref. 3). This is illustrated by Fig. 5 which is a photomicrograph 100 times full size taken of an etched cross-cut at the groove.

In any event it seems doubtful whether pressure surge cavitation can have the same surface damaging effect as cavitation in pumps. In the case of the latter vapour bubbles have to make room for themselves when the pressure drops to vapour pressure and they then collapse violently due to an increase in pressure. In pressure surge cavitation in a rising pipe the velocity gradient set up by the surge creates a vacuous cavity (in the form of one or more bubbles) into which the vapour evaporates; this cavity is then closed during a reversal of the velocity gradient and results in a rise in pressure in the pipe which is calculable. There does not seem to be any question of damage to the pipe surface as occurs in pumps, turbines and control valves.

However, if the vapour cavity is in the form of a long thin bubble at the top segment of the pipe as observed by Baltzer (Ref. 4) and Fox (Ref. 5) the repeated collapse of this cavity presumably would affect the top of the pipe to some extent in due course. It was felt that this alone would not have caused the severe damage that occurred at Witbank and that there must be other contributing causes.

It seemed probable that the other cause or causes of the damage would be connected with the peculiarities of the Witbank area, particularly in regard to the water pumped through the pipeline.

Witbank is a fairly large industrial town and is the centre of an extensive coal mining district. Water is supplied from a dam about 10 km away. The catchment area of the dam contains a large number of coal mines which pump water from underground and there is also runoff from waste dumps. Such mine water is usually polluted with sulphur compounds (formed by the action of bacteria on the iron pyrites in the coal) carbonates, methane gas, carbon dioxide gas and sulphur cycle bacteria.

The drainage from coal mines is at present strictly controlled by legislation (since 1976) but during the years 1966 to 1974 when the damage to the pipeline occurred there was much less control (Ref. 6). Routine monthly tests of the water in the Witbank dam taken from 1969 to 1974 reveal the following relevant particulars:

(a) the water was generally alkaline ranging from pH 6.7 (slightly acid) to pH 8.4 (fairly alkaline).

(b) the sulphate content was fairly high (as expected from the coal mine drainage) varying from 36 to 170 mg/ℓ (maximum desirable limit for domestic supplies is 200 mg/ℓ).

In the light of these considerations two probable causes of the pipeline damage present themselves, namely, gaseous cavitation and sulphate reducing bacteria, possibly in combination. (The latter were found to be present in the bitumen lining by ESCOM in 1974 but it was considered that cavitation erosion was the predominant cause).

5. GASEOUS CAVITATION

This subject has been extensively investigated in recent years mainly in regard to the effect on transient pressures (Ref. 7 & 8). The term refers to the release of dissolved gases during cavitation, in contrast with vapourous cavitation which refers to the vapour from a liquid.

The main gases which can be expected to become dissolved in the water supplying or stored in the Witbank dam are as follows: (Ref. 6 & 9).

(a) nitrogen (N_2), oxygen (O_2), carbon dioxide (CO_2) and sulphur dioxide (SO_2) - from the atmosphere.

(b) SO_2, CO_2 and methane (CH_4) - from coal mine drainage.

(c) SO_2, CO_2, CH_4, hydrogen (H_2), hydrogen sulphide (H_2S) and ammonia (NH4) - from decaying organic matter.

H_2S and SO_2 are rapidly oxidised in the presence of dissolved oxygen which can normally be expected to be present. It is not known whether dissolved oxygen was always present in the past but a recent analysis indicated that it was present in the dam water although to a relatively small extent. Ammonia and hydrogen are not likely to be present in significant amounts.

The only gases likely to have been dissolved in the water pumped through the pipeline in appreciable amounts are accordingly nitrogen, oxygen, methane and carbon dioxide. Each of these has been found to occur in natural waters up to a concentration of at least 10 mg/ℓ but not all together; CO_2 for example occurs in air-saturated water at a concentration of only 0.6 mg/ℓ but can reach more than 10 mg/ℓ at the expense of oxygen. (Ref. 9). Air-saturated water contains about 25 mg per litre of water and it accordingly seems reasonable to assume that the Witbank water could have contained a mixture of the 4 gases amounting to 20 mg/ℓ. If it is assumed that there was 5 mg/ℓ of each gas, the total volume of gas when released from the water at 15°C and sea level atmospheric pressure would be 1.80% of the water, as shown below (the maximum solubility of each gas is given for comparison):

	Density (kg/m³)	Volume of gas in water equivalent to 5 mg/ℓ each (% of water)	Maximum solubility by volume (% of water)
	(all figures at 15°C and 100 kPa pressure)		
Nitrogen	1.19	0.42	1.8
Oxygen	1.36	0.37	3.4
Methane	0.68	0.74	3.7
Carbon Dioxide	1.88	0.27	102
	Total	1.80%	

If all the oxygen or methane was replaced by carbon dioxide the total volume would be reduced to 1.70 and 1.33 respectively.

Some of these dissolved gases would be released from solution during each cavitation period, the amount being dependent on the duration of the cavitation and the size of the cavity. In the Appendix it is shown that for the wide range of 10 to 100% release of gases the width of the resulting cavity would vary on the average from 75 to 165 mm.

As the gas became re-dissolved the gas cavity would be reduced in size but over a longer period than that in which the gas bubbles were released owing to the greatly reduced gas/water surface. On the other hand the gas cavity would be enlarged by the gases carried up from the ascending section of the pipe marked CD in Fig. 1.

The actual width of the damage at the grooved section varies between 50 and 200 mm which is sufficiently close to the possible range of gas cavity widths to indicate that there should be a close relationship between the extent of the damage and the extent of the gas.

In the first few years after installation of the pipeline the pump would have been operated intermittently for short periods in view of the relatively large capacity of the pump and the need to maintain full reservoirs. Thus the crown of the pipe would be exposed to gases for relatively long periods at this stage.

Such gases could be expected to have a minor cushioning effect on the rise in pressure which occurs on closure of a cavity; they would also reduce the acoustic velocity of the pressure wave as pointed out by Fox (Ref. 5). This indeed appears to be reflected by certain pressure gauge observations taken in 1974 of the after-wave on tripping a pump; the observations, however, were not sufficiently accurate and further tests are to be undertaken.

Severe cavitation damage to turbines from gaseous cavitation has been reported as a result apparently of liquefaction of CO_2 (Ref. 10) but this does not seem likely in the present case unless excessive amounts of CO_2 were prevalent at some stage; this has yet to be investigated further.

Corrosion due to the presence of oxygen in the released gases could also possibly be responsible for the damage but the unusual nature of the damage, particularly the groove in the crown and the complete removal of the bitumen lining on either side calls for some other explanation.

The most likely cause of the removal of the bitumen at the crown of the pipe and the formation of the groove in the steel appears to be the presence of sulphate reducing bacteria.

6. SULPHATE REDUCING BACTERIA

Extensive damage in the form of pitting of the interior (and exterior) of steel pipes has been traced to the presence of certain sulphate reducing bacteria (Sporovibrio desulphuricans). They have been found to occur in alkaline coal mine drainage (Ref. 11) and thrive in water containing little or no dissolved oxygen provided the pH is between 5.5 and 8.5. Their energy requirements are met from inorganic sulphates which they reduce to sulphides with the aid of hydrogen or organic matter (Ref. 12).

The reason why these bacteria are found on steel pipes appears to be because they can conveniently obtain a continuous supply of hydrogen gas adsorbed on cathodic areas of the steel surface. By removing this hydrogen they effectively depolarise the corrosion cells which invariably exist and thus allow electrochemical corrosion to continue unabated. Hydrogen is then formed as fast as it is removed and at the same time iron goes into solution at the anodic areas resulting in deep pits. The cathodic area around each pit remains intact and thus forms a convenient stable surface to which the bacteria can adhere.

These bacteria also require a source of carbon (for proteins) which they normally obtain from carbon dioxide or organic matter; it seems possible that they may be able to work in conjunction with other bacteria which utilise methane for this purpose; they have been known to feed on bitumen (Ref. 13 & 14 & 15).

The crown of the Witbank pipeline, at those sections where gaseous cavitation takes place, would accordingly form an ideal home for these bacteria. The reasons are summarised as follows:

(a) a regular supply of CO_2 would be available for carbon requirements.

(b) the bitumen lining would conveniently harbour the bacteria, protecting them from the force of the water while pumping was taking place and providing them with a supplementary supply of carbon.

(c) the steel surface would provide a continuous supply of the necessary hydrogen gas.

(d) the water would be at a suitable pH(6,9 to 8,5) and would provide the necessary sulphates and small quantities of organic matter.

(e) conditions would be more or less anaerobic as required by the bacteria particularly at the lining/steel interface.

There would be a tendency in the first instance for the bacteria to concentrate in the top centre of the crown as this is where the last vestiges of the gases would finally dissolve after each tripping of the pump. Thus a series of pits would form along the centre line of the pipe which would eventually join up and form a groove; the bacteria would then congregate on either side of the groove and suppress any pits which may have started to form there. The groove would tend to be self-centering as any deviation from a straight line would cause an imbalance in the rates of corrosion on the two faces of the groove.

In course of time the adhesion between the steel and the lining at the crown of the pipe would fail and the bitumen would probably fall off exposing the steel to the corrosive attack of any oxygen dissolved in the water.

No damage appears to have been caused in the first Witbank pumping main which was laid more or less along the same route. This may have been due to the fact that it was lined with concrete and/or to the fact that the pumping velocity is much less than in the second main.

7. SURGE TANK

A small experimental one-way surge tank has been installed at the section where severe cavitation was taking place and this appears to have slowed down the rate of corrosion. Further investigations into the problem are being carried out but pumping is now almost continuous making it difficult to carry out certain tests. A full investigation will probably not be possible until a third pumping main, now being planned, has been installed.

8. CONCLUSIONS

In designing a steel pipeline for conveying water the question of cavitation resulting from pressure surges deserves careful consideration, especially in the case of raw water from a river or lake which may be polluted with industrial effluents.

The experience at Witbank described in this paper indicates that in certain circumstances a steel pipeline can become severely corroded if it is subjected to cavitation even where bitumen lining has been provided.

The special circumstance basically responsible for the corrosion damage at Witbank appears to have been the ingress of polluted coal mine drainage into the waters of the Witbank dam. This drainage contains in particular an abundance of sulphates and carbon dioxide both of which are required by sulphate-reducing bacteria which can also be expected to be present. These bacteria are known to cause severe corrosion of steel pipelines.

A suitable habitat for these bacteria appears to be provided at a particular section of the Witbank pumping main where gaseous cavitation invariably accompanies the shutting down of pumps. It is concluded that such bacteria are largely responsible for the removal of the bitumen lining at the crown of the pipe and the severe corrosion of the steel which led to failure of the pipeline. Where such circumstances prevail it would clearly be necessary to adopt measures such as surge tanks or air vessels to prevent cavitation; it might also be advisable to provide a different type of internal lining.

Further investigations and laboratory tests are being undertaken in an attempt to establish beyond doubt the cause or causes of the corrosion.

9. ACKNOWLEDGEMENTS

The author wishes to acknowledge the assistance given him by the Town Engineer, Witbank (Mr. P.K. Odendaal) and his departmental officials particularly the Water Engineer (Mr. F.J. Calitz). He also wishes to thank Dr. D. Stephenson, Professor of Hydraulic Engineering and Dr. F.P.A. Robinson, Professor of Corrosion Science, both of the Witwatersrand University for their helpful advice; the latter kindly provided the photomicrograph shown in the paper.

Finally, thanks are due to the Electricity Supply Commission, in particular to Mr. D.G. Hazleton for information on the Commission's pipeline and many helpful discussions.

10. REFERENCES

1. Institution of Mechanical Engineers. "Cavitation". Proc. conference organised by Fluid Machinery Group at Heriot-Watt University, Edinburgh. (3-5 Sept. 1974).

2. Provoost, C.A.: "Investigation into cavitation in a prototype pipeline caused by waterhammer". Proc. 2nd International Conference on Pressure Surges. Paper D2-13. Organised by BHRA at City University, London (22-24 Sept. 1976).

3. Uhlig, H.H.: "The Corrosion Handbook". John Wiley & Sons. (1948).

4. Baltzer, R.A.: "Column separation accompanying liquid transients in pipes". Jnl. of Basic Engineering, Trans. ASME, Series D, 89 No. 4, pp. 837-846. (Dec. 1967).

5. Fox, J.A.: "Hydraulic analysis of unsteady flow in pipe networks". Macmillan. (1977).

6. Henzen, M.R. and Pieterse, M.J.: "Acidic mine drainage in the Republic of South Africa". 8th Int. Conv. organised by Int. Assoc. of Water Pollution Research, Sydney. (Oct. 1976). Pub. in Progress in Water Technology, 9, No. 4, pp 981-992. Pergamon Press. (1978).

7. Wiggert, D.C. and Sundquist, M.J.: "The effect of gaseous cavitation on fluid transients". ASME Journal of Fluids Engineering, 101, pp. 79-86, March 1979.

8. Kranenburg, C.: "Gas release during transient cavitation in pipes". ASCE Journal of Hyd. Div., 100, HY10, pp 1383-1398, Oct. 1974.

9. Camp, T.R.: "Water and its impurities". Reinhold Pub. Corp., New York. (1963).

10. Knapp, R.T., Daily, J.W. and Hammitt, F.G.: "Cavitation". McGraw-Hill (1970).

11. Olson, G.J., Turbank, S.C. and McFeters, G.A.: "Impact of western coal mining II. Microbiological studies". Water Research. Journal of Int. Assoc. on Water Pollution Research, 13, No. 11. (1979).

12. Uhlig, H.H.: "Corrosion and corrosion control". John Wiley & Sons. (1971).

13. Peabody, A.W.: "Control of pipeline corrosion." National Assoc. of Corrosion Engineers, Texas. (1967).

14. Hawken, L.E. and Linton, A.H. (Editors): "Micro-organisms". Edward Arnold. (1979).

15. Costello, J.A.: "The role of sulphate-reducing bacteria in metallic corrosion." Conf. on Corrosion organised by South African Council for Scientific & Industrial Research, on board M.S. Europa. (12-17 Mar. 1972).

APPENDIX

SIZE OF GASEOUS CAVITIES AT GROOVED SECTION

Computer analyses of the surge pressures have been carried out assuming 11 discrete points along the pipeline. These show that at the middle of the grooved section a vapour pocket of 1.8 m^3 would be formed over a length of 687 m or 0.00262 m^3 per metre length of pipe; the volume capacity of the pipe is 0.268 m^3 per metre length so the void fraction is nearly 1%. Reference to Fig. 6 shows that a void fraction of 1% is equivalent to a segmental vapour cavity 208 mm wide.

During the period of cavitation part of the dissolved gases would be released in a multitude of fine bubbles which would collect in the vapour cavity and raise the pressure above the vapour pressure of the water. On damping out of the pressure surge the pressure would rise to the normal static pressure, which is about 220 kPa (305 kPa absolute).

In part 5 of the paper it is shown that the water could contain dissolved gases which if released from solution at 15°C and 100 kPa pressure would occupy between 1.33 and 1.80% of the volume of the water, say an average of 1.5%. Thus at the static pressure of 305 kPa (assuming pressure x volume is constant) the volume of these gases would be 1.5 x 100/305 = 0.5% of that of the water i.e. the void fraction would be 0.5%. The corresponding width of a segmental cavity would be 165 mm (see Fig. 6).

The computed cavitation time is about 45 seconds (see Fig. 7) but this does not allow for the release of any gases; the actual gaseous cavitation period is expected to be longer but Kranenburg points out that some observations have shown that it might be less. In either case a fairly large proportion of the dissolved gases should be released, particularly if CO_2 is predominant as it probably is (Ref.8).

The void fractions and corresponding widths of gas cavities are as follows for various amounts of released gases: (See Fig 6).

Amount of dissolved gas released (%)	Void Fraction at 305 kPa (%)	Width of Cavity (mm)
10	0.05	75
50	0.25	132
75	0.38	150
100	0.5	165

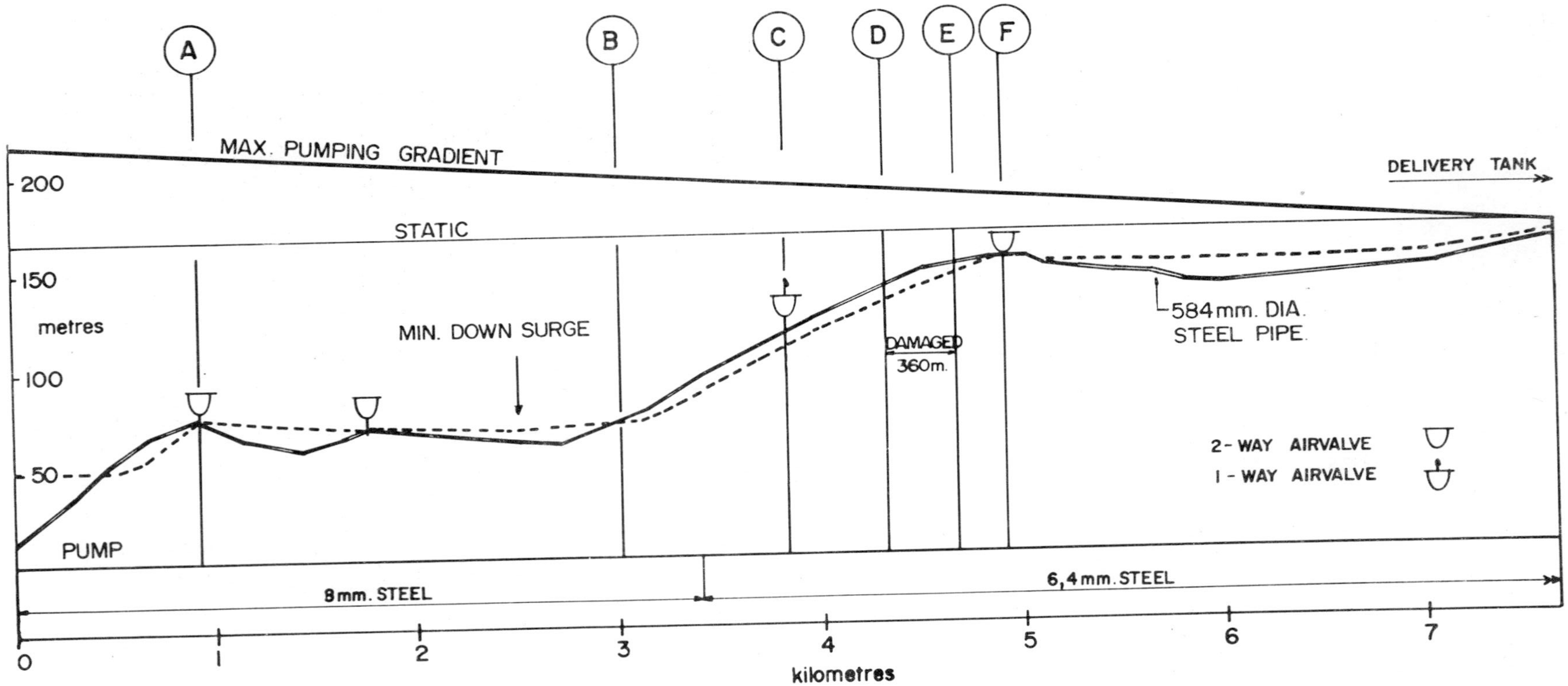

Fig. 1 Longitudinal section of 584 mm dia. steel pumping main at Witbank.

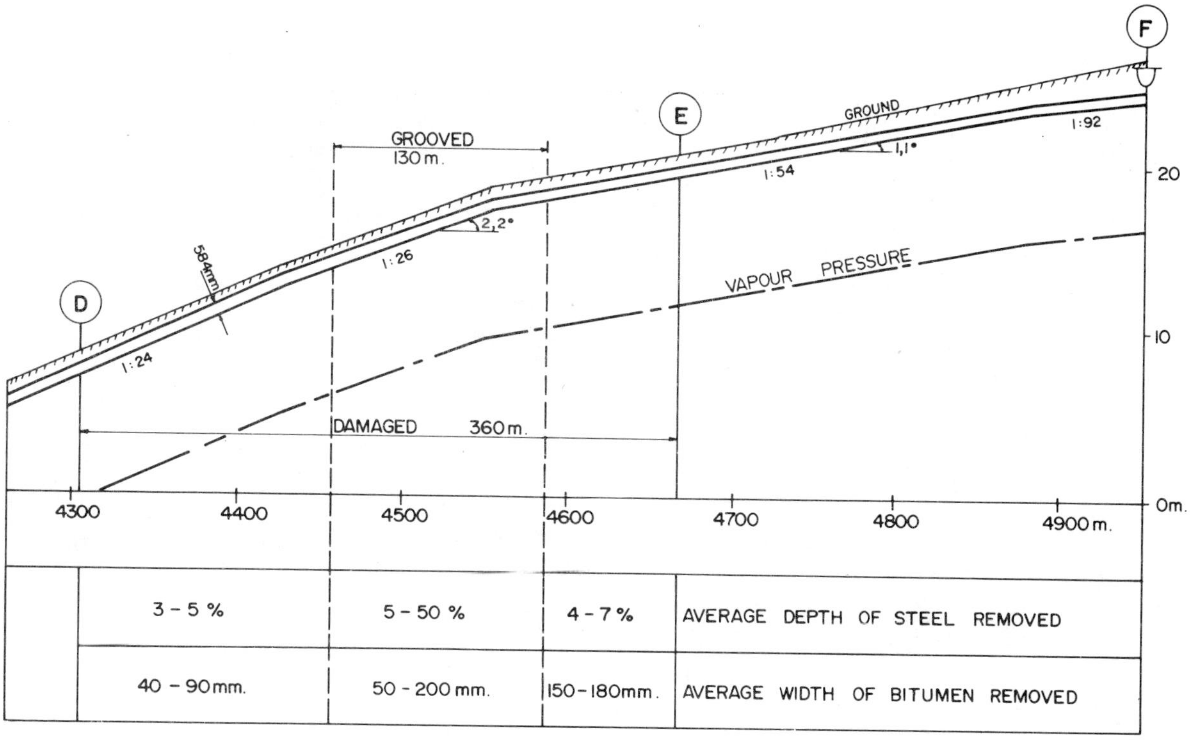

Fig. 2 Detail longitudinal section at damaged portion of pumping main.

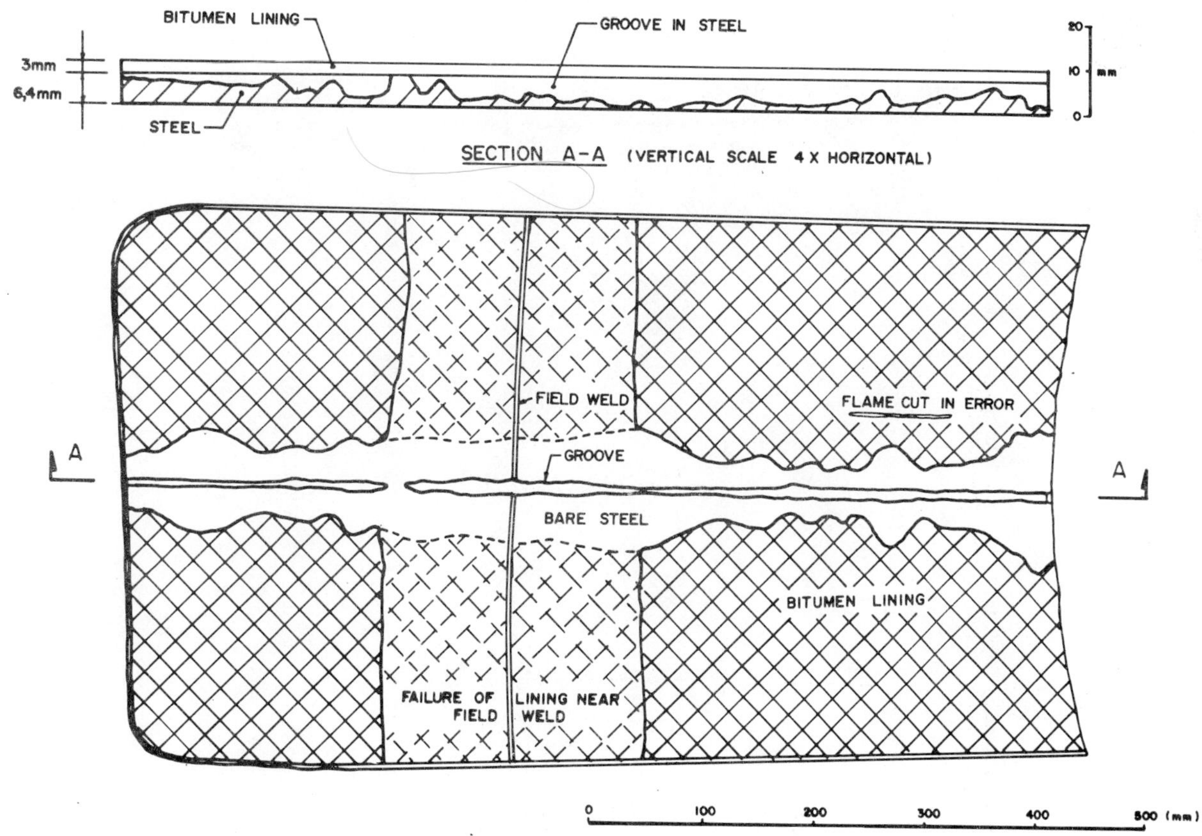

Fig. 3 Plan and section of sample cut from crown of damaged pipe (lying upside down).

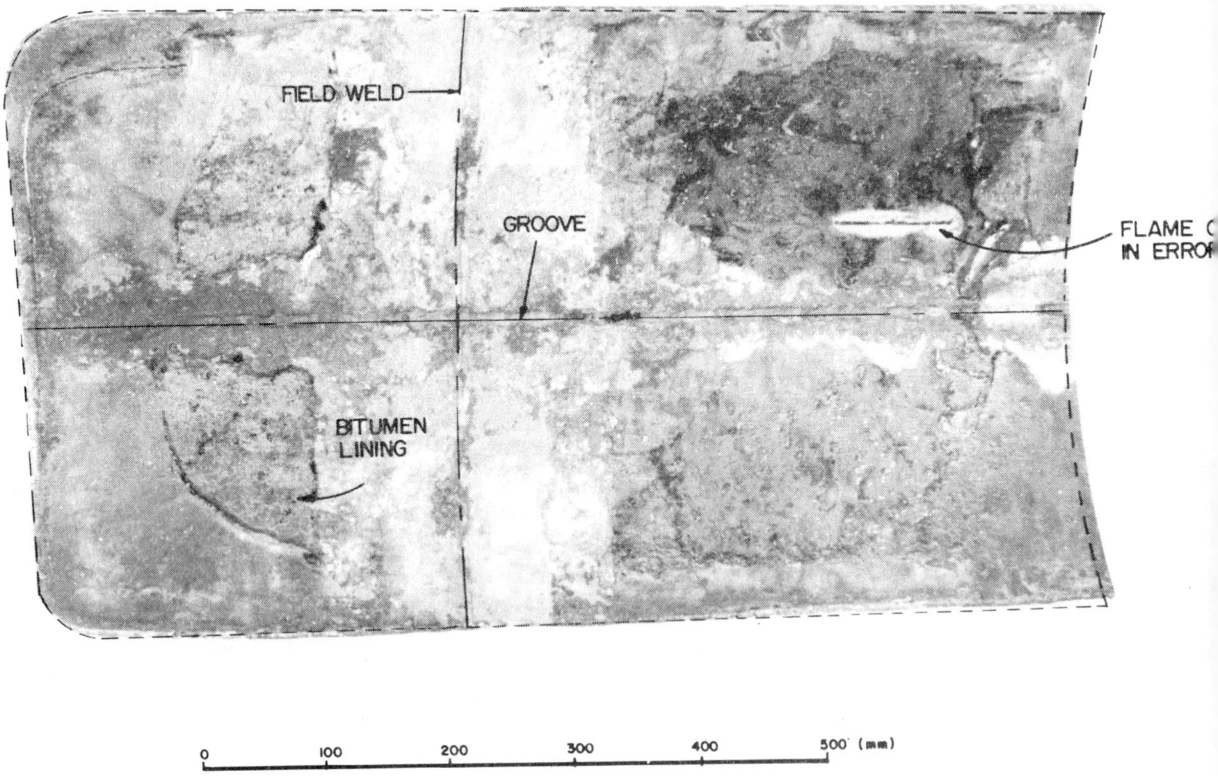

Fig. 4 Photograph of sample cut from crown of damaged pipe.

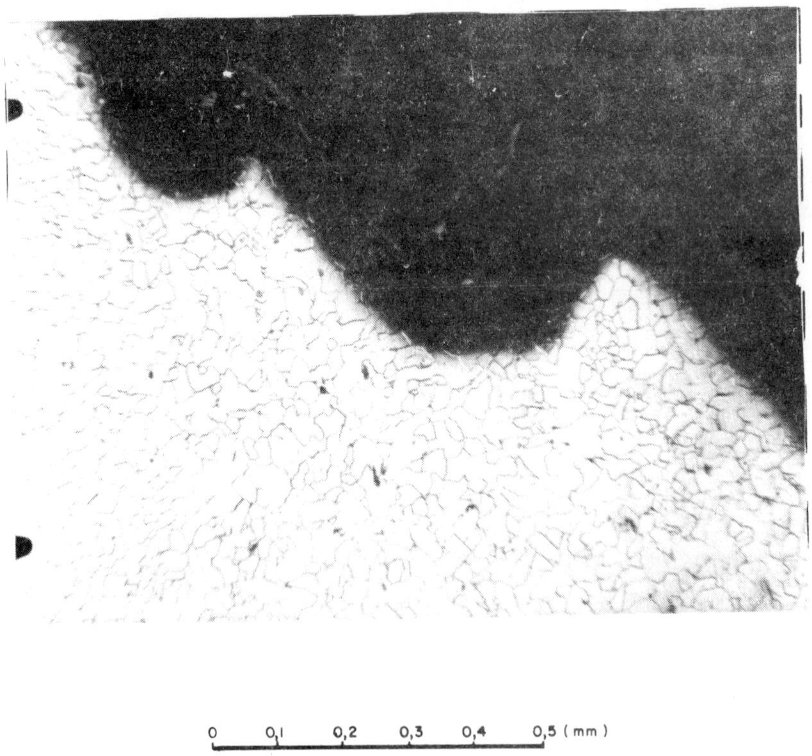

Fig. 5 Photomicrograph of cross-cut at groove.

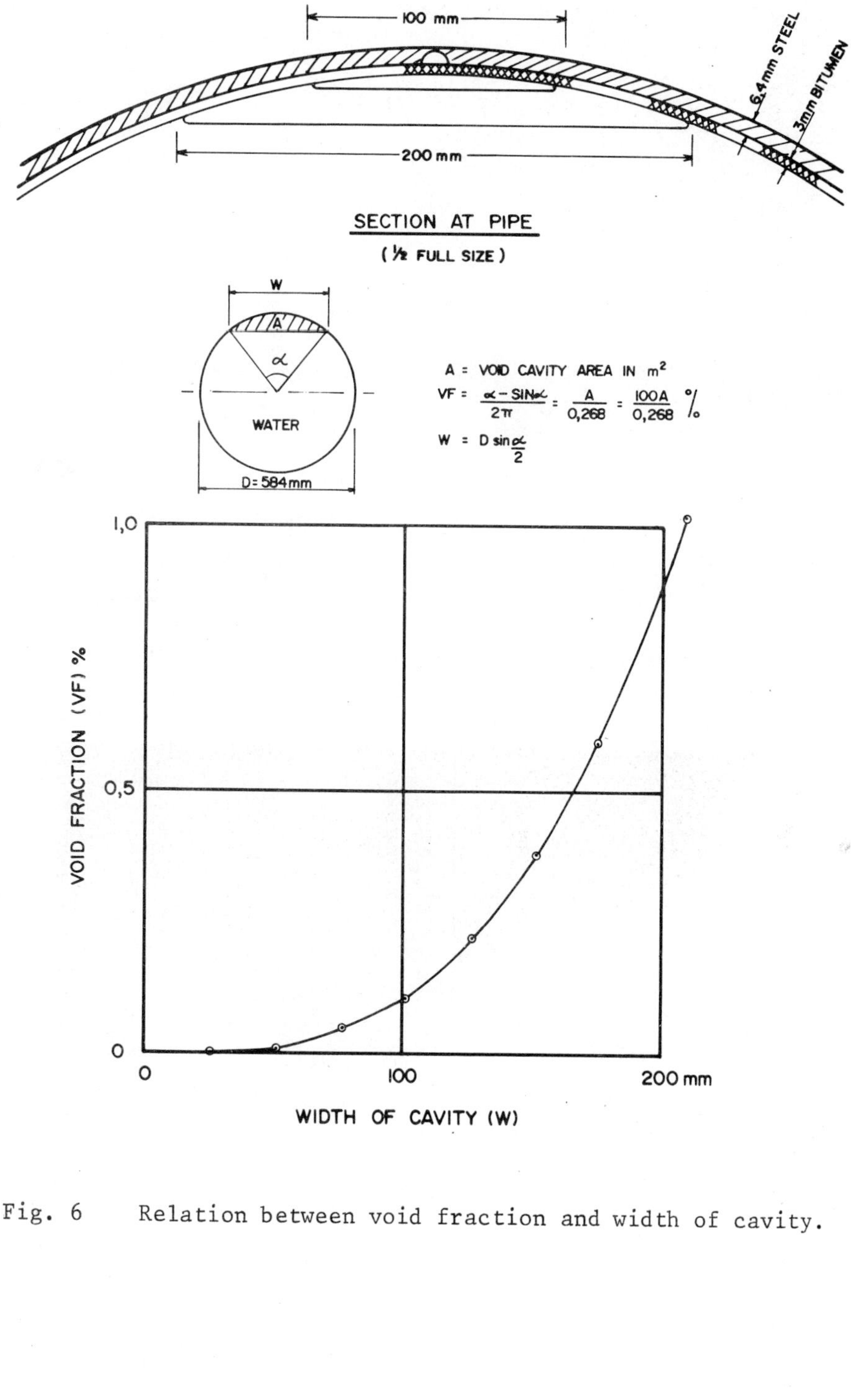

Fig. 6 Relation between void fraction and width of cavity.

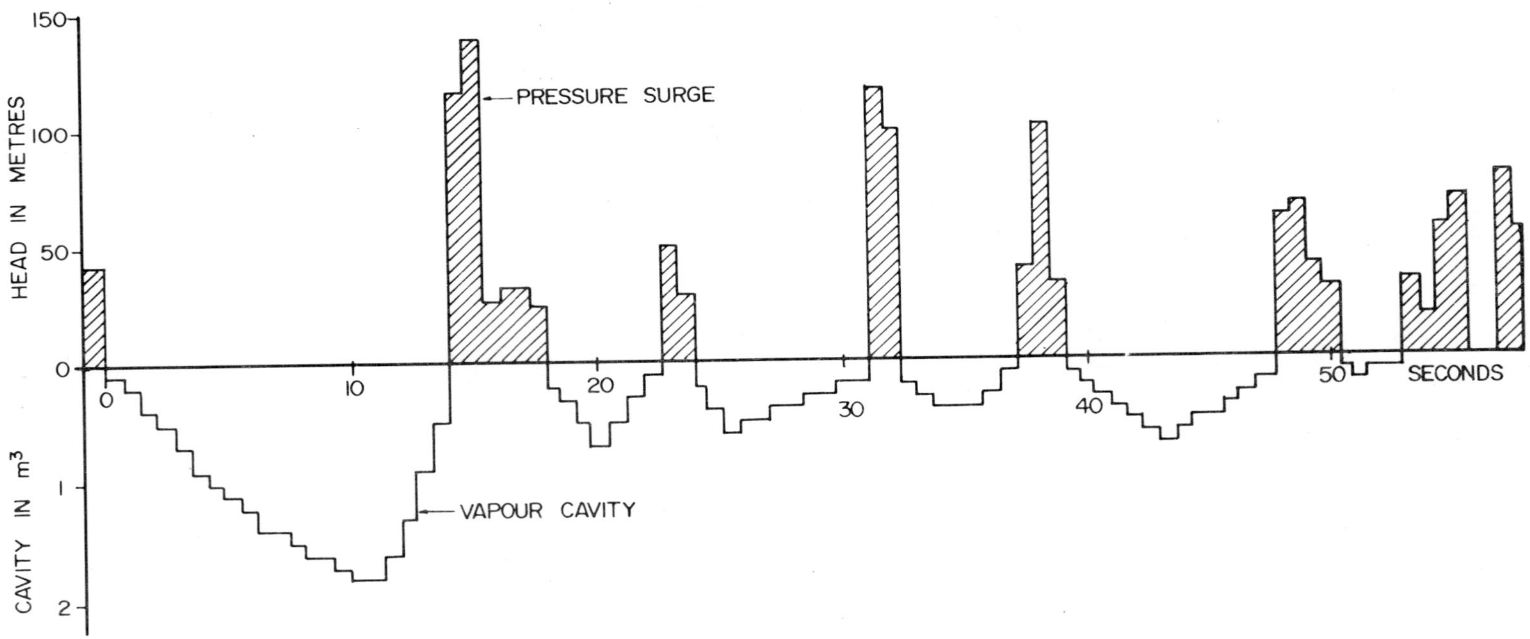

Fig. 7 Computed surge pressure diagram, showing vapour cavity volumes at middle of section DE after pump trip.

Third International Conference on
PRESSURE SURGES
Canterbury, England
March 25-27, 1980

PAPER J1

THE DYNAMIC BEHAVIOUR OF NON-RETURN VALVES

G.A. Provoost
Delft Hydraulics Laboratory, The Netherlands.

Summary

Non-return valves serve to prevent the reversal of the flow. If, however, the reversal of the flow occurs in a very short time the valve may close when the flow already has been reversed. Depending on the type of non-return valve a sudden decrease of the reversal flow will occur, possibly resulting in unallowable pressure variations and slamming of the valve.

To theoretically predict whether slamming of a valve is to be expected,data under dynamic conditions of this valve have to be known. Since in most cases these data are not known the behaviour of a non-return valve during transient flow may be predicted by means of dynamic tests for several conditions. These tests can be carried out in a circuit which allows a sudden reversal of the flow. A description is given of such a test-rig. With a non-return valve installed in this test-rig,flow rate and upstream and downstream pressures can be measured during decreasing and possibly reversed flow under various conditions.

The results of tests carried out for two types of non-return valves are shown. The results of the measurements are explained and discussed. Tests have shown that especially the fluid velocity gradient with time is of decisive importance when considering the dynamic behaviour of non-return valves.

Held at the University of Kent at Canterbury, U.K.

Conference Organised by BHRA Fluid Engineering, Cranfield, Bedford, England.
In conjunction with The City University, London.

NOMENCLATURE

a	torque coefficient
c	velocity of propagation of pressure waves
D	pipe diameter
f	friction coefficient according to White-Colebrook
g	acceleration due to gravity
H	pressure head
ΔH	pressure head loss
I	mass moment of inertia
Q	flow rate
t	time-coordinate
T	total torque exerted on valve body
T_e	torque exerted on valve body due to external forces
T_f	torque exerted on valve body due to friction
T_h	torque exerted on valve body due to hydrodynamic pressures
T_w	torque exerted on valve body due to weight of valve body
v	fluid velocity
x	coordinate along axis of pipeline
α	angle of rotation of valve body
ξ	pressure head loss coefficient

1 INTRODUCTION

A non-return valve is a valve which serves to prevent reverse flow. Non-return valves are often used in combination with pumps. When these pumps are switched off the flow through these pumps, and consequently the flow through the non-return valves, will decelerate to zero. If the non-return valves are not closed at the instant of zero flow, the flow will reverse and the non-return valve will close during reverse flow, possibly resulting in unexpected high pressure variations and slamming of the valve.

It is important to know the dynamic behaviour of non-return valves so that in the design-stage of a hydraulic project it is possible to predict whether the chosen type of non-return valve will satisfy. For the purpose a theoretical description of the behaviour of non-return valves during transient flow was set up. Since up to now insufficient data are available only dynamic tests under several conditions may lead to acceptable predictions. The aim of this study was to find the determinants on which the dynamic behaviour of non-return valves depends.

2 THEORY

A non-return valve consists of a valve body, whose position is influenced by the flow (closed, partly open or open position). Generally the valve body will be in the closed position if a pressure drop exists in the direction in which the flow has to be prevented. In this situation no reverse flow occurs. If, however, the direction of the flow is changing from the permitted one into reverse direction in a very short time, it is possible that the valve body is still in a partly open position while reverse flow occurs. Due to the reverse flow the valve body will move fast into the closed position, thus causing a sudden zero flow condition. Two phenomena can be observed at the instant of closure of the non-return valve, viz.

(1) a sudden stop of the movement of the valve body
(2) a sudden stop of the reversed flow through the non-return valve.

The first phenomenon can be noted by a more or less heavy sound depending on the velocity of the valve body at the instant of closure. This phenomenon is often referred to as the slamming of the valve.

The second phenomenon causes a pressure rise at the initial downstream side of the valve. At the same time a pressure drop occurs at the initial upstream side of the valve. This pressure drop may be so large that the vapour pressure of the fluid is reached. In that case a vapour bubble will grow at the initial upstream side of the valve. The disappearance of such a vapour bubble causes a sudden pressure rise which may lead to a re-opening of the valve. The second closure of the valve may again cause reaching of the vapour pressure thus leading to a repeating phenomenon.

To avoid slamming of the valve it may be required that the valve-body reaches its closed position at the instant the flow becomes zero. Whether this occurs may be predicted by comparing the calculated time interval in which the flow in the relevant pipe system decreases to zero and some minimum possible closing time of the valve. Such a method was proposed by Lewinsky-Kesslitz (Ref. 1) who has also given theoretical minimum possible closing times for rectangular swing type non-return valves. However, the use of this method will not always result in acceptable predictions since the motion of the water during the closing procedure is neglected.

Only a more detailed analysis of the motion of the valve body will lead to a good prediction whether slamming will occur or not. Hereto all forces acting on the valve body have to be taken into account. The following consideration is made for valves with rotating valve bodies but will be equivalent for valves with translating valve bodies. The motion of the valve body will depend on the flow rate through the valve and the upstream and downstream pressures. The relationship between these variables is given in the equations given below:

- The motion of a rotating valve body is described by Newton's Second Law of Motion:

$$T = I \frac{d^2\alpha}{dt^2}. \qquad (1)$$

where: T = torque of the valve body,
I = mass moment of inertia of rotating parts and a part of the surrounding water, and
α = angle of rotation of the valve body.

The effect of inertia of the fluid surrounding the valve body will be appreciable, especially for fast motion of the valve body, and has to be added to the mass moment of inertia of the moving parts of the valve. As a simplification it has been assumed in equation (1) that added mass is independent of the motion of the valve body and flow around the valve body.

The torque exerted on the valve body is due to the weight of the valve body (T_w), external forces (T_e), friction (T_f) and hydrodynamic pressures (T_h):

$$T = T_w + T_e + T_f + T_h. \qquad (2)$$

The torque caused by the weight of the valve body (gravity) depends on the dimensions and geometry of the valve. The torque due to external forces may be caused by a ballast weight or a spring and depends also on the geometry of the valve. The torque caused by hydrodynamic pressures does not depend on the geometry of the valve but rather on the flow around the valve body. Since the motion of the valve body disturbs the steady state flow around the valve body the flow pattern around the valve body will depend on the flow passing the valve, on the position of the valve body, and on the angular speed of the valve body:

$$T_h = f(Q, \alpha, \frac{d\alpha}{dt}). \qquad (3)$$

- The pressure loss created by the valve also depends on the flow pattern around the valve body. Written in terms of pressure head we obtain:

$$\Delta H = g(Q, \alpha, \frac{d\alpha}{dt}). \qquad (4)$$

- The equations describing unsteady one-dimensional pipe flow in both adjacent pipes are obtained from the consideration of the mass conservation of a control-volume for a pipe segment (5) and from the momentum equation along the axis of the flow (6) as given by Bergeron (Ref. 2):

$$\frac{\partial H}{\partial t} + v\frac{\partial H}{\partial x} + \frac{c^2}{g}\frac{\partial v}{\partial x} = 0. \qquad (5)$$

$$\frac{\partial v}{\partial t} + v\frac{\partial v}{\partial x} + g\frac{\partial H}{\partial x} = -\frac{f}{2D}v|v|. \qquad (6)$$

These equations hold as long as the actual pressures are higher than the vapour pressure of the liquid. If, however, the actual pressure equals the vapour pressure further expansion of the liquid only leads to the appearance and increase of vapour bubbles, a phenomenon known as cavitation of the liquid. The equations describing this kind of two-phase flow are described elsewhere (Ref. 3).

With equations (1) to (6) and the boundary conditions of the adjacent pipe ends (such as a pump or an air vessel) the motion of the valve body can be predicted.

However, up to know the relationships (3) and (4) are not known. Only by means of dynamic tests the relationships for a certain type of valve can be determined.

This is the reason why the behaviour of a non-return valve under dynamic conditions cannot be predicted accurately.

Only for slow motion of the valve body a prediction of the behaviour of the valve can be made. If the influence of the motion of the valve body on the flow pattern around the valve body may be neglected both the torque exerted on the valve body caused by hydrodynamic pressures and the pressure loss created by the valve will depend only on the flow passing the valve and the position of the valve body.

With the assumption that the flow pattern is independent of the flow rate (no Reynolds effects) the torque and pressure head loss will be proportional to the flow rate squared:

$$T_h = a(\alpha)\ Q|Q|. \tag{7}$$

$$\Delta H = \xi(\alpha)\ Q|Q|. \tag{8}$$

The coefficients a and ξ which depend on the position of the valve body have different values for the two possible directions of the flow. These coefficients can be determined experimentally by means of steady state flow tests or theoretically estimated with the assumption of potential flow.

3 DESCRIPTION OF THE TEST-RIG FOR NON-RETURN VALVES

To determine the dynamic behaviour of a certain type of non-return valve the flow through this valve has to be reversed in a very short time. For this purpose a special test-rig for non-return valves was set up in which the initial flow can be reversed in a controlled time interval (see Fig. 1). This test-rig consists of a by-pass on the test-rig for valves of the hydraulic machinery laboratory which forms part of the facilities of the Delft Hydraulics Laboratory. This test-rig for valves is a closed loop circuit equiped with pumps and a pressure control tank and has been described in detail by Oldenziel and Teijema (Ref. 4). The pressure level in the pipe system is adjusted by the pressure control tank in the test-rig for valves. During the tests a certain flow rate is pumped through the test-rig for valves. The flow rate through the non-return valve to be tested in the initially steady state condition can be adjusted by the control valve in the test-rig for valves. The initial upstream side of the test-section of the test-rig for non-return valves is provided with two combined fast acting valves. With this kind of three-way valve the initial upstream side of the test-section can be disconnected from the test-rig for valves and connected with a small air vessel. The pressure in this vessel can be set at a predetermined low value.

With the aid of the combined fast acting valves the pressure at the initial upstream side of the test-section can be lowered in a controlled time to the predetermined value present in the small air vessel. This causes a sudden pressure difference over the test-section leading to a fast, almost linearly decreasing flow and possibly to a reversal of the flow until the non-return valve closes.

The time-dependent flow rate in the test-rig is measured with a fast acting flow-meter. This flow-meter is situated at the initial upstream side of the valve to be tested. The pressures at both sides of the valve are measured with piezo-electric transducers. Both flow rate and pressures are recorded on a UV-recorder.

4 EXPERIMENTS

Experiments were carried out for two types of valves having a diameter of 100 mm each:

(a) a moving ball type non-return valve (see Fig. 2). In this case the valve body consists of a ball which is pushed from its seat by the flow into a slot under normal flow condition.
(b) a swing type non-return valve (see Fig. 3). In this case the valve body consists of a rotating steel plate. At the axis an arm provided with a ballast-weight was installed.

The experiments were carried out for steady state velocities of 1 m/s, 2 m/s and 3 m/s. For each velocity tests were carried out with different pressures in the small air vessel resulting in different gradients for the decrease of velocity in the test section. Moreover for the swing type non-return valve a test was carried out for a smaller distance between ballast-weight and centre of rotation. This test is referred to as experiment No. 7. Time histories of velocity, initially upstream and downstream pressures were recorded for each test. For the swing type non-return valve also the position of the valve body was recorded.

A summary of the experiments shown in this paper is given in table 1.

Experiment No.	Type of non-return valve	Initial steady state velocity	Created pressure drop at initial upstream side
1	moving ball type	1 m/s	15 kPa
2	moving ball type	3 m/s	30 kPa
3	moving ball type	3 m/s	40 kPa
4	moving ball type	3 m/s	50 kPa
5	swing type	1 m/s	60 kPa
6	swing type	2 m/s	70 kPa
7	swing type	2 m/s	70 kPa
8	swing type	3 m/s	75 kPa

Table 1: Summary of experiments.

5 RESULTS

In Figs. 4 to 11 the results of the experiments are shown. These results show that, in each experiment, reversal of the flow occurs and that the valve closes in reverse flow. In table 2 a summary is made of the time interval needed for reversal of the flow, the measured fluid velocity gradients and velocities of the reversed flow at the instant of closure. The positive direction of the flow is taken in the allowed direction of the flow through the non-return valve.

Experiment No.	Time interval for reversal of flow	Fluid velocity gradient	Fluid velocity at closure
1	0,28 s	$-3{,}6$ m/s^2	$-1{,}0$ m/s
2	1,15 s	$-2{,}6$ m/s^2	$-0{,}7$ m/s
3	0,88 s	$-3{,}4$ m/s^2	$-1{,}0$ m/s
4	0,56 s	$-5{,}4$ m/s^2	$-1{,}6$ m/s
5	0,20 s	$-5{,}0$ m/s^2	$-1{,}05$ m/s
6	0,30 s	$-6{,}7$ m/s^2	$-1{,}45$ m/s
7	0,30 s	$-6{,}7$ m/s^2	$-2{,}0$ m/s
8	0,40 s	$-7{,}5$ m/s^2	$-1{,}9$ m/s

Table 2: Summary of measured fluid velocities.

The experiments show that, with the described test-rig, the initial flow can be reversed in a controlled time interval with a nearly constant gradient with time.

In all experiments the valve closed in reversed flow. At the instant of closure a pressure rise occurred at the initial downstream side of the valve and a pressure drop occurred at the initial upstream side of the valve.

From experiments Nos. 2, 3 and 4 (see Figs. 5 to 7) it follows that, for a given steady state flow rate, a larger fluid velocity gradient - which gives a shorter time interval needed for reversal of the flow - leads to larger negative velocities at the instant the valve body (the ball) reaches its seat. And thus with a larger fluid velocity gradients larger pressure fluctuations occur. In experiments Nos. 2 and 3 (see Figs. 5 and 6) vapour pressure is not reached at the initial upstream side of the valve. However, in experiment No. 4 (see Fig. 7) vapour pressure is reached at the initial upstream side of the valve leading to the formation of a concentrated cavity (column-separation). The disappearance of this cavity causes a sudden pressure rise leading to re-opening of the valve and to a repetitive slamming of the valve. Also in experiments Nos. 2 and 3 the valve re-opens after the first closure but no repetitive slamming occurred. The phenomena after closure of the valve in experiments Nos. 1 and 3 (see Figs. 4 and 6) are comparable with each other although the initial fluid velocities were different. The fluid velocity gradients, however, were of the same order. This implies that the fluid velocity gradient dominates the dynamic phenomena during valve closure and that the time interval needed for reversal of the flow is of minor importance.

In all experiments for the swing type non-return valve (see Figs. 8 to 11) the valve body began to move round about the instant of reversal of the flow. This means that friction plays an important role. The valve is in more open position for larger flow rates. In experiment No. 8 (fluid velocity 3 m/s) the valve was in complete open position. A comparison of experiments Nos. 6 and 7 shows that the position of the ballast-weight influences the valve-opening. This means that the position of the valve body during steady state flow depends on the steady state flow rate and the position of the ballast-weight. As the valve body began to move round about the instant of reversal of flow the dynamic phenomena will only depend on the fluid velocity gradient and the position of the valve body during steady state flow. This implies that experiments for steady state fluid velocities larger than 3 m/s will give the same results as an experiment for 3 m/s and the same fluid velocity gradient, since for 3 m/s the valve is already in a fully open position. This is the reason why for large steady state fluid velocities and consequently for large steady state flow rates, the fluid velocity gradient is of decisive importance. For velocities smaller than 3 m/s both fluid velocity gradient and position of the valve body during steady state flow are important. A comparison of experiments Nos. 6 and 7 (see Figs. 9 and 10) shows the influence of the position of the valve body during steady state flow. In both experiments the steady state velocities were 2 m/s and the fluid velocity gradients were - 6,7 m/s. Due to a smaller distance between ballast weight and centre of rotation the valve in experiment No. 7 was in more open position than in experiment No. 6. This implies that in experiment No. 7 the distance the valve body had to travel to closed position was larger than in experiment No. 6. As a result, in experiment No. 7 the time interval between reversal of flow and closure of the valve was larger than in case 6 which is the reason why the absolute value of the fluid velocity at the instant of closure in experiment No. 7 was larger than in experiment No. 6 (velocities were respectively - 2,0 m/s and - 1,45 m/s). For a given geometry of a valve the position of the valve body during steady state flow is only influenced by the steady state flow rate. This is the reason why besides the fluid velocity gradient the steady state flow rate is important when choosing several conditions for dynamic tests of non-return valves.

6 CONCLUSIONS

When theoretically predicting the behaviour of a certain non-return valve, data of this valve under dynamic conditions must be available. Since in most cases these data are not known the behaviour of the non-return valve during transient flow may be predicted by means of dynamic tests for several conditions. When choosing these conditions both steady state flow rate and fluid velocity gradient with time have to be considered. Especially for large flow rates the fluid velocity gradient is of decisive importance. The time interval needed for reversal of flow seems to be of minor importance. The dynamic tests can be performed with the test-rig for valves as described in this paper.

Experiments carried out for two types of non-return valves show that in many practical situations these non-return valves will close in reverse flow. At the instant of closure pressure variations will be initiated. At the initial downstream side of the valve a pressure rise will occur and at the initial upstream side a pressure drop will occur. Repetitive slamming may occur if vapour pressure is reached at the initial upstream side of the valve.

7 ACKNOWLEDGEMENTS

The measurements concerning the moving ball type non-return valve form part of the measurements carried out for such a valve provided with and without a special anti-slamming device. These tests were carried out on request of Hillen de Lelie B.V. The author thanks Hillen de Lelie B.V. for permission to publish some results of these measurements in this paper.

8 REFERENCES

1. Lewinsky-Kesslitz, H.P., Über die Dynamik der Rückschlagklappe, Österreichische Ingenieurs-Zeitschrift, 110. Jahrgang, Juni 1965, pp. 185-191.

2. Bergeron, L., Water-hammer in hydraulics and wave surges in electricity, John Wiley and Sons, Inc., New York, 1961.

3. Provoost, G.A., Investigation into cavitation in a prototype pipeline caused by water-hammer, Proc. 2nd International Conference on Pressure Surges, London, Sept. 22-24, 1976.

4. Oldenziel, D.M. and Teijema, J., Cavitation on valves in correlation to liquid properties, Proc. IAHR Symp. Problems of hydraulic machine - hydraulic structure interaction, Leningrad, Sept. 6-9, 1976.

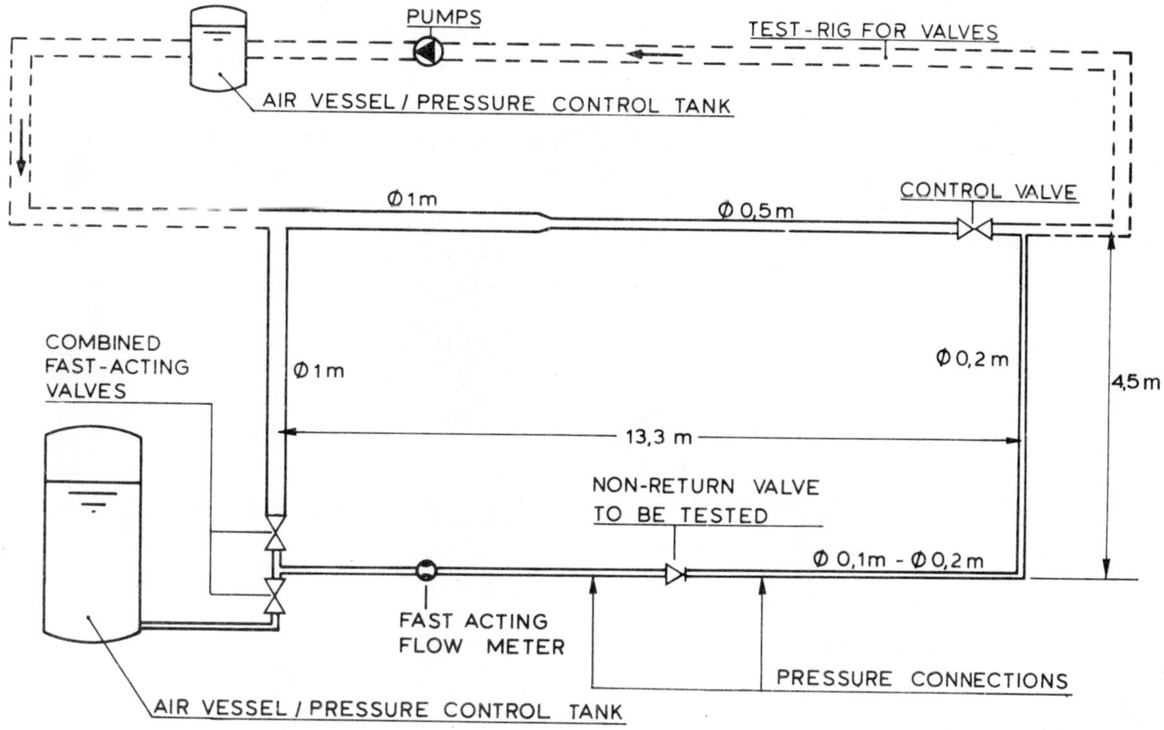

Figure 1 Schematic diagram of test-rig for non-return valves

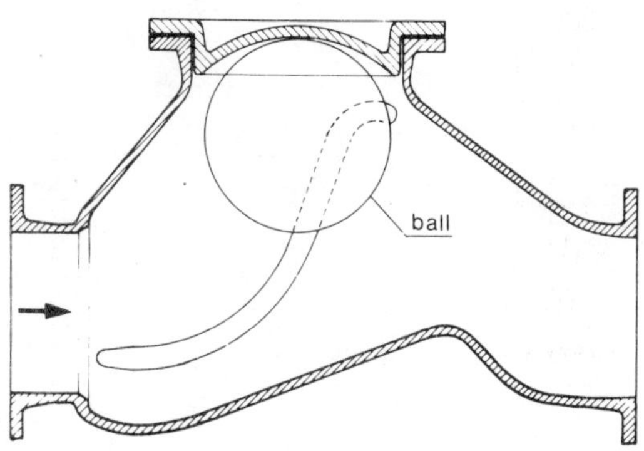

Figure 2 A moving ball type non-return valve

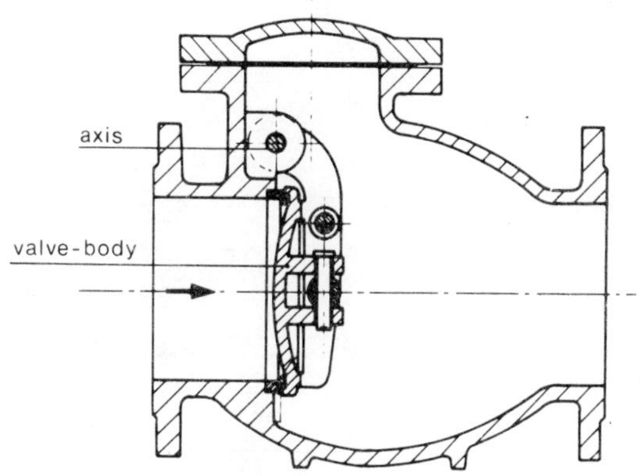

Figure 3 A swing type non-return valve

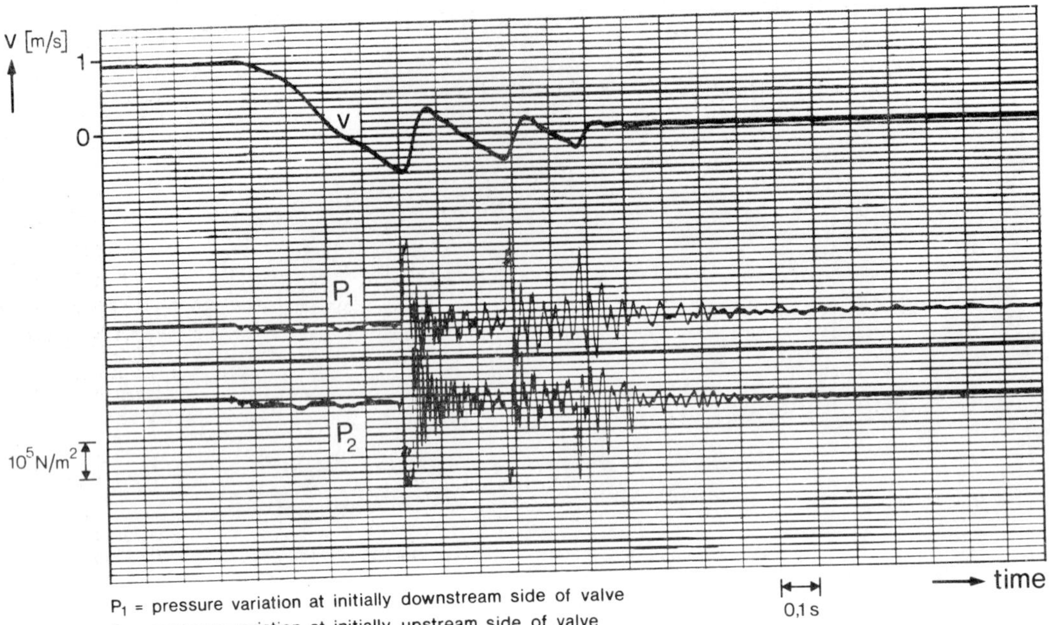

P_1 = pressure variation at initially downstream side of valve
P_2 = pressure variation at initially upstream side of valve
v = velocity of fluid in initially upstream pipe

Figure 4 Recording of experiment No. 1
(moving ball type non- return valve)

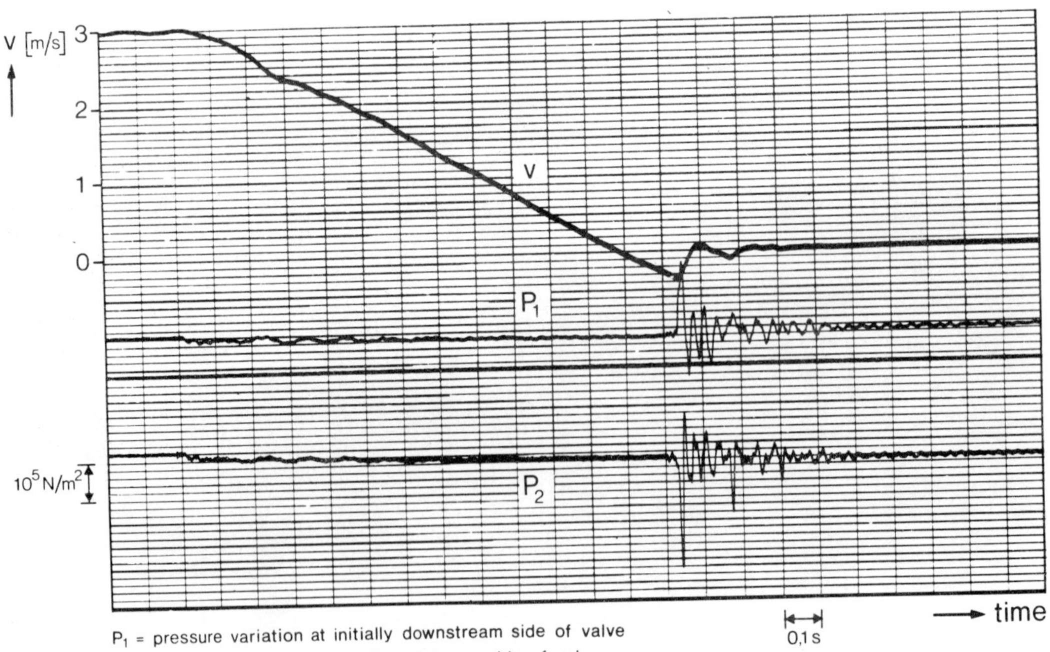

P_1 = pressure variation at initially downstream side of valve
P_2 = pressure variation at initially upstream side of valve
v = velocity of fluid in initially upstream pipe

Figure 5 Recording of experiment No. 2
(moving ball type non-return valve)

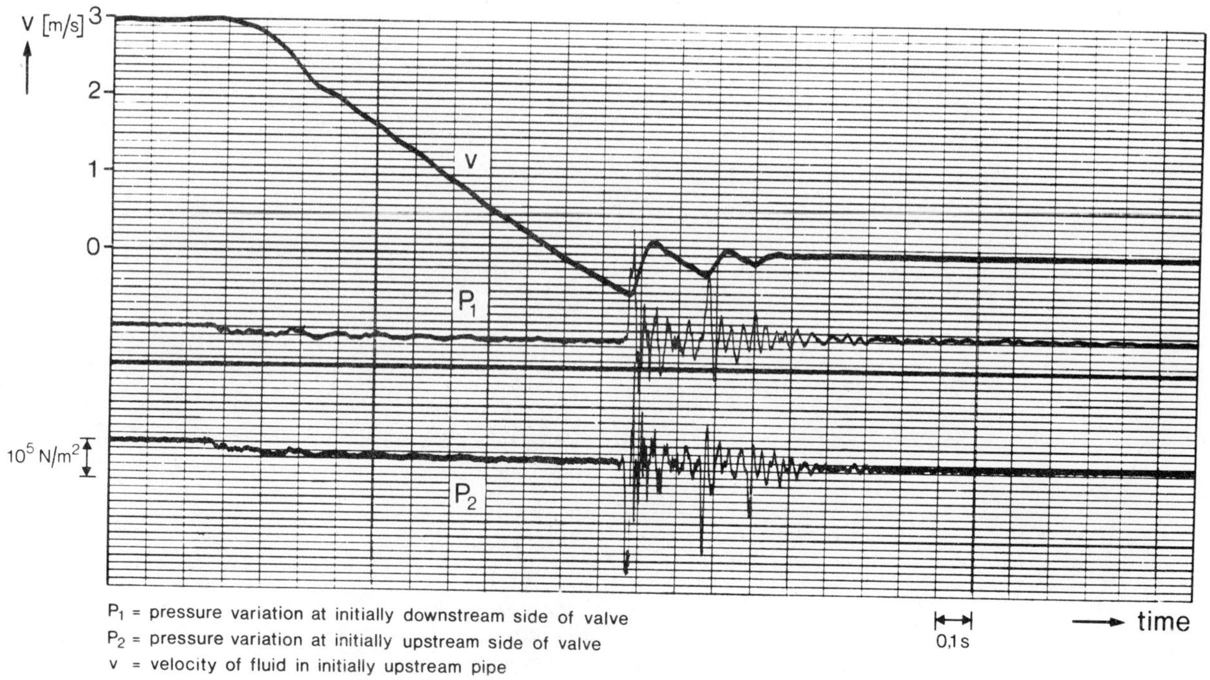

Figure 6 Recording of experiment No. 3 (moving ball type non-return valve)

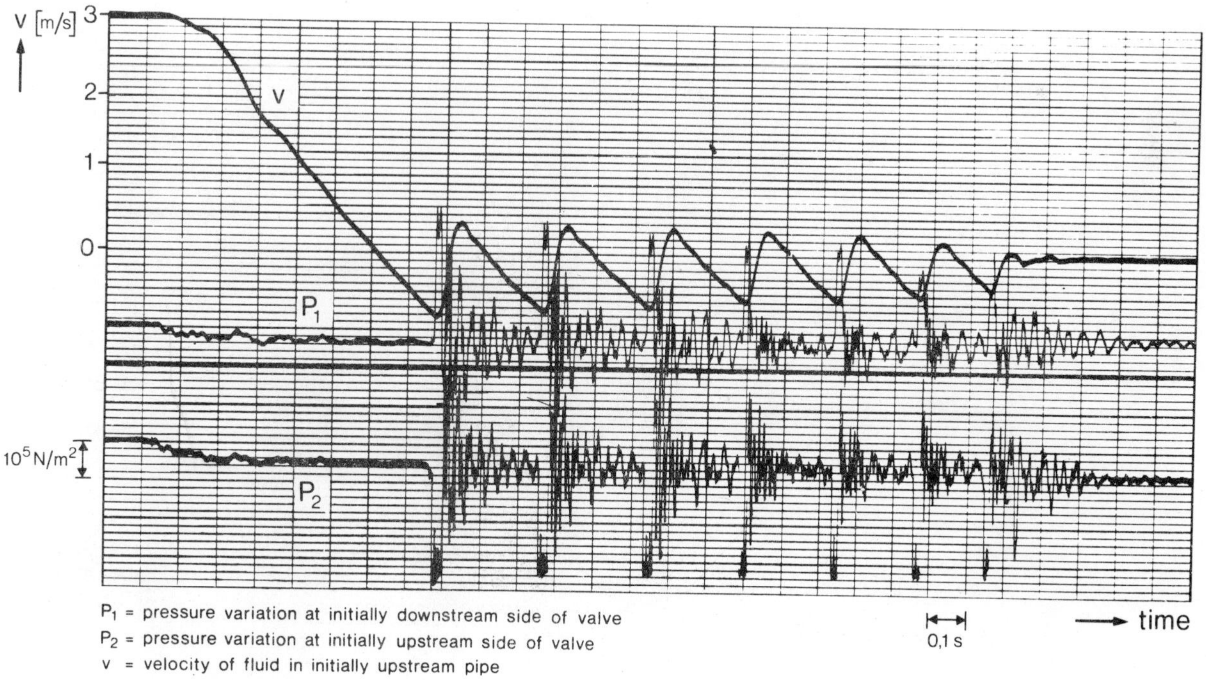

Figure 7 Recording of experiment No. 4 (moving ball type non-return valve)

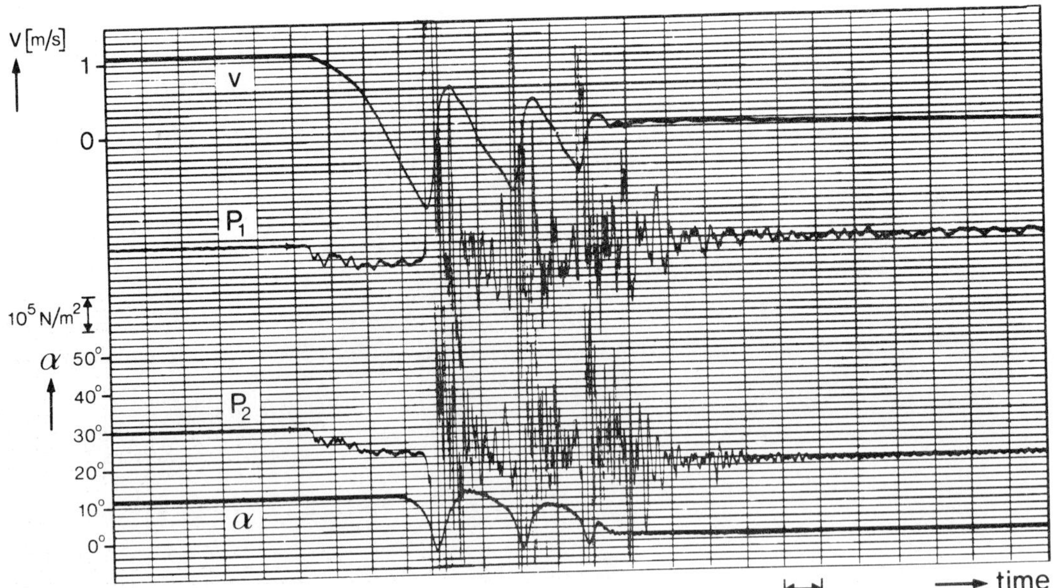

P_1 = pressure variation at initially downstream side of valve
P_2 = pressure variation at initially upstream side of valve
v = velocity of fluid in initially upstream pipe
α = rotation of valvebody to closed position

Figure 8 Recording of experiment No. 5 (swing type non-return valve)

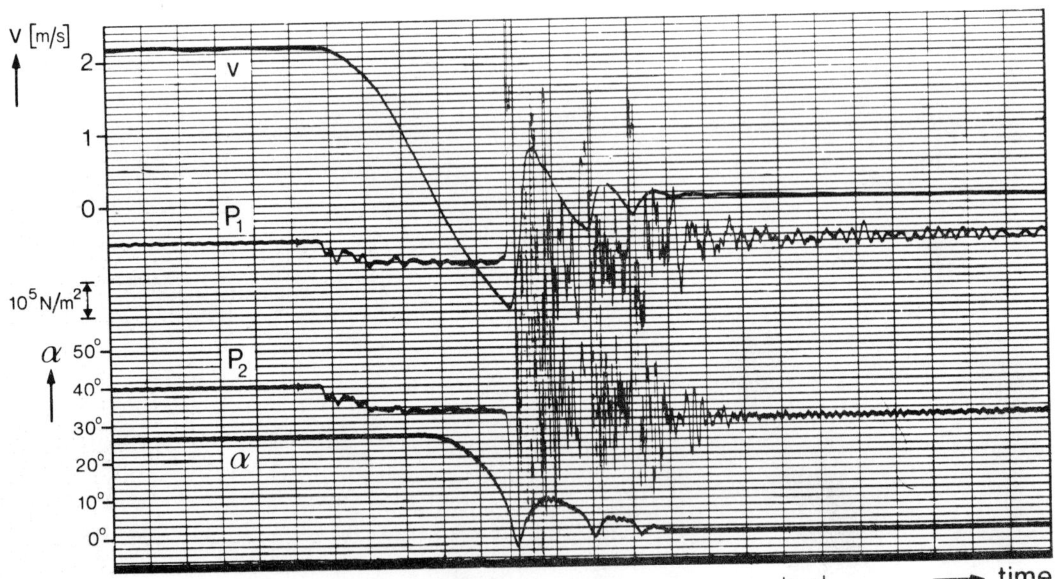

P_1 = pressure variation at initially downstream side of valve
P_2 = pressure variation at initially upstream side of valve
v = velocity of fluid in initially upstream pipe
α = rotation of valvebody to closed position

Figure 9 Recording of experiment No. 6 (swing type non-return valve)

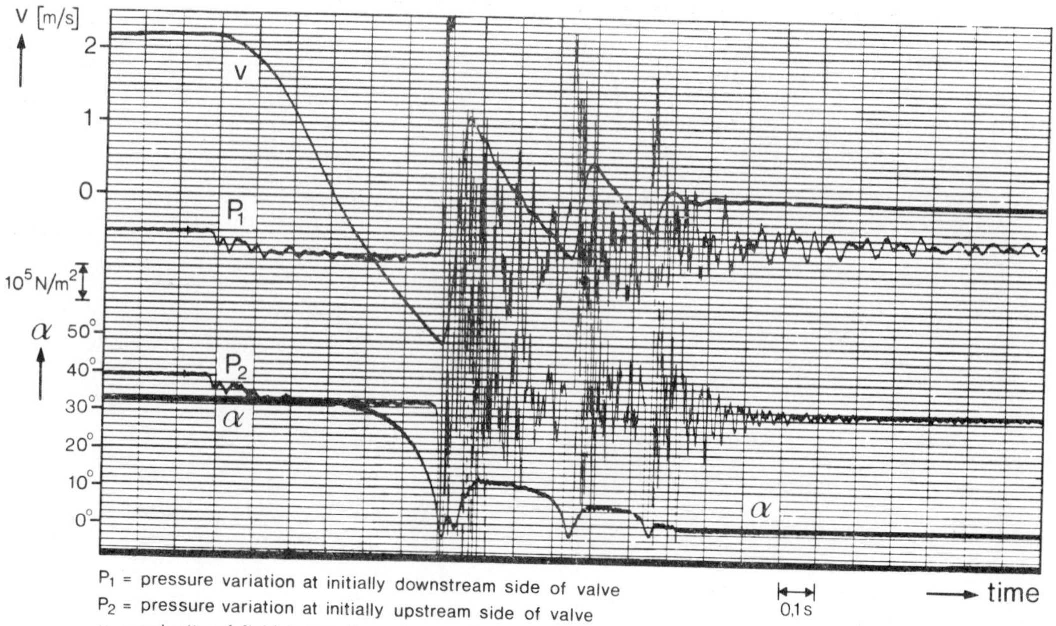

P₁ = pressure variation at initially downstream side of valve
P₂ = pressure variation at initially upstream side of valve
v = velocity of fluid in initially upstream pipe
α = rotation of valvebody to closed position

Figure 10 Recording of experiment No. 7 (swing type non-return valve)

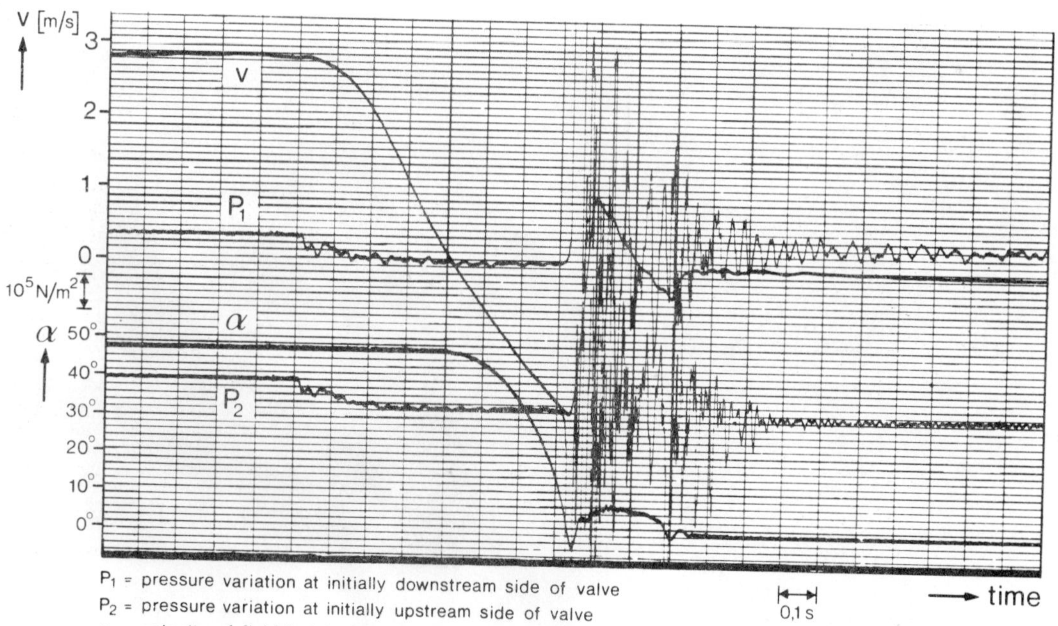

P₁ = pressure variation at initially downstream side of valve
P₂ = pressure variation at initially upstream side of valve
v = velocity of fluid in initially upstream pipe
α = rotation of valvebody to closed position

Figure 11 Recording of experiment No. 8 (swing type non-return valve)

Third International Conference on

PRESSURE SURGES

Canterbury, England
March 25-27, 1980

PAPER J2

THE INFLUENCE OF PRESSURE SURGES ON THE FUNCTIONING OF SAFETY VALVES

B. Föllmer and H. Zeller

Aerodynamisches Inst. der RWTH Aachen,
Federal Republic of Germany.

Summary

The functioning of a spring-loaded safety valve depends mainly on the magnitude of the pressure forces in the flow and on the force of the spring at the respective stroke. The functioning can be impaired by variations of the pressure in the valve inlet. Variation of the pressure can be caused by the stagnation-pressure loss in the feed line. The stagnation-pressure loss increases with the mass flux during the opening of the safety valve. A decrease of stagnation-pressure can be originated by the transient flow in the feed line at fast opening of the valve. In addition, oscillations of the stagnation-pressure in front of the valve disk may occur due to variations of the extent of the supersonic field, formed through jet contraction in a sharp edged inlet at supercritical pressure ratios; the variation of the extent of the supersonic field is fast. It is caused by the rapidly varying mass flux through the contracted cross section of the flow in the valve inlet and between the valve seat and disk. The influence of the stagnation pressure losses and the variations on the flow in the inlet depend on the geometry of the feed line. By suitable choice of the length and the diameter, malfunction of the valve can be avoided. The paper discusses further oscillation of the stagnation pressure in front of the valve disk. Mach-Zehnder-Interferometer pictures and a film obtained with a high speed camera in a two-dimensional model of a valve exhibit clearly the characteristic parameters of the time depended flow. The amplitudes and the frequencies of the stagnation-pressure oscillations measured in experiments are discussed in conjunction with flow observations.

Held at the University of Kent at Canterbury, U.K.

Conference Organised by BHRA Fluid Engineering, Cranfield, Bedford, England.
In conjunction with The City University, London.

Nomenclature

a	speed of sound
a_o	speed of sound at stagnation condition
a_K	speed of sound in the vessel
A_E, d_E	cross-sectional area, diameter in the entrance of the valve
A_o, d_o	narrowest cross-section and diameter ahead of the valve disk
b	height of the valve connection in the two-dimensional model
c_F	flow force coefficient
c_S	spring force coefficient
c_λ	spring stiffness coefficient
F	flow force
h	lift
H	dimensionless lift
K_{DM}	ratio of the pressure force to the inertial force (see eq. 2-23)
l_E	length of the feed line
m	mass of the moved parts of the valve
Ma	Mach number
p	pressure
p_o	total (stagnation) pressure
p_K	pressure in the vessel
p_a	back pressure
S	spring force
t	time
t_{op}	opening time of the valve
u	velocity
x	local position
Z	dimensionless time (see eq. 2-21)
α	discharge coefficient (see eq. 2-1)
$\varkappa$	isentropic exponent
λ	spring constant
ζ	drag coefficient

ϱ density

ψ isentropic discharge function

1. Introduction

Safety valves, which have been tested according to national regulations under normed test conditions, i.e. without connecting pipes directly connected to the test vessel may not work properly under real conditions since their installation differs from the test conditions. Especially in spring loaded safety valves the disk can fall into oscillating motion. In the pipe system oscillatory forces may arise which can destroy parts of the whole system. In order to avoid such dangerous conditions the German technical regulations (Ref. 1) for example prescribe a maximum permissible total pressure drop in the connecting line, which is only conditioned by the steady flow resistance. The maximum permissible total pressure loss must be smaller by a fixed amount than the reseat pressure difference of the valve. (The reseat pressure difference is the difference between the start-to-leak pressure and the seat-off-pressure of a safety valve.) The functioning of valves may also be influenced by pressure surges of large amplitudes in long pipes, which can be caused by fast opening of valves. The pressure surges may generate a large decrease of the total pressure in the entrance or a large increase of the back pressure in the exit of the valves for an interval of the time.

In this paper only the influence of the flow in the feeding line on the functioning of spring loaded valves will be investigated for gases. It will further be shown that also the shape of the entrance of the valve can cause oscillating functioning and pressure surges in the feed line.

2. Pressure surges in the feeding line

2.1 Total pressure loss caused by flow resistance in the feeding line (Ref. 5)

For steady flow conditions the total pressure loss depends on the flow resistance in the feeding line and on the state of the fluid in the vessel. For adiabatic flow conditions the mass flux per unit area $\varrho_E u_E$ in the entrance of the valve or the entrance Mach number Ma_E is determined by the pressure ratio p_a/p_{0E} and the effective ratio of the cross-sectional areas $A_E/\alpha A_0$ which is the ratio of the area of the pipe A_E and of the narrowest cross-section A_0 of the valve multiplied by the discharge coefficient α. The mass flux in the entrance of the valve can be expressed by the isentropic mass flux per unit area at the exit, the narrowest cross-sectional area of the valve and the discharge coefficient α.

$$\varrho_E u_E A_E = \alpha (\varrho_a u_a)_{th} A_0 \tag{2-1}$$

The Mach number Ma_E in the entrance of the valve can be calculated from the following relation

$$\frac{\varrho_E\, u_E}{\varrho_{OE}\, a_{OE}} = Ma_E \left(1 + \frac{\varkappa-1}{2} Ma_E^2\right)^{-\frac{\varkappa+1}{2(\varkappa-1)}} . \qquad (2\text{-}2)$$

For the theoretical mass flux density in the exit of the valve one obtains

$$\frac{\varrho_a\, u_a}{\varrho_{OE}\, a_{OE}} = \sqrt{\frac{2}{\varkappa}}\, \psi \qquad (2\text{-}3)$$

with
$$\psi = \sqrt{\frac{\varkappa}{\varkappa-1}\left[\left(\frac{p_a}{p_{OE}}\right)^{\frac{2}{\varkappa}} - \left(\frac{p_a}{p_{OE}}\right)^{\frac{\varkappa+1}{\varkappa}}\right]}$$

for pressure ratios
$$\frac{p_a}{p_{OE}} > \frac{p^*}{p_{OE}} = \left(\frac{2}{\varkappa+1}\right)^{\frac{\varkappa}{\varkappa-1}}$$

and
$$\psi = \psi_{max} = \left(\frac{2}{\varkappa+1}\right)^{\frac{1}{\varkappa-1}} \left(\frac{\varkappa}{\varkappa+1}\right)^{\frac{1}{2}} \quad \text{for} \quad \frac{p_a}{p_{OE}} < \frac{p^*}{p_{OE}} .$$

In adiabatic flow of an ideal gas with constant specific heats the drag coefficient ζ_{12} of a cylindrical pipe of length l_{12} is

$$\zeta_{12} = \zeta_{max,1}(Ma_1, \varkappa) - \zeta_{max,2}(Ma_2, \varkappa) \qquad (2\text{-}4)$$

In this expression $\zeta_{max,1}$ and $\zeta_{max,2}$ are the drag coefficients of fictitious pipes of length $l_{max,1}$ and $l_{max,2}$ with critical conditions at their ends $[Ma = 1, p = p^*, T = T^*]$.

$$\zeta_{max} = \frac{1-Ma^2}{\varkappa\, Ma^2} + \frac{\varkappa+1}{2\varkappa} \ln \frac{(\varkappa+1)\, Ma^2}{2\left(1 + \frac{\varkappa-1}{2} Ma^2\right)} \qquad (2\text{-}5)$$

Since ζ_{max} depends only on the Mach number at the entrance of the pipe they are state variables in adiabatic flow. Values of ζ_{max} are tabulated for example in (Ref.2 and 3). The tables also contain the ratio of the stagnation pressure p_o at the entrance and the stagnation pressure p_o^* at the exit of the fictitious pipes which for adiabatic flow depend only on the Mach number at the pipe entrance. The ratio of the stagnation pressures at the entrance and at the end or exit of the pipe with length l_{12} is therefore for known ζ_{12}

$$\frac{p_{02}}{p_{01}} = \frac{p_{02}/p_0^*}{p_{01}/p_0^*} = f(Ma_1, Ma_2, \varkappa) \; . \tag{2-6}$$

Since in adiabatic flow ζ_{max} is a state variable one can imagine every feeding pipe with arbitrary cross-section distribution and known total drag coefficient ζ_Z as an isentropic nozzle flow followed by an adiabatic pipe flow of cross-sectional area A_E whose drag coefficient generates the same total pressure loss and therefore the same state E in the entrance of the valve as the actual feeding line. According to eq. (2-4) and (2-6) the ratio between the pressure in the vessel p_K and the stagnation pressure in the entrance p_{oE} is

$$\frac{p_K}{p_{oE}} = f(\zeta_Z, Ma_E, \varkappa) \; . \tag{2-7}$$

It follows from the equations (2-1), (2-2), (2-3), and (2-7) that the pressure loss nondimensionalized by the difference $p_K - p_a$ depends on the following variables

$$\frac{p_K - p_{oE}}{p_K - p_a} = f\left(\zeta_Z, \frac{A_E}{\alpha A_o}, \frac{p_a}{p_K}, \varkappa\right) . \tag{2-8}$$

Fig. 2 shows the maximum permissible drag coefficient ζ_Z as a function of $A_E/\alpha A_o$ with the pressure ratio p_a/p_K as a parameter for a maximum total pressure loss $(p_K-p_{oE})/(p_K-p_a) = 0.03$ and $\varkappa = 1.3$. This dependence has been accepted in the German regulations (Ref. 1). The drag coefficient for gases with different isentropic exponents can be obtained from Fig. 3.

2.2 Stagnation pressure distribution in the entrance of fast opening valves with long feeding lines

Fast opening of the valve causes a steep expansion wave with a large amplitude, which is connected with a large decrease of the stagnation pressure ahead of the valve. The stagnation pressure rises again, when the reflected compression wave reaches the valve. In order to get an estimate of the dependence of the stagnation pressure p_{oE} on the time for a safety valve the time dependent lift of the disk during the opening phase is approximated by

$$h = h_{op} \sin^2\left(\frac{t}{t_{op}}\right) \; . \tag{2-9}$$

The quantity h_{op} is the lift for the fully opened valve and t_{op} is the corresponding opening time. The dependence of the discharge coefficient α on the lift h is approximated by

$$\alpha = \alpha_{op} \frac{h}{h_{op}} \tag{2-10}$$

with the discharge coefficient α_{op} at h_{op} determined in experiments. For this estimate it is assumed that the unsteady flow in the feeding line between vessel and entrance of the valve is onedimensional and isentropic. Then for an ideal gas with constant specific heats along the characteristics

$$\frac{dx}{dt} = u \pm a \tag{2-11}$$

the compatibility conditions are

$$du = \mp \frac{2}{\varkappa - 1} da \tag{2-12}$$

With the following dimensionless variables

$$\bar{x} = \frac{x}{l_E} \quad ; \quad \bar{t} = \frac{t}{t_{op}} \frac{\alpha_{op} A_0}{A_E}$$

$$\bar{u} = \frac{u}{a_K} \frac{t_{op} a_K}{l_E} \frac{A_E}{\alpha_{op} A_0} \quad ; \quad \bar{a} = \frac{a}{a_K} \frac{t_{op} a_K}{l_E} \frac{A_E}{\alpha_{op} A_0}$$

one obtains for the characteristic directions

$$\frac{d\bar{x}}{d\bar{t}} = \bar{u} \pm \bar{a} \tag{2-13}$$

and for the compatibility conditions

$$d\bar{u} = \mp \frac{2}{\varkappa - 1} d\bar{a} \tag{2-14}$$

In the entrance the stagnation pressure p_{oE} is connected with the flow velocity and the speed of sound by the following relations

$$\left(\frac{p_{oE}}{p_K}\right)^{\frac{\varkappa-1}{\varkappa}} = \frac{a_E^2}{a_K^2} + \frac{\varkappa-1}{2} \frac{u_E^2}{a_K^2} \tag{2-15}$$

$$\left(\frac{p_{oE}}{p_K}\right)^{\frac{\varkappa-1}{\varkappa}} = \left(\bar{a}_E^2 + \frac{\varkappa-1}{2} \bar{u}_E^2\right) \cdot \frac{1}{\left(\frac{t_{op} a_K}{l_E} \cdot \frac{A_E}{\alpha_{op} A_0}\right)^2} \quad . \tag{2-16}$$

For the stagnation pressure in the entrance of the valve the following dependence is valid

$$\frac{p_{oE}}{p_K} = f\left(t, \frac{A_0}{A_E}, \alpha_{op}, t_{op}, a_K, l_E, p_a\right) \tag{2-17}$$

or in terms of dimensionless parameters

$$\frac{p_{oE} - p_a}{p_K - p_a} = f\left(\bar{t}, \frac{t_{op} a_K}{l_E} \frac{A_E}{\alpha_{op} A_0}, \frac{p_a}{p_K}\right) \quad . \tag{2-18}$$

Fig. 4 shows the calculated distribution of the dimensionless pressure difference $(p_{oE}-p_a)/(p_K-p_a)$ for $p_a/p_K = 0.2$ and $(t_{op}a_K/l_E)(A_E/\alpha_{op}A_o) = 9$ as a function of time. The values chosen for the dimensionless parameters correspond to an actual case with $A_E/(\alpha_{op}A_o) = 3$, $t_{op} = 0.08$ s, $a_K = 340$ m/s and $l_E = 9$ m. It is seen in Fig. 4 that after the beginning of the opening the stagnation pressure ahead of the valve p_{oE} drops as a consequence of the expansion wave entering the feeding line. When the pressure wave reflected at the vessel arrives at the valve, the stagnation pressure p_{oE} increases beyond the pressure in the vessel p_K. The maximum dimensionless decrease of the total pressure is $[(p_K-p_{oE})/(p_K-p_a)]_{max} = 0.145$ in this example. This value is much larger than the maximum permissible stagnation pressure loss caused by the flow resistance in the pipe and also much larger than the reseat pressure difference of many valves. Fig. 5 shows the maximum stagnation pressure drop $[(p_K-p_{oE})/(p_K-p_a)]_{max}$ as a function of $(t_{op}a_K/l_E)(A_E/\alpha_{op}A_o)$ with p_a/p_K as parameter. If it is requested that during the opening phase a certain value of the dimensionless total pressure difference $[(p_K-p_{oE})/(p_K-p_a)]_{max}$ is not exceeded, then for given values $A_E/(\alpha_{op}A_o)$, a_K and t_{op}, the permissible length of the feeding pipe l_E can be calculated from the values given in Fig. 5 for the assumed distribution h(t) according to eq. (2-9) and the dependence $\alpha(h)$ according to eq.(2-10). Whether the decrease of the stagnation pressure during the unsteady flow phase influences the functioning of the valve depends mainly on the dependence between lift and time. In an actual valve this relation may deviate from the one given by eq. (2-9). In one hand the lift time relation is determined by the mass of the parts of the valve to be accelerated and on the other hand by the length and the cross-sectional area of the feeding pipe.

In order to demonstrate this point the equation of motion for the disk was solved simultaniously with the equations of motion for unsteady flow in the pipe. The motion of the disk is determined by

$$m\ddot{h} = F(h, p_{oE}) - S(h) \tag{2-19}$$

with m the mass of the moved parts of the valve, h the lift, F the force of the flow and S the force of the spring. The distribution of the flow force $F = F(h,p_{oE})$ can be determined experimentally in a quasisteady flow without feeding pipe with the stagnation pressure p_{oE} being equal to the pressure in the vessel p_K. The force of the spring then is

$$S = S_o + \lambda h \tag{2-20}$$

where λ is the constant of the spring stiffness and $S_o = (p_K-p_a)\,A_o$.

With the dimensionless variables

$$H = \frac{h}{d_o} \quad \text{and} \quad Z = \frac{t\,a_K}{l_E},$$

the coefficient of the flow force

$$c_F = f(H)\cdot\left[1 + \frac{p_{oE}-p_K}{p_K-p_a}\right] \tag{2-21}$$

and the coefficient of the force of the spring

$$c_S = 1 + \frac{\lambda\, d_0}{(p_K - p_a) A_0}\, H = 1 + c_\lambda H \qquad (2\text{-}22)$$

is obtained from eq.(2-19)

$$\frac{d^2 H}{d Z^2} = (c_F - c_S)\, \frac{(p_K - p_a)\, A_0 \cdot l_E^2}{m\, d_0\, a_K^2} \quad . \qquad (2\text{-}23)$$

In the last equation the expression

$$\frac{(p_K - p_a) A_0}{m \dfrac{d_0 a_K^2}{l_E^2}} \equiv K_{DM}$$

is a similarity parameter, which gives the ratio of the pressure force acting on the disk to the adapted inertia force of the system. Measured flow force coefficients $c_F(H)$ may be found in (Ref.4). In Fig. 6 the flow force coefficient of a safety valve and the spring force coefficient for $c_\lambda = 1.0$ are shown. These values are used in the following calculations. Fig. 7 shows the influence of the similarity parameter K_{DM} on the dimensionless pressure difference and the lift, respectively, for $c_\lambda = 1.0$, $A_E/\alpha A_0 = 3$ and $p_a/p_K = 0.2$. For $K_{DM} = 0.5$ the amplitude of the oscillation of the disk decreases despite of the assumptions of inviscid flow in the feeding pipe and undamped motion of the disk. For $K_{DM} = 1.1$ the amplitude of the oscillation increases with time. It is quite natural that the parameters c_λ, $A_E/\alpha A_0$, p_a/p_K and especially the dependence of the flow force on the lift of the valve influence the motion. The influence of the spring constant c_λ on the motion is shown in Fig. 8 for $K_{DM} = 1.0$. For this value of the similarity parameter a soft spring ($c_\lambda = 0.8$) as well as a hard spring ($c_\lambda = 1.2$) lead to an unfavourable functioning of valves. The amplitude of the oscillation of the lift remains almost constant for the soft spring, while for the hard spring the initially small amplitude is increased rapidly since it causes a slow opening of the valve. On the whole, a hard spring increases the sensibility of the system against initiation of oscillations through wave motion in the feeding pipe. On the contrary the influence of the parameters $A_E/\alpha A_0$ and p_a/p_K on the functioning of the valve is small. Since the possibility to vary the stiffness of the spring is small, it is mainly the parameter K_{DM}, which decides on a safe functioning of the valve.

3. Oscillation of the stagnation pressure ahead of the disk in a sharp edged valve entrance

In some valves oscillations of the stagnation pressure upstream of the disk have been observed when the valve is directly connected to the pressure vessel without a connecting pipe. This happens when the entrance is not rounded but sharp edged such that the flow separates at the edge and a contraction is formed. The Mach-Zehnder-Interferogram shown in Fig. 9 was taken at an exposure time of approximately one microsecond. It shows the flow through a plane model of such a sharp edged valve with a disk at lift position h/b = 0.5 and for a pressure ratio $p_a/p_K = 0.35$. The cross-sectional area of the entering flow is contracted to about 3/4 of the cross-section of the duct. The evaluation of the interferogram shows that the Mach number along the axis of the duct increases to Ma = 0.965 in the narrowest cross-section. Despite of the supercritical pressure ratio sonic conditions are not reached in the entrance of the valve and the mass flow is still determined by the sonic cross-sectional area of the separated flow between the valve seat and the disk. For a larger lift the Mach number in the valve inlet is increased since the mass flux per unit area is increased. The picture of Fig. 10 shows that for h/b = 0.61 and p_a/p_K = 0.35 the flow in the valve inlet reaches sonic condition in the narrowest cross-section and expands to supersonic flow. The Mach number in the supersonic flow upstream of the recompression shock has been increased to Ma = 1.3. The entropy rise across the shock has decreased the critical mass flux per unit area only by such an amount that the cross-sectional area between seat and disk for a lift h/b = 0.61 is just large enough for the mass flux passing the narrowest cross-section of the duct.

The flow becomes unstable, if the lift is increased to h/b = 0.62 for the same pressure ratio p_a/p_K = 0.35. The flow pattern is then changed periodically with a frequency of about 300 Hz. For a lift of h/b = 0.66 and p_a/p_K = 0.35 Fig. 11 shows one period of this oscillation in pictures of a movie taken with a Mach-Zehnder-Interferometer and a high speed camera with 5800 pictures per second. It can be recognized clearly that the upstream and the downstream motion of the normal shock is connected with a change of the flow contraction in the valve inlet. The 9th picture in Fig. 11 shows the shock in his furthest downstream position. The narrowest critical cross-sectional area of the flow in the valve inlet has increased to about 85 % of the cross-sectional area of the valve inlet, and the Mach number ahead of the shock is about Ma = 1.59.

Thereby the mass flow through the valve has increased. Because of the increased shock intensity and the increased entropy rise across the shock the critical mass flux per unit area has diminuished. Under such conditions the mass flow through the effective cross-sectional area in the valve inlet can no longer be the same as the mass flow through the cross-sectional area between the seat and the disk. Thereby a compression wave is generated, which is travelling upstream and moving the shock upstream into his position shown in picture 1 of Fig. 11. The narrowest cross-sectional area of the flow in the valve inlet, the Mach number ahead of the shock and hence also the shock intensity and the losses across the shock are diminuished. The mass flux through the cross-sectional area near the disk is now larger than the mass flux through the narrowest cross-section in the duct, and an expansion wave travels upstream and retracts the shock to a position shown in the 9th picture. Thereby the sonic cross-sectional area of the flow in the valve inlet, the Mach number ahead of the shock, the shock intensity and the losses are increased such that again a compression wave travelling upstream is generated.

The variation of the stagnation pressure corresponding to the flow shown in the interferograms of Fig. 11 can be measured in the stagnation point of the disk. Fig. 12 shows that the stagnation pressure can vary by about 11 % of the pressure in the vessel. Therefore the flow force acting on the disk can vary by the same amount.

The discussion of the flow pictures of Figures 9, 10 and 11 has shown that in the duct chosen for the experiment different flow pattern can appear. The limiting curves of these have been determined experimentally. They depend on h/b and p_a/p_K shown in Fig. 13. For flow conditions below the curve A the flow in the inlet never reaching sonic conditions while for all conditions above the curve A sonic conditions are reached in the contracted flow. Locally supersonic flow may occure, which is decelerated to subsonic flow again through a recompression shock. For flow conditions between the curves B and C the flow is no longer steady. In parts of this region it oscillates with a frequency determined by the geometry of the valve.

4. Acknowledgements

The authors wish to thank Prof. E. Krause, Ph.D. for his critical help.

The work has been sponsored by the "Deutsche Forschungsgemeinschaft" within the "Sonderforschungsbereich Strömungsmechanik und Thermogasdynamik an der RWTH Aachen".

5. References

1. AD-Merkblatt A 2: "Safety valves" (Sicherheitsventile), (sketch, March 1979). (In German).

2. Hall, N.: "Thermodynamics of fluid flow." Englewood Cliffs, N.J. (1951).

3. Shapiro, A. H.: "The dynamics and thermodynamics of compressible fluid flow." New York, (1953).

4. Wilhelmsson, T.: "Safety valves for steams and gases." (Säkerhetsventiler för anger och gaser). Technish Tidskrift 18, pp. 1-6. (Nov. 1960). (In Swedish).

5. Zeller, H., Föllmer, B.: "The influence of the flow resistance in the feed line on the functioning of safety valves." (Der Einfluß des Strömungswiderstandes der Zuführungsleitung auf die Funktion von Sicherheitsventilen). Abhandlungen aus dem Aerodynamischen Institut der RWTH Aachen, Heft 23, pp. 18-23. (1978). (In German).

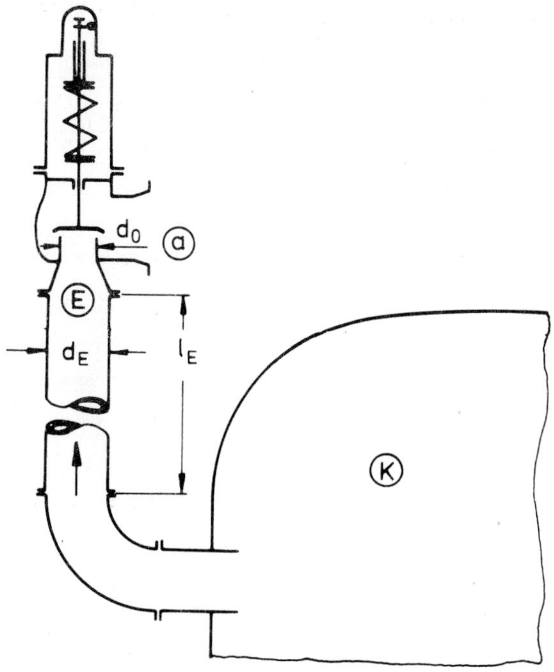

Fig. 1 Safety valve with feed line

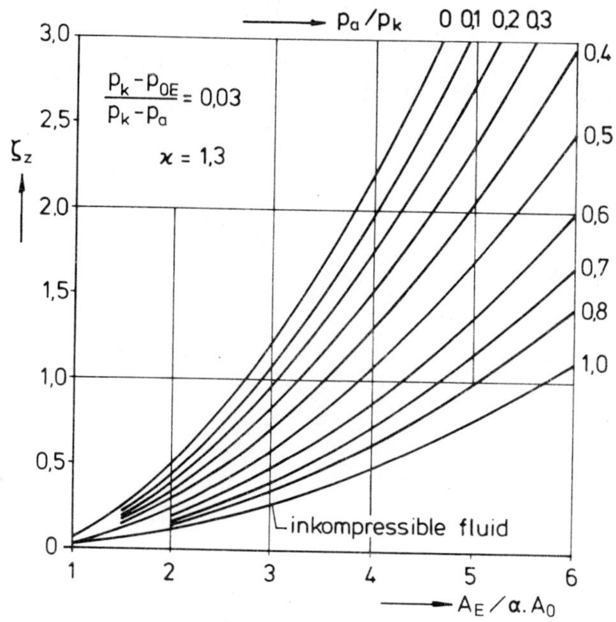

Fig. 2 Drag coefficient ζ_Z of the feed line

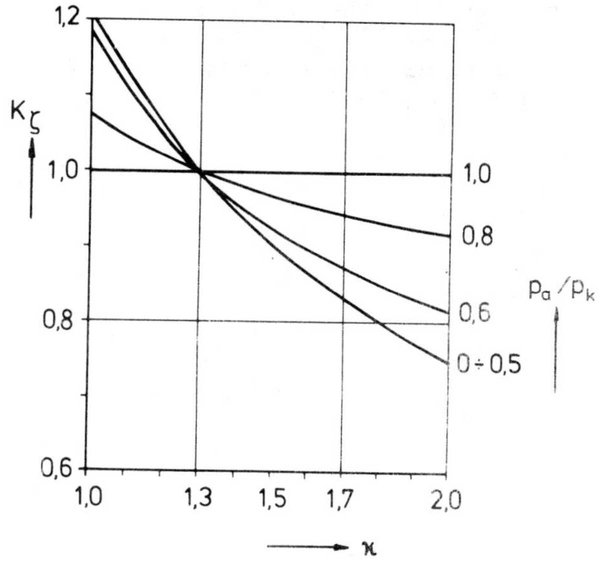

$$\zeta_z = K_\zeta \cdot \zeta_{z(\varkappa = 1,3)}$$

Fig. 3 Correction factor for the drag coefficient ζ_Z

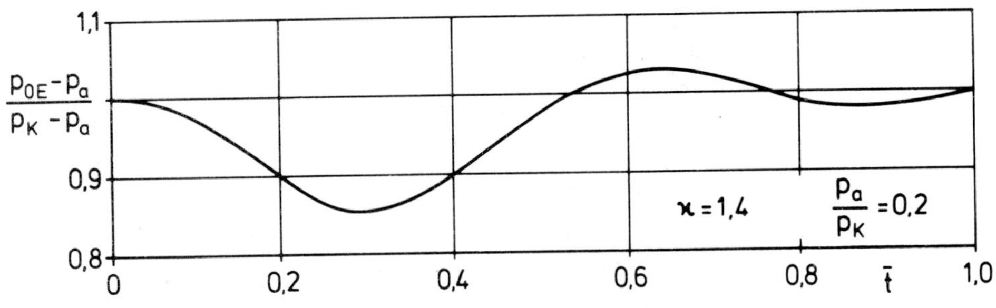

Fig. 4 Stagnation pressure ahead the valve for $(t_{op}a_K/l_E)(A_E/\alpha A_o) = 9$

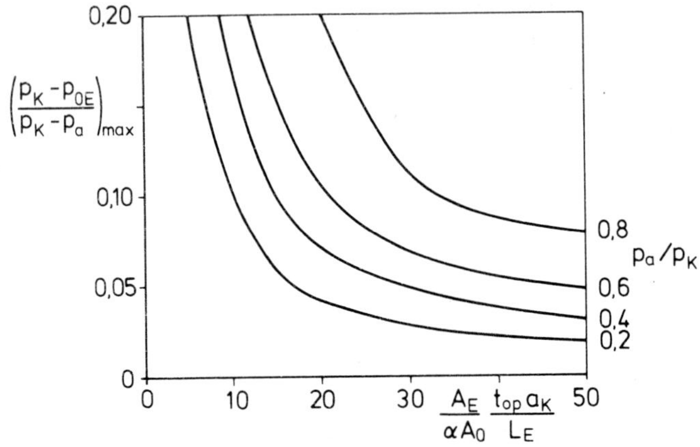

Fig. 5 Maximum stagnation pressure loss ahead the valve

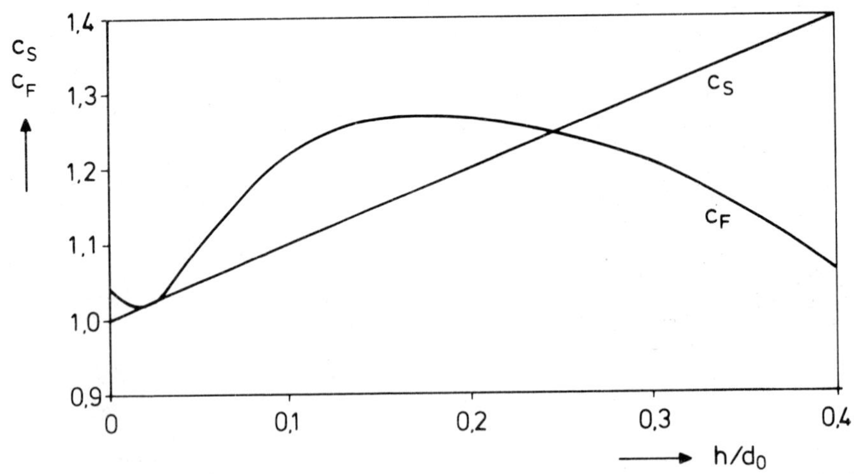

Fig. 6 Flow force and spring force coefficients

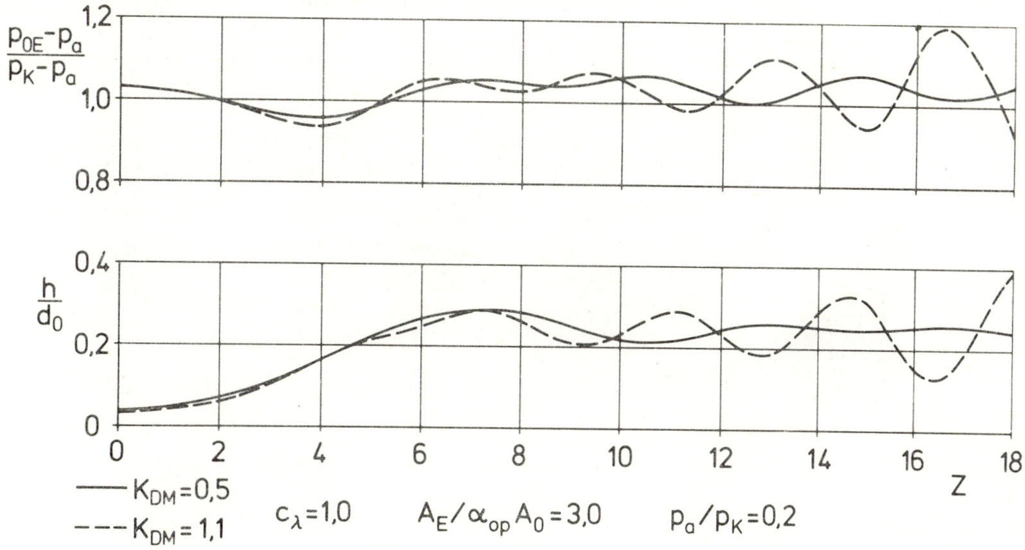

Fig. 7 Influence of the parameter K_{DM} on lift and stagnation pressure

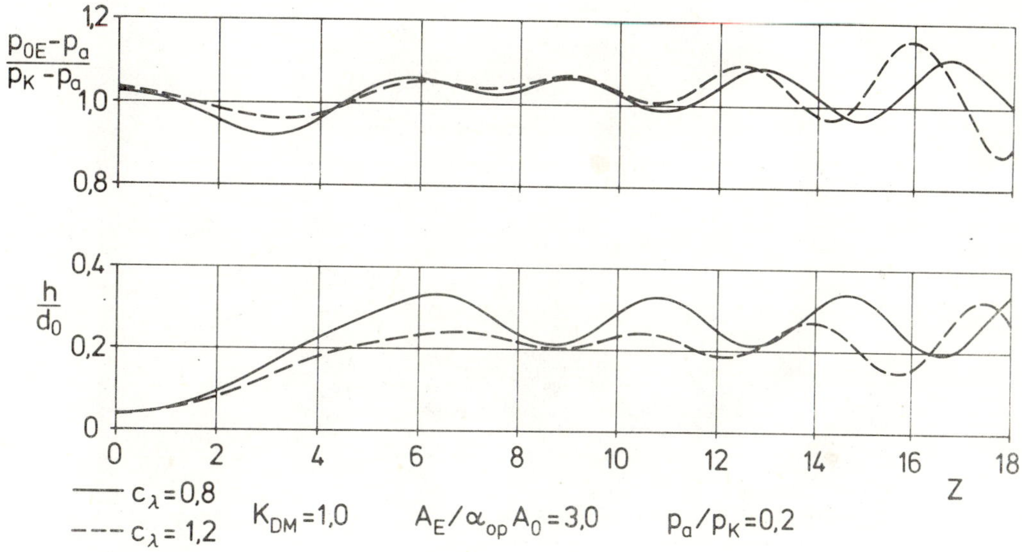

Fig. 8 Influence of spring stiffness on lift and stagnation pressure

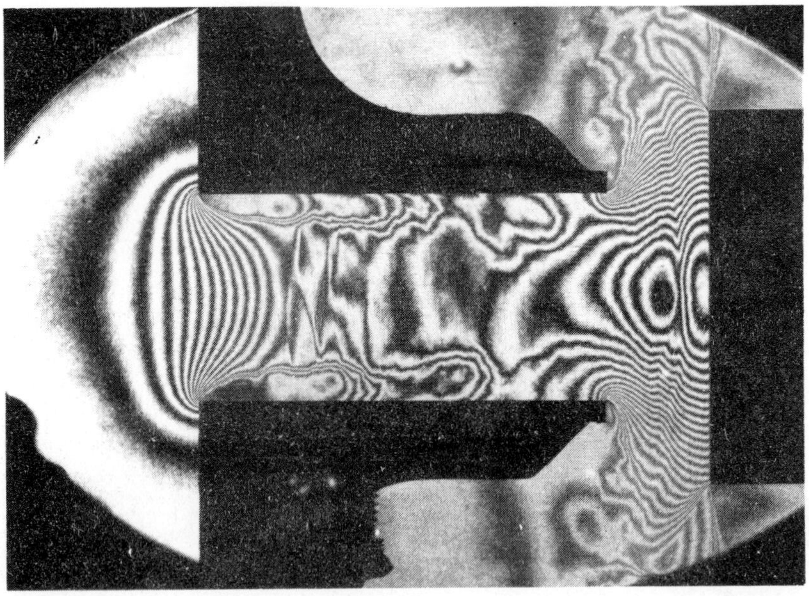

Fig. 9 Mach-Zehnder-Interferogram of the flow through a plane valve with $h/b = 0.5$ and $p_a/p_K = 0.35$

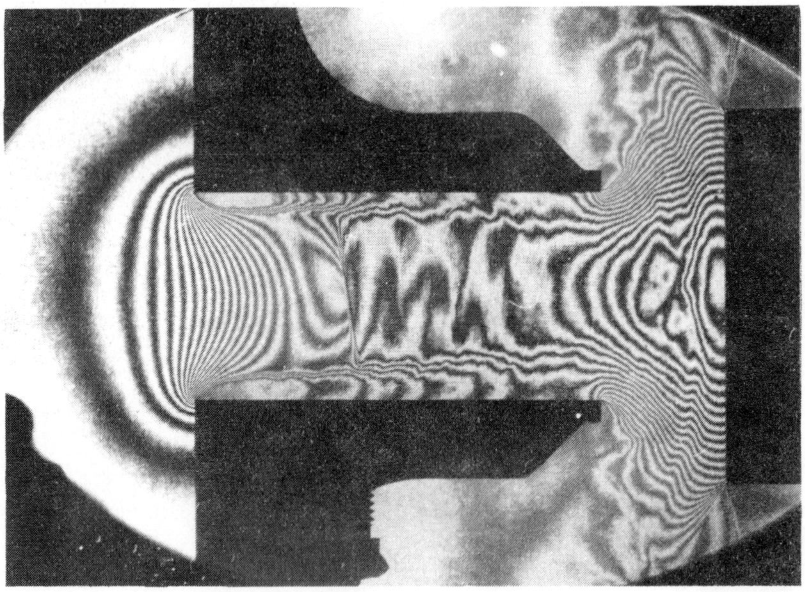

Fig. 10 Mach-Zehnder-Interferogram of the flow through a plane valve with $h/b = 0.61$ and $p_a/p_K = 0.35$

Fig. 11 Part of a Mach-Zehnder-Interferometer movie of the flow through a plane valve with $h/b = 0.66$ and $p_a/p_K = 0.35$

(density in the vessel 1.196 kg/m^3, density difference for one step in the interferogram 0.0341 kg/m^3)

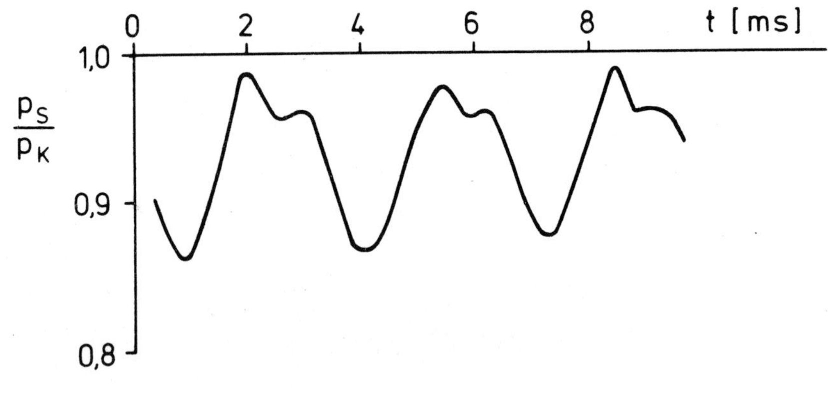

Fig. 12 Pressure at the stagnation point of the disk (plane model)

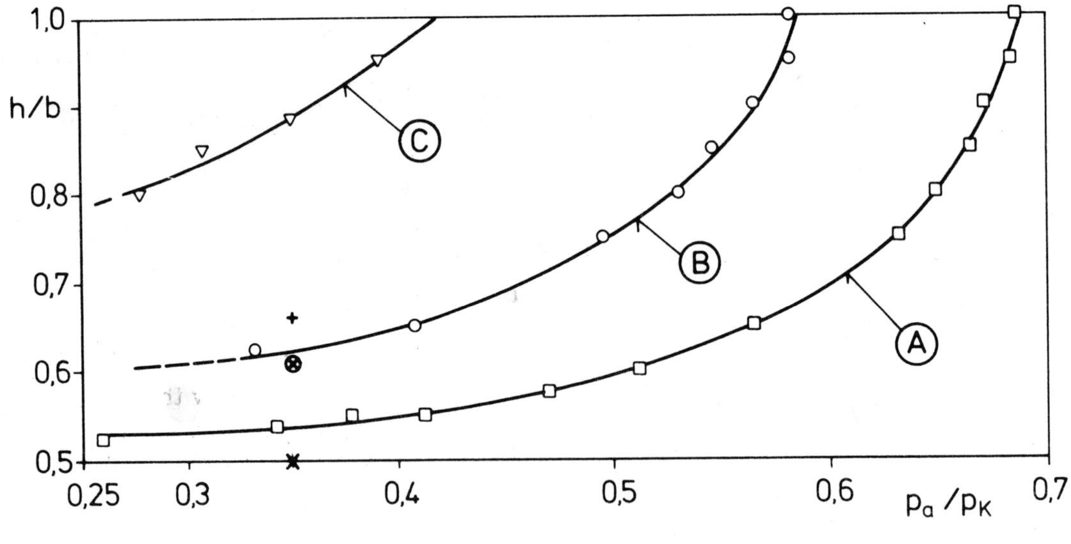

Fig. 13 Limiting curves for the different flow pattern in the sharp edged inlet of the plane model
× flow conditions of Fig. 9
⊗ flow conditions of Fig. 10
+ flow conditions of Fig. 11

Third International Conference on

PRESSURE SURGES

Canterbury, England
March 25-27, 1980

PAPER J3

DESIGN CHARTS FOR SURGE TANKS ON PUMP DISCHARGE LINES

A.T.K. Fok

Proctor & Redfern Group, Canada.

Summary

A set of comprehensive charts is presented for surge tanks designed against power failure. The charts are in a dimensionless form and were developed using a computer program based on the method of characteristics. Pipeline friction loss and the different ratios of in and out throttlings at the differential orifice are considered. The extreme surge pressures in the tank can be calculated. The time of occurrence of critical transients, such as maximum and minimum pressures, and the maximum reverse flow, can be estimated. In addition, the time to resume the pump operation after power failure can be determined.

The charts are applicable to the design of simple surge tanks, one-way surge tanks, and by-passing check valves between suction and discharge lines.

Examples are presented to demonstrate the application of the charts.

Held at the University of Kent at Canterbury, U.K.

Conference Organised by BHRA Fluid Engineering, Cranfield, Bedford, England.
In conjunction with The City University, London.

Nomenclature

A. Dimensional Variables

a	m/s	Velocity of pressure wave
A_p	m^2	Cross-sectional area of pipe
A_s	m^2	Cross-sectional area of surge tank
c	m^3	Maximum water volume leaving one-way surge tank due to downsurge
g	m/s^2	Acceleration of gravity
H_{do}	m	Maximum pressure drop below normal due to downsurge
H_f	m	Total friction loss of the pipeline
H_o	m	Head difference between the water surface of one-way surge tank and the normal pressure head
H_{or}	m	Throttling loss for normal flow, Q_o, leaving simple surge tank
H'_{or}	m	Throttling loss for normal flow, Q_o, entering simple surge tank
H_v	m	Velocity head of normal condition, $V_o^2/2g$
H_{f+v}	m	Sum of H_f and H_v
H_{up}	m	Maximum pressure rise above static level due to upsurge in simple surge tank
L	m	Length of pipeline
Q_o	m^3/s	Normal flow in pipe
Q_{re}	m^3/s	Maximum reverse flow entering simple surge tank
V_o	m/s	Normal flow velocity in pipe
T	sec	Time
T_{do}	sec	Time for maximum downsurge to occur in surge tank
T_o	sec	Normal flow travel time as defined L/V_o
T_{re}	sec	Time for maximum reverse flow to occur in simple surge tank
$(T_{re})_{0.3}$	sec	Time for 30% of normal flow reversed into simple surge tank
T_{up}	sec	Time for maximum upsurge to occur in simple surge tank
$(T_{up})_{1.0}$	sec	Time for upsurge pressure equal to 1 H_{f+v} to occur in simple surge tank

Nomenclature (cont'd)

B. Dimensionless Variables

$b_o = \frac{H_{f+v}}{Q_o}\sqrt{\frac{A_s\, g\, A_p}{L}}$

$c = \frac{C}{LA_p}$

$h_{do} = \frac{H_{do}}{H_{f+v}}$

$h_f = \frac{H_f}{H_o}$

$h_v = \frac{H_v}{H_o}$

$h_{f/v} = \frac{H_f}{H_v}$

$h_o = \frac{H_o}{H_{f+v}}$

$h_{or} = \frac{H_{or}}{H_{f+v}}$

$h'_{or} = \frac{H'_{or}}{H_{f+v}}$

$h_{up} = \frac{H_{up}}{H_{f+v}}$

$k_{or} = \frac{H'_{or}}{H_{or}}$

$q_{re} = \frac{Q_{re}}{Q_o}$

$t_{do} = \frac{T_{do}}{T_o}$

$t_{re} = \frac{T_{re}}{T_o}$

$(t_{re})_{0.3} = \frac{(T_{re})_{0.3}}{T_o}$

$t_{up} = \frac{T_{up}}{T_o}$

$(t_{up})_{1.0} = \frac{(T_{up})_{1.0}}{T_o}$

$\rho = \frac{a\, V_o}{g\, H_o}$

C. Subscripts

do	Downsurge
f	Friction
o	Normal condition
or	Orifice
re	Reverse flow
s	Surge tank
up	Upsurge
v	velocity

1. INTRODUCTION

In a large pumping-discharge system, a simple surge tank (Fig. 1) can be used to effectively control the transients resulting from the shut-down or start-up of a pump. Following a power failure or sudden shut-down of a pumping system, the water surface in the surge tank drops as the potential energy of the water in the tank is converted into kinetic energy of the flow entering the system. Such a conversion of energy reduces the rate of change of flow and consequently the magnitude of the transients in the system. At start-up, part of the initial flow from the pump enters the surge tank, thereby reducing the effects on the system induced by the sudden increase in flow. In a pumping-discharge system, a power failure usually causes a more serious adverse effect than the start-up of pumps. In this paper, only the surge transients induced by power failure are discussed.

In order to use a simple surge tank more effectively, a differential orifice is usually installed at the riser of the tank. This differential orifice throttles the reverse flow of water from the system into the tank, but causes very little throttling for the flow leaving the tank.

In a high head pumping system a simple surge tank may not be feasible or economical because of its height. In such a case, a one-way surge tank may be preferable. In general, a one-way surge tank (left-hand side of Fig. 1) is a relatively small conventional surge tank with a check valve in the riser permitting flow out of the tank only. The tank level is maintained by an altitude valve by-passing the check valve.

Recently the application of both simple and one-way surge tanks has been increased substantially because of their effectiveness in reducing critical transient pressures. Considerable effort has been devoted by Jacobson (Ref.1), Parmakian (Ref.2), Ruus (Ref.3), and many others to develop design charts for sizing simple surge tanks. With recent developments in computational technique, and the application of digital computers to analysis of transient phenomena, these charts can be modified or extended to provide more versatile design tools.

In this paper, a set of comprehensive charts is presented which attempts to incorporate the best aspects of the previous publications along with the use of modern computer techniques. This set of charts expands the selection and range of parameters, improves the accuracy and completeness of simple surge tank designs, and more importantly, it provides the means for designing one-way surge tanks. A review of the literature* indicates that there are no existing design aids for one-way surge tanks.

2. ASSUMPTIONS

The following assumptions were made in deriving the design charts:

1. The pump discharge system is a single uniform pipeline connecting the surge tank and the receiving reservoir.

2. The cross-sectional area inside the surge tank is constant over the range of fluctuating water levels.

3. Water column separation does not occur in the system.

4. Surge transients directed upstream (from the pump to the tank) do not affect the surge tank.

---

* Journal of Fluid Mechanics (from 1967 to 1979); Proceedings of ASCE Journal of the Hydraulics Division (from 1964 to 1979); Journal of Hydraulic Research (from 1963 to 1979); Proceedings of Canadian Journal of Civil Engineering (1977-1979); Proceedings of the 1st (1972) and of the 2nd (1976) Conference on Pressure Surges, BHRA.

The first assumption permits the simplication of a complex piping system. For a pipeline network, an equivalent single-line hydraulic condition can be estimated by considering the kinetic energy of the system, looping effect of the network distribution, and other hydraulic conditions in the system.

The second assumption is valid for most common surge tank designs.

The third assumption is a realistic simplifying assumption. If water column separation occurs, a new boundary condition, the existence of vapour pressure, would have to be considered. The resulting complexity would prevent the developing of charts in the present form. Additionally, one of the fundamental principles of surge tank design and installation is the prevention of water column separation. This tends to minimize the need for including the consideration of water column separation in the design charts.

The fourth assumption needs to be studied for the requirement of two different situations. The first situation is that of a surge tank located near the pump. Following pump shut-down, the high pressure head in the surge tank will usually cause the check valve at the pump to close very rapidly and hence the pump characteristics can be eliminated. However, for a simple surge tank with considerable orifice throttling, or for a one-way surge tank with a very low water level, the elimination of the effect of pump and motor inertia may result in over-estimating the size requirements of the surge tank. The second situation is that of a surge tank located at the high elevation of the system, far from the pump. In this case, the surge tank would be designed to protect the downstream part of the system while the pipeline between pumps and the surge tank would be protected by other means (see Page 124, Ref.2).

3. METHOD OF ANALYSIS

Considering the previous assumptions, any property M, of the surge transients in a simple surge tank may be expressed as a function of the following parameters:

$$M = F_1 \; (Q_o, A_p, L, H_f, A_s, H_{or}, H'_{or}, a, g) \qquad (1)$$

where Q_o = normal flow in pipe; A_p = cross-sectional area of pipe; L = Length of pipeline; H_f = friction loss of pipeline; A_s = cross-sectional area of surge tank; H_{or} and H'_{or} = throttling losses for normal flow, Q_o, out and in the surge tank respectively; a = velocity of pressure wave; and g = gravitational acceleration.

The following dimensional variables of practical importance are also introduced: $V_o = Q_o/A_p$ = normal flow velocity in pipe; $H_v = V_o^2/2g$ velocity head in pipe; $H_{f+v} = H_f + H_v$ = sum of friction head and velocity head.

The results of preliminary dimensional analysis* and the presentation in the previous publications can be combined to derive variables which are useful in engineering applications. Equation 1 can be re-written in dimensionless form as:

$$m = f_2 \; (b_o, h_{f/v}, h_{or}, k_{or}) \qquad (2)$$

where $b_o = \dfrac{H_{f+v}}{Q_o}\sqrt{\dfrac{A_s \, g \, A_p}{L}}$; $h_{f/v} = \dfrac{H_f}{H_v}$; $h_{or} = \dfrac{H_{or}}{H_{f+v}}$ and $k_{or} = \dfrac{H'_{or}}{H_{or}}$

* Dimensional analysis requires that each parameter in Equation 1 is tested against all the transient properties induced by power failure. This can be accomplished by varying the tested parameter while all other parameters being kept constant. As the main purpose of this paper is to develop a design tool, detailed methodology and procedure of the dimensional analysis in arriving at the Equations 2 to 11 will not be presented.

The variable, b_o reflects the most important characteristics of the pipeline and surge tank; $h_{f/v}$ reflects the relationship between the line friction loss and the kinetic energy in the pipeline; h_{or} and k_{or} reflect the characteristics of out and in orifice throttling respectively. Any property of the transients in a simple surge tank can be expressed as a function of some or all of these four dimensionless variables.

The surge transients that govern the design of a simple surge tank are: H_{do}, H_{up}, Q_{re}, T_{do}, T_{up}, T_{re}, where H_{do} and H_{up} = maximum downsurge and upsurge pressure respectively in the tank; Q_{re} = maximum reverse flow into the surge tank; T_{do}, T_{up}, and T_{re} = time from power failure for the maximum downsurge H_{do}, the maximum upsurge H_{up}, and the maximum reverse flow, Q_{re}, to occur respectively. The effect of surge transients in a simple surge tank can be formulated as follows:

$$h_{do} = f_3 (b_o, h_{or}) \tag{3}$$

$$h_{up} = f_4 (b_o, h_{or}, k_{or}) \tag{4}$$

$$q_{re} = f_5 (b_o, h_{or}, k_{or}) \tag{5}$$

$$t_{do} = f_6 (b_o, h_{or}, h_{f/v}) \tag{6}$$

$$t_{up} = f_7 (b_o, h_{or}, k_{or}, h_{f/v}) \tag{7}$$

$$t_{re} = f_8 (b_o, h_{or}, k_{or}, h_{f/v}) \tag{8}$$

In the case of a one-way surge tank, the new variable, H_o = head difference between the water surface of one-way surge tank and the normal head, reflects the difference between a simple and a one-way surge tank. For the case of a one-way surge tank only, the following downsurge transients are applicable since the check valve is closed upon flow reversal:

$$h_{do} = f_9 (b_o, h_o) \tag{9}$$

$$t_{do} = f_{10} (b_o, h_{f/v}, h_o) \tag{10}$$

where the new parameter, $h_o = H_o/H_{f+v}$ reflects the one-way surge tank characteristic H_o.

For a one-way surge tank with large cross-sectional area or for a suction reservoir connected to the discharge line by the by-passing check valve, the drop of water surface, H_{do} is very small and hence h_{do} is unimportant. Under these conditions a new parameter, C = maximum water volume leaving the tank, or its dimensionless form, $c = C/(LA_p)$, may be more appropriate to represent the downsurge transients. On this basis, the equations 9 and 10 can be re-written as:

$$c = f_{11} (h_v, h_f) \tag{11}$$

$$t_{do} = f_{12} (h_v, h_f) \tag{12}$$

where $h_v = H_v/H_o$ reflects the relationship between kinetic energy and potential energy and $h_f = H_f/H_o$ reflects the friction loss of the pipeline.

Preliminary analysis indicates that the speed of the pressure wave, a, or its dimensionless form, $\rho = (a V_o)/(g H_o)$, does not affect the transients in surge tanks under most hydraulic conditions. The effect of the speed of pressure waves will be discussed further in later sections of this paper (see the discussion of Fig. 3).

4. DISCUSSION OF DESIGN CHARTS

4.1 Maximum Downsurge

The variation of the maximum downsurge pressure, h_{do}, for a simple surge tank is shown in Fig. 2 based on Equation 3. The results of the present analysis for a simple surge tank (dashed lines in Fig. 2) were found to be consistent with Jacobson's results (Refs. 1 and 2). For a highly restricted orifice (high h_{or}), the downsurge pressure outside the tank experiences two peaks (see Fig. 4 and Table 2 of Ref. 4). The first peak usually occurs within 2 L/a time interval and when the outside tank pressure is lower than the inside tank pressure. The second peak occurs when the flow leaving the tank decreases to zero and the pressures inside and outside of the tank are equal. The time of occurrence of the second peak depends on the pipeline and surge tank characteristics as expressed by Equation 6. For values of orifice, h_{or}, larger than that indicated in the Balanced Design Curve (dot-semi-dashed line in Fig. 2), the first peak of the downsurge transient is greater than the second and the pipeline is subjected to a more serious impact of the downsurge transient than obtained from Fig. 2. However, as mentioned in the discussion of assumption 4, the momentum developed by the motor and pump upon power failure may decrease this impact substantially.

The maximum downsurge pressure of a one-way surge tank, h_{do}, is plotted in Fig. 2 (solid lines) to illustrate the relationship given by Equation 9. The importance of the pressure head difference, h_o, in the one-way surge tank is clearly identified.

In normal design practice, a surge tank (simple or one-way) should never drain so as to admit air into pump discharge lines. Therefore, the bottom of the tank should be lower than the minimum surge level (maximum downsurge) as calculated from Fig. 2.

As shown in Fig. 3 which depicts the relationship given by Equation 11, the maximum volume of water, c, leaving the one-way surge tank under a constant water level is proportional to the increase in kinetic energy, h_v, and line friction, h_f.

In order to assess the effect of the speed of the pressure wave, a, the relationship between its dimensionless form, ρ , and the maximum volume of drained water, c, is also shown in Fig. 3. For values of ρ less than 1, no water drains out of the tank because the downsurge pressure in the pipeline never drops below the tank operational level. For ρ greater than 1.5, the water leaving the tank, c, is independent of ρ . Hence, the effect of pressure wave speed, a, can only influence the surge transients of a one-way surge tank within the small range of $1.0<\rho<1.5$. In practice, the design of a one-way surge tank with $\rho < 1.5$ is considered to be uncommon. For a simple surge tank, the value of ρ is infinity (i.e. $H_o = 0$); hence the speed of the pressure wave, a, will not affect the transients in a simple surge tank.

4.2 Maximum Upsurge

In Fig. 4, the maximum upsurge in a simple surge tank, h_{up}, is plotted against the three independent variables given by Equation 4. As shown in the Figure, the in and out throttlings have significant effects on decreasing the upsurge pressure in the tank. The importance of the ratio of differential throttlings, k_{or}, on reducing the upsurge transients is also shown in the Figure.

The upsurge pressure calculated from Fig. 4 can assist the designer in the determination of the height of a simple surge tank for preventing overflow during the upsurge conditions.

4.3 Maximum Reverse Flow

In Fig. 5, the maximum reverse flow into the simple surge tank, q_{re}, is shown as a function expressed by Equation 5. Similar to the upsurge pressure, the maximum reverse flow, q_{re}, is significantly reduced by an increase of throttlings, H_{or} and k_{or}.

As part of a simple surge tank design, it is important to consider the turbulence created by the reverse flow with high velocity (usually for high orifice value, h'_{or}). The maximum reverse flow obtained from Fig. 5 can aid in the estimation of the maximum flow velocity through the orifice into the tank.

4.4 Time of Maximum Downsurge

Fig. 6 shows the time, t_{do}, of the maximum downsurge in surge tanks (Equations 6 and 10). For both simple and one-way surge tanks, the time, t_{do}, decreases significantly with the increasing of $h_{f/v}$. For a one-way surge tank, the pressure head difference, h_o, drastically reduces the time duration of the downsurge transient. The family of curves (upper half of the Figure), h_o versus t_{do}-reduction, can be applied to each line of $h_{f/v}$. The reduction of t_{do} due to the effect of h_o should be measured in terms of absolute distance (see the demonstration of Design 4 in the Practical Application Section).

In a simple surge tank, the orifice, h_{or}, reduces slightly the time of surge transients, t_{do}, t_{up}, and t_{re}. However, this reduction is usually not significant and hence for simplicity, the effects of orifice, h_{or} and k_{or}, on the time variables, t_{do}, t_{up}, and t_{re}, are not shown in Figs. 6, 8 and 9.

In Fig. 7, the time, t_{do}, for maximum downsurge pressure in a constant water level one-way surge tank is plotted to show the relationship given by Equation 12. In the same way as in Fig. 3, the variable t_{do} increases with the increasing h_v, and h_f.

For ρ greater than 1.0, the time, t_{do}, of maximum downsurge pressure for both simple and one-way surge tanks will not be less than 2 L/a.

4.5 Time of Maximum Upsurge and Maximum Reverse Flow

In Figs. 8 and 9, the times t_{up} and t_{re} for the maximum upsurge pressure and the maximum reverse flow in a simple surge tank are plotted to express the relationship given by Equations 7 and 8 respectively. Similar to Fig. 6, the time variables, t_{up} and t_{re} increase linearly with increasing b_o, but decrease with the increasing of $h_{f/v}$.

4.6 Resumption of Normal Pump Operation

As time elapses, surge transients in the simple surge tank will decay. In practice, the pump operation should not be resumed until the surge transients induced by pump shut-down are substantially dampened. In Figs. 8 and 9, the time, $(t_{up})_{1.0}$ and $(t_{re})_{0.3}$ for the upsurge pressure diminishing to its normal head, i.e. $h_{up} = 1$, and for only 30% of normal flow reversed into the simple surge tank are presented respectively. The significant effect of line friction, $h_{f/v}$, and orifice, h_{or} are evident. For simplicity, the effect of the ratio of the differential throttlings is not presented herein. In Figs. 8 and 9 the times, $(t_{up})_{1.0}$ and $(t_{re})_{0.3}$, for a simple surge tank with a differential orifice can be established by assuming that the orifice has the outgoing throttling also equal to h'_{or} ($h_{or} \times k_{or}$). This assumption should be reasonable since the in-going throttling affects the upsurge transients (both upsurge pressure and reversal flow) much more significantly than the same outgoing throttling.

5. PRACTICAL APPLICATIONS

Consider a pumping system that discharges water into a reservoir (Fig. 1). The data relevant to the transient analysis of this system are given as: $Q_o = 0.567$ m^3/s; L = 982 m; $A_p = 0.372$ m^2; $V_o = 1.52$ m/s; $H_v = 0.118$ m; $H_f = 0.237$ m; $H_{f+v} = 0.355$ m; $h_{f/v} = 2$; $T_o = 644$ sec; a = 982 m/s; g = 9.82 m/s. Estimate the surge transients induced by power failure with various types of surge tank protections.

5.1 Design 1

A simple surge tank is to be installed with $A_s = 6.85\ m^2$ and hence $b_o = 0.1$; no orifice, i.e. $H_{or} = 0$ m ($h_{or} = 0$). From Fig. 2 the maximum downsurge in the tank is found to be $H_{do} = 3.73$ m ($h_{do} = 10.5$); from Fig. 4, the maximum upsurge $H_{up} = 3.12$ m ($h_{up} = 8.8$); from Fig. 5, the maximum reverse flow $Q_{re} = 0.510\ m^3/s$ ($q_{re} = 0.9$); from Figs. 6, 8 and 9, the times for the maximum transients to occur can be estimated as $T_{do} = 64.4$ sec ($t_{do} = 0.1$), $T_{up} = 193$ sec ($t_{up} = 0.3$) and $T_{re} = 129$ sec ($t_{re} = 0.2$); from Figs. 8 and 9, it can be found that after $(T_{up})_{1.0} = 5925$ sec ($(t_{up})_{1.0} = 9.2$); less than 0.355 m upsurge pressure (1 unit of H_{f+v}) and after $(T_{re})_{0.3} = 4190$ sec ($(t_{re})_{0.3} = 6.5$), less than $0.170\ m^3/s$ (30% of Q_o) reverse flow will occur in the tank.

5.2 Design 2

Install an additional orifice $H_{or} = 1.42$ m ($h_{or} = 4$) to the simple surge tank of Design 1. From the design charts, the magnitude and time occurrence of the following transients in the tank can be obtained: $H_{do} = 2.84$ m ($h_{do} = 8.0$); $H_{up} = 1.78$ m ($h_{up} = 5.0$); $Q_{re} = 0.352\ m^3/s$ ($q_{re} = 0.62$); $T_{do} \approx 64.4$ sec ($t_{do} \approx 0.1$); $T_{up} \approx 193$ sec ($t_{up} \approx 0.3$); $T_{re} \approx 129$ sec ($t_{re} \approx 0.2$); $(T_{up})_{1.0} = 900$ sec ($(t_{up})_{1.0} = 1.4$); and $(T_{re})_{0.3} = 450$ sec ($(t_{re})_{0.3} = 0.7$).

5.3 Design 3

Change the orifice of Design 2 to a differential orifice with $k_{or} = 10.0$ and $h_{or} = 4$. Only the transients affected by the modification are presented here: $H_{up} = 0.50$ m ($h_{up} = 1.4$), $Q_{re} = 0.187\ m^3/s$ ($q_{re} = 0.33$); $(T_{up})_{1.0} \approx 220$ sec ($(t_{up})_{1.0} \approx 0.34$); and $(T_{re})_{0.3} \approx 129$ sec ($(t_{re})_{0.3} \approx 0.2$).

5.4 Design 4

A one-way surge is designed to protect the system with $A_s = 6.85\ m^3$ (same as Design 1) and $H_o = 3.55$ m ($h_o = 10$). From Figs. 2 and 6, the maximum downsurge $H_{do} = 4.83$ m ($h_{do} = 13.6$) will occur at $T_{do} = 27.7$ sec ($t_{do} = .043$).

5.5 Design 5

If a reservoir (at suction side) has the same water level as the one-way surge tank in Design 4, i.e. $H_o = 3.55$ m ($h_o = 10$), a by-passing check valve can be installed between the suction reservoir and the discharge line. The aforementioned variables can be converted into: $h_v = 0.0333$ and $h_f = 0.0667$. Upon power failure, a maximum volume of water $C = 12.24\ m^3$ ($c = 0.0335$ from Fig. 3) will by-pass the pumps during the period of $T_{do} = 43.1$ sec ($t_{do} = 0.067$ from Fig. 7).

6. ACKNOWLEDGEMENT

The writer would like to express his gratitude to I. Butler, D. Fisher and A. Perks of Proctor & Redfern Limited for their encouragement during the preparation of this publication. Comments from S. Vatagodakumbura, C. Bishop and S.H. Zemelle are deeply appreciated. In particular, the writer would like to thank S.H. Tan and Miss B. Shoofey for drafting and typing the difficult drawings and manuscripts within very limited time constraints.

7. REFERENCES

1. Jacobson, R.S., "Charts for Analysis of Surge Tanks in Turbine or Pump Installations", Special Report 104, Bureau of Reclamation, Denver, Colorado, February, 1952.

2. Parmakian, J., "Water Hammer Analysis", Dover Publications Inc., New York, N.Y., 1963.

3. Ruus, E., and El-Fitiany, F., "Maximum Surges in Simple Surge Tanks", Canadian Journal of Civil Engineering, March 1977.

4. Fok, A.T.K., "Design Charts for Air Chamber on Pump Pipe Lines", Journal of the Hydraulic Division, ASCE, September 1978.

APPENDICES

APPENDIX A - DIMENSIONAL ANALYSIS

Dimensional analysis was employed to investigate surge phenomena during development of the subject design charts. The techniques involved are well documented, and detailed general discussions on the principles of dimensional analysis theory are available in Refs. 5,6,7,8 and 9. This appendix is included to illustrate the dimensional analyses which were made during the development of the design charts presented in this paper; a prior general knowledge of dimensional analysis is assumed.

Within the limitation and assumptions noted in the main text (Section 2.0), any property, M, of the surge transients in a simple surge tank can be expressed by the following functional relation:

$$M = \phi_{13} (\rho', f', V, L, D, a, g, D_s, D_{or}, D'_{or}) \qquad (13)$$

where ρ' = fluid density; f' = friction factor; V = pipe flow velocity; L = pipe length; D = pipe diameter; a = surge wave speed; g = gravitational acceleration; D_s = surge tank diameter; D_{or} = diameter of outgoing orifice and D'_{or} = diameter of incoming orifice.

Choosing ρ', V and D as basic quantities and using the π theorem, the parameters in Eq. 13 can be replaced by a reduced number of equivalent dimensionless groups:

$$m = \phi_{14} \left(f', \frac{L}{D}, \frac{a}{V}, \frac{V^2}{Dg}, \frac{D_s}{D}, \frac{D_{or}}{D}, \frac{D'_{or}}{D}\right) \qquad (14)$$

The dimensionless parameters of Eq. 14 reflect the characteristics of f', L, a, g, D_s, D_{or} and D'_{or} respectively.

Certain combinations of parameters may be derived which are suitable for experimental planning. For example, given $H_f = f'\frac{L}{D}\frac{V^2}{2g}$ and $H_v = \frac{V^2}{2g}$, a dimensionless variable $h_{f/v}$ can be defined:

$$h_{f/v} = \frac{H_f}{H_v} = f'\frac{L}{D} = \phi_{15} \left(f', \frac{L}{D}\right) \qquad (15)$$

Similarly Jacobson's recommended variable b_o can be expressed as:

$$b_o = \frac{H_{f+v}}{Q_o} \sqrt{\frac{A_s\ g\ A_p}{L}} = \left[\frac{1}{2}\left(\frac{D_s}{D}\right)^2 \left(\frac{V^2}{Dg}\right)\left(1+f'\frac{L}{D}\right)^2 \Big/ \frac{L}{D}\right]^{0.5}$$

$$= \phi_{16} \left[f', \frac{L}{D}, \frac{V^2}{Dg}, \frac{D_s}{D}\right] \qquad (16)$$

where $H_{f+v} = H_v + H_f = \frac{V^2}{2g}\left(1+f'\frac{L}{D}\right)$, $Q_o = V\frac{\pi D^2}{4}$, $A_s = \frac{\pi D_s^2}{4}$ and $A_p = \frac{\pi D^2}{4}$

Rewriting Eqs. 15 and 16, new functions may be defined:

$$\frac{L}{D} = \phi_{17} (f', h_{f/v}) \qquad (17)$$

$$\frac{D_s}{D} = \phi_{18} \left(f', \frac{L}{D}, \frac{V^2}{Dg}, b_o\right)$$

$$= \phi_{18} \left(f', \phi_{17} (f', h_{f/v}), \frac{V^2}{Dg}, b_o\right)$$

$$= \phi_{18} \left(f', h_{f/v}, \frac{V^2}{Dg}, b_o\right) \qquad (18)$$

To incorporate various coefficients (orifice losses, etc.) into the derived relations, a number of techniques are used.

For outgoing flow, the throttling loss through an orifice is $H_{or} = C\, V_{or}^2/2g$ where C is an orifice throttling coefficient and can be considered as a constant value (say, C = 2.8). V_{or} is the outgoing flow velocity through the orifice and can be expressed as $V_{or} = V\,(D_{or}/D)^2$. Hence $H_{or} = C\,V^2/2g\,(D_{or}/D)^4$.

The dimensionless variable, h_{or}, which is applied in developing the design charts, can thus be expressed as:

$$h_{or} = \frac{H_{or}}{H_{f+v}} = \frac{C\,V^2}{2g}\left(\frac{D_{or}}{D}\right)^4 \Big/ \frac{V^2}{2g}\left(1+f'\frac{L}{D}\right)$$

$$= C\left(\frac{D_{or}}{D}\right)^4 \Big/ \left(1+f'\frac{L}{D}\right)$$

$$= \phi_{19}\left(\frac{D_{or}}{D}, f', \frac{L}{D}\right) \qquad (19)$$

in which

$$\frac{D_{or}}{D} = \phi_{20}\left(h_{or}, f', \frac{L}{D}\right)$$

$$= \phi_{20}\left(h_{or}, f', \phi_{17}\left(f', h_{f/v}\right)\right)$$

$$= \phi_{20}\left(h_{or}, f', h_{f/v}\right) \qquad (20)$$

The ratio between throttling losses from inward and outward flow can be considered for two different orifice designs:

(i) same inflow and outflow orifice diameter but with different values of orifice coefficients (C).

(ii) same value of C, but with different inflow and outflow orifice diameters.

In the present analysis, the latter is assumed. Hence:

$$K_{or} = \frac{H_{or}}{H'_{or}} = \frac{C\,V^2}{2g}\left(\frac{D_{or}}{D}\right)^4 \Big/ \frac{C\,V^2}{2g}\left(\frac{D'_{or}}{D}\right)^4$$

$$= \left(\frac{D_{or}}{D}\right)^4 \Big/ \left(\frac{D'_{or}}{D}\right)^4$$

$$= \phi_{21}\left(\frac{D_{or}}{D}, \frac{D'_{or}}{D}\right) \qquad (21)$$

Substituting Eq. 20 into 21:

$$K_{or} = \phi_{22}\left(\phi_{20}\left(h_{or}, f', h_{f/v}\right), \frac{D'_{or}}{D}\right)$$

$$= \phi_{22}\left(f', h_{f/v}, h_{or}, \frac{D'_{or}}{D}\right) \qquad (22)$$

Eq. 22 can be rewritten as:

$$\frac{D'_{or}}{D} = \phi_{23}\ (f',\ h_{f/v},\ h_{or},\ k_{or}) \qquad (23)$$

Substituting Eqs. 17, 18, 20 and 23 into 14:

$$m = \phi_{24}\ (f',\ \phi_{17}\ (f',\ h_{f/v}),\ \frac{a}{v},\ \frac{V^2}{Dg},\ \phi_{18}\ (f',\ h_{f/v},\ \frac{V^2}{Dg},\ b_o),$$

$$\phi_{20}\ (h_{or},\ f',\ h_{f/v}),\ \phi_{23}\ (f',\ h_{f/v},\ h_{or},\ k_{or}))$$

$$= \phi_{24}\ (f',\ h_{f/v},\ \frac{a}{v},\ \frac{V^2}{Dg},\ b_o,\ h_{or},\ k_{or}) \qquad (24)$$

In general design practice, the friction factor, f', and gravitational acceleration, g, do not vary greatly, (say $f' = \pm 0.012$ and $g = \pm 9.8\ m/s^2$). Hence, the dimensionless parameters, f', and V^2/Dg may not be very significant as variables. Also, preliminary analysis indicated that surge wave velocity, "a", does not affect the surge inside a simple surge tank (see discussions in Section 4.1). The dimensionless variable, a/v, can also be eliminated for simple surge tank analysis. In light of the above, Eq. 24 can be simplified to become:

$$m = \phi_{25}\ (h_{f/v},\ b_o,\ h_{or},\ k_{or}) \qquad (25)$$

This is the same expression as Eq. 2 of Section 3.0.

Each of the dimensionless parameters in Eq. 25 may or may not be a governing parameter for any particular phenomenon. For example, the dimensionless variable, k_{or}, is not important in determining downsurge. In addition, the dimensionless variable, $h_{f/v}$ was found during the surge analysis to be important only in determining the surge time duration. Based on the above considerations, Eqs. 3 to 8 in Section 3.0 can be established to relate the surge transients under investigation.

For the case of a one-way surge tank, it appears that orifice parameters, h_{or}, k_{or} should be eliminated while the new variable, H_o = head difference between water surface of one-way surge tank and the normal head, should be included. Accordingly, Eq. 14 can be rewritten as:

$$m = \phi_{26}\ (f',\ \frac{L}{D},\ \frac{a}{v},\ \frac{V^2}{Dg},\ \frac{D_s}{D},\ \frac{H_o}{D}) \qquad (26)$$

The dimensionless variable, h_o, used in the chart formation can be expressed as

$$h_o = \frac{H_o}{H_{f+v}} = \frac{H_o}{\frac{V^2}{2g}\ (1+f'\frac{L}{D})} = 2\ \left(\frac{H_o}{D}\right) \Big/ \left[\frac{V^2}{Dg}\ (1+f'\frac{L}{D})\right]$$

$$= \phi_{27}\ (\frac{H_o}{D},\ \frac{V^2}{Dg},\ f',\ \frac{L}{D}) \qquad (27)$$

Hence:

$$\frac{H_o}{D} = \phi_{28}\ (h_o,\ \frac{V^2}{Dg},\ f',\ \frac{L}{D})$$

$$= \phi_{28}\ (h_o,\ \frac{V^2}{Dg},\ f',\ \phi_{17}\ (f',\ h_{f/v})) = \phi_{28}\ (h_o,\ \frac{V^2}{Dg},\ f',\ h_{f/v}) \qquad (28)$$

The famous Allievi's variable, ρ, can be expressed as:

$$\rho = \frac{av}{gH_o} = \frac{a}{v}\frac{V^2}{Dg} \Big/ \frac{H_o}{D} = \phi_{29}\left(\frac{a}{v}, \frac{V^2}{Dg}, \frac{H_o}{D}\right) = \phi_{29}\left(\frac{a}{v}, \frac{V^2}{Dg}, \phi_{28}\left(h_o, \frac{V^2}{Dg}, f', h_{f/v}\right)\right)$$

$$= \phi_{29}\left(\frac{a}{v}, \frac{V^2}{Dg}, h_o, f', h_{f/v}\right) \quad (29)$$

Hence:

$$\frac{a}{v} = \phi_{30}\left(\rho, \frac{V^2}{Dg}, h_o, f', h_{f/v}\right) \quad (30)$$

Substituting Eqs. 17, 18, 28 and 30 into 26:

$$m = \phi_{31}\left(f', \phi_{17}(f', h_{f/v}), \phi_{30}\left(\rho, \frac{V^2}{Dg}, h_o, f', h_{f/v}\right), \frac{V^2}{Dg}\right.$$

$$\left.\phi_{18}\left(f', h_{f/v}, \frac{V^2}{Dg}, b_o\right), \phi_{28}\left(h_o, \frac{V^2}{Dg}, f', h_{f/v}\right)\right)$$

$$= \phi_{31}\left(f', h_{f/v}, \rho, \frac{V^2}{Dg}, b_o, h_o\right) \quad (31)$$

As discussed previously, f' and V^2/Dg are relatively unimportant, hence:

$$m = \phi_{32}\left(h_{f/v}, \rho, b_o, h_o\right) \quad (32)$$

As discussed in Section 4.1, ρ becomes important only when $1.5 > \rho > 1.0$ which is not a usual one-way surge tank design and $h_{f/v}$ is important related to the time duration. Eqs. 9 and 10 in Section 3.0 can thus be formulated as shown.

For a very large diameter surge tank, D_s/D approaches infinity and its variation becomes unimportant. Hence, Eq. 26 can be simplified as:

$$m = \phi_{33}\left(f', \frac{L}{D}, \frac{a}{v}, \frac{V^2}{Dg}, \frac{H_o}{D}\right) \quad (33)$$

The dimensionless variables used for charts, h_v and h_f can be expressed as

$$h_v = \frac{H_v}{H_o} = \frac{1}{2}\frac{V^2}{Dg} \Big/ \frac{H_o}{D} = \phi_{34}\left(\frac{V^2}{Dg}, \frac{H_o}{D}\right) \quad (34)$$

Hence:

$$\frac{H_o}{D} = \phi_{35}\left(\frac{V^2}{Dg}, h_v\right) \quad (35)$$

and

$$h_f = \frac{H_f}{H_o} = \frac{1}{2} f' \frac{L}{D}\frac{V^2}{Dg} \Big/ \frac{H_o}{D}$$

$$= \phi_{36}\left(\frac{V^2}{Dg}, f', \frac{L}{D}, \frac{H_o}{D}\right) \quad (36)$$

Hence:

$$\frac{L}{D} = \phi_{37}\left(\frac{V^2}{Dg}, f', h_f, \frac{H_o}{D}\right)$$

$$= \phi_{37}\left(\frac{V^2}{Dg}, f', h_f, \phi_{35}\left(\frac{V^2}{Dg}, h_v\right)\right)$$

$$= \phi_{37}\left(\frac{V^2}{Dg}, f', h_f, h_v\right) \quad (37)$$

Substituting Eqs. 40*, 35 and 37 into 33:

$$m = \emptyset_{38}\left(f', \emptyset_{37}\left(\frac{V^2}{Dg}, f', h_f, h_v\right), \emptyset_{40}(\rho, h_v), \frac{V^2}{Dg}, \emptyset_{35}\left(\frac{V^2}{Dg}, h_v\right)\right)$$

$$= \emptyset_{38}\left(f', h_f, \rho, \frac{V^2}{Dg}, h_v\right) \qquad (38)$$

Finally, as discussed previously, f', V^2/Dg and ρ (for $\rho > 1.5$) are relatively unimportant. Hence Eqs. 11 and 12 in Section 3.0 can be formulated.

REFERENCES (continued):

5. Yalin, M.S., "Mechanics of Sediment Transport", Pergamon Press, Oxford, 1972.

6. Yalin, M.S., "Theory of Hydraulic Models", MacMillan, London, 1971.

7. Sedov, L.I., "Similarity and Dimensional Methods in Mechanics", Academic Press, New York, 1959.

8. Langhaar, H.L., "Dimensional Analysis and Theory of Models", John Wiley and Sons, New York, London, 1962.

9. Birkhoff, G., "Hydrodynamics, a Study in Logic, Fact and Similitude", Princeton, Harvard University Press, 1950, Dover, 1955.

*As $\rho = \frac{av}{gH_o} = \frac{a}{v}\left[\frac{v^2}{2g}\Big/H_o\right] = \frac{a}{v} h_v = \emptyset_{39}\left(\frac{a}{v}, h_v\right)$ (39)

Hence $\frac{a}{v} = \emptyset_{40}(\rho, h_v)$ (40)

APPENDIX B - ILLUSTRATION OF DESIGN CHART APPLICATION

The purpose of this section of Appendix is to demonstrate the technique for extracting the dependent parameters from the design charts. The illustrations cover the common ranges of each independent variable. For practical design consideration, the reader should also refer to Section 5.0.

B.1 ILLUSTRATION OF FIGURE 2

TYPE OF SURGE TANK	INDEPENDENT VARIABLES (GIVEN) h_{or}	h_o	b_o	DEPENDENT VARIABLE (FROM CHART) h_{do}	ILLUSTRATED IN SECTION 5.0
Simple	0	0	0.02	50.0*	
	0	0	0.10	10.5	Design 1
	0	0	0.50	2.4	
	4	0	0.02	48	
	4	0	0.10	8.0	Design 2
	4	0	0.50	1.5	
	40	0	0.02	34.2	
	40	0	0.10	37.0	
	40	0	0.50	<1.1	

*obtained from dashed-lines

TYPE OF SURGE TANK	h_{or}	h_o	b_o	h_{do}	ILLUSTRATED IN SECTION 5.0
One-way	0	5.0	0.02	50.0**	
	0	5.0	0.10	11.0	
	0	5.0	0.50	5.4	
	0	10.0	0.02	52.0	
	0	10.0	0.10	13.6	Design 3
	0	10.0	0.50	10.0	
	0	50.0	0.02	60.0	
	0	50.0	0.10	50.0	
	0	50.0	0.50	50.0	

**obtained from solid lines

B.2 ILLUSTRATION OF FIGURE 3

TYPE OF SURGE TANK	INDEPENDENT VARIABLES (GIVEN) h_f	h_v	DEPENDENT VARIABLE (FROM CHART) C	ILLUSTRATED IN SECTION 5.0
One-way	0.0	0.002	0.002	
(constant tank	0.0	0.033	0.033	Design 5
water level)	0.0	0.5	0.5	
	0.25	0.002	0.0025	
	0.25	0.033	0.039	
	0.25	0.5	0.6	
	0.75	0.002	0.0038	
	0.75	0.033	0.062	
	0.75	0.5	0.92	

B.3 ILLUSTRATION OF FIGURE 4

TYPE OF SURGE TANK	INDEPENDENT VARIABLES (GIVEN) h_{or}	k_{or}	b_o	DEPENDENT VARIABLE (FROM CHART) h_{up}	ILLUSTRATED IN SECTION 5.0
Simple	0	1	0.02	50*	
	0	1	0.1	8.8	Design 1
	0	1	0.5	0.96	
	4	1	0.02	41	
	4	1	0.1	5	Design 2
	4	1	0.5	2.7	
	40	1	0.02	20	
	40	1	0.1	1.16	
	40	1	0.5	<0.1	

*obtained from right half of the figure

TYPE OF SURGE TANK	h_{or}	k_{or}	b_o	h_{up}	ILLUSTRATED IN SECTION 5.0
Simple	4	10	0.02	22**	
	4	10	0.1	1.4	Design 3
	4	10	0.5	<0.1	
	40	10	0.02	3.3***	
	40	10	0.1	0.13	
	40	10	0.5	<0.1	

** obtained from the solid-line in the left-half of the figure
*** obtained from the dashed-dot-line in the left half of the figure

B.4 ILLUSTRATION OF FIGURE 5

TYPE OF SURGE TANK	INDEPENDENT VARIABLES (GIVEN) h_{or}	k_{or}	b_o	DEPENDENT VARIABLE (FROM CHART) q_{re}	ILLUSTRATED IN SECTION 5.0
Simple	0	1	0.02	1.0*	
	0	1	0.1	0.9	Design 1
	0	1	0.5	0.6	
	4	1	0.02	0.88	
	4	1	0.1	0.62	Design 2
	4	1	0.5	0.18	
	40	1	0.02	0.52	
	40	1	0.1	0.21	
	40	1	0.5	0.01	

*obtained from right half of the figure

TYPE OF SURGE TANK	h_{or}	k_{or}	b_o	q_{re}	ILLUSTRATED IN SECTION 5.0
Simple	4	10	0.02	0.65**	
	4	10	0.1	0.33	Design 3
	4	10	0.5	0.07	
	40	10	0.02	0.26	
	40	10	0.1	0.09	
	40	10	0.5	<0.01	

** obtained from left half of the figure

B.5 ILLUSTRATION OF FIGURE 6

TYPE OF SURGE TANK	INDEPENDENT VARIABLES (GIVEN) $h_{f/v}$	h_o	b_o	DEPENDENT VARIABLE (FROM CHART) t_{do}	ILLUSTRATED IN SECTION 5.0
Simple	0	0	0.02	.068*	
	0	0	0.1	.34	
	0	0	0.5	1.7	
	2	0	0.02	.02	
	2	0	0.1	.10	Design 1
	2	0	0.5	.50	
	10	0	0.02	$<$0.01	
	10	0	0.1	.03	
	10	0	0.5	.15	

*obtained from lower half of the Figure

TYPE OF SURGE TANK	$h_{f/v}$	h_o	b_o	t_{do}	ILLUSTRATED IN SECTION 5.0
One-way	0	10	0.02	.52**	
	0	10	0.1	.14	
	0	10	0.5	.18	
	2	10	0.02	.015	
	2	10	0.1	.43	Design 4
	2	10	0.5	.54	
	10	10	0.02	0.01	
	10	10	0.1	.013	
	10	10	0.5	.017	

**included the reduction obtained from the upper half of the Figure

B.6 ILLUSTRATION OF FIGURE 7

TYPE OF SURGE TANK	INDEPENDENT VARIABLES (GIVEN) h_f	h_v	DEPENDENT VARIABLE (FROM CHART) t_{do}	ILLUSTRATED IN SECTION 5.0
One-way (Constant tank water level)	0.0	0.002	.004	
	0.0	0.033	.066	Design 5
	0.0	0.5	1.0	
	0.25	0.002	.005	
	0.25	0.033	.084	
	0.25	0.5	$>$ 1.0	
	0.75	0.002	.01	
	0.75	0.033	.17	
	0.75	0.5	$>$ 1.0	

B.7 ILLUSTRATION OF FIGURE 8

TYPE OF SURGE TANK	INDEPENDENT VARIABLES (GIVEN) $h_{f/v}$	h_{or}	b_o	DEPENDENT VARIABLES (FROM CHART) t_{up}	$(t_{up})_{1.0}$	ILLUSTRATED IN SECTION 5.0
Simple	2	0	0.02	.06*	>10**	
	2	0	0.1	.3	9.2	Design 1
	2	0	0.5	1.5	N.O	
	20	0	0.02	.01	>10	
	20	0	0.1	.048	4.0	
	20	0	0.5	.24	N.O	

* obtained from solid lines
**obtained from dashed-dot lines

TYPE OF SURGE TANK	$h_{f/v}$	h_{or}	b_o	t_{up}	$(t_{up})_{1.0}$	ILLUSTRATED IN SECTION 5.0
Simple	2	4	0.02	approx.	10***	
	2	4	0.1	same as	1.4	Design 2
	2	4	0.5	h_{or} = 0	N.O	
	20	4	0.02	"	4.0	
	20	4	0.1	"	.40	
	20	4	0.5	"	N.O	

***obtained from dash-lines

TYPE OF SURGE TANK	$h_{f/v}$	h_{or}	b_o	t_{up}	$(t_{up})_{1.0}$	ILLUSTRATED IN SECTION 5.0
Simple	2	40	0.02	"	1.0****	
	2	40	0.1	"	.34	Design 3
	2	40	0.5	"	N.O	
	20	40	0.02	"	.21	
	20	40	0.1	"	.52	
	20	40	0.5	"	N.O	

****obtained from long-short dash lines N.O Never Occur

B.8 ILLUSTRATION OF FIGURE 9

TYPE OF SURGE TANK	INDEPENDENT VARIABLES (GIVEN) $h_{f/v}$	h_{or}	b_o	DEPENDENT VARIABLES (FROM CHART) t_{re}	$(t_{re})_{0.3}$	ILLUSTRATED IN SECTION 5.0
Simple	2	0	0.02	.04*	> 7.0**	
	2	0	0.1	.2	6.5	Design 1
	2	0	0.5	1.0	5.7	
	20	0	0.02	<0.01	> 1.0	
	20	0	0.1	.032	.80	
	20	0	0.5	.16	.56	

*obtained from solid lines
**obtained from dashed-dot lines

TYPE OF SURGE TANK	$h_{f/v}$	h_{or}	b_o	t_{re}	$(t_{re})_{0.3}$	ILLUSTRATED IN SECTION 5.0
Simple	2	4	0.02	approx.	.80***	
	2	4	0.1	same as	.70	Design 2
	2	4	0.5	h_{or} = 0	N.O	
	20	4	0.02	"	.12	
	20	4	0.1	"	.11	
	20	4	0.5	"	N.O	

***obtained from dashed lines

TYPE OF SURGE TANK	$h_{f/v}$	h_{or}	b_o	t_{re}	$(t_{re})_{0.3}$	ILLUSTRATED IN SECTION 5.0
Simple	2	40	0.02	"	.12	
	2	40	0.1	"	N.O	Design 3
	2	40	0.5	"	N.O	
	20	40	0.02	"	.02	
	20	40	0.1	"	N.O	
	20	40	0.5	"	N.O	

****obtained from long-short dashed lines N.O Never Occur

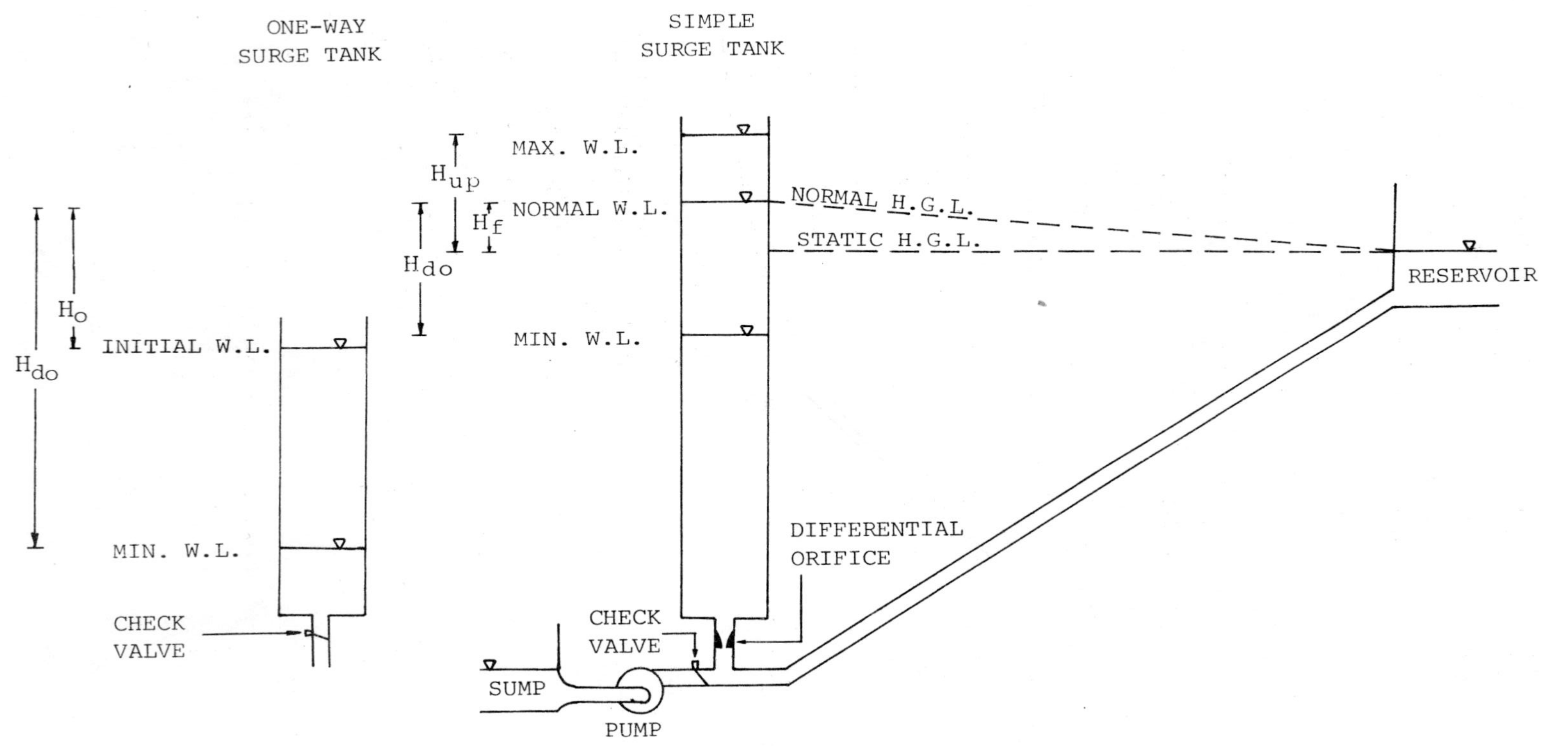

FIG. 1. SCHEMATIC OF SURGE TANK INSTALLATIONS ON PUMP DISCHARGE LINE

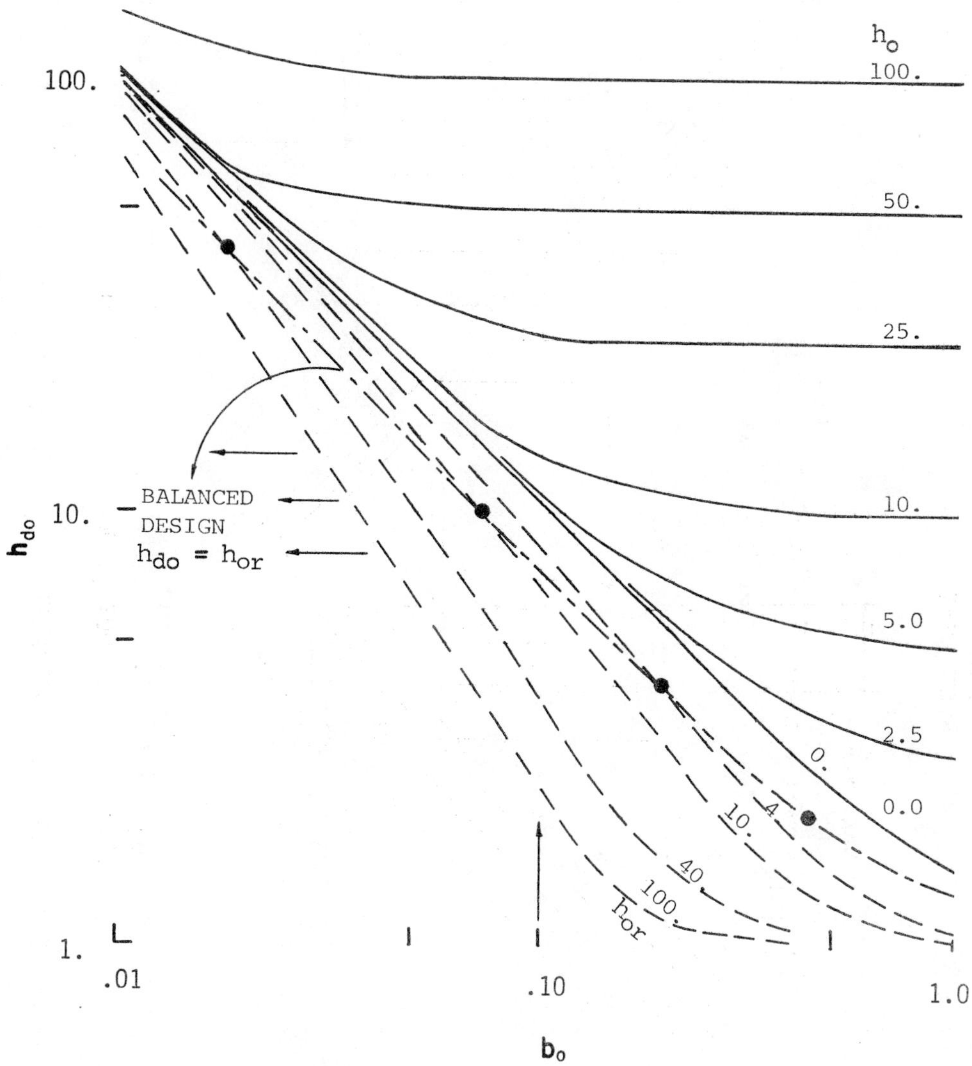

FIG. 2. MAXIMUM DOWNSURGE IN SURGE TANK

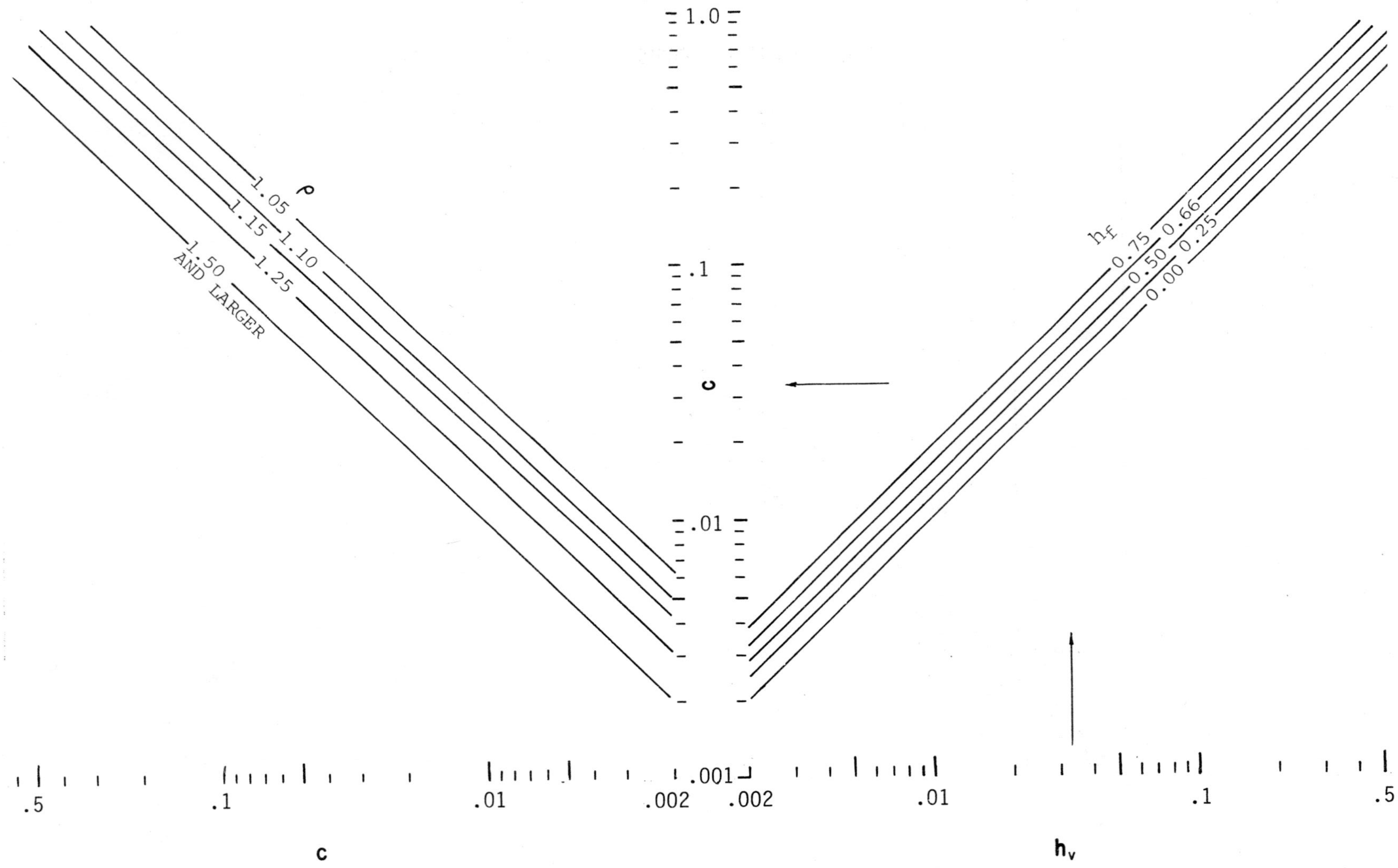

FIG. 3. MAXIMUM VOLUME OF WATER LEAVING ONE-WAY SURGE TANK (CONSTANT WATER LEVEL)

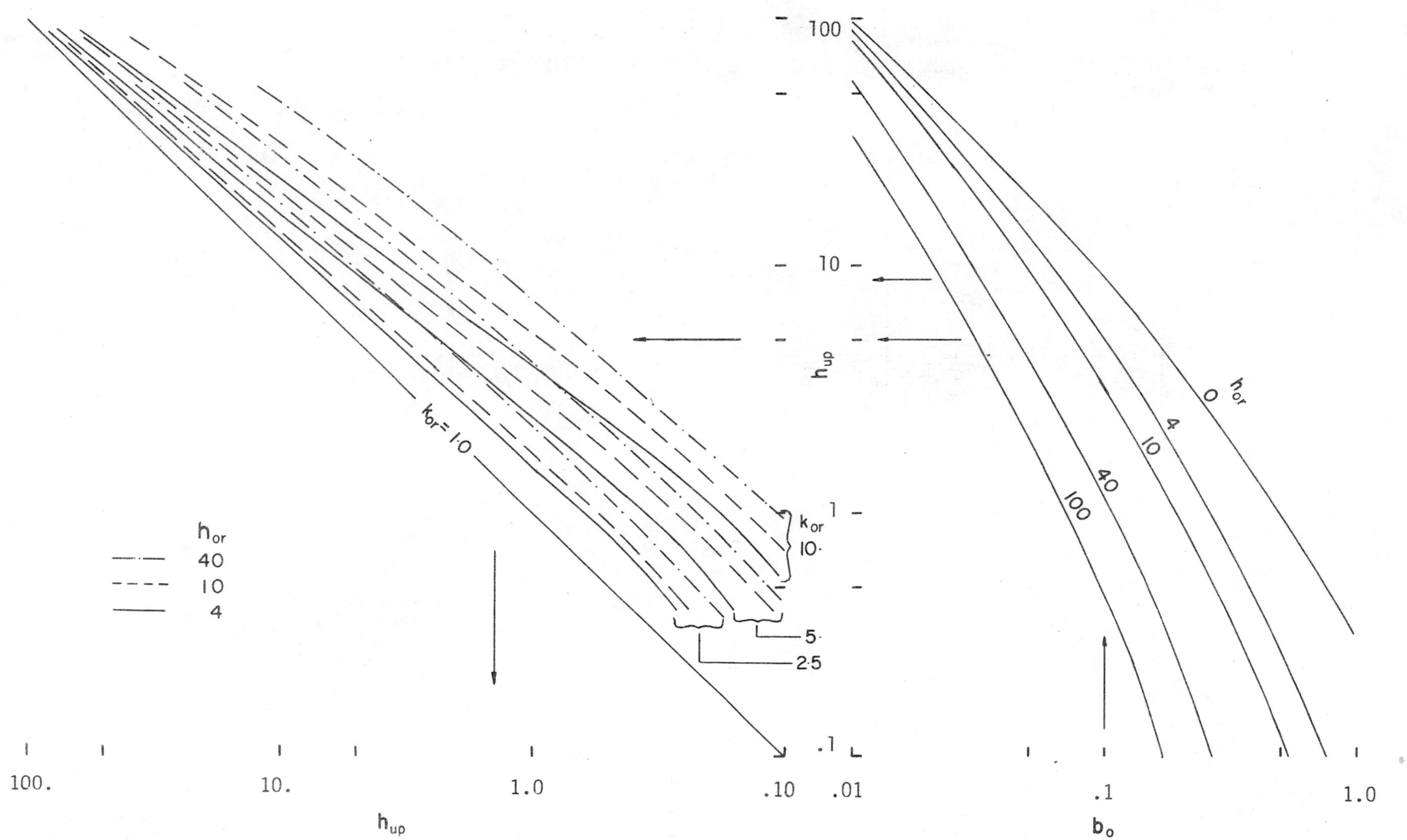

FIG. 4. MAXIMUM UPSURGE IN SIMPLE SURGE TANK

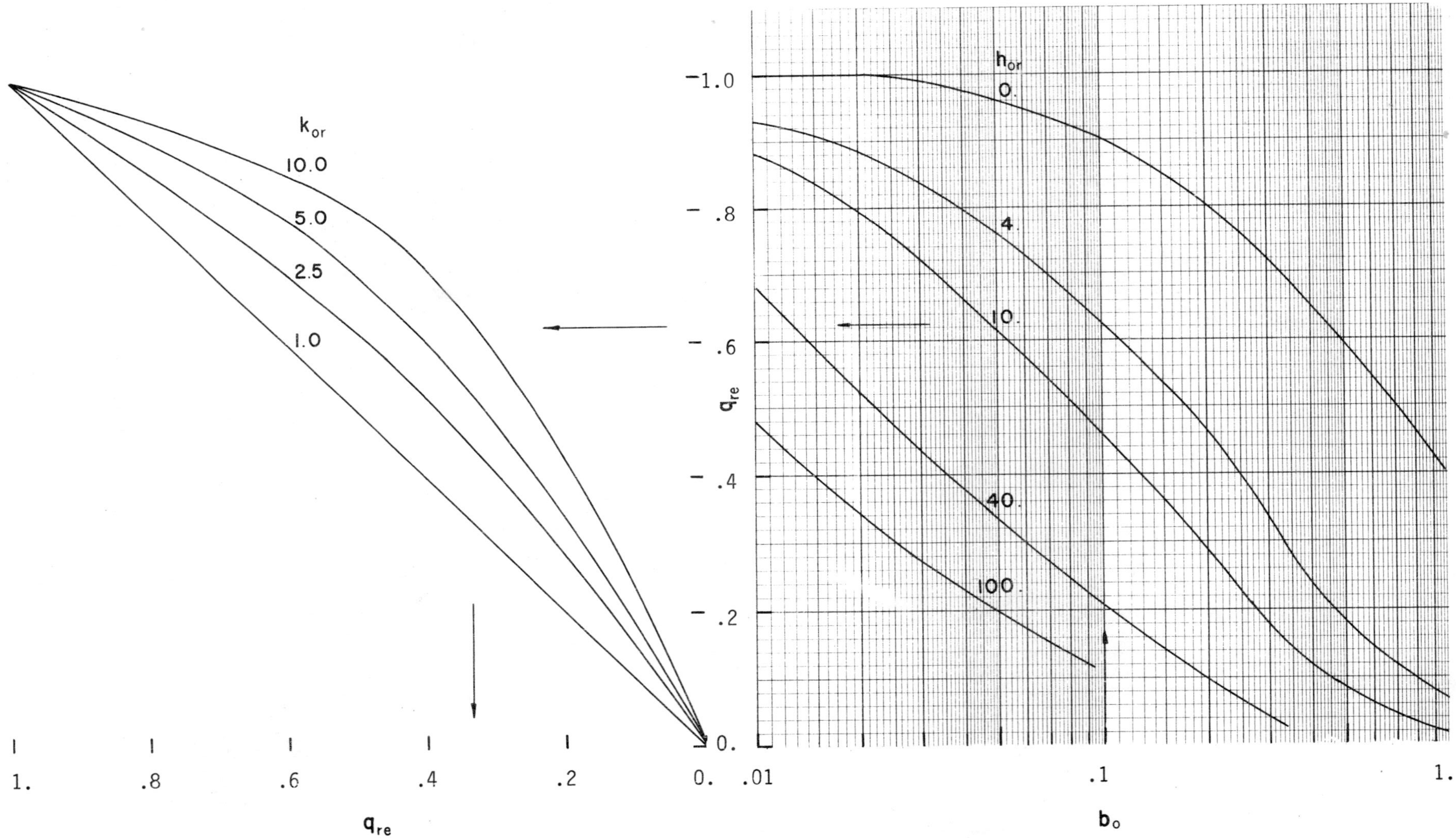

FIG. 5. MAXIMUM REVERSE FLOW INTO SIMPLE SURGE TANK

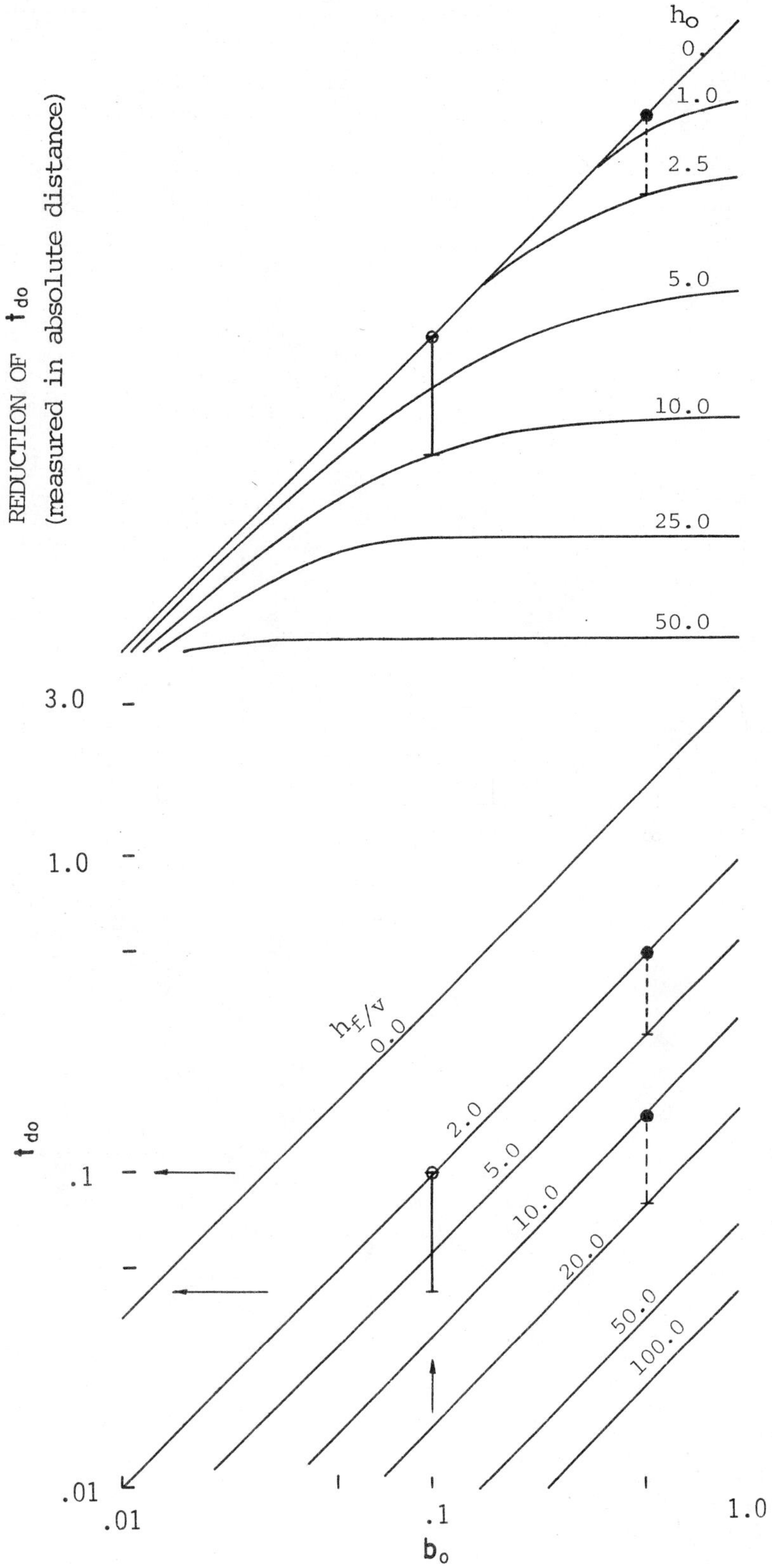

FIG. 6. TIME OF MAXIMUM DOWNSURGE IN SURGE TANK

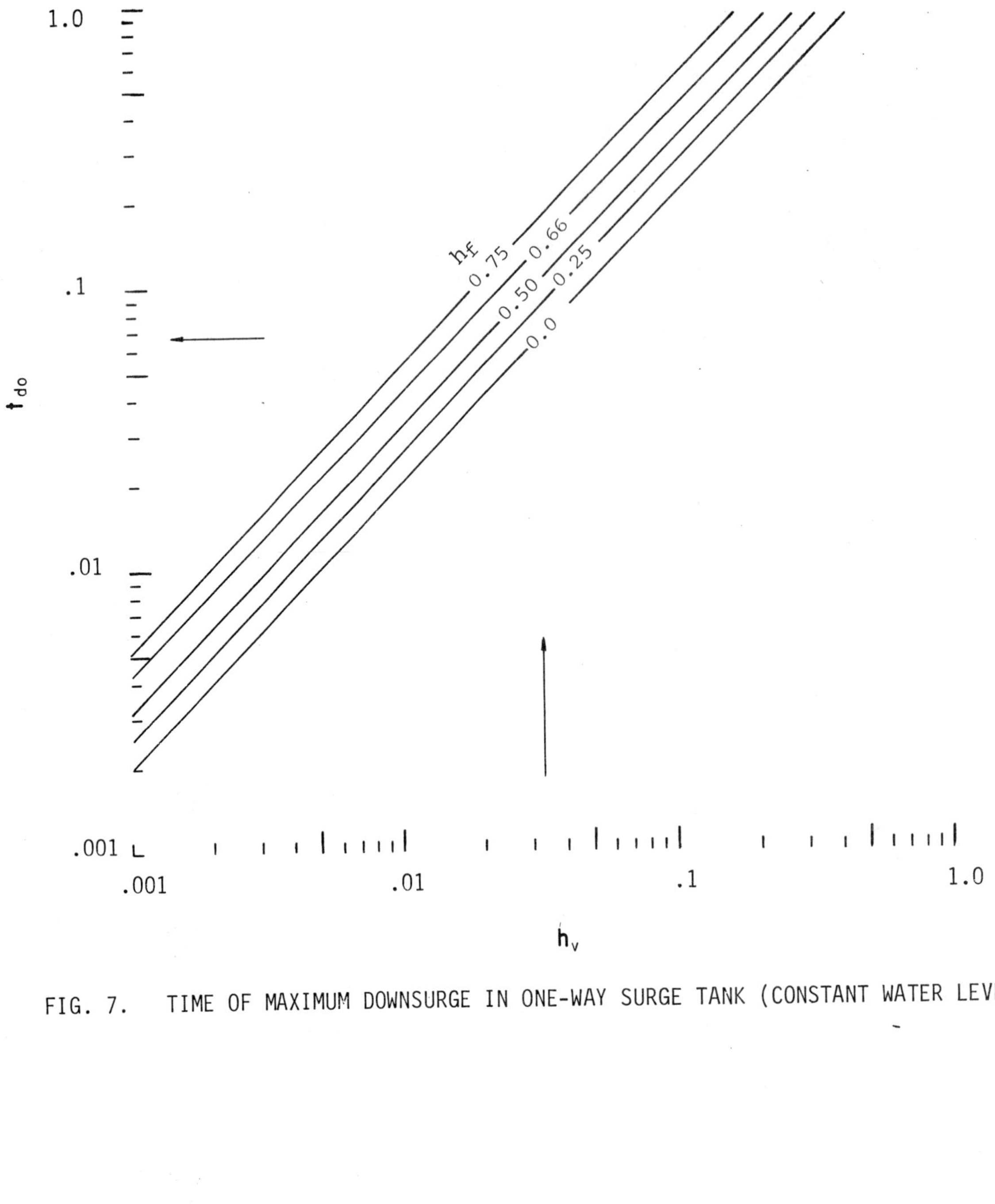

FIG. 7. TIME OF MAXIMUM DOWNSURGE IN ONE-WAY SURGE TANK (CONSTANT WATER LEVEL)

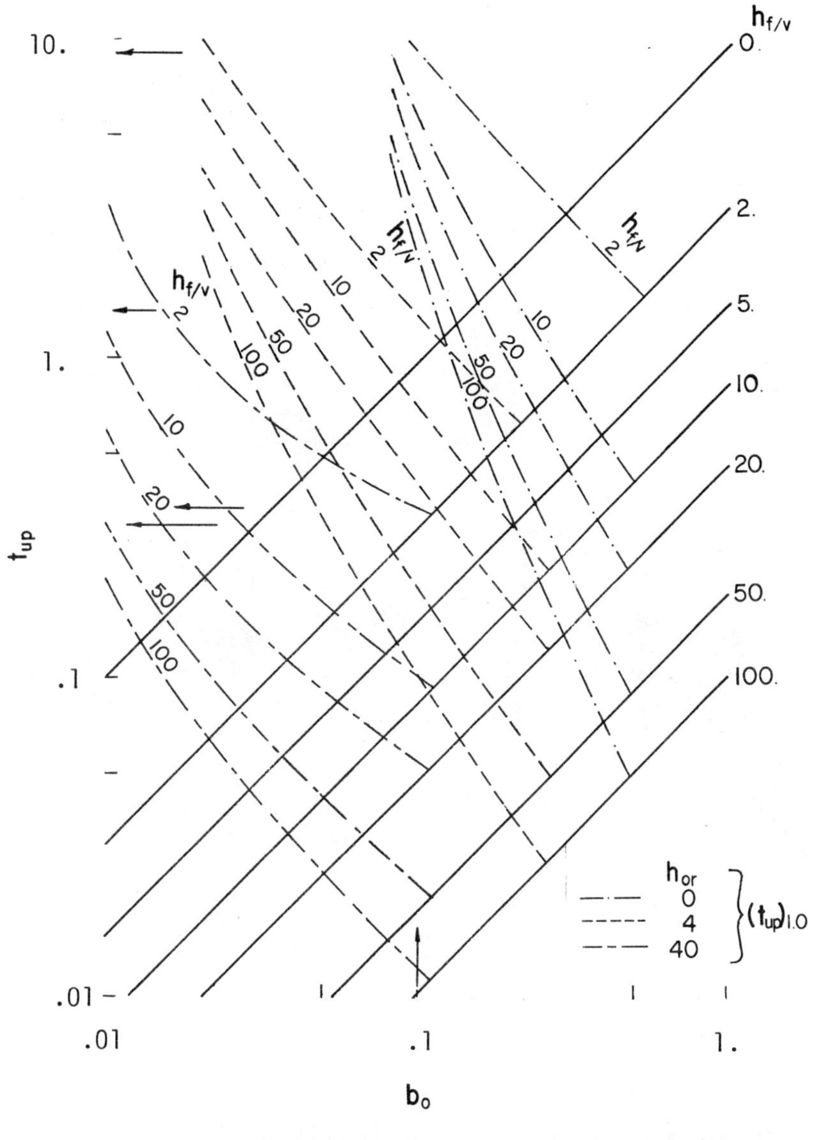

FIG. 8. TIME OF MAXIMUM UPSURGE IN SIMPLE SURGE TANK

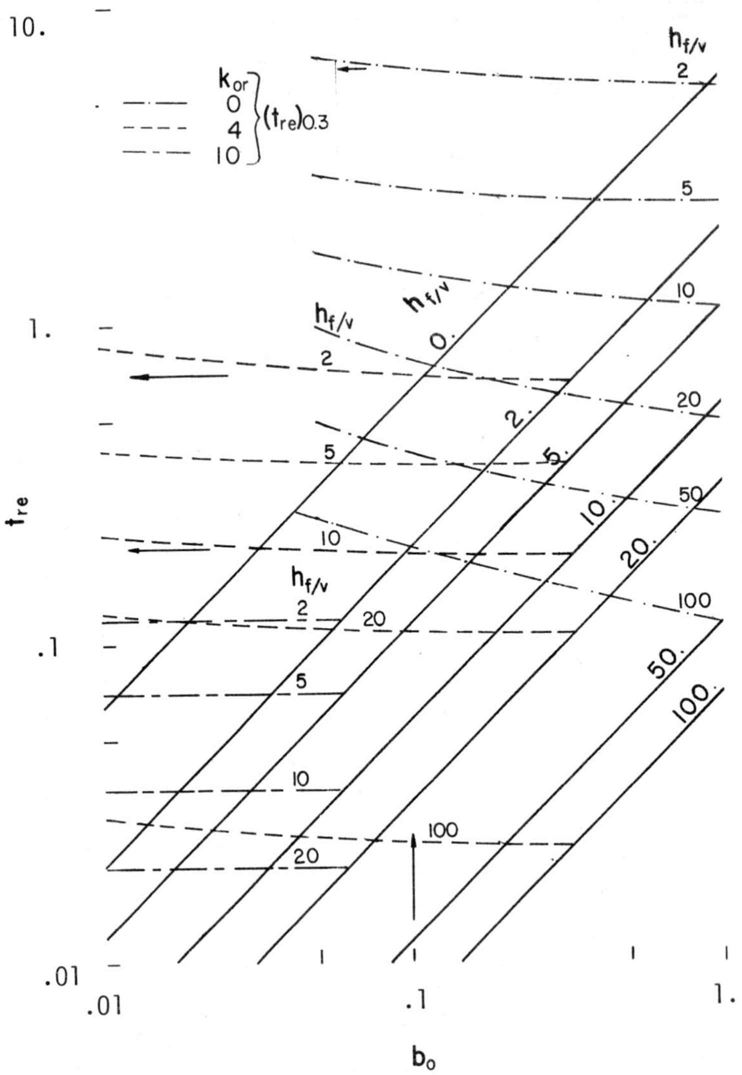

FIG. 9. TIME OF MAXIMUM REVERSE FLOW IN SIMPLE SURGE TANK

PROCEEDINGS OF THE

THIRD INTERNATIONAL CONFERENCE ON

PRESSURE SURGES

Canterbury, England
March 25-27, 1980

VOLUME 2

Organised by BHRA Fluid Engineering
in conjunction with The City University, London

bhra fluid engineering

THE CITY UNIVERSITY

TO SERVE MANKIND

EDITORIAL NOTE

Volume 1 contains the papers presented at the 3 rd International Conference on Pressure Surges held at Canterbury, England, March 25-27, 1980. An edited record of the discussion and contributions to these papers will be produced in a separate volume - Volume 2.

The Organisers cannot be held responsible for statements or opinions made in the papers in Volume 1 or in the discussion recorded in Volume 2.

These papers have been reproduced by offset printing from author's original typescript to minimise delay.

When citing papers from this volume the following reference should be used:- Title, Author(s), Paper No., Pages, 3 rd International Conference on Pressure Surges, BHRA Fluid Engineering, Cranfield, Bedford, U.K. (March 1980)

Editors: H.S. Stephens
A.R.D. Thorley
J.A. Hanson

British Library Cataloguing in Publication Data

International Conference on Pressure Surges,
3rd, Canterbury, 1980
Proceedings of the Third International Conference on Pressure Surges, Canterbury, March 1980.
1. Hydrodynamics - Congresses
2. Hydraulic transients - Congresses
532'.56 QC150

ISBN 0-906085-24-1

Printed and Published by:

BHRA FLUID ENGINEERING
CRANFIELD, BEDFORD, U.K.

ORGANISING COMMITTEE

Dr. K.J. Enever (Chairman)	Cranfield Institute of Technology
Mr. J.A. Fox	University of Leeds
Mr. A. Lupton	Binnie & Partners
Mr. M. Papworth	BHRA Fluid Engineering
Mr. H.S. Stephens	BHRA Fluid Engineering
Dr. A.R.D. Thorley	The City University, London
Prof. A.E. Vardy	University of Dundee
Mr. J.A. Hanson (Secretary)	BHRA Fluid Engineering

ACKNOWLEDGEMENTS

BHRA wishes to thank Dr A.R.D. Thorley for compiling and editing the discussion published in this volume.

His efforts should have made an earlier publishing date possible but, due to printing difficulties, the publication date had to be postponed.

We apologise for this delay to the delegates who were expecting to receive this volume before the end of 1980.

PROCEEDINGS OF THE THIRD INTERNATIONAL CONFERENCE ON

PRESSURE SURGES

CANTERBURY, ENGLAND

March 25-27, 1980

Volume II

CONTENTS

Third International Conference on

PRESSURE SURGES

Canterbury, England
March 25-27, 1980

PAPER E6

WAVE PROPAGATION IN PLASTICALLY DEFORMING DUCTS

J.W.R. Twyman & A.R.D. Thorley

The City University, London.

and

R. Hewavitarne,

British Gas Corporation

Summary

An experimental study of the deformation of hexagonal section pipe subjected to large amplitude pressures is described. The results demonstrate that the corners yield plastically first, followed by the mid-points of the sides. This behaviour is used as the basis for an idealised deformation procedure for the theoretical determination of the propagation speed of transient pressure waves in regular polygonal ducts when plastic deformation occurs.

Held at the University of Kent at Canterbury, U.K.

Conference Organised by BHRA Fluid Engineering, Cranfield, Bedford, England.
In conjunction with The City University, London.

NOMENCLATURE

A	duct cross-sectional area
B	see equation (30)
c	speed of propagation of a pressure wave
C	constant of integration in equation (36)
D	distance across the flats of a regular polygon
e	wall thickness (e_o = initial value of)
E_p	tangent modulus
F	see equation (24)
G	see equation (25)
H	see equation (26)
K	bulk modulus
L	width of the flats of a regular polygon
M	a bending moment; abbreviation for equation (27)
$M_{A,B}$	bending moment at A, B
M_o	bending moment at which plastic yielding occurs
n	number of sides in a regular polygon
$N_{A,B}$	normal force at A, B.
$N_{y,o}$	normal force associated with plastic yielding
P, P_y	pressure, pressure to cause plastic yielding
R	instantaneous value of a half diagonal of a polygon
R_o	initial value of a half diagonal of a polygon
t	time
T	see equation (29)
α	geometric parameter $= e_o/L$
β	$= \pi/n$ for a regular polygon
δp	change in internal pressure in a duct
$\Delta_{A,B}$	stretch of a side wall of a duct at A, B.
Δ_T	total stretch of a half side wall of a polygon
ε	hoop strain
ρ, ρ_s	density of fluid, and of wall material
σ, σ_y	stress, yield stress
θ	angle change of a side wall of a polygon

INTRODUCTION

Rapid pressure changes having a large amplitude can occur due to thermo-chemical reactions, plant maloperation or system faults. If these large pressure changes are contained within a duct the limit of elastic behaviour of the duct wall may be reached, or even exceeded so that plastic deformation takes place.

This problem has been encountered principally in connection with sodium-cooled fast reactors having the added complication that the flow passages are initially of hexagonal cross-section. It has been amply demonstrated (1) that a pressure pulse emanating from the site of a thermo-chemical reaction is rapidly attenuated to a pressure level corresponding to that at which plastic yielding of the duct would just start to occur. The attenuation is caused by virtue of the fact that pressure levels above that to cause plastic yielding are propagated at velocities much lower than pressure waves causing only elastic deformation (2). This means that the "unloading" wave, travelling at the higher speed associated with elastic deformation, overtakes the pressure levels associated with plastic deformation, as shown schematically in Fig.1.

As in conventional pressure surge problems the speed of propagation of pressure waves in ducts of arbitrary cross-sectional shape can be predicted from a solution of

$$c = \left\{ \rho \left[\frac{1}{K} + \frac{1}{A}\frac{\delta A}{\delta p} \right] \right\}^{-1/2} \tag{1}$$

The term representing area strain per unit pressure change $\frac{1}{A}\frac{\delta A}{\delta p}$ must be evaluated in terms of the behaviour of the duct wall material and the cross-sectional shape of the duct itself. In the case of transient pressure waves in elastically deforming circular section pipes the results have been well known for many years. Similarly, wave propagation in elastically deforming ducts of a regular n-sided polygonal section have been analysed (3) including the hexagonal section encountered in nuclear systems.

Fox and Merckx (4) were the first to consider a theorectical study of pressure wave transmission in a plastically deforming pipe, subsequently utilised and developed by Thorley and Twyman (5). However, although dealing with the propagation of large amplitude pressure waves in an initially hexagonal duct, it was assumed that wave propagation would be governed by one of two modes of pipe deformation, i.e.

i) elastic deformation due primarily to the elastic bending of the 'flats' of the hexagon, and

ii) plastic hoop straining of a circular section into which it was assumed that the hexagon would deform instantaneously when the yield point of the wall material was reached.

A recent experimental study (6) has indicated that as the pressure in a hexagon section is increased a point is reached at which the corners deform plastically. Subsequent deformation lends support to the hypothesis that the plastically deforming hexagon deforms from a 6 to a 12 sided shape, with six additional plastic hinges forming at the mid-points of the sides.

EXPERIMENTAL STUDY

The experiments were carried out on a 2.13 m length of mild steel hexagonal section pipe. The cross-sectional dimensions were 127 mm. across the flats of the hexagon and the walls were 3.43 mm thick.

Both ends of the tube were closed off by welding steel plates onto them, with provision in them for filling and pressurising with water and for air release. The test-pipe was mounted vertically.

Ten strain gauges, arranged in two banks of five, were arranged symmetrically across two opposite flats in the mid-length of the pipe. The gauges were of the post-yield type, having a gauge length of 5mm. Data output was via a multichannel digital recorder which could be set to scan all channels automatically. Initial calibration and zeroing was done manually.

The test procedure was to check the zero readings of the strain gauges, then increase the pressure in steps to a target pressure, monitoring the strains on the way, then releasing the pressure and re-checking the zero-reading of the strain gauges. The initial target pressure was 13.8 kPa (2 lbf/sq.in) and this was increased in steps of 13.8 kPa. At higher pressures the magnitude of the increments was increased so that for target pressures of around 7MPa, for example, the increments were 138 kPa. This procedure was repeated until total failure of the tube occurred.

DISCUSSION OF EXPERIMENTAL RESULTS

As the tests proceeded graphs of pressure versus strain were plotted. These provided a check that no anomalies in the results were developing and also enabled predictions to be made of the pressures at which plastic yielding might occur. For mild steel of the type from which the pipe was made, yielding was expected at 1200 microstrain. Fig.2 is a typical example of the experimental results.

Before any yielding occurred at all the pressure-strain curves were straight lines. For the corner gauges the pressure-strain relationships before yielding occurred were in the range

$$1.10 \times \varepsilon < p < 1.182 \times \varepsilon$$

where ε is the microstrain and p is the corresponding pressure in kPa. Hence assuming yield occurred at a microstrain of 1200 led to the prediction that the corners of the hexagon section would deform plastically at a pressure in the region of 1.32 - 1.42 MPa.

Similarly, for the mid point of the originally flat sides

$$1.476 \times \varepsilon < p < 1.723 \times \varepsilon$$

and the yield pressure would be in the range 1.77 - 2.07 MPa. With all these results an encouraging level of consistency was noted, not only in terms of the numerical values but also in the fact that the behaviour of the hexagonal section was symmetrical.

As implied by the above relationships the strain at the mid-point of the sides was lower than that experienced at the corners for a given pressure. The commencement of plastic deformation at the corners, as indicated by the non-linearity of the pressure-strain graphs, was found to occur at 1.38 MPa. This was within range predicted above. Similarly, plastic yielding of the mid-points of the two opposite flats which had been strain gauged appeared to occur at a pressure of 1.79 MPa. This is at the lower end of the range at which yielding was predicted and it could be a slightly misleading result on the grounds that, as the corners had yielded by this time, the hexagon section, as a structure, was no longer truly elastic and hence linear in its behaviour. However, yielding clearly occurred at the mid-point at a higher pressure, and the behaviour of the hexagon was symmetrical.

As the tests proceeded further into the zone of plastic behaviour of the material, the pressure-strain path was linearly elastic as the pressure was relieved. Then when re-pressuring for the next text, to take the target pressure one increment higher, a linear p-ε path was followed until it reached the plastic curve, after which the plastic curve was extended. See, for example, Fig.2. From this it was deduced that the shape of the pipe did not affect the material characteristics.

Fig.3 provides a comparison of the pressure-strain behaviour of the corners and mid-points of the pipe wall. One interpretation from this is that in the region of pressures between 2 and 3.5 MPa the corners continued to open up considerably, and more significant radial deformation took place at the higher pressure levels. Meanwhile at the mid-point plastic hinges formed as the radial deformation increased, but then further plastic deformation at this point was small due to there being greater flexibility at other points along the sides.

A plaster cast of the deformed pipe at a pressure of 8.8 MPa was taken and the result is shown in Fig.4. Failure of the pipe occurred at a pressure of 9.13 MPa when the welded seams down two of the sides parted. One interesting feature was that a section of the pipe removed from close to the stopped ends after failure fitted the plaster cast almost exactly, indicating that the end affects were small.

THEORETICAL ANALYSIS

In the interests of generality the theory is developed for an arbitrary regular polygon having initially n sides. A detailed example refers to an initially hexagonal duct for comparison with previously published results (5).

Fig. 5 is a cross-sectional view of an pressurised n-sided duct of initial wall-thickness e_o and side width L. It is assumed that the duct is made of a "rigid" but perfectly plastic material having a yield stress σ_y, that elastic deformation is small compared with plastic deformation, and that strain hardening can be neglected compared with the large scale geometric effects.

It is also assumed that each side of the polygons may be treated as a fixed-end beam of uniform cross-section carrying a uniformly distributed load. As the load is increased on the 'beam', yielding first occurs at the ends, and after a further small increase plastic hinges form at these sections. Prior to this no rotation is assumed, but axial deformation is permitted.

The formation of the plastic hinges at the corners does not cause collapse, as the central bending moment at this load is still less than the value which causes yield, so that the beam has not by this stage been reduced to a mechanism. Thus, if the load is increased still further the plastic hinges at the ends of the beam undergo rotation while the bending moment at each hinge, remains constant at the fully plastic value. The central bending moment increases first to the yield value and soon after to the fully plastic value and additional 'hinges' form.

The shape of the deformed beam may be characterised by a straight line between any pair of hinges, as shown in Fig.6.

LIMIT LOAD ANALYSIS

The free-body diagram for the left half of the beam section AB is shown in Fig.7 and a consideration of the force balance yields the following relations for the situation representing the limit of elastic deformation i.e. the yield point of the material has just been reached at the centre

$$\underline{\Sigma F_x = 0} \qquad N_A = N_B = N_y \tag{2}$$

$$\underline{\Sigma F_y = 0} \qquad R_A - p_y L/2 = 0 \tag{3}$$

and since by symmetry for an n-sided duct

$$N_y = R_A/\tan\beta \tag{4}$$

$$\therefore \quad N_y = p_y L/2\tan\beta \tag{5}$$

$$\underline{\Sigma M_A = 0} \qquad M_y = p_y L^2/16 \tag{6}$$

Substitution of equations (5) and (6) into the 'yield condition' equation

$$\frac{M}{M_o} + \left(\frac{N}{N_o}\right)^2 = 1 \tag{7}$$

and noting that a) $N_o = \sigma_y e_o$ (8)

b) $M_o = \sigma_y e_o^2/4$ (9)

the result is a quadratic in p_y/σ_y

$$\frac{1}{\tan^2\beta}\left(\frac{p_y}{\sigma_y}\right)^2 + \frac{p_y}{\sigma_y} - 4\left(\frac{e_o}{L}\right)^2 = 0 \tag{10}$$

from which the positive root gives

$$p_y = \frac{1}{2}\sigma_y \tan^2\beta\left(\sqrt{1 + 16\alpha^2/\tan^2\beta} - 1\right) \tag{11}$$

where p_y is the limit load and the geometric parameter α is defined by

$$\alpha = e_o/L \tag{12}$$

A further small increase above the limit load leads to plastic deformation.

PLASTIC DEFORMATION

The distortion caused by the increase of internal pressure results in the formation of plastic hinges at the corners and mid-points of the sides, the hinges being connected by rigid sections of width L/2. Fig.8 shows the free body diagram of the typical half-side AB. The stretch of the half-side at each hinge is Δ_A and Δ_B and the angle formed due to the rotation of the half-side at each hinge is θ

Noting equation (4) and applying Newton's Second Law of Motion for the half-side AB gives (8):-

$$-M_A - M_B + \frac{L}{4}\left[N_A(\tan\beta\cos\theta - \sin\theta) - N_B\sin\theta\right] = \rho_s \frac{e_o L^3}{96}\frac{d\omega}{dt} \tag{13}$$

$$\delta p.\frac{L}{2} - N_B\sin\theta + N_A(\sin\theta - \tan\beta\cos\theta) = \rho_s \frac{e_o L}{2}\frac{dV_R}{dt} \tag{14}$$

$$N_A(\cos\theta + \tan\beta\sin\theta) - N_B\cos\theta = \rho_s \frac{e_o L}{2}\frac{dV_T}{dt} \tag{15}$$

where ω is the angular velocity and V_R and V_T are the radial and tangential velocity components of the centre of mass of the half-side. ρ_s is the density of the duct wall material.

The yield conditions for the plastic hinges at A and B are, from equation (7)

$$\frac{M_A}{M_o} + \left(\frac{N_A}{N_o}\right)^2 = 1 \tag{16}$$

$$\frac{M_B}{M_o} + \left(\frac{N_B}{N_o}\right)^2 = 1 \tag{17}$$

and the "flow rule" for these hinges can be expressed (7) in the form:

$$d\Delta_A = \frac{N_A}{2\sigma_y} . d\theta \qquad (18)$$

$$d\Delta_B = \frac{N_B}{2\sigma_y} . d\theta \qquad (19)$$

Neglecting inertial effects as a first approximation for thin-walled tubes and combining equations (16) and (17) with equation (13) leads to:

$$4M_o\left[2-\left(\frac{N_A}{N_o}\right)^2-\left(\frac{N_B}{N_o}\right)^2\right] + N_B L \sin\theta + N_A L\left(\sin\theta - \tan\beta\cos\theta\right) = 0 \qquad (20)$$

Solving equations (14) and (15) simultaneously gives:

$$N_A = \frac{\delta p \; L\cos\theta}{2\tan\beta} \qquad (21)$$

$$N_B = \frac{\delta p L}{2\tan\beta}\left(\cos\theta + \tan\beta\sin\theta\right) \qquad (22)$$

Combining these results with equation (20) and recalling equations (8) and (9) yields, after some re-arrangement:

$$F\tan^2\theta + G\tan\theta + H = 0 \qquad (23)$$

where

$$F = M\alpha^2\tan^2\beta\,(1 - M\tan\beta) \qquad (24)$$

$$G = M\alpha^2\,(1 - 2M.\tan\beta) \qquad (25)$$

$$H = 2 - 2M^2\alpha^2 - M\alpha^2 \qquad (26)$$

$$M = \delta p/(2\sigma_y\tan\beta) \qquad (27)$$

and α has been defined previously by equation (12)

Solving for the positive root gives

$$\theta = \tan^{-1}(T/B) \qquad (28)$$

where

$$T = \left(M^2\alpha^4(1-2M\tan\beta)^2 - 4\alpha^2 M\tan\beta(1-M\tan\beta)(2-2M\alpha^2-M\alpha^2)\right]^{1/2} - M\alpha^2(1-2M\tan\beta) \qquad (29)$$

$$B = 2\alpha^2 M\tan^2\beta\,(1 - M\tan\beta) \qquad (30)$$

AREA STRAIN DURING PLASTIC DEFORMATION

It is assumed in this section that as the internal pressure acting on the walls of the duct increases the deformation proceeds as follows:

i) plastic hinges form at the corners

ii) plastic hinges are then formed at the mid-point of each side

iii) plastic stretching of the walls of the duct occurs.

Under conditions of elastic deformation the area strain is made up of three components. These are due to bending and shear deflections and also stretching of the side walls, but the principal component will be the bending deflection if the polygon is initially of eight sides or less (3). However, in the case of plastic deformation, as the cross-sectional shape tends towards a circle the area strain will become primarily due to the (hoop) stretching of the side walls, hence

$$\frac{\delta A}{A} = 2\varepsilon \tag{31}$$

where the hoop strain

$$\varepsilon = \frac{2\Delta_T}{L} = \frac{\Delta_T}{R_o \sin\beta} \tag{32}$$

now

$$\Delta_T = \Delta_A + \Delta_B \tag{33}$$

hence, from equations (18), (19), (21) and (22) with N_A and N_B in terms of the angle β

$$\frac{d\Delta_T}{d\theta} = \frac{\delta p . R_o \sin\beta}{\sigma_y}\left(2\cos\theta\cos\beta + \sin\theta\sin\beta\right) \tag{34}$$

As β takes on discrete values as deformation progresses this equation can be integrated to give

$$\Delta_T = \frac{\delta p\, R_o \sin\beta}{\sigma_y}\left(2\sin\theta\cos\beta - \cos\theta\sin\beta + C\right) \tag{35}$$

The hoop strain therefore becomes

$$\varepsilon = \frac{\delta p}{\sigma_y}\left[2\sin\theta\cos\beta - \cos\theta\sin\beta + C\right] \tag{36}$$

For the case of a duct having been deformed into a circle it has been shown previously (5) that the hoop strain is given by

$$\varepsilon = \frac{R}{e}\left(\frac{\delta p}{E_p - 2\sigma}\right) \tag{37}$$

Hence by combining equations (36) and (37) and setting θ to zero for a circular section the integration constant is

$$C = \frac{R}{e}\left(\frac{\sigma_y}{E_p - 2\sigma}\right) + \sin\beta \tag{38}$$

and so

$$\varepsilon = \frac{\delta p}{\sigma_y}\left[2\sin\theta\cos\beta - \cos\theta\sin\beta + \sin\beta + \frac{R}{e}\left(\frac{\sigma_y}{E_p - 2\sigma}\right)\right] \tag{39}$$

Combining equations (1) (31) and (39) the speed of propagation may be expressed as

$$c = \left\{\rho\left(\frac{1}{K} + \frac{2}{\sigma_y}\left[2\sin\theta\cos\beta - \cos\theta\sin\beta + \sin\beta + \frac{R}{e}\left(\frac{\sigma_y}{E_p - 2\sigma}\right)\right]\right)\right\}^{-1/2} \tag{40}$$

In the case of a duct which is of circular section equation (40) reduces to the earlier result (2,5) of

$$c = \left\{ \rho \left(\frac{1}{K} + \frac{2R/e}{(E_p - 2\sigma)} \right) \right\}^{-1/2} \tag{41}$$

SPECIMEN RESULTS

Brief comparision is made with previously published data (5) which dealt with the deformation of a hexagonal duct. The following data apply:

$$L = R_o = 79\,mm$$

$$e_o = 2.6\,mm$$

$$\alpha = e_o/R_o = 0.03291$$

$$\left.\begin{array}{l} K = 4.86\ GPa \\ \rho = 856.2\ kg/m^3 \end{array}\right\}\ \text{liquid sodium at } 400\,°C$$

$$\beta = 30°$$

$$p_y = 15\ MPa$$

From equation (11) $\sigma_y = 3506.86\ MPa$

From equation (28) for an increase in pressure to 17.63 MPa

$$\theta = 2.48°$$

and

$$\frac{1}{A}\frac{\delta A}{\delta p} = \frac{2}{\sigma_y}\left[2\sin\theta\cos\beta - \cos\theta\sin\beta + \sin\beta + \frac{R}{e}\left(\frac{\sigma_y}{E_p - 2\sigma}\right)\right]$$

$$= \frac{1}{1.753 \cdot 10^9}\left(0.07504 - 0.49953 + 0.5 + 2.82644\right)$$

$$= 1.6554 \cdot 10^{-9}\ m^2/N$$

From equation (40) C = 792.17 m/s

Whereas from eqn (41) C = 801.59 m/s.

There is a 1.2% disparity between these two results due to the influence of the additional terms in equation (40). Further comparisons are made in Fig. (9) which is a graph of the wave speed according to equation (40) compared with earlier results (5) based on equation (41).

DISCUSSION OF THE THEORETICAL RESULTS

Further results are shown in Tables 1 and 2 and the graph in Fig.9. It will be noted that the angle θ increases to a maximum of the order of 6° which implies that although the hexagon section is deforming into a twelve sided polygon it is no longer a regular polygon.

The numerical values of wave speed are only marginally different from those predicted by assuming a truly circular section duct undergoing plastic deformation. This is illustrated in Fig.9, however it must be remembered that equations (40) and (41) are concerned with a duct that has been pressurised to the point that plastic deformation has been initiated i.e. in the region where wave pressures δp are in excess of around 15MPa. Below this value is a region of wave propagation associated with relatively large scale elastic deformation, indicated by a dashed line in Fig.9. This region has not been explored analytically, but at its left hand end must merge into a curve for wave speeds for small amplitude pressure waves. For the numerical example presented in this paper a wave speed of 867 m/s applies (2) if the amplitude of the pressure waves is small.

CONCLUSIONS

a) An experimental study has indicated that when an initially hexagonal pipe section is subjected to large amplitude internal pressures plastic yielding occurs first at the corners of the hexagon, followed at a slightly higher pressure, by plastic yielding at the mid-points of the side walls.

b) A revised expression for wave propagation speeds in plastically deforming ducts of an initially regular polygonal section has been developed.

c) Further work is required to develop a suitable relationship for wave speeds in regular polygonal ducts undergoing large amplitude elastic deformation.

REFERENCES

1. Florence A L and Abrahamson G R "Simulation of a hypothetical core disruptive accident in a fast flux test facility" Stanford Research Institute, Menlo Park, California HEDL-SRI-1 (May 1973)

2. Fox G L and Stepnewski D D "Pressure wave transmission in a fluid contained in a plastically deforming pipe" ASME Paper 74-PVP-26 Pressure Vessel and Piping Conference, (June 1974)

3. Thorley A R D and Buttigieg G L G "Surge propagation speeds inducts of regular polygonal cross-section" Procs. 6th Australasian Hydraulics and Fluid Mechanics Conference, Adelaide, (1977).

4. Fox G L and Merckx R R "Pressure wave transmission in a fluid contained in a plastically deforming pipe" Hanford Engineering Development Laboratory. HEDL-TME-73-61; (1973)

5. Thorley A R D and Twyman J W R "Propagation of Transient pressure waves in a sodium-cooled fast reacter" Procs.Second International Conference on Pressure Surges, Publ. British Hydromechanics Research Association, (Sept.1976).

6. Wicklen V E "Deformation under internal pressure of hexagon section steel pipe" Final Year Project Report. Dept. of Mech.Eng. The City University, London (July 1979)

7. Youngdahl C K "Large deformation plastic response of hexagons to pressure pulses" ANL Report 798, Argonne National Laboratory, (1973).

8. Hewavitarne R "Propagation of transient pressure waves in the plastic deformation region for hexagonal-section ducts". Thesis submitted in partial fulfilment of the requirements of the M.Sc.degree. The City University, London 1978.

ACKNOWLEDGEMENTS

The experimental work was conducted in the laboratories of The City University and the authors wish to acknowledge the assistance of the laboratory technicians. Also Mrs.V.E.Maginnis (nee Wicklen), a former student, contributed to this paper through her undergraduate project and her painstaking work in determining experimentally the behaviour of the hexagon section during pressure tests. This is much appreciated.

TRUE STRAIN ξ (%)	TRUE STRESS σ (MPa)	TANGENT MODULUS Ep (MPa)	$\frac{R}{e} = \frac{R_o}{e_o} \exp(2\xi)$	Ep − 2σ (MPa)	WAVE SPEEDS C (m/s)	WAVE PRESSURE δp (MPa)
0.1	88.6	88600	29.0966			3.04
0.2	177.0	88600	29.1548			6.07
0.3	356.0	89000	29.2717	88288	1157.70	12.16
0.6	518.0	37500	29.3890	36464	801.01	17.63
1.0	575.0	7216	29.6250	6066	342.21	19.41
2.0	617.0	3448	30.2235	2214	206.04	20.41
3.0	649.0	3279	30.8341	1981	193.05	21.05
4.0	684.0	3061	31.4570	1693	176.79	21.74
5.0	713.0	2727	32.0925	1301	153.54	22.22
6.0	740.0	2521	32.7408	1041	136.04	22.60
7.0	764.0	2362	33.4022	834	120.59	22.87
8.0	785.0	2317	34.0769	747	113.01	23.04
9.0	811.0	2255	34.7655	633	102.82	23.33
10.0	832.0	2166	35.4677	502	90.85	23.46
15.0	933.0	1973	39.1978	107	39.92	23.80
20.0	1026.0	1371	43.3203	-	-	23.68
25.0	1050.0	-	47.8763	-	-	21.93

Table 1 **Materials Data,** Wave Pressure and Wave Speeds (from **Eqn** 41) in the Plastically Deforming Duct.

WAVE PRESSURE δp (MPa)	θ^{o}	β^{o}	1/k (x 10^{-9})	$1/A(\delta A/\delta\ )$ (x 10^{-9})	WAVE SPEED C (M/S)
17.63	2.483	30	0.20576	1.65540	792.17
19.41	3.747	30	0.20576	9.83272	341.10
20.41	4.354	30	0.20576	27.37798	205.77
21.05	4.711	30	0.20576	31.21119	192.81
21.74	5.071	30	0.20576	37.24968	176.58
22.22	5.307	30	0.20576	49.427713	153.40
22.60	5.488	30	0.20576	62.998372	135.94
22.87	5.611	30	0.20576	80.199147	120.52
23.04	5.688	30	0.20576	91.335987	112.95
23.33	5.816	30	0.20576	109.945169	102.95
23.46	5.872	30	0.20576	141.408133	90.81
23.80	6.016	30	0.20576	732.774258	39.92
23.68					
21.93					

TABLE 2 Wave Pressure, Deformation Angle and Wave Speeds Calculated Using Equations (28) and (40.)

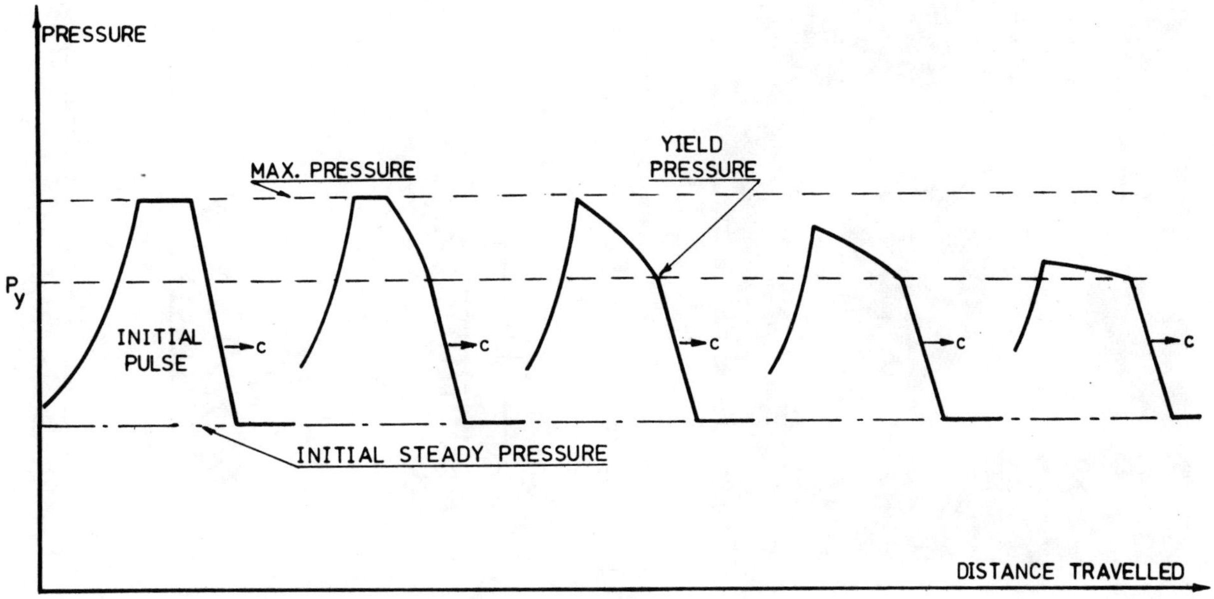

Fig.1. Schematic representation of the attenuation of a large amplitude pressure wave.

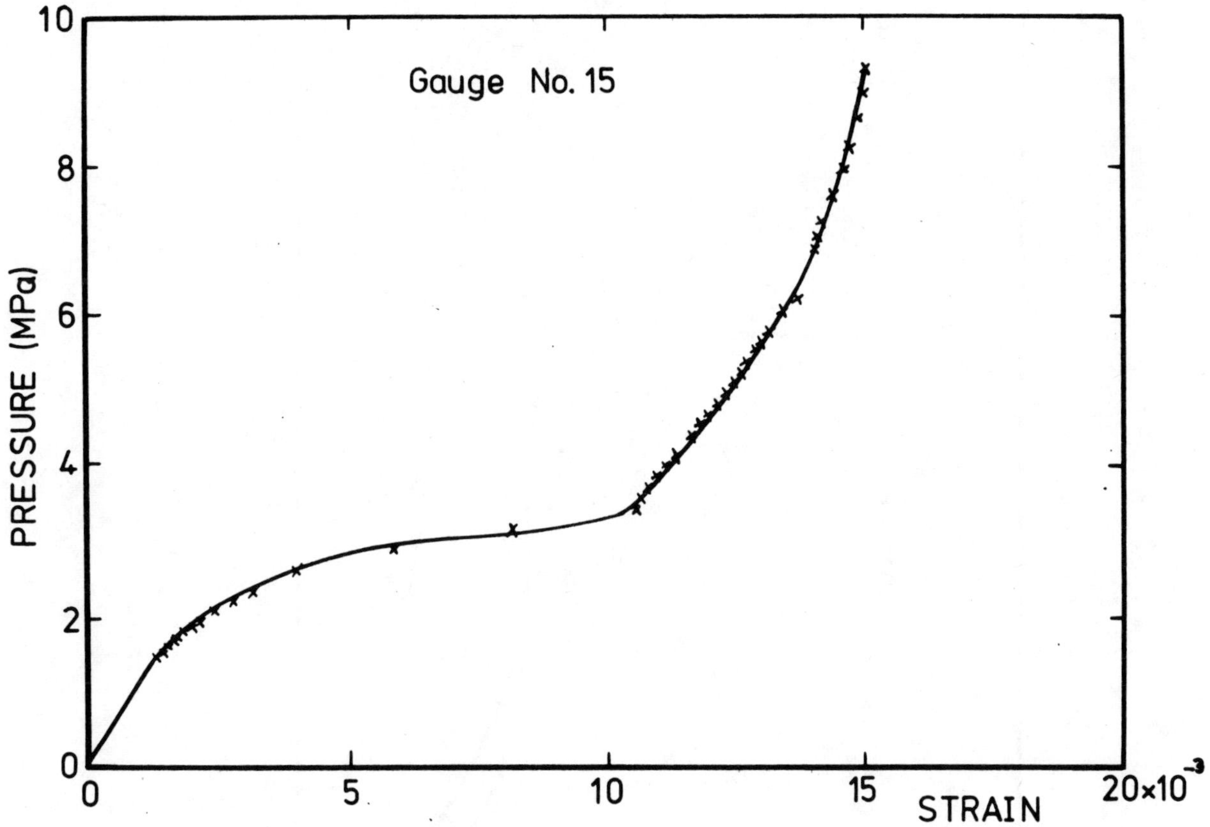

Fig.2. Typical example of a pressure-strain curve.

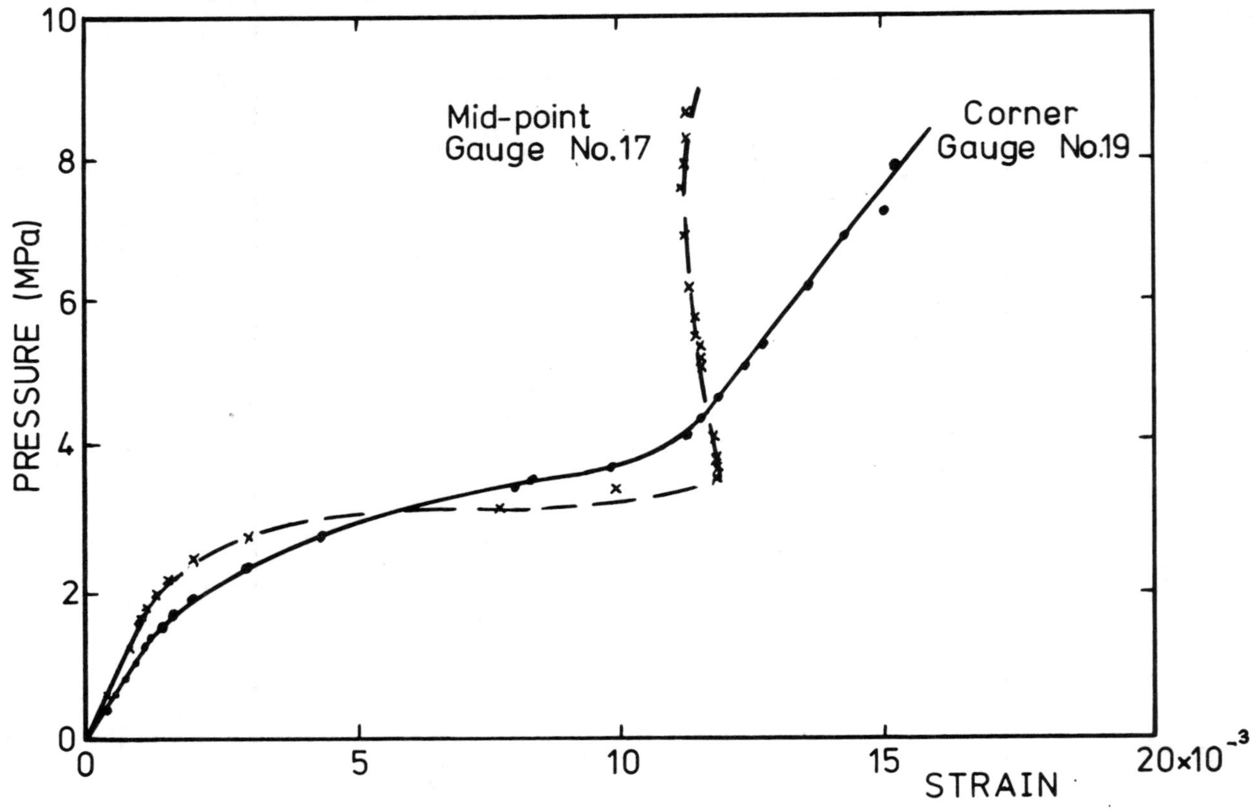

Fig.3. Comparison of pressure-strain behaviour at the mid-point and corners of the hexagon.

Fig.4.
Plaster-cast of the hexagon section taken just before tube failure occurred.

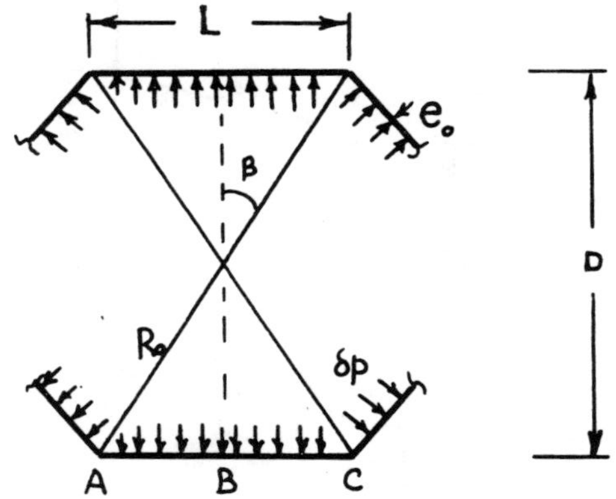

Fig.5.
Cross-sectional view of an internally pressurised n-sided duct.

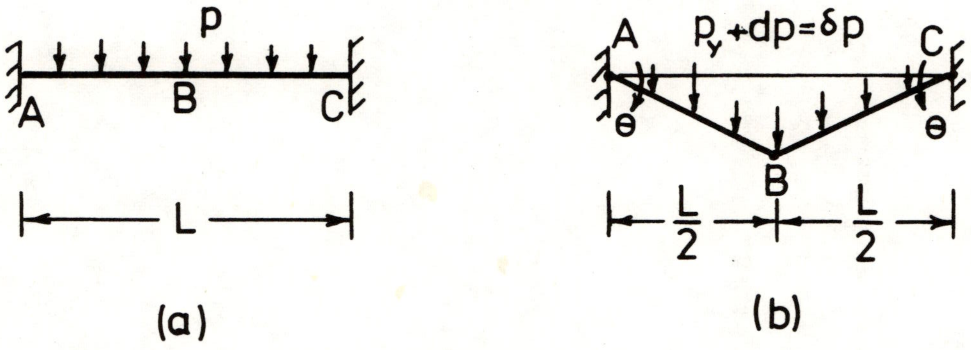

Fig.6. Deformation of a fixed-end beam from an initially straight section (a) to two separate 'rigid' links having plastic hinges at the ends and at the mid-point (b).

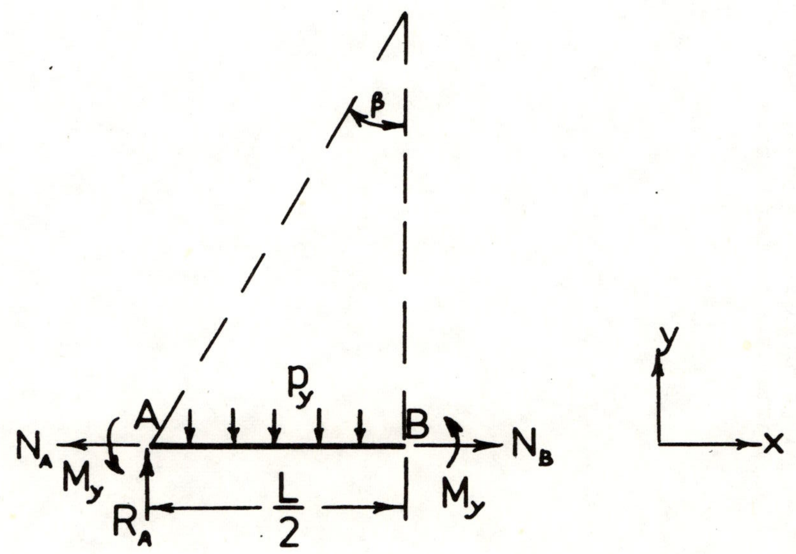

Fig.7. Free body diagram for one half of a 'beam'.

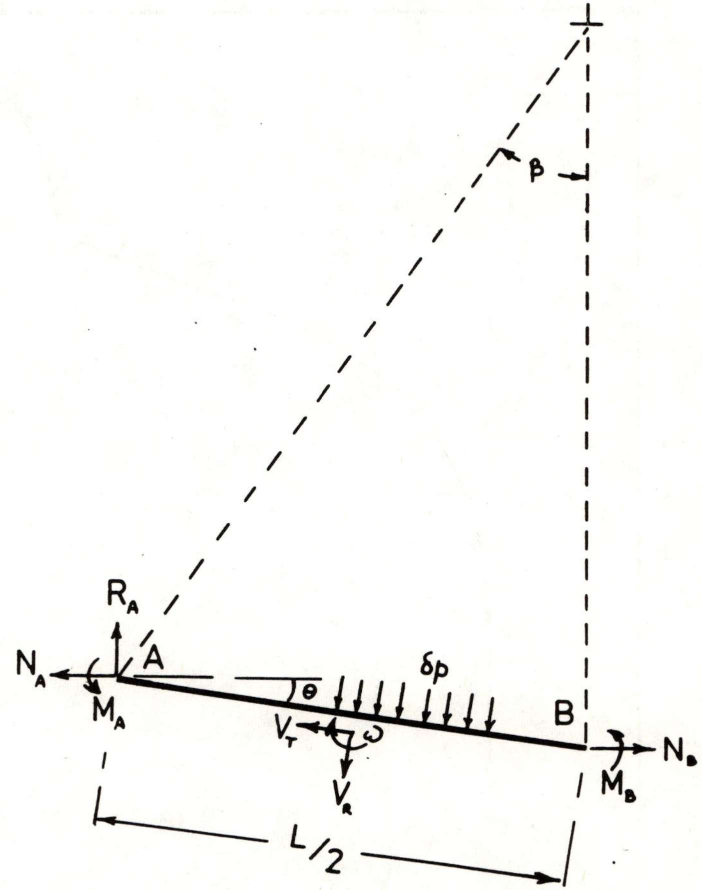

Fig.8.
Free body diagram for one half of a 'beam' subject to plastic deformation at its ends.

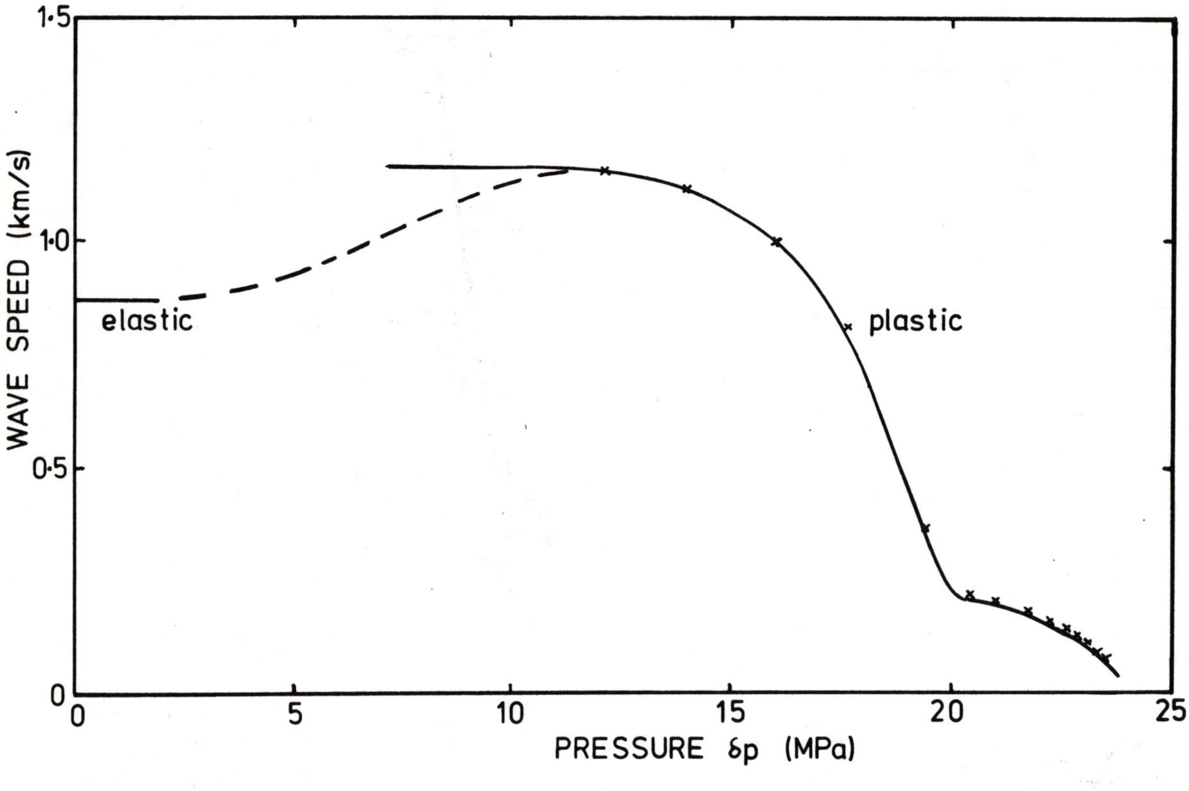

Fig.9. Graph of wave speed versus wave pressure for a duct having initially a hexagonal cross-section.

Third International Conference on

PRESSURE SURGES

Canterbury, England
March 25-27, 1980

SPECIAL CONTRIBUTION

MODELLING OF TWO-COMPONENT FLOWS

D.C. Wiggert
Michigan State University, USA

P.V. Rao and C.S. Martin
Georgia Institute of Technology, USA

During the past several years a significant degree of interest has been demonstrated to analyze unsteady pipeline flows in which free gas is present either by entrainment or by release mechanisms. The numerical modelling of these flows has posed special problems due to the unavoidable presence of numerical desperse and attenuation in the employed finite-differencing schemes. Recently we have been assessing several solution techniques as part of a study to evaluate transient flows in large cooling water systems. The objective is to devise a scheme which would minimize the numerical error, provide sufficient detail in the condenser water boxes, and yet be rather straightforward and uncumbersome when constructing the complete system model and conducting subsequent analysis.

Held at the University of Kent at Canterbury, U.K.

Conference Organised by BHRA Fluid Engineering, Cranfield, Bedford, England.
In conjunction with The City University, London.

MODELLING OF TWO-COMPONENT FLOWS

D. C. Wiggert, Michigan State University, USA
P. V. Rao and C. S. Martin, Georgia Institute of Technology, USA

During the past several years a significant degree of interest has been demonstrated to analyze unsteady pipeline flows in which free gas is present either by entrainment or by release mechanisms. The numerical modelling of these flows has posed special problems due to the unavoidable presence of numerical dispersion and attenuation in the employed finite-differencing schemes. Recently we have been assessing several solution techniques as part of a study to evaluate transient flows in large cooling water systems. The objective is to devise a scheme which would minimize the numerical error, provide sufficient detail in the condenser and water boxes, and yet be rather straightforward and uncumbersome when constructing the complete system model and conducting subsequent analysis.

Consider two-component bubbly flow in a pipe in which the fluid may be experiencing gas release. The conservation equations can be expressed in the manner [1]

$$\frac{\partial \alpha}{\partial t} + V\frac{\partial \alpha}{\partial x} - c_1\frac{\partial V}{\partial x} = b_1 \tag{1}$$

$$\frac{\partial p}{\partial t} + V\frac{\partial p}{\partial x} + c_2\left(\frac{\partial \alpha}{\partial t} + V\frac{\partial \alpha}{\partial x}\right) = b_2 \tag{2}$$

$$\frac{\partial V}{\partial t} + V\frac{\partial V}{\partial x} + c_3\frac{\partial p}{\partial x} = b_3 \tag{3}$$

The dependent variables are α = volumetric void fraction, V = mixture velocity and p = pressure. The b and c terms account for gas release, elastic properties of the fluids and pipe, and force terms [1]. The compatibility and characteristic relations derived from equations (1) to (3) are

$$c^{\pm}\begin{cases}\dfrac{dp}{dt} \pm \dfrac{a}{c_3}\dfrac{dV}{dt} \mp \dfrac{ab_3}{c_3} + b_1c_2 - b_2 = 0 & (4)\\[2ex] \dfrac{dx}{dt} = V \pm a \mathrel{\tilde{=}} \pm a & (5)\end{cases}$$

$$\text{pathline}\begin{cases}\dfrac{dp}{dt} + c_2\dfrac{d\alpha}{dt} - b_2 = 0 & (6)\\[2ex] \dfrac{dx}{dt} = V \mathrel{\tilde{=}} 0 & (7)\end{cases}$$

$$a^2 = c_1 c_2 c_3 \quad \text{(mixture celerity)} \tag{8}$$

Method of characteristics finite-difference solutions can proceed by employing either a fixed grid with spatial or temporal interpolations for the c^+ or c^- characteristics [2,3] or a characteristics grid with interpolations for the pathline characteristic [4]. Finite-differencing can be first or second order, or a combination of the two. An alternate scheme can be formulated by combining equations (1) and (2) to eliminate the derivatives with respect to α:

$$\frac{\partial p}{\partial t} + V\frac{\partial p}{\partial x} + c_2\left(b_1 + c_1\frac{\partial V}{\partial x}\right) = b_2 \tag{9}$$

Along with equation (3), equation (9) can be solved for p and V using a four-point implicit scheme which has found widespread use in unsteady open channel flow analysis [5]. The void fraction is subsequently evaluated by integration of equation (6) along the pathline characteristic.

As a means of comparing these two techniques, consider the following application. The Tracy pumping system has previously been analyzed by Wylie and Streeter [5]. It consists of a pipeline 1,564 m long and 4.57 m in diameter with a pump located at the upstream end and a reservoir downstream. Under steady flow conditions the pump loses power, creating transient flows and pressures in the line with eventual flow reversal. In the present analysis, the pipe is assumed to be horizontal, no gas release takes place, and the void fraction in the pipe is varied parametrically. Figures 1 and 2 show a comparison of the characteristic and implicit solutions for initial void fractions at the pump outlet of $\alpha_o = 1\text{x}10^{-6}$ and 0.01, respectively. The pipe was discretized into twenty sections and the Courant time step was employed. The method of characteristics solutions are the fixed-grid with spatial interpolations and the characteristics grid. In Figure 1, the low void fraction characterizes a fluid which is nearly identical to pure liquid with no free gas, hence the predicted transient is indistinguishable from the original analysis in which no void is present. All three numerical solutions correlate well. With the larger void fraction, Figure 2, some discrepancies are evident when rapid changes in pressure or flow take place; most notably the fixed-grid characteristics solution shows typical smoothing because of interpolations and the implicit solution underestimates peak values.

In Figure 3 the effect of varying α_o from $1\text{x}10^{-6}$ to 0.05 is shown. Here one can observe the pronounced changes in response due to the presence of free air; the pressure resurge tends to increase in magnitude as the void fraction increases. One useful feature of the implicit method is the possibility for one to increase the time step significantly beyond that dictated by the Courant condition and yet maintain numerical stability; this is shown in Figure 4 for the present application, where time steps of ten and twenty times the Courant condition (.0855 sec) yield reasonable solutions.

With two-component gas-liquid flows, the initial conditions can be formulated by setting the time derivatives to zero in equations (1) to (3). By recombining, one can solve for the total derivatives of the dependent variables:

$$\frac{d\alpha}{dx} = (-b_1V + b_2c_1c_3/V - b_3c_1)/(a^2 - V^2) \tag{10}$$

$$\frac{dp}{dx} = (b_1c_2V - b_2V + b_3c_1c_2)/(a^2 - V^2) \tag{11}$$

$$\frac{dV}{dx} = (-b_1c_2c_3 + b_2c_3 - b_3V)/(a^2 - V^2) \tag{12}$$

These relations define a steady nonuniform flow condition, and they can be solved numerically by conventional means, e.g., Runge-Kutta. An example is shown in Figure 5 for a pipeline similar in size to the previous illustration, and in which the void fraction at the pipe inlet is $\alpha_o = 1\text{x}10^{-6}$. The saturation pressure was set equal to the inlet pressure for demonstrative purposes. When gas release is allowed to take place using an experimentally determined gas release rate coefficient [2], the void fraction distributed along the pipe varies by three orders of magnitude. Such a variation would profoundly affect the mixture celerity and subsequent pressure transient.

In summary, favorable comparisons are made between method of characteristic and implicit solutions. The numerical results show the effects of compressibility and the highly variable wave celerity on system response. In order to more fully assess their accuracy, the models should be correlated with experimental data and additional numerical sensitivity studies should be conducted. The method of time line interpolations, as proposed by Wylie at this conference [3], demonstrates an alternate characteristics scheme, one which may improve numerical accuracy. The techniques presented herein may be useful in application to condenser flows and to two-component flows in general.

The work presented herein was supported by the Electric Power Research Institute. R. Otwell assisted in computer programming.

REFERENCES

1. Martin, C. S., Padmanabhan, M., and Wiggert, D. C., "Pressure wave propagation in two-phase bubbly air-water mixture," Proc. 2nd Intl. Conf. on Pressure Surges, London (1976).

2. Wiggert, D. C. and Sundquist, M. J., "The effect of gaseous cavitation on fluid transients," Journal of Fluids Engineering, Vol. 101 (March 1979), pp. 79-86.

3. Wylie, E. B., "Free air in liquid transient flow," Proc. 3rd Intl. Conf. on Pressure Surges, Canterbury (1980).

4. Ferch, R. L., "Method of characteristics solutions for non-equilibrium transient flow-boiling," Intl. Journal Multiphase Flow, Vol. 5 (1979), pp. 265-279.

5. Wylie, E. B., and Streeter, V. L., Fluid Transients, McGraw-Hill Book Co. (1979).

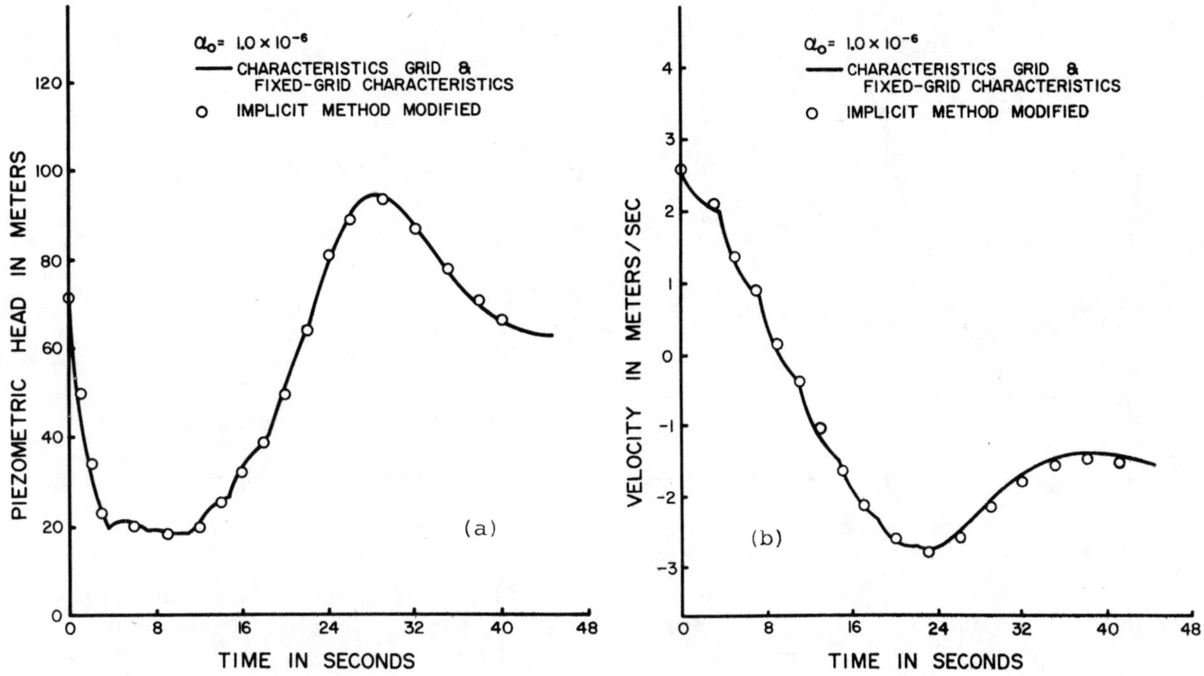

Figure 1. Piezometric head (a) and velocity (b) at upstream end of pipe for $\alpha_o = 1.0x10^{-6}$.

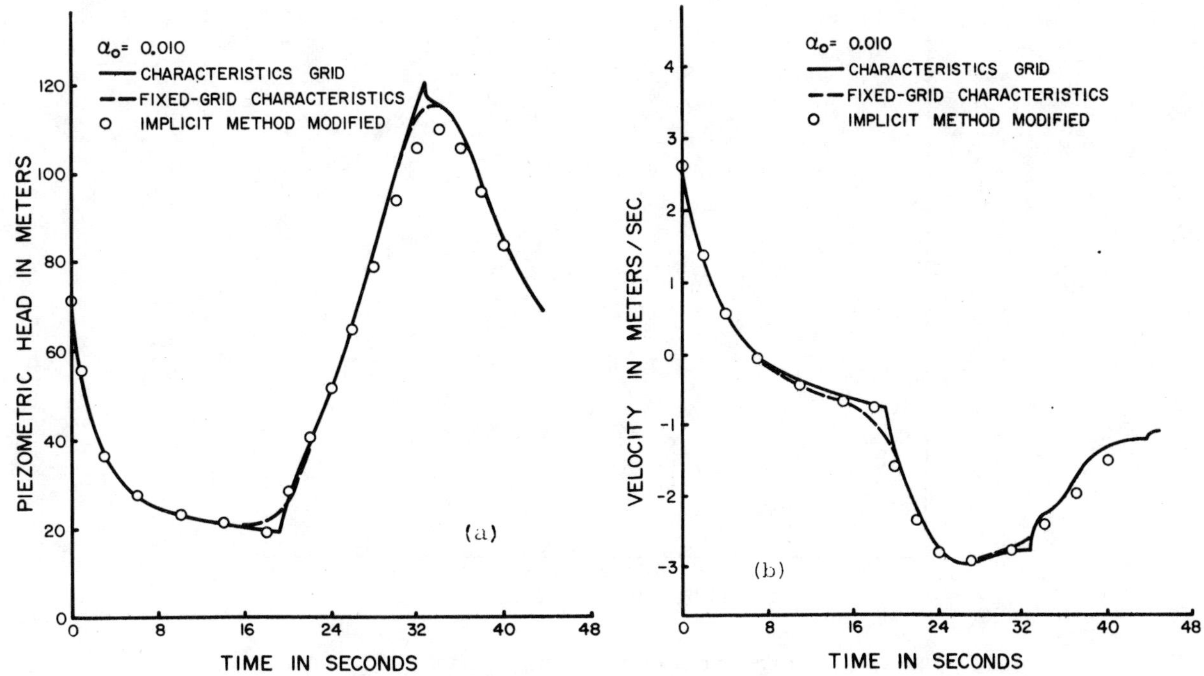

Figure 2. Piezoetric head (a) and velocity (b) at upstream end of pipe for $\alpha_o = 0.010$.

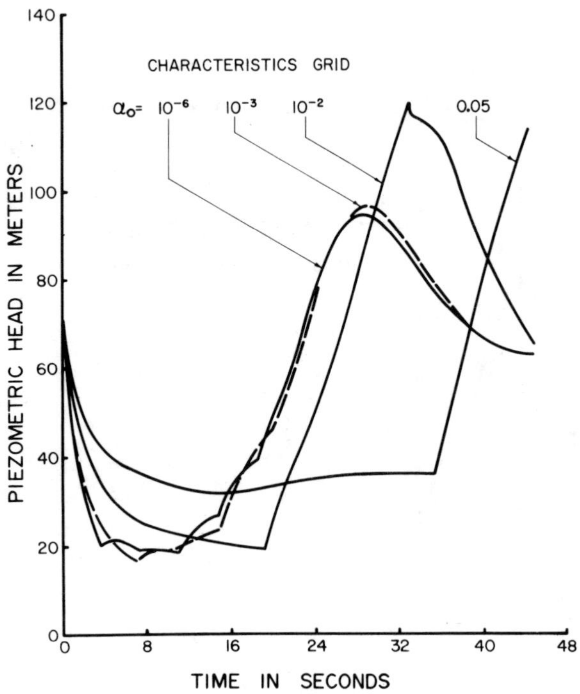

Figure 3. Piezometric head at upstream end of pipe with α_o variable.

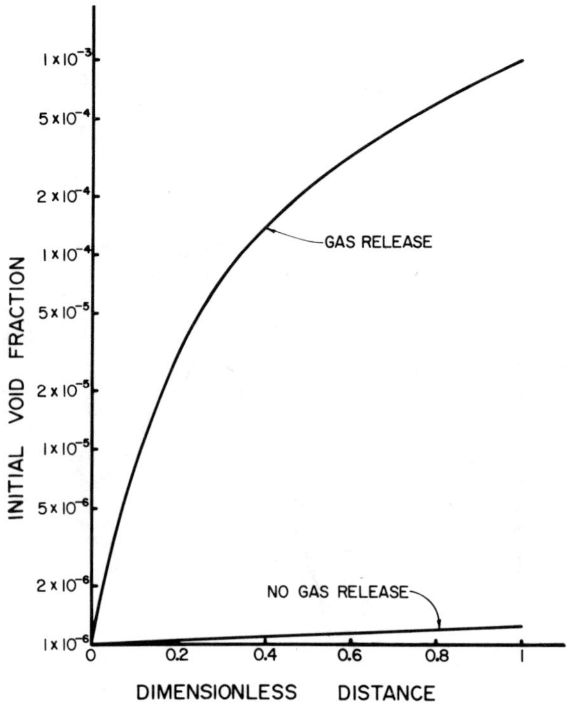

Figure 5. Initial value solution showing distribution of void fraction along length of pipe.

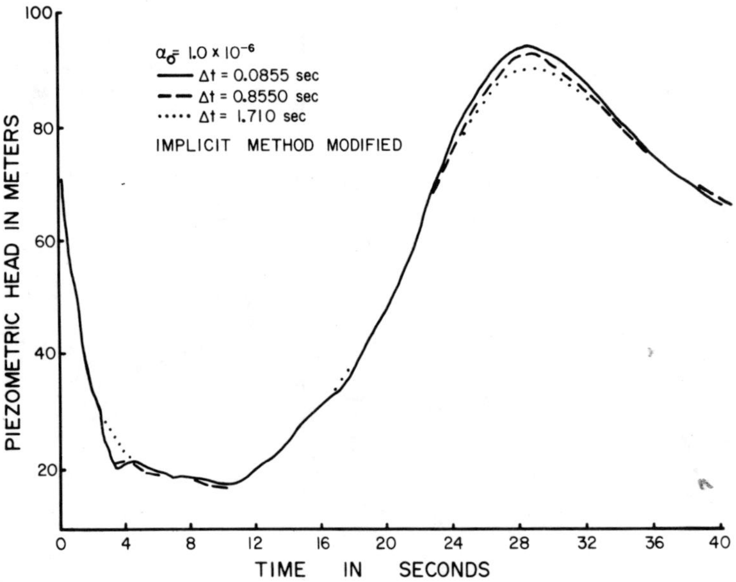

Figure 4. Piezometric head at upstream end of pipe with time increment variable.

CLOSING ADDRESS

C.S. Martin, Georgia Insitute of Technology

It is a special honor to me to be designated by the organizing committee to deliver the closing speech to the Third International Conference on Pressure Surges. I think I can speak for many in the audience that it has been very enjoyable and pleasant to return to the site of the initial conference in 1972. First, let us acknowledge BHRA, the organizing committee, the authors, the discussors, and contributors for their efforts in making this meeting an outstanding success. Everyone I have spoken with has been well pleased with the contributions in general, and especially the lively and vigorous discussions that have concluded each lengthy session. Let us not forget how important and valuable the discussion period is to the worth of the meeting. If each paper were delivered without discussion we might as well stay home.

Regarding an assessment of the success of the conference and the ultimate future of this area and similar conferences I have prepared a table summarizing by category the contributions at the three BHRA Pressure Surge Conferences. It is interesting that there is still considerable activity in hydroelectric schemes and surge tanks. For some reason there were not as many papers on two-phase flow this year as in 1976, at which time there were several contributions from the nuclear industry. I am pleased to see the activity in an area frequently neglected -- fluid-structure interaction and plastic detormation. Five papers were devoted to fluid-structure interaction and check-valve behavior. This is probably only the tip of the iceberg. It is an important field that needs more research concentration.

Eight years ago Professor Streeter closed by stating that the acoustic velocity prediction is our greatest problem in math models of transient flow. It may not be our greatest problem today but it certainly is not our least. We have witnessed the introduction of several interesting methods at these sessions for dealing with free air in water hammer analysis. There was no lasting contribution that allows one to predict the amount of gas release, however. There is a definite need for experimental studies that yield an understanding of the physical process of gas release and a reliable constitutive relationship. These appear to be a somewhat diminishing effort concerning column separation and water hammer in cooling water systems. Perhaps the problem has been over emphasized to some degree. A novel device for controlling pressure transients was introduced and commissioned to be reliable and predictable. For the first time at these conferences there have been two serious attempts in understanding the fluid mechanics of unsteady flow friction. While the effect may not be significant in many engineering problems it is always (or should be) a nagging worry to analysts. Two papers addressed the problem of design -- one by design charts and the other by doubt in the design process.

What is the future of pressure surge, or more directly Pressure Surge Conferences. To be pragmatic I feel that we as individuals and mini-organizations do not have much control over the matter. Most of the participants with a direct or indirect tie with industry have no choice in the types of problems that we will encounter. Advances in technology, developments in energy resources, new designes, etc. will dictate the activity, not so much our wishes -- with the exception, perhaps, of a few of us in the academic world. I personally believe that the events will take care of themselves but predict, and hope, that we will all return in four years.

TABLE SUMMARIZING CONTRIBUTIONS AT THE
THREE BHRA PRESSURE SURGE CONFERENCES

	'72	'76	'80
COMPUTATIONAL METHODS AND PIPE NETWORKS	7	3	4
PLASTIC DEFORMATION AND FLUID STRUCTURE INTERACTION	1	2	5
RESONANCE, VIBRATION AND STABILITY	4	3	0
TWO-PHASE FLOW AND COLUMN SEPARATION	5	11	5
VALVES AND VALVE CLOSURE	3	1	2
AIR CHAMBERS AND LINERS	4	2	1
HYDROELECTRIC SCHEMES AND SURGE TANKS	8	5	6
COOLING WATER SYSTEMS	3	2	2
DESIGN	0	0	2
OTHER	2	0	3
	37	29	30

Discussion & Contributions

SESSION A

UNSTEADY FRICTION AND VELOCITY PROFILES

Chairman: Dr. K.J.Enever
Cranfield Institute of Technology, U.K.

Papers:

A1	Velocity profiles in accelerated pipe flows started from rest	E.van de Sande, A.P.Belde and B.J.G.Hamer KEMA, Netherlands.
A2	Unsteady flows: Fact and Friction	A.E.Vardy University of Dundee,U.K.

<u>PAPER A1</u>

Velocity profiles in accelerating pipe flows started from rest.

E. van de Sande, A.P. Belde and B.J.G. Hamer
KEMA, Netherlands

<u>J.F.Robbie</u>, University of Aberdeen, U.K.

The strength of this paper lies in the convincing demonstration of contemporary visualisation and velocity measuring techniques in relation to accelerating flow - an updating and extension of Reynolds classic demonstration.

In relation to turbulence, the influence of local secondary effects can be considerable. It is a little surprising therefore to find, in Fig.1, the air-controller valve located at the upstream end of the pipe - in marked contrast to Reynolds carefully designed pipe entrance.

What considerations led to its location at this point and what checks were made on its influence on the onset of turbulence?

Past experience in monitoring unsteady flow effects in hydraulic models of closed conducts systems with open surface shafts, has shown that secondary effects can, on occasion, influence the detailed response of the system - e.g. constant level tanks are rarely constant when monitored closely.

What influence, if any, would the Authors expect the increasingly adverse secondary pressure gradient produced from this source in their apparatus to have

a) on the onset of the turbulence detected by them and

b) in producing the velocity profile at time t = 40s in Fig.2

The importance of studies in accelerating flows to the practical designer may be assessed when it is realised that accelerations in excess of $0.12 ms^{-2}$ are required in water flows in tunnels, in excess of 10m diameter, if design criteria of certain contemporary pumped storage schemes were to be met.

<u>D. Goldberg</u>, Stoner Associates, Inc., U.S.A.

The author are to be congratulated on their fine experimental study and flow visualisation. The results of this study help advance general knowledge of unsteady friction effects.

We have seen the results of this phenomenon since 1975, during startups of a 105 mile, 20" diameter, partially looped pipeline which carries crude oil beneath the Gulf of Mexico. The delayed transition from laminar to turbulent flow first became apparent during tests of a transient model leak detection system. Transition Reynold's numbers were of the same unusual magnitude as the authors'.
In modeling this phenomenon; we are concerned with correlating the transition point. Have the authors made any progress in identifying this point in different sized pipes?

Dr.Ing.Hans-Burkhard Horlacher, Akad, Oberrat

In the paper the authors mentioned that they intend to investigate accelerating and decelerating flows. They have shown very interesting results for the case of accelerating flows, initially at rest.

Have the authors carried out investigations for decelerating flows? Perhaps they can give some results or presumptions: Transition from turbulent to laminar flow, frictional losses etc.

Author's Closure

J.F. Robbie, University of Aberdeen, U.K.

The location of the valve seems to be insignificant. It was placed alternatively upstream as well as downstream of the pipe and the velocity profiles were both identical. Accelerating flows are essentially stable as long as the velocity is still increasing.

The second question about the onset to turbulence is more difficult to answer. In the experimental set-up the unsteady pressures were measured but level variations were not observed. In a system in which these variations are present, the onset to turbulence can take place somewhat earlier but it still will be at the end of the acceleration.

D. Goldberg, Stoner Associates, Inc., U.S.A.

The investigations were performed in two pipes of different diameters. Only at the end of the acceleration, turbulence is originated. In pipes of larger diameter the same result can be expected so the value of the transition Reynolds Number will increase with increasing diameter. The transition point is determined by a critical decrement of the acceleration.

H.B. Horlacher, Akad, Oberrat

We are now studying decelerating flows. A transition from turbulent to laminar flow does not take place because the life time of the eddies is greater than the deceleration time in our experiments.
The velocity profiles show an adverse boundary layer flow and this effect enhances the dissipation processes in unsteady flows. The transient friction factor in decelerating flows will be much higher than the steady state values.

PAPER A2

Unsteady flows: Fact and friction.

A.E. Vardy
University of Dundee, U.K.

C.S. Martin, Georgia Institute of Technology, U.S.A.

What is the relative magnitude of unsteady friction in the model compared to steady-state friction? Furthermore, how was the unsteady friction parameter B chosen in the analysis?

J.A. Fox, Leeds University, U.K.

It may be thought that your arguments are relevant to trains in tunnels and so concern bodies which are shorter than a dx length. In fact, in an ordinary pipeline we are always looking at points a dx length apart and are not concerned with anything shorter than a dx length. With a control volume of constant length do your comments require modification?

A.Anderson, University of Newcastle Upon Tyne, U.K.

I apologise for not raising this point during the Conference but I did not have the necessary material with me to refer to.

As Prof.Vardy commented in his presentation, formulas of the type of his Eq.(9) have frequently been quoted in the literature but unfortunately with a very wide range of experimental values for the constant B - see Table 3 of Paper G4 for some examples in addition to Safwat and Polder (Ref.5). It is possible, however, that the "theoretical inadequacies" arise not only from the modelling of the friction term in this sense but also from its discretisation in the numerical solution process and this can be shown by a very simple example.

Consider the case of a single uniform pipe loading from a constant head reservoir and terminated by a valve (Fig.A), this being commonly used as an elementary teaching example in e.g.Streeter(Ref.A pp599-610) or Watters (Ref.B pp.56-59) et al. The pipe is usually divided into a number of elements, typically 5 say, but in such a simple case with a constant wave velocity c it should be possible to treat the pipe as a single element with only 2 nodes, one at each end at each boundary condition, and then the solution grid with the two characteristic lines is as sketched in Fig.A. The solution in either case is trivial. Now consider, however, the friction term. It has been usual in the past (e.g.Ref.A and B) to approximate the friction head F along the characteristic by evaluating U/U/ only at the beginning of the characteristic at time t:

$$F \approx (fL/2gD) \begin{cases} U_R/U_R/ & \text{along } +C \\ U_V/U_V/ & \text{along } -C \end{cases} \qquad (A)$$

Some papers have suggested that this representation can be improved by taking the mean velocity head along each characteristic (superscript dash' denotes value at time $t+\Delta t$):

$$F \simeq (fL/2gD) \begin{cases} \frac{1}{2}\left[U_R|U_R| + U'_V|U'_V|\right] & \text{along } +C \\ \frac{1}{2}\left[U_V|U_V| + U'_R|U'_R|\right] & \text{along } -C \end{cases} \qquad \text{(B)}$$

which still gives an explicit formulation but requires solution of a quadratic in place of a linear equation.

Both these formulations are completely satisfactory for "slowly varied" flows and give effectively identical solutions for any number of discrete pipe elements. They also give good answers for "rapidly varied" flows (where valve closure time $\leqslant 4L/c$) but only as long as a reasonable number of pipe elements are taken (say 3 or more) because for a small number of pipe elements accuracy begins to deteriorate. This is shown quite clearly in Watters' Table 5.2 (Ref.B pp.75-76). Ultimately taking the limit and reducing the number of pipe elements to one as suggested in Fig.A still makes little difference to solutions for "slowly varied" flows but with "rapidly varied" flows the tendency hinted at by Watters' Table 5.2 is confirmed and the numerical solution bears little resemblance to the correct (experimental) solution, whichever of the above methods of formulating the friction term is adopted.

It is possible, however, to get good solutions for the "rapidly varied" case with a single pipe element simply by modifying the discrete representation of the friction loss term, e.g. it was found that using the following weighting about the grid points still allowed explicit solution by means of quadratic equations but gave excellent results:

$$F \simeq (fL/2gD) \begin{cases} \left[\frac{1}{2} U'_V|U'_V| + \frac{1}{4}(U'_R|U'_R| + U_R|U_R|)\right] & \text{along } +C \\ \left[\frac{1}{2} U_V|U_V| + \frac{1}{4}(U'_R|U'_R| + U_R|U_R|)\right] & \text{along } -C \end{cases} \qquad \text{(C)}$$

In "slowly varied" flow this gives effectively identical solutions to the simpler methods of Eq.A and B above but in "rapidly varying" flows it seems to give far superior results. It therefore allows the most economical pipe representation to be used even for "rapidly varying" flow.

Finally, then, the general point arising out of the above particular example seems to be that for "rapidly varying" flows (i.e. where the "step change" occurs entirely within a single element of the grid spacing) the form of discretisation of the friction term alone can have a marked influence on the amplitudes of the pressure variations at that element. This effect, of course, does not contribute to the change of phase of the solution in the way that Prof.Vardy's expression does, so it clearly remains the case that further development of that expression is of central importance. Nevertheless it is possible that the question of discretisation may also yield improvements in amplitude prediction if pursued further.

Additional references:

A. Streeter, Victor L.: "Fluid Mechanics",4th edn.,McGraw-Hill,New York 1966.

B. Watters,Gary Z.: "Modern Analysis and Control of Unsteady Flow in Pipelines", Ann Arbor Science Publishers,Ann Arbor Mich., 1979.

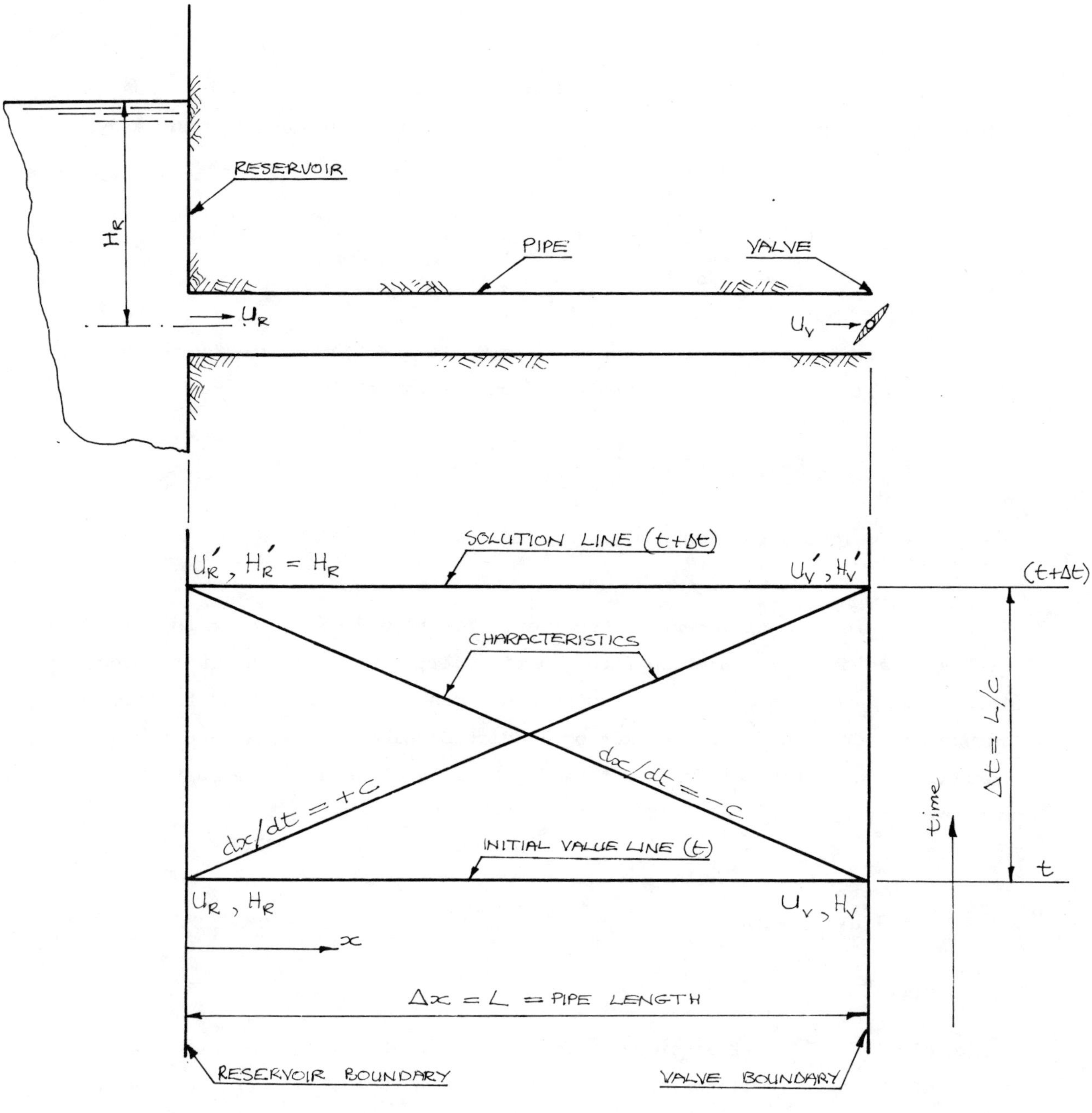

FIG. A. SIMPLE PIPE WITH ELEMENTARY SINGLE ELEMENT SOLUTION GRID

D.C. Wiggert, Michigan State University, U.S.A.

Mention is made of Zielke's analysis in which a time-dependent wall shear stress term for unsteady laminar flows is developed. If one assumes laminar flow in the present study, then from Zielke's development the wall shear stress would be

$$\tau_w = \tau_{ws} + \tau_{wu}$$

$$= 2\rho\nu \frac{\ell}{a} \bar{U} + \rho\nu \frac{\ell}{a} \int_0^t \frac{\partial \bar{U}}{\partial t}(u) W(t-u)\, du$$

in which ν = kinematic viscosity and W = weighting for past velocity changes. Zielke [1] has suggested that a similar relation could account for unsteady shear stress in turbulent flow; one could postulate a term such as

$$\tau_w = \frac{1}{2}\rho C_F \bar{U}|\bar{U}| + \rho C \int_0^t \frac{\partial \bar{U}}{\partial t}(u) W_t (t-u)\, du$$

where C is a constant of dimension LT^{-1} and W_t is a weighting for turbulent flows. Placing this relation into Equation (2) one finds

$$\frac{\partial p}{\partial x} + \frac{1}{2}\rho \frac{\ell}{a} C_F \bar{U}|\bar{U}| + \rho \frac{\ell}{a} C \int_0^t \frac{\partial \bar{U}}{\partial t}(u) W_t (t-u)\, du + \frac{\partial}{\partial t}(\rho\bar{U}) + \frac{\partial}{\partial x}(\beta\rho\bar{U}^2) = 0$$

which is an integro-differential equation. Zielke did not attempt to solve such a relation in conjunction with Equation (1), but instead he introduced the integral term after the continuity and momentum equations were transformed into characteristic form. The method predicted phase lag and attenuation which correlated well with carefully obtained experimental data. Note that the dimension of the integral

$$\int_0^t \frac{\partial \bar{U}}{\partial t}(u) W_t (t-u)\, du$$

is merely LT^{-1}, i.e., that of velocity.

For laminar oscillatory flow it can be shown that τ_w is dependent upon the frequency of oscillation and the mean velocity and acceleration [1]. In the limit as the frequency goes to zero, Zielke demonstrated that the shear stress could be given in a form identical to Equation (7) and continued on to show a modification in the characteristic directions, an effect equivalent to increasing the fluid inertance by one-third. Use of the above-mentioned integral term, however, did not alter the characteristic directions.

The wave patterns shown in Figure 2 contain a range of Fourier components, some of which may be high frequency. If this were true, then the integral form of the shear stress might be more appropriate to use, rather than the inertance approximation which (for laminar flow) is valid for low or zero frequency components. In this light a physical interpretation of the term $(1+B\ell/a)^{-1/2}$ would be instructive.

1 Zielke, W., "Frequency Dependent Friction in Transient Pipe Flow," Ph.D. Thesis, University of Michigan (1966).

Author's Closure

These questions are, in practice, closely related. As a preamble, I would like to emphasise again the inability of numerical schemes to simulate the detailed shape of rapidly varying pressure pulses. In presenting the paper, I illustrated this point by considering a shallow water wave gradually steepening into a surge. During the steepening process, the back of the wave gradually catches up with the front. Our difficulty is to decide whether to regard the wave as broken (e.g. conservation schemes) or continuous (e.g. MOC) when simulating it numerically. When the whole of the wavefront exists within a single δx length, numerical tests cannot provide the answer.

In some cases, for example in short railway tunnels, it is known *a priori* that the air pressure waves do not develop into shocks. Even when a complete wavefront exists within one δx length, the MOC type solution is therefore valid. However, an analogous difficulty presents itself when an attempt is made to simulate the unsteady skin friction in the flow because (by stipulation) no information is known about the manner in which the velocity varies along the wavefront. All that can be found is the magnitude of the net velocity change. Moreover, although it is known that this change exists within a single δx length, there is no way of deciding whether the change actually occurs within, say, $\delta x/2$ or $\delta x/4$.

The inevitable conclusion is that even if a precise relationship was known between the skin friction and the velocity history, little use could be made of it in a numerical solution with a coarse grid size.

With this in mind, it should be easy to see why I was willing to produce the equations (6) and (7) "out of thin air". I am not trying to simulate the actual phenomenon of unsteady skin friction, but rather to show how its overall affect might be approximated in a numerical solution when the unsteadiness is very rapid indeed.

In developing the argument, I accepted that there must be some dependence of skin friction on acceleration as well as upon velocity. As shown by Zielke (Ref. 11) and others, there is also a dependence on higher order derivatives, but this cannot be included in a model that must accommodate complete wavefronts in a single δx length. By simple reasoning, it is shown in the paper that, in addition to influencing the magnitudes of pressure fluctuations, unsteady skin friction must affect the speed of wave propagation.

Having restated these points, I am in a position to consider the specific questions raised in the discussion. Firstly, as explained in Section 5 of the paper, the value of the dimensionless product $B\ell/a$ was selected by trial and error to give good agreement with the phase of the experimentally measured pressure fluctuations. By using a value of 0.05, the wavespeed was "altered" by $2\frac{1}{2}\%$. Using equation (11), it can be seen that the value of the ratio $\Delta c/\Delta U$ is typically altered by a similar amount. Unfortunately, an important difficulty is hidden within this simple statement. The average values of $\Delta c/\Delta t$ and $\Delta U/\Delta t$ reduce when Δt increases, but the right hand side of (11) does not. The answer to Prof. Martin's query about the relative magnitudes of unsteady and steady skin friction in the model is therefore a function of the chosen grid size.

Some light can be thrown on the question by inserting typical values into equations (3) and (8). Using $B\ell/a = 0.05$ and $C_F = 0.005$, we obtain

$$\frac{\tau_{wu}}{\tau_{ws}} = 20\,\frac{a}{\ell}\,\frac{1}{U^2}\,\frac{dU}{dt} \qquad (13)$$

Naturally, the numerical value of the ratio is wholly dependent on the choice of a, ℓ, U and dU/dt.

Mr Fox's question has been covered in the preliminary remarks. He has correctly associated the existence of the problem with the need to use finite δx lengths. However, it is worth stressing that trains have no special role in the current discussion. They are simply the method used to produce rapid pressure fluctuations in the particular example chosen.

Dr Anderson's contribution is very interesting. I was not aware of the improvement to be gained from the stated choice of weighting factors and I have not yet managed to obtain a copy of Watters' work. The first of the pair of equations (C) includes values at two points at the solution time level. Presumably, the C- equations must be solved to give values at R' before the C+ equations are solved for the values at V'. It is interesting to note that the difference between the expressions (B) and (C) is simply

$$\frac{fL}{8gD}\ (U_R' - U_R) \tag{14}$$

which is proportional to the average acceleration during the time interval Δt. A possible improvement of (14) in comparison with (8) is that the former implicitly provides a numerical value for the parameter B.

Professor Wiggert has extended the discussion to include a more detailed description of the behaviour of unsteady friction. In order to make use of expressions involving the weighted contributions of the past velocity history, it is necessary to consider the class of unsteady flows in which wavefronts are spread over several grid lengths. This implies either that the flows vary less rapidly or that many more grid points are included in both space and time. I would prefer not to move too far in this direction in this discussion in case the overall objective is clouded. In particular, it would be inappropriate to attempt to give physical meaning to the term $(1 + B\ell/a)$. However, the contribution raises some interesting points which are worthy of attention.

The effective increase of the fluid inertance by one third in laminar flows is not really caused by skin friction directly. It is simply the result of the existence of a paraboloidal velocity distribution. In other words, the momentum flux correction coefficient β is equal to 4/3 in steady laminar flows in circular pipes. In principle, an identical value would be obtained in an inviscid flow with the same velocity distribution. Of course, the velocity distribution exists in practice only because of the shear stresses in the fluid. However, I deliberately neglected this phenomenon in the paper (see Section 2) in order to avoid confusion. Brown (1) gives useful comments.

Professor Wiggert points out that phase changes can be caused by skin friction even without the need for an assumption that there is a dependence on acceleration. I agree with this statement, but wish to warn against the possibility of drawing erroneous conclusions. Suppose that a wave is generated in a pipeline. In the absence of an acceleration dependence, skin friction will influence the shape of the wave as it propagates along the pipeline, but it will not alter the speed of the leading edge of the wavefront. The times and positions of the peaks and troughs in the pressure histories will be influenced by the changed waveforms.

Nevertheless, the local speed of propagation of each elemental part of the wavefront relative to the fluid will always be equal to c.

If the skin friction has an acceleration dependence, the above effect will be further modified because the local relative wavespeed will be, say, $c(1 + B\ell/a)^{-\frac{1}{2}}$. Of course, this in no way implies that the normal mechanics of wave propagation are altered. Within the cross-section, the wavespeed is still equal to c, but the cross-section average wavespeed is reduced (as it is when a pipe wall is distensible or a liquid contains bubbles, for example).

Discussion & Contributions

SESSION B(i)

MATHEMATICAL TECHNIQUES

Chairman:	D.H.Wilkinson C.E.G.B., U.K.	
Papers:		
B1	Free air in liquid transient flow	E.B. Wylie University of Michigan, U.S.A.
B2	Combination of finite difference & finite element techniques in hydraulic transient problems.	C.S.Watt Merz & McLellan, U.K., A.P.Boldy University of Warwick,U.K. J.M.Hobbs Thames Polytechnic,U.K.
B3	Calculation of pressures surge in complex networks.	K.G.Priddin BHRA Fluid Engineering,U.K.

PAPER B1

Free air in liquid transient flow.

E.B. Wylie
University of Michigan, U.S.A.

A.P. Boldy, Warwick University, U.K.

What is the effect on computer storage and time when using the time-line interpolations?

S.Whittle, British Aerospace, U.K.

Prof.Wylie considers in his paper what effect void fraction has on the wave celerity in homogeneous two phase flow; and shows that this may reduce to approximately 1% of its value in the liquid only case. In many problems this may mean that the acoustic velocity in the two phase fluid will be of the same order of magnitude as the flow velocity. With this in mind I would suggest that the non-linear convective terms ($V\,\partial V/\partial x$ & $V\,\partial P/\partial x$) should be included in the mass and momentum continuity equations, resulting in a slight change in the characteristic directions. I assume Prof.Wylie considered this, and if so, what effect did he think this would have on the results obtained.

K.J.Enever, C.I.T.

I am very interested in the results presented in this paper as, I am sure, will be all those concerned with pressure surge analysis in systems containing small quantities of gas.

I should like to make one small point. Any gas surrounded by a liquid is bound to contain vapour regardless of the pressure and, by Dalton's Law of Partial Pressure, the total pressure is the absolute pressure of the gas plus the vapour pressure of the liquid, whereas it is of course just the gas which expands according to equation (3). Thus strictly speaking, as H should be related to the pressure of the gas rather than the total pressure of the gas/vapour mixture I believe equation (9) should read:

$$H = p/\rho g + \overline{H} - Z - Hv$$

where Hv is vapour pressure head.

This is unimportant in many cases, but when H becomes small, or if Hv is large it can be seen that this would have a appreciable effect on a' in equation (9) and, presumably, change the period and to some extent the amplitude of the pressure fluctuation.

Does the author agree with this analysis of the situation, and if so does he consider that the pressure in his system dropped close enough to vapour pressure to have any effect on his results?

A.E.Vardy, Dundee

1. Your suggestion that we should incorporate simple checks into our computer programs is very well timed. I agree that we should include a continuity balance check whenever possible and I admit to having failed to do so on many occasions. All the same, this check must not be regarded as a "sufficient" test; it is merely "necessary".

2. Your time-line interpolation scheme has given impressive results. How sensitive will it be to rapidly changing boundary conditions? I am wondering about the accuracy with which the point R_o can be established in Fig.4 of your paper. This is not very important in comparison with the errors associated with a simple interpolation at the point R_M. However, might there be an economic limit to the number of time levels that could usefully be retained in care? If you retained only three additional levels, say, it would be possible to interpolate at the point R_1. This would overcome the occasional need to retain a very large number of time levels when the local wavespeed was exceptionally small.

G.A.Provoost, Delft Hydraulics Laboratory, Netherlands

Due to the presence of free air in a fluid the wave speed becomes pressure-dependant. The presented time-line interpolation method is very interesting for this kind of phenomena. The example of Figure 9 of your paper shows that the pressure rise in case of the smallest air contents occurs earlier than in the vapour only case.

My question is whether there is a physical reason causing this pressure rise to occur earlier in such cases. In my opinion each free air content will lead to lower shock wave velocities and thus leading to a delay in the occurrence of the pressure rise.

This may be concluded from the equations as derived by Kalkwijk and Kranenburg (Refs. 1 and 2) which describe a shock wave.

The laws of conservation of mass and momentum are applied to a small section of the pipeline containing a shock wave which moves with a velocity a'_s with respect ro the pipe wall. Disregarding the momentum of the air the results are:

$$(a'_s - v_1)\rho_1 A_1 (1 - k_1) - (a'_s - v_2)\rho_2 A_2 (1 - k_2) = 0$$

$$(a'_s - v_1)\rho_1 A_1 v_1 (1 - k_1) - (a'_s - v_2)\rho_2 A_2 v_2 (1 - k_2) = A_2(p_1 - p_2)$$

substituting:

$$\rho_1 A_1 = (1 + \frac{p_1 - p_2}{\rho_o c_o^2}) \rho_2 A_2$$

$$\Delta p = p_1 - p_2$$

$$\Delta v = v_1 - v_2$$

$$\rho_1 \approx \rho_2 \approx \rho_o$$

$$a_s = a'_s - v_1$$

and assuming $k << 1$ and $\frac{\Delta p}{\rho_o c_o^2} << 1$, the following equation for the shock wave velocity as with respect to the fluid particles in the high pressure region is obtained:

$$a_s = c_o \left[1 + \Delta k \frac{\rho_o c_o}{\Delta p} \right]^{-\frac{1}{2}}$$

where: a_s = shock wave velocity
A = cross-sectional area
c = wave velocity
k = void fraction
p = absolute pressure
v = fluid velocity
ρ = density
Δ = difference across a shock wave
subscripts:
0 = absence of free air
1,2 = indication of two cross-sections at both sides of a shock wave.

Since the void fraction and thus the difference across a shock wave in case of free air is larger than without free air the shock wave velocity will be smaller. This result does not seem to be present in the presented time-line method.
Since also Mr. Enever mentioned in the discussion that an advance of the pressure peak for small free air contents was found when the free air was lumped in a number of points of the pipeline, some calculations were carried out with the mathematical model as used by the Delft Hydraulics Laboratory. In this model the free air is concentrated in a number of nodes and the continuity equation is fullfilled exactly. The air was assumed to behave isothermal. The wave speed in the sections between the nodes was kept constant at the value for the case without free air. This method will be referred to as the concentrated air method. Calculations were carried out for the same example as presented in Figure 9 of your paper.
The data of the example are:
length 3048 m, internal diameter 610 mm, friction factor 0,02, upstream head 50,38 m, downstream head 3,05 m, wave speed without free air 981,5 m/s, steady state velocity 3,0483 m/s, level of horizontal pipe 0 m and head at vapour pressure - 10,06 m.
A transient is introduced by lowering instantaneously the upstream head to zero, being the level of the pipe.
The number of computing sections chosen was 60 and consequently the

number of internal nodes 59. The case without free air (the vapour only case) reproduced with this method, which is also referred to as the concentrated cavitation method, is shown in Figure 1. In this Figure profiles of head (relative to the head at vapour pressure) at three points of the pipeline are plotted against time after the start of the transient together with the analytic solution as given by Wylie and Streeter (Ref. 3) as a comparison. This analytic solution was mentioned in your paper as the vaporous cavitation result.

In this analytic solution the continuity was fullfilled exactly thus leading to a good prediction of the shock wave velocity and can be compared with the afore mentioned equation for the shock wave velocity. The agreement between the concentrated cavitation method and the analytic solution is obvious. The occurring pressure peaks which occur in the concentrated cavitation method after passage of the shock wave are caused by the collapse of the several cavities. These spikes are not realistic but caused by the numerical method. This method, however, predicts the shock wave velocity well and consequently it maintains the proper timing of the transient. This causes that the average pressure rise after disappearance of the cavitation region is a good prediction of the physical pressure rise. Since the concentrated cavitation method only gives a good prediction of the value and the timing of this pressure rise this model gives a macroscopic view of the phenomenon. This was verified by both laboratory and prototype measurements (Refs. 4 and 5). With a smoothing technique to reduce the numerical oscillations behind the shock wave this model will also lead to a good microscopic view. Smoothing techniques which can be used in such cases are mentioned by Kranenburg (Ref. 6).

A comparison of this analytic solution and a calculation with an initial free gas content of 2.10^{-6} at atmospheric pressure was already presented as a discussion of the article of Wylie and Streeter (Ref. 5) but is given here to show the importance of discretisation.

Figure 2 shows the results of the calculations for 0 (vapour only case), 0,002%, 0,004% and 0,01% free air at standard conditions (= at atmospheric pressure).

In Figures 3, 4 and 5 these results are shown together with the presented results of the time-line method for the same free air content.

As a conclusion of these comparisons it is noted that the pressure peak after disappearance of the cavitation region occurs earlier with the time-line method than with the concentrated air method.

In my opinion this time shift is due to a less accurate prediction of the shock-wave velocity in the time-line method.

This may be caused by fullfilling the continuity equation not enough accurate. Fullfilling this equation very accurately is required to obtain a good prediction of the shock wave velocity. When using interpolating techniques (either with the specified time interval method or with the presented time-line method) attention has to be paid to this requirement in order to obtain reliable results, especially in case of the occurrence of shock waves.

REFERENCES

1. Kalkwijk, J.P.Th. and Kranenburg, C., Cavitation in Horizontal Pipelines Due to Water Hammer, Journal of the Hydraulics Division, ASCE, Vol. 97, No. HY 10, Oct., 1971, pp. 1585-1605.
2. Kalkwijk, J.P.Th. and Kranenburg, C., Cavitation in Horizontal Pipelines Due to Water Hammer, Report No. B/71/3, Laboratory of Fluid Mechanics, Dept. of Civil engineering, Delft University of Technology, Delft, the Netherlands, 1971.
3. Wylie, E.B. and Streeter, V.L., Column Separation in Horizontal Pipelines, Proc. Joint Symposium Design and Operation of Fluid Machinery, Fort Collins, Colorado, June 12-14, 1978.
4. Kalkwijk, J.P.Th., Kranenburg, C., Vreugdenhil, C.B. and Vries, A.H. de, Investigations into Cavitation in Long Horizontal Pipelines caused by Water Hammer, Trans. of the 6th I.A.H.R. Symposium, Section for Hydraulic Machinery, Equipment and Cavitation, Rome, 1972.
5. Provoost, G.A., Investigation into Cavitation in a Prototype Pipeline Caused by Water Hammer, Proceedings of the Second International Conference of Pressure Surges, BHRA, London, Sept. 22-24, 1976.
6. Kranenburg, C., Transient Cavitation in Pipelines, Ph.D. Thesis, report No. 73-2, Laboratory of Fluid Mechanics, Dept. of Civil Engineering, Delft University of Technology, Delft, the Netherlands, 1973.
7. Provoost, G.A., Discussion on: Wylie, E.B. and Streeter, V.L., Column Separation in Horizontal Pipelines, Proc. Joint Symposium Design and Operation of Fluid Machinery, Fort Collins, Colorado, June 12-14, 1978.

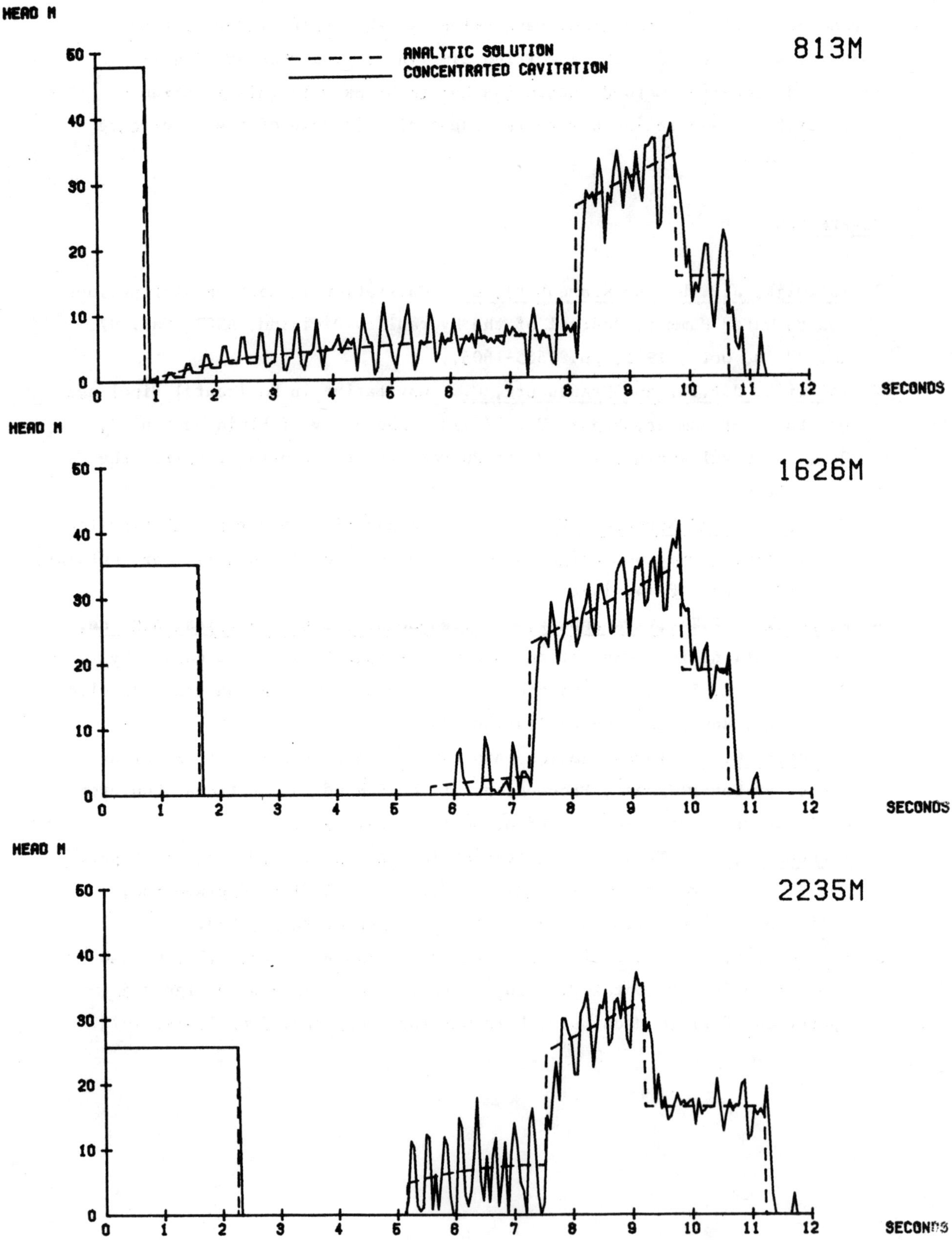

Figure 1 Comparison of an analytic solution and the concentrated cavitation method for the vapour only case (air content $\alpha_o = 0$)

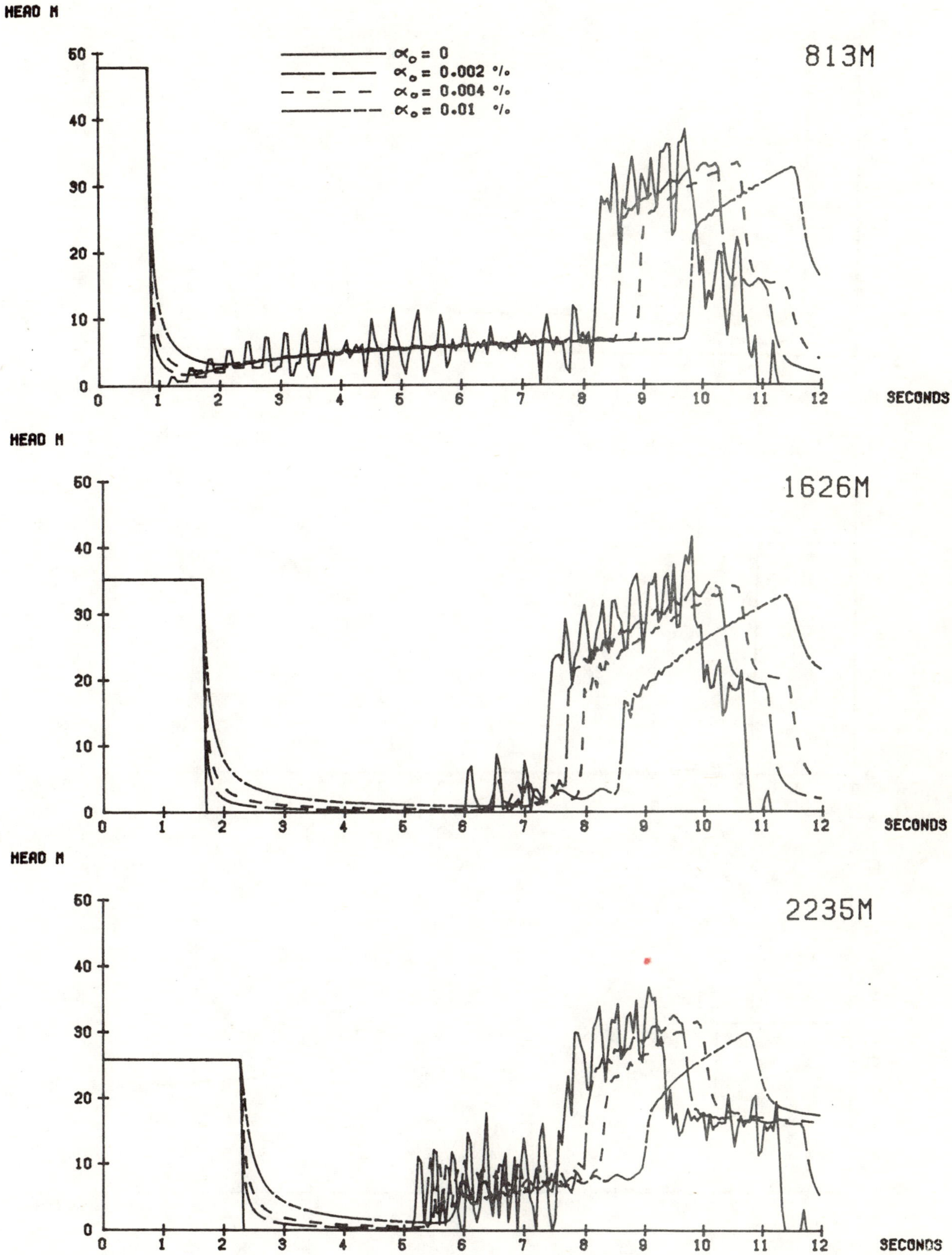

Figure 2 Comparison of the concentrated cavitation method (α_o = 0) and the concentrated air method for three different air contents (α_o = 0,002%, 0,004% and 0,01%)

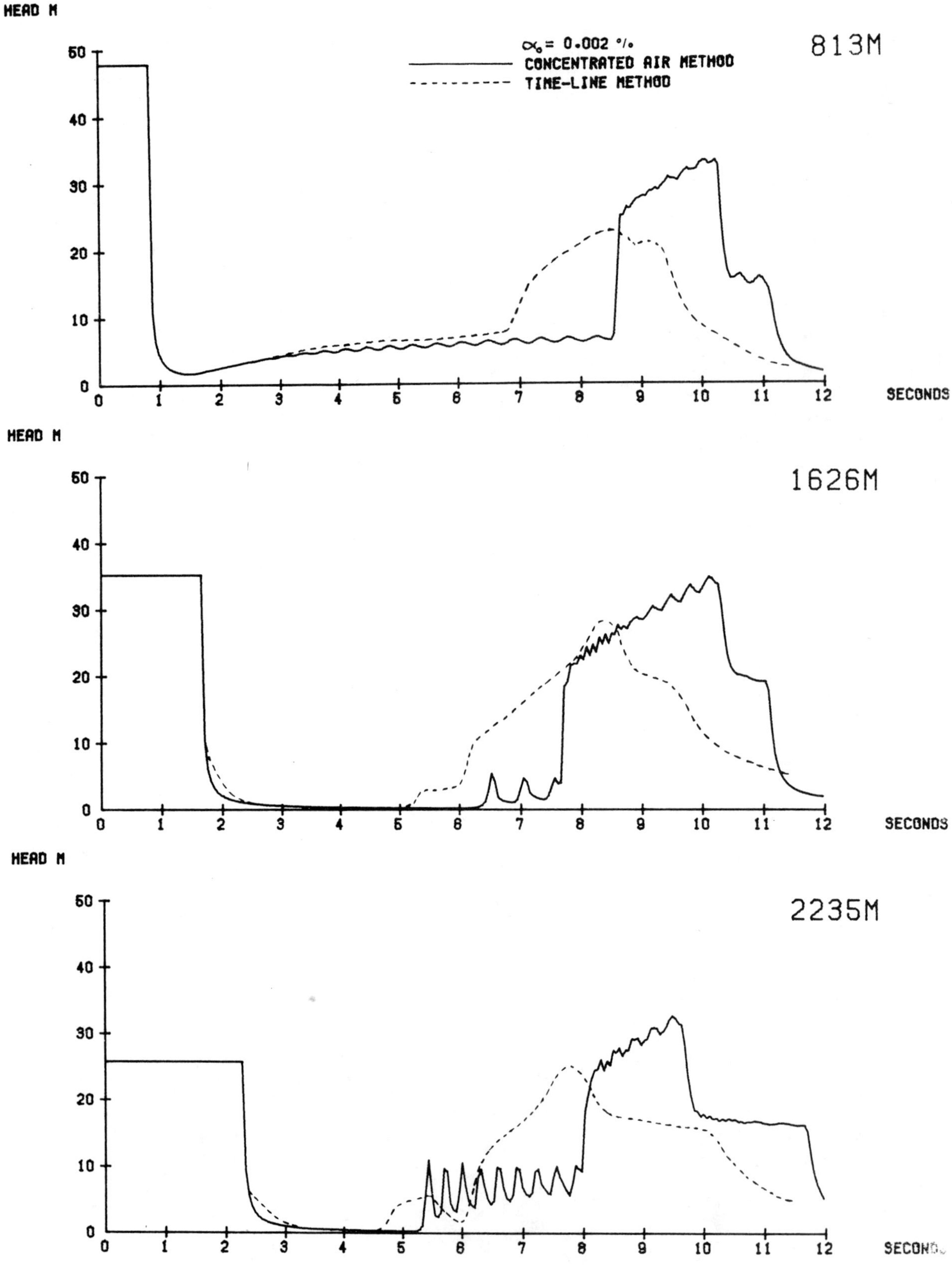

Figure 3 Comparison of the concentrated air method and the time-line method for the air content α_o = 0,002%

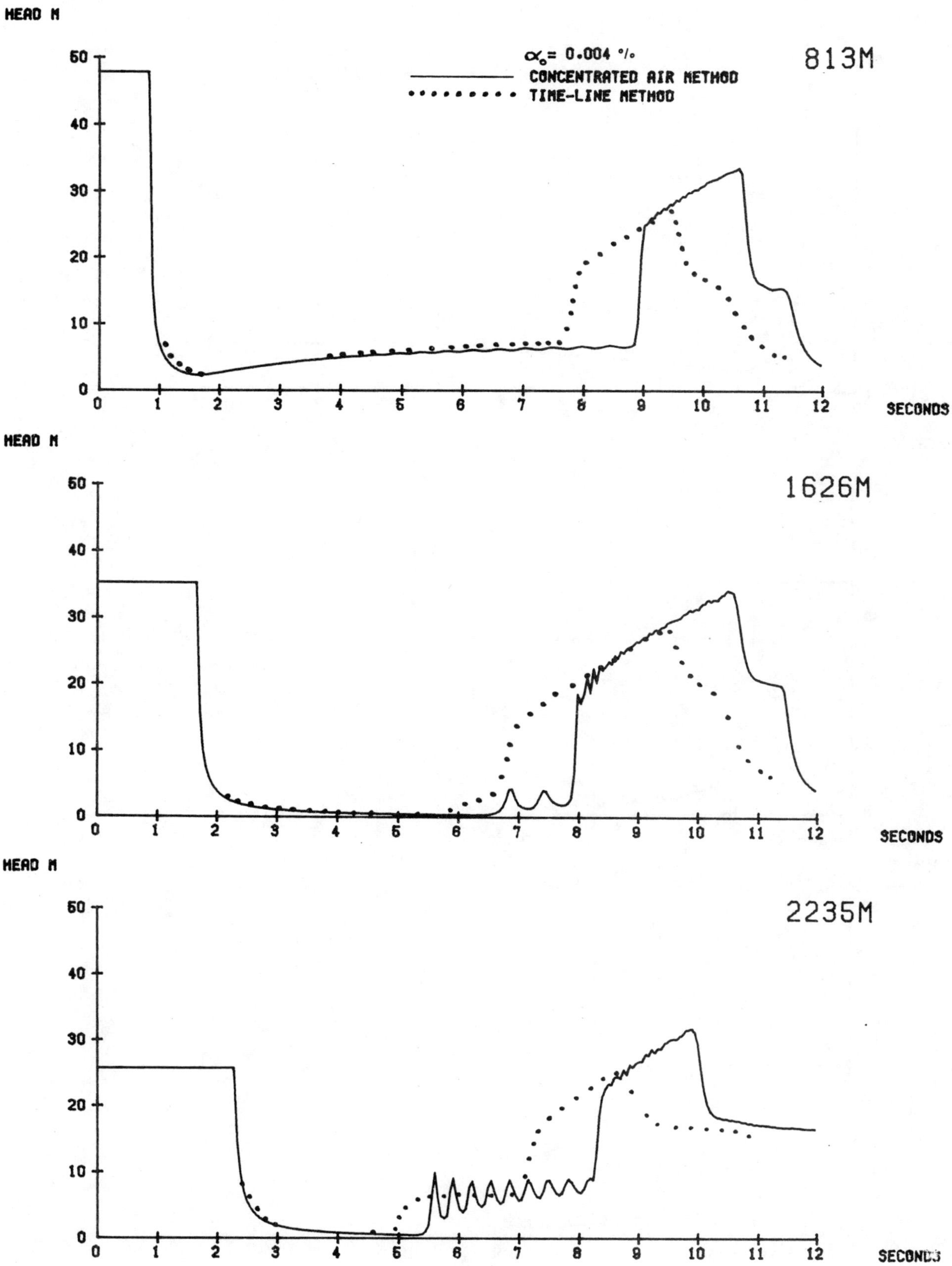

Figure 4 Comparison of the concentrated air method and the time-line method for the air content α_o = 0,004%

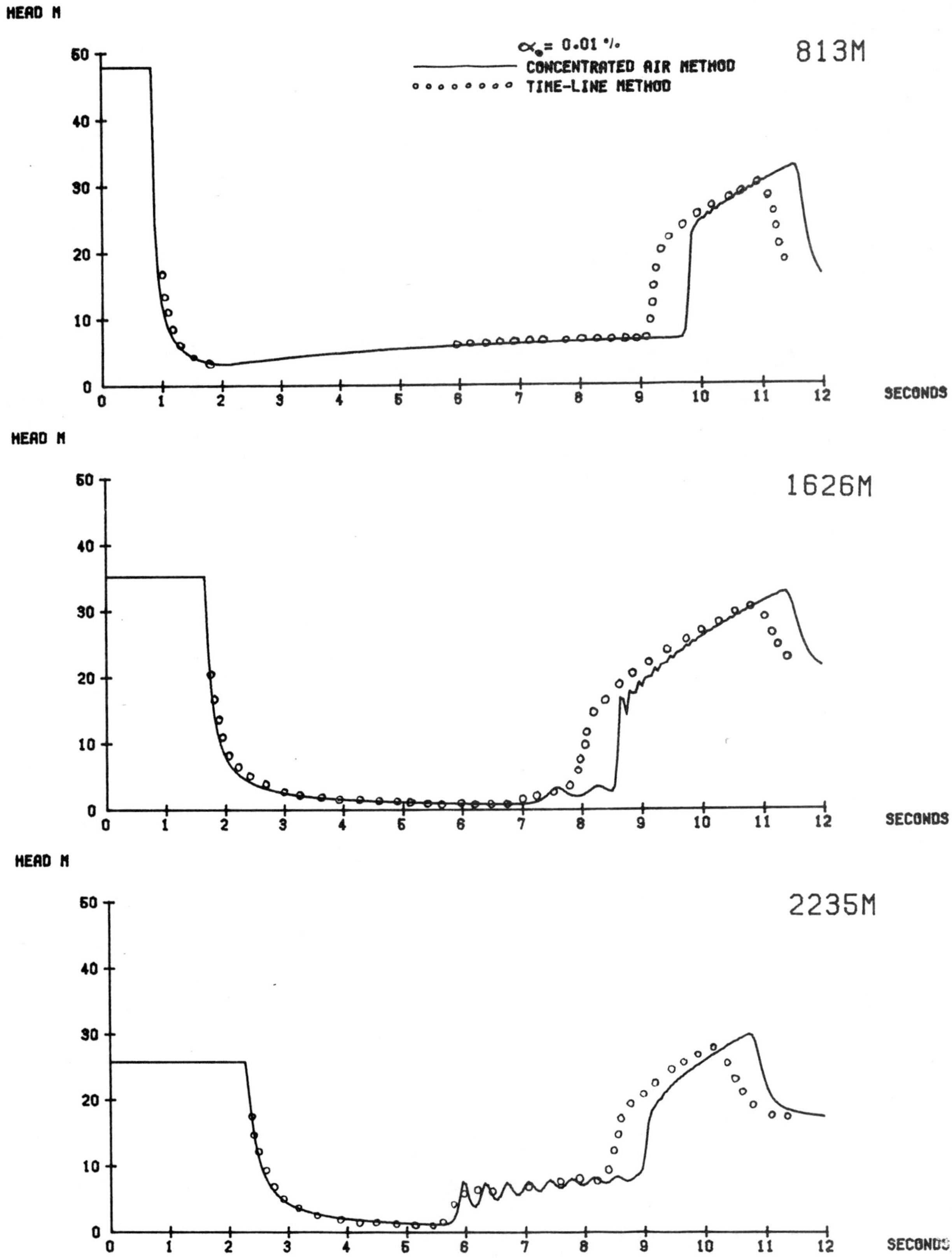

Figure 5 Comparison of the concentrated air method and the time-line method for the air content α_o = 0,01%

FREE AIR IN LIQUID TRANSIENT FLOW

E. B. Wylie

University of Michigan, U.S.A.

CLOSING DISCUSSION

There is an increase in both computer storage and time in the time-line method. Storage requirements increase by a factor of m - 1 for each of the dependent variables H and Q, where m is the maximum number of time steps that the characteristics may extend. There is little CPU time disadvantage in this part of the problem since a continuous storage scheme can be adopted so that the variables never have to be relocated in storage, as they do in the standard rectangular grid method. The time disadvantage arises in the iterative solution. As is the case with any iterative technique CPU time probably increases by a factor of 2 to 3.

The influence of the transport terms $V\partial V/\partial x$ and $V\partial p/\partial x$ will certainly be more significant during the low wave speed portion of the transient. However during this condition the unsteady behavior is generally quite uniform, i.e., more of a surging response and, as a result, the terms in their respective equations are of lesser importance. During the high pressure, rapid transient portion of the problem the terms are unimportant. Although a formal study was not completed in an attempt to evaluate their influence it is felt they are not the significant quantitities influencing the dynamic behavior in these sorts of problems.

The author agrees with Dr. Enever that the air pressure used in Eq. (3) should be the partial pressure of the air, p_a. If p is the total absolute pressure then

$$p = p_a + p_v$$

and Eq (3) should be

$$\alpha = mRT/(p - p_v)$$

The equation in the line after Eq.(8) should read $H = p/\rho g$, in which H is the absolute pressure head. If H_a represents the absolute partial pressure head of air then Eq. (9), and each subsequent use of this equation, would read

$$a' = \frac{a\,H_a}{\sqrt{C_3 + H_a^2}}$$

This has no influence on the integration of the differential equations since $dH = dH_a$, assuming constant vapor pressure. In the first example this change would likely make little difference; however in the second case in which the pressure reduces nearly to vapor pressure, depending upon the air content, this is likely to have some influence on the timing of the events.

In response to Professor Vardy, it is agreed that if the built-in continuity check indicates a perfect mass balance it does not necessarily guarantee an accurate solution. The benefit of the check is more direct in the cases in which there is a continuity imbalance; then one can be reasonably assured that the results are open to question.

Interpolations in the time-line direction converge efficiently in the presence of rapid pressure changes when the characteristics are not extended too far back. Normally when the characteristics do extend several levels to earlier time the fluid is "soft" and sharp transients do not exist. The exception to this is when the liquid vaporizes at low pressures. Then rapid transients persist, at least in our typical models, and the location of the "correct" interpolation point becomes more difficult. Undoubtedly there are acceptable alternatives to extending back beyond perhaps 5 or 6 time steps. An investigation was not conducted into other schemes.

The post-conference discussion submitted by Mr. Provoost was most appreciated. He raised some excellent questions primarily relating to Figure 9. The results in the Figure were questioned by the author, and discussed by others, during the presentation of the paper. There is no physical reason known to the author for the response with a free air content to occur earlier than the vapor-only case in this particular problem. However the time-line interpolation procedure cannot be lightly dismissed. It does, after all, model distributed parameter equations which describe the physical system. It also approximates the characteristics grid method, which seems to be universally recognized as a method by which others may be judged when dealing with wave propagation problems.

Historically most different approaches to problems go through various levels of refinement before the "ultimate" scheme is adopted. If the time-line approach survives as a viable method for variable wave speed unsteady flow problems, and I suspect it will, then it too will be improved upon.

Further evaluation of the time-line method has been conducted, leading to a better evaluation of its strengths and weaknesses. Various alterations in the scheme as proposed in the paper were investigated. It was found that a simpler integration of Eq. (12) produced just as good, if not better, results than the integration shown in Eq. (24). By referring to Fig. 4, the location of point R_o is found exactly in the manner described in the paper. However Eq. (24) is replaced by

$$H_P - H_{R_o} + B\frac{\bar{a}}{a}(Q_P - Q_{R_o}) + \Delta z + R\,Q_{R_o}^2 = 0$$

in which $\bar{a}$ is an average wavespeed defined by the slope of the characteristic line between R_o and P. The same approximation in the integration of the dQ term was used in the C- compatibility equation. As stated above the changes in results from this different integration were not dramatic, but it does represent a simplification since it removes the need to interpolate in the x-direction to find Q at points such as R_1, R_2, R_3, R_M, S_1, and S_2 in Fig. 4.

A concentrated air model was also implimented in which an equivalent amount of free air is lumped at computing sections. The air is assumed to expand and contract isothermally and a continuity balance at the node relates the volume change to the net liquid inflow. Continuous liquid is assumed between the lumped air masses, with a wave speed equal to the propogation velocity in the liquid without free gas in the pipeline. A number of investigators have used a discrete bubble model to simulate transient problems in gaseous liquids. Most have found the results to be plagued with significant pressure oscillations, either physical or numerical. This has been the situation at Michigan. However it appears as if the Delft model satisfactorily smoothes out the oscillations, at least in the high friction example that is being compared.

In order to compare the time-line interpolation model with the discrete bubble model, and with the analytic solution, an example is presented. An imposed pressure history is used to excite an infinite pipeline of the following characteristics:

$D = 0.61$ m, $f = 0.0$, $Q_o = 0.89\ m^3/s$, $a = 1000$ m/s, $\rho = 1000 kg/m^3$,

$H_o = 20$ m, $\Delta x = 25$ m, and $\alpha = 0.2\%$.

The forcing function at x = 0 is shown in Fig. 1 along with the response at x = 200 m. The spreading of the wave of pressure reduction is apparent and agreement amongst the three methods is excellent. For the pressure rise portion of the problem the analytic method, Eq. (19), shows that a shock wave would form at x = 45 m when H = 10 m and at x = 191 m when H = 20 m. Thus the resulting pressure rise at x = 200 m is clearly downstream of the formation of a shock wave. The simplified analytics cannot provide the exact timing of the shock wave at x = 200 m; however the two numerical procedures provide close agreement on the timing. Following the passage of the shock wave the discrete volume method shows a significant pressure oscillation about the actual pressure of 20 m. The frequency of this oscillation is a function of the distance interval between discrete cavities that contain a specified mass of free gas at a given pressure level. In the figure, which results from $\Delta x = 25$ m, the period is approximately 0.2 sec. At a distance interval of 50 m in this problem the oscillation was at a period of approximately 0.4 sec at about the same amplitude. This is not a numerical oscillation but a physical response of the lumped concentrated air cavity model. Since it is physical, in order for the discrete model to simulate the actual distributed system response, selected frequencies may be filtered out of the numerical predictions, the results may be smoothed, or some other means needs to be used to get rid of the apparent oscillations. In many situations this is an entirely acceptable solution, but it must be realized what is being done since it seems likely that in some examples a problem may arise when, unknowingly, the real response may be filtered or smoothed out.

The time-line method and discrete model are compared again in Fig. 2. This is the same frictionless example as in Figure 6 of the original paper. In both models $\Delta x = 200$ m, $\alpha = 0.2\%$. Once again considerable smoothing would be necessary to accept the discrete model results.

The discussion of Mr. Provoost focused on Fig. 9 of the original paper. As indicated in his discussion the timing of events is open to some question, and the author agrees, as stated earlier. The revised formulation of the time-line method was used to simulate this same problem, with results presented in Fig. 3. The no-air case, $\alpha = 0$, is the same analytic solution that was presented in the original Fig. 9 from reference 9. The two air contents $\alpha = 0.004\%$, and $\alpha = 0.01\%$ provide a delayed response which seems physically reasonable. The simulation of the lowest air content, $\alpha = 0.002\%$, produced vapor column separation over a portion of the pipeline with the revised time-line method. Inasmuch as this study did not address the issue of vaporous cavitation, only gaseous, this case is not presented in Fig. 3 of this closing discussion.

The author is most grateful to the discussors for their contributions. As a result the study has been broadened and enhanced considerably.

ERRATA

Page	
30	line 1 after Eq. (8) should read $H = H_P + \bar{H} - z$, $H = p/\rho g$, in which H_P = height of hydraulic grade line.
30	line 1 after Eq. (9) should read $C_3 = C_2\, m/(\rho g)^2$
36	The last paragraph before the conclusion should be deleted.
42	Fig. 3 of the closing discussion should replace Fig. 9.

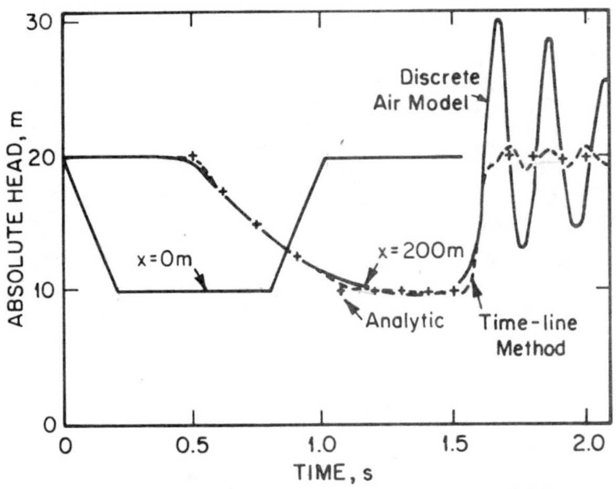

Figure 1. Pressure head response at x=200m

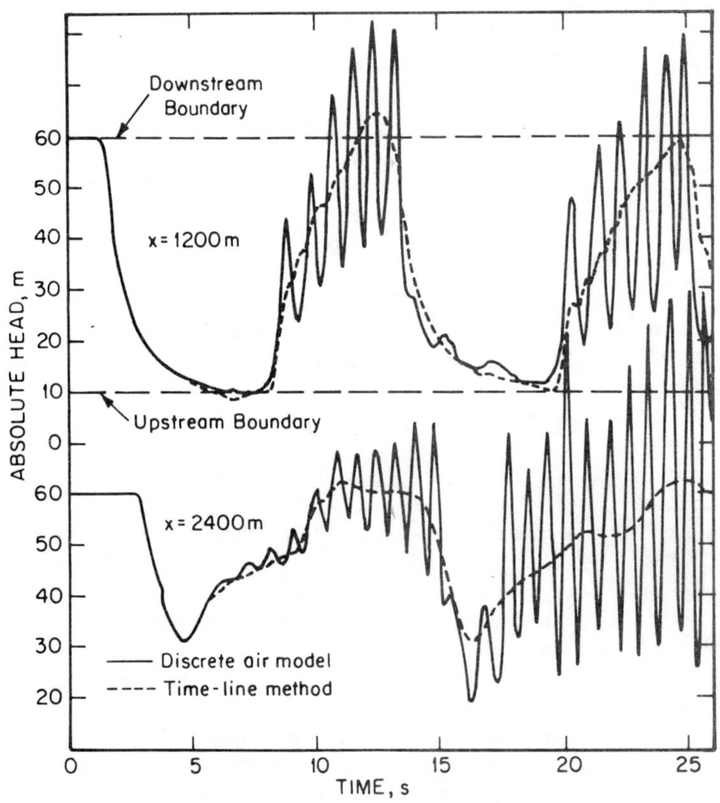

Figure 2. Pressure head response at 1200m and 2400m

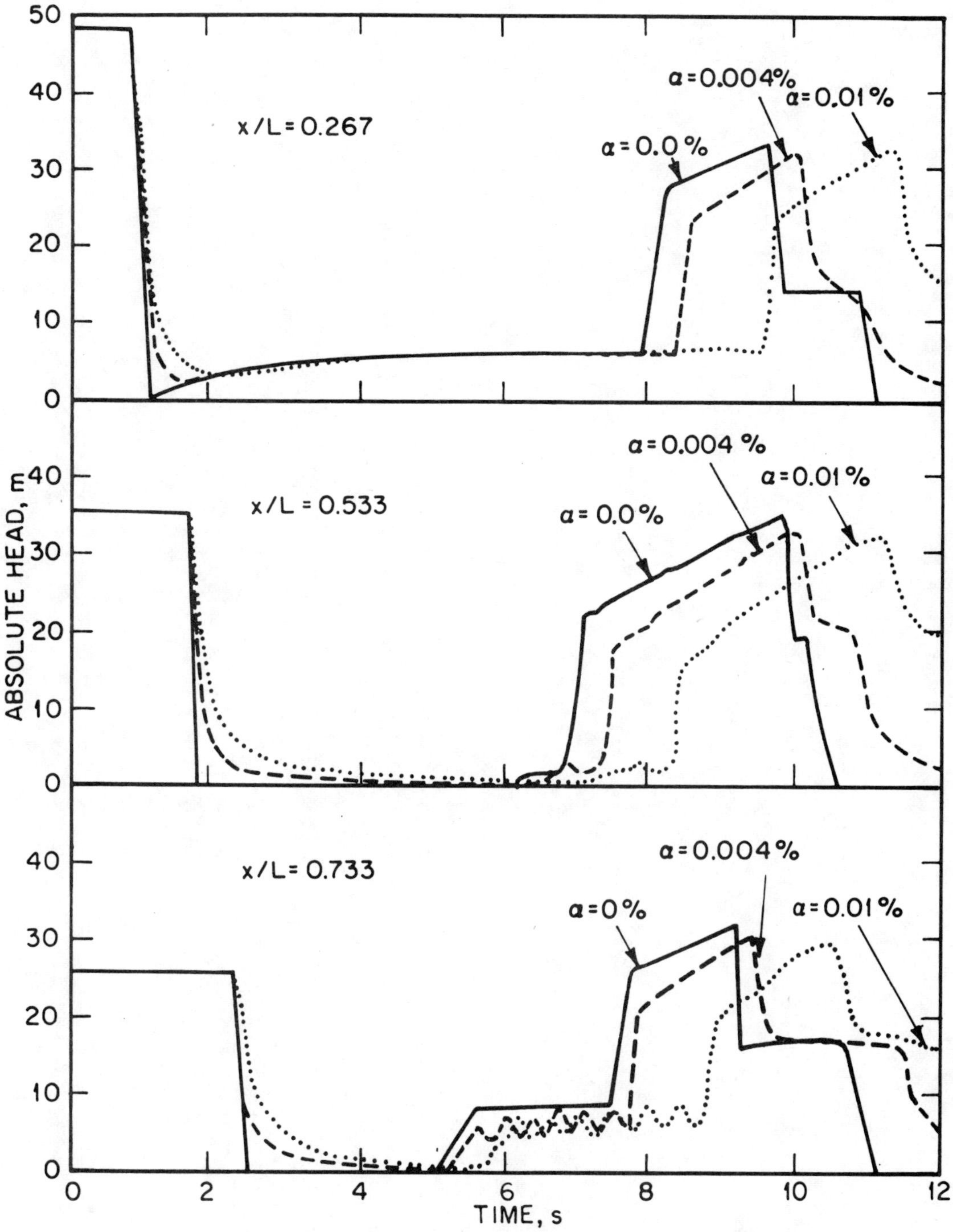

Figure 3. Example. Pressure response; time-line method.

PAPER B2

Combination of finite difference and finite element techniques in hydraulic transient problems.

C.S. Watt
Merz & McLellan, U.K.

A.P. Boldy
University of Warwick, U.K.

J.M. Hobbs
Thames Polytechnic, U.K.

D.C. Wiggert, Michigan State University, U.S.A.

With reference to Fig.5, it appears that incompressible unsteady flow is taking place up to approximately one second; thereafter a compressible water hammer-like transient is evident. With the numerical predictions, both finite-element and characteristic methods give comparable results up to one second, but in the waterhammer zone, the method of characteristics shows a better prediction. It is necessary to assess the numerical accuracy of various schemes - whether they are based on the characteristic method, or the finite element or another implicit scheme. One suggested manner is to perform a Von Neuman analysis, which would show what degree of attenuation and dispersion is present due to the numerical discretization; such effects have been noted for characteristic schemes which employ spatial interpolations.

A.E.Vardy, Dundee

Have you considered using the results in the part of the pipeline that is wholly upstream of the valve to compare theory with experiment? By using the measured pressure history just upstream of the valve as the pipe downstream boundary condition prior to closure, you could eliminate uncertainties resulting from the unknown behaviour of the flow through the valve. This will be particularly useful as a means of ensuring accurate conditions immediately prior to the water-hammer phase of the results.

Author's Closure

Authors' reply to D.C. Wiggert

The authors have found that in general the finite difference schemes (both method of characteristics (MOC) and implicit finite difference scheme (IFD)) seem able to model compressible flow slightly better than the finite element method (FEM). A possible cause is the choice of the high order polynomials which provide benefits of larger mesh lengths as shown in Fig. 8. We noted that the numerical attenuations on IFD and FEM are both more severe than the MOC (as seen in Fig. 5 and 8) as the time discretization is progressively refined. Although no similar numerical discretization errors were reported elsewhere, there are suggestions that the FEM will be subjected to similar stable criterion as the IFD in the application to open channel flow (e.g. Ref. 15). The need to carry out a more detailed error analysis is noted.

Reference

(15) Cooley, R.L. and Moin, S.A: "Finite element solution of Saint-Venant equation". J. of Hyd. Div., ASCE, Vol. 102, No. HY6, pp.759-775. (June 1976).

Authors' reply to A.E. Vardy

We welcome the suggestion by Prof. Vardy to use the measured pressure history just upstream of the valve as the downstream boundary condition prior to closure. This of course will by-pass all the uncertainties of the valve at the final stage of the closure. But an accurate account of a valve, especially near its final closure position, can be of practical interest to larger hydraulic systems. We found it necessary to calibrate the valve rather precisely in order to predict the transient pressure accurately (Fig. 10). A perfect closure (watertight and the valve behaves like a dead end condition) can rarely be achieved by mechanically operated valves and a slight leakage can contribute a significant damping to the waterhammer waves as seen in Fig. 11.

Errata

1) In equation (4.1) the terms with $\sin\phi$ should read

$$\left(-\frac{a^2}{g}-\frac{1}{2}\Delta x\sin\phi\right) \text{ instead of } \left(-\frac{a^2}{g}+\frac{1}{2}\Delta x\sin\phi\right)$$

and

$$\left(\frac{a^2}{g}-\frac{1}{2}\Delta x\sin\phi\right) \text{ instead of } \left(\frac{a^2}{g}+\frac{1}{2}\Delta x\sin\phi\right)$$

2) In table 1 the second and third rows should read

Finite element method alone (at 50 times MOC Δt)	11.6	0.06
Implicit finite difference alone (at 50 times MOC Δt)	4.25	0.02

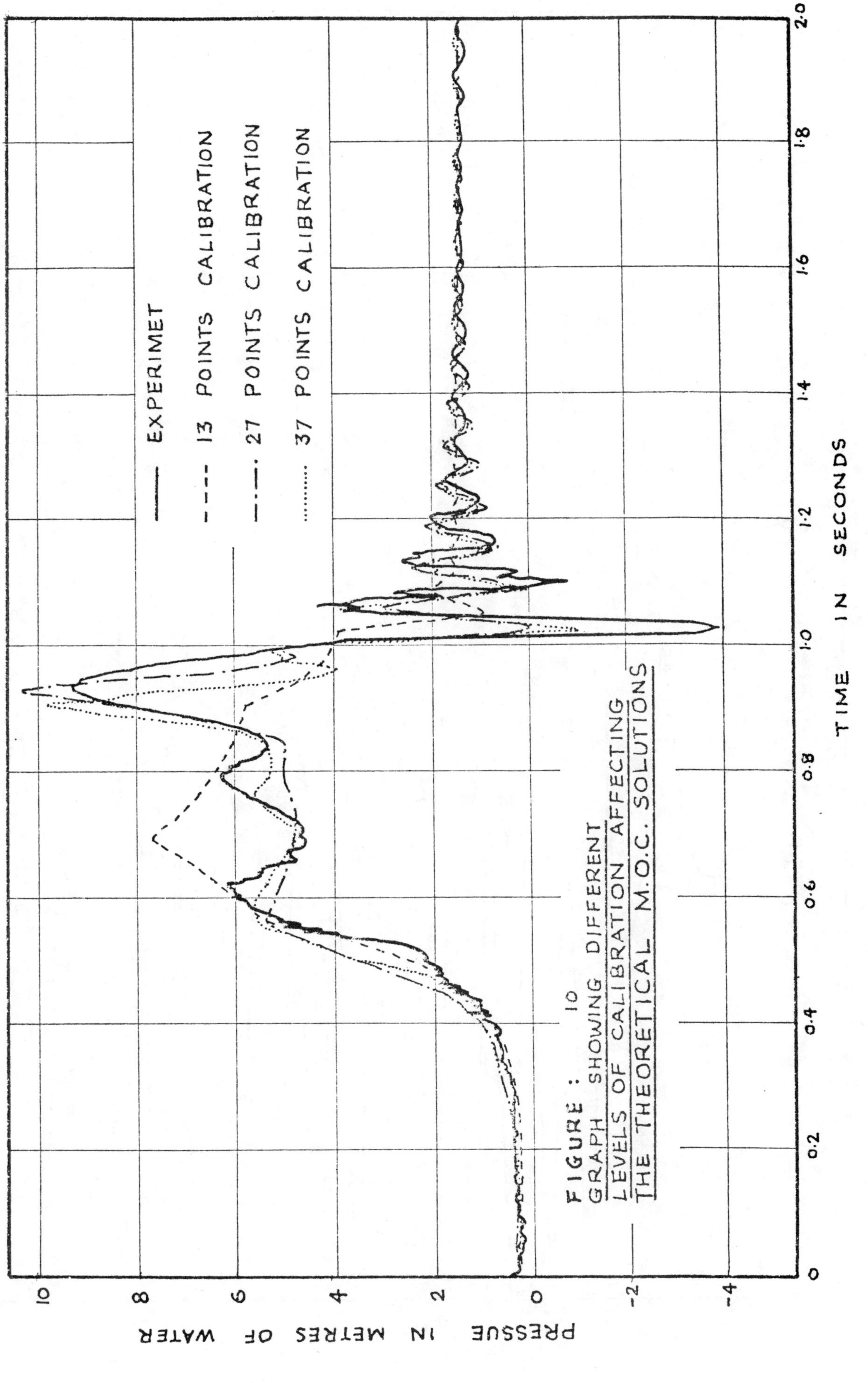

FIGURE : 10
GRAPH SHOWING DIFFERENT LEVELS OF CALIBRATION AFFECTING THE THEORETICAL M.O.C. SOLUTIONS

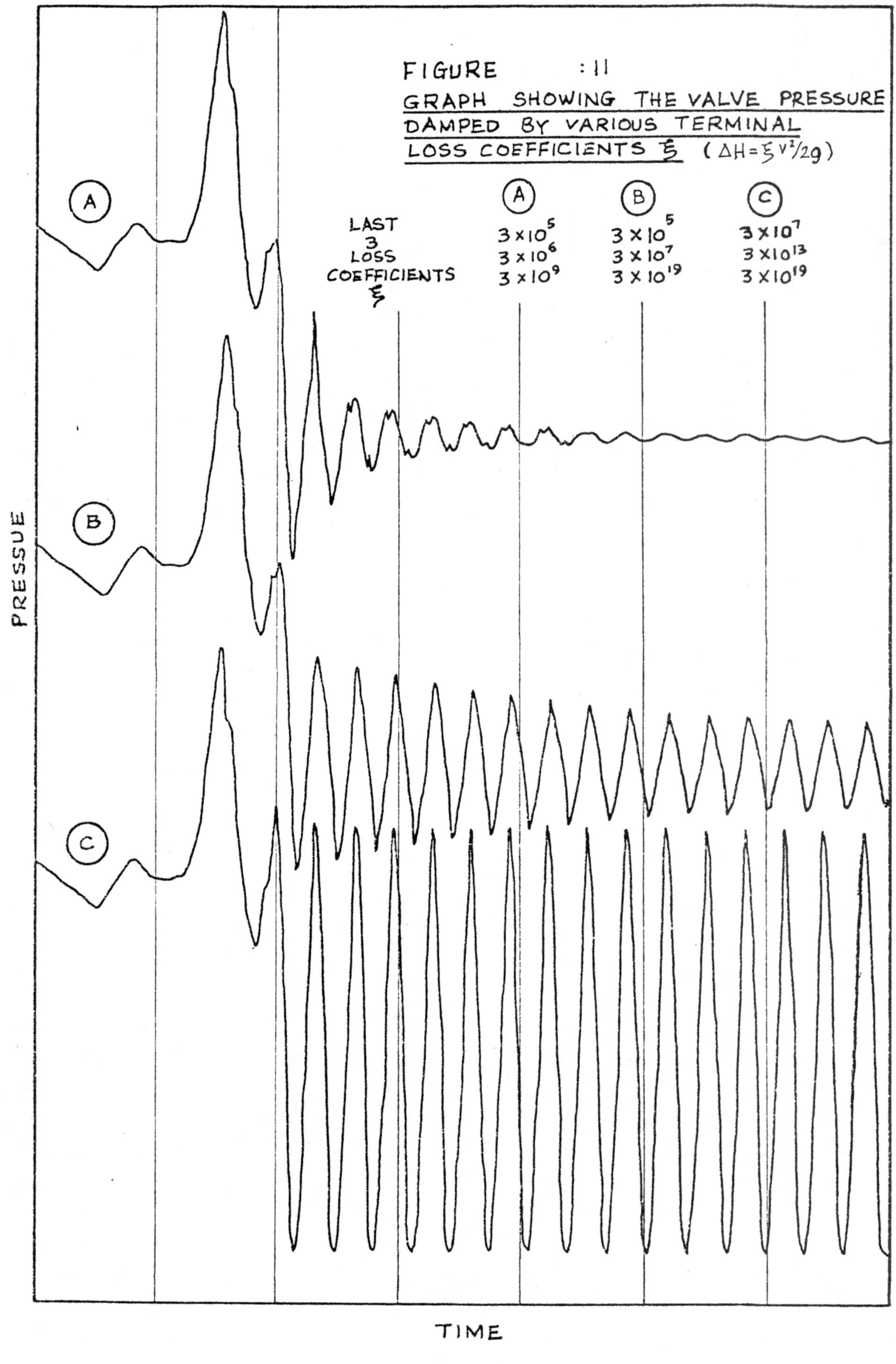
FIGURE : 11
GRAPH SHOWING THE VALVE PRESSURE DAMPED BY VARIOUS TERMINAL LOSS COEFFICIENTS ξ (ΔH=ξV²/2g)
A
B
C
LAST 3 LOSS COEFFICIENTS ξ
A
3 x 10^5
3 x 10^6
3 x 10^9
B
3 x 10^5
3 x 10^7
3 x 10^19
C
3 x 10^7
3 x 10^13
3 x 10^19
PRESSUE
TIME

PAPER B3

Calculation of pressure surge in complex networks.

K.G. Priddin
BHRA Fluid Engineering, U.K.

D.J.F.Ewing, C.E.G.B.

I followed this paper with great interest. Really good modularisation of a computer program is more easily said than done, and the author's approach is certainly very effective for the class of systems considered.

One hesitates to suggest modifying a program which is clearly working well, but there is one aspect - the integration in time - where a more powerful method may be worth considering, especially if the program is to be enhanced. If the implicit method with weighted timesteps is used, the problem of numerical stability disappears and one can use unrestricted timesteps. The method is outlined, for example, on pp 294-295 of Wylie and Streeter's book (Ref. 1). However the method - and more particularly the reasons why it works so well - are perhaps not as widely known as they might be: so it may be worth explaining the technique.

In one unknown, when the differential equation to be solved is $dx/dt = f(x,t)$, the method requires one to solve the implicit equation

$$x - x_{old} = \alpha\Delta t\ f(x,t) + (1 - \alpha)\ \Delta t\ f(x_{old}, t_{old}) \qquad \ldots (1)$$

where α is the weight parameter. The equation is called 'implicit' because x appears on both sides. α must be between $\frac{1}{2}$ and 1 for numerical stability (as will be seen later). Where there are n unknowns $(x_1, \ldots x_n)$ and a set of equations $dx_i/dt = f_i(x_1, \ldots x_n,t)$ to solve, the method consists in solving simultaneously the n equivalents of equation (1), with x and f replaced by x_i and f_i as appropriate. These equations are usually nonlinear and so must be solved by iteration. The generalised Newton-Raphson method is normally suitable, as follows. To solve the n simultaneous equations $F_i(x_1, \ldots x_n) = a_i$, for example, starting from an initial estimate $(y_1, \ldots y_n)$ one first calculates the residuals $r_i = a_i - F_i(y_1, \ldots y_n)$ and the gradient matrix $F_{ij} = \partial F_i/\partial x_j$ (evaluated at $\underline{x} = \underline{y}$) and then solves the n linear equations $\sum_{j=1}^{n} F_{ij}\ h_j = r_i$. The new estimate is $\underline{x} = \underline{y} + \underline{h}$. The method converges quadratically and by examining the sum of the squares of the residuals r_i one has an easy test for convergence.

A pitfall to avoid is trying to solve equation (1) by direct iteration - that is by substituting for x on the right-hand side, calculating a new x directly and looping round and round. This will not converge unless Δt is small enough: so one gains nothing over an explicit method. Incidentally, $\alpha = 0$ gives the orthodox 'explicit Euler' method - the simplest scheme of all.

The reason why the implicit method works where the explicit method breaks down can be seen by considering the case $f(x,t) = - Nx$. This has solution $x = x_0 \exp(- Nt)$. Ordinary explicit methods are equivalent to replacing $\exp(- N\Delta t)$ by some polynomial in Δt chosen to match the first few terms of its Taylor series. But any such polynomial grows large in absolute magnitude as $N\Delta t$ increases: so it becomes a poorer and poorer representation of $\exp(- N\Delta t)$. Once this polynomial exceeds unity in absolute magnitude (i.e. once Δt exceeds a critical value) the calculation goes numerically unstable. Similarly, when integrating a system of equations, trouble can be expected if there is a rapidly-decaying transient (this arises, for instance, if a particular physical quantity can adjust itself much faster than the other terms in the physical balance - e.g. if one inertia is very small compared to the others). Numerical instability therefore occurs unless the timestep is sufficiently small, and the equations are said to be 'stiff'.

The implicit method, in contrast, replaces $\exp(- N\Delta t)$ by the fraction $(1 - N\Delta t + \alpha N\Delta t)/(1 + \alpha N\Delta t)$. This fraction lies between -1 and +1 for all positive N and all Δt, if α is in the range $\frac{1}{2} \leqslant \alpha \leqslant 1$.

For $\frac{1}{2} < \alpha \leqslant 1$ the term tends to die out as it should: so the method is stable. The case $\alpha = \frac{1}{2}$ is neutrally stable. Thus the implicit method represents systems with rapidly-decaying transients more effectively than an explicit method can. In general one has to trade stability for accuracy: $\alpha = \frac{1}{2}$ gives best accuracy but oscillations once introduced cannot be suppressed. However these oscillations can only arise from physical discontinuities - e.g. the sudden powering-up of a pump - so if these are known not to happen, one can get away with $\alpha = \frac{1}{2}$. Conversely, $\alpha = 1$ gives maximum stability. However, the internal bookwork involved in changing α during a calculation complicates matters, and fixing α somewhere around 0.6 to 0.75 gives a reasonable compromise between accuracy and stability.

I can recommend the whole approach from practical experience in a variety of applications - ranging from heat transfer to creep buckling - over the last ten years. In theory the Newton-Raphson method need not converge if the starting guess is poor but in practice this has never been a problem. Incidentally there is no need to program the algebra of the individual matrix partial derivatives - if one uses double precision FORTRAN then the derivatives can be calculated to sufficient accuracy by central differences.

The method can easily be adapted to cover equations involving $(n + m)$ unknowns $(x_1, \dots x_n)$ and $(z_1, \dots z_m)$ where in addition to the n differential equations $dx_i/dt = F_i$ one has m constraint equations of form $g_j(x_1, \dots x_n, z_1, \dots z_m, t) = 0$. Such 'mixed' sub-systems of equations continually arise in hydraulic systems at system boundaries (pumps, surge chambers, etc.)

Reference: Wylie, E.B. and Streeter, V.L.: "Fluid Transients". McGraw-Hill (1978).

A.E.Vardy, Dundee University

1. I would like to congratulate the author on developing a delightfully straightforward technique. By linearising the equations, unstabilities have been found in some cases such as low flows and closely spaced boundaries. Linearising might also lead to inaccuracies especially if the coefficients in, say, (2.6.2) are not centred in time. Overcoming this problem by reducing the mesh (Δx and/or Δt) may lead to an unacceptable increase in computing costs.

2. How well does the scheme simulate the passage of step wavefronts? I refer here to the numerical representation as well as to the linearisation technique.

Author's Closure

Reply to Dr. Ewing's contribution

I would like to thank Dr. Ewing for his contribution concerning the integration in time and linearisation techniques currently used. I am aware that there are more powerful techniques than the one we are using, and his method certainly seems general enough for our purpose. I have not yet considered in detail how much extra complexity it would require to use it, but there are probably some standard routines for parts of it, and so the extra work involved may not be as high as I first imagined. At present we have found the current technique to be suitable for our needs, in that the results agree well with physical measurements and do not require excessive computer time. However, we are always looking at ways of increasing the program efficiency and this suggestion will certainly be evaluated to see whether the modification can be justified in terms of accuracy, stability and computer time.

Reply to Professor Vardy's contribution

In reply to Professor Vardy's comments, I would agree that the problems he mentions in point 1 are very real and I do not think that we have yet achieved the ideal solution. Reducing the timestep has been one of the main weapons used against these problems, but it cannot always be used. Where the timestep has needed to be unrealistically small we have sometimes found the cause of the problem to be a small volume changing pressure rapidly. It has been argued that these very small volumes will not alter the macroscopic behaviour very much and so they have sometimes been ignored. This limits the accuracy of the results in favour of a more stable program. I rather think that a method similar to that outlined by Dr. Ewing would help this sort of problem greatly.

In reply to the second point, I think I only need to say that when the timestep and number of interior sections match up, the implicit method reduces to the characteristic method for pipe flow and so the method should be quite accurate for steep wave fronts. Clearly the accuracy is less when using a larger timestep, but a coarse timestep should not be used when events are occurring quickly anyway. The linearisation at losses does have a little problem with steep wave fronts, but it quickly sorts itself out in practice.

Discussion & Contributions

SESSION B(ii)

MATHEMATICAL TECHNIQUES (cont'd)

Chairman: Professor E.B.Wylie
University of Michigan, U.S.A.

Papers:

B4	Simulation of hydraulic transients in heating networks.	A.Serwach Ind.Inst.of Automation & Measurements,Poland.
B5	Limits of simplification of strongly looped pipe networks in waterhammer Analysis.	O.Fuzy & L.Kullman Sci.Soc.of Mechanical Engineers, Hungary.
B6	Combined analysis of surge and waterhammer pressures.	H.N.F.Pells & M.J.Shand Ninham Shand Inc., South Africa.

Special Contribution:

Modelling of two-component flows.	D.C.Wiggert, Michigan State Union. P.V.Rao & C.S.Martin, Georgia Inst.of Tech., U.S.A.

PAPER B5

Limits of simplification of strongly looped pipe networks in waterhammer analysis.

O.Fuzy and L.Kullman
Sci.Soc. of Mechanical Engineers, Hungary

H. Safwat, Bechtel Power Corporation, U.S.A.

What were the criteria used in introducing the additional links or flow paths between the beginning and end points for case a of Fig.1?

A.E.Vardy, Dundee University, U.K.

Equation 3 in this paper appears to be inconsistent. The relationship

$$a^2 = \frac{dp}{d\rho} \tag{7}$$

applies only in rigid pipes because it takes no account of the elasticity of the pipe. In contrast, the relationship

$$a^2 = \frac{E_r}{\rho} \tag{8}$$

takes account of the elasticity of the pipe. It corresponds to, say, equation 7 in paper E1 and it implies that

$$a^2 = \frac{Adp}{d(\rho A)} \tag{9}$$

in which A denotes the cross sectional area of the pipe.

The continuity equation for unsteady flow in a pipe is

$$(\rho A)_t + (\rho Av)_s = 0 \tag{10}$$

In order to replace the term $(\rho A)_t$ by Ap_t/a^2, it is necessary to use equation (9). Using the same substitution, the term $(\rho Av)_s$ can be developed as

$$(\rho Av)_s = v(\rho A)_s + \rho A\, v_s \tag{11}$$

i.e.

$$(\rho Av)_s = \frac{A}{a^2}\, v\, p_s + \rho A\, v_s \tag{12}$$

and so the continuity equation (10) is

$$\frac{A}{a^2}(p_t + v\, p_s) + \rho A\, v_s = 0 \tag{13}$$

i.e. $$p_t + v\,p_s + \rho\,a^2 v_s = 0 \qquad (14)$$

This equation differs from the authors' continuity equation. Their equation (2) can be obtained using (9) to replace $(\rho A)_t$ as above, but (7) to replace $(\rho Av)_s$.

I have not seen the authors' reference 5 and would like to know which of (7) or (9) is needed to develop the equations (5) and (6) of the present paper.

Author's Closure

Author's reply to the question of Dr.H. Safwat:

The criteria Dr. Safwat is asking for are listed in paragraph 3 /"Structure of networks"/ of our paper on page 91 in Volume 1. We disregard of listing them here again.

Author's reply to the comments of Prof.A.E. Vardy:

While writing our computer program we have always had in mind to make a simulator for the praxis as generally applicable as possible. That is why we assume the cross-sectional area A of the pipe to be an explicit function of the length s i.e. A = A(s).
The wave speed a is calculated by the formula $a^2 = E_r/\rho$ taking the elasticity of the pipe into consideration through the reduced modulus of elasticity E_r. Having done this one need not assume an explicit dependence of the cross-sectional area of the model pipe on the pressure. So we may write

$$a^2 = \left(\frac{d\rho}{dp}\right)^{-1} .$$

In his comment Professor Vardy has based on the two functions $\rho=\rho(p)$ and $A = A(p)$. We take the functions $\rho=\rho(p)$ and $A = A(s)$ as starting points. Turning to the continuity equation

$$(\rho A)_t + (\rho Av)_s = 0 ,$$

we must calculate the partial derivatives:

$$(\rho A)_t = A\rho_t = A\,\frac{d\rho}{dp}\,p_t = \frac{A}{a^2}\,p_t$$

and $$(\rho Av)_s = Av\rho_s + \rho(Av)_s = Av\,\frac{d\rho}{dp}\,p_s + \rho(Av)_s = \frac{Av}{a^2}\,p_s + \rho(Av)_s .$$

Substituting these derivatives into the continuity equation we get

$$\frac{A}{a^2}\,p_t + \frac{A}{a^2}\,vp_s + \rho(Av)_s = 0 ,$$

or for a cylindrical pipe

$$p_t + vp_s + \frac{\varrho a^2}{d^2}(d^2 v)_s = 0$$

as it appeares in our paper /Eq. (2) /.

We think it necessary to remark, that Eq.(2) in our paper becomes identical with Eq.(14) of the Comment. This proves that there is no need to assume $A = A(p)$ if calculating the wave speed from E_r as we do.

If we differentiate a function $A = A(s,p)$ in Eq.(11) of the Comment, we receive $(\varrho Av)_s = \frac{A}{a^2} vp_s + \underbrace{v\varrho A_s + \varrho Av_s}_{\varrho(Av)_s}$ instead of Eq.(12).

We developed Eqs.(5) and (6) of our paper using Eq.(2) which is consistent as we have just shown.

PAPER B6

Combined analysis of surge and waterhammer pressures.

H.N.F.Pells and M.J.Shand
Ninham Shand Inc., South Africa

J.F. Robbie, University of Aberdeen, U.K.

The Authors rightly remind us of the significant effect surge tank orifices may have on the net head across the pump/turbines. In the studies undertaken, was this effect allowed for on the behaviour of the machines and what form of machine representation was made?

Machine operation associated with the case of Hydraulic change over was found to be the most onerous in relation to surge amplitude with the upper reservoir full.

What factors influenced the times of application of the particular flow sequences and what effect, if any, would you expect the residual surges from the original pump start up to be?

If such a sequence were applied when the upper reservoir is near maximum drawdown would the resulting minimum surge be contained in the chamber?

Were other sequences of load changes considered of practical importance?

The response of the system to machine load changes presented in Figs.5 - 10 appears to be predominantly mass surge. In the light of the findings of the Forrest/Robbie paper (G3)[1], would the Authors elect to carry out waterhammer and surge analysis as a single operation by the method of characteristics in future schemes?

I should like to thank the Authors for sharing their experiences on this interesting scheme with us and congratulate them on the satisfactory corelation they obtained with observed values. Such exchange of information is essential in meeting the increasingly rigerous design specifications for pumped storage schemes satisfactorily.

Mass Oscillation Prediction - a Comparative Study of Mass Surge and Waterhammer Methods - Paper G3, proceedings Third International Conf. on Pressure Surges, March, 1980, BHRA Fluid Engineering, Canterbury.

H. Brekke, Kvarner Brug A/S, Norway

Comments to the paper of Mr.H.N.F.Pells and M.J.Shand. During the discussion of your paper a question was raised about the methods of analysis used by the turbine manufacturers.

As I am representing a turbine manufacturer I would inform that we as many other turbine manufacturers are using the characteristic methods together with a turbine characteristic program.

However, for Reversible Pump turbines numerical instability may occur in the analysis as well as instability at site if the shut down of the guide vanes is too fast at a shut down test.

This instability occurs if the speed is so high and/or the pressure is so low compared with the guide vane position that the zero efficiency line is passed and the area of unstable guide vane characteristics is reached. To avoid this a slow closure of the guide vanes is preferred. The same thing may also occur if a runaway test at site is carried out when reaching high speed even at large guide vane opening if a drop in pressure occurs due to pressure surges.

Unstable conditions may also occur after syncronising if the guide vanes are opened too fast causing a drop in pressure while the turbine speed is constant.

M.H. Chaudhry, Old Dominion University, U.S.A.

As indicated in the paper, the physical model was not built according to dynamic similitude relationships to predict the prototype behaviour. However, three figures are included in the paper showing comparison of the mathematical and hydraulic model results. Could the author give the purpose of these comparisons?

Dr. A.P.Boldy, Warwick University.

With reference to your Fig.11 showing the pressure rise on load rejection, I suggest that the difference between observed and computed values may be caused by the shape of the machine characteristics and the method used to interpolate between the guide vanes in the region encountered during a load rejection.

C.S. Martin, Georgia Institute of Technology, U.S.A.

Would you please comment on the origin of high frequency pressure fluctuations. Could they be due to (i) use of steady flow model results in a dynamic situation, (2) complex two or three dimensional wave interaction in the scroll case, or (3) fluid-structure interaction.

During commissioning tests was reverse pumping in fact actually observed.

A.Anderson, University of Newcastle Upon Tyne, U.K.

1. Site recordings:

In his presentation Mr.Pells showed some most interesting recordings of transients obtained at site after the Conference paper had been prepared. Such observations are extremely valuable to all engineers involved in this type of work and it is hoped that Mr.Pells will be able to publish some of these in the Discussion Volume. Mr.Pells has already indicated that it appears that the turbine manufacturers' predicted pressure rise does not always appear to correspond well with the site observations available and it would again be very valuable to the practising engineer if he could expand on the comparisons, especially regarding the following:

(a) In some schemes it has seemed that prediction of machine speed rise can be more reliable than prediction of pressure transients. Is this the case for Steenbras and if so is it consistently thus or only under certain operating conditions?

(b) In some schemes cavity collapse or "column separation" phenomena in the draft tube have a significant effect on penstock pressures, as demonstrated by Prof.Chaudhry in relation to Fig.9 of Paper G1. Is there any evidence from the Steenbras site recordings of pressure peaks from the draft tube passing into the penstock and disturbing the pressure traces there? If so, then what sort of minimum pressures occurred in the draft tube? Were these near normal vapour pressure, i.e. -8 to -10 m WC gauge, or were they higher, e.g. approx. -4 to -6 m WC gauge? Experience on other schemes has suggested that the effective vapour pressure under a turbine runner can appear to be higher than might be expected and that consequently column separation effects occur much more often than might be predicted.

(c) During his presentation Mr.Pells suggested that the site recordings showed that little or no waterhammer pressure passed beyond the surge tank into the low pressure conduit. Is this generally the case for all Steenbras site recordings? Obviously if it is (and it would be expected to be so on the basis of Paper K3 by H.H.Bernhart at the 1976 Pressure Surge Conf.) then this has an important bearing not only on the design of low pressure tunnels but also on the choice between elastic column and the much cheaper rigid column (mass oscillation) methods of surge shaft level prediction. In view of the extensive experience gained on Steenbras would Mr.Pells now have greater confidence in the inelastic mass oscillation theory for surge chamber design than was the case at the start of the project?

2. Mathematical model:

In view of the very extensive range of operational possibilities modelled in the course of the design and reported in the Paper it would also be most useful if Mr.Pells could amplify the information given on the mathematical model itself, in particular the following:

(a) How was the surge chamber headpond modelled? Usually in simple mass oscillation calculations it is assumed that the horizontal water surface in the shaft remains horizontal across the full shaft cross-section and that the value of this cross-sectional area changes abruptly as the surge level passes from the riser into the headpond and vice versa. Was this simple model the one adopted for Steenbras? If so at what level was the headpond assumed to begin, at 373.45 m or at 371.45 m? Also have there been any site observations made of the actual surge behaviour in the headpond and if so how does this seem to compare with the usual simple model?

(b) Some information is given on the representation of the machines in the first paragraph of Section 7 of the Paper but it would be most interesting if this could be expanded. Was the spherical valve treated independently of the pump-turbine or were these combined together in some way? Were the 4 machines "lumped" together as a single representative machine on a single representative conduit or were all 4 machines modelled separately on their own pipe branches? Were full 4-quadrant machine characteristics used or only simple fixed speed relationships between head and flow at certain guide vane settings? Was the machine behaviour (speed rise) modelled or was the machine treated simply as a hydraulic device, e.g. a valve? How were the "variations in the surges ......allowed for automatically in the calculations"? Was the same machine model used to obtain Fig.11 and 12 as for Fig.5-8?

(c) Finally, could Mr.Pells give an indication of the number of pipe nodes taken and the wavespeeds and time step sizes used in the analyses? Were various "mesh sizes" (i.e. numbers of elements) used and if so was there any detectable effect on the solutions? Were the calculations sensitive to the values of wavespeed used?

PAPER B 6 - COMBINED ANALYSES OF SURGE AND WATERHAMMER PRESSURES

H.N.F. PELLS AND M.J. SHAND - NINHAM SHAND Inc. SOUTH AFRICA

AUTHORS ANSWERS TO QUESTIONS

J.F. Robbie - University of Aberdeen:

The object of the combined analysis was to determine the waterhammer pressure both for the design of the spiral casing and for the design of the penstock and to achieve maximum damping of surge without an unacceptable increase in penstock pressures.

In the analysis for the penstock design the machine was represented as a variable orifice which simulated the flow versus time characteristics of the actual machine.

The factors influencing the times of application of particular flow sequences were determined by analysing a number of situations and selecting those which gave the worst surge conditions. The residual surges from the original pump start were disregarded as they were found to be minor.

The maximum drawdown can be contained in the surge shaft when the upper reservoir is at its minimum level. This was another of the design criteria for the surge shaft.

The other sequences of load changes considered of practical importance were the consecutive pump starts shown in Figure 8 of the paper. The timing of the consecutive starts of each pair of pumps was selected in order to minimise upsurge and reduce the possibility of the head on the pumps moving into the unstable region of the pump characteristics.

The response of the system shown in Figures 5 to 10 of the paper is predominantly mass surge as is rightly pointed out in the question. This must be so since the orifice was sized to avoid excessive waterhammer, thus minimising pressure in the low pressure tunnel. This is the very reason why combined analysis was done and why in future schemes it is considered that the combined analysis should be done in order to achieve efficient and economical orifice and surge tank design at minimum cost. The cost of computation which seems to concern the questioner and was clearly of concern to others at the conference is quite negligible compared to the potential savings that can be made by an optimisation of the orifice, surge shaft and penstock design.

H. Brekke:

H. Brekke comments on the use of the characteristic method together with a turbine characteristic programme and the possibility that numerical instabilities may occur. Discussion with others at the conference and post conference discussion with manufacturers of the turbines does highlight two major problems in reconciling the discrepancy between predicted and actual turbine behaviour. The one is the numerical instability which Brekke points out and the other is the fact that the four quadrant diagrams are determined in a hydraulic laboratory on scale models of the machine tested under a quasi-steady state condition. Whether the machine will behave during transient conditions as indicated by the quasi-steady state model tests and the majoration formula used in extrapolating the model performance to actual machine performance, are both uncertainties.

The turbine manufacturers are interested in achieving the guarantees of performance given during tender and are not unduly concerned whether the actual form of the transients corresponds with the predictions based on the four quadrant diagrams.

M.H. Chaudry:

Mr M.H. Chaudry asks the purpose of comparison between the hydraulic model and the mathematical results.

It must be emphasised that the hydraulic model was built as a surge model and not as a waterhammer model and its purpose was to check that the mathematical model would accurately predict the surges measured in the hydraulic model. This was achieved as shown in Figure 4.

The discrepancies between the predictions of waterhammer by the mathematical model and the hydraulic model indicate that the hydraulic model was not suitable for this purpose. It was never intended to check the mathematical model predictions of waterhammer with the hydraulic model.

Dr A.P. Boldy, Warwick University:

Dr Boldy suggests that the difference between observed and computed values of pressure on load rejection are caused by the machine characteristic and the interpolation between the guide vanes in the region encountered during a load rejection. This could well be the case as the four quadrant diagram extrapolated from the quasi-steady state hydraulic machine model analyses is particularly inaccurate in the region which Dr Boldy mentions.

C.S. Martin, Georgia Institute of Technology:

The origin of the high frequency pressure fluctuations on the pressure rise due to turbine operation is of very considerable interest. Martin mentions three possible causes and the authors have considered a fourth possibility, namely that the pressure fluctuations could be due to incipient resonant conditions or in other words that they are caused by higher order harmonics superimposed on the fundamental pressure rise.

This subject is being investigated further and will be reported on in the future.

It is not possible to confirm whether reverse pumping occurred during commissioning as no flow measurements were made. The good agreement between the predicted pressure waves assuming reversal of flow and the measured pressure waves indicated that flow reversal did occur. Furthermore, pressure measurements taken on the spiral casing of the machine and in the penstock show that the pressure in the spiral casing exceeded that in the penstock over a period of about four seconds. We believe therefore, that reverse pumping took place.

A. Anderson, University of Newcastle Upon Tyne:

1. Site Recordings

   (a) Experience at Steenbras appears to confirm that predictions of machine speed rise are more reliable than prediction of pressure transients.

   (b) There was no evidence at Steenbras that cavity collapse or column separation in the draft tube had a significant effect on pressure variations in the penstock.

(c) Dr Anderson remarks on the slide shown during the presentation which showed negligible transmission of waterhammer pressure past the surge tank. It must be borne in mind that the presentation showed the results of load rejection of only one machine. With a full four machine load rejection the transmission of waterhammer pressure past the orifice would be considerably greater.

He also suggests that the experience gained at Steenbras should give greater confidence in the mass oscillation method of surge chamber design. It has never been suggested that the mass oscillation theory is less accurate or less satisfactory than the combined analysis using the characteristic method. The objection to the mass oscillation is that it does not show transmission of waterhammer pressure past the surge shaft and cannot predict waterhammer pressures in the low pressure tunnel. The design of the orifice and the surge shaft is an optimisation process in which alternative sizes of orifice and surge shaft must be considered and by use of the more sophisticated analysis considerable savings in construction costs can be achieved. As mentioned above, the cost of analysis is insignificant compared with the possible savings in construction costs which can be achieved by sophisticated design.

2. Mathematical model

(a) Dr Anderson asks for details of the modelling of the surge chamber headpond which was modelled as a differential surge tank as described below.

When upsurge in the shaft occurs flow into the headpond first commences through the orifices at the base of the riser wall and depends on the differential between the riser and the headpond. When the level of the water within the riser reaches the overflow level flow into the headpond is modelled according to weir formula until such time as the level in the headpond drowns out the weir flow. Thereafter it is assumed that the level rises horizontally. Drawdown is modelled similarly except that flow from the headpond into the riser is assumed without restriction until the level in the riser drops below that of the headpond when reverse flow through the orifices occurs. The accuracy of this modelling is indicated in Figure 10 which shows the excellent agreement between computed and observed water levels in the riser.

In common with the mass oscillation method of analysis the upward velocity in the surge shaft was ignored.

(b) For the purposes of the surge analysis shown in Figures 5 to 9 the spherical valve was not treated independantly of the pumps turbines. The behaviour of the system was modelled as an equivalent valve, the four machines and spherical valves being lumped together as a single representative machine on a single representative conduit. Only simple relations between head and flow were used.

The reference to variations due to surges being allowed for automatically in the calculation was an inellegant way of describing variations in penstock flows caused by variations in surge head, in contrast to rigid column surge analysis where the penstock flows are usually taken as independant of surge head.

The mathematical model used to obtain the results in Figures 11 and 12 was the four quadrant analysis. The inaccuracy still inherent in this method are emphasised in the poor agreement that is obtained.

(c) The number of pipe nodes used in the analysis was 22. The wave speeds varied between 1 325 metres per second for the steel pipe to a maximum of 1 438 metres per second for the concrete lined tunnel sections. These rather high wave velocities were obtained from the formula given in Streeter and Wylie, Hydraulic Transients 1967, but correllated well with the observed period of the pressure wave in the conduit of 2½ seconds.

The normal time step of 0,0767 used in the calculation was determined by the 105,7 m lenth of the first reach and the wave speed of 1 370 m/s. Varying mesh sizes were used particularly when it was desired to carry out fairly long periods of surge analysis. The penstock instead of being represented by 5 sections of varying diameter and length was represented by a single equivalent pipe with a coarser grid. This coarse mesh gave no appreciable variations in the surge results but considerable differences in waterhammer pressures were obtained.

Calculations were carried out using lower values of wave speed to assess the effects of entrained air. These showed that waterhammer pressures were not significantly different from those for the higher wave speeds discussed above. Agreement between observed and calculated wave speeds indicates that the water in the system is substantially free of dissolved gas.

DISCUSSION ON SPECIAL CONTRIBUTION

MODELLING OF TWO-COMPONENT FLOWS

D. C. Wiggert, Michigan State University, USA

P. V. Rao and C. S. Martin, Georgia Institute of Technology, USA

J.Fox, Leeds University, U.K.

In view of your comment that it is possible to use large dts using implicit methods when analysing two phase flows and the behaviour almost seemed to be that of mass oscillation have you considered the possibility of using mass oscillation methods i.e. rigid column theory, and do you think there is any possibility of such methods working adequately well.

C.S. Martin, Georgia Institute of Technology, U.S.A.

In response to Professor Vardy's and Fox's comments the analysis can not be performed using a rigid-column approach. As shown by Dr.Wiggert's slide illustrating pressure variation at the pump, rigid column theory would only predict the case of the lowest void fraction. In fact, a rigid column analysis would not be completely valid for higher void fractions because of mixture elasticity.

Discussion & Contributions

SESSION C

CAVITATION AND TWO-PHASE FLOW

Chairman: Prof. M.Fanelli
ENEL-DRS-CRIS, Italy.

Paper:

C1	Testing of transient flow and column separation in crude oil pipelines	J.Aga, Norconsult AS, Norway T.J.Karterud & T.K.Nielsen, River & Harbour Lab., Norway.
C2	Allowing for free air in waterhammer analysis	D.J.F.Ewing, CEGB, U.K.
C3	Analysis of water clearing loads in steam quenching systems	H.H.Safwat, Bechtel Power Corp.U.S.A.

PAPER C1

Testing of transient flow and column separation in crude oil pipelines.

J. Aga
Norconsult AS, Norway

T.J.Karterud & T.K.Nielsen
River and Harbour Lab., Norway

D.J.F.Ewing, C.E.G.B.

The authors show very convincing overall agreement between their column-separation model and their experiments. In particular they have managed to bracket the experimental response.

Nevertheless, as the cavity spreads, the pressure has to be reset to vapour pressure every time the cavity spreads to a new node. Is there any danger of such resetting resulting in a cumulative buildup of errors?

J.J.van deu Berf, Groutmÿ N.V., Netherlands.

1. What are the limitations of using the model you presented? For example: What happens when ΔX is very small and cavity is growing very big? I think that you can get an error. Do you have an indication about this error?

2. In the same time interval does the temperature remain constant?

E.Benjamin Wylie, University of Michigan, U.S.A.

A concern was raised about an accumulative effect introduced by continuously setting the pressure to vapor pressure during calculations of cavity formation in long pipelines by use of individual cavities at computing sections.

In fact, the pressure at a vapor cavity is set to vapor pressure when the vapor pocket is initially formed, and the calculations are immediately repeated so that the momentum and continuity equations are satisfied. Thereafter, as long as the vapor pocket exists, the pressure is held at vapor pressure. Conservation of momentum and mass are continuously maintained within the normal limits of the numerical procedure.

D.C.Wiggert, Michigan State University, U.S.A.

The experimental data in Figures 2-9 clearly show the effect of gas release taking place during the transient. The authors assume a discrete gaseous cavity opening once the pressure at an interior location reaches the liquid vapor pressure (-0.7 bar), and thereafter the mass of the released gas is calculated by assuming its growth to be proportional to the volume of the cavity developed during the separation period, i.e., while vapor pressure conditions remain. Gas release is most likely to begin when the system pressure p falls below the saturation pressure p_s. An approximate relation to estimate the rate of release Γ of dissolved gases has been used by several investigators [1, 2, 3]:

$$\Gamma = \begin{cases} K\beta(p_s - p) & , p<p_s \\ 0 & , p>p_s \end{cases}$$

The parameter β is Henry's proportionality constant, and K is an experimentally determined coefficient which accounts for the gas bubble population density, diffusion of gas in the liquid, and the relative motion between the gas bubbles and surrounding liquid.

The flow phenomena in the present study likely has distributed voids because of the large amount of gas release taking place. The writer has conducted tests similar to the authors', but instead has employed CO_2-H_2O mixtures [2]. A homogeneous bubbly flow model which includes gas release was developed and successfully used to predict the transients. In addition to pressure and mean mixture velocity, the volumetric void fraction becomes a third dependent variable. The wave celerity developed from a method of characteristics analysis is not constant, but varies significantly with the void fraction and system pressure. The damping effect mentioned by the authors (page 121) may be due not only to the complex velocity distributions, but additionally and perhaps more significantly, to the wave dispersion phenomenon taking place.

1. Kranenburg, C., "Gas release during transient cavitation in pipes," Jour. Hydraulics Div., ASCE, Vol. 100, No. HY10 (October 1974), pp. 1383-1398.

2. Wiggert, D. C. and Sundquist, M. J., "The effect of gaseous cavitation on fluid transients," Jour. Fluids Engineering, ASME, Vol. 101 (March 1979, pp. 79-86.

3. Wylie, E. B. and Streeter, V. L., Fluid Transients, McGraw-Hill Book Company, (1979).

Author's Closure

Reply to D.J.F. Ewing:

When the cavity increases the pressure is assumed to be constant and equal to the vapour pressure. Therefore there is no "resetting" in our model.

Reply to J.J. van deu Berf:

1. The cavity may extend over several Δx's. For smaller Δx's the cavity can be located more accurately.

2. In our numerical model the temperature is held constant overall during the whole similation.

   We believe that there is a considerable heat transfer to the surroundings such that this assumption is reasonable.

Reply to E.B. Wylie:

In Streeter & Wylie's model the pressure is equal to the vapour pressure as long as the cavity exists. This, however, is not the case in our model. Here the pressure is assumed to be constant and equal to the vapour pressure for an increasing cavity. When the calculated pressure increases, the mass of the developed gases will remain as gas and not condense. Then the pressure in the vapour pocket will be calculated by the equation of state $pV = z \cdot m R T$.

We had no collapsing of the cavity like Streeter & Wylie's model.

Our numerical model shows relatively good agreement with the experimental results only for our actual crude oil. This particular crude showed relatively little effervescence. We have not tested other crudes with higher vapour pressure. Crude oil is a very complex fluid and it will be necessary to perform further experiments in order to visualize the dynamic behaviour and gas release.

We thank you for your interest in this paper.

PAPER C2

Allowing for free air in waterhammer analysis.

D.J.F. Ewing
CEGB, U.K.

Dr.K.J.Enever, Cranfield Inst. of Tech.

I was very interested to read Dr.Ewing's equation (5). In my paper (Ref.5) I used pseudo-viscosity purely to give stability and I am pleased to see that there may be some physical justification for this.

In fact, I decided to use a non-linear pseudo-viscosity and a later paper ('Further investigation of the theory of shock waves in bubbly liquids' - K.J.Kneser, Bulletin de la Direction des Etudes et Recherches, E.D.F., Serie A, No.2, pp. 29-46, 1977) confirmed that this gave greater stability over a wider range of conditions.

I am pleased to see that Dr.Ewing, like myself, is continuing to use the method of air lumping suggested by myself (Ref.5) and developed further by Tullis, Streeter & Wylie (Ref.14). Though there is some danger of numerical instability I believe this can be avoided if suitable numerical integration techniques are used as suggested both in Ref.14 and by Dr.Ewing in this paper.

Author's Closure

Author's reply to Dr.K.J. Enever

Dr Enever's interest and comments are appreciated. Unfortunately the parallel with pseudoviscosity seems to be qualitative rather than quantitative. If for example one has a physical diffusivity of 300 $m^2\ s^{-1}$ (which is representative of those calculated in the paper) and a wave speed of 300 $m\ s^{-1}$, then a shock will be around 1 m thick. Numerical pseudoviscosities are normally chosen so as to spread shocks over adjacent grid points. So, unless the numerical grid also happens to be about 1 m wide, there will be no correspondence between the magnitudes of the pseudoviscosity and physical effective viscosity.

Dr Enever also quite rightly points out that he pioneered the lumped-air method (Ref. 5): the essential improvement of Ref. 14 was of course the use of a numerically stable integration technique, sidestepping the need for a pseudoviscosity. This should certainly have been made clear in my paper.

Erratum

Equation (B14) is incorrect and should read

$$U/\rho_o g A\bar{H}^2 = (4gL/\pi^2 a_o^2)G(Y) - gL/2a_o^2$$

but equation (28) is unaffected.

PAPER C3

Analysis of water clearing loads in steam quenching systems.

H.H. Safwat
Bechtel Power Corp., U.S.A.

A.H.Fox, Nuclear Power Co.Ltd.(Whetstone), U.K.

1. Since the pressure transducer Pw, situated in the Wetwell(Fig.3), was only six downcomer diameters from the outlet, would not a better agreement be forthcoming by simply considering stagnation conditions of the submerged jet, rather than using the 1-D analysis shown?
2. His present model underpredicts the measured pressures and as a designer would he be better using the more pessimistic model during the water clearing phase?

Author's Closure

1. The approach suggested by Mr. Fox has been tried. It leads to very high loads on a submerged object below the downcomer. The test results as well as the two dimensional analysis show that the velocities outside a distance two diameters below the downcomer outlet are significantly lower than those velocities within the downcomer. The physical explanation is that the emerging jet from the downcomer presses on the water which is confined, as a result there is a build-up in the pressure outside that region, and the velocities outside the direct vicinity below the downcomer outlet do not reach high values.

   If one uses the direct jet approach, he calculates excessive loads on an object below the downcomer outlet. He does not calculate any effect on objects that are not directly below the downcomer outlet. In reality, such objects do experience some small loads (due to induced flow effects).

2. A clarification should be made in regard to Mr. Fox's comment and question. Examination of the phenomena that take place reveals that as the water clearing ends (water originally in the downcomer is expelled), air starts to enter into the tank. This phase is referred to as charging air bubble. During this phase, a pressure build up for contents inside the air bubble occurs because the velocity of the air bubble surface is not moving fast enough with respect to the air flow rate into the air bubble. The sharp increase exhibited by the measured wetwell pressure occurs after air starts flowing in the tank. This paper only deals with the water clearing phase of the transient. The charging air bubble loads are calculated using different procedures which have not been presented in this paper.

   Interesting enough that the direct jet approach suggested by Mr. Fox in Question 1 gave higher loads than the charging air bubble loads. The experimental data show that the charging air bubble loads are higher.

Discussion & Contributions

SESSION D

DISCUSSION PAPER

Chairman: Dr. A.R.D.Thorley,
The City University, London, U.K.

Paper:

D	Pressure Surge: Doubts in the design process for real systems.	M.U.Papworth BHRA Fluid Engineering, U.K.

PAPER D

Pressure Surge: Doubts in the design process for real systems.

M.U. Papworth
BHRA Fluid Engineering, U.K.

D.H.Wilkinson, CEGB Marchwood Eng. Lab.

Although there was much I agreed with in the paper I have a number of comments which are probably best brought out by discussion of the paper's conclusions. On a) I would have thought that for conventional Hydro schemes the mechanisms causing surge, starting and stopping pumps, exercising valves etc., were well understood. I would instead refer the conference to fields of waterhammer which were largely neglected in the papers in general but were of interest to the Board, namely w.h. in L.P. feed systems and in water and sodium cooled reactors. In many cases these involved collapse of a cavity due to steam condensation following boiling off or poor priming or to expulsion of an air pocket. Origins and magnitudes of pulses required investigation. Propagation through thin walled flexible pipework systems to find pipe stresses and deflections and loads on components was another rather new area. It was necessary to avoid both pipe rupture and disablement of valve operating mechanisms, which could have safety implications. On b) I think that on the contrary, a study of computer predictions for a range of cases from simple working up to complex can give a very good 'feel' for the fluid dynamics of hydraulic systems, particularly with an interactive computer.

On c) we had an initial problem with lack of data on bursting pressures and bending moments for cast iron and steel valves and for steel pipes but further searching and perseverance revealed quite a lot of data some of which has been published in an I.Mech.E. paper.

On d) there is great concern in the Generating Board, EPRI and no doubt elsewhere to avoid potential waterhammer problems at the design stage. The Board consider it necessary to have a number of people with enough knowledge about w.h. to attain this position and we are frequently asked on new proposals for fossil and nuclear fired plant to assess the possible occurance and consequences of w.h. and either design out of it or to withstand it.

In general the 'classical' waterhammer field showed diminishing scientific returns with a concentration on applications and fine tuning of methods. The new challenges would seem to lie in the problems discussed above. Two papers to be given tomorrow were a start towards a flexible pipe pulse propagation theory. They did not by any means give complete methods but they attempted to describe the main phenomena with simple expressions for estimating magnitudes. Otherwise a recent I.Mech.E. paper gave a more general survey of the state of knowledge on Feed systems and Water Reactors, its occurance, magnitudes and prevention.

T.G.F.Welshman, Thames Water Authority, U.K.

I was delighted to receive this paper, which gives a very practical general outlook on pressure surge problems. Mr. Papworth's presentation was clear and direct, but also lightly amusing.

As a civil engineer experienced in water supply systems, I am well acquainted with pressure surge problems, albeit that the solution of these problems is not always easy. However, having become more involved recently in pumped sewerage systems it has been made all too clear to me that many of the design and operational engineers involved are barely aware of the problems which can arise due to pressure surges and water hammer conditions.

I fully endorse Mr. Papworth's plea for the specific education of engineers, designers and operators into surge and water hammer phenomena. Perhaps there could be, also, a more conscious effort to pull together the excellent studies being pursued by the researchers and mathematicians with a view to producing investigation and design manuals which could be understood readily by practising engineers less versed and less experienced in wrestling with complex mathematics.

There might be benefit in dividing future International Pressure Surge Conferences into two sections - one section to deal with the more academic aspects, and the other to cover practical aspects. Delegates could then attend either, or both, of the sections according to their knowledge and interest.

<u>Dr. K.J.Enever</u>, Cranfield Institute of Technology

I feel that Mr.Papworth's paper is very useful and I agree with almost all the points he has made in Sections 1,3,4 and 5 but there are a few specific points on which I should like to express some disagreement.

In his introduction he states that in the days of graphical analysis the rigours of the method forced the user to understand the basic phenomena. I do not believe that this is true. Many people learned rules for carrying out graphical analysis without any real understanding of the fluid mechanics on which it was based. As a result, when faced with an unusual problem they were at a loss or analysed it incorrectly just as they might do using computational methods.

In Section 2 Mr.Papworth states that pressure surge experts have failed to educate design engineers in the perils of certain situations. Clearly there is always room for improvement in the education of engineers, but I do not think that the author has put forward evidence that there is any less knowledge of pressure surges than of other phenomena of importance to design engineers.

In his conclusions I think that he is rather unfair both to design engineers and to researchers. Many of the former have a very good knowledge of fluid mechanics and are perfectly capable of understanding the mathematics in a book such as Wylie & Streeter's Fluid Transients, whilst I believe that most of the latter have a better knowledge of the engineering fluid dynamics than he suggests and regard the mathematic involved as a tool to solve a problem.

My own summary of what is required for good pressure surge analysis is as follows:

i) A good understanding of the basic physics which may include applications of fluid mechanics, thermodynamics, strength of materials and structural dynamics.

ii) A good understanding of the engineering situations encountered including a knowledge of limitation of data and a knowledge of available surge protection measures.

iii) The mathematical formulation of problems arising under ii) making use of the knowledge from i).

iv) The development of numerical (or graphical) methodsto solve the equationsdeveloped in iii).

v) Interpretation of the results using the understanding acquired in i) and ii).

vi) Site tests which will, it is hoped, confirm the analysis and also add to the understanding in i) and ii) for future reference.

I think I agree with the author's implied criticism that too much time is currently spent on item iv) in view of the limitation in data, but I think that there are still a number of physical phenomena involved in pressure surge that we do not fully understand and research under item i) is still vital.

With regard to data from real systems I entirely agree that insufficient is available and hope that we may see more papers in the future on this important part of the cycle I have suggested above.

Author's Closure

Before replying to the detailed points raised by the contributors, I should perhaps admit that the main purpose of the paper was to provide a forum for a wide-ranging discussion on the doubts concerning the success of applying surge knowledge to the design of piping systems. Unfortunately this debate did not take place during the formal discussion period, although the informal sessions of subsequent evenings did contain useful exchanges of information between researchers and designers.

D. W. Wilkinson

a) The test of success in surge prediction is the ability to predict accurately the maximum and minimum pressures and flows in systems before commissioning. Thus, although it is well known that air evolvement during transient events in high level condenser systems modifies the subsequent pressures considerably, it is still not possible to predict the evolvement well enough to yield accurate pressure predictions. A second example of a known phenomenon, which does not not yet yield to precise analysis, is transient pipe friction. Here it is known that frictional losses are dependent on rate variables as well as pipe geometry and flowrate, but there are few attempts so far to actually include transient friction in the standard analysis.

I would not disagree with the general point that there are areas of interest which are not yet fully understood.

b) Computer models can be used as educational tools to give a feel for pressure surge, but all too often they are used to predict performance of engineering systems without the necessary engineering feel. Most university students can write a pipe/valve/pump program and get answers: the danger is that the lack of knowledge of typical systems will lead to proposals which take insufficient account of doubts in the data, method of operation and maintenance (see item 1 on page 160). We have found that graphical displays along the lines demonstrated by Provost in J1 to be an excellent method of understanding the influence of parameters on system performance.

c) I am not aware of any data on system design pressures which includes rate terms. In steel these rate terms may be irrelevant because the basic failure mechanism is understood well enough to use the maximum design stress. For other materials, such as uPVC, GRP and possibly cast iron, there are considerably more doubts and it is certain that user confidence will increase if rate dependent tests on such materials are performed. It will be extremely interesting to see whether the rate dependent tests on GRP pipe currently underway at BHRA do show different results to steady state tests.

d) It is encouraging to note that EPRI have a sensible approach to pressure surge avoidance at the design stage. BHRA encounters a much wider section of the fluids industry and there is no doubt whatsoever that this approach is not successfully attained in general.

General:- I fully concur with the comments that classical waterhammer is showing limited scientific returns, but this only seems to emphasise the need to evangelise the lessons among technical as well as scientific engineers.

T. G. F. Welshman

It is always a pleasure to have practising engineers agree with one's opinions and I take particular encouragement over the education aspects of the comments.

The problem with separating the practical and theoretical sides of the conference is that it reduces the interaction between researchers and users: personally, I find the after-hours' discussion the most useful part of these conferences. Possibly the attempt to put the practical aspects on one or two days and the academic work on another day could be continued, or even expanded. It should be pointed out that both City University and BHRA/CIT run practical courses on pressure surge annually.

Dr. K. J. Enever

I find Dr. Enever's comments on graphical analysis interesting and bow to his greater experience.

The evidence for our failure to educate design engineers is an opinion based on discussions with a large number of such designers in a wide variety of industries. These industries are often in the production field where fluid flow is a small part of the problems. The fact that a number of firms make a good living out of suggesting modifications to design proposals indicates that the problem does exist. Whether other areas of design also lack an efficient education system does not affect the need in pressure surge.

It took me a year to understand Wylie and Streeter's excellent book when I first became involved in pressure surge seven years ago. This amount of time is just not available to most design engineers, who spend only a small fraction of their time on pressure surge problems. In larger firms specialist teams exist, but in most medium sized firms the teams are built up on a project basis and do not have room for 'experts'. I'm afraid Dr. Enever underestimates the communication gap between university and practising engineers.

The summary of requirements for good pressure analysis is very much the traditional one and for unusual problems I would not dissent. It is, however, a costly and time-consuming approach, which cannot be justified for the standard problems which make up the bulk of surge analyses. In most cases items (i), (iii) and (iv) can be condensed under the general heading of experience.

In fact, there is little difference between developing a surge program and producing a piece of technical hardware. A prototype is generated based on known techniques leading to a production model. This production model is supported and improved gradually in the light of practical experience, but not radically modified for each new sale. At a later stage a Mk2 version may be produced which incorporates lessons and research since the prototype production. Finally, the technical salesman must understand both the product and the user requirements if the product is to be successful.

Discussion & Contributions

SESSION E (i)

PIPE-LIQUID INTERACTION

Chairman: Dr.H.H.Safwat
Bechtel Power Corporation, U.S.A.

Papers:

E1	Transient cavitation phenomena in fluid-structure interactions.	C.A.Kot, B.J.Hsieh, C.K.Youngdahl & R.A.Valentin, Argonne National Lab., U.S.A.
E2	Dynamic response of pipework systems to water hammer.	D.H.Wilkinson, CEGB, U.K.
E3	Study of pipe-liquid interaction following pump trip and check-valve closure in a pumping station.	J.Ellis, University of Strathclyde, U.K.

PAPER E1

Transient cavitation phenomena in fluid-structure interactions.

C.A.Kot, B.J.Hsieh, C.K.Youngdahl and R.A.Valentin
Argonne National Lab., U.S.A.

D.J.F.Ewing, C.E.G.B.

The snapthrough behaviour of the thin relief disc is extremely complex to analyse. For example, for certain geometries the snapthrough need not remain axisymmetric - bifurcation buckling will occur.

Have you tried uncoupling the problem - i.e. determing the response of the relief disc independently and feeding this information (as a static load - displacement curve) into your model?

Does this give any useful correlation with the more rigorously formulated but more complex method given in the paper?

Author's Closure

Yes, we have considered solving the structural problem separately from the fluid and then using the result in the fluid-hammer analysis. We find however that the flurel-structure coupling is too strong to use this approach and we are forced to solve the coupled problem to obtain reasonable results.

The objective of our effort was to develop a computational procedure which treats fluid-structure interactions in the context of a general pressure-transient analysis in piping systems. It was found that it is not possible to uncouple the fluid calculations from the structural response predictions. However, coupled calculations can be carried out even when there is a large disparity in the required time steps for each of the computations. While the fluid is treated in one dimension the structural response calculations can be multi-dimensional. Finally, both the pressure loading and the structural response are drastically altered when fluid cavitation is allowed for in the computations. Hence the effect of cavitation cannot be neglected in fluid-structure interaction problems.

PAPER E2

Dynamic response of pipework systems to water hammer.

D.H. Wilkinson
C.E.G.B., U.K.

M.Sandberg, Swedish Institute for Building Research

In your paper you mention two types of waves, strong and weak waves. From a mathematical standpoint all types of shock-waves are weak waves because they are so called weak solutions.

How do you define the difference between the two types of waves?

A.E.Vardy, Dundee University

To what extent is it permissible to simplify the analysis of the fluid/structure interactions by neglecting some of the longitudinal, shear and bending waves in (a) freely supported networks and (b) more firmly supported networks? If it is not possible to neglect both shear and bending waves, will it sometimes be possible to neglect either shear or bending waves? Transverse movements in some pipes will, of course, be directly linked to longitudinal movements in others.

Author's Closure

To M. Sandberg

Weak and strong shock waves in gas dynamics have a rather different meaning to that used here. I have assumed that for axial waves in straight pipes there will be two composite pressure/stress waves, where the relative magnitudes of the pressure and stress components and the wave speeds, c_s and c_ℓ are initially unknown. Solution of sets of equations for these parameters, among others, indicates that the faster moving, c_s wave has a high 'stress' and low 'pressure' component while the slower, c_ℓ, wave has a low 'stress' and a high 'pressure' component. Paper E4 gives expressions for the relative magnitudes of the components. These then are the 'strong' and 'weak' waves and each composite pressure/stress wave is made up of one strong wave and one weak one travelling at the same speed. The assumption made in the 'basic' method of this paper is, among other things, that the 'weak' waves can be neglected. The solutions can then be obtained in simple analytical form and trends seen with an error of less than 10% in most variables.

To Prof. Vardy

Straight pipes involve only axial deflections. As soon as a bend or T-junction of any angle is introduced then the axial deflection of one pipe imposes a transverse deflection component for one or more of the other pipes. The shear and bending waves then represent the transverse 'stiffness' of the other pipes and contribute to the generation, transmission and reflection phenomena at the bend or T-junction. The effect of varying the shear/bending stiffness is shown in Fig. 7 and, over a wider range can be found from Equations (22) and (23) by putting in values of f from 0 to ∞. Similarly the effect of neglecting longitudinal waves can be seen by varying A in Equation (7) from zero for a very weak pipe to infinity for a rigid pipe with results given by Equations (8) or (23). In very simple terms the magnitude of the flexible pipe effects relative to a rigid pipe solution are given by the expression for V_1/V_o in Equation (8), $1/(1+A)$. In that case, with $A = 1.853$, $V_1/V_o = 0.350$. A rigid pipe solution would have $V_1/V_o = 0$ giving a 35% error in p. Thicker pipes will have smaller errors and it is for the user to decide whether flexibility effects are important or not.

Shear and bending waves are inseparable, as an initial shear wave at a bend turns after a short time into a combined shear/bending wave.

PAPER E3

Study of pipe-liquid interaction following pump trip and check-valve closure in a pumping station.

J. Ellis
University of Strathclyde, U.K.

W.L. Glass, Stewart, Sviridov & Oliver

It is noted that the non-return valve is not the quick closing type and that there was a considerable return flow before it closed. It is possible that there was some re-bound on closing similar to that which is sometimes experienced even with quick closing valves?

Dr.A.P.Boldy, Warwick University, U.K.

In connection with the comparison of computed and observed values, Figs. 13,14, 15 and 16, have you considered the accuracy of the observed values bearing in mind a time of 1/10th sec. of the transient effects.

M.H. Chaudhry, Old Dominion University, Norfolk, VA

It appears that computational procedures, similar to that of the Author, for the closure of the valve have been reported earlier in the literature (Ref.1). In addition, extensive series of tests were conducted by Parmley (Ref.2, 3) at Glenfield and Kennedy Ltd., Britain for determining coefficients to describe the closure of check valve.

Is the author aware of the above works and if so, how does his investigations differ from the above mentioned researches.

References:

1. Pool, E.B., Porwit, A.J., and Carlton, J.L. "Prediction of Surge Pressures from Check Valves for Nuclear Loops" Paper No.62-WA-219, presented at Winter Annual Meeting Amer.Soc. of Mech.Engrs., Nov.1962, pp.8.
2. Parmley, L.J., "The Behaviour of Check Valve During Closure", Research Report, Glenfield and Kennedy Ltd., Kilmarnock, England, Oct. 1965.
3. Chaudhry, M.H., "Applied Hydraulic Transients", Van Nostrand Reinhold Co., 1978, pp.105- 106.

A.H.Lupton, Binnie & Partners

A comment: I have had my hand on the valve lever during closure and there was no bounce.

A.E.Vardy, Dundee University

To what extent is it permissible to simplify the analysis of the fluid/structure interactions by neglecting some of the longitudinal, shear and bending waves in (a) freely supported networks and (b) more firmly supported networks? If it is not possible to neglect both shear and bending waves, will it sometimes be possible to neglect either shear or bending waves? Transverse movements in some pipes will, of course, be directly linked to longitudinal movements in others.

Dr.Ing.Hans-Burkhard Horlacher, Akad. Oberrat

In this paper Mr.Ellis describes that he uses an interpolation to calculate the unknown velocity v_3 (equation 1,2,3,4). Perhaps he can give further details about the type of interpolation.

Interpolations often cause a numerical damping to the solution.

In the equations 19, 25, 30 Mr.Ellis introduce the expression M - lumped mass at a point. What is the exact value of M?

Author's Closure

Mr. Glass has raised an interesting point in referring to the possibility of elastic rebound of the valve flap. Such an occurrence was actively considered since an event of this sort would necessarily have influenced the pressure surge in the liquid. The computation did not predict any rebound and eye witnesses present during the tests did not observe any rebound. I am very pleased to note Mr. Lupton's confirmation that there was no rebound. The phenomenon of elastic rebound is the result of wave motion in the pipe wall producing the necessary acceleration to overcome the fluid forces tending to hold the valve on its seat. The explanation for the absence of rebound in this instance must be attributed to the liquid forces being greater than those tending to displace the valve flap.

With reference to the question of valve closure computations raised by Prof. Chaudhry, when initially faced with the task of assessing the magnitude of the reversed velocity at the time of closure computations were set up which equated the out of balance torque exerted on the valve door to the inertial torque of the door. In this way the time of closure and the conditions of flow in the line were established. While carrying out this exercise the writer also conducted a literature search which revealed that swing check valve calculations had been carried out by other investigators. Having obtained a copy of the confidential report of Messrs Glenfield and Kennedy it was gratifying to learn that similarities existed between the two approaches and since the Glenfield and Kennedy calculations compared favourably with laboratory measurements this in turn gives some confidence in the computed time of closure of the present paper. In respect of how the present work differs from that of these previous researches the accuracy of Glenfield's work must be greater since they were able to determine precisely the moment of inertia of their valve doors while the present study relied only on manufacturer's drawings to make this assessment.

Professor Vardy has touched on a most important aspect of the transient behaviour in mentioning the extremes of fixity. The extent to which it will be possible to neglect transverse effects such as shear, bending and torsion will very much depend upon the ability of the pipes' supports to carry such loads. In the case of a freely supported network it would be necessary to consider these effects since they represent the only manner of absorbing the thrust at bends valves and branch connections. Although having said this it is likely that there will be secondary effects which will be of lesser importance, although further research will be necessary before an assessment can be made.

The ultimate in fixed systems is where all bends, valves and branches are rigidly anchored in which case there is no scope for the type of coupling defined in the paper and hence an analysis of coupling effects would be solely restricted to consideration of Poisson coupling. Fixed anchorages would eliminate transverse effects. The question then arises of how "fixed" are the types of support in common usage today. As far as displacements discernible to the eye are concerned there is little doubt that there was a considerable degree of restraint offered by the supports employed in the present study. Considering small movements such as occur during high-frequency vibrations however it is the writer's opinion that these supports offer little restraint, and that the primary means by which strain energy is dissipated is at the place where pipes pass through the walls of the pumping station. Thus in cases where even apparently substantial supports are provided it may be necessary to consider these transverse effects, although common methods of pipe jointing which offer some flexibility will probably have an appreciable effect on the progress of these transverse waves.

Dr. Boldy has raised the matter of accuracy of the observations. Since the writer was not present at the original tests it is not possible to comment on the type of gauge employed or its method of employment. An oscilloscope was used to record the transient and additional pipe strain measurements were made which did indicate that pressures within the pipe were of the recorded magnitude.

Discussion & Contributions

SESSION E (ii)

PIPE-LIQUID INTERACTION (cont'd)

Chairman: Professor D.C.Wiggert
Michigan State University, U.S.A.

Papers:

E4	Water hammer in a thin-walled pipe.	D.H.Wilkinson & E.M.Curtis CEGB, U.K.
E5	Effect of compressible pipe-liners on pressure transient generation.	A.K.Grundy, Shell Intel.Petroleum, Netherlands, J.A.Fox, University of Leeds,U.K.
E6	Wave propagation in plastically deforming ducts.	J.W.R.Twyman & A.R.D.Thorley, The City University,U.K. R.Hewavitarne, British Gas Corp.

PAPER E4

Water hammer in a thin-walled pipe.

D.H.Wilkinson and E.M.Curtis
C.E.G.B., U.K.

J.A.Fox, Leeds University, U.K.

The authors are to be greatly congratulated upon a most detailed and careful investigation. It is possible that a somewhat less detailed approach may be almost as good, as some of the terms included in the analysis must be very small.

H.Safwat, Bechtel Power Corp., U.S.A.

How was the velocity V_D measured?
What was the value of the velocity V_o in the test whose results are shown in Fig.9. Was any air trapped at the flap? If so what would be its influence on the results?
Have there been any tests performed on bends?

C.A.Kot, A.N.L., U.S.A.

Question for Dr.Wilkinson.

Have you compared your complex theory with experimental data.

We have found that a simple ring-hoop deformation approach works quite well in the plastic regime for thin wall pipes. Agreement is good with experimental data both for the pressures and strains.

C.S.Martin, Georgia Institute of Technology

Would the Author care to comment on any effect of the possible instability of the free surface on the initial rounding of the pressure trace at the top of the pipe?

Also, can you please comment on the effect of flexibility of the flap on the results?

Author's Closure

To Mr. J. A. Fox

We would like to thank Prof. Fox for his appreciative comments on our paper. As far as the detail in the approach is concerned we have given three levels of the method, the 'basic' theory with about 4 equations in paper E2 and, from this paper the 'simplified' theory with 11 equations and the 'exact' theory with 21 equations. For elastic deformation Fig. 9 shows that the 'simplified' method gives significantly better agreement with experiment than the 'basic' theory. However, it is for the user to decide how accurate he wants to be and in many cases the 'basic' or even rigid pipe methods, may be quite adequate. The 'basic' method is in fact used in paper E2 as the starting point for extending the theory to bends and T-junctions.

Going the other way the 'exact' method has no advantage over the 'simplified' one for elastic deformation the solutions being almost identical numerically. For plastic deformation the 'exact' theory is, at least in one sense, necessary in that by trying to include everything that might be significant it should give a good solution against which more approximate ones can be compared. An approach to simplifying it might then be to look at the results in Fig. 2 and 3 and see which variables, if any, could be approximated to and hence deleted as unknown. In all regions R_1/R_0' and t_1/t_0' are almost identically 1.000 so this could be assumed. c_s also seems to remain constant over most of the range but changes in the plastic-plastic zone. V_1/V_0 is also always 1.000 but that is about all as the other variables change in a non-linear and, to us, unpredictable way. Assumptions about these 3 or 4 variables would still leave 17 or 18 simultaneous highly non-linear equations to be solved. There is still no chance of analytical or calculator solutions. Once the library non-linear equations method has been found and put to work it makes no practical difference whether there are 10, 20 or 30 equations, only a few lines of FORTRAN being involved either way.

Looking at the plastic case on a rather more approximate basis more sweeping assumptions, like the neglect of axial stresses and Poisson's ratio effects, could be, and may have been, made. We have not investigated this approach and so do not know how accurate such simplified assumptions might be. However, we would suggest comparison of the predictions of such methods with those from our 'exact' solution over a range of cases to assess their accuracy.

To H. Safwat

The water velocity V_0 was measured by timing the passage of the free surface between wires stretched across the tube at locations A, B and C as indicated in Fig. 5 of the paper. At each location two parallel wires were used and the change in conductivity between them used to provide a signal. Some preliminary experiments with a thick walled tube having wires at eight different locations plus some calculations of the rig dynamics enabled a small correction of about 1% to be derived to account for the slowing of the liquid column between the wires at A and B and the end of the tube.

We consider that in the tests reported there was no significant effect of trapped air. However, as mentioned in Section 4 of the paper, there had been some initial difficulties which resulted in ragged, unrepeatable, pulses and these were attributed to the presence of air although this could not be definitely established.

We have not yet performed any tests on bends although this will be done as the next stage of our experimental programme.

To C. A. Kot

No, comparisons await the completion of our planned experimental programme on water hammer with plastic deformation. We have not seen your test results or simple theory and so cannot comment on them or on the degree of confirmation established. Onepoint we have made however is in a question to Dr. Thorley on paper E6 which quotes a wave speed equation of yours which can lead to apparently anomalous predictions. We would therefore query whether a simple method can be generally applicable over a wide range of geometry, materials and impact velocity unless it has been compared with a method such as ours or with a comprehensive series of experiments covering a wide range of cases.

To C. S. Martin

The effect of free surface geometry is probably not important since the technique of free surface impact produces sharply rising water-hammer pulses in horizontal pipes also. Presumably the water-hammer pressure wave does not appear until complete liquid contact with the tube end has occurred and no voidage remains. In our experiments the pulse rise time is about 0.3 ms which was satisfactory for our purposes. Our initial tests were with a thickwalled test section so that any effects due to causes such as surface instability or flap flexibility could be seen and not confused with pipe flexibility effects. We were able to achieve satisfactorily square pulses (Fig. A) with this thick walled test section and concluded that secondary rig related effects were not significant. Flap flexibility will be important if the flap is so stiff that closure is slow, or so flexible that it closes prematurely. We found that our first metal shim flap was satisfactory so it does not seem critical. There is generally a blip about 40% of the way up the leading edge of the pulse which we think is the flap closing, about 0.1 ms after column impact. Our first flap was made of rubber and this did not work because it was too flexible and closed on airflow.

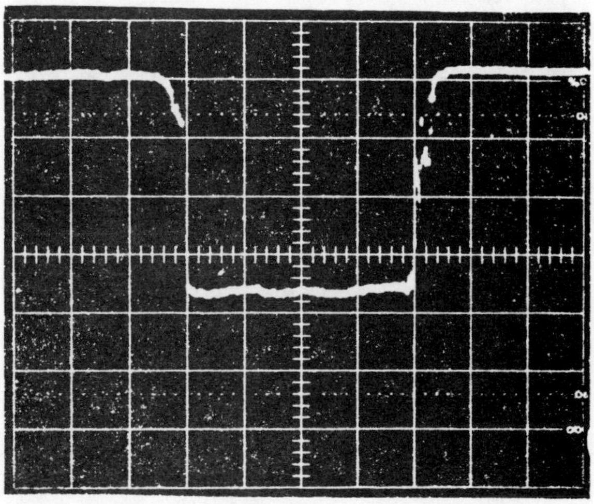

Fig. A

PAPER E5

Effect of compressible pipe-liners on pressure transient generation

A.K. Grundy

Shell Internationale Petroleum Maatschappij B.V. The Netherlands

J.A. Fox

University of Leeds, U.K.

A.H. Fox, Nuclear Power Co. (Whetstone)

What was the % open area of the perforated tube used in the attenuator and what is the effect of open area upon attenuator performance? In the limit as the number of holes is reduced the perforated tube tends to a unperforated simple tube, where no attenuation occurs. Thus the open area would be expected to have an effect on attenuator performance.

How would the attenuator design be optimised for a particular design application i.e. diameters, number of tubes, pressures etc.? Since the tests performed have not covered a wide range of variables.

Reply:

The perforated core had 6 mm bore perforations with a 15% open area. At this stage of the work the intention was to study only the effects of a pipe-liner on pressure transients. Therefore, a relatively large perforation size was chosen to minimize the head loss across the core wall as fluid entered or left the attenuator interspace. Certainly the percentage open area and hole size will effect the attenuator performance, and it may prove interesting to investigate this aspect in future development work.

The most effective attenuator design can be determined by carrying out an unsteady flow analysis incorporating one of the mathematical models described. With experience the analyst should be able to estimate the approximate storage capacity and overall dimensions of the device, and then, quickly arrive at an optimum design after a number of computer runs. In practice the design of the attenuator may be influenced by the constraints of a particular application. For instance, space may be at a premium, or perhaps the engineer may wish to build a certain amount of redundancy into his system; e.g. by installing additional tubes. From this point of view, it is felt that the practical experience gained from field tests and monitoring of the device in operational installations will be invaluable.

A.E. Vardy, Dundee University

1. In what ways does your device differ from a pressurised surge vessel?

2. What other methods of introducing high flexibility have you considered? Do any exhibit hysteresis?

Reply:

1. Both the attenuator described and a pressurised surge vessel are a means of introducing capacitance into a pipeline system. The fundamental difference between the two lies in the way in which this storage is provided, and the effect this has on the transient response of the system. A pressurised surge vessel provides capacitance at a single point in a pipeline, via a branched connection. Hence, the transient

wavespeed in the pipeline is unaffected by the vessel. Conversely the attenuator occupies a certain length of the pipeline, reducing the wavespeed along that length by means of a compressible pipe-liner. During unsteady flow conditions the pipe-liner is compressed and distended, and hence the local wavespeed is continually changing. Therefore, in addition to providing capacitance, the attenuator also modifies the pipe period of the pipeline altering the response of the system.

This fundamental difference between the two devices is highlighted when one closely examines the methods of analysis presented in the paper. The 'boundary condition' model is in essence the analysis of a type of air-vessel, where flow passes through the device, but no account is taken of the time it takes a wave to propagate through the device. In general this analysis is valid when the ratio of the length of the attenuator to that of the pipeline is small. However, when this is not the case the local changes in wavespeed may become significant. The 'transient analysis' takes account of waves propagating through the attenuator and simulates the varying transient response of the system, including the interaction between waves reflected at the boundaries of the attenuator. If one considers the extreme theoretical case where an attenuator occupies the whole length of a pipeline; the 'boundary condition' model will compute the wavespeed incorrectly at every point along its length, whereas, the 'transient' analysis will accurately simulate the unsteady flow conditions.

2. Experimental work has also been carried out using a compressible pipe-liner of polystyrene balls. The balls, approximately 4 mm dia., were held in the interspace by means of a fine nylon mesh sheath slipped over the perforated core. The attenuator proved very effective in this form during initial tests; the maximum surge pressure with the attenuator installed was less than 25% of the value generated in the unprotected pipeline. However, over a prolonged period of testing the polystyrene balls suffered permanent deformation drastically reducing the performance of the attenuator. After one series of tests, spanning a week, the diameter of the balls had reduced by 50%. For this reason, polystyrene was rejected as a suitable material for a compressible liner of this type. Another interesting feature of these results was that the physical properties of the polystyrene were highly time dependent, the stress: strain curve exhibiting hysteresis. It is felt that significant advances in the design and operation of the device may be made by investigating the properties and practical suitability of other such materials.

D. Goldberg, Stoner Associates, Inc.

As the Authors are using a storage mechanism for surge suppression which relies on an enclosed air volume, would there be any significant difference in response between the authors' device and a standard air bottle of similar enclosed volume? The lack of difference between the authors' "transient model" and "boundary condition" model seem to affirm the negative.

On this light, would the authors reflect on the practical commercial viability of their device as well as any maintenance difficulties they foresee?

Reply:

For the experimental investigation described it would be true to say that the response of the attenuator was similar to that of an air bottle. This was due to the dimensions of the attenuator chosen relative to the test pipeline, but is not necessarily true for all pipeline/attenuator configurations. In this instance the storage effects of the attenuator dominated the response of the system. For further discussion on this, the reader is referred to the question raised by Prof. Vardy.

On the question of maintenance, the attenuator, in the form presented, may have to be removed from the pipeline to effect internal maintenance. However, this is not considered a major problem and no other significant difficulties are envisaged.

It is proposed that the attenuator should be considered as an alternative method of surge control for particular applications where its design offers definate advantages over the more conventional methods. The commercial viability of such a device has not been examined in detail, yet, if one considers the major industries where surge protection is a troublesome problem, and often a very expensive item, then the attenuator must have an optimistic future.

A.R.D. Thorley, The City University, London

This is an interesting development, partly because a similar type of device was conceived by Dr. Johal a former research assistant working at City University. After a series of numerical studies and some experimental work (unpublished) which yielded results similar to those presented here a provisional patent was taken out by Dr. Johal. Nevertheless would it not be true to say that a surge attenuator having the form described is really nothing more than a somewhat refined air vessel under another name?

Reply:

The authors are obliged to Dr. Thorley for raising this point at the conference.

For a comparison between the attenuator and an air-vessel the reader is again referred to the question posed by Prof. Vardy.

K.S. Johal, Pipe Line Technologists Ltd.

The authors are to be commended for developing this Pressure Surge attenuator for use on relatively short pipelines. However, the authors should be aware that a patent has been taken out on a similar device by myself following the work undertaken at the City University during the period 1977 - 1979. An experimental demonstration of this method of pressure surge suppression was presented at the short course on "Pressure Surge" run at the City University in April 1979 to provide the participants on this course with an alternative method of pressure surge suppression. This demonstration showed that if the ratio of the volume of gas contained in the Surge Suppression device to the total volume contained in the pipeline to be protected was approximately 1% at NTP, then the pressure amplitude due to an instantaneous downstream valve closure was reduced by 60%. Computer simulations of this surge protection device via the method of characteristics taking into account the effects of reflections and transmissions at the boundaries of this device (due to sudden changes in wave velocity) showed that this was an effective surge suppression device which could be utilised even on relatively long pipes when a number of these devices were placed in parallel to each other. The computer output of this simulation model also showed that effects of the disturbance in pressure and flow velocity, due to the attenuator boundaries, propagated towards the upstream pipeline boundary causing additional agitation in the main flow along the complete length of the pipe, causing additional pressure loss due to this agitation. Did the authors come across this phenomena? Further effect of the flow agitation due to pressure wave reflections and transmissions at the boundaries of this device was that the flow velocity oscillations were generated before the return of the main reflected wave from the upstream pipeline boundary. It was felt that this effect could be used to reduce knocking of reflux valves, situated on the downstream side of pump, by installing the attenuator on the downstream side of the reflux valve, serving to protect the pump from power failure. The results of these studies can be obtained either from myself or the City University.

Reply:

Firstly, it must be stated that this device was not developed specifically for use on 'short' pipelines as K.S. Johal implies in his discussion. Nonetheless, it is the authors' opinion that its most attractive applications will lie in this area. This comment was made purely on practical grounds. Indeed, the larger the attenuator, or the more attenuators there are installed in parallel or in series, the longer the pipeline which can be protected, but there will be a certain point at which this type of equipment becomes too large and expensive in comparison to other methods of surge control.

The authors would like to thank Dr. Johal for his comments which back-up the results presented in this paper. Dr. Johal's work appears to have been along very similar lines to that presented, and it is unfortunate that the research carried out at Leeds University from 1975, and presented as a Ph.D thesis in 1978, was not published earlier to avoid duplication of effort.

The question about the reflection of pressure waves from the boundaries of the attenuator has been covered in the paper. This phenomenon was observed in the results of the 'transient' analysis and is seen as pressure fluctuations on the theoretical traces in Figs. 8 to 10. Reference to this is made in 'Results of the Experimental Tests and Theoretical Analyses', where this phenomenon is cited to illustrate the fundamental differences between the two methods of analysis. However, these pressure fluctuations were not detected in the experimental tests because of the strong damping effect of the attenuator.

T.G.F. Welshmen, Thames Water Authority, U.K.

This type of paper is of especial interest to experienced engineers involved in a variety of pressure surge problems. New ideas like this one, which has been modelled and then tested under experimental conditions, give me real hope that there may be tangible end products resulting from the extensive theoretical and practical research being carried out, world-wide, into pressure surge problems.

The publication and presentation of this paper at the International Conference was most welcomed by me and, I feel sure, by many of the practical engineering delegates who were present.

Dr. Grundy gave a clear explanation and description of this attenuation device. He also defined his considered limitations on its application- i.e. the protection of "short" pipelines.

It is to be hoped that B.H.R.A. Fluid Engineering will maintain a close interest in the development of this device, particularly the results of field monitoring of any such installations on real systems. I am sure that many engineers, including me, would like to be appraised of any such "field" results.

PAPER E6.

Wave propagation in plastically deforming ducts.

J.W.R.Twyman and A.R.D.Thorley
The City University, London

R.Hawavitarne
British Gas Corporation

Pells, Ninham Shand Inc.

I wish to ask Dr.Thorley how the hexagonal pipe was manufactured?

Was it welded out of flat plate with a weld on each corner or was it rolled with a single weld.

D.H.Wilkinson, C.E.G.B., Marchwood Engineering Lab.

An expression for the wave speed in the liquid is given in equation (41), which is also given as equation (5) in paper E1 presented by Dr.Kot. This expression has the term $(\frac{d\alpha}{d\varepsilon} - 2\sigma)$ in the denominator. My comment is that this term could be zero or negative for some stress-strain curves. If it becomes zero then the predicted wave speed is zero and there is no propagation of the pressure pulse, an unlikely result physically and not supported by my own analysis. The stress-strain curve shown in my paper E4, in Fig.1, a bi-linear type, has σ_y = 34,000 psi, σ_n = 84,000 psi and ε_n = 0.25. This therefore gives $d\sigma/d\varepsilon$ = 200,000 psi in the plastic region, so at σ = 100,000 psi, above ultimate stress, the term $(d\sigma/d\varepsilon - 2\sigma)$ would become zero and the predicted c_ℓ from equation (41) would be zero.

It might be argued that the bi-linear stress-strain curve is unrealistic and different curve between the yield and ultimate points such as one that was convex upwards would shift the stress for zero c_ℓ from equation (41) downwards below the ultimate stress point. In particular Table 1 of paper E6 gives experimental material properties with σ and $d\alpha/d\varepsilon$ values as a function of ε. For the last two points, ε = 20 and 25% the quantity $(d\sigma/d\varepsilon - 2\alpha)$ is not entered but is negative, going through zero between ε = 15 and 20%. The equation (41)may therefore be incorrect and it is necessary to reconsider its derivation in the light of the analysis given in paper E4 to see which of the assumptions leading to equation (41) are not supported by the more exact analysis.

Dr.K.J.Enever, Cranfield Institute of Technology

I was interested in Mr.Wilkinson's contribution to the discussion of this paper and in particular in his doubt over the validity of equation (41) because of the fact that imaginary wave velocities can apparently be obtained. Unfortunately, I do not have the reference to hand in which this formula was proved so I have re-derived it for myself.

In the case of plastic deformation, Poisson's ratio (μ) is $\frac{1}{2}$ and hence we get the longitudinal stress equal to half the hoop stress whether we regard the pipe as being fixed at one end or fixed throughout its length (because $\mu/2 = \mu^2$). As one or other of these cases is likely to apply we can safely work on this basis.

Now σ_θ (using subscript θ for circumferential) is given by pR/e. Normally changes in R and e are regarded as small, but for large deformations:

$$\frac{d\sigma_\theta}{dR} = \frac{p}{e} + \frac{R}{e} \cdot \frac{dp}{dR} - \frac{pR}{e^2} \frac{de}{dR}$$

Further calculation gives:

$$\frac{1}{A} \cdot \frac{dA}{dp} = \frac{2}{R} \cdot \frac{dR}{dp} = \frac{2R}{e\left\{\frac{\partial\sigma_\theta}{\partial\varepsilon} - K_1\sigma_\theta\right\}}$$

where ε_θ is dR/R and K_1 is a constant which is 2 for plastic deformation, but depends upon the degree of restraint for elastic deformation.

For elastic deformation $\frac{\partial\sigma_\theta}{\partial\varepsilon} = E/K_2$ where K_2 again depends on pipe restraint. Also E $\ggg \sigma_\theta$ so the term becomes:

$$\frac{2RK_2}{eE}$$

For plastic deformation however it becomes:

$$\frac{2R}{e\left\{\frac{\partial\sigma_\theta}{\partial\varepsilon_\theta} - 2\sigma_\theta\right\}}$$

Thus the equation is indeed correct if σ and ε are circumferential values. The physical significance of this is that when $\frac{\partial\sigma_\theta}{\partial\varepsilon_\theta} = 2\sigma_\theta$ the graph of pressure against radial extension has become horizontal and any increase in pressure would cause an indefinite increase in radius. Thus no wave could propagate forward and a wave speed of zero seems to be physically correct. A value of $\sigma_\theta > \frac{\partial\sigma_\theta}{\partial\varepsilon_\theta}/2$ should not be possible as this involves stressing the pipe beyond this limit.

However, the data for stress/strain shown in Table 1 does go beyond this limit which suggests that it is possibly not a true plot of σ against ε_θ.

A bar under a tensile uni-axial stress will extend in theory up to the point at which $\frac{\partial\sigma}{\partial\varepsilon} = \sigma$ before failure and I wonder if the data come from such a test, as this is about the value of δ at which the tabulated results cease. Can the authors say whether this is true? If so would this not lead to difficulties in using this data? I appreciate that the authors were in fact dealing with an hexagonal duct and of course this might lead to differences which I have not appreciated.

<u>D.H. Wilkinson and E.M.Curtis</u>, C.E.G.B., Marchwood Engineering Lab.

We would like to comment on the expression for the area strain.

$$\frac{1}{A} \cdot \frac{dA}{dp} = \frac{2R}{e\left\{\frac{\partial\sigma_\theta}{\partial\varepsilon_\theta} - 2\sigma_\theta\right\}} \qquad (1)$$

used in the general expression for the speed of propagation of pressure waves

$$C = \left\{\rho\left(\frac{1}{K} + \frac{1}{A} \cdot \frac{dA}{dp}\right)\right\}^{-\frac{1}{2}} \qquad (2)$$

which gives equation (41) in the paper.

It seems to us that while initially plausible this expression is actually wrong both numerically and in concept and we shall try to give a generally valid expression to replace it.

Considering numbers we compare with our results in Figs. 2 and 3 of our paper E4. The bilinear stress strain curve is made up of an elastic line with slope $E = 1.756_{10}11\ N/m^2$ reaching a yield point at $\sigma_y = 2.44_{10}8\ N/m^2$ with $\varepsilon_y = 1.39_{10}{-3}$ and an in-elastic line joining this yield point to an ultimate point with $\sigma_u = 5.946_{10}8\ N/m^2$ and $\varepsilon_u = 0.25$. The slopes $\partial\sigma_\theta/\partial\varepsilon_\theta$, if interpreted as $\partial\bar{\sigma}/\partial\bar{\varepsilon}$ are $1.756_{10}11\ N/m^2$ elastic

and $1.41_{10}9\ N/m^2$ plastic

We also had $R/e = 45.24$, $K = 2.146_{10}9\ N/m^2$, $\rho = 10^3 kg/m^3$

The expressions (1) and (2) give the following values, using the plastic stress-strain slope and stress at the yield point, i.e. just after yield.

$$\frac{1}{A}\frac{dA}{dp} = 9.811_{10}{-8}\ \frac{m^2}{N}$$

$$C = 100.7\ m/s$$

For the value at ultimate stress we take $\ell n\ R/R_o = 0.25$, $R/R_o = 1.284$ $\ell n\ e/e_o = -0.25$, $e/e_o = 0.779$.

Therefore

$$\frac{1}{A}\frac{dA}{dp} = 6.756_{10}{-7}\ \frac{m^2}{N}$$

$$C = 38.5\ m/s$$

The 'true' values, from our Fig. 2 are 1009 m/s at yield and 126 m/s at ultimate. These are so different from the values from equations (1) and (2) as to suggest something is fundamentally wrong. In particular the expressions give a discontinuity in C at the yield point due to the change in slope of the stress-strain curve while our solution gives a continuous variation through the point. This indicates that the slope used is the problem and brings us to the fundamental assumption leading to your result. This is that expressions like $\sigma_\theta = pR/e$ can be differentiated to give the $\partial R/\partial p$ value required. This is only true for small perturbations about the plastically strained condition. What you have derived is the speed of Mach waves of zero strength at the strained condition since your relation between changes in area and pressure are only appropriate for small perturbations. The real water hammer pressure pulse is a large perturbation taking the area right through the elastic region and into the plastic region. The slope that matters then is $\bar{\sigma}/\bar{\varepsilon}$, that of the line joining a point on the stress-strain curve to the origin. The slope of this line is continuous through yield and we give below a derivation of an alternative equation for strong shock waves.

It is necessary, as in your expression, to specify a point on the stress-strain curve $(\bar{\sigma},\bar{\varepsilon})$ and work from there to the wave speed.

In general

$$\sigma_\theta = p\frac{R}{e}$$

and, from the end load condition for an unrestrained pipe,

$$\sigma_z = \frac{\sigma_\theta}{2}$$

For elastic or plastic regions, neglecting σ_r as small,

$$\varepsilon_\theta = \ell n\,\frac{R}{R_o} = \frac{\bar{\varepsilon}}{\bar{\sigma}}\,\{\sigma_\theta - \nu\sigma_z\}$$

where

$$\bar{\sigma} = \frac{1}{\sqrt{2}}\,\{(\sigma_\theta - \sigma_z)^2 + \sigma_\theta^2 + \sigma_z^2\}^{\frac{1}{2}}$$

From these

$$\bar{\sigma} = \frac{\sqrt{3}}{2}\,\sigma_\theta$$

$$\ell n\,\frac{R}{R_o} = \frac{\bar{\varepsilon}}{\bar{\sigma}}\,(1 - \frac{\nu}{2})\sigma_\theta$$

and, eliminating $\sigma_\theta/\bar{\sigma}$,

$$\ell n\,\frac{R}{R_o} = \frac{2}{\sqrt{3}}\,(1 - \frac{\nu}{2})\bar{\varepsilon}$$

or

$$\frac{R}{R_o} = e^{\frac{2}{\sqrt{3}}(1 - \frac{\nu}{2})\bar{\varepsilon}} \qquad (3)$$

To find p we note that R and e seem to vary inversely from our solution so put

$$\frac{e}{e_o} \simeq \frac{R_o}{R}$$

Then

$$\frac{R}{e} = \frac{R}{R_o}\cdot\frac{e_o}{e}\cdot\frac{R_o}{e_o} \simeq \left(\frac{R}{R_o}\right)^2\cdot\frac{R_o}{e_o}$$

Therefore $$p = \sigma_\theta \frac{e}{R} = \frac{2}{\sqrt{3}} \bar{\sigma} \cdot \frac{e_o}{R_o} \cdot \left(\frac{R_o}{R}\right)^2$$

The quantity required in equation (2) is not actually 1/A dA/dp but 1/A ΔA/p and this is given by

$$\frac{1}{p}\frac{\Delta A}{A} = \frac{\sqrt{3}}{2}\frac{R_o}{e_o}\left(\frac{R}{R_o}\right)^2 \frac{1}{\bar{\sigma}} \left\{\left(\frac{R}{R_o}\right)^2 - 1\right\} \tag{4}$$

Combining equations (2), (3) and (4) into one expression gives finally

$$C = \left\{\rho\left(\frac{1}{K} + \frac{\sqrt{3}}{2}\frac{R_o}{e_o}\frac{1}{\bar{\sigma}} e^{\frac{4}{\sqrt{3}}(1-\frac{\nu}{2})\bar{\varepsilon}} \left(e^{\frac{4}{\sqrt{3}}(1-\frac{\nu}{2})\bar{\varepsilon}} - 1\right)\right)\right\}^{\frac{1}{2}} \tag{5}$$

Note that for small deflections $e^x \simeq 1 + x$, $x<<1$, and

$$C \simeq \left\{\rho\left(\frac{1}{K} + 2\frac{R_o}{e_o}\frac{\bar{\varepsilon}}{\bar{\sigma}}\left(1 - \frac{\nu}{2}\right)\right)\right\}^{\frac{1}{2}} \tag{6}$$

which is the standard elastic pipe expression with $\bar{\sigma}/\bar{\varepsilon} = E$

For the 'yield' case equation (5) gives C = 1048 m/s and for the 'ultimate' case 134 m/s. These are 3.8% and 6.5% above the 'true' values from out paper E4. This small remaining difference is probably due to neglect of weak waves which we showed in paper E2 to give rise to 4 to 8% errors and to neglect of the external pressure p_a which has a slight effect in a non-linear case. Equation (5) is therefore a complete replacement for yours, Thorley and Kot's expression from equations (1) and (2).

$$C = \left\{\rho\left(\frac{1}{K} + \frac{R}{e} \cdot \frac{1}{\left(\frac{\partial\bar{\sigma}}{\partial\bar{\varepsilon}} - 2\bar{\sigma}\right)}\right)\right\}^{\frac{1}{2}} \tag{7}$$

which represents only the speed of Mach waves at the strained condition. While equation (7) can apparently give a zero or imaginary Mach wave speed for some materials when $2\bar{\sigma} \geq \partial\bar{\sigma}/\partial\bar{\varepsilon}$ this is not so for the main pressure wave speed. Equation (5) always gives a positive, non-zero, finite wave speed for any point on any conceivable stress-strain curve. As is usual shock waves travel a good deal faster than Mach Waves. Although Equation (5) replaces Equation (7) both give C as a function of $\bar{\sigma}$ which does not seem all that helpful. The connection between $\bar{\sigma}$ or p and the impact velocity V_o is still not known and solution of a set of non-linear simultaneous equations still seems to be required.

Author's Closure

In answer to Mr.Pels, each length of the hexagonal pipe was manufactured from two pieces of flat plate. Each piece was deformed into a half-section and the two sections then butt-welded along axial seams which were mid-way across two opposing flats.

Mr. Wilkinson and Dr.Enever both refer to the wave speed equation, No.(41), as also does the additional written contribution by Messrs. Wilkinson and Curtis. The source to which Dr.Enever refers for a derivation of equation (41) for a circular section pipe is probably reference (2) in the paper as this is more readily available than ref.(4) which is much more detailed. Ref.(1) is a complementary report which describes some parallel experimental studies. However, Dr.Enever's re-derivation is substantially similar to that in refs. (2) and (4).

As he has correctly deduced the graph of true stress versus true strain does become horizontal near the point of ultimate failure for the specimen material for which our numerical results of wavespeed apply. The experimental stress-strain data were supplied by the material manufacturers in the form of a load-extension curve for a tensile test on a bar. The data were converted to the true stress-true strain curve of Fig.A1.

Turning to the physical nature of the problem, it is quite clear from published experimental data, e.g. Ref.1, that following a rapid increase in pressure at a location in a pipe, such that the pipe is stressed beyond the yield point, a pressure wave propagates ina manner similar to that shown schematically in Fig.1 in the paper. The initial portion of the wave propagates at a speed associated with the elastic modulus of the pipe wall material, but plastic deformation at the higher pressure levels considerably reduces the speed with which these higher pressure levels are propagated. This leads to a considerable dispersion of the pressure wave and if, as in the case shown, an unloading wave is also introduced, the net effect is an attenuation of the pressure pulse, since the whole of the pressure decrease is propagated at the 'elastic' wave speed. It is necessary therefore to distinguish between a pressure rise and a pressure drop when in the 'plastic' region of the material behaviour. This also reinforces the argument that if one thinks in terms of a large pressure change being made up of a group of small but discrete steps it is the slope of the appropriate point on the true stress-true strain curve which determines the speed with which the various pressure levels are propagated.

If the slope of a line from the origin to a point (σ ,ε) in the plastic region of the stress-strain curve is used it is not clear to us to what point on the wave front this would apply.

We do not believe that it is unreasonable or illogical for there to be a rapid change in the wave propagation speed for the various pressure levels as one passes through the yield condition. The discontinuity referred to by Wilkinson and Curtis is due to the idealised form of their stress-strain curve of Fig.1 in Paper E4. The fact that an equation gives a 'continuous solution' through this point is not of itself a validation. It may, however, be noted that equation (41) of the present paper does give such a solution (albeit rapidly varying) through this point for a real material as was shown in Fig.9 in the Paper.

Although we stand by the basis upon which equation (41) is developed it may be worth recalling some of the assumptions upon which it, as well as the Wilkinson-Curtis equation is developed. In particular, it is assumed that bending and shear effects are small, that the inertia of the duct wall is negligible and that the flow may be treated as one-dimensional. All of these assumptions become dubious at the higher strain levels, e.g. 20-25%, and detract from any theories based upon them when considering unsteady behaviour under these conditions.

The quantity $(1/A)(\delta A/\delta p)$ in equation (1) of the paper has its origin in the equation describing conservation of mass. In particular it relates the mass flow-rate outwards from a control volume to the radial deformation of the pipe wall. There is a change δA of area caused by a change δp in the net pressure across the pipe wall.

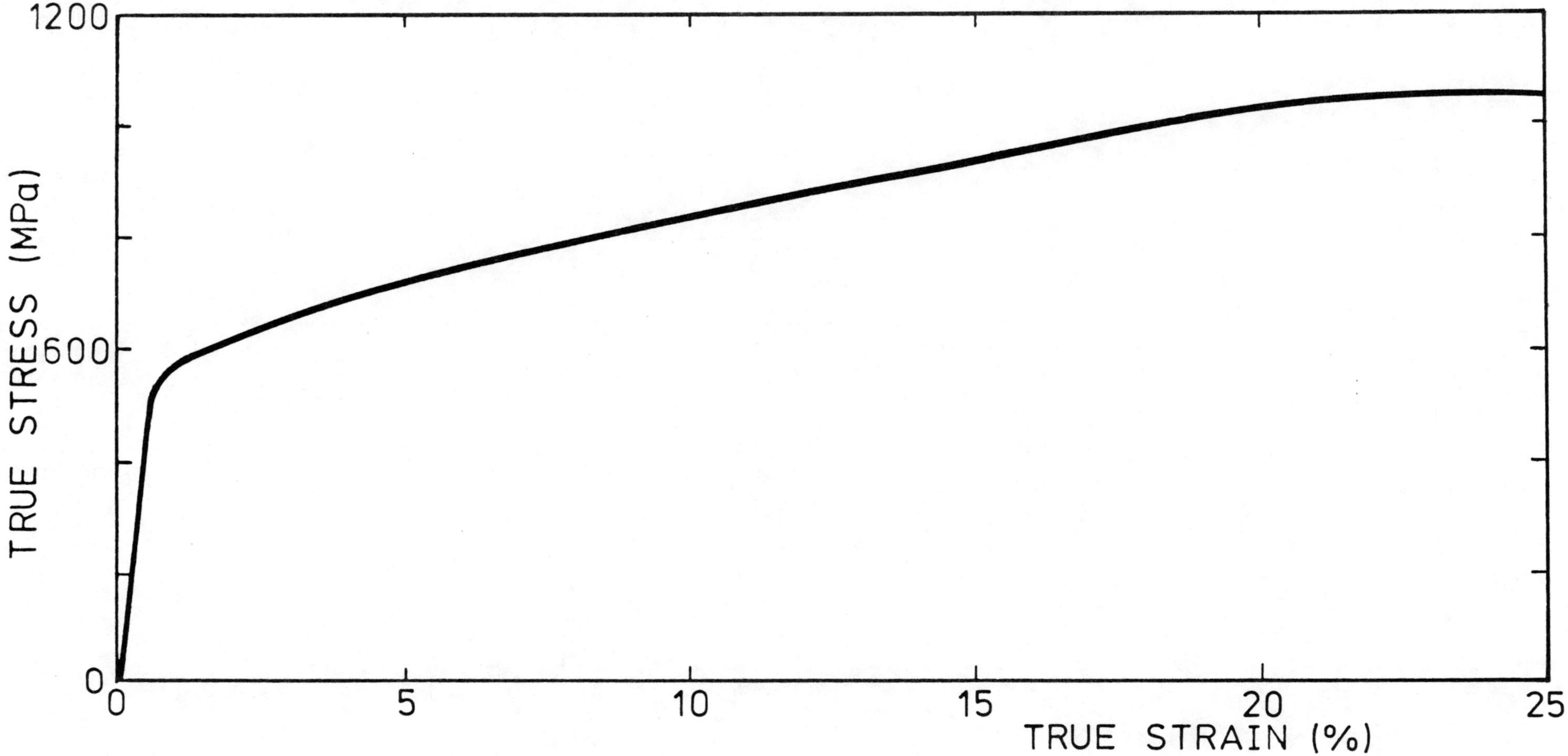

Fig. A1: Graph of true-stress against true-strain for the pipe wall material.

Discussion & Contributions

SESSION F

COOLING WATER SYSTEMS

Chairman: Dr. D.J.F.Ewing
CEGB, U.K.

Papers:

F1	Recent developments of an existing transient-simulation programme for cooling water circuits.	G.Fenini and M.Fanelli ENEL-DSR-CRIS, Italy.
F2	Level surges in the circulating water downshaft of large generating stations.	B.T.Goldring and W.T.Mawer, C.E.G.B., U.K. N.Thomas, University of Cambridge, U.K.

PAPER F1

Recent developments of an existing transient-simulation programme for cooling water circuits.

G.Fenini and M.Fanelli
ENEL-DSR-CRIS, Italy

Dr.A.Boldy, University of Warwick

In connection with Fig.2, the representation of the pump characteristics; have you considered using the generally accepted 'Suter' method which gives a continuous curve?

Author's Closure

"Suter" [1] and other similar methods [2] refer to continuous curves based on dimensionless - homologous equations like :

$$\frac{h}{a^2} = \text{const.} \quad \text{and} \quad \frac{v}{a} = \text{const.} \quad \text{or} \quad \omega = \frac{h}{a^2 + v^2} \quad \text{and}$$

$\theta = \tan^{-1}\frac{v}{a}$, where the ratio of H (piezometric heads), Q (flow discharge) and N (pump revolutions) to their rated values H_o , Q_o and N_o (nominals) are used to yeld a dimensionless presentation like : $h = \frac{H}{H_o}$, $v = \frac{Q}{Q_o}$ and $a = \frac{N}{N_o}$.

These equations, for small speed ratio a, are not satisfactory and then substituted by $\frac{h}{v^2}$ = const.and $\frac{a}{v}$ = const. or similar, since a is proportional to v for homologous situations.

In any case both set of equations are used, in computer programs, by storing a few points and making then parabolic interpolations between groups of three consecutive stored points.

Our method is quite similar and follows a representation of pump curves which is generally more used by pump manufacturers [3] . Also in this case we have continuous curves discretized by points for an easy use with computer programs.

But our representation allow to use an interpolation criteria (based on radial interceptors) which avoids singularities both on piezometric head and torque curves.

[1] Marchal M., G.Flesch and P. Suter : "The calculation of waterhammer problems by means of the digital computer", Proc. Int. Symp. Waterhammer pumped storage projects, ASME, Chicago, November 1965.

[2] Victor L. Streeter, E.Benjamin Wylie : "Hydraulic transients" Mc Graw Hill Book Company.

[3] M.Fanelli : " Il regime transitorio nei circuiti di raffreddamento dei condensatori per le grandi unità termoelettriche. Studi di carattere analitico"
L'Energia Elettrica n° 12 , Dic. 1971

PAPER F2

Level surges in the circulating water downshaft of large generating stations.

B.T.Goldring and W.T.Mawer
CEGB, U.K.

N.Thomas
University of Cambridge, U.K.

M.H.Chandhry, Old Dominion University, U.S.A.

To the writer's knowledge, no satisfactory similitude relationships have been reported in the literature to model air entrainment in physical hydraulic models. What simulitude relationships were used by the Authors during operation of their models and are they satisfied by them?

The Authors had two models built at different scales. Were any attempts made to determine the "scale effects" by comparing results of these two models?

Author's Closure

As stated on p. 281 of the paper, the models were operated with Froude scale flows, since the aim of the hydraulic modelling was to reproduce the surface flow patterns giving rise to air entrainment, not to measure the rates of air entrainment; no model simulation of these was intended. The authors are indeed confident that Froude scaling reproduces the essential features of the free-surface flows.

Again as stated on p. 281, the smaller model was constructed (quickly) to evaluate the bulk flow patterns and was modified in the light of observations of the flow stability and air-entrainment potential of the free-surface flows. The larger model, whose design incorporated the modifications to the smaller model, was constructed to verify the earlier findings, and was of sufficient scale to provide guidelines on the relative (not absolute) air transport capacities of the various modes of internal flow.

Any attempts, based on the hydraulic model data alone, to determine the 'scale effects' on air entrainment would have been naive, in view of its unknown parametric structure; on this point the questioner is quite correct - there are no valid similitude relationships in the literature. However, the fundamental studies at Cambridge have not only identified the essential characteristics of the mechanisms by which air is entrained and transported (see Section 6.3), but have also provided a theoretical framework for the observations. For example, the theoretical curves in Fig. 13, which compare most favourably with experimental siphon data, were calculated from equation 6.2.2(c), that is,

$$\text{Net air entrainment } \Phi(B) = \frac{(1 - B)^2}{(1 - \frac{1}{3}B)} \, 0.025(F_r - 1)$$

where B = effective bubble escape speed divided by r.m.s. velocity fluctuation in submerged shear layer.

Note that this is equivalent to:

B = effective entrainment inception speed divided by mean velocity at impingement

since the ratio mean to r.m.s. turbulence is a universal coefficient (about 20) in the submerged jet shear layer region (Bradshaw, P. (ed), 1978, 'Turbulence', Springer-Verlag). Thus, observed entrainment inception speeds of 1 to 1.5 m/s (e.g. Ervine, 1976, loc. cit.) correspond to effective bubble escape speeds of about 5 to 8 cm/s, i.e. a reduction by a factor 3 to 4 from the still water values; this is attributed to the mechanism described in Section 6.3.

This theoretical scaling has since been used at C.E.R.L. to correlate siphon priming curves. The shape of the curves was accurately reproduced, but the entrainment 'constant', 0.025 in the above equation, was found to be very sensitive to flow conditions. All the correlations available in the literature consider net air entrainment as an entity, and not as a combination of entrainment and detrainment. We are satisfied that the latter approach is more fundamental, but further work is needed to evaluate the main factors influencing the entrainment constant.

There was no attempt to determine scaling relationships from the downshaft model tests due to the lack of data on the quantities of air entrained by the plunging weir flows. However, the scale effects for the siphon seal weirs, located upstream of the downshaft, have been investigated and the results published in Ref. 12.

12. Goldring, B.T., 'The Use of Small-scale Siphon Models', Proc. Inst. Civil Engrs., Part 2, December 1979.

Discussion & Contributions

SESSION G

HYDROELECTRIC SYSTEMS

Chairman: G.A.J.Young
formerly BHRA Fluid Engineering, U.K.

Papers:

G1	A nonlinear mathematical model for analysis of transients caused by a governed Francis turbine.	H.Chaudry, Old Dominion University, U.S.A.
G2	Chibro Khodri tandem control.	J.L.Khosa, Bharat Heavy Electricals Ltd., India.
G3	Mass oscillation prediction - A comparative study of mass surge and water-hammer methods.	J.A.Forrest and J.F.Robbie, University of Aberdeen, U.K.
G4	Effect of equation formulation on the prediction of mass oscillations in closed conduits and surge tanks.	A.Anderson, University of Newcastle-Upon-Tyne, U.K. J.F.Robbie, University of Aberdeen, U.K

PAPER G1

A non-linear mathematical model for analysis of transients caused by a governed Francis turbine.

H.Chaudry,
Old Dominion University, U.S.A.

K.L. Ariyananda, Kennedy & Donkin, U.K.

The assumption that effective gate closing/opening time is twice the time taken to close/open the wicket gates between 25 per cent and 75 per cent is not valid in general, (e.g. two stage gate closures).

Also, for gate closing times in excess of 6s the total gate closing time is taken as effective gate closing time + 1.6s.

N.Dahlback, Swedish State Power Board.

What is the influence n the results when introducing the turbine characteristics compared to a calculation with the turbine simulated just with a valve?

Especially how much does the maximum (minimum) pressure change in the calculation? This can be shown in a diagram where the flow related to the gale opening is compared with your resulting flow. Note that the pressure is essentially proportional to dQ/dt.

Dr.A.Boldy, University of Warwick, U.K.

In connection with the turbine boundary condition, I suggest it is better to extrapolate previous calculated values of machine discharge instead of the machine head. The discharge will follow a smooth curve whereas the head is unpredictable.

The extrapolated discharge may then be substituted into the c+ and c- (or fixed head) characteristic equations for the end of the pipes connected to the turbine producing a good approximation for the net head across the turbine.

The difference in extrapolated technique may not be important for a Francis turbine which has a 'flat'characteristic curve but would be very significant for a reversible pump turbine which has a very variable shaped characteristic.

A.Anderson, University of Newcastle Upon Tyne, U.K.

Prof.Chaudry is to be congratulated on his paper which should be of considerable value to the practising engineer and this comment is intended as a contribution to, rather than a question on, the material which he presented.

Referring to Fig.9 and the unusual "spike" in the measured pressure rise at approx. 7 sec.(possibly followed by lesser spikes at approx. 9 and 11 sec.), though in the Paper itself Prof.Chaudry stated that "the cause of the spike in unknown" he was able during his presentation to show pressure traces indicating that its origin was probably due to a sharp pressure peak in the draft tube following vapour cavity formation below the runner .

I would entirely agree with this explanation. It may be a relatively common phenomenon - other examples are shown in Ref.A and in Fig.A - and as shown in these and in Fig.9 it is of considerable interest because it causes the maximum pressure rise in both penstock and draft tube, though as shown in both cases it not unexpectedly appears to have little effect on the turbine speed rise.

There are, however, some difficulties in modelling this phenomenon compared with column separation effects in other pipelines. Fig.A indicates that cavity formation occurs at an apparent vapour pressure head of only -4 to -6 m (WC gauge) instead of the usual value of -8 to -10 m. This is quite simply because, though waterhammer theory is one-dimensional, assuming pressures uniform right across the conduit cross-section, beneath a turbine runner the flow will always be swirling, with a variable pressure distribution across the draft-tube and generally with lower pressures in the core of the flow than at the walls. Consequently vapour cavities may form in the draft tube long before a waterhammer calculation would normally predict they should.

In practice this effect can be rather crudely allowed for in calculations by artificially increasing the vapour pressure head to around -4 to -6 m depending on the situation and in this connection it would be most valuable if Prof.Chaudhry could indicate what sort of values were obtained in the tests at Mica.

This problem in one-dimensional modelling may also occur in connection with dissolved air release in power stations with long tail race pressure tunnels, especially pumped-storage schemes which may be loaded from a spinning-in-air mode.

Additional reference:

A. Sono,M., Fukumasu, K., Watanabe, T. and Mutaguchi, K.:
"Design and manufacture of Francis type reversible pump turbines", Mitsubishi Technical Review, February 1977, pp.285-293, especially Fig.4, and
Yamaguchi, Y. and Takoh, T.:
"An experimental research on the transient phenomena of a pumped-storage power station", American Society of Mechanical Engineers, Paper 76-FE-17, 1976, especially Fig.6 and 7.

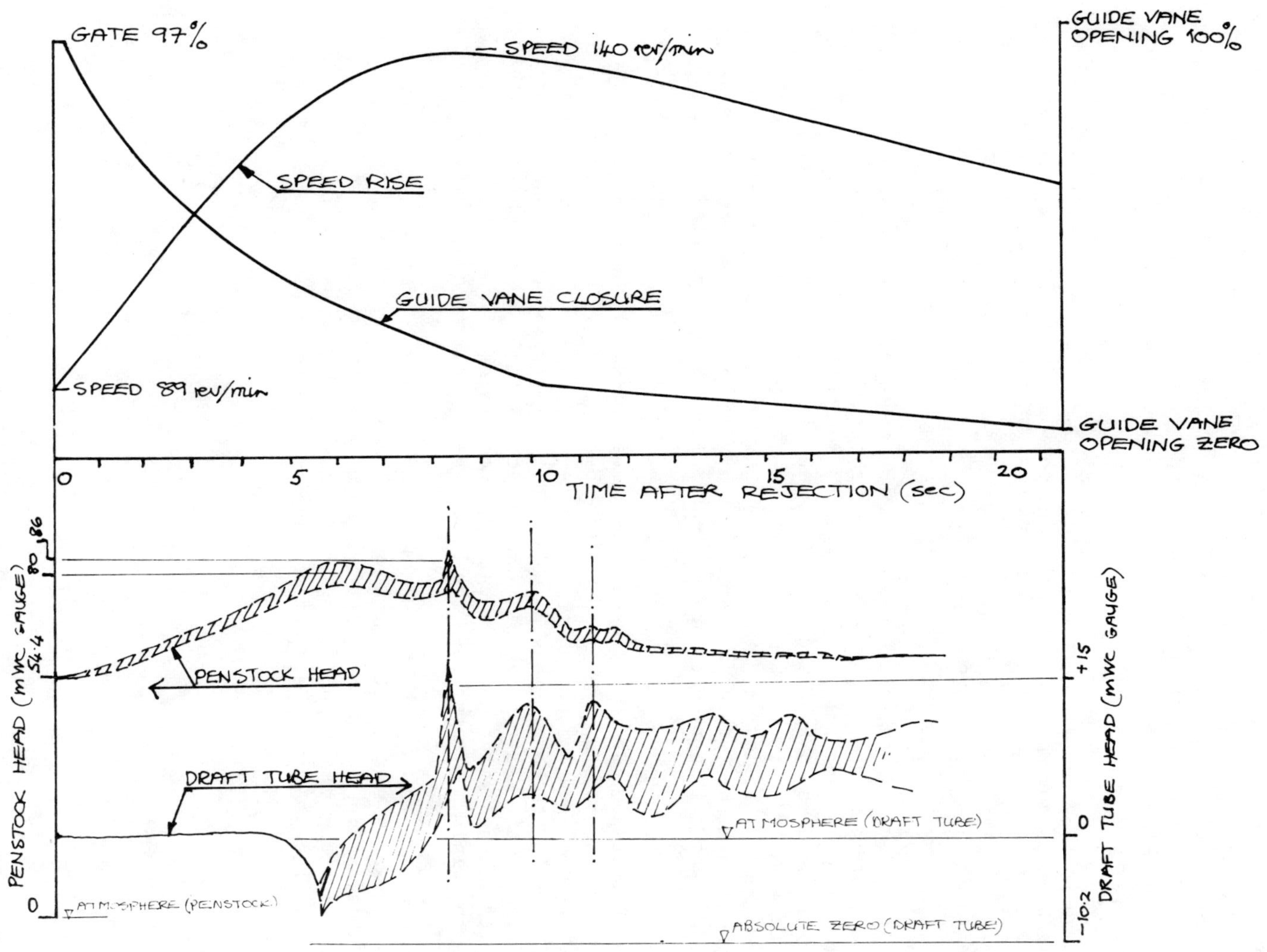

FIG. A. TRACING FROM UV RECORDER FOR 220MW LOAD REJECTION SHOWING CAVITY COLLAPSE IN DRAFT TUBE WITH CORRESPONDING PRESSURE PEAKS IN PENSTOCK

Author's Closure

The author is thankful to Ariyananda, Boldy, Dehlback, and Anderson for their useful comments.

It is correct that the definition of effective gate closing/opening time is not valid for cases having more than one rate of closure or opening. The total closing time is equal to the effective closing time plus the time for the cushioning stroke. The time of the cushioning stroke (generally 1 to 5 seconds) varies from project to project and is not always equal to 1.6 seconds as implied by Ariyananda.

The author agrees with Dr. Boldy that it would be better to extrapolate discharge rather than head for the turbine boundary condition because the discharge varies more gradually than the head during transient-state conditions. This is especially true for a reverse pump turbine although it may not be important for a Francis turbine.

The inclusion of turbine characteristics while calculating the maximum pressure following load rejection is very important since the discharge is reduced due to wicket-gate closure and due to choking effects caused by a rise in the turbine speed especially in the case of Francis turbines. The influence of choking effect depends upon the specific speed of the turbine and it is difficult to quantify this influence. In one case we analyzed, the pressure rise in the penstock following total load rejection was about 25 per cent of the gross head when the turbine characteristics were included in the analysis as compared to about 15 per cent when the turbine boundary condition was assumed as a valve.

The prototype test data presented by Anderson should be welcomed by the practicing engineers. Similar results have been reported in the literature by others. As outlined in the paper and discussed during the presentation of the paper, no definite conclusions have been drawn regarding the cause of the pressure spike in the draft tubes of Mica Project; water-column separation may be considered as one of the possible causes. Concerning Mr. Anderson's querry, no data on the minimum pressures in the Mica draft tubes is readily available.

The author agrees with Mr. Anderson's suggestion to use a lower value of minimum pressures to predict column separation in the draft tube of a turbine. This assumption may be justified to account for the actual flow conditions below the runners not being one dimensional and the flow not being over the full cross-sectional area of the draft tube due to presence of an air core.

PAPER G3

Mass oscillation prediction - A comparative study of mass surge and water-hammer methods.

J.A.Forrest and J.F.Robbie,
University of Aberdeen, U.K.

A.E.Vardy, Dundee University

It is disconcerting to find that the method of characteristics solution is less able than the mass oscillation solution to handle the compressed air installation. Since the period and magnitude of the oscillations predicted by the two techniques are different, it is clear that the air pressure varies differently for the two analyses. How certain are the authors that this result is an inevitable consequence of the use of the method of characteristics and that it is not the consequence of a particularly well-hidden bug in the computer program?

C.S.Martin, Georgia Institute of Technology, U.S.A.

I presume everyone expects me to respond to the stand taken by Dr.Robbie. He has made a very eloquent delivery and should be congratulated on an excellent paper. I do not intend to question the record of my statements at the 1972 and 1976 conferences, but at no time have I personally felt that entirely one analysis - waterhammer or surge alone - is adequate for engineering purposes. My intent in preparing the 1972 paper, and in discussing Dr.Robbie's 1976 paper, was to take the position that final analyses, for which both waterhammer pressures and surge levels were desired, would be best served by a technique that yielded both - the method of characteristics.

I would like to ask Dr.Robbie if he has an explanation for the differences in the results of the two analyses, surging and waterhammer, for the case of the presence of an air chamber.

B.B.Queen, Merz and McLellan

Following the last Conference we included a surge chamber representation in our Waterhammer programmes which use the method of characteristics not to study mass surge as such, (which would be very expensive on our programmes as the time interval used for the Dinorwic studies is 0.05 sec.) but to ensure that any effect of the surge chamber on the waterhammer studies and particularly the rapid loading requirement for Dinorwic was properly taken into account. We accordingly carried out two studies which are shown on the attached diagram. The first study represented an infinite free surface at the location of the surge chamber and the second included fully the low pressure tunnel and surge chamber with its throttle losses. It will be observed that in the waterhammer regime the effect of the surge chamber is negligible and only takes small effect towards the end of the loading, slightly reducing the output.

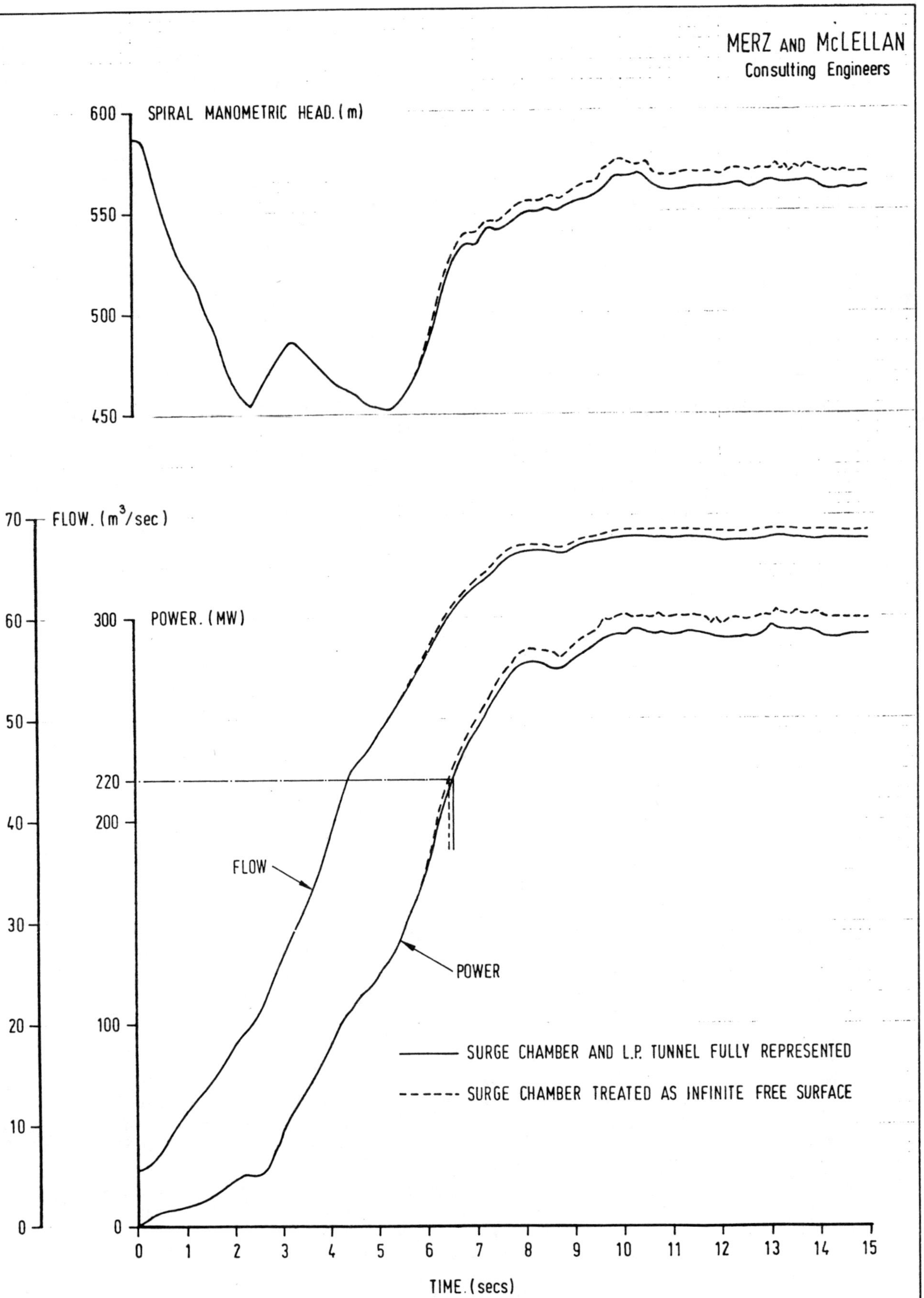

DINORWIC POWER STATION

TURBINE RAPID LOADING FROM NO-LOAD

The surge chamber would go on reducing level for about 70 sec., when it would reach a minimum 18 m approx. below the steady state level, with corresponding temporary reduction in output, but still well over the target figure of 220 MW per machine.

J.A.Fox, Leeds University

With the rapidly reducing costs of computer time with the advent of the micro computer I wonder whether the costs of computing time are really a factor in deciding whether to use compressible or incompressible methods. To analyse a mass oscillation problem using incompressible theory is undoubtedly correct but if a surge problem exists in a network in which there is incidentally a surge tank one would surely have to perform an elastic analysis anyway and the presence of the surge tank would make trivial additions to the cost of the run.

Author's Closure

Reply to Professors Martin and Vardy

Professor Martins's contribution alludes to background material from earlier conferences which formed part of the oral presentation of the paper.

This material is repeated here in order to clarify his discussion contribution and set the paper in context.

Somewhat inadvertently perhaps, Martin (2,3) and Robbie (4,5) through papers and discussion contributions at the 1972 and 1976 Conferences, have represented the differing schools of thought on what is the best numerical approach to the calculation of surge tank oscillations in practice.

In 1972, Martin had "no reservations in stating that, as far as accuracy is concerned, the method of characteristics will be just as good as, if not better than, any method based on mass oscillation theory" and concluded realistically and fairly that "the decision to employ either of the methods must then rest on time, the nature of the particular problem, the availability of computer programs, economics and personal preference".

The 1972 Robbie/Robson (5) and the 1976 Anderson/Robbie (4) papers dealt with aspects of mass surge prediction related to the design of the Dinorwic Pumped Storage Scheme. For this, the practical necessity of avoiding unnecessary waste or misuse of computer time, while maintining adequate accuracy, was of prime importance - and hence the adoption of the incompressible approach for mass surge studies, complemented by waterhammer studies as necessary.

The work of Bullough (6), reported in 1972 and 1976, was directed specifically at assisting the practicing designer select the most appropriate numerical method to be used in relation to mass surge equations.

Professor Martin's re-opening, in 1976, of the issue of preferred method of analysis by enquiring "whether it is still too costly to analyse both phenomena one stage" and seeking "comment on the merits of a simultaneous analysis of surging and waterhammer in pumped storage schemes" provoked our response that "the trend, which had emerged over a number of years, of attempting to advance 'elastic column' methods at the expense of often highly developed incompressible methods could be economically and technically myopic" and we reaffirmed "our long held view that it is much preferable to devise and apply, without prejudice, appropriate tools for appropriate jobs in the analysis of pressure surges".

Clearly Professor Novak's 1972 suggestion, during the discussion of the Robbie/Robson paper (5), "that it would be interesting to compare the results of the method of characteristics with conventional method based on mass oscillation theory" was sound and forms the basis of the comparability study reported in this paper, with the intention of enabling practicing designers to reach their own conclusions on which method is appropriate in a given circumstance, by examining the application of both methods to the prediction of mass surge in a representative selection of hydro power installations.

Now to deal with specific points, the differences between observed values and values predicted by incompressible and elastic methods for enclosed air chambers (see Figs. 13-15), commented upon by Professors Vardy and Martin are essentially an intriguing by-product of the main objectives of the study stated above.

Accordingly the Authors remain diffident in leading the investigation of the phenomenon and indeed had hoped that a cogent explanation would have emerged from authorative participants at the Conference.

We would suggest that the following points may be relevant.

Deficiencies in the rational heat transfer (RHT) equation (37), such as the omission of latent heat terms, are known to exist and would apply equally to both methods. Dr. Graze (8, 9, 10, 11), the originator of the RHT approach, is known to be continuing his refinement of his model and it is hoped that he will remedy the above deficiency soon. It is our guess that the mass surge analysis would then coincide with observed values but not the characteristic analysis.

A second contributory cause, which has greater significance in realtion to the waterhammer approach, is in the <u>parallel solution</u> of the ordinary differential equation (equ. 37) representation of conditions within the air vessel and the partial differential waterhammer equations.

Examination of the predicted responses for Driva (Fig. 12) and more particularly for Villingen (Figs. 13-15) reveals high frequency discontinuities in the variables when modelled elastically.

The effect of these on the parallel numerical solution of the RHT equation (37) could be severe because, as Bullough (6) has shown, disproportionately large errors can result from disconuities occurring during the numerical solution of ODE's unless the solution is constrained and pass exactly through the discontinuity.

Unfortunately little can be done to overcome this effect in this instance other than adopting extremely small time intervals, since frequency differences have been found (11) to exist between the major variables concerned (heat, volume and temperature). Tentative reductions in time step had little effect on the response but did increase the time of computation significantly.

The development of the high frequency discontinuities may of course be caused by the interaction of the numerical solutions.

Some credence may be given to this possibility by the fact that a solution based on polytropic simulation of the thermodynamic effects (PV^n = Const. with n = 1.4) yielded a smooth solution of frequency higher than that of the mass surge. Exclusion of thermodynamic effects by setting $\frac{dHc}{dt} = 0$, also yielded smooth responses and fairly coincident solutions.

This would suggest that Professor Vardy's suggested explanation of a well hidden bug in the basic computer program does not apply.

The Authors understand that site observations of the response of the Driva Air Chamber installation have been taken and are currently being processed. These could be of considerable interest and value in resolving the current dilemma and it is to be hoped that they may be made generally available soon.

Reply to Mr. B. B. Queen

Mr. Queen's contribution is a most welcome demonstration of the relative coarseness of boundary condition representation in the method of characteristics permissible under certain circumstances. In so doing, he lends support to the desirability of using the right tool for the right job.

In the context of mass surge prediction however, the paper equally demonstrates the sensitivity of solution to fine detailed considerations which cannot be overlooked.

Reply to Mr. J.A. Fox

In due course Mr. Fox's view on the relative insignificance of cost of computation may well be realised but it is significant to note that 90% of the Authors presenting papers based on computer studies at this Conference, specifically mentioned the importance of cost of computation at this time, during presentation.

The network problem involving an incidental surge tank envisaged by Mr. Fox seems far removed from the problems facing the designers of the hydraulic system and hydraulic machines with a hydro or pumped storage scheme subjected to various plant operation sequences. The latter undoubtedly are faced with the need to apply both elastic and incompressible methods to specific aspects of the design. It is hoped that the present paper enables them to make a more informed choice in relation to mass oscillation prediction.

In conclusion it seems appropriate that Professor Martin has the last word - "the decision to employ either of the methods must then rest on time, the nature of the particular problem, the availability of computer programs, economies and personal preference - to name a few".

PAPER G4

Effect of equation formulation on the prediction of mass oscillations in closed conduits and surge tanks.

A.Anderson,
University of Newcastle-Upon-Tyne, U.K.

J.F.Robbie,
University of Aberdeen, U.K.

A.E.Vardy, Dundee University

1. The authors helpfully draw attention to the difficulties experienced in determining local loss coefficients. Is it really necessary, however, to worry about local changes in the static head that are not accompanied by changes in the Bernoulli Sum (i.e. "recoverable losses" in the authors' terminology)?

2. It is interesting to note that mass oscillation problems are more sensitive to local losses than are waterhammer problems. In the latter, the rapid pressure changes are much greater than those due to local effects.

J.A.Fox, Leeds University

In the diagram for the general case for flow through junctions (Fig.1) an error exists in that the boundary layer separation on the "straight through" limb of the junction has been omitted. I trust that this omission has not been duplicated in the theoretical development in the paper.

H.Safwat, Bechtel Power Corporation, U.S.A.

This is a comment rather than a question, reviewing the references cited at the end of the paper, I would like to draw the attention to a discussion by E.B.Wylie to an ASME paper by Contractor published in the Journal of Basic Engineering, in 1965. In this discussion Professor Wylie addresses the inclusion of secondary losses (minor losses) in connection with the application of method of characteristics for the analysis of waterhammer problems.

Dr.A.Boldy, University of Warwick, U.K.

In connection with the Tee junction loss coefficients, the research undertaken at BHRA (documented in D.S.Miller's book on Internal Flow) showed that the coefficients were effected significantly by the 'sharpness' of the edge at the branch. Is this taken into account or is it a case of correct data input?

Author's Closure

Reply to Professor Vardy

1. It was argued that junction "losses" are unlike other local losses in that, far from being merely measures of dissipation, they also ensure that the "Bernoulli Sum" is

indeed satisfied by including the "recoverable" effects. The essential point which the paper was seeking to make is that, if the junction loss terms are omitted, the "Bernoulli Sum" is satisfied only for the two limiting cases discussed in Section 2.2 and not throughout the rest of the transient cycle. Also, because junction losses are sensitive to the shape of the junction (as emphasised by Dr. Boldy), the paper sought to show that it is unrealistic to represent a wide range of types of surge shaft/conduit junction configuration by any single formulation which neglects or simplifies junction losses, as has been attempted in the past (Eq. 19). The examples and literature references in Section 4 show that junction losses have already been seen to have a significant influence on predicted surge levels. Attention was drawn, however, to two general exceptions in the last paragraph of Section 4, but it was not emphasised that these are of declining importance for the following reasons:

i) The "reservoir" or "tank" variety of surge chamber in Fig. 4(b) is becoming rare and for economic reasons most surge chambers today are of the throttled type, e.g. the pumped storage schemes of Paper B6 and Ref. 6, 10, 28.

ii) In the past, rapid load rejection was regarded as being of fundamental importance but with modern pumpted storage schemes there is now developing an emphasis on rapid turbine loading and on sequential operations, e.g. frequency regulation duty, consecutive pump starts, hydraulic reversal (pumping to turbining), etc. (see e.g. Ref. 43, 48 and Paper B6), which are much more subject to the influence of junction losses.

Thus though junction losses may reasonably have been regarded as a "secondary effect" in the past, they are now assuming greater practical significance.

2. Professor Vardy draws attention to an important practical aspect. As Mr. B.B. Queen showed in his discussion of Paper G3, elastic wave transient effects are usually completely damped out before the mass oscillation translational wave effects begin to be of significance, so that the small changes occurring in junction loss coefficients will consequently have little influence on the elastic wave response. If, however, a waterhammer solution is used over a longer period to obtain surge levels (as in Papers B6 and G3) these will be sensitive to junction losses in exactly the same way as a mass oscillation solution (see also reply to Dr. Safwat). Junction or combined junction and throttle loss coefficients vary substantially throughout the transient cycle depending on the flow rates at the branches. Thus damping effects may be poorly represented by assuming a single valued "orifice" type local loss. As mentioned in the paper and also by Idel'chik (Ref A) and others, across any junction as a whole there is net dissipation of energy but this is not necessarily evenly divided between the branches and individual branches may have "negative losses" for certain flow configurations.

Reply to Mr. Fox

Equations 3-6 indicate that there are no omissions from the junction losses in the theoretical development, these being well established standard definitions for junction losses as quoted in Ref. 23 or Ref A and B. Mr. Fox, however, has kindly drawn attention to some failings of the sketches in Fig. 1. He is quite correct in stating that illustration of the separation on the "straight through" limb of the junction was inadvertently omitted (generally it would be expected to occur just opposite the lateral branch). Also, where there is flow in the lateral branch, separation would generally be expected on both sides of the lateral branch,as sketched in Fig. 4(c), rather than just on one side as in Fig. 1.

Reply to Dr. Safwat

We would like to thank Dr. Safwat for reminding us about the contributions of Contractor and Wylie and in that connection Ref. C is also interesting. These papers suggest that at a junction there will be wave reflection due to two causes:

a) Changes in flow cross-sectional area
   - formulas for this have been established for many years (e.g. Ref. 14, 31, etc.) and the method of characteristics solution automatically implements this effect.
b) Occurrence of loss mechanisms
   - if the loss coefficient values are omitted from the problem data then wave reflection effects due to these will obviously not appear in a solution.

It is, perhaps, worth emphasising that though the paper was restricted to incompressible flows, the formulations generally used for the surge shaft boundary condition in waterhammer programs have exactly the same limitations as those indicated for traditional mass oscillation formulations (Eq. 19), i.e. the surge shaft is generally represented as "measuring" either the hydraulic grade line or energy grade line in the conduit as though conditions were steady (honourable exceptions which incorporate junction losses include Ref. 29 and Ref D). The comment by Professor Wylie (in his discussion of Contractor's paper) that velocity head effects may not always be negligible at minor losses assumes considerable significance at a junction where, it has been argued in the paper (see also Reply to Professor Vardy), these "recoverable" effects are combined with the dissipative mechanisms in a single junction "loss" term.

Reply to Dr. Boldy

For the two limiting regimes discussed in Section 2.2 and the corresponding loss coefficients quoted in Eq. 8 and 10, a survey of experimental results in the literature, including those by BHRA (Ref. E), shows that these extremes are influenced remarkably little either by branch area ratio or by branch edge sharpness (except obviously for the "elbow" loss J_{ab} in Eq. 10). Thus the inferences drawn from these particular results (which apply only for the two particular junction flow regimes indicated) remain valid for a wide range of junction types. Dr. Boldy, however, stresses an important feature of junction losses, ie that away from these limiting conditions they generally vary substantially with changes in junction geometry, including changes in the "sharpness" of the edge of the branch. Thus, as Dr. Boldy suggests, for any particular junction configuration it is necessary to provide the appropriate junction loss coefficient data as surge response may be influenced by the form of the junction, e.g. Ref.28. Unfortunately, as commented in the Conclusions, little appropriate data for surge shafts is available as yet (Ref. F and G provide some information additional to that referenced in the paper).

Additional references

A. Idel'chik, I.Y. "Pressure losses in flow of a real liquid (or gas) in pressure systems". Fluid Mechanics - Soviet Research, 5,3, pp 44-49. (May-June 1976).

B. Villemonte, J.R. "Some basic concepts on flow in branch conduits". Journal of the Hydraulics Division, Proc.ASCE, 103, HY 7, pp 685-697, (July 1977).

C. Swaffield, J.A. "The influence of bends on fluid transients propagated in incompressible pipe flow". Proceedings, Institution of Mechanical Engineers, 183, Part I, 29, pp 603-614, (1968-69).

D. Perkins, F.E., et al "Hydro-power plant transients. Part 2. Response to load rejection". Report No. 71, Hydrodynamics Laboratory, Dept. of Civil Engineering, Massachusetts Institute of Technology. (Sept. 1964).

E. Miller, D.S. "Internal flow. A guide to losses in pipe and duct systems". Brit. Hydromech. Res. Assoc., Cranfield. (1971).

F. Wisner, P. and Bratosin, D. "On dimensioning throttles for surge chambers. Preliminary notes". (Asupra dimensionarii diafragmelor la castelele de echilibru). La Revista Hidrotehnica, 3, 6, pp 230-233, (1958). (In Roumanian).

G. Wisner, P. and Bratosin, D. "Studies on head losses in surge chamber throttles". (Cercetari asupra pierderii de sarcina in diafragmele castelelor de echilibru). Hidrotehnica, 5, 1/2, pp 9-12, (1960). (In Roumanian).

Discussion & Contributions

SESSION H

CASE STUDIES

Chairman: M.U.Papworth (BHRA Fluid Engineering, U.K.)

Papers:

H1	Fire main failure due to rapid priming of dry lines.	P.Hope & M.U.Papworth BHRA Fluid Engineering, U.K.
H2	Kaplan turbine accidents and reverse water hammer.	S.Pejovic, L.Krsmanovic & A.Gajic, University of Belgrade, Yugoslavia & D.Obradovic, "Energoprojekt", Yugoslavia.
H3	Cavitation and corrosion in a pumping main.	W.L.Glass, Stewart, Sviridov & Oliver, South Africa.

PAPER H1

Fire main failure due to rapid priming of dry lines.

P.Hope & M.U.Papworth
BHRA Fluid Engineering, U.K.

H.Safwat, Bechtel Power Corporation, U.S.A.

Would you please provide some details on how the air/water interface is modelled during filling of the line and how the trapped air was modelled?

A.R.Pearson, National Engineering Laboratory,U.K.

Question is related to System B: Rapid priming against a partial flow obstruction.

During the computer analysis three conditions were considered; pump start-up, pump failure and priming of an empty line. What relationship was used for the pump speed with time until full speed was reached during start-up?

C.Papadakis, Bechtel Inc., Ann Arbor, U.S.A.

I would like to comment that it is common practice to burry fire protection pipes 4 to 5 feet below ground to prevent freezing. In such case the pipes can be kept full of water and if other peripheral systems (such as sprinkler system) with different pressure requirements are not present, no priming is needed.

Author's Closure

Response to Mr. H.H. Safwat

Water entering the canvas suction hose was considered to have the form of a variable length incompressible cylindrical slug, of uniform cross sectional area, entering a rigid duct. The trapped air in the hose was considered to be concentrated at the end of the duct in the form of an air slug having an initial volume consistent with the assumed mass of trapped air at atmospheric pressure. The piston formed by the water slug was advanced along the cylinder representing the hose in time in accordance with the flowrate through the hydrant valve. No air compression was considered to occur until the piston had reached the air slug at the end of the cylinder. This approach was justified because in practice the canvas hose was observed to be virtually flat prior to priming, such that the water column did not encounter air until the hose was nearly full. The equation of motion of the water column was obtained using Newtons second law, in a finite difference form with a variable mass term to account for the increasing length of the column as water entered the hose. The upstream pressure acting on the column and the flowrate into the hose were the interface variable with the rest of the system and the downstream pressure acting on the column was of course the trapped air pressure-initally considered to be atmospheric. Once the water piston reached the air pocket, polytropic compression of index 1.2 was assured. The constant ($\rho V^{1.2}$) was obtained using the perfect gas equation and the initial conditions.

Response to Mr. A.R. Rearson

The full pump/motor characteristics were used to simulate startup, and because the system compiresed a diesel driven pump and a electrically driven pump, the anaysis was based on the diesel pump as this had a much lower inertia and would therefore be a "worst case" regarding transient problems.

Response to Mr. C. Papadakis

Thank you for your comment, it is of course one of the best ways of overcoming the freezing problem for pipework that has to be located out of doors, providing that the buried pipework is laid in such a way that it is not prone to fracture. Some plastic and GRP pipework can cause problems when buried underground. Priming of the fire fighting device above gound it still necessary in many cases however.

PAPER H2

Kaplan turbine accidents and reverse water hammer.

S.Pejovic, L.Krsmanovic and A. Gajic
University of Belgrade, Yugoslavia

D.Obradovic
Energoprojekt, Yugoslavia

M.Papworth, BHRA

It is important to recognise failures as a means of learning: does the incident reported in the paper teach us anything about sensible design or operational methods.

A.Anderson, University of Newcastle Upon Tyne, U.K.

In view of the reluctance of many utilities and manufacturers to publicise operating accidents Prof.Pejovic and his colleagues are to be congratulated for making the information in their Paper available, and particularly for presenting it in such a practical and easily used form. I would be particularly interested if the authors could provide additional information on the following points:

(a) Prof.Pejovic implied in his presentation that he knew of a number of similar accidents not only in Yugoslavia but also elsewhere in the world. Would it be possible for the authors to prepare a small table of the countries where such accidents have been known to take place? Also in this type of accident has damage to the thrust bearing due to lifting and subsequent dropping of the rotating parts also been a problem?

(b) How was the value established for the additional criterion. C_{rez} = 0.5 m/s? Presumably this was based on a calculation of the runner blade strength against a blow from the water column beneath it. If so then at what section of the blade was failure assumed to occur in arriving at the criterion? Is this the same section at which the blade in Fig.1.1 and 1.2 actually failed? In other accidents have the blades tended to fail at similar sections?

(c) Air supply to the turbine should alleviate the pressure rise at the top cover but it is not clear whether it will have any influence on the loading of the turbine blades. Do the authors have an opinion on this?

(d) In non-cavitating conditions the pressure distribution across the turbine top cover can be measured and predicted but when the pressure falls the situation is less certain. Does vapour pressure occur across the full top cover area or only in limited zones? Is the effect of the runner pumping water out from under the top cover as the guide vanes close important? Also, when the returning water column strikes the runner blades (leading to possible breakage) do these absorb most of the energy or is the top cover also subjected to a large impact or pressure rise?

(e) In Francis turbines the situation is rather different and damage to the runner itself due to column separation beneath it seems much less likely, but very high draft tube pressure rises have been recorded following accidents with very rapidly closing guide vanes (Ref.A), causing damage throughout the turbine, spiral casing and suction cone. Have the authors come across any similar problems in their investigations? If so, then what amplitudes of pressure rise have they come to expect in such cases?

Additional reference:

A. Bernhard, H.J. and Mathews, R.V.: "The design of guide apparatus for large turbines and pump turbines", proceedings, IAHR Symposium, Vienna, 1974, Session XI, Paper 4.

Editors' Note - No Closure has been received, July, 1980.

PAPER H3

Cavitation and corrosion in a pumping main

W.L. Glass
Stewart, Sviridov & Oliver, South Africa

A.H. LUPTON, Binnie & Partners, U.K.

What explanation have you for the extraordinary symmetry of the attack on the bitumen?

Author's reply

One possible explanation is that the groove in the steel consists of a series of pits which have joined together; each of these pits would be surrounded initially by a circular colony of bacteria with the pit in the centre, some pits being larger than others and having correspondingly larger circular colonies. This theory seems to be reasonably consistent with the shape of the exposed bare steel shown on Fig. 3. Another possibility is that the heat given off during the corrosion process would tend to soften the bitumen equally on each side of the groove.

J.A. FOX, Leeds University

Is it possible that the overburden pressures could cause some flexing of the pipe which would explain the straightness of the crack. If the overburden were sufficiently large it might even place the soffit into such a large tension as to cause cracking once the bacterial attack had progressed sufficiently, so producing a self reinforcing mechanism.

Author's reply

The maximum overburden on the pipeline was 4,7 m but where the groove occurred it was less than 0,8 m and the sides were well padded and consolidated; nevertheless it does seem possible that overburden pressures did cause some flexing thus helping the groove to form along the crown.

H. SAFWAT, Bechtel Power Corporation, U.S.A.

Was the fracture of the pipe examined to determine whether the failure did occur due to under pressure in the line or due to the subsequent over pressure that followed the collapse of the vapour cavity?

Author's reply

It might indeed be expected that the groove formed in the crown of the pipe would lead to fracture of pipe along the groove either by the initial underpressure after pump trip or the subsequent overpressure. This, however, does not appear to have occurred, the failure being due to small holes penetrating through the pipe wall and resulting in sufficient leakage to be eventually observed.

C. PAPADAKIS, Bechtel Inc., U.S.A.

I would like to congratulate Mr. Glass for his excellent presentation. Without having studied the paper it seems to me that waterhammer cavitation should have little to do with the formation of the groove and the astonishingly symmetrical removal of the bitumen lining along the groove. Pump start-up and shutdown transients may have assisted the groove formation process. However, the majority of the corrosion damage should be the result of a rather steady state process, i.e. by bubbles trapped at the pipe crown for long periods of time. The release of dissolved gases during cavitation may not be the only cause of corrosion because of the short duration of negative pressures and the ability of these gases to get back into solution soon after negative pressures are over. After the groove has progressed in depth, failure could be caused by internal underpressures developed in the pipe after pump failure. I have seen 2 inch deep grout being removed from 4 ft. diameter concrete pipe joints because of negative pressure in the pipe.

Author's reply

If by 'cavitation' you are referring to 'cavitation erosion' I entirely agree with you. I also agree that the majority of the corrosion damage should be the result of a rather steady state process. It will be seen in Fig. 1 that for about 1.5 km upstream of the grooved section the pipeline is subjected to a sub-atmospheric pressure and that it rises fairly steadily all the way (point B to beyond point D). Thus any gases released during the short low pressure period (about 7 seconds) would tend to rise slowly up towards point D. At point C the gases should be released through the one-way air valve but the remainder would proceed up to the groove where the static pressure from the air valve at F is extremely low (the air valve at point F may not close for several hours due to the water level at the tank being very little higher than the air valve).

There are presumably many similar bitumen lined pipelines where bubbles tend to be trapped for long periods of time without causing any severe damage. In the present case I feel that gas bubbles alone are an inadequate explanation whereas the presence of sulphate-reducing bacteria does give a reasonably satisfactory explanation for this very unusual failure.

T.G.F. WELSHMAN, (Thames Water Authority, U.K.)

This paper was, to me, most refreshing in that it dealt with, what would appear to be, a rather unique problem. Although the ensuing verbal discussion left me with the feeling that the causes of the mode of failure of the pipeline might not be as definitive as Mr. Glass suggests, there is no doubt that an extensive 'post-mortem' was carried out on this failure and very plausible, if not unusual, conclusions were drawn.

Considerable benefit arises from the inclusion of papers such as this one, whereby engineers gain knowledge of phenomena which may not have come to their attention before. By being aware of such potential dangers, and having the advantage of previous knowledge, they are better equipped to investigate, and to keep a wary eye on, their own systems which might be subject to the same problems.
Although the Conference included several papers which dealt with real system failures, I single out this one as being of particular interest, by virtue of the somewhat novel nature of the identified problem.

Author's reply

I agree that there is not sufficient evidence to fully substantiate the theory put forward; the matter is being followed up, however, and I hope to be able to write another paper confirming or modifying the theory of bacterial corrosion in the light of further evidence. Since the paper was written a single hole, also exactly at the crown, has developed in the ESCOM pipeline at a section where cavitation occurs but where no trouble had previously been experienced. Whatever the cause of the corrosion it seems clear that cavitation contributes towards it and should therefore be prevented in pipelines in the Witbank area, particularly if bitumen lined. It is hoped shortly to examine the concrete lined main mentioned in the paper to see if any corrosion has occurred there.

A.K. GRUNDY, Shell International Petroleum Mij., B.V.

Has the author been able to obtain any information about this phenomenon from the oil-industry, as they have experienced various problems because of the activities of sulphate reducing bacteria.

Author's reply

Yes, comparison with the problems of the oil industry have been made and were most fruitful. Their problems, however, are related to external corrosion whereas the paper deals with internal corrosion. Peabody (Ref. 13) refers to oil and gas pipeline corrosion problems in some detail and reports that sulphate-reducing bacteria have been responsible for the disintegration of ashphalt coating to pipes.

Author's request

Cavitation erosion is a common occurrence in pumps where considerable damage can be done to the interior surface of an impeller as a result of very high pin-point pressures set up on collapse of vapour bubbles; if a pump is run continuously, several millimetres of metal can be removed in a few weeks. Cavitation due to waterhammer downsurge with vapour bubble collapse on subsequent upsurge, presumably results in similar high local pressures; no damage however, appears to have been reported in the literature. Is this because the mechanism is different and no damage in fact occurs, or is it because the damage is relatively small due to the relatively short duration of the cavitation erosion during the intermittent periods of waterhammer? If the latter is correct then cavitation erosion could be quite serious in the case of an installation where a pump is stopped and started several times a day; it might at least result in failure of the internal lining of a pipe in a few years. I would be most interested to learn whether any such damage has been observed.

R.H. FALCONER, Babtie, Shaw & Morton, U.K.

I am sure that I must be one of many who would wish to congratulate Mr. Glass for presenting such a well researched paper describing the detective work involved in tracking down the culprits responsible for this unique type of pipe failure.

The nature of a similar form of 'bacterial' attack in gravity sewers and sewage pumping mains is well reported (Ref. 'The Problem of Hydrogen Sulphide in Sewers' R.D. Pomeroy, Clay Pipe Development Association Publication, April 1977). Concrete and cast iron pipes have deteriorated and failed as a result of sulphuric acid produced from H_2S by close relatives of Mr. Glass's bacteria namely, 'Thiobacillus Concretivorous'. These bacteria survive in similar anaerobic conditions to those described by Mr. Glass. A slimey mass tends to be produced where the sulphuric acid formed by the bacteria attacks the pipe material. This is usually above the liquid surface.

Bearing this similar type of 'bacterial' attack in mind I should like Mr. Glass to comment on the following points:-

a) Was there any evidence of bacterial slime at the point of attack and was the existence of Sporovibrio Desulphuricans inferred from the conditions pertaining or specifically isolated under the microscope?

b) Was there any evidence of attack by sulphuric acid or H_2S itself i.e. the formation of iron sulphide?

c) In the first Witbank pumping main have there been any indications of the concrete lining being attacked? It might be expected that with a lower velocity there might be less aeration and more H_2S produced from organic material in the water.

d) Has Mr. Glass considered any methods for controlling the occurrence of these bacteria by methods such as oxygen or merely air entrainment into the pipeline near the failure site to create aerobic conditions under which these bacteria should not survive whilst reducing surge and cavitation effects by cushioning and raising sub-atmospheric pressure levels? Air entrainment would also inhibit H_2S production from organic material in the water. Perhaps nothing more would be required that a two-way air valve with vented non-return valve at location C to provide adequate air inflow. The bitumen lining would however, have to be replaced to prevent straightforward corrosion of the bared metal.

Oxygenation is a recognised technique in sewage pumping mains for maintaining aerobic conditions to prevent the build up of H_2S whilst giving a measure of sewage purification in the pumping main itself where the capacity of the treatment works is overloaded.
(Ref. 'BOC Blows Bubbles in Dirty Bath Water' - article in New Civil Engineer, February 1977).

Author's reply

Thank you for your remarks. I am glad that you have raised the question of corrosion due to hydrogen sulphide in sewers and sewage pumping mains as I should have made some reference to this in the paper. Hydrogen sulphide is known to be produced in sewage pumping mains by sulphate-reducing bacteria acting on sulphates under anaerobic conditions. Being very soluble in water, however, the H_2S remains in solution and very little if any corrosion normally occurs in the pumping main. When the sewage enters a sewer, however, it frequently becomes aerated, the H_2S then comes out of solution and re-dissolves in the water condensed on the crown of the sewer (which seldom flows full). It is there that your Thiobacillus oxidises the H_2S to sulphuric acid under aerobic conditions (such oxidation would apparently occur in the presence of oxygen in any case).

The existence of Sporovibrio (now apparently called Desulphovibrio) desulphuricans was inferred in the Witbank pipeline. A type of sulphate-reducing bacteria was reported to have been isolated from the bitumen taken from the ESCOM pipeline.

Regarding question (b) tests for sulphides must be carried out on fresh samples and this was unfortunately not done when the failure occurred.

Examination of the concrete-lined main will be carried out in due course and I will then be able to reply to your question (c).

In reply to question (d), air valves and injection of air or oxygen have been considered but could lead to further corrosion even if the bitumen lining was repaired. The best solution will probably be to arrange for the pumps to be shut down gradually to avoid cavitation; if this is impracticable two or three small one-way surge tanks similar to the existing one will probably be the best solution. It is hoped to definitely establish the exact cause of the corrosion before making a final decision on preventive measures.

Discussion & Contributions

SESSION J

SURGE CONTROL AND SUPPRESSION

Chairman: Prof.C.S.Martin
Georgia Inst. of Technol., U.S.A.

Papers:

J1	Dynamic behaviour of non-return valves.	G.A.Provoost, Delft Hydraulics Lab., Netherlands.
J2	Influence of pressure surges on the functioning of safety valves.	B.Foellmer and H.Zeller, Aerodynamische Inst., Federal Republic of Germany
J3	Design charts for surge tanks on pump discharge lines.	A.T.K.Fok, Proctor & Redfern Group, Canada.

PAPER J1

Dynamic behaviour of non-return valves.

G.A. Provoost
Delft Hydraulics Lab., Netherlands

Dr.J.Ellis, Strathclyde University

Regarding the addition of a weight to the lever of the spring check valve, this valve bears a considerable resemblance to the valve which was discussed in paper E3. One of the suggestions which was made to improve the behaviour of this valve was to add weight to the lever of the valve; but the effects on the pressure pulses following closure were negligible. My own view on this matter is that while the added weight gives an additional closing moment it also gives an increased moment of inertia which may be counterproductive. I would like to learn of the speaker's experiences regarding the effectiveness of adding weight to the lever of swing - check valves and whether alteration of the angle between the lever and the valve door has a significant bearing on performance.

C.Papadakis, Bechtel Inc., Ann Arbor, U.S.A.

I hope that Mr.Provoost's work will be completed in the near future since it should have quite an impact in check valve design. In power plant large pipes (such as steam lines) check valve closures in 20-30 miliseconds and valve velocities of 120-150 ft/sec. upon impact are not uncommon. For fast closing large check valves a dashpot mechanism may be more desirable than springs or counterweights which may shear upon impact. Manufacturers' information on check valves usually uncouples the flow dynamics from the valve motion which makes difficult any transient analysis of the system incorporating the check valve response. These facts should make Mr.Provoost's contribution quite valuable in design.

E.M.Curtis, CEGB, Marchwood

I should like to know what type of flowmeter was used to measure the transient flows in the non-return valve tests.

W.L.Glass, Stewart, Sviridov & Oliver

The ingenious animated pressure surge diagrams were indeed fascinating to watch and beautifully illustrated the pressure waves set up as a result of cavitation in a horizontal pipeline. Could the author state whether any evidence has been found of damage to the interior surface of the pipe as a result of vapour pocket collapse?

In regard to the swing type non-return valve illustrated in Fig.3 the writer has observed that similar valves sometimes remain open when the flow reverses very rapidly ; this has happened for example when two pumps were running in parallel and one pump was tripped out; in this case the water flowed backwards through the idle pump. Does the author think that the gate could be held in the open position during backward flow by means of hydraulic forces in the sort of way in which a ping-pong ball moves up into the neck of a funnel when one blows down on it.

D.J.F.Ewing, C.E.G.B.

Thank you for your excellent presentation. I have two questions and a comment.

(i) In your equation (2), do you know anything yet about the relative importance of the five quantities?

(ii) Normally least is known about the hydraulic moment T_h. For a tilting disc, Worster (Ref.1) suggested putting

$$T_h = a(\alpha) Q^*/Q^*/$$

(in your notation) in place of your equation (7), where

$$Q^* = Q + AR\dot{\alpha}$$

(A = area, R = distance to centre of pressure, α = angle turned through from horizontal). Kane and Cho (Ref.2) use a similar idea except that $Q^* = Q + AR\dot{\alpha}/\sin\alpha$. Thus, one is using flow relative to the moving disc rather than absolute flow. How does this simplified treatment compare with the theories that you are developing?

(iii) The dynamic effects you discuss apply when a butterfly valve is closing down at a known rate, as well as to a check valve. For example using Worster's approach one could use your equation (8) with Q replaced by Q*.

This dynamic correction sometimes improves agreement between calculation and experiment even though Worster's theory is only approximate.

References:

1. Worster, R.C. The use of linearised equations in calculating the motion of reflux valves, BHRA report RR676 (1960).

2. Kane,R.S. and Cho, S.M., Hydraulic performance of tilting-disc check valves, ASCE J, Hydr.Div.102, HYI,pp.57-72 (1976).

Answer to Dr. J. Ellis:

The addition of a weight to the lever of a swing type non-return valve gives indeed both an additional closing moment and an increased moment of inertia. Whether the effect is an improvement of the dynamic behaviour of the non-return valve depends on the system in which the valve is incorporated.

In most cases the effect will be small. As long as the valve remains in full open position during normal flow, the effect will generally be negative. If, however, the valve body gets a more closed position during normal flow by the addition of a weight to the lever the effect may be positive.

The addition of a weight to the lever has about the same effect as the increase of the distance between a given weight and the pivot. This effect was studies in the Experiments Nos. 6 and 7 in my paper. In case of the large distance between weight and pivot (Experiment No. 6) the position of the valve body changed into more closed position. The angle of the valve body related to closed position was changed from 33 ^{o} into 28 ^{o} (see Fig. 10 respectively Fig. 9). As shown in the paper the smaller angle gives an improvement of the dynamic behaviour.
Instead of the addition of a weight to the lever the addition of a spring to the lever will improve the dynamic behaviour, as it gives an additionals closing moment and does not increase the moment of inertia.
The best improvement of the dynamic behaviour of each non-return valve is to decrease the displacement of the valve body. For a swing type non-return valve, however, a decrease of the displacement is always attended with a decrease of the cross-sectional area of the valve and thus with an increased pressure loss during normal steady flow. So a good dynamic behaviour and a small pressure loss are contradictory.
A swing type non-return valve is relatively cheap, gives only a small pressure loss but has a moderate dynamic behaviour. Valves which have a valve body consisting of a spring loaded disc or ring are more expensive, create a larger pressure drop but have a much better dynamic behaviour.
The alteration of the angle between the lever and the valve door will in general have insignificant bearing on the dynamic behaviour.
At least attention has to be paid to the behaviour of a non-return valve once it has closed. The valve may reopen due to a pressure difference across the valve in the original flow direction, perhaps leading to a repetitive slamming of the valve. Whether reopening occurs is influenced by the presence of a weight on the lever (or spring).

Answer to C. Papadakis:

Thank you for your valuable comments. The aim of my paper was not to give a complete view of the phenomena but to show the importance of taking into account the interaction between flow dynamics and valve motion when predicting the valve motion. Since manufacturers do not make this combination no sufficient data are available to perform surge calculations when the behaviour of the non-return valve dominates the surge phenomena. Therefore the only possibility up to now to examine these phenomena is to measure the non-return valve response in a test circuit for the same steady state flow and same fluid velocity gradient as calculated in a mathematical model without non-return valve.
Meanwhile theories will be developed which can be verified by means of experiments in the described test-circuit.

Answer to E.M. Curtis

The transient flows were measured by means of an electromagnetic flowmeter of which the frequency of the magnetic field was increased to 1 kHz allowing to measure the time-dependent flow up to frequencies of 50 Hz.

Answer to W.L. Glass:

Thank you for your remark over the motion picture I showed before my presentation. The animation showed the computed occurrence and disappearance of a cavitation region in a horizontal pipeline. In real systems of large diameters (from 0,5 m up to 2,00 m) in which cavitation was accepted no damage to the interior surface of the pipes was found since the water contained dissolved air which is released at low pressures and certainly at vapour pressure. Therefore in practice no collapse of cavities (which causes sudden pressure rises) but a more or less slow compression of the released gas will occur.
The observations you mention in the second part of your question are rather strange in my opinion. It must be a very bad design of a non-return valve if the valve body is pushed in open position due to backward flow. My opinion about the observed cases is that friction must have played an important role there. As long as the backward flow does not exert a moment larger than friction minus gravity forces the valve body will not move. But since the moment exerted by the backward flow increases with increasing flow at a certain instant the valve body will begin to move and close in a very high backward flow thus causing probably a serious damage to valve and surrounding pipes.

Answer to D.J.F. Ewing:

Thank you for your useful contribution, which definitely supllements my paper.
The total moment exerted on a valve body was devided into four quantities. The relative importance of these four quantities will depend on the geometry and dimensions of the valve. However, the hydraulic moment T_h is influence by the valve motion itself and therefore T_h is the most important quantity of the total moment exerted on a valve body.
This hydraulic moment as well as the pressure loss over the valve depend on the Reynolds number and the flow pattern around the valve body. Neglecting Reynolds effects this flow pattern will depend on the position of valve body and the ration between the flow induced by the motion of the valve body and the flow passing the valve. The flow in-

duced by the motion of the valve body is proportional to the angular velocity $\dot{\alpha} = \frac{d\alpha}{dt}$ of the valve body so that the equations (3) and (4) of my paper may be replaced by:

$$T_h = a' \left(\alpha, \frac{\dot{\alpha}}{Q}\right) Q|Q|, \text{ and} \tag{9}$$

$$\Delta H = \xi' \left(\alpha, \frac{\dot{\alpha}}{Q}\right) Q|Q| \tag{10}$$

Written in terms of flow relative to the moving valve body these equations should read:

$$T_h = a'' \left(\alpha, \frac{\dot{\alpha}}{Q}\right) Q^*|Q^*|, \tag{11}$$

$$\Delta H = \xi'' \left(\alpha, \frac{\dot{\alpha}}{Q}\right) Q^*|Q^*|, \text{ and} \tag{12}$$

$$Q^* = Q - AR\dot{\alpha}, \tag{13}$$

where: A = area of valve body (m^2),

R = resistance of centre of valve body to pivot

α = angle relative to closed position

$\dot{\alpha} = \frac{d\alpha}{dt}$ = angular velocity (radians/s)

The simplification suggested by Worster assumes that:

$$a''\left(\alpha, \frac{\dot{\alpha}}{Q}\right) = a(\alpha), \text{ and} \tag{14}$$

$$\xi''\left(\alpha, \frac{\dot{\alpha}}{Q}\right) = \xi(\alpha), \tag{15}$$

where $a(\alpha)$ and $\xi(\alpha)$ are respectively torque coefficient and pressure head loss coefficient for steady state flow as used in equtions (7) and (8) of my paper.

In other words: hydraulic moment and pressure head loss are determined by the flow relative to the moving valve body and coefficients related to steady state flow. I do not know what the validity is of this simplification, but I suppose that in case of a fast moving valve body it will certainly give better results that the use of equations (7) and (8) of my paper which are only valid for slow motion of the valve body. All mentioned torque coefficients and pressure head loss coefficients depend on the flow pattern in the valve.

When the valve opening is small compared with the cross-sectional area of the surrounding pipes, the flow pattern will be nearly independent of the angular velocity of the valve body and thus equations (14) and (15) will be satisfied.

In the past I have carried out calculations for consultancy purpose using the flow relative to the moving valve body as mentioned by Worster for determining both hydraulic moment and pressure head loss. From these calculations it appeared that the relative flow never reversed. Only when a dashpot was used to lower the valve velocity upon impact the relative flow reversed in the last part of the closure.
Also Dr. Ellis used in his paper the approsimation suggested by Worster, but he did not use physical data to obtain the coefficients $a(\alpha)$ and $\xi(\alpha)$. His coefficients were obtained for a rectangular valve and a contraction coefficient equal 1. In the notation as mentioned before we get:

$$\xi(\alpha) = \frac{1}{2gA^2(1-\sin\alpha)^2} \quad \text{, and} \tag{16}$$

$$a(\alpha) = AR\rho g\xi(\alpha) \tag{17}$$

The results of Ellis' calculations also show that the relative flow hardly reverses (see Fig. 7 if paper E3, page 215 of volume 1 of the proceedings: "Flow velocity in pipe due to through valve").
In my further investiation I will determine the coefficients a' and ξ' as a function of Q and $\frac{\dot{\alpha}}{Q}$. Measurements have been carried out recently for a rectangular valve but results are not yet available. This study should show the validity of the simplification as mentioned by Worster. This simplification can also be written as:

$$a'(\alpha, \frac{\dot{\alpha}}{Q}) = a(\alpha)\ (1 + AR\,\frac{\dot{\alpha}}{Q})^2, \text{ and} \tag{18}$$

$$\xi'(\alpha, \frac{\dot{\alpha}}{Q}) = \xi(\alpha)\ (1 + AR\,\frac{\dot{\alpha}}{Q})^2. \tag{19}$$

As soon as results are available a comparison will be made between the measured coefficients a' and ξ' and the coefficients determined according equations (18) and (19).
The equation mentioned by Kane and Cho differs somewhat from the one given by you. Kane and Cho used:

$$Q^* = Q - AR\dot{\alpha}/\cos\alpha, \tag{20}$$

where: α = angle of disc from seat.
For a valve with a vertical closed position this equation corresponds with the one given by you.
The approach of Kane and Cho gives me some doubts. They use:

$$a'(\alpha, \frac{\dot{\alpha}}{Q}) = a(\alpha)\ (1 + \frac{AR}{\cos\alpha}\,\frac{\dot{\alpha}}{Q})^2, \text{ and} \tag{21}$$

$$\xi'(\alpha, \frac{\dot{\alpha}}{Q}) = \xi(\alpha)\ (1 + \frac{AR}{\cos\alpha}\frac{\dot{\alpha}}{Q})^2. \tag{22}$$

For a valve whose body can turn over $\frac{\pi}{2}$ radians (90 ^{o}) both a' and ξ' would be infinite for full open position, which is far from realistic. Since there is no physical reason to introduce the cos-term this approach seems to be suitable only for the typical valve they examined.

Your comment that the above discussed method of using relative flow rather than absolute flow is not only valid for non-return valves but also for butterfly valves is quite true. For quick acting butterfly valves the water displacement induced by the valve motion may cause additional pressure losses or closing moments, or perhaps smaller closing times than expected from dry tests. For centric buttterfly valves, however, the mentioned simplification will not give any improvement since then R = 0 and according to (13) $Q^* = Q$. Only the equations (9) and (10) are then useful. Also this aspect, which can be important for large butterfly valves with small closing times, needs further investigation.

PAPER J2

Influence of pressure surges on the functioning of safety valves.

B. Foellmer and H. Zeller
Aerodynamisches Institut, Federal Republic of Germany

G.W. Bleasdale, British Aerospace, U.K.

Has the Author tried to reduce the oscillations by re-design of the disc valve, say, by changing to a conical shape. In this case, if a shock wave develops, it would attach to the nose of the cone and would have much less tendency to oscillate. The problem stems mainly from the separation at the front of the Borda tube and rounding the leading edge of this would be benefical.

Author's reply

Indeed the observed flow oscillation depends mainly on the contraction coefficient of the flow in the pipe. When the edge of the inlet is rounded (see Fig. 1) the flow contracts less and no flow oscillations are observed.

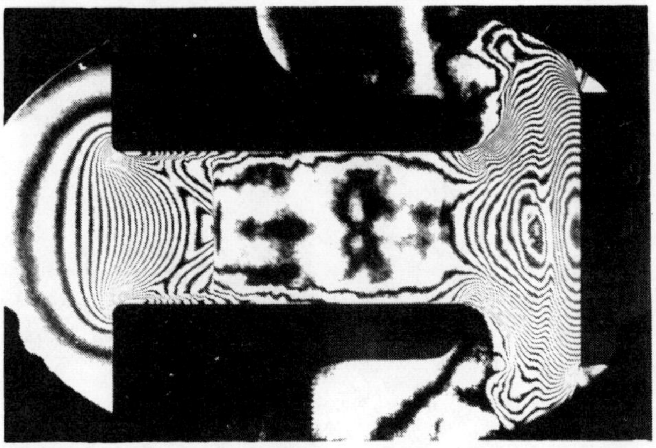

Fig. 1 Mach-Zehnder Interferogramm of the flow through a plane valve with $h/b = 0.7$ and $p_a/p_k = 0.3$

Changing the disk to a conical shape can reduce the oscillations when the cone diving in the duct is long enough for the given flow conditions. But for such a conical disk the inlet must be divergent.

H. Safwat, Bechtel Power Corp., U.S.A.

Can Mr. Foellmer comment on what parameters would affect the frequency of the observed frequency? Is the 300 HZ that he observed mainly a function of geometry?

Author's reply

The frequency depends mainly on the length of the duct and the temperature of the gas flowing through the valve.

PAPER J3

Design charts for surge tanks on pump discharge lines.

A.T.K.Fok
Proctor & Redfern Group, Canada

W.L.Glass, Stewart Sviridov & Oliver

The charts presented should prove invaluable in the design of surge tanks. In regard to the one-way surge tanks, could the author comment on the possibility of using them on sewage pumping mains? Presumably it would be necessary to use a supply of clean water. There is also the problem of insuring that the tanks are full before pump trip.

Author's Closure

One-way surge tanks can be used effectively for surge control on sewage pumping mains. Because of the odour of the sewage water, it would be necessary to supply clean water in filling the tank. From the author's past experience on one-way surge tank design, special considerations should be given to the two areas, namely, the design of the check valve at the tank's riser and the operation of pump restart-up following the pump trip-out.

A swing check valve is more preferable than a tilting disc check valve in order to solve debris jam problems. However, for a low-head system, it may require an additional counterweight to close the swing check valve tightly. The counterweight would be adjusted to just prevent leakage under the normal operating pressure.

In cold weather area, the one-way surge tank may have to be installed inside an enclosed heated chamber with a continuous flow into the tank and drain from the tank. Very often, two check valves in parallel are recommended to increase the reliability.

A probe control should be installed inside the tank so that the pump will not resume its normal operation until the tank is full. In order to minimize this interruption during normal pump operation, variable speed pumps are desirable in most cases.